Contents

Credits

The authors and publisher would like to thank the following individuals and organisations for permission to reproduce figures and extracts from tables and articles:

p.52 (BSI certification mark and Kitemark) BSI, **p.131** (Findings of public attitudes to science report) Crown Copyright Click use licence C2008002221. Department for Business Innovation and Skills (BIS) and RCUK: Findings of public attitudes to science Report 2008, **p.217** (Plate showing a balanced diet) Human Physiology and Health 2e by David Wright/Pearson Education, **p.309** (Representation of Baltimore classification system for viruses) Professor Vincent Racaniello, PhD, **p.427** (Findings from the British Crime Survey and police recorded crime) Crown Copyright Click use licence C2008002221. Home Office Official Statistics Crime in England and Wales 2007/08 Findings from the British Crime Survey and police recorded crime.

The authors and publisher would like to thank the following individuals and organisations for permission to reproduce photographs:

p.1 Dr Jurgen Scriba/Science Photo Library, **p.3** Tracy Whiteside/Shutterstock, **p.13** lightpoet/Shutterstock, **p.16** Thomas Deerinck, NCMIR/Science Photo Library, **p.24** Ray Tang/Rex Features, **p.26** Julie Grimshaw, **p.29** TEK Image/Science Photo Library, **p.31** PT Images/Shutterstock, **p.34** Digital Stock, **p.39** Photodisc/Kim Steele, **p.43** Image100/Alamy, **p.45** Thomas M Perkins/Shutterstock, **p.49** left (**l**) Justin Kase z03z/Alamy, right (**r**) Photodisc/Photolink, **p.50** Crown copyright, **p.54** P. Goldsmith/Eli Lilly and Company Ltd, **p.57** CERN Geneva, **p.59** Sergej Khakimullin/Shutterstock, **p.64 l** Pearson Education Ltd/Trevor Clifford, **r** Andrew Lambert Photography/Science Photo Library, **p.67** Mary Altaffer/AP/Press Association Images, **p.68** Falconia/Shutterstock, **p.71 l** middle (**m**) Glue Stock/Shutterstock, **l** bottom (**b**) Jozsef Szasz-Fabian/Shutterstock, **r** David J. Green/Alamy, **p.77** Obianuju Ekeocha, **p.79** Adam Hart-Davis/Science Photo Library, **p.81** kolosigor/Shutterstock, **p.83** Pixmann/Alamy, **p.98** Photodisc/Russell Illig, **p.101** Andrew Lambert Photography/Science Photo Library, **p.103** top (**t**), m J. Irvine, Antonine Education, **p.106** Ilya Akinshin/Shutterstock, STILLFX/Shutterstock, Bragin Alexey/Shutterstock, **p.108** Pearson Education Ltd/Trevor Clifford, **p.109** Camlab, serving science since 1950, **p.110** Martyn F. Chillmaid/Science Photo Library, **p.111** Pearson Education Ltd/Trevor Clifford, **p.119** AstraZeneca/Helen Marie Bristow, **p.121** Gustoimages/Science Photo Library, **p.123** Jason Stitt/Shutterstock, **p.124** Getty Images/Popperfoto, **p.129** Getty Images/Stone, **p.133** James Fraser/Rex Features, **p.134 t** CERN Geneva, **b** Sam Ogden/Science Photo Library, **p.137** Alex Macnaughton/Rex Features, **p.140** NASA, **p.141 l** Rex Features, **r** Corbis/Bettman, **p.142** Lucy Goodchild, Imperial College London, www.imperial.ac.uk/media by Colin Smith 2010, **p.145** Lawrence Manning/Corbis, **p.147** Robert Sanderson, **p.152** Andreas Reh/iStockphoto, **p.153** NASA/Stöckli, Nelson, Hasler, **p.155** medicalpicture/Alamy, **p.157 l** Peter Gould, **r** NASA, **p.159** Photodisc/Karl Weatherly, **p.164** concinnitās ltd/Dr M. J. Hatcher, **p.167** Jeff Metzger/Shutterstock, **p.169** Robert Sanderson, **p.173** Pasieka/Science Photo Library, **p.191** Hannah Victoria Hewson, **p.193** Pasieka/Science Photo Library, **p.195** Pedro Vidal/Shutterstock, **p.198** Phototake Inc./Photolibrary.com, **p.210** CNRI/Science Photo Library, **p.219 l** R. Bodine, Custom Medical Stock Photo/Science Photo Library, **r** Michael Donne/Science Photo Library, **p.226** Biophoto Associates/Science Photo Library, **p.232** Tom Warrender, www.classroommedics.co.uk, 01902 565457, **p.235** ynse/Shutterstock, **p.237** Thinkstock, **p.273** moshimochi/Shutterstock, **p.275** SCIMAT/Science Photo Library, **p.277** omkar.a.v/Shutterstock, **p.279** Michael Abbey/Science Photo Library, **p.280** Science Source/Science Photo Library, **p.283 l** Eye of Science/Science Photo Library, **r**, **t** BSIP, PR Bouree/Science Photo Library, **r**, **m** Michael Abbey/Science Photo Library, **p.285 m** Eye of Science/Science Photo Library, **b** Steve Gschmeissner/Science Photo Library, **p.288** Will & Deni McIntyre/Science Photo Library, **p.291** CNRI/Science Photo Library, **p.292** John Durham/Science Photo Library, **p.297** Eye of Science/Science Photo Library, **p.299** Sascha Burkard/Shutterstock, **p.301** Jim Varney/Science Photo Library, **p.305** Agricultural Research Service/U.S. Department of Agriculture/Science Photo Library, **p.309** Pasieka/Science Photo Library, **p.310** Eye of Science/Science Photo Library, **p.313** E. Gueho/Science Photo Library, **p.316** Amber Lansley/J.Kane, Health Protection Agency, **p.319** Massimo Brega/Eurelios/Science Photo Library, **p.321** Dan Brandenburg/iStockphoto, **p.329** Adrian T. Sumner/Science Photo Library, **p.330** CNRI/Science Photo Library, **p.332** Ed Reschke, Peter Arnold Inc./Science Photo Library, **p.334** Science Pictures Ltd/Science Photo Library, **p.336** James King-Holmes/Science Photo Library, **p.337** Joy M. Prescott/Shutterstock, **p.340 l** Studiotouch/Shutterstock, **r** SIPA Press/Rex Features, **p.341** Andy Clarke/Science Photo Library, **p.344 l** Eye of Science/Science Photo Library, **r** Sue Ford/Science Photo Library, **p.346 l** Library of Congress/Science Photo Library, **r** Joti/Science Photo Library, **p.347** Look at Sciences/Science Photo Library, **p.350** Patrick Landmann/Science Photo Library, **p.353** Dr L. Caro/Science Photo Library, **p.361** Gillian Hamilton/The Alzheimer's Society/The Alzheimer's Research Trust/Alzheimer Scotland, **p.363** Getty Images/Universal Images Group, **p.365** Alamy/Image Source, **p.369** Photodisc/Jim Wehtje, **p.371** Suzanne Tucker/Shutterstock, **p.376 t** RVI Medical Physics, Newcastle/Simon Fraser/Science Photo Library, **b** Sovereign, ISM/Science Photo Library, **p.378** Gonul Kokal/Shutterstock, **p.381 l** OSF/Photolibrary, **r** Basov Mikhail/Shutterstock, **p.383** Getty Images/Hans Neleman, **p.386** Dmitriy Shironosov/iStockphoto, **p.389** Maximilian Stock Ltd/Science Photo Library, **p.391** Ümit Erdem/Shutterstock, **p.393** Martyn F. Chillmaid/Science Photo Library, **p.396** Maximilian Stock Ltd/Science Photo Library, **p.412** all Pearson Education Ltd, **p.413** Sheila Terry/Science Photo Library, **p.420** Rachel Slater, www.almacgroup.com, **p.423** Nicholas Bailey/Rex Features, **p.425** Ajay Bhaskar/Shutterstock, **p.432 t l** photobank.ch/Shutterstock, **t r** ostill/Shutterstock, **b** Thinkstock, **p.446** Monkey Business Images/Shutterstock, **p.449** kilukilu/Shutterstock, **p.451** Photos.com, **p.457** Jim Varney/Science Photo Library, **p.465** Mauro Fermariello/Science Photo Library, **p.474** tbkmedia.de/Alamy, **p.476 l** FirearmsID.com, **r** Valery Kraynov/Shutterstock, **p.476** Dr Gary Settles/Science Photo Library, **p.481 l** Photos.com, **r** Thinkstock, **p.484** Sebastian Kaulitzki/Shutterstock, **p.499** Lipik/Shutterstock

About the authors

Frances Annets has taught vocational qualifications for over 10 years and is a GCE chemistry moderator, GCSE science moderator and GCSE chemistry and applied science examiner. She is also an examiner for an international teacher and trainer diploma.

Shirley Foale is a lecturer in a FE college. Having worked in laboratories over a number of years she has first-hand Level 2 and 3 knowledge.

Jo Hartley is a science teacher, GCSE biology examiner and science BTEC coordinator at a secondary school in Merseyside.

Sue Hocking has been an examiner for almost 30 years. She has delivered BTEC science and health studies courses in FE colleges, as well as GCSE and A level biology courses in secondary schools and a sixth form college. Her specialist fields are biology, biochemistry and health promotion. Sue is a series editor for Pearson and has written many books and teacher support resources.

Lee Hudson is a chemistry teacher with experience of teaching all levels of science.

Tony Kelly went into management in industry after achieving a doctorate in chemistry. He teaches criminology and psychology for the OU and works as an independent inspector at a prison.

Roy Llewellyn is a senior science teacher, GCE physics examiner and head of vocational studies at a secondary school in South Wales.

Ismail Musa is an examiner for A level physics. Ismail has been teaching vocational applied science courses for over 10 years. As well as teaching and examining he has been working as a Subject Learning Coach for Science (SLC), coaching students and staff and organising and running teaching and learning sessions.

Joanna Sorensen is a course manager for the BTEC National Diploma in Applied Science (Forensic Science) at a FE college and teaches on a foundation degree in crime scene and forensic investigation. She has also worked for the Forensic Science Service.

About your BTEC Level 3 National Applied Science

Choosing to study for a BTEC Level 3 National Applied Science qualification is a great decision to make for lots of reasons. More and more employers are looking for well-qualified people to work within the fields of science, technology, engineering and maths. The applied sciences offer a wide variety of careers, such as forensic scientist, drug researcher, medical physics technician, science technician and many more. Your BTEC will sharpen your skills for employment or further study.

Your BTEC Level 3 National Applied Science is a **vocational** or **work-related** qualification. This doesn't mean that it will give you *all* the skills you need to do a job, but it does mean that you'll have the opportunity to gain specific knowledge, understanding and skills that are relevant to your chosen subject or area of work.

What will you be doing?

The qualification is structured into mandatory units (ones you must do) and optional units (ones that you can choose to do). This book contains 15 units and includes all Level 3 mandatory units. How many, and which, units you do depends on the type of qualification you are working towards. See the specification for a full list of unit and pathway opinions.

- BTEC Level 3 National **Certificate** in Applied Science: 3 mandatory (M) units that provide a combined total of 30 credits.
- BTEC Level 3 National **Subsidiary Diploma** in Applied Science: 3 mandatory units plus optional (O) units that provide a combined total of 60 credits.
- BTEC Level 3 National **Diploma** in Applied Science: 6 mandatory units plus optional units that provide a combined total of 120 credits.
- BTEC Level 3 National **Extended Diploma** in Applied Science: 6 mandatory units plus optional units that provide a combined total of 180 credits. To achieve the Extended Diploma you must complete additional units to those covered in this book.

Unit number	Credit value	Unit name	Cert	Sub Dip	Dip
1	10	Fundamentals of science	M	M	M
2	10	Working in the science industry	M	M	M
3	10	Scientific investigation			M
4	10	Scientific practical techniques	M	M	M
5	10	Perceptions of science		O	M
6	5	Using mathematical tools for science		O	M
8	5	Using statistics for science		O	O
11	10	Physiology of human body systems		[O]	[O]
13	10	Biochemistry and biochemical techniques		[O]	[O]
15	10	Microbiological techniques		[O]	[O]
18	10	Genetics and genetic engineering		O	O
20	10	Medical physics techniques		O	O
22	10	Chemical laboratory techniques		(O)	(O)
31	10	Criminology		O*	O*
32	10	Forensic evidence collection and analysis		O*	O*

*Units only available for the Forensic Science pathway
[ ] Units not applicable to the Forensic Science pathway
( ) Units not applicable to the Medical Science pathway

How to use this book

This book is designed to help you through your BTEC Level 3 National Applied Science course. It is divided into 15 units and provides enough coverage to achieve a Certificate, Subsidiary Diploma or Diploma.

This book contains many features that will help you use your skills and knowledge in work-related situations and assist you in getting the most from your course.

Introduction

These introductions give you a snapshot of what to expect from each unit – and what you should be aiming for by the time you finish it!

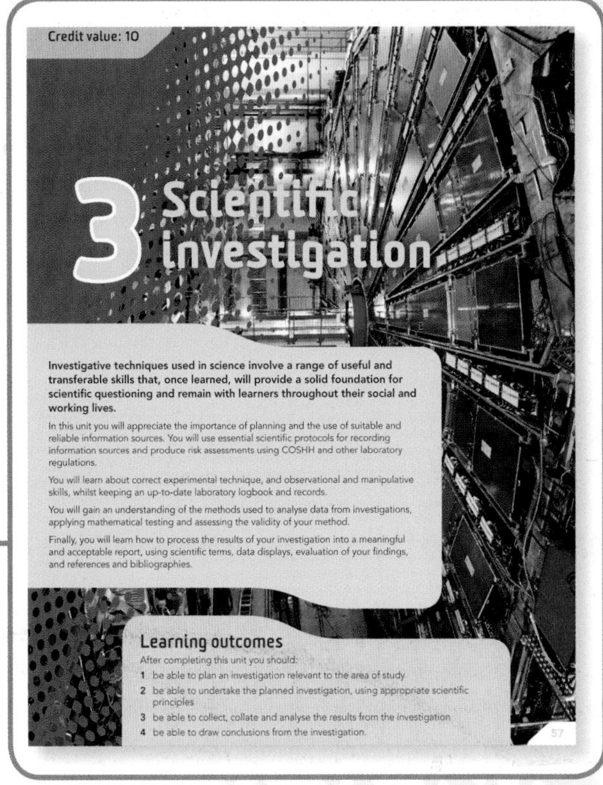

Credit value: 10

3 Scientific investigation

Investigative techniques used in science involve a range of useful and transferable skills that, once learned, will provide a solid foundation for scientific questioning and remain with learners throughout their social and working lives.

In this unit you will appreciate the importance of planning and the use of suitable and reliable information sources. You will use essential scientific protocols for recording information sources and produce risk assessments using COSHH and other laboratory regulations.

You will learn about correct experimental technique, and observational and manipulative skills, whilst keeping an up-to-date laboratory logbook and records.

You will gain an understanding of the methods used to analyse data from investigations, applying mathematical testing and assessing the validity of your method.

Finally, you will learn how to process the results of your investigation into a meaningful and acceptable report, using scientific terms, data displays, evaluation of your findings, and references and bibliographies.

Learning outcomes

After completing this unit you should:

1 be able to plan an investigation relevant to the area of study
2 be able to undertake the planned investigation, using appropriate scientific principles
3 be able to collect, collate and analyse the results from the investigation
4 be able to draw conclusions from the investigation.

57

Assessment and grading criteria

This table explains what you must do to achieve each of the assessment criteria for each unit. Each unit contains a number of assessment activities to help you with the assessment criterion, shown by the grade button P1.

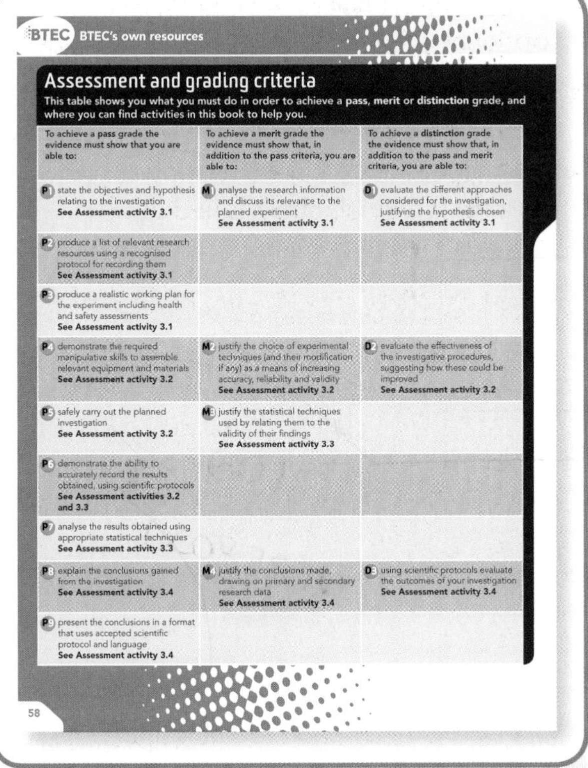

BTEC BTEC's own resources

Assessment and grading criteria

This table shows you what you must do in order to achieve a pass, merit or distinction grade, and where you can find activities in this book to help you.

To achieve a pass grade the evidence must show that you are able to:	To achieve a merit grade the evidence must show that, in addition to the pass criteria, you are able to:	To achieve a distinction grade the evidence must show that, in addition to the pass and merit criteria, you are able to:
P1 state the objectives and hypothesis relating to the investigation **See Assessment activity 3.1**	M1 analyse the research information and discuss its relevance to the planned experiment **See Assessment activity 3.1**	D1 evaluate the different approaches considered for the investigation, justifying the hypothesis chosen **See Assessment activity 3.1**
P2 produce a list of relevant research resources using a recognised protocol for recording them **See Assessment activity 3.1**		
P3 produce a realistic working plan for the experiment including health and safety assessments **See Assessment activity 3.1**		
P4 demonstrate the required manipulative skills to assemble relevant equipment and materials **See Assessment activity 3.2**	M2 justify the choice of experimental techniques (and their modification if any) as a means of increasing accuracy, reliability and validity **See Assessment activity 3.2**	D2 evaluate the effectiveness of the investigative procedures, suggesting how these could be improved **See Assessment activity 3.2**
P5 safely carry out the planned investigation **See Assessment activity 3.2**	M3 justify the statistical techniques used by relating them to the validity of their findings **See Assessment activity 3.3**	
P6 demonstrate the ability to accurately record the results obtained, using scientific protocols **See Assessment activities 3.2 and 3.3**		
P7 analyse the results obtained using appropriate statistical techniques **See Assessment activity 3.3**		
P8 explain the conclusions gained from the investigation **See Assessment activity 3.4**	M4 justify the conclusions made, drawing on primary and secondary research data **See Assessment activity 3.4**	D3 using scientific protocols evaluate the outcomes of your investigation **See Assessment activity 3.4**
P9 present the conclusions in a format that uses accepted scientific protocol and language **See Assessment activity 3.4**		

58

Assessment

Your tutor will set **assignments** throughout your course for you to complete. These may take the form of projects where you research, plan, prepare, make and evaluate a piece of work or activity, case studies and presentations. The important thing is that you collect evidence of your skills and knowledge to date.

Stuck for ideas? Daunted by your first assignment? These learners have all been through it before…

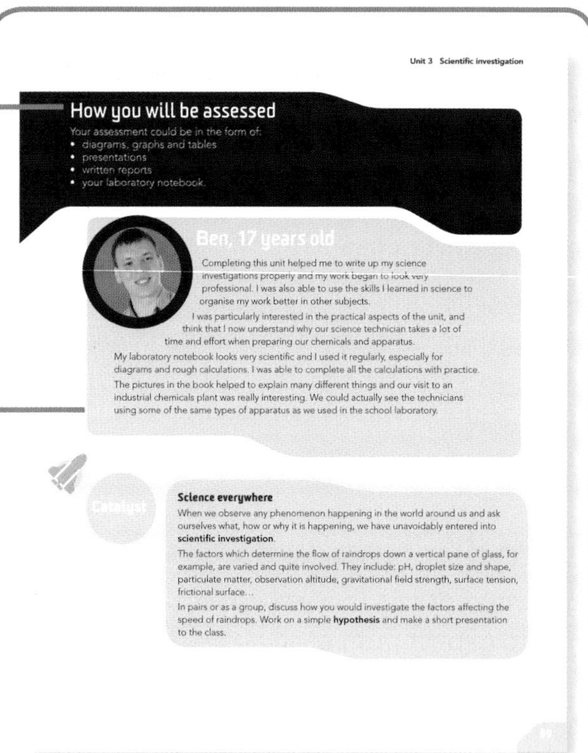

Unit 3 Scientific investigation

How you will be assessed

Your assessment could be in the form of:
- diagrams, graphs and tables
- presentations
- written reports
- your laboratory notebook.

Ben, 17 years old

Completing this unit helped me to write up my science investigations properly and my work began to look very professional. I was also able to use the skills I learned in science to organise my work better in other subjects.

I was particularly interested in the practical aspects of the unit, and think that I now understand why our science technician takes a lot of time and effort when preparing our chemicals and apparatus.

My laboratory notebook looks very scientific and I used it regularly, especially for diagrams and rough calculations. I was able to complete all the calculations with practice.

The pictures in the book helped to explain many different things and our visit to an industrial chemicals plant was really interesting. We could actually see the technicians using some of the same types of apparatus as we used in the school laboratory.

Science everywhere

When we observe any phenomenon happening in the world around us and ask ourselves what, how or why it is happening, we have unavoidably entered into **scientific investigation**.

The factors which determine the flow of raindrops down a vertical pane of glass, for example, are varied and quite involved. They include: pH, droplet size and shape, particulate matter, observation altitude, gravitational field strength, surface tension, frictional surface…

In pairs or as a group, discuss how you would investigate the factors affecting the speed of raindrops. Work on a simple **hypothesis** and make a short presentation to the class.

Activities

There are different types of activities for you to do: **assessment activities** are suggestions for tasks that you might do as part of your assignment and will help you develop your knowledge, skills and understanding. Each one has **grading tips** that clearly explain what you need to do in order to achieve a pass, merit or distinction grade.

Assessment activity 3.1

P1 M1 D1 P2 P3 BTEC

Your work as a junior research technician in the science department of a renowned national university allows you to develop your investigation procedures at every opportunity. You must set up an investigation and demonstrate your activity to visiting sixth formers from a local secondary school.

1 In your chosen investigation, state what you plan to achieve. P1 Use research notes, and list your notes in rough according to the suggested protocol ready to reproduce them for your final report. P2 Produce a concise hypothesis from your research. P1

2 Draw a table of your research references showing the information from each concerning your chosen topic of investigation. Analyse which ones are relevant to your work and which ones are not and discuss giving reasons. M1

3 Write out a method using information from this section. P3 Outline any other methods you may have considered and give the reasons to justify your final choice. D1

Grading tips

To achieve P1, P2 and M1 make sure you provide good points of reference and background information from many sources. List each source as you work through them.

To achieve P3 include descriptions of the main aspects covered under 'Method' above.

It is essential that you choose an investigation which has variations of approach so that you can evaluate them to achieve D1.

There are also suggestions for activities that will give you a broader grasp of the industry, stretch your imagination and deepen your skills.

Activity 3.1B

A student records the rate of transpiration of a sunflower over a 24-hour period. Identify the dependent and independent variables.

Key terms

Technical words and phrases are easy to spot. There is also a glossary at the back of the book with additional terms.

Key terms

Accuracy – closeness of readings to actual value.

Precision – the degree of uncertainty of a measurement; usually the size of the unit of measurement used.

Worked examples

Worked examples provide a clear idea of what is required for calculations.

Worked example

1 How many moles are there in 5 g of magnesium oxide?

2 How many moles are there in 10 g of sodium chloride?

(Hint: use the periodic table to check.)

Answers

M_r of MgO: 24.0 + 16.0 = 40.0 g mol^{-1}

moles in 5 g of MgO = $\frac{5}{40.0}$ = 0.125 moles of MgO

M_r of NaCl: 23.0 + 35.5 = 58.5 g mol^{-1}

moles in 10 g of NaCl = $\frac{10}{58.5}$ = 0.17 moles of NaCl

Personal, learning and thinking skills

Throughout your BTEC Level 3 National Applied Science course, there are lots of opportunities to develop your personal, learning and thinking skills. Look out for these as you progress.

PLTS

Self-manager

You will have the opportunity to practise this skill when organising laboratory time, dealing with the pressure of experimental procedures and time/equipment constraints, and seeking advice from your tutor.

Functional skills

It's important that you have good English, maths and ICT skills – you never know when you'll need them, and employers will be looking for evidence that you've got these skills too.

Functional skills

English

You will be developing your English skills when carrying out research (safety in the lab).

Did you know?

Where you see these boxes, look out for interesting snippets of information related to the science topics.

Did you know?

Many of Sir Isaac Newton's laboratory notes have still not been fully interpreted. Scientists of that time often used unusual jargon in their works and Newton also used his own unique symbols.

Case studies

Case studies show you examples of specific scientific topics applied in the workplace.

Case study: Risk assessment

David works as a production operative for a large chemical manufacturer. His work involves weighing substances with precision and accuracy and brings him into regular contact with dangerous chemicals which need very careful handling.

David is fully aware of the dangers in using these substances and how to dispose of them safely: 'Preparation is fundamental in preventing accidents and reducing risks. I attend many short courses on risk assessment and how to deal with incidents in and around the laboratory'.

WorkSpace

Case studies provide snapshots of real workplace issues, and show how the skills and knowledge you develop during your course can help you in your career. Where you see the STEM ambassadors logo, the person featured is a working scientist who is part of the Science, Technology, Engineering and Mathematics Network. See the STEMNET website for more information.

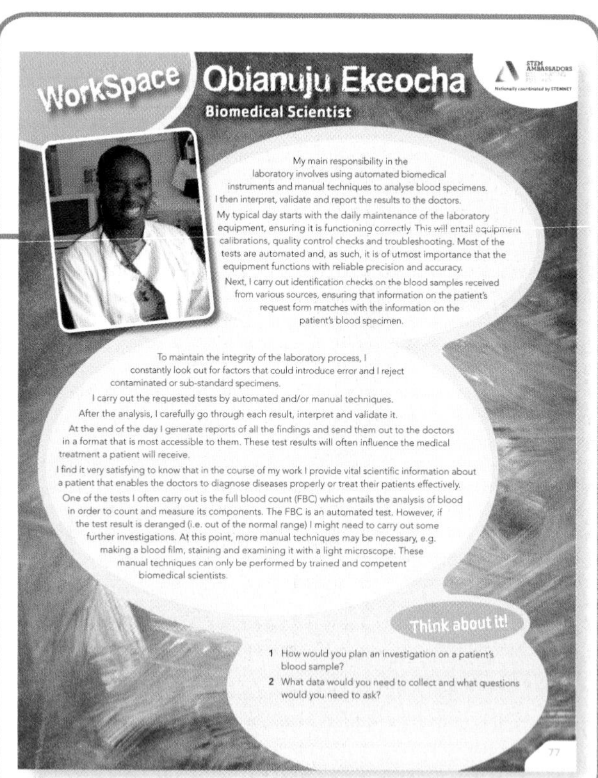

WorkSpace

Obianuju Ekeocha

Biomedical Scientist

STEM AMBASSADORS

Nationally coordinated by STEMNET

My main responsibility in the laboratory involves using automated biomedical instruments and manual techniques to analyse blood specimens. I then interpret, validate and report the results to the doctors.

My typical day starts with the daily maintenance of the laboratory equipment, ensuring it is functioning correctly. This will entail equipment calibrations, quality control checks and troubleshooting. Most of the tests are automated and, as such, it is of utmost importance that the equipment functions with reliable precision and accuracy.

Next, I carry out identification checks on the blood samples received from various sources, ensuring that information on the patient's request form matches with the information on the patient's blood specimen.

To maintain the integrity of the laboratory process, I constantly look out for factors that could introduce error and I reject contaminated or sub-standard specimens.

I carry out the requested tests by automated and/or manual techniques.

After the analysis, I carefully go through each result, interpret and validate it.

At the end of the day I generate reports of all the findings and send them out to the doctors in a format that is most accessible to them. These test results will often influence the medical treatment a patient will receive.

I find it very satisfying to know that in the course of my work I provide vital scientific information about a patient that enables the doctors to diagnose diseases properly or treat their patients effectively.

One of the tests I often carry out is the full blood count (FBC) which entails the analysis of blood in order to count and measure its components. The FBC is an automated test. However, if the test result is deranged (i.e. out of the normal range) I might need to carry out some further investigations. At this point, more manual techniques may be necessary, e.g. making a blood film, staining and examining it with a light microscope. These manual techniques can only be performed by trained and competent biomedical scientists.

Think about it!

1 How would you plan an investigation on a patient's blood sample?
2 What data would you need to collect and what questions would you need to ask?

77

Just checking

When you see this sort of activity, take stock! These quick activities and questions are there to check your knowledge. You can use them to see how much progress you've made or as a revision tool.

Edexcel's assignment tips

At the end of each unit, you'll find hints and tips to help you get the best mark you can, such as the best websites to go to, checklists to help you remember processes and useful facts and figures.

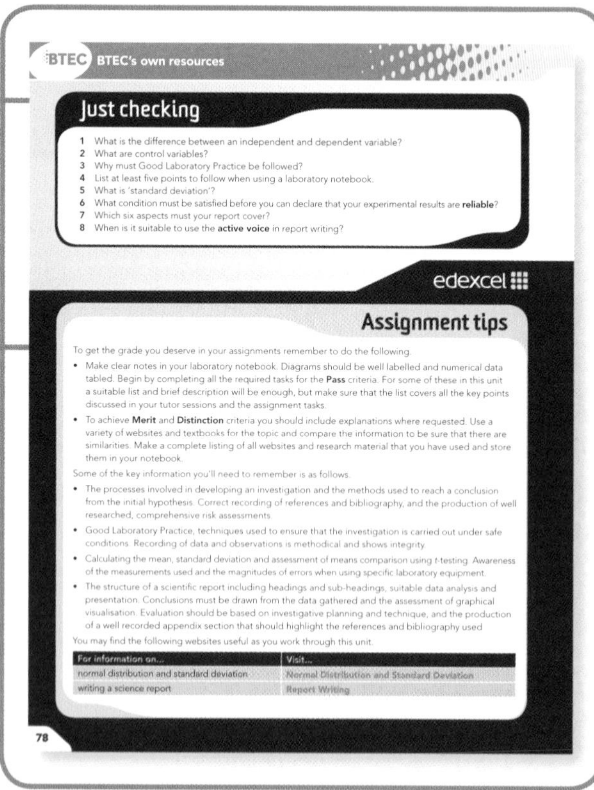

BTEC BTEC's own resources

Just checking

1 What is the difference between an independent and dependent variable?
2 What are control variables?
3 Why must Good Laboratory Practice be followed?
4 List at least five points to follow when using a laboratory notebook.
5 What is 'standard deviation'?
6 What condition must be satisfied before you can declare that your experimental results are **reliable**?
7 Which six aspects must your report cover?
8 When is it suitable to use the **active voice** in report writing?

edexcel

Assignment tips

To get the grade you deserve in your assignments remember to do the following.

- Make clear notes in your laboratory notebook. Diagrams should be well labelled and numerical data tabled. Begin by completing all the required tasks for the **Pass** criteria. For some of these in this unit a suitable list and brief description will be enough, but make sure that the list covers all the key points discussed in your tutor sessions and the assignment tasks.
- To achieve **Merit** and **Distinction** criteria you should include explanations where requested. Use a variety of websites and textbooks for the topic and compare the information to be sure that there are similarities. Make a complete listing of all websites and research material that you have used and store them in your notebook.

Some of the key information you'll need to remember is as follows.

- The processes involved in developing an investigation and the methods used to reach a conclusion from the initial hypothesis. Correct recording of references and bibliography, and the production of well researched, comprehensive risk assessments.
- Good Laboratory Practice, techniques used to ensure that the investigation is carried out under safe conditions. Recording of data and observations is methodical and shows integrity.
- Calculating the mean, standard deviation and assessment of means comparison using t-testing. Awareness of the measurements used and the magnitudes of errors when using specific laboratory equipment.
- The structure of a scientific report including headings and sub-headings, suitable data analysis and presentation. Conclusions must be drawn from the data gathered and the assessment of graphical visualisation. Evaluation should be based on investigative planning and technique, and the production of a well recorded appendix section that should highlight the references and bibliography used

You may find the following websites useful as you work through this unit.

For information on...	Visit...
normal distribution and standard deviation	Normal Distribution and Standard Deviation
writing a science report	Report Writing

78

Have you read your **BTEC Level 3 National Study Skills Guide**? It's full of advice on study skills, putting your assignments together and making the most of being a BTEC Applied Science student.

Your book is just part of the exciting resources from Edexcel to help you succeed in your BTEC course. Visit www. Edexcel.com/BTEC or www.pearsonfe.co.uk 2010 for more details.

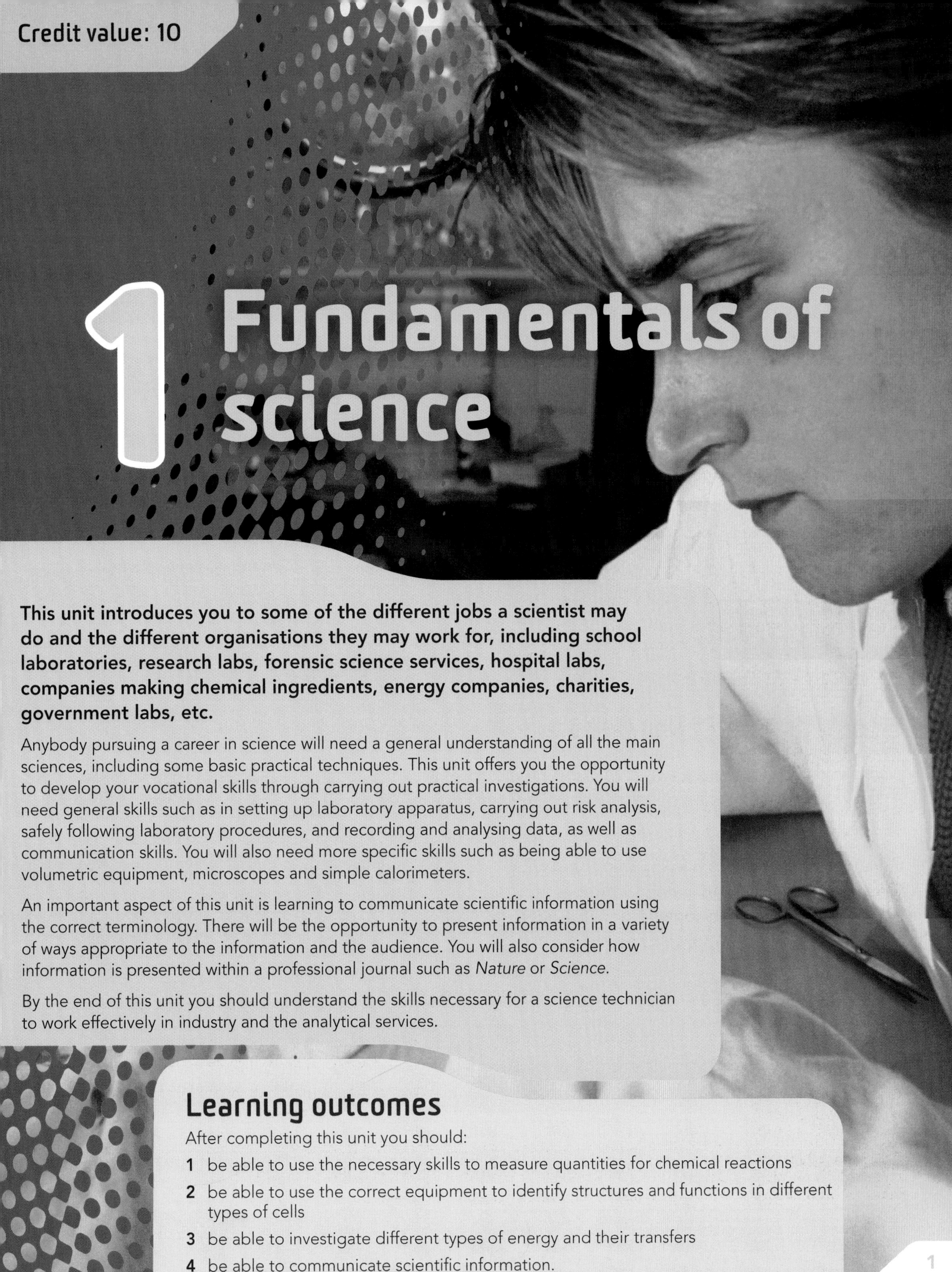

Credit value: 10

1 Fundamentals of science

This unit introduces you to some of the different jobs a scientist may do and the different organisations they may work for, including school laboratories, research labs, forensic science services, hospital labs, companies making chemical ingredients, energy companies, charities, government labs, etc.

Anybody pursuing a career in science will need a general understanding of all the main sciences, including some basic practical techniques. This unit offers you the opportunity to develop your vocational skills through carrying out practical investigations. You will need general skills such as in setting up laboratory apparatus, carrying out risk analysis, safely following laboratory procedures, and recording and analysing data, as well as communication skills. You will also need more specific skills such as being able to use volumetric equipment, microscopes and simple calorimeters.

An important aspect of this unit is learning to communicate scientific information using the correct terminology. There will be the opportunity to present information in a variety of ways appropriate to the information and the audience. You will also consider how information is presented within a professional journal such as *Nature* or *Science*.

By the end of this unit you should understand the skills necessary for a science technician to work effectively in industry and the analytical services.

Learning outcomes

After completing this unit you should:

1 be able to use the necessary skills to measure quantities for chemical reactions
2 be able to use the correct equipment to identify structures and functions in different types of cells
3 be able to investigate different types of energy and their transfers
4 be able to communicate scientific information.

Assessment and grading criteria

This table shows you what you must do in order to achieve a **pass, merit or distinction** grade, and where you can find activities in this book to help you.

To achieve a **pass** grade the evidence must show that you are able to:	To achieve a **merit** grade the evidence must show that, in addition to the pass criteria, you are able to:	To achieve a **distinction** grade the evidence must show that, in addition to the pass and merit criteria, you are able to:
P1 outline the key features of the periodic table, atomic structure and chemical bonding **See Assessment activity 1.1**	**M1** relate the key features of the periodic table to the conclusions drawn from the practical activities **See Assessment activity 1.1**	**D1** explain how standard solutions and titrations are prepared in industry **See Assessment activity 1.1**
P2 demonstrate practically the ability to prepare chemical solutions and test their accuracy **See Assessment activity 1.1**		
P3 record accurately observations of different types of tissues from a light microscope **See Assessment activity 1.2**	**M2** explain how the relative presence of different cell components influences the function of tissues **See Assessment activity 1.2**	**D2** compare different tissues with similar functions in terms of their structure and functions **See Assessment activity 1.2**
P4 interpret electron micrographs of different types of tissues **See Assessment activity 1.2**		
P5 describe the key structures and functions of a eukaryotic and prokaryotic cell **See Assessment activity 1.2**		
P6 describe different types of energy transfer **See Assessment activity 1.3**		
P7 carry out a practical investigation into the calorific value of different fuels **See Assessment activity 1.3**	**M3** carry out a practical demonstration of a range of energy interconversions with appropriate explanations of the systems investigated **See Assessment activity 1.3**	**D3** evaluate the efficiencies of energy conversion systems **See Assessment activity 1.3**
P8 outline the methods by which scientific information is communicated **See Assessment activity 1.4**	**M4** produce a detailed, correctly structured report which demonstrates a high level of presentation **See Assessment activity 1.4**	**D4** compare and contrast the report with a similar report from a professional journal **See Assessment activity 1.4**
P9 report on a scientific investigation that has been carried out **See Assessment activity 1.4**		

How you will be assessed

Your assessment could be in the form of:

- presentations
- case studies
- practical investigations
- practical demonstrations
- scientific reports
- scientific posters.

Nuala, 17 years old

I didn't realise how many different opportunities there were for working in science. This unit has helped me understand how I can use this course in different jobs. It helped me focus on some of the basic skills I am going to need when I start work.

I enjoyed looking at the different roles of a lab technician. It was good to explore the different skills and techniques needed to carry out that role. I also understand more clearly how scientists need basic chemistry, biology and physics skills. I can now apply these skills and knowledge to a range of different activities that will be useful in the course and in the world of work.

There were lots of practical tasks and activities for this unit so that made it more exciting for me. The bit I enjoyed most was investigating the energy value of fuels and looking at energy efficiency. I'm particularly concerned about saving energy and the environment so it was good to find out more about these things.

I found some of the maths hard, but reading the examples and following the methods in practice made it a lot less daunting.

Catalyst

Working as a laboratory technician

Laboratory technicians have many different roles and work in many different industries. They can be chemists, biologists, physicists or a mixture of all three. They may work in research and development or they may be involved in testing or quality control. They could work in industry or in a service area like a hospital laboratory. They could prepare chemicals and equipment for other scientists, and may help write reports or make presentations on their work.

Work in groups to think about the different industries and services a laboratory technician may work in. List all these industries and services. Discuss the roles and jobs they may carry out within each of these areas.

1.1 Measuring quantities for chemical reactions

In this section:    

Key terms

Periodicity – the repeating pattern seen in the periodic table of the elements.

Relative atomic mass – the ratio of the average mass per atom of an element to 1/12th of the mass of a carbon-12 atom.

Mass number – the number of protons plus the number of neutrons in the nucleus of an atom.

Ionic bonding – bonds formed by the transfer of electrons.

Covalent bonding – bonds formed by the sharing of electrons.

Relative molecular mass – the ratio of the average mass per molecule of an element or compound to 1/12th of the mass of a carbon-12 atom. It is equal to the sum of all the relative atomic masses of all the atoms in that molecule or compound.

Mole – a unit of substance equivalent to the number of atoms in 0.012 kg of carbon 12. One mole of a compound has a mass equal to its relative atomic mass expressed in grams.

Molar mass – the mass of one mole of a substance.

Standard solution – a solution of known concentration used in volumetric analysis.

Titration – a method of volumetric analysis used to calculate the concentration of a solution.

The periodic table

The periodic table shows all the chemical elements arranged in order of increasing atomic number. Chemists can use it to predict how elements will behave, or what the physical or chemical properties of the element may be. The repeating pattern of behaviour shown by the elements in the periodic table is called **periodicity**.

A laboratory technician needs to be very familiar with the periodic table. It is an information sheet on all the elements and their properties.

It was first published in 1869 by a Russian chemist called Dmitri Mendeleev. At the time, not all the elements we know of now had been discovered, but Mendeleev was still able to organise the periodic table by leaving gaps for the missing elements.

The elements in the periodic table are organised into groups (vertical columns) and periods (horizontal rows). Chemical properties are similar for elements in the same group. The atomic number increases as you move from left to right across a period. This is because each successive element has one more proton than the one before.

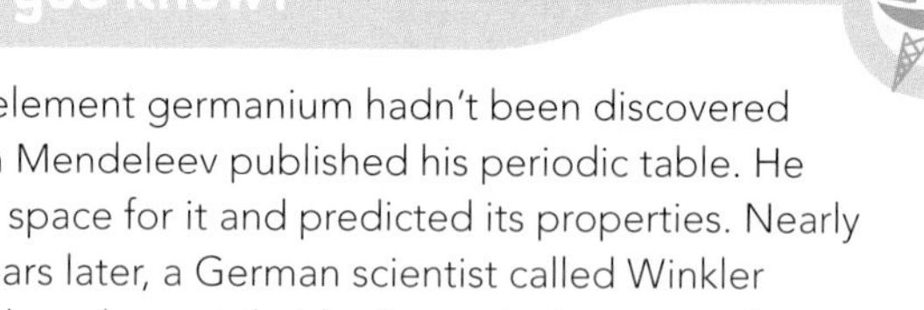

Did you know?

The element germanium hadn't been discovered when Mendeleev published his periodic table. He left a space for it and predicted its properties. Nearly 20 years later, a German scientist called Winkler found an element that had very similar properties to those predicted by Mendeleev and which fitted into the space left in the periodic table. He named it germanium after his country, Germany.

Activity 1.1A

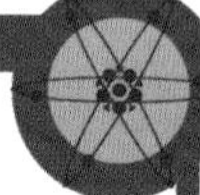

Look at the periodic table and write down five key features. You should consider what you learnt in Key Stage 4.

Work in pairs and try to list the names of as many groups in the periodic table as you can. Discuss any facts you know about the elements in the groups you have listed. These may be properties of the elements or trends within the groups. You may even be able to discuss the electronic structure of some of the elements.

PLTS

Team worker and Effective participator

Working in a group will help you develop these skills.

Trends in the periodic table

Elements in the periodic table are either metals or non-metals. There are more metals than non-metals, and they are all on the left-hand side of the table. So, by looking at the table, a lab technician can see if an element is a metal or a non-metal. This is just one

Key

relative atomic mass **atomic symbol** name atomic (proton) number

1	2											3	4	5	6	7	8
							1 **H** hydrogen 1										4 **He** helium 2
7 **Li** lithium 3	9 **Be** beryllium 4											11 **B** boron 5	12 **C** carbon 6	14 **N** nitrogen 7	16 **O** oxygen 8	19 **F** fluorine 9	20 **Ne** neon 10
23 **Na** sodium 11	24 **Mg** magnesium 12											27 **Al** aluminium 13	28 **Si** silicon 14	31 **P** phosphorus 15	32 **S** sulfur 16	35.5 **Cl** chlorine 17	40 **Ar** argon 18
39 **K** potassium 19	40 **Ca** calcium 20	45 **Sc** scandium 21	48 **Ti** titanium 22	51 **V** vanadium 23	52 **Cr** chromium 24	55 **Mn** manganese 25	56 **Fe** iron 26	59 **Co** cobalt 27	59 **Ni** nickel 28	64 **Cu** copper 29	65 **Zn** zinc 30	70 **Ga** gallium 31	73 **Ge** germanium 32	75 **As** arsenic 33	79 **Se** selenium 34	80 **Br** bromine 35	84 **Kr** krypton 36
85 **Rb** rubidium 37	88 **Sr** strontium 38	89 **Y** yttrium 39	91 **Zr** zirconium 40	93 **Nb** niobium 41	96 **Mo** molybdenum 42	[98] **Tc** technetium 43	101 **Ru** ruthenium 44	103 **Rh** rhodium 45	106 **Pd** palladium 46	108 **Ag** silver 47	112 **Cd** cadmium 48	115 **In** indium 49	119 **Sn** tin 50	122 **Sb** antimony 51	128 **Te** tellurium 52	127 **I** iodine 53	131 **Xe** xenon 54
133 **Cs** caesium 55	137 **Ba** barium 56	139 **La*** lanthanum 57	178 **Hf** hafnium 72	181 **Ta** tantalum 73	184 **W** tungsten 74	186 **Re** rhenium 75	190 **Os** osmium 76	192 **Ir** iridium 77	195 **Pt** platinum 78	197 **Au** gold 79	201 **Hg** mercury 80	204 **Tl** thallium 81	207 **Pb** lead 82	209 **Bi** bismuth 83	[209] **Po** polonium 84	[210] **At** astatine 85	[222] **Rn** radon 86
[223] **Fr** francium 87	[226] **Ra** radium 88	[227] **Ac*** actinium 89															

** The lanthanoids (atomic numbers 58–71) and the actinoids (atomic numbers 90-103) have been omitted.*

The periodic table of the elements.

example of the information the periodic table provides to help the lab technician.

Elements in the same group will behave in a similar way. For example, all Group 1 metals react with water to form an alkaline solution and hydrogen gas. As you go down a group the elements may get more or less reactive depending on which group it is.

Activity 1.1B

A lab technician reacted three Group 1 metals, lithium, sodium and potassium, with water. She made the following observations.

lithium	fizzed, moved slowly on surface of water
sodium	fizzed, moved quickly on surface of water
potassium	fizzed, moved quickly on surface of water, caught alight with a purple flame

1. Why did the lab technician not react caesium with water to make more observations on Group 1 metals?
2. What might she have observed if she had reacted caesium with water?
3. What can you say about the reactivity of the elements as you go down Group 1?

Atomic structure

The periodic table also tells you the atomic number and **relative atomic mass** for each element.

Atomic number = the number of protons in an atom; this is the same as the number of electrons in the atom.

Relative atomic mass = the ratio of the average mass per atom of an element to 1/12th of the mass of a carbon-12 atom.

When chemists write the full chemical symbol for an element they show the mass number (top) and the atomic number (bottom).

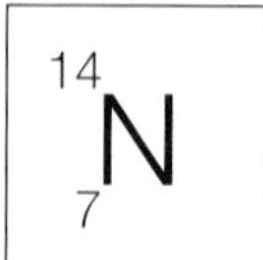

Nitrogen.

Mass number = the number of particles in the nucleus; this is the number of protons plus the number of neutrons.

If an element has only one isotope, its mass number and relative atomic mass will be the same value. If an element occurs as more than one isotope (atoms with different numbers of neutrons in the nucleus), the relative atomic mass will be different from the mass number.

From the periodic table we can see that nitrogen has an atomic number of 7 and a mass number of 14. This means nitrogen has seven protons, seven electrons and seven neutrons.

Activity 1.1C

Find the following elements in the periodic table:

lithium, copper, manganese.

Fill in the table to show the number of each type of particle in these elements.

	Number of protons	Number of electrons	Number of neutrons
lithium ^{7}Li			
copper ^{65}Cu			
manganese ^{55}Mn			

The protons and the neutrons are found in the nucleus at the centre of an atom. The electrons are in shells or energy levels surrounding the nucleus. Each shell can hold electrons up to a maximum number. When the first shell is full, electrons go into the second shell, and so on.

Electron shell	Max. no. of electrons
1	2
2	8
3	18
4	32
5	50

A sodium atom containing 11 electrons has an electron arrangement of 2, 8, 1. This can be represented by a Bohr diagram.

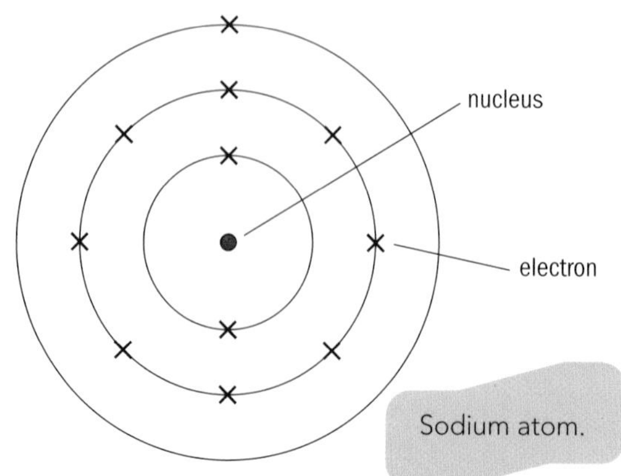

Sodium atom.

This is the simple version of electron structure you will have seen at Key Stage 4.

Under Bohr's theory, an electron's energy levels (also called electron shells) can be imagined as concentric circles around the nucleus. However, electrons within each shell will not have the same amount of energy, and so the energy levels or shells can be broken down into sub-shells called orbitals.

When writing out electron structures you should follow these rules.

- The electrons sit in orbitals within the shell. Each orbital can hold up to two electrons.
- The first shell can hold two electrons in an *s*-type orbital.
- The second shell consists of one *s*-type orbital and three *p*-type orbitals
- The third shell consists of one *s*-type orbital, three *p*-type orbitals and five *d*-type orbitals.
- Electrons fill the lowest energy level orbitals first.
- Where there are several orbitals of exactly the same energy, e.g. the three 2*p* orbitals in the second shell, then the electrons will occupy different orbitals wherever possible.

So the electronic structure of a sodium atom becomes:

$1s^2\ 2s^2\ 2p^2\ 2p^2\ 2p^2\ 3s^1$

Activity 1.1D

Complete the electronic structures for the elements in the table. Three have been done for you.

Element	No. of electrons	Electron structure
hydrogen	1	$1s^1$
helium		
lithium		
boron		
carbon	6	$1s^2\ 2s^2\ 2p^1\ 2p^1$
oxygen	8	$1s^2\ 2s^2\ 2p^2\ 2p^1\ 2p^1$
magnesium		
chlorine		
calcium		

Chemical bonding

One of the tasks of a lab technician is to make up solutions ready for experiments or for making products. Different types of compounds dissolve in different types of solvents depending on what type of bonding is in the compound. The lab technician must know what type of compound they are using in order to use the correct solvent.

Ionic bonding. The bonding in sodium chloride is ionic. This means that the sodium atom loses the electron in its outer shell to become the positively charged sodium ion, Na^+, with the same electron configuration as neon. Chlorine gains an electron to become the negatively charged chloride ion, Cl^-, with the same electron configuration as argon. This means that both the sodium ion and the chloride ion have a full outer shell and become stable.

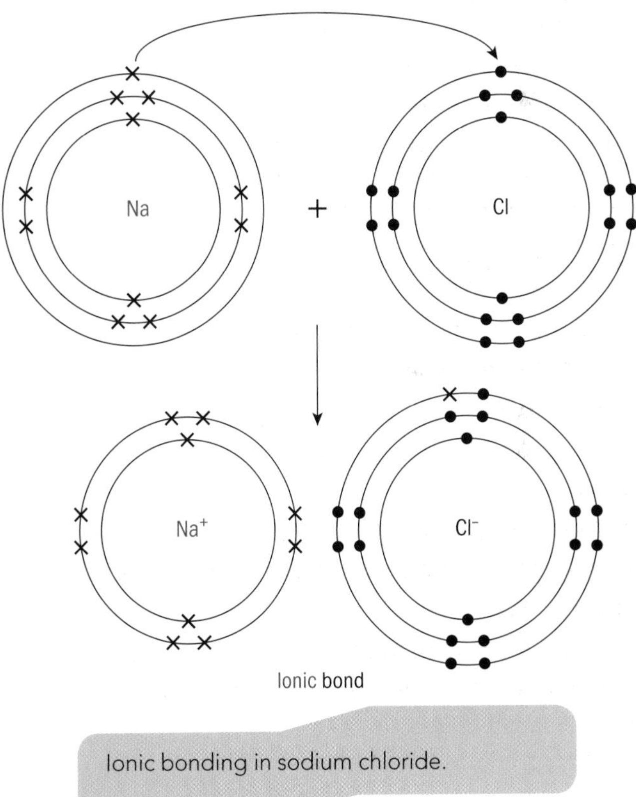

Ionic bonding in sodium chloride.

The positive charge on the sodium ion and the negative charge on the chloride ion are attracted and this is what holds the ions strongly together. The oppositely charged ions form a giant ionic lattice.

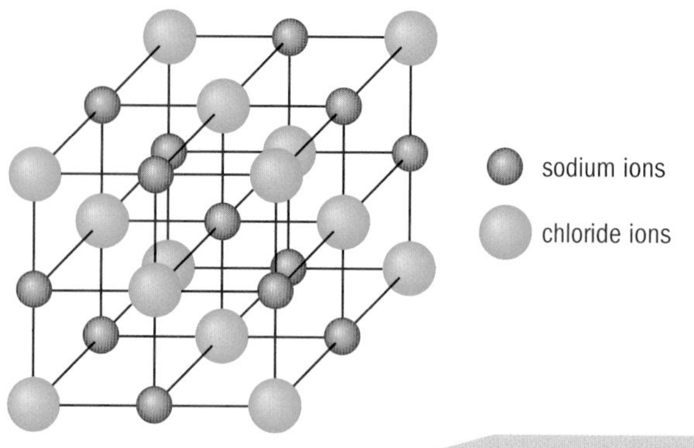

Lattice structure of sodium chloride.

Covalent bonding. Covalent bonding occurs between atoms of two non-metals. A covalent bond forms when an electron is shared between the atoms. These electrons come from the top energy level of the atoms.

A chlorine molecule has a covalent bond. The highest shell in each chlorine atom contains seven electrons. One electron from the highest shell in each atom is shared to give each chlorine atom the electron configuration of argon with a stable full outer shell.

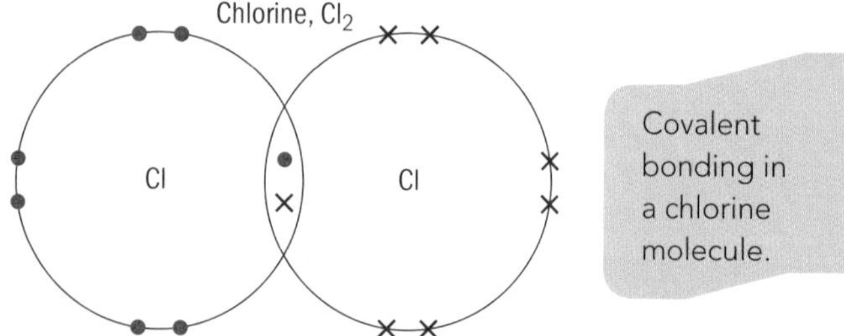

Covalent bonding in a chlorine molecule.

Activity 1.1E

Research ionic and covalent compounds in textbooks and on the Internet. Discuss with a partner the differences between ionic and covalent compounds.

You should consider:

- melting point
- boiling point
- strength of bonds
- physical state at room temperature
- solubility in water
- solubility in other solvents
- conductivity when in solution.

You may want to put your findings into a table.

Organic compounds also bond covalently. Carbon makes four covalent bonds.

Methane has the formula CH_4. Each carbon atom bonds covalently with four hydrogen atoms. The carbon gains the stable electron structure of neon and hydrogen gains the stable electron structure of helium.

These four bonds mean that methane has a tetrahedral structure. This is because the bonds are as separated from each other as possible with each bond angle being 109.5°.

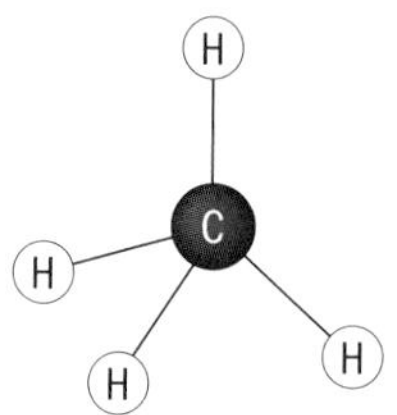

Tetrahedral structure of methane.

Activity 1.1F

1 Use molecular model kits to build models of the following organic compounds.

methane	CH_4
ethane	CH_3CH_3
propane	$CH_3CH_2CH_3$

2 What do you notice about the structure of these molecules?

3 Look at one of the carbons in each molecule and the atoms bonded to it. What do you notice about the shape?

Organic compounds with three or more carbons in a chain cannot be linear because of the tetrahedral structure around each central carbon.

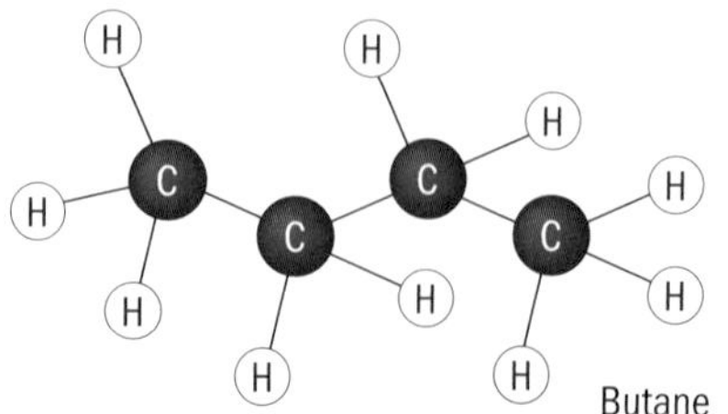

Structural diagram of a butane molecule.

Preparing chemical solutions and testing their accuracy

When carrying out experiments and investigations it is important that a chemist knows the concentration of the chemical solutions that they are using. The best way to make sure the solution is the correct concentration is to prepare a fresh solution and then test its accuracy by **titration**.

Formulas and balancing equations

All chemical reactions can be written as a balanced equation using the chemical formulas for the reactants and the products involved in the reaction. The symbols for the elements can be found in the periodic table and the numbers in the formulas show how many atoms of each element there are. Carbon dioxide has the formula CO_2 so it contains one carbon atom and two oxygen atoms. (Chemists don't write the 1 in a formula or an equation. If the symbol is present that means there must be at least one atom for that element.)

Carbon reacts with oxygen to make carbon dioxide.

The equation for this reaction is:

$C + O_2 \rightarrow CO_2$

The equation must balance like a maths equation. There should be the same number and types of atoms on both sides of the equation. There is one carbon and two oxygen atoms on both sides of this equation so it does balance.

A chemist needs to follow these steps to write a balanced equation.

1. Write the equation as a word equation including all the reactants and all the products:

   sodium hydroxide + hydrochloric acid → sodium chloride + water

2. Write out the formula for each substance in the reaction:

   $NaOH + HCl \rightarrow NaCl + H_2O$

3. Balance the equation, checking there is the same number of each element on both sides:

left-hand side	right-hand side
Na 1	Na 1
O 1	O 1
H 2	H 2
Cl 1	Cl 1

This equation is already balanced.

Worked example

Write a balanced equation for the following reaction.

$C_2H_5OH + O_2 \rightarrow CO_2 + H_2O$

1. Write out the number of atoms of each element on both sides:

left-hand side	right-hand side
C 2	C 1
H 6	H 2
O 3	O 3

2. Multiply each type of atom as necessary to make the two sides equal. After increasing the number of atoms by increasing the total number of molecules, make sure you count the atoms again.

   In this case, put a 2 in front of the carbon dioxide to equal out the carbons. This will also add two more oxygens to the right-hand side:

   $C_2H_5OH + O_2 \rightarrow 2CO_2 + H_2O$

left-hand side	right-hand side
C 2	C ~~1~~ 2
H 6	H 2
O 3	O ~~3~~ 5

Put a 3 in front of the water to balance the hydrogens. Remember to add to the oxygens again:

$C_2H_5OH + O_2 \rightarrow 2CO_2 + 3H_2O$

left-hand side	right-hand side
C 2	C ~~1~~ 2
H 6	H ~~2~~ 6
O 3	O ~~3~~ ~~5~~ 7

The carbons and hydrogens are now equal on both sides so you must multiply the oxygens on the left-hand side to finish balancing the equation:

$C_2H_5OH + 3O_2 \rightarrow 2CO_2 + 3H_2O$

left-hand side	right-hand side
C 2	C ~~1~~ 2
H 6	H ~~2~~ 6
O ~~3~~ 7	O ~~3~~ ~~5~~ 7

This equation is now balanced.

Activity 1.1G

Write out a balanced equation for each of the following word equations:

1 methane + oxygen → carbon dioxide + water
2 calcium carbonate + hydrochloric acid → calcium chloride + carbon dioxide + water
3 calcium hydroxide and hydrochloric acid → calcium chloride + water

Functional skills

Mathematics

Balancing these equations will improve your maths skills.

Quantities in reactions

Chemical equations allow us to work out the masses of the reactants we need to use in order to get a specific mass of product. Chemists never use one molecule of a substance because that would be too small. Even 0.1 g of hydrochloric acid will contain millions of molecules of the acid. These numbers are very big and difficult to work with so chemists use a quantity called a **mole** with the symbol mol. In the same way that we know that one dozen eggs means 12 eggs, chemists know that one mole of a chemical means that there are 6.023×10^{23} particles (the Avogadro constant). This is defined in reference to carbon 12 as follows.

'A mole is the amount of a substance which has the same number of particles as there are atoms in 12 g of carbon 12.'

So one mole of carbon dioxide has the same number of particles as one mole of gold.

The **molar mass** of a substance is equal to the mass of one mole of a substance. It is useful to be able to convert masses into moles and moles into masses. The relative atomic mass (A_r) of an element in the periodic table tells us how much mass there is in one mole of the element.

'Relative atomic mass is the average mass of an atom of an element compared to one twelfth of the mass of an atom of carbon 12.'

The relative atomic mass of hydrogen is 1.0. The relative atomic mass of oxygen is 16.0.

Molecules have a **relative molecular mass** (M_r).

'Relative molecular mass is the sum of all the relative atomic masses of all the atoms in a molecule.'

The relative molecular mass of water, H_2O, is $(1 \times 2) + 16 = 18$.

Relative atomic and molecular masses do not have any units as they are only relative to carbon 12.

Did you know?

The Avogadro constant is named after a 19th century Italian physicist. He was investigating gases and found that under the same conditions the same volume of gas always contains the same number of particles. Later it was found that 1 mole of a gas always contains 6.023×10^{23} particles.

Activity 1.1H

What is the relative molecular mass for these molecules?

1 CO_2
2 NaOH
3 $Ca(OH)_2$

Worked example

Masses are converted to moles as follows.

1 What is the number of moles in 136.5 g of potassium?

Number of moles of an element = mass ÷ A_r

For potassium, $A_r = 39.1$

Number of moles = 136.5 ÷ 39.1
= 3.5 mol

2 What is the number of moles in 20 g of sodium hydroxide, NaOH?

Number of moles of a molecule = mass ÷ M_r

For NaOH, $M_r = 23.0 + 16.0 + 1.0 = 40.0$

Number of moles = 20 ÷ 40.0
= 0.5 mol

Preparing standard solutions

When carrying out titrations a chemist has to use solutions of a known concentration. These are called **standard solutions**. They have been prepared and tested to ensure they are of the specific concentration needed.

The number of moles of **solute** in a given volume of **solvent** tells us how concentrated the solution is. When one mole of solute is dissolved in one cubic decimetre of solvent, its concentration is written as 1 mol dm^{-3}. This is written as 1 M for short, and is known as the molarity of the solution.

- One mole of HCl has a mass of 1 + 35.5 = 36.5 g
- 36.5 g of HCl in 1 dm^3 of solution has a concentration of 1 mol dm^{-3} or 1 M.

Did you know?

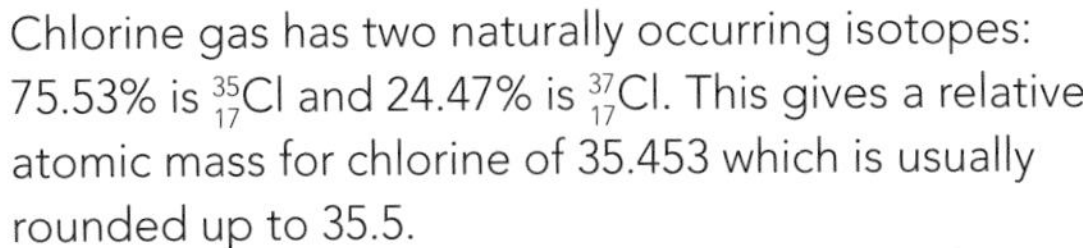

Chlorine gas has two naturally occurring isotopes: 75.53% is $^{35}_{17}Cl$ and 24.47% is $^{37}_{17}Cl$. This gives a relative atomic mass for chlorine of 35.453 which is usually rounded up to 35.5.

Worked example

How many moles of hydrochloric acid are there in 100 cm^3 of 1 M hydrochloric acid solution?

Number of moles = volume of solution in dm^3 × molarity

$$= \frac{100}{1000} \times 1$$

$$= 0.1 \text{ mol}$$

(remember 1 dm^3 = 1000 cm^3).

Assessing accuracy of prepared solutions

It is important that the accuracy of any solution is assessed in order to get accurate results from an investigation. This can be done with acids and bases using a procedure called titration. This involves reacting the prepared solution with another solution with a known concentration. An indicator is used that will change colour when the prepared solution has completely reacted. This is called the end point. The end point is reached when the acid has neutralised the base. You can then use the volumes reacted to calculate the concentration of the prepared solution.

Worked example

In a titration 25.0 cm^3 of 0.100 mol dm^{-3} sodium hydroxide solution is neutralised by 30.0 cm^3 hydrochloric acid. Calculate the concentration of the hydrochloric acid.

1 Find the number of moles of sodium hydroxide in 25 cm^3.

Moles of NaOH = volume (dm^3) × concentration (mol dm^{-3})

$$= 25.0 \times 10^{-3} \times 0.1$$

$$= 2.50 \times 10^{-3} \text{ mol}$$

2 Use a balanced equation to find out how many moles of hydrochloric acid react with this amount of sodium hydroxide.

$HCl(aq) + NaOH(aq) \rightarrow NaCl(aq) + H_2O(l)$

So, 1 mol of HCl reacts with 1 mol of NaOH. Therefore, 2.50×10^{-3} mol of HCl react with 2.50×10^{-3} mol NaOH as this is a 1 : 1 ratio.

3 Find the concentration of hydrochloric acid.

Concentration = mol dm^{-3} = mol ÷ dm^3

$$= \frac{2.50 \times 10^{-3}}{30.0 \times 10^{-3}}$$

$$= 0.083 \text{ mol dm}^{-3}$$

Assessment activity 1.1

1 The Royal Society of Chemistry produces posters to go on classroom walls that give key information to learners about different ideas in chemistry. You have been asked to design a poster on the key features of the periodic table. P1

The poster should include information on:

- periods
- groups (chemical and physical properties)
- trends
- atomic structure.

2 Ionic or covalent compounds can be used for making up standard solutions or carrying out titrations. These can include acids, alkalis and salts. It is important for a lab technician to know what type of compound is being used in order to choose the correct solvent.

Produce a leaflet to be used as a reference by other lab technicians that outlines the key features of ionic and covalent bonding using the following examples:

- one inorganic acid
- one organic acid
- one salt.

Add a safety note to show what the hazards are for each of the acids and the salt. P1

3 Companies that make cleaning products have to make sure the products are at the correct concentration. As a quality control assistant you need to make up a range of standard solutions to be used to check the concentrations of the cleaning products.

a Prepare a 1 M solution of sodium hydroxide (CORROSIVE) and a 2 M solution of sodium hydrogen sulfate (HARMFUL). Wear your eye protection and take care not to splash your skin with the solutions.

b Test the accuracy of the two solutions you have made. P2

c Write a report showing the calculations you have used to make and test the solutions.

d Explain how you used the periodic table in the calculations needed before you could prepare your standard solutions. M1

e Explain how these procedures are carried out in industry. D1

Extra safety note for question 3a:

If learners are to use solid sodium hydroxide to make the solution, rather than diluting a stronger molar solution, they will also need to wear goggles, handle the sodium hydroxide pellets with a spatula or similar and only add the pellets to the water a few at a time – stirring regularly – as so much heat is generated that the solution may boil. Never add the water to the pellets.

Grading tips

In question 1 to achieve P1 pick two or three groups to focus on to explain the key features. When discussing atomic structure use examples of elements to show electron arrangements.

In question 2 discuss the tetrahedral basis of the organic salt to ensure you have achieved P1. Use the information you have researched for previous activities to help you.

In question 3 to achieve M1 make sure you relate your work to the key features of the periodic table. Check all your calculations and that you are using the correct units.

PLTS

Independent enquirer, Self-manager and Reflective learner

Researching the information should show your skills as an independent enquirer. How you design your poster will help develop your skills as a self-manager. When you use your knowledge to write the report for M1 then you will be using and developing your skills as a reflective learner.

Functional skills

Mathematics and English

The calculations in this activity will improve your maths skills. Producing the poster, leaflet and writing the report will develop your English skills.

1.2 Identifying structures and functions in different cells

In this section:     M2 D2 P3 P4 P5

Key terms

Eukaryotic cell – a cell which contains a distinct membrane-bound nucleus. Most also contain membrane-bound organelles. They are found in plants, animals, fungi and protists.

Prokaryotic cell – a cell characterised by the absence of a membrane-bound nucleus or membrane-bound organelles. Bacterial cells are prokaryotic.

Organelle – a structure within a shell that performs a specific function.

Tissue – a group of cells that perform specific functions.

When someone is ill in hospital **tissue** samples are often sent to the lab in the cytology department to be analysed. A light (optical) microscope will be used to look at the cells in these tissues to see if they are healthy or if there is any disease. It is important that the lab technician can use the light microscope properly, can record what they see accurately and can also recognise what they see.

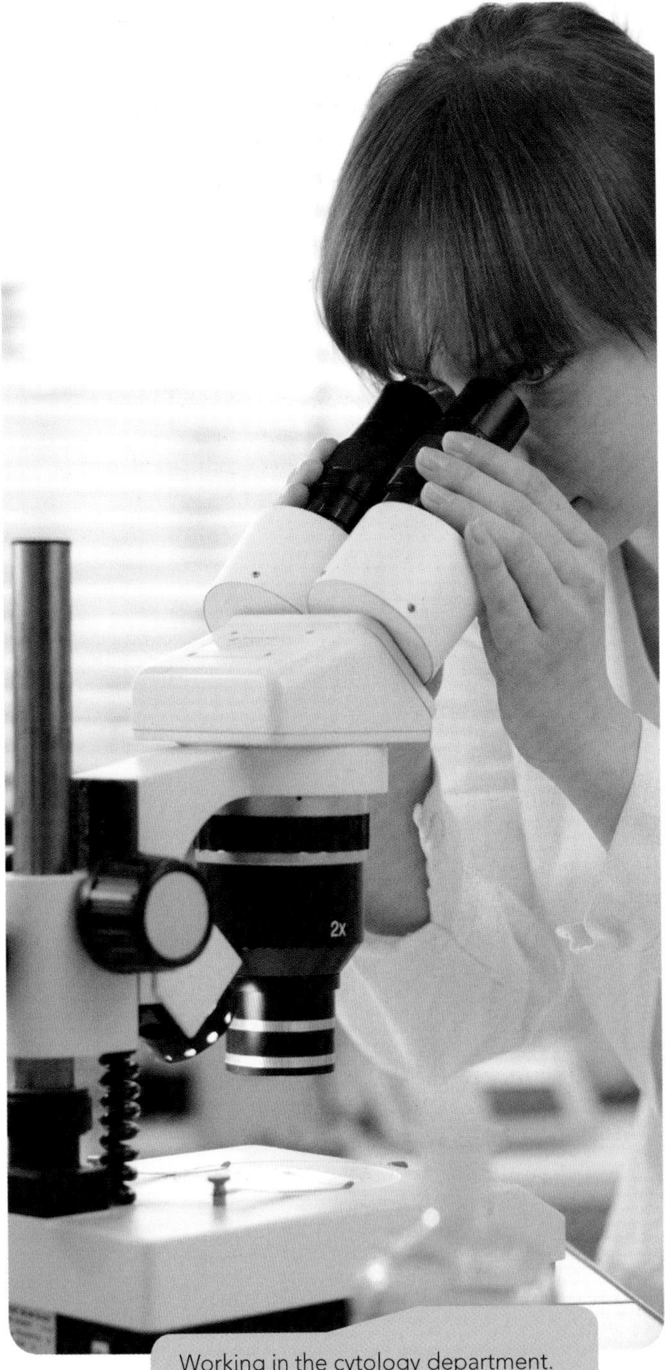

Working in the cytology department.

Did you know?

Early simple microscopes only had one lens and were really just magnifying glasses. Scientists found it interesting to look at insects and fleas through these lenses and so they were called 'flea glasses'.

Microscopic structures of cells

The technician will need to recognise different types of cells such as plant cells, animal cells and bacteria. A bacterial cell is **prokaryotic**. These cells have a simple construction with no nucleus or membrane-bound **organelles**. A cell from a plant or an animal is called **eukaryotic**. These cells are more complex with a nucleus and membrane-bound organelles.

Below are diagrams of a generic plant cell, animal cell and bacterial cell.

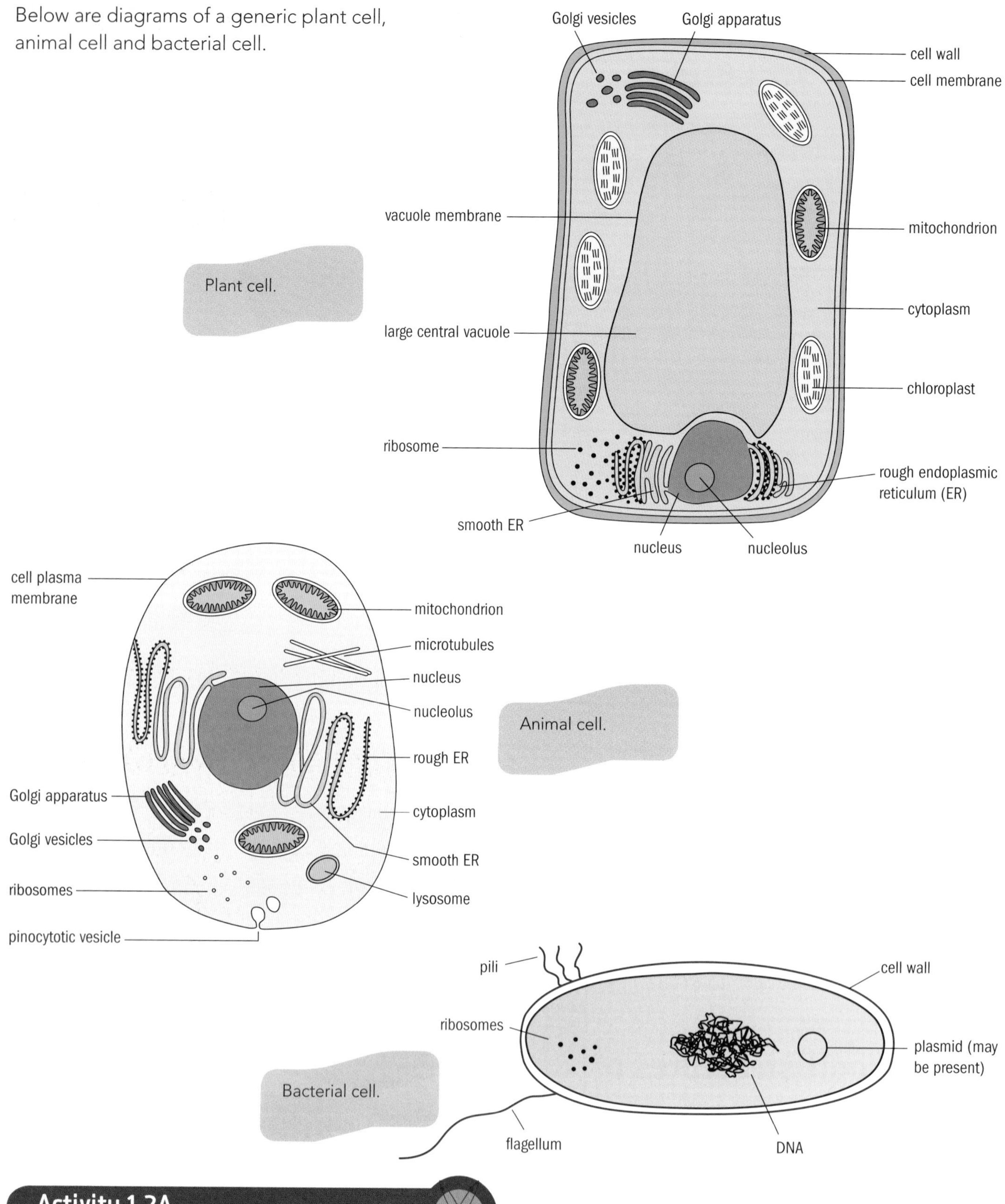

Plant cell.

Animal cell.

Bacterial cell.

Activity 1.2A

Look at the diagrams of the animal and plant cells. List the main features of each. What do they have in common? What are the differences?

Cell organelle structure and function

Technicians also need to be able to recognise the different organelles within the cell as well as the type of tissue the cell comes from. An organelle is a specialised subunit within the cell that has a specific function. The organelles can be seen, using light and electron microscopy.

Organelle	Function
cell membrane	• serves as a boundary between the outside environment and the inside of the cell • gives form and shape to the cells • connects one cell to two or more adjacent cell(s) • may serve as an organ of locomotion like that of an amoeba, a protozoan that utilises its membrane as pseudopodia or false feet for movement
cell wall	• protects and supports
nucleus	• regulates cell activity
nucleolus	• instructions in DNA are copied here • works with ribosomes in the synthesis of protein
cytoplasm	• maintains cell shape • stores chemicals needed by the cell
mitochondria	• site of cellular respiration
ribosome	• site of protein synthesis
endoplasmic reticulum (ER)	• transports chemicals between cells and within cells • provides a large surface area for the organisation of chemical reactions and synthesis • there are two types of ER: rough ER is covered with ribosomes and this is where proteins are assembled; smooth ER does not have ribosomes and this is where steroids and lipids are assembled
Golgi body	• modifies chemicals to make them functional • secretes chemicals in tiny vesicles • stores chemicals
lysosome	• breaks large molecules into small molecules
vesicles	• for transportation in and out of and within the cell

Unit 11: See page 199 for more information on organelles in eukaryotic cells.

Activity 1.2B

In pairs find a diagram of a cell in a textbook or on the Internet. Can you recognise the following organelles in the diagram:

cell membrane, cell wall, nucleus, nucleolus, cytoplasm, mitochondria, ribosome, endoplasmic reticulum, Golgi body, lysosome, vesicles?

PLTS

Independent enquirer

Using the Internet to research will develop your skills as an independent enquirer.

Tissues and their functions

Cells are grouped together to form tissues.

Tissue	Function/location
epithelial	• covers surface of the body • lines body cavities and major organs • filters, absorbs and diffuses various substances • sensory perception • bodily secretions • skin
connective	• structure • support • connects two types of tissue to each other
nerve	• carries information to and from the brain • controls how the body works • central nervous system • peripheral nervous system
muscular	• movement

Unit 11: See page 201 for more information on tissues and their functions.

Activity 1.2C

You are working in a hospital laboratory in the cytology department. This week, you have two work experience students with you. You decide to produce a leaflet for the students to help them recognise different tissues. The leaflet must include:

- a diagram of the structure of a type of cell that makes up each of the tissue types
- a description of the tissue and its function.

Using electron micrographs

Sometimes it is difficult to get enough detail from looking at a slide under a light microscope. In this case a cytologist may use an electron microscope, which uses a beam of electrons to produce an image on a fluorescent screen. The image produced is called an electron micrograph.

An electron microscope is much more powerful than a light microscope and can magnify objects up to half a million times bigger. Think about this in comparison to how much the light microscope you have used will magnify an object.

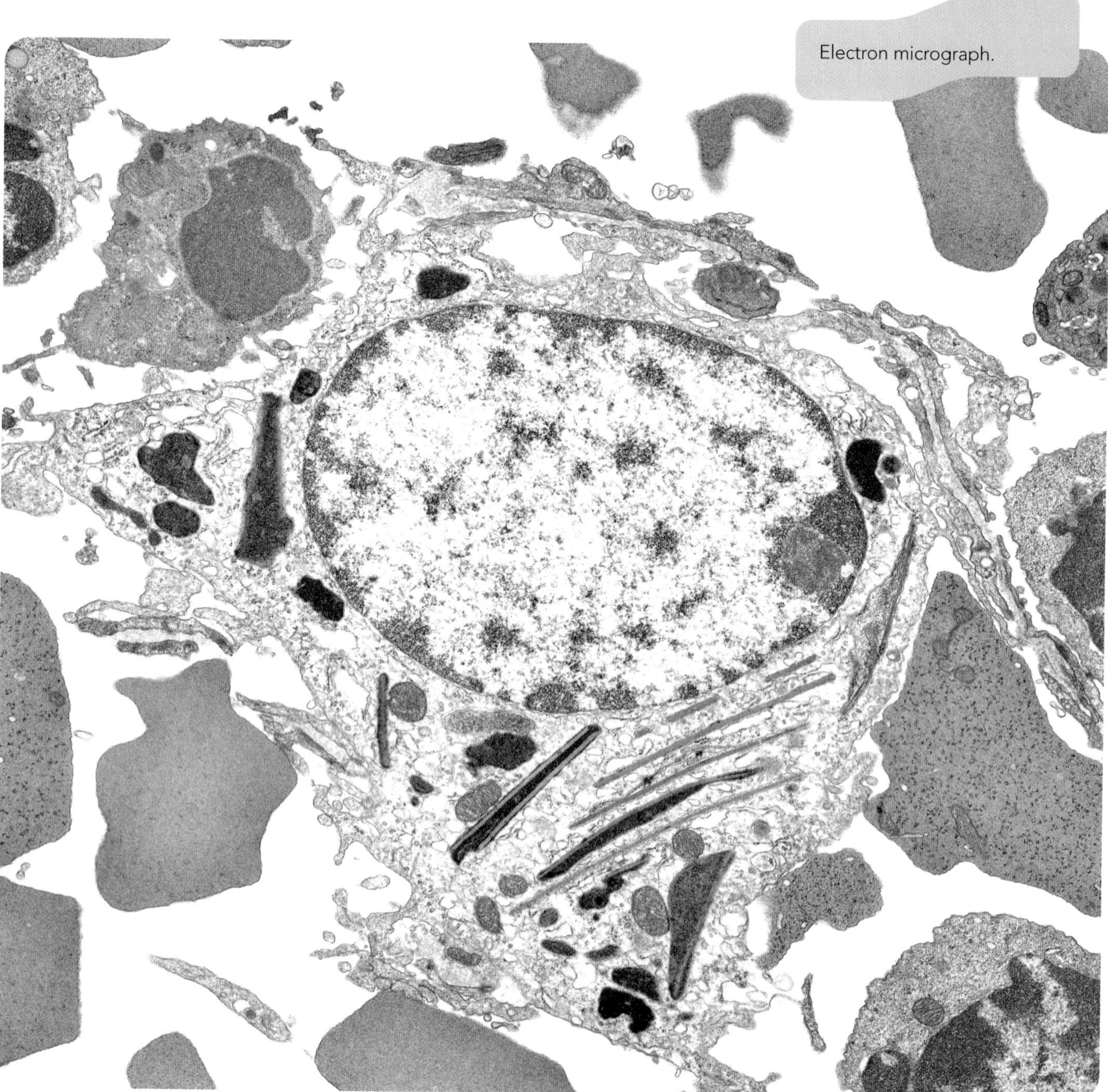

Electron micrograph.

Case study: Hospital technician

Raul has just started working in the cytology lab in his local hospital. Samples of blood and tissues come to the lab to be analysed for diseases or other problems. Raul has to check blood samples for unusual numbers of white blood cells. This can show that there may be an infection present or even a disease such as cancer. He uses a light microscope for his initial checks.

1 What other equipment might Raul use to examine the blood samples?

2 Why do you think it would be important to check his results more than once?

Assessment activity 1.2

BTEC

1 You are working in the cytology department of a hospital and you have been given some slides of different types of cells and tissues. By mistake, the slides have not been labelled.

   a Look at each slide given to you by your tutor using a microscope fitted with its own light source.

   b Draw and label what you see on each slide.

   c Write a label for each slide so that it is clear what type of cell or tissue each has on it. P3

   Choose one slide containing a eukaryotic cell and one containing a prokaryotic cell. Describe the key structures and functions for each of these. P5

   Write a report for your supervisor explaining the function of the cells on each slide. M2 Include in your report a comparison of some of the tissues that have similar functions. D2

   Safety note: If you have to use a microscope with a mirror that uses daylight illumination, never use it where the Sun's rays can hit the mirror directly, as the magnification will cause the light to permanently damage your retina. Choose a spot near a window that is not in direct sunlight.

2 You have been given some electron micrographs to interpret. Label the cells and organelles on the micrograph. Write a report describing what each micrograph is showing. P4

Grading tips

To achieve P3 remember to draw what you see, not what you expect to see. You may need to look at your answer to Activity 1.2B to help you label the organelles.

To achieve M2 you need to consider the different organelles to help you decide the function of the cell.

PLTS

Self-manager and Independent enquirer

In answering question 1, you will be able to show some self-manager skills if you manage your work carefully. Interpreting electron micrographs for question 2 will help develop your skills as an independent enquirer.

Functional skills

English

Writing the reports will help develop your English skills.

1.3 Transfer of different types of energy

In this section:    M3 D3 P6 P7

Key terms

Kinetic energy (KE) – amount of energy an object has due to its motion.

Potential energy (PE) – energy stored by an object.

Metabolism – the chemical reactions in a living organism that maintain life.

Heat capacity – the amount of heat required to raise the temperature of a substance by a given amount.

Calorimeter – a device used to measure the quantity of heat flow.

Energy and how we use it and waste it is often in the news. With more people concerned about global warming and how to save energy, it has become even more important for us to understand how energy is used.

Did you know?

There are two types of kinetic energy, internal and external. When an object is moving it has kinetic energy, but even when it is still the atoms inside it will be vibrating. This is called internal kinetic energy.

The main types of energy we need to consider are as follows.

Mechanical energy – the sum of a system's kinetic and potential energy. For example, a swinging pendulum has **kinetic energy** due to its movement and **potential energy** due to its position. Its potential energy will be greatest at the top of its swing and its kinetic energy will be greatest at the lowest point of its swing.

Chemical energy – the potential energy stored in chemical bonds. It is released due to a chemical reaction. When sodium reacts with chlorine to make sodium chloride (table salt) the chemical energy in the bonds is released as heat.

Thermal energy – total internal kinetic energy of an object due to the random motion of its particles. Heat transfer is the flow of thermal energy from one object to another. A hot iron has a lot of thermal energy and if you touch it the thermal energy will flow to your hand making it feel hot.

Electrical energy – the presence and flow of electric charge. For example, electrical energy flows though conducting wires.

Nuclear energy – the potential energy stored in the nucleus.

Light and **sound** are also types of energy.

Activity 1.3A

Work in groups to discuss the main types of energy.

- What produces these types of energy?
- What is the energy used for?
- How can the energy be transferred from one type to another?
- What are the units used for each type of energy?

Produce a poster or a leaflet explaining the different energy types.

PLTS

Team worker and Effective participator

You will develop these skills by taking part in this group activity.

Types of energy transfer

Energy exists in many different forms. For example, energy is stored as chemical energy in fuels, food and batteries, and can be converted into heat, light and sound.

Did you know?

Energy is usually measured in joules. Energy in food is often measured in calories or kilocalories.

Energy cannot be created or destroyed; it can just be converted from one form to another. So when you watch television all the electrical energy is converted into heat, light and sound. This is called the law of conservation of energy.

Even though none of the energy is destroyed, some of it is converted into energy that is not useful. In the case of the television, the light and sound energy is useful, but the heat energy is wasted. An efficient television is one where most of the energy is converted to light and sound and only a little to heat. Such energy transfer can be shown using diagrams.

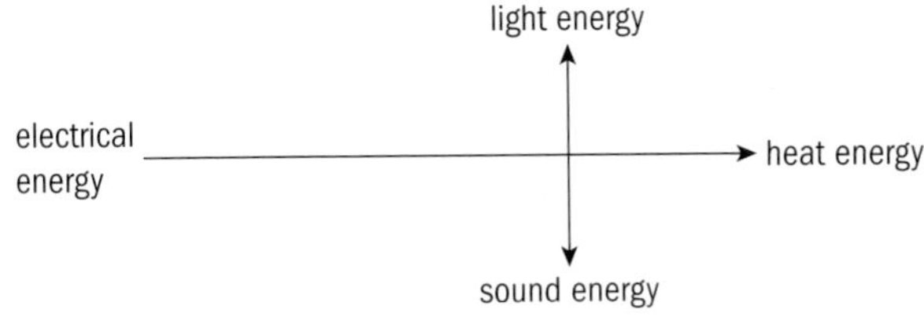

Energy transfer in a television.

This diagram shows the types of energy transfer that occur when a television is used. A Sankey diagram can be used to give more information. The widths of the arrows are proportional to the amounts of energy, so you can see how much energy is wasted and how much is useful.

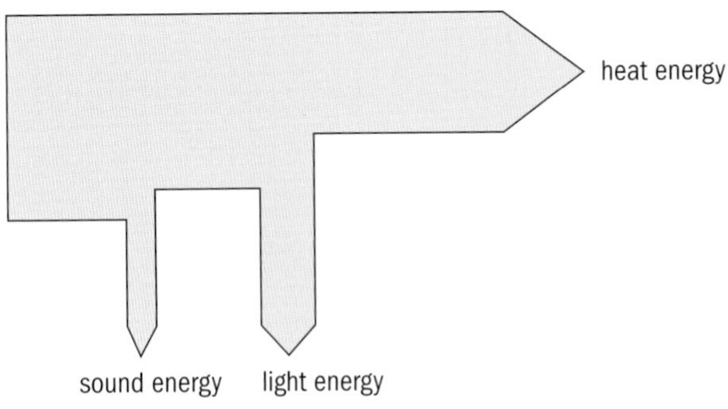

Sankey diagram showing energy transfer in a television.

Activity 1.3B

You work for a company that makes household appliances. The company wants to make their appliances as energy efficient as possible. In order to do this they need to understand the energy transfers that happen when each appliance is used.

For each of these appliances draw a diagram to show the energy transfers that occur when the appliance is used:

- gas hob
- computer
- food mixer
- washing machine.

Make sure you show which is useful and which is wasted energy. Now draw a Sankey diagram for two of the appliances.

PLTS

Independent enquirer

The research you carry out for this activity will help you develop this skill.

We all use a large amount of electrical energy in our homes. This can be produced in a variety of ways. However, in the UK most electrical energy is produced by burning fossil fuels.

Fossil fuels contain chemical energy. When they burn they give off heat and light energy. The heat energy is used to heat water. The water turns to steam and has kinetic energy. The steam moves a turbine. This is also kinetic (mechanical) energy. The movement of the turbine powers a generator. This produces electrical energy.

Hydroelectric power stations convert the potential energy of water stored high in a reservoir into kinetic energy as the water flows out through the dam. This kinetic energy then turns the turbine.

A nuclear power station uses nuclear energy in the form of uranium as its primary energy source. This undergoes nuclear fission that generates the thermal energy required to heat water in a manner similar to a conventional power station.

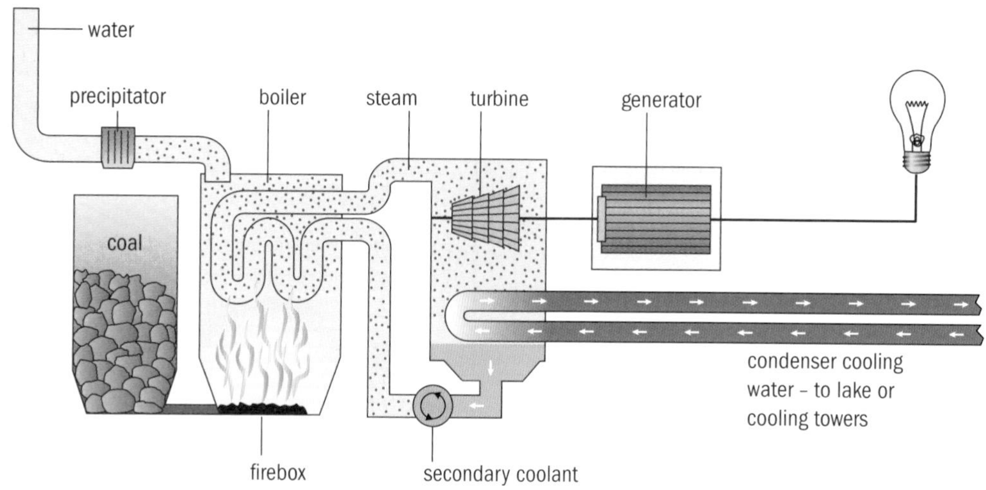

Using fossil fuel to produce electricity.

Activity 1.3C

Draw an energy diagram showing all the energy interconversions when a fossil fuel is burned to produce electricity. Label useful and wasted energy.

We also need energy in our bodies to keep us warm and do work. This energy comes from the food we eat. The energy in the food is transferred in our bodies through a process called **metabolism**: chemical reactions break down complex molecules to release energy and to use energy to build up other complex molecules. So when we eat, the molecules in the food are broken down to release energy that we can then use to grow, respire, move, etc.

Measuring the calorific value of different fuels

A **calorimeter** is a device used to measure heat of reaction. In industry a sophisticated and expensive calorimeter is often used. In the school lab you might use a Styrofoam cup as a calorimeter, because styrofoam is a good insulator that prevents heat exchange with the environment. In order to measure the heat of reaction, the reaction is carried out in the Styrofoam cup and the temperature difference before and after the reaction is measured. The temperature difference enables us to evaluate the heat released in the reaction.

A beaker of water could also be used as a calorimeter.

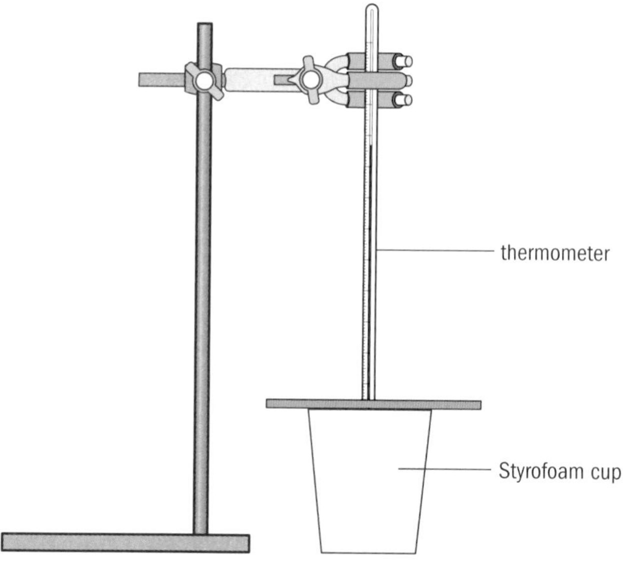

Styrofoam calorimeter.

We first need to know the **heat capacity** of the calorimeter. The heat capacity is the amount of heat required to raise the temperature of the entire calorimeter by 1 K, and it is usually determined experimentally before or after the actual measurements of heat of reaction.

A bomb calorimeter is sometimes used in industrial laboratories.

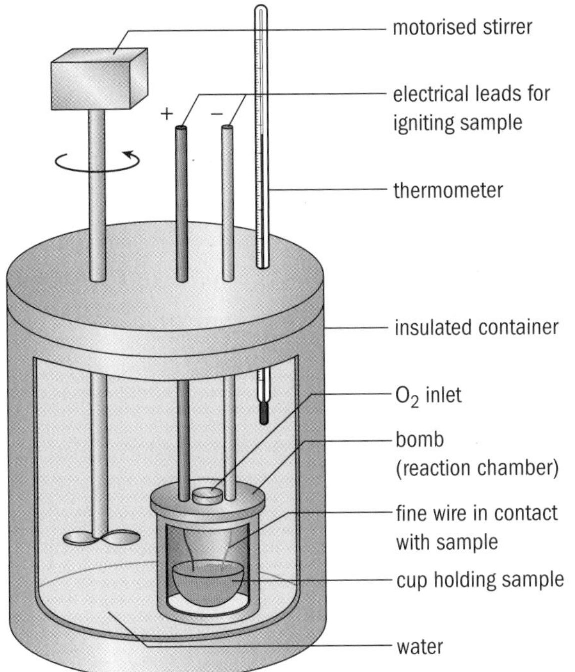

Bomb calorimeter.

Activity 1.3D

Use the Internet and textbooks to research how a bomb calorimeter is used. The diagram will also help you. Discuss your findings with a partner.

PLTS

Independent enquirer

The research you carry out for this activity will help you develop this skill.

Different fuels will contain different amounts of energy. The best fuels will give off a large amount of energy very easily. An energy company needs to know which fuel will give off the most energy.

If you burn a known amount of a fuel under a known volume of water and the temperature change of the water is measured, then you can work out how much energy the fuel has produced. The amount of heat energy (Q) gained or lost by a substance is equal to the mass of the substance (*m*) multiplied by its specific heat capacity (*c*) multiplied by the change in temperature (final temperature – initial temperature):

$$Q = m \times c \times (T_f - T_i)$$

The specific heat capacity of water is $4.18\,\text{J}\,\text{g}^{-1}\,\text{K}^{-1}$.

Worked example

Ethanol, 3 g, was burnt under 300 g of water. The temperature of the water increased from 20 °C to 77 °C.

Calculate how much energy was gained by the water. Use this answer to calculate how much energy was provided by 1 g of ethanol.

$Q = m \times c \times (T_f - T_i)$

$m = 300\,\text{g}$

$c = 4.18\,\text{J}\,\text{g}^{-1}\,\text{K}^{-1}$

$T_f = 87\,°\text{C}$

$T_i = 20\,°\text{C}$

$Q = 300 \times 4.18 \times (77 - 20)$

$Q = 300 \times 4.18 \times 57$

$Q = 71478\,\text{J} = 71\,\text{kJ}$

Three grams of ethanol produce 71 kJ. Therefore, 1 g of ethanol produces 71 ÷ 3 = 24 kJ of energy.

Published data on ethanol show that it has an energy content of $29\,\text{kJ}\,\text{g}^{-1}$. What reasons could there be to explain why the value found experimentally is lower than the published value?

Calculating the efficiencies of energy conversion systems

It is important that when energy is converted in a system as much of the energy it is converted to is useful and as little as possible is wasted. The efficiency of a device can be calculated using the following equation:

$$\text{efficiency} = \frac{\text{useful energy output}}{\text{total enegy input}}$$

To calculate % efficiency the following equation is used:

$$\%\ \text{efficiency} = \frac{\text{useful energy output}}{\text{total enegy input}} \times 100$$

Assessment activity 1.3

1 You are working for an energy company and are investigating the energy transfers for different sources of energy.

Produce a presentation describing the energy transfers that occur when the following sources are used to provide energy to our homes:

- solar power
- coal
- wind power
- tidal power
- nuclear power. P6

Carry out a practical demonstration of a range of energy interconversions to show your supervisor. Explain each of the systems you are demonstrating. If fuels are used in your practical demonstration, you will need to read the safety note for question 3. M3

2 You work as a nutritionist at a health spa. You want to explain to the guests at the spa how they get energy from food to help them understand how to have a healthy diet.

Produce a leaflet describing how food is metabolised in our bodies showing the energy transfers that take place. Include a list of low energy foods that might be good to eat when trying to lose weight. P6

3 You are working for an energy company and need to investigate a selection of fuels to see which one gives out the most energy per gram.

Carry out an investigation into a range of fuels. Use your results to calculate how much energy each fuel provides. Write a report for your supervisor explaining which fuel provides the most energy per gram, using your results and calculations to back up your conclusions. P7

Safety note: Many fuels are highly flammable and small quantities will need to be used in a controlled way. Your tutor will need to check your plan and may not allow you to use certain fuels. You will need to wear eye protection.

4 You work for a company that makes household appliances. The company wants to make their appliances as energy efficient as possible. In order to do this you must calculate the energy efficiencies of the appliances.

Investigate different household appliances and work out their efficiency and their percentage efficiency. You may compare different types of appliance or you may compare the same appliance made by different manufacturers.

Write a report evaluating the appliances in terms of efficiency. D3

Grading tips

In question 1 to achieve P6 it is useful to explain which is useful energy and which is wasted. You should also give the units of the energy types. You could draw a normal energy diagram, but remember that a Sankey diagram will give more information. To achieve M3 add information to your presentation to explain the energy systems you have demonstrated.

In question 2 to achieve P6 make sure you give the energy content of the foods with the correct units.

In question 3 to achieve P7 put the fuels in order of how much energy they provide. Make sure you use the right units in the calculations.

In question 4 to achieve D3 consider what the purpose of the appliance is and how you know it is efficient.

PLTS

Independent enquirer and Reflective learner

The research you carry out for questions 2, 3 and 4 will help you develop your skills as an independent enquirer. In question 4 you will also practise your reflective learner skills.

Functional skills

ICT, Mathematics and English

Preparing a presentation will help you develop your skills in ICT. Questions 3 and 4 will develop your maths and English skills.

1.4 Communicating scientific information

In this section:

Key terms

Abstract – a brief summary of a report or paper.

Bibliography – a list of books and sometimes other resources that have been used as references when writing a report or article.

Audience – the type of people you expect to read a report, watch a presentation, etc.

Journal – a periodical that publishes articles and reports in a particular area or field, and is read by other professionals in that field

Did you know?

On April Fools' day in 1957 the BBC showed a news report on how spaghetti was grown on trees. Spaghetti was not a common food in Britain in the '50s and so most people did not know much about it. Many people believed the report, as it was presented in a scientific manner. This shows how important it is to question scientific reports, to ask for evidence and to look at the references used.

At the moment most of the scientific information you learn comes from your tutors and textbooks. People who work in the science industry will probably still use textbooks but are unlikely to have a tutor. They need to make sure they are up to date with relevant scientific developments and so must get their information from other sources.

Activity 1.4A

Work in groups to make a list of all the ways in which scientific information is communicated to you. For each method discuss how useful it is, and how valid it is.

- Is the method reliable?
- Is the method prone to bias?
- Which method do you think is best?

PLTS

Team worker and Effective participator

This activity will help you to develop these skills.

Methods by which scientific information is communicated

The method by which scientific information is communicated is very important. It must be accessible to the **audience**, clear to understand and in a suitable format. Often scientists will expect to see a report written using scientific terminology and structured in a specific way. Think of your practical write-ups, which should always include an aim, a method, results and a conclusion, in that order. The conclusion cannot come before the results as that would not make sense.

Activity 1.4B

Visit a science museum to study how they communicate scientific information.

1. Who is the audience?
2. What type of information is being communicated?
3. What is the purpose of the communication?
4. What methods are used to communicate the information? Is there just one method or a variety?
5. Why are the different methods used?
6. Which methods work the best?

PLTS

Reflective learner

Answering the questions in this activity will develop this skill.

Scientific information is not always communicated by scientists. Often it comes from sources on the Internet, and sometimes we get information from posters, presentations, newspapers, or the TV or radio. Sometimes it comes from conversations with friends or family. What do you think about scientific communication that comes from these sources?

Activity 1.4C

Find some articles in newspapers that are communicating scientific information. In groups, discuss the articles.

1 How has the information been presented?
2 Do you think the information is correct and reliable?
3 Why do you think the newspaper printed an article on these topics?
4 Research the topic on the Internet and in textbooks; how was the information presented there?
5 Did your research agree with the article?

If there was any inaccurate or incorrect science in the articles, discuss why you think this might have happened.

Science in the news.

The correct structure of a detailed scientific report

When a scientist has made a new discovery or investigated and found evidence for a new theory they need to share their findings with the rest of the scientific community. They will write a report and send it to a scientific **journal** to be published. Most journals would expect the report to be written in a standard format. This could include the following sections: title, **abstract**, introduction, method, results, accuracy, discussion, conclusions, references, **bibliography**.

This makes the report easier to read and understand. If the report is very long it also means that other scientists reading it will know where to find the summary and conclusions, and other main points fairly easily.

Records have to be kept of all investigations carried out within organisations that use science. Reports will be written about these investigations and will be read by a variety of audiences. They may be read by the directors and managers of the company. They may also be read by other scientists within the company and possibly also by scientists that do not work for the company.

Activity 1.4D

Look at a selection of scientific reports from journals and textbooks. Can you find the following sections of each report:

title, abstract, introduction, method, results, accuracy, discussion, conclusions, references, bibliography?

Produce a poster showing the meaning of each of the above terms. You may need to research what each term means.

Case study: Government physicist

Alex works for a government science team looking at how energy is produced and used in this country. The team's most recent brief has been to look at how energy can be most efficiently used in industry and to make suggestions on how energy can be saved. The team has to gather data from a range of industries around the country, and it is Alex's job to compile these data and use the results to write a report to give to the government.

1 In what ways could Alex compile and present these data in the report?
2 Why do you think the government would think this an important project?

PLTS

Independent enquirer

Carrying out the research for this activity will develop this skill.

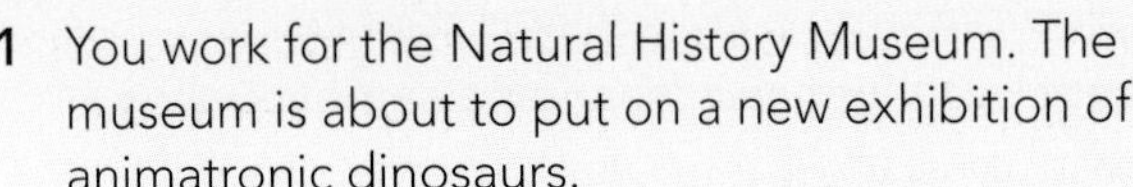

Assessment activity 1.4

M4 D4 P8 P9 BTEC

1 You work for the Natural History Museum. The museum is about to put on a new exhibition of animatronic dinosaurs.

The curator wants to make sure that the visitors to the museum can find out as much information about the dinosaurs as possible. She has asked you to investigate the different methods of communicating the information to the visitors and to produce a report for her on these methods. P8

2 You are applying for a job as a writer on a scientific journal. As part of the interview process you have been asked to write a scientific report.

Write a report on one of the following topics.

a The use of standard solutions in analytical chemistry.

b A comparison between a light microscope and an electron microscope.

c The use of fuels to produce electricity. M4

Write an evaluation comparing your report with a similar one in a professional journal. D4

3 You are the Science Information Officer for a large research and development company. You have to write a report on a scientific investigation. This will be read by the directors of the company and may be used in a scientific journal.

You may choose any investigation either from this unit or any other unit you have studied to write about. P9

Grading tips

In question 1 to achieve P8 describe each method in your report briefly. Who would the audience be? Why is it suitable?

In question 2 to achieve M4 the report you produce must be detailed and it must be correctly structured. Consider the different sections you learnt about for Activity 1.4D. Do demonstrate a high level of presentation: remember the report must be of a standard that could be published in a scientific journal.

In question 2 the report you compare with does not have to be on the same topic, but to achieve D4 it should be similar in its layout and its purpose.

In question 3 to achieve P9 make sure you use the appropriate standard format to write the report. Consider your audience. Ensure the report contains all the necessary sections and information.

Remember to use spell-check and to read through your reports for any grammatical or scientific errors before handing it in.

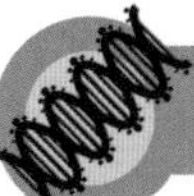

Functional skills

English

The questions in this activity will develop your English skills.

WorkSpace

Jacky Chaplow

Informatics Liaison Officer, Centre for Ecology & Hydrology (CEH), Natural Environment Research Council (NERC)

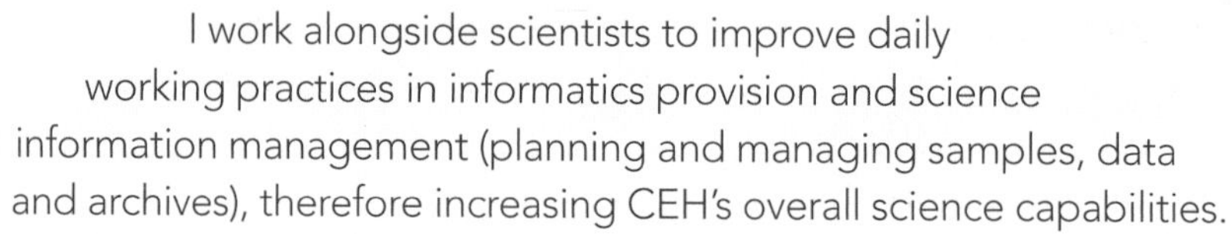

I work alongside scientists to improve daily working practices in informatics provision and science information management (planning and managing samples, data and archives), therefore increasing CEH's overall science capabilities.

My typical day begins by logging onto the network, and checking and responding to emails. I check the intranet for pages that require approval and assist staff with new working procedures around science information management. I may present and demonstrate new tools, policies and procedures to staff and manage data for scientists.

I work within a science department studying environmental chemistry – we have different groups working on a range of disciplines such as radioecology, soil ecology, water chemistry modelling, analytical chemistry, stable isotope ecology and the Predatory Bird Monitoring Scheme, among others.

I very much enjoy working with scientists at the forefront of knowledge discovery as my own background is in soil ecology. I was originally employed to set up and run laboratory and field experiments, to collect and prepare samples, and collate and present data. I now focus on data management. I love the variety and my career has developed over the years.

Someone within CEH may have an idea or a science question that could be answered by a research experiment. They talk to their colleagues and other experts within the field and write a science proposal which is then sent to funding bodies (e.g. NERC, DEFRA, EA) who decide whether to allocate funding. If the project is funded the scientist then needs help to manage samples, data and archives.

A project may run for a number of years. Experiments produce masses of samples, data and archives, which are dealt with under new policies and procedures. It is my role to guide the scientist through this complicated process.

Think about it!

1 Why do you think science information management is a useful tool?

2 Would you like to use your science background to work in a role where you are supporting scientists?

Just checking

1. State the meaning of atomic number.
2. Give the electron configuration of a chlorine atom in terms of energy levels and sub-shells.
3. Define the term 'relative molecular mass'.
4. What is the relative molecular mass of sodium carbonate?
5. Describe the trends seen in Group 1 and Group 7 of the periodic table.
6. Define covalent bonding.
7. What are the expected properties of an ionic compound?
8. Balance the equation: $Al + NaOH \rightarrow Na_3AlO_3 + H_2$.
9. How many moles of sodium hydrogen carbonate are there in 200 cm^3 of 1 M sodium hydrogen carbonate solution?
10. Define prokaryote and eukaryote cells.
11. State the functions of the following organelles: Golgi body, cell wall, vesicles, mitochondria.
12. Describe the following tissues: epithelial, connective, nerve, muscular.
13. Give an example of each of the following types of energy: mechanical, chemical, thermal, electrical.
14. Describe the energy transfers in metabolism.
15. Draw an energy transfer diagram for nuclear energy to electrical energy.
16. Define a calorimeter.
17. A food mixer has an input energy of 5 kJ and 4 kJ becomes the kinetic energy of the blades. Calculate its effciency.
18. What are the main sections you would expect to see in a professional scientific report?

Assignment tips

There are standard procedures used when carrying out the investigations in this unit.

- Make sure you know what the procedures are and carry them out carefully.
- It is important to do a risk assessment for all practical investigations.
- Try to be as precise and accurate as possible when carrying out practical investigations. Making a mistake like spilling some solvent can affect the final result. If this happens you will probably have to start again.
- Make sure you are using the correct equipment. For example, when measuring temperature changes make sure the thermometer you are using has a big enough temperature range.
- When writing a report it is important that it is written to a professional standard.
- Make sure you structure your report correctly.
- If you are word-processing the report remember to use spell-check.
- Read through the report carefully once it is complete to ensure there are no errors and that the report makes sense.

You may find the following websites useful as you work through this unit.

For information on...	Visit...
laboratory techniques	Science buddies
cells	CELLS alive
energy	Think Quest
academic writing	The Writing Center
writing in the biological sciences	The Scientific Paper

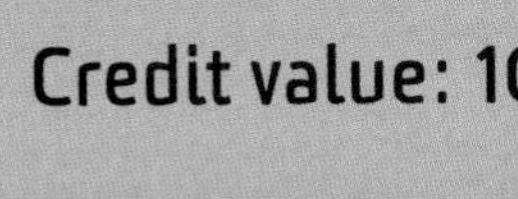

2 Working in the science industry

You might know already which branch of science you are most interested in, or perhaps you are using this course to find out about the different branches of science and the hundreds of careers that could be open to you. You may even change your mind as you come across the different aspects of this fascinating subject.

This unit gives you the chance to explore what happens in many science workplaces. Some of the procedures are used in all science workplaces, while others are much more specialised. This unit will familiarise you with the skills and knowledge that an employee in the science industry needs to have to be an effective, efficient and safe member of a team.

Learning outcomes

After completing this unit you should:

1 know how procedures are followed and communicated in the scientific workplace
2 be able to design a scientific laboratory
3 know about laboratory information management systems
4 be able to demonstrate safe working practices in the scientific workplace.

Assessment and grading criteria

This table shows you what you must do in order to achieve a pass, merit or distinction grade, and where you can find activities in this book to help you.

To achieve a **pass** grade the evidence must show that you are able to:	To achieve a **merit** grade the evidence must show that, in addition to the pass criteria, you are able to:	To achieve a **distinction** grade the evidence must show that, in addition to the pass and merit criteria, you are able to:
P1 outline procedures in the scientific workplace **See Assessment activity 2.1**	**M1** explain why procedures and practices are followed in the scientific workplace **See Assessment activity 2.1**	**D1** analyse why laboratory procedures and practices must be clearly communicated **See Assessment activity 2.1**
P2 identify how information is communicated in the scientific workplace **See Assessment activity 2.1**	**M2** explain how information is communicated in the scientific workplace **See Assessment activity 2.1**	
P3 design a scientific laboratory, identifying its individual key features **See Assessment activity 2.2**	**M3** justify key features in the non-specialist and specialist laboratory **See Assessment activity 2.2**	**D2** analyse why good laboratory design is important for efficiency, effectiveness and safety **See Assessment activity 2.2**
P4 describe the procedure for storing scientific information in a laboratory information management system **See Assessment activity 2.3**	**M4** explain the processes involved in storing information in a scientific workplace **See Assessment activity 2.3**	**D3** discuss the advantages gained by keeping data and records on a laboratory information management system **See Assessment activity 2.3**
P5 demonstrate safe working practices in a scientific workplace **See Assessment activity 2.4**	**M5** explain the need for current regulations and legislation in safe working practices **See Assessment activity 2.4**	**D4** evaluate the regulation of safe working practices in a scientific workplace **See Assessment activity 2.4**

How you will be assessed

Your assessment could be in the form of:

- a table showing procedures and practices in a scientific workplace
- a leaflet detailing the importance of communication in a scientific workplace
- a design plan for your own laboratory
- an article on laboratory safety.

Stephen, 18 years old

The information in this unit helped me to decide that I wanted to be a technician in education, perhaps in a college or university.

The unit covered all aspects of the work in general but also looked at the specialist tasks used in some types of laboratory. Although, at that stage, I was not sure what some of the information might be used for, I did the tasks and assignments as well as I could.

I finished the course with some good grades and started looking for work. There was a technician job in a college of a London university advertised. I applied for the job and was asked to go for an interview.

When I went for the interview a lot of the questions that I was asked had been covered in 'Working in the science industry' and this made me comfortable answering the questions. I was told afterwards that I had shown a good understanding of the work and organisational aspects and that had got me the job. I've a lot to learn so I hope to go on to do a degree part time and to get promotion.

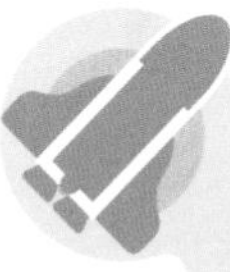

Catalyst

Where is science in the workplace?

Write down any six organisations that you know of and then think about whether they use science to carry out their work.

Get into small groups and compare what you have written with others in the group.

What conclusions have you come to about science in the workplace?

You should have included all types of workplace which involve computers in your list as well as science as it is understood by most people.

2.1 Procedures and practices

In this section: 

M2

Key terms

Science environment – any location where science is carried out as part of the day-to-day routine.

Procedure – method of conducting the business.

Practice – established method used regularly.

Calibration – checking that the equipment is working to a set standard.

Hierarchy – a number of people arranged in order of rank.

Whatever type of science you go into, either straight from this course or after higher education, there will be a number of **procedures** that are found in all parts of the **science environment**.

In any laboratory it is important to have:

- knowledge of equipment
- ability to use equipment
- ability to keep equipment in good working order
- knowledge of laboratory procedures
- communication skills and information on how people work.

Knowledge of equipment

You may already be familiar with different pieces of equipment that you have used in science. Some of these pieces of equipment might be new to you.

Equipment	Use
centrifuge	separates solids and liquids, particles of different sizes or liquids of different densities
microscope	optical instrument used for magnifying very small things so that they are visible to the naked eye
balance	measures the mass of objects, powder or liquids
pH meter	measures the pH of a liquid
oven	heats substances, often to high temperatures
incubator	maintains objects or substances at a particular temperature, often human body temperature
fume cupboard	protects the operator and others from hazardous fumes and powders
graduated pipette	measures liquids for use
heating mantle	a source of heat when a naked flame could be dangerous
burette	graduated glass tube with a tap at one end used to perform titrations
Bunsen burner	a source of heat where a naked flame is required or where it would not be dangerous
desiccator	a vessel containing a substance that attracts moisture so keeping the contents drier than in the open atmosphere
oscilloscope	used to measure voltage and frequency in electric circuits

Activity 2.1A

Think about the science you are most interested in and find out about some of the equipment that is used there. Make bullet point notes on what the equipment is and what it is used for.

Glassware

Glassware is commonly used in science. You are probably already familiar with some of the glassware which is used in most scientific workplaces. Why do you think glass is used so much?

The glassware used in science is not the same as the borosilicate glass used at home. There are several different types of glass used in science and it is important to ensure that the correct type of glassware is used, and is kept in good condition. Glassware is expensive and chips and cracks can be repaired. Borosilicate glass is soft and can be melted easily using a Bunsen burner. You may be familiar with the name Pyrex®. This type of glass is stronger, particularly when subject to sudden changes of temperature, and does not melt easily. It requires additional oxygen to make the flame hot enough to melt it. This is important to know if your job is to mend or make safe cracked or chipped glassware.

Activity 2.1B

Complete the following sentences:

When finding broken glassware you should...

When heating glassware you should...

When moving hot glassware you should...

When washing glassware you should...

When storing glassware you should...

Maintenance of equipment

It is very important that equipment is kept in good, safe working order as other people using it could be put in danger and results produced from work using the equipment could be wrong. For example, broken or wrongly **calibrated** pH meters can give incorrect results, making the results useless. In a research laboratory this could lead to wrong conclusions, wasted time and wasted resources. Broken or damaged wires on equipment can cause electric shock or fire.

To make sure that this does not happen all scientists and technicians look for faults all the time. This could be as simple as chipped or cracked glassware or it could mean looking out for problems with pieces of sophisticated equipment, or being aware of their maintenance schedules. Some types of equipment will draw the user's attention to a problem with warning lights but others will need to be carefully watched.

Modern equipment, for example high specification spectrophotometers, is so sophisticated that only the manufacturers and their specially trained technicians can perform maintenance and servicing on them. Other pieces of equipment can be maintained, serviced and calibrated by trained technical staff in the laboratory. When equipment is being cleaned the power should be switched off to prevent electric shock and the manufacturers' instructions should be followed.

Calibration, maintenance and servicing can, in some cases, be carried out by the laboratory technical staff and appropriate records should be kept of the work done and dates.

If the maintenance, calibration and servicing has to be carried out by specialist personnel from the manufacturer, either because the equipment is very technical or the laboratory staff have not been trained to do it, a contract will be set up with the manufacturer. The contract will state how often the maintenance will be carried out and what will be done. This will have a cost implication for the laboratory. This type of maintenance and calibration will be required for the validation of the data produced by the laboratory and, perhaps, for the laboratory to be allowed to do particular work.

Equipment	Maintenance required
microscope	General cleaning, taking particular care to use lens tissues on the objectives and eye pieces. Immersion oil can cause a particular problem if left to dry.
pH meter and electrode	These should be rinsed clean and stored with the protective cap to prevent drying out. The machine itself should be calibrated using standard solutions of known pH.
spectrophotometer	Spillage must be avoided but if it does occur the machine should be switched off before cleaning can take place. The light path in the machine should be cleaned, as should the carriage which holds the cuvettes with the test solutions inside.
centrifuge	The inside, including buckets, must be cleaned regularly to prevent the build up of spilt material and dust which may be hazardous to the user and the machine. Seals around lids may need changing at times.
burette	Burettes will need regular cleaning and checking of the glass body, tap and spout. Normal cleaning procedures will be sufficient in most cases. The tap should be taken apart, cleaned, checked for damage, greased and reassembled. The spout will need checking for damage as a break or chip will affect the accuracy of the amount of liquid dispensed.
Bunsen or similar type of burner	Burners will have a gas tube which must be checked for cracks or breaks that could cause a gas leak. Any connections should be checked for leakage which could cause explosions. The moving parts, such as the collar, should be freely moving to allow the air content to be easily controlled. If there is a possibility of drips falling down the chimney of the burner this should be checked for blockages.

Equipment	Maintenance required
fume cupboard	Used to protect the worker and other staff from potentially hazardous substances such as chemicals, fumes or microorganisms. It works like an extractor fan pulling air from inside the laboratory into the fume cupboard, across the work area then up a chimney and outside into the atmosphere. Maintenance is required on the motor and filters to ensure that air is moving fast enough to remove any dangerous substances. The movement of air can be checked by the technical staff using a basic piece of equipment. If further work is required a specialist company may need to be called in.
oscilloscope	Used to measure the value of ac voltages and to make measurements so that their frequencies can be calculated. Modern instruments are very reliable but in situations where a high level of accuracy is needed their calibration should be checked routinely. This could usually be carried out by technical staff in the workplace but if adjustments are needed then a specialist would be required.

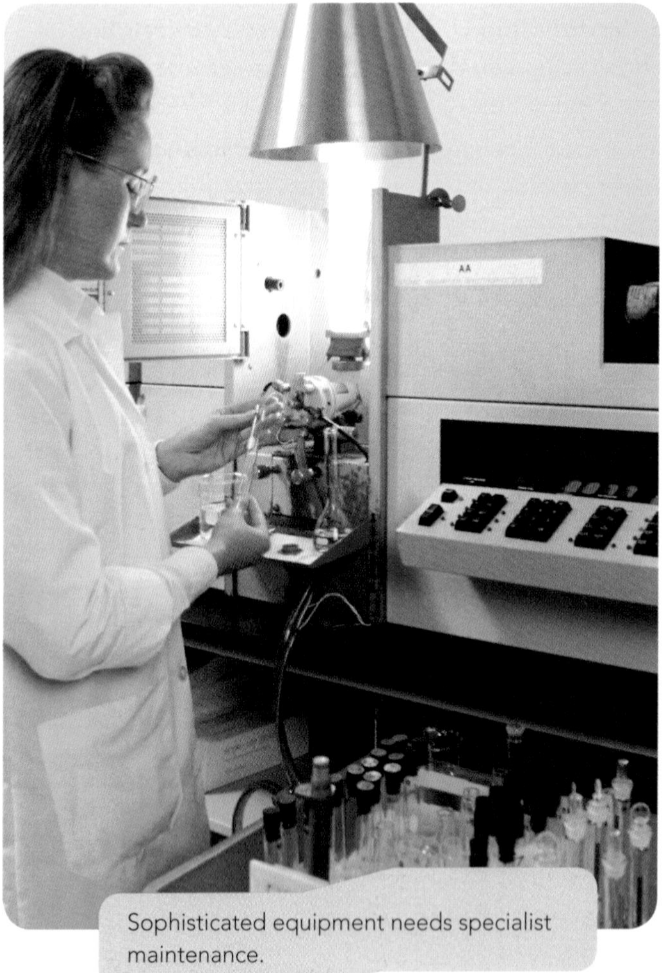
Sophisticated equipment needs specialist maintenance.

Activity 2.1C

List reasons why equipment must be maintained.

Moving equipment

Sometimes you will need to move equipment, either for maintenance or for disposal. When equipment is to be moved for maintenance purposes the maker's instructions should be followed to prevent damage. Moving within the laboratory might just mean providing access for the maintenance engineer to be able to work on the equipment or gathering the pieces of equipment together in one place for the engineer to work on them. If this is the case then good manual handling techniques should be used to prevent injury to staff and damage to the equipment.

For example, training can be organised for staff who might be required to lift or move equipment or stock. This could involve the best way to lift using a straight back and bent knees or the use of ladders, etc. for accessing higher level shelves. Technicians should also be aware of the fact that equipment can be very heavy and, as well as needing more than one person to move it, there might be a need to use a trolley or wheels.

When a piece of equipment has become obsolete, due to age and/or damage, it should be disposed of in a suitable way.

Some companies will buy old but working equipment for resale to organisations who can still use it. For example, obsolete blood testing equipment can be sold by an NHS trust and bought by colleges for training.

If equipment is to be disposed of there are several procedures that need to be observed. Some analysers, for example, contain hazardous substances such as mercury, radioactive material or gases. Local council disposal regulations should be followed before disposal can take place. Mercury should be drained and collected for disposal. Radioactive sources can be sent to the relevant authorities for safe disposal. Gases, such as refrigeration gases, can be removed by an authorised organisation.

Laboratory procedures and practices

Depending on what type of work the laboratory is carrying out the procedures and **practices** in use will be different. However, some procedures will be carried out in some form in all laboratories. These procedures and practices could be documented in a Laboratory Information Management system (see page 47).

Standard Operating Procedures (SOP) are in place in many laboratories. These are documented procedures that are carried out regularly as part of the routine and could cover many different aspects of the work. For example, how tests are carried out, the procedures for handling the chemicals or samples involved and the disposal of samples or waste. This could also cover the way different pieces of equipment must be used and maintained, including the correct solutions for use in them.

A typical standard operating procedure

At the start of the day:

- Switch on the analyser. Allow machine to run its initial set up procedure.
- Check internal standards have been met by consulting the screen or print out. Address any problems that have been highlighted.
- Check any solutions required for the running of the equipment. Replace or refill as required.
- During the day monitor the correct running of the machine and the level of solutions.

At the end of the day:

- Clean and shut down the machine according to the manufacturer's instruction.
- Switch off or leave safe.

Store management and ordering

In all laboratories there will be one or more stores of chemicals and equipment. A procedure which must always be under strict control (see below) is the ordering, storage and use of these day-to-day requirements. The senior technician or laboratory manager is likely to be in control of ordering, and will have to manage a budget. Normal procedures will include:

- stock rotation – using the oldest chemicals first to make sure they do not become too old; some chemicals can become dangerous if kept too long
- correct storage procedures – taking into account any special storage requirements
- how stock is ordered and moved for immediate use.

Orders may be placed automatically on a regular basis for chemicals or equipment, called consumables, or stock may be ordered as and when it becomes necessary. The senior technician or scientist will probably be the person who has the authority to place orders, but they will need to be told when stocks become low. The order will need to be authorised and this might include agreement from the finance department of the company.

In industry, stocks from the supplier will usually be delivered in large quantities, even by road tanker. The chemicals could automatically be taken from the storage place for direct industrial use. It will be the responsibility of the technicians or store person to deliver what is required for use in the laboratory. In some laboratories a system using bar codes and a bar code reader is used to control stock, a bit like on the shelves of a supermarket.

There are various categories of materials that may be used in the laboratory. Some can be bought normally while others might need a special licence to be bought.

Material	Storage requirements
flammable material: large quantities and explosive material	Stored in a special brick-built store usually away from the main building with a roof that can detach in case of an explosion rather than blowing out walls.
flammable material: small quantities for immediate use	Stored in a metal flammables cabinet.
drugs and poisons	Stored in a locked cupboard.
dry chemicals	Stored in a cool dry place.
biological materials: bacteria, plants and animals, including clinical test samples	Live material will need to be stored correctly so that it stays viable. This might involve keeping it cool or warm, in the dark or light.
radioactive materials	Store in a locked metal cupboard away from where flammable materials are stored. Removal from the cupboard and details of use should be recorded in a log book.

Transfer of materials

It is important that, as a technician or scientist, you know how materials should be transferred from place to place and what quantity is safe to carry. Flammable, biological and radioactive materials are commonly moved around and need special handling. When materials are ordered from a supplier, the supplier will be responsible for delivering them safely to the laboratory. Once they arrive, a technician will be responsible for the safe handling of any hazardous material.

Dry chemicals should be kept in a suitable container if taken from their original container and relevant labelling should be transferred as well. You may need to use a fume cupboard to prevent dust particles contaminating the surrounding area and other workers.

Wet chemicals must be carried in a bottle carrier in a suitable container.

Biological material, such as living organisms, must be treated with care to keep them in good condition and, in the case of animals, to prevent suffering.

In general, if substances are heavy and are to be transferred from the store to the area of use they must be carried on a trolley whenever possible.

Disposal of waste

In every science laboratory there will be some form of waste produced and it must be disposed of safely.

Waste	What it might include
chemical	waste from school laboratories waste from manufacturing
clinical	waste from hospital wards and clinics sharps bins waste test samples from hospital laboratories waste reagents from testing in hospital laboratories waste from veterinary surgeries
pharmaceutical	waste from pharmaceutical manufacturers out of date drugs from local pharmacies and chemist shops
biohazardous	waste from research laboratories samples from patients
radioactive	waste from hospitals waste from the energy industry

Options for disposal

Some non-toxic, non-polluting waste products can be flushed away into the drain in the laboratory using water, while others must be disposed of by mixing with other chemicals to make them safe (for example, sodium carbonate added to acids).

Disposal of waste may involve storing it until it can be disposed of. This could pose a problem as large amounts of mixed waste can create a new hazard, e.g. carcinogens or mutagens.

The storage of waste material will be specific to each laboratory and local regulations will detail what needs to be done. The technicians in charge will have detailed knowledge of substances in use in the laboratory and the requirements for their safe disposal.

Some waste must be taken away for expert disposal. Some can be taken away, cleaned and recycled, such as mercury. If this forms part of your job you will be trained in which process you should use. A technician will need to label and log all waste products ready for collection. Disposal of large quantities of waste must only be carried out by a reputable company, a registered waste carrier, to prevent illegal fly tipping of potentially dangerous substances into the environment.

Processes for disposing of waste

Chemical waste

Flammable solids:

Large amounts (or even small amounts) will require collection. Small quantities of water-reactive solids, such as calcium dicarbide, may be added carefully to a large volume of water in a fume cupboard. Metals such as sodium, lithium and calcium should be destroyed chemically before disposal via a drain with great dilution. Sodium should be dissolved in ethanol or propan-2-ol in a fume cupboard. Lithium and calcium may safely be dissolved in excess water.

Toxic chemicals:

Most of these should be stored for collection or made safe chemically if in large amounts. Small amounts (10 g or less) of toxic salts may be dissolved, diluted and flushed away.

Corrosive liquids:

These should be diluted and neutralised using sodium carbonate (for acids) or ethanoic (acetic) acid (for alkalis) before washing to waste with large amounts of water.

Water-reactive corrosives:

These should be added cautiously to a large excess of water in a bucket or bowl in a fume cupboard before washing to waste with large amounts of water.

Corrosive solids:

These should be dissolved carefully, diluted greatly and preferably neutralised with sodium carbonate as above before washing to waste.

Oxidising agents:

These should be dissolved in water and diluted greatly before washing to waste. Care should be taken that wood, paper or cloth does not become contaminated with the solution.

General chemicals:

Low-hazard inorganic chemicals may safely be disposed of via the refuse or drains but large amounts of metal compounds such as copper or zinc salts should be kept for collection. Small amounts of copper or zinc salts may be dissolved and flushed away.

Special cases

Asbestos:

Asbestos can only be collected by a carrier specifically licensed for that purpose. However, if a small amount of asbestos is found, it should be sealed in a tough plastic bag and a special collection arranged by contacting the Environmental Services department of your local council.

Mercury:

Mercury recovered from spills or otherwise considered 'dirty' could be kept and sent for purification when enough has been collected. Lime/sulfur is used to help clean up small spills of mercury and the mix contaminated with dust and small droplets of mercury from broken thermometers, etc. should be stored in a strong bottle and kept for collection by a registered waste carrier.

Clinical waste

Clinical waste is placed in special yellow plastic bags and collected separately by the refuse service for incineration. Sometimes it will need to be double bagged, i.e. putting a full bag inside another empty bag to prevent leakage. Sharps that might be contaminated with biologically hazardous material, such as used needles, are collected in special reinforced bins and are usually also taken away for incineration.

Pharmaceutical waste

Pharmaceutical waste from manufacturing and quality control processes will be treated as chemical waste. Other controls might be used for particularly dangerous drugs. Out of date or returned medicines are waste from the local pharmacy or chemist shop.

Did you know?

Old medicines should not be put into household waste or flushed down the sink or toilet but returned to your local chemist shop or pharmacy.

Radioactive waste

This waste can come from the energy industry where large amounts of radioactive material will be generated. Some medical tests require the use of very small amounts of radioactive material. This can be mixed with clinical waste and disposed of in the same way.

Regulations about waste disposal

There are many rules and regulations about waste and its disposal, for example, the Hazardous Waste Regulations 2005.

Below are outlined two quite complicted regulations that are covered by European legislation.

Trade Effluent (Prescribed Processes and Substances) Regulation 1989:

Companies who dispose of their waste into rivers and waterways need to seek the water authority's permission if they wish to dispose of chemicals such as mercury, cadmium and their compounds, tetrachloromethane and pesticides. It advises that the disposal of other substances in river water such as ammonia and metals like zinc, nickel, lead and copper is kept to a minimum.

Radioactive Substances Act 1993:

Organisations that use radioactive material, and therefore produce radioactive waste, must be registered so that their activities can be monitored.

Communication in the workplace

The work carried out by scientists and technicians relies heavily on the structure of the team they work in, and the way each team member acts. In most work places there is a **hierarchy**.

This means the most senior person will have various levels of personnel reporting to them. How this is organised depends on:

- how large the team is
- the particular routines that are carried out in the workplace
- whether the team is spread over a large area or different sites
- if the team is split into smaller groups carrying out a particular job or at particular times of the day or night.

No matter how people are organised, the way they communicate within their team or outside of it is crucial to the safe and smooth running of the organisation.

Type of organisational structure

Manager

Team leader(s)

Senior technician(s)

A number of technicians or scientific officers

A number of laboratory assistants

Reporting of results

Results generated in a workplace will be specific to that workplace. They may be results of research performed by colleagues or results generated for the use of outside agencies. Whatever the results are they must only be communicated to those who need to know them.

Internal day-to-day results will probably be reported via the laboratory notebooks, printouts from the laboratory equipment and at team meetings. These results may be gathered together to produce a report on completion of the research.

Unless there are reasons for urgent results to be communicated directly to another person, results will normally go through an office procedure where they are written up and copied to the recipient, for example a GP. In some cases results, such as scans, can be viewed via a computer screen along with test results.

Scientific terminology

It is essential that scientific terminology is used and understood by all members of the team if effective communication is to take place. This is particularly important where research work or production is being carried out in different countries where language may cause confusion if standard terminology is not used.

Security

Keeping information secure means being sure only certain people can access it. This is important as industrial espionage can cost a company large sums of money. For example, a pharmaceutical company could lose money if a competitor reaches drug production first.

Individuals also want their personal records (e.g. medical records) kept private. A patient can be badly

Communication in the workplace.

affected if their personal clinical information is lost or reaches the wrong person. The Data Protection Act is in place to prevent personal information falling into the wrong hands.

Roles and responsibilities

It is important that each member of the team knows their role and responsibilities. This helps the team work well together but lets each member know and feel comfortable with what is expected of them.

Your place in the team

Most workplaces will give a new person a diagram showing how the team is organised. This shows the new person in the team who they should report to and also who they can go to for help or guidance. This also affects how the work is organised.

Work schedules or rotas will be written so that staff know when they should be working and when their days off will be. This might include working at other sites. This differs depending on the organisation. For

Case study: The importance of communication

The importance of communication in the workplace is highlighted in this fictitious case study.

In a pharmaceutical laboratory the research and development technicians had been working on a new drug. They had reached a critical stage of the work but had changed one of the chemicals in the mix. The next day, one of the technicians was due to be on holiday and the other was on day release. They left work on Tuesday evening leaving a process running that needed 12 hours to complete. Their laboratory note book, noting their findings and changes to formulation, had not been updated recently and was left on the bench.

The last person in the laboratory saw the equipment still on and turned everything off. The following day a laboratory assistant arrived to find the process apparently complete but with no information about what to do next. Not realising that the process had not been completed and results had not been documented, they cleared away the samples and equipment before asking anyone else. The senior technician was also away on a management course.

Whilst being disposed of, the chemicals were mixed with others and toxic fumes were created. These unknown fumes caused the assistant to become unconscious and, when she was found, the emergency services were called and the assistant taken to hospital. When an emergency contact number for the assistant was called, the number was found to be no longer in use. The emergency services could be given no clue as to the cause of the collapse.

Meanwhile, the R&D technician on holiday talked at length to some casual acquaintances in technical detail using technical language to impress them about what he had been doing at work.

The technician on day release returned to find a considerable amount of research lost and had to start it again. The senior technician returned to find one of their staff in hospital and the process being repeated. Shortly afterwards news on the Internet informed them that the drug was being developed by a rival company.

Who is responsible for this situation arising?

instance, if there is a continuous process running, such as a production line in a pharmaceutical company producing tablets, there may be a need for staff to work day and night in shifts. This means working for a number of hours and then having time off. This could be over an extended period.

At busy times, a day worker may need to work night shifts to keep up with the work. If cover is needed for emergencies then some qualified or experienced people will be 'On call'. This means they could be called in to work at short notice. If handover meetings are required they will be scheduled so that team members can attend.

Holidays, time off, sickness and day release to attend training can be added to the schedules to ensure that enough people are available to cover the work and that the staff working together are qualified and experienced.

Assessment activity 2.1

BTEC

1 Produce a table listing all of the procedures and practices that you have investigated and in a second column give the reasons why they are carried out. P1 M1

2 You have been asked to produce a leaflet for trainee technicians covering communication in the scientific workplace. In the introduction say how information is communicated in the workplace. Then, in a table list the ways communication can happen in a workplace, the reasons for communication and why any communication must be clear. Finally, use case studies (real or imagined) to show how important it is that procedures and practices are communicated clearly and what might happen if they are not. Display the information as a leaflet for distribution to prospective technicians at a job fair. D1 P2 M2

Grading tips

In order to achieve P1 you are only asked to list procedures and practices in the table. Don't do more than is required. For M1 you may have to think about why these procedures and practices are carried out so make sure you understand what each one is.

For P2, you need to list the ways in which information is communicated in the workplace, whereas for M2 you need to give more in-depth explanations of the points listed for P2.

Think about leaflets that you have picked up in the past. What attracted you to them, why did you want to read them and how did they hold your attention? Along with the information required to show that you understand why communication of procedures and practices is necessary in a scientific workplace, add in these factors to produce an interesting and informative leaflet. D1

PLTS

Self-manager

Completing work to standard and deadline will use your skills as a self-manager.

2.2 Laboratory design

In this section: 

Key terms

Key features – anything that is vital for the work of the organisation.

Services – utilities such as gas, electricity and water. In some laboratories there might also be vacuum lines available.

At some stage in your career in a scientific workplace you might be involved in the planning, building, rebuilding or refurbishment of a laboratory.

This is a complex task requiring specialised knowledge and experience. However, practical day-to-day knowledge of the work and the workplace may prevent costly mistakes in time, money and personnel problems.

What makes a workplace scientific?

You have already investigated the use of science in the workplace and you found that science in some form or another is present in many workplaces.

In this section of the unit you will be looking at what makes a workplace really scientific.

Let's consider what you would expect to find inside a scientific workplace.

Think about laboratories that you are familiar with at school, during work experience or visits.

The **key features** of a laboratory are those that make it a scientific workplace, for example, building materials that will not be affected by substances in use in the laboratory. These could include floor, ceiling and wall coverings, such as special tiles or paint, types of specialist furniture such as benches with clear surfaces and surfaces on furniture that can withstand the types of substances in use. There will also be greater access to more **services** in the laboratory than in many other workplaces. Ventilation will be an important part of the laboratory, as there must be adequate fresh air, but this must not allow substances inside the laboratory to escape into the environment unchecked.

The services that are required in the laboratory are: water, gas, electricity and you could include in this list a **vacuum line**.

The furniture will depend on the work being carried out but essential pieces might include benches and stools, sinks, storage cupboards, drawers and fume cupboards.

Activity 2.2A

List the key features that you would expect to see in a laboratory. What furniture and services should be there?

Furniture

Furniture in laboratories can be free standing or fixed. It is usual for benches to be fixed because they will have services, water and waste outlets, gas and electric fitted to them and it would be dangerous if the benches were able to be moved. In some classrooms, however, modular furniture can be moved around and

Case study: Poor design

A new suite of laboratories was built as part of the General Hospital. The designs were approved and the laboratory was built. When the staff came to move in they found that the bench where the haematologists use microscopes to look at blood smears was positioned underneath the outlet to the air conditioning. They spent all day with cold air blowing down their necks and this resulted in many stiff necks and aching shoulders. The air conditioning could not be altered, as this had an effect on several areas of the laboratory which became uncomfortably hot. The laboratory had to be rearranged so that the microscopes were located away from the outlet. This meant that urgent work was held up until the relocation was completed.

How could these problems have been prevented?

linked to fixed units to change the configuration of the teaching area. There might, in some laboratories, be requirements for teaching equipment such as whiteboards.

In some old laboratories wood is still in use for work surfaces. It is a very good material but requires continuous care to keep it in good condition. More and more synthetic surfaces are being introduced into laboratories which are easier to keep in good condition.

The surfaces in a laboratory will be different from those in your home kitchen as they have to be chemical proof as well as heat proof. Benches may have storage below them but there should be space for people to sit and work at them. Some laboratories, microbiology for instance, require that there are no gaps between separate lengths of work surface to prevent the build up of chemical substances or microorganisms.

Storage is a vital part of the laboratory. If there is not enough storage space, things that should be put away safely can get left out. This can be dangerous as they can block access and create a hazard for people moving about in the laboratory. Other hazards include paper, which could be a fire risk if left lying around the laboratory.

In all laboratories there will be safety equipment. This could include fire extinguishers, a fire blanket, safety solutions and eye wash. All laboratories will have a first aid box in the room or close by. Sometimes a safety shower will be available to wash away large spills of substances from a person's body.

There may be one or a row of fume cupboards. Fume cupboards need maintenance to make sure that they are removing the correct volume of air to protect staff both working at the cupboard and in the surrounding area. The speed of flow depends on the substance in the fume cupboard. For instance, chemicals and radioactive substances require a flow of 0.4 and 0.7 m/s. The fume cupboard acts as an extractor fan pulling air into itself, across the work space and then up a chimney to be vented into the atmosphere. It is important that air contaminated with harmful organisms or fumes is cleaned (filtered) before being released into the atmosphere. Air might also be filtered before going into the cabinet to prevent contamination of what is being worked on.

Fume cupboards can also contain gas, water and electric supplies, as well as a sink.

There are also cabinets called laminar flow cabinets. These are designed to blow cleaned air across the work surface. Any hazards being produced in the cabinet will be blown out into the laboratory and over the workers. Therefore, laminar flow cabinets should only be used for non-hazardous work.

Unit 15: See page 288 for more information on laminar flow cabinets.

Activity 2.2B

As a preparation for the assignment task make a sketch of the laboratory you are working in or that you are most familiar with.

Sketch in the furniture and any large pieces of equipment and make a note of where services, exits and store rooms are located. A plan in this case is a view of the lab looking down onto it as if you were a spider on the ceiling.

Access and workspace

Have you left space for people to move around? There are rules and regulations stating how much space is required in secondary schools for movement between benches and cupboards. They are published in a document called Building Bulletin 80 Science Accommodation in Secondary Schools. This can also be found in a CLEAPSS document called G14 Designing and Planning Laboratories.

One of the biggest problems in any workplace is the lack of space for people to move around and to work. This can have dangerous consequences.

When scientists and technicians are moving around the laboratory they may need to carry samples or equipment in their hands, or move them on trolleys if

A dangerous situation developing in an overcrowded laboratory.

A modern clean and ordered laboratory.

they are heavy. If a technician is working at a laboratory bench, other people need to move past them without hitting them. If the bench space is insufficient, samples or equipment might be knocked off the bench or things could be put on the floor, which would add to the risk of tripping.

Equipment will need space around it for access should it need repair but also for air circulation to prevent overheating.

Has this made a difference to your plan? Redraw your plan if you have to make changes.

The specialist laboratory

The work that is carried out in a laboratory will influence the type of equipment that is used there. Most laboratories will have several large pieces of equipment, numerous smaller pieces and several computers linked in.

The storage required will vary, with special facilities for radioactive, flammable and toxic materials.

Medical

In a medical laboratory machines will be used to diagnose illness, as well as monitor patients' treatment. Many and varied tests can be carried out on many different types of sample. In particular, work in a microbiology laboratory is carefully controlled to prevent the spread of disease-causing microorganisms into the surrounding area. Any sample may be hazardous and so it must be treated accordingly.

Forensic

In a forensic laboratory there will be areas for checking and testing evidence. It is important that evidence does not become contaminated as this can lead to a criminal case being thrown out of court. Different areas of the laboratory will be used for specific types of testing. Firearms, ballistic and explosives tests will be carried out in an area away from the main work area for obvious safety reasons.

Pharmaceutical

In a pharmaceutical research laboratory there will be equipment for producing and testing drugs. This could include testing and producing new drugs or testing drugs where patients have highlighted a problem, for example strange side effects. In a pharmaceutical dispensing laboratory there will be equipment for weighing and measuring out medicines and, possibly, for making pills. There will be an area for this activity and an area for actually handing over prescriptions to patients.

Preventing contamination

In some laboratories, especially the forensic and pharmaceutical labs mentioned above, there will be a definite difference between clean areas and general areas. In clean areas special protective clothing must be worn, such as suits covering the whole body, masks and shoe covers, hats and, in some cases, beard covers. This is to protect work from contamination by workers' clothes, hair and skin. In clean areas the air coming in is filtered to prevent contamination from outside. On the way into these clean areas there are lockers for clothes, supplies of the right protective clothing and bins for used protective clothing. There is often a barrier of some sort, such as a low wall or a door with labelling, to remind people that they are entering a special area.

Food science

In a food science laboratory, usually associated with large food producing companies, there will be kitchens producing new food samples or maybe improving an existing product. There will also be laboratories checking the quality of products and investigating complaints made by customers.

Physical sciences

In physical laboratories, physical tests might be carried out, for example to check the strength of a metal component. This could involve stretching a piece of metal until it breaks. Special equipment is required, as are special precautions to protect the workers.

Chemical

Chemical laboratories could be producing small amounts of chemicals for specialist work or they could be researching chemical reactions in preparation for much larger scale production in a factory situation. In each case there will be chemicals which must be stored correctly and waste chemicals which must be disposed of correctly.

The health and safety issues will be different for different types of laboratory.

Laboratory design will need to address all of the issues relating to health and safety to comply with current regulations in a new building. In an older building that is being refurbished or updated, such as a listed building, it may not be possible to comply with all of the rules and regulations.

Assessment activity 2.2

1. Using all of the information given and work you have done in class produce a simple design for a laboratory identifying the key features that must be present to be able to call it a scientific workplace. P3
2. Your school or college laboratory will be a non-specialist laboratory as it can be used for most types of science, even though you might only carry out one of the sciences in it. If it has been designated as a microbiology suite then it will be a specialist laboratory.

   Look at the facilities and equipment that are required in a specialist laboratory to carry out their work and justify the inclusion of the key features in a non-specialist and a specialist laboratory. M3
3. You are the senior technician in a laboratory and you have been told that there is a grant available for the refurbishment of your workplace. There have been problems over the years with the increase in size and complexity of the equipment that you are using in the laboratory and the number of staff that have to be accommodated as well as the amount of work expected.

   To be given the grant you have to make an application with an analysis of the changes that you want to make and how they impact on the efficiency, effectiveness and safety aspects of your workplace. You need to write a report outlining the changes you would like to make and should cover the design and the content of the laboratory. D2

Grading tips

Gather together all of the information that you may have learned in class or from your own research so that you can draw a simple plan of a scientific workplace. As a practice for the assignment you might like to draw a sketch plan of the laboratory you are familiar with and mark on this plan all of the services, furniture and large pieces of equipment that are there. Add on to this all exits and other doors to preparation or store rooms and then locate safety items such as fire blankets, eyewash bottles, etc. that may be there. This will help you to attain P3.

Find out what different pieces of equipment are required for a named type of specialist laboratory and what they are used for to help you reach M3.

Putting all of the information you gathered for M3 into a professional report that could be shown to the management of your company to encourage them to give you the grant that you require will help you reach D2.

PLTS

Independent enquirer, Creative thinker, Reflective learner and Self-manager

You will develop these skills by researching, creating a scenario, using several sources to reach a conclusion and completing work to standard and deadline

Functional skills

ICT

You will develop your ICT skills by finding and saving information for later use, bringing together information to suit a purpose, presentation of information

2.3 Storage and management of information in the laboratory

In this section:

Key terms

Data protection – the Data Protection Act 1998 was passed by Parliament to control the way information is handled and to give legal rights to people who have information stored about them.

In the course of work being carried out in a laboratory, large amounts of information will be produced. It has become increasingly important for data to be stored so that they can be retrieved at a later date.

Can you think of some reasons why data might be needed, possibly some time after the work has been completed?

Storage of records has changed enormously since the computer arrived in the workplace. In most organisations large boxes or filing cabinets of documents have been replaced by hard drives and disks containing the vast majority of data.

Data storage nightmare!

Benefits of computer storage

- The amount of space required for computer storage is much smaller than the space required for paper storage.
- Computer storage is less of a fire risk than large amounts of paper.
- Computer records can be searched very quickly.
- If records have to be accessed from a number of sites, a computer system allows this without paper records having to be copied and sent to different locations. This saves time and removes the risk of them getting lost.
- Records can be updated quickly and there is less chance of technicians using out of date information.

The data that might need to be stored in a laboratory could fall into one of several categories:

- data produced from research carried out in the laboratory
- data about staffing levels
- personal data about members of staff, etc.
- data about resources/equipment.

All types of data will need careful management if they are to be kept safe, secure and be available at a later date.

Did you know?

In health-related organisations information stored from different outbreaks of disease such as 'flu (influenza) could help to prevent pandemics.

Many thousands of people died from 'flu in Britain following the First World War in 1919. Scientists have been trying to compare this virus with those that have caused the disease in subsequent years, going as far as trying to get lung tissue from people who died and were buried at the time. If records had been available this could have been easier. However, DNA information on the virus was not known at that time.

Data storage

Type of data	Reason to keep data	Who should record it, have access and be able to make changes
COSHH records	to ensure awareness of health and safety issues with substances being used in the organisation (see page 50 for more information on COSHH)	stores technicians and whoever is involved in ordering, storing and use of the substances
scientific data	in any scientific workplace it is vital to be able to safely store and then retrieve scientific data generated by that workplace and also data from other sources (scientific literature, for example)	heads of department, deputies, and those working in the laboratories
scientific apparatus	data such as date of purchase, maintenance data and schedules for maintenance	heads of department, deputies and those involved in the schedules
waste disposal	to show what and how much waste is produced and how it is disposed of	stores technicians and those involved in disposal; heads of department may need to authorise costs of disposal
health and safety checks	to show that health and safety is being monitored and to hold accident reports if necessary	heads of department, health and safety officers and possibly others who have special responsibilities
training records	to know the level of training or qualification of members of staff, and to keep and maintain a record of training required and completed by staff	training officer, heads of department, supervisors, human resource department and individual members of staff
quality assurance	to be able to show that quality procedures are being carried out (for audit purposes)	head of department, quality officers and those with special responsibility
report records	reports following tests for GPs or hospital records, or for use in developing new medicines, etc.	office support personnel will usually be responsible for recording results, with access needed by clinical staff (in a clinical environment); report records in this setting wouldn't usually be subject to change by anyone
specification levels	this could be the level at which the organisation is allowed to work, for example the danger levels of microorganisms in use	the head of department and organisation management
sample throughput	this gives information about the number of samples going through processes in the laboratory in a given time and could be an indicator of the efficiency and effectiveness of the organisation	head of department and deputies, organisation management
management	this could cover the management hierarchy and their roles	organisation management and human resources department
security	different types of laboratory might need different levels of security depending on the work being carried out	head of department, security staff, health and safety officer and all staff

With more organisations using computers to store and process personal information, there is a danger that the information could be misused or get into the wrong hands.

- Who could access this information?
- How accurate is the information?
- Could it be easily copied?

Activity 2.3A

When an organisation keeps certain types of records they will have to comply with the Data Protection Act.

What does this cover and how might it be relevant in the science laboratory?

Use the website detailed below to produce a summary.

ICO – Data Protection Act

- Is it possible to store information about a person without the individual's knowledge or permission?
- Is a record kept of any changes made to information?

How might the storage of data be different on a computerised system and a manual system?

Look again at the list of the different types of data that might be stored and consider how a computer system might be used to store the same information. Think about the benefits of using a computer compared with a manual system.

Many laboratories have bought a Laboratory Information Management System (LIMS) which is like an electronic filing cabinet. The system allows laboratories to input data in the form that they need in order to use it and they can customise the system so that they can input information relevant to their organisation.

The LIMS can store text and graphical documents and can use the data to produce relevant information such as investigation results. It can also be used to monitor good laboratory practice by, for example, monitoring sample collection, testing, quality assurance and outgoing results.

The system can alert the laboratory of incoming samples so that when they are received into the laboratory they can be bar coded and devices can be used to generate labels for quick error-free processing. A hand-held device can be used to enter the samples onto the LIMS. The sample can then be put through the testing procedure with minimal work for the technical staff.

The LIMS can also be used to monitor stock levels so that ingredients or products do not fall below safe levels for the company to continue working.

Depending on the organisation and how sophisticated the LIMS is, much of the laboratory documentation can be taken over by the system.

Assessment activity 2.3

BTEC

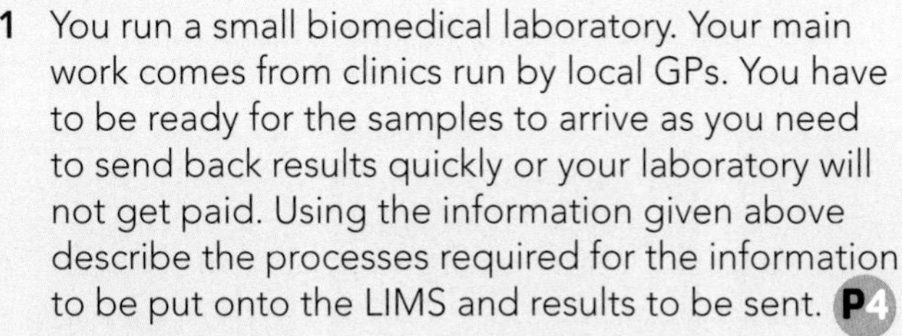

1 You run a small biomedical laboratory. Your main work comes from clinics run by local GPs. You have to be ready for the samples to arrive as you need to send back results quickly or your laboratory will not get paid. Using the information given above describe the processes required for the information to be put onto the LIMS and results to be sent. **P4**

2 Explain the different processes involved in storing information. Why are these processes necessary? **M4**

3 You are the senior technician in an expanding laboratory. You have been using a manual system for holding information and data but you now believe it is time to update and you have been looking into the use of LIMS.

Research a LIMS and using this information produce a presentation to give to the management of your company to encourage them to give you the funds to buy and run this system. **D3**

Grading tips

For **P4** put yourself in the place of the person in charge in the scientific workplace and suggest what type of procedures could be used.

For **M4** show a process that you would follow to store various types of information. Think about who should have access.

For **D3** the scenario gives you the opportunity to research a real LIMS and you could use the information made available by the companies who produce a LIMS in your presentation. Remember not to copy the website, but you can use some of the information as quotes as long as you don't forget to show your sources.

2.4 Safe working practices in the laboratory

In this section:   

Key terms

Risk – a situation involving exposure to danger.

Risk assessment – an investigation of the risks involved with a certain procedure and whether there is a safer alternative.

Laws – a collection of rules according to which a country is governed.

Regulations – rules or orders.

Carcinogen – an agent or substance that has been suspected of causing or increasing the risk of cancer.

Mutagen – an agent such as a chemical, ultraviolet light or a radioactive element that can induce or increase the risk of genetic mutation in an organism.

Teratogen – a drug or other substance capable of interfering with the development of the fetus (unborn child), therefore causing birth defects.

You have already seen in the laboratory there are practices and procedures that are carried out on a regular basis as the work of the laboratory requires. These will have been built up over a period of time, changed and perhaps changed again as equipment and techniques have evolved. They will have been assessed for their **risks** each time a change has occurred.

The practical work that you carry out in school or college will have been assessed, using a **risk assessment**, to see if it is safe for you to carry out with the level of knowledge and experience you have at the moment. This may have been done by a member of the technical team before the practical requirements are prepared for the class. The aim of the assessment is to minimise the risks. However, accidents can still happen, even in the most safety conscious workplace.

As part of your scientific practical work in class you will have produced your own risk assessment before commencing your work.

Labelling

It is important that common hazard labels are recognised and used when handling chemicals and other substances.

Tankers carrying toxic chemicals have large labels on the side giving details of the substance being carried and the way to deal with it in case of accident or spillage.

When substances are transferred from their original container and put into smaller containers for laboratory use, all labelling must be carried over, including the substance name and its hazard if it has one. Orange sticky labels are available for putting on containers with labels stating the content of the container.

Common hazard labels found in the laboratory.

Hazard sign on a tanker carrying hazardous chemicals.

Activity 2.4A

Look at the hazard labels shown on page 48.
The hazard labels shown are:

- Toxic
- Oxidising
- Extremely flammable
- Harmful
- Very toxic
- Corrosive
- Highly flammable
- Irritant
- Dangerous to the environment
- Explosive.

Draw the symbol, add the correct label and explain what the symbol means in more detail. Give a chemical example for each category.

Sources of information

The data required for the labelling of substances can be taken from several sources of information.

In schools and colleges, CLEAPSS Hazcards will give the potential hazards, as well as information for the technical staff who might be using more concentrated chemicals to prepare dilute solutions for the class.

In the workplace, chemical manufacturers will produce data sheets giving all of the relevant information for their products' use. These can be found on websites for the manufacturers or suppliers or alternatively on discs that the manufacturer supplies that can be also used for ordering supplies.

Hazard data sheets will give details such as what to avoid when using the substance, exposure limits and any specific hazards, such as the substance being a **carcinogen**, a **teratogen** or a **mutagen**.

Safe working practices

A lot of the procedures and practices that were investigated in section 2.1 of this unit will have health and safety implications.

Risks can be minimised by using personal protective equipment (PPE) and/or other pieces of equipment such as fume cupboards or laminar flow cabinets, as well as ensuring that staff use safe procedures when handling substances.

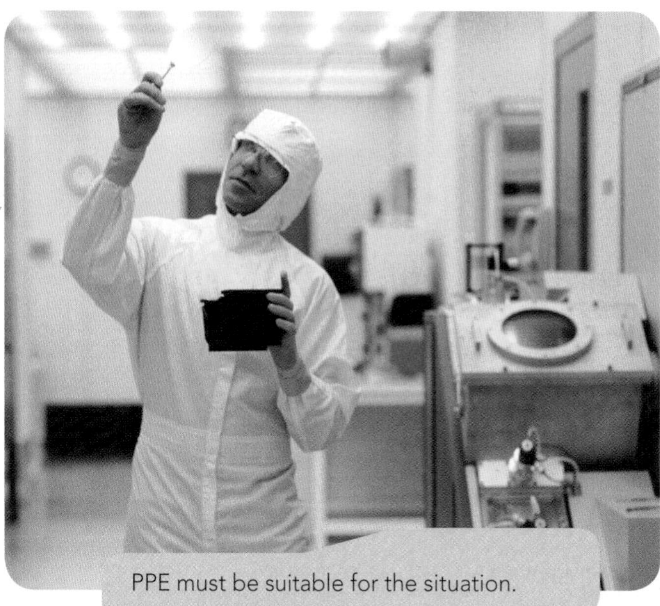
PPE must be suitable for the situation.

Activity 2.4B

Use the list of PPE below to say what each one might be used for in various types of laboratory.

Laboratory coat	
Protective gloves	
Goggles	
Visor	
Protective shoes	
Protective glasses	
Fume cupboard	

Regulations and legislation

As well as laboratories' own safety standards, there are many organisations that keep a check on laboratories to make sure they are maintaining the standards required for their particular scientific area and that the staff who work there are not put at risk.

The people who use the products or services supplied by the scientific workplace are also protected by other organisations, as is the environment around the workplace.

There are **laws** and **regulations** in place to monitor all types of scientific work in factories, in the field and in laboratories themselves.

Health and Safety at Work Act

This is the law that most people have heard of. It covers all aspects and areas of the workplace. It is important when investigating or using this law that you look at the most up-to-date version, as parts of it are updated to meet current standards.

The Health and Safety at Work poster.

All workers are made aware of this law in posters which explain the basics. The 'Health and Safety at Work' poster should be displayed somewhere in the workplace.

There are many regulations and laws that have to be obeyed in the workplace.

Some of these are made by the Health and Safety Executive (HSE). Their purpose is to 'prevent death, injury and ill health to those at work and those affected by work activities'.

Organisations with more than five employees have to produce a 'Health and Safety Policy' document. The HSE oversees this document which sets out the general approach, objectives and the management of health and safety in the business. To do this they oversee the following processes:

1 Writing the policy
2 Assessing risks
3 Providing facilities
4 Training workers
5 Consulting workers
6 Displaying Health and Safety at Work posters.

The HSE work in conjunction with many other agencies to ensure that all aspects of health and safety are covered in a common way.

The HSE have their own laboratories with scientists and technicians researching the problems seen in different types of workplace.

Did you know?

The HSE's main laboratory is set in 550 acres, with over 350 highly qualified staff.

For example, 'Health and Safety and the Healthcare Sector' is a part of the HSE where, althought clients work in different areas of the health service, they all have one common aim. This is to improve patient and worker safety. They work closely with hundreds of occupational health companies providing support and a wide range of analytical services. For instance, they monitor the survival and spread of resistant microorganisms in the hospital environment.

HSE inspectors carry out visits to fact find, monitor and advise on situations concerning health and safety.

COSHH Regulations 2002

COSHH stands for Control of Substances Hazardous to Health.

These regulations aim to control exposure to chemicals and protect workers' health.

COSHH research and publish data giving the amounts of chemicals or length of time workers can be exposed to chemicals before they are endangering their health.

Action to be taken when incidents do occur

However well an organisation is run, illness and accidents can still occur. This is where the Reporting of Injuries, Diseases and Dangerous Occurrences Regulations 1995 (RIDDOR) are used.

Employers have a legal duty to report ill health, accidents and also accidents which did not quite happen (or a near miss as it is sometimes called). As part of this procedure the employer must keep a register of all accidents and near misses so that they can look at the way they work to see if changes can be made to make the workplace safer. HSE inspectors will also need to see this documentation if an incident occurs.

Activity 2.4C

Using this government website, access the form for reporting accidents:

Health and Safety Executive – forms

Print the form and complete the details required as if you were the person who was a witness to an incident. Use your imagination and the information you have about laboratory work.

In the event of an accident, details must be logged in the organisation's accident book. On the government website of the Health and Safety Executive (HSE) there is a set of forms that can be used in the event of an accident. Further information on RIDDOR can also be found there.

Standards

As well as laws there are regulations and standards. Laws must be enforced in all workplaces, however, each type of laboratory will have laws, regulations and standards that are specific to them.

UKAS regulations – United Kingdom Accreditation Service

UKAS is 'the sole national accreditation body recognised by the government to assess, against internationally agreed standards, organisations that provide certification, testing, inspection and calibration services. Accreditation by UKAS demonstrates the competence, impartiality and performance capability of these evaluations.'

This means that, if UKAS agree, the organisation can show that it is working to internationally agreed standards.

UKAS is a non-profit distributing company and it deals with testing, calibration laboratories, certification bodies, proficiency-testing schemes and medical laboratories.

Quality standards

Many of these standards are monitored by the British Standards Institution (BSI). Certification demonstrates to customers, competitors, suppliers, staff and investors that industry-respected practices are in use.

Many of the quality standards are prefixed by 'ISO'. Other commonly used prefixes include:

BS – this is a British standard and is mostly used in the UK

EN – this is a European standard and is used throughout Europe

ISO – this is an international standard and may be used throughout the world.

BS compliance shows that the organisation is working to the standards set by the British Standards Institution.

What is a 'standard'?

The BSI (British Standards Institution) website states that:

'A Standard is an agreed, repeatable way of doing something. It is a published document that contains a technical specification or other precise criteria designed to be used consistently, as a rule, a guideline, or a definition.

Standards help to make life simpler, and increase the reliability and the effectiveness of many goods and services we use.'

Did you know?

Standards are designed for voluntary use and do not impose any regulations. However, laws and regulations may refer to certain standards and make compliance with them compulsory.

An example is the size and shape and layout of credit cards. Standard BS EN ISO/IEC 7810:2003 defines their physical characteristics. Adhering to this standard is one of the reasons why the cards can be used worldwide.

A further example is ISO 14001. The ISO 14000 family covers issues relating to environmental management and BS EN ISO 14001 helps an organisation to:

- understand the impacts on the environment caused by its activities, and to
- achieve continual improvement of its environmental performance.

BS EN ISO IEC 17025 is a standard that deals with testing and calibration. It contains all the requirements that testing and calibration laboratories have to meet if they wish to demonstrate that they operate a quality system, are technically competent and are able to generate technically valid results.

Standards for international trade

As international science cooperation, trade, travel and communications increase, international standards are used to enable products to be traded across Europe and worldwide. Standards are an important tool in removing barriers to international trade.

The BSI Certification mark is awarded to businesses demonstrating best practice in management systems, e.g. quality management (ISO 9001) or environmental management (ISO 14001). The Kitemark is awarded by BSI to products and services that have been tested against the requirements of a standard to show that they are safe and fit for purpose. The Kitemark is owned by BSI which means that only BSI can award it.

Standards can cover:

- quality
- environment
- sustainability
- information security and governance
- risk
- food safety
- occupational health and safety
- energy management
- testing and calibrating
- manufacturing
- services
- materials and chemicals
- building and construction
- electrotechnical matters
- protective equipment
- information and communications technology.

Good Laboratory Practice and Good Management Practice (GLP and GMP)

This is part of a quality assurance procedure which is aimed at ensuring that products are consistently manufactured to a quality appropriate to their intended use. They provide guidelines for quality control and assurance in testing laboratories.

Particular standards are required here as, unlike other goods that can be taken back if faulty, if medicines or drugs are not of good quality, they could have already caused damage to the patient before the problem is detected.

Compliance regulations

Medicine and Healthcare products Regulatory Agency (MHRA)

A government agency with responsibility for standards of safety, quality and performance. Involved with clinical trials and approval of products. Good Clinical Practice (GCP) is part of the Inspection and Standards division of MHRA.

European Medicines Agency (EMEA)

Relates to medicines for human use and deals with regulation of medicines based on objective, scientific assessment of their quality, safety and efficiency.

Herbal Medicine Products (HMPC)

This is a part of the EMEA that deals with herbal and traditional ingredients and products.

Food and Drug Administration (FDA) US

As many of the medicines and drugs we use are manufactured in the US by US companies, these regulations will have an impact on the medicines and drugs we can use.

Assessment activity 2.4

BTEC

1 Investigate the safe working practices in a particular scientific workplace and make bullet point notes on them. P5

2 Explain the reasons why current regulations and legislation are required in the scientific workplace to ensure safe working practice. Use examples to show how things can go wrong if these regulations are not followed. Examples must be real and can be taken from the Internet, journals or the news. All sources and quotes must be acknowledged. M5

3 You are to take the role of a scientific journalist who has been given the task of writing a report for a national newspaper following accidents and allegations that health and safety are not being taken seriously in Britain's scientific workplaces. You should choose a scientific workplace and find out all of the legislation that is relevant for that workplace. D4

Grading tips

For P5 you might like to look at a few health and safety documents from organisations carrying out scientific work. Don't forget to use bullet points and keep to the requirements of the task.

For M5 give brief details of the regulation or legislation you are talking about. Think carefully about what could go wrong in a particular situation.

You will have to have an understanding of each of these types of legislation if you are going to reach D4. It would be useful to use headings such as 'Why is risk assessment carried out?' and then go on to explain how the organisation you are investigating is or is not doing as it should. Consider the consequences of not following regulations in your evaluation. Use your imagination. It is essential that, when considering legislation, rules and regulations, the most up to date information is used.

PLTS

Independent enquirer, Creative thinker, Reflective learner and Self-manager

You will develop these skills by researching, creating a scenario, using several sources to reach a conclusion and completing work to standard and deadline

Functional skills

ICT

You will develop your ICT skills by finding and saving information for later use, bringing together information to suit a purpose, presentation of information

WorkSpace

Marie Cribb

Analytical Chemist, Lilly

I work as an analytical chemist in the Analytical Technologies group at Lilly, which is part of Discovery Chemistry. I develop analytical and purification methods using chromatography to purify compounds before they go through to biological testing. I've been at Lilly since I was 18.

On a typical day I have to check the submission system to see which samples need to be run, and talk to the chemist about their submission. Then I will either start developing a method so a sample can be purified, or I will start purifying a sample. I have to multi-task so I can get as many samples run during the day as possible, without losing any quality.

Sometimes I have meetings or courses with the rest of my department, or with external visitors, or I could be involved with projects that are to do with something other than chromatography.

The things I like best about my job include the fast pace of work, the variety, and the opportunities I have been given. I completed a chemistry degree part time through the company. Also, knowing I may contribute to a project that creates a new drug to combat a disease state is very rewarding.

Yesterday, one of the samples didn't behave as expected when I started purifying it. I realised there was a problem with the chromatography instrument. I had to work out what the problem was by checking parts of the instrument, such as the pumps, and then speak to an engineer from the instrument manufacturer to order a spare part so I could fix the fault. I needed to make sure the manufacturer had the right information so they could send the correct part and minimise downtime, and also tell the chemist that the purification of their sample would be delayed.

Think about it!

- Why is chromatography such an important part of analytical chemistry?
- Why is it important for an analytical chemist to know something about the instruments they use?

Just checking

1 Why are some procedures and practices appropriate and important in some areas and not in others?
2 Why is communication in many forms in the workplace essential for effective, efficient and safe working?
3 What has a great impact on the efficiency and safety of the people working there?
4 Do you remember which rules and regulations in the workplace are laws and which are not?
5 Who is responsible for safety in the workplace?

Assignment tips

To get the grade you deserve in your assignments remember to do the following:

- Work on the tasks required for the pass criterion. You should be given lots of information to help with the pass tasks. If the merit and distinction work follows on, work to complete to the highest level of which you are capable. Read the information that you are given and do what is required.
- To achieve merit and distinction criteria you will be expected to do more of the work by yourself. Research is fundamental and a variety of websites and text books for the topic is essential to provide further information. Make a note of websites you have used and include them in your references section. Many sites provide valuable links which should be explored. Specific scientific sites and journals are important and can give some in-depth information for the merit and distinction grades.
- When researching websites only use information that you understand and do not just copy and paste information from specialist sites.

Some of the key information you'll need to remember includes the following:

- Specialist laboratories will have their own set of procedures and practices which they will have to observe to be allowed to continue their work.
- Communication is critical in the laboratory and between laboratories and outside agencies to allow the vital flow of information. Communication can help prevent serious breaches of health and safety rules.
- The design of the scientific workplace can have a positive or negative effect on the people who work there. Consideration of many aspects must be given to make the workplace a safe and pleasant place to work.
- Laboratory information, test results and staff information must be managed and stored correctly, safely and securely to avoid the information being lost or corrupted. Modern methods will include computer storage often using purpose-designed programmes.
- Each scientific workplace will have their own hazards and the safety rules will be designed to prevent damage to the work force, buildings and equipment and the outside environment.

For more information on ...	Visit:
Equipment in use in school and college laboratories Consumables These websites will show up-to-date supplier websites:	CLEAPSS website. Contact your tutors for access details. The Association for Science Education Laboratory News MedLab News Laboratorytalk Supplier websites
Procedures and practices Different branches of science have their own websites outlining the work carried out by their members	The Institute of Biomedical Science The Royal Society of Chemistry The Institute of Physics The Society of Biology
Disposal of waste	CLEAPSS website. Contact your tutors for access details.
Laboratory design	Computer packages for design CLEAPSS website. Contact your tutors for access details. CLEAPSS G14 Designing and Planning Laboratories The Association for Science Education
Laboratory Information Management Systems	STARLIMS Corporation Accelerated Technology Laboratories Inc.
Health and safety	Government websites HSE – incident forms
Quality control British Standards	British Standards Institution

Credit value: 10

3 Scientific investigation

Investigative techniques used in science involve a range of useful and transferable skills that, once learned, will provide a solid foundation for scientific questioning and remain with learners throughout their social and working lives.

In this unit you will appreciate the importance of planning and the use of suitable and reliable information sources. You will use essential scientific protocols for recording information sources and produce risk assessments using COSHH and other laboratory regulations.

You will learn about correct experimental technique, and observational and manipulative skills, whilst keeping an up-to-date laboratory logbook and records.

You will gain an understanding of the methods used to analyse data from investigations, applying mathematical testing and assessing the validity of your method.

Finally, you will learn how to process the results of your investigation into a meaningful and acceptable report, using scientific terms, data displays, evaluation of your findings, and references and bibliographies.

Learning outcomes

After completing this unit you should:

1 be able to plan an investigation relevant to the area of study
2 be able to undertake the planned investigation, using appropriate scientific principles
3 be able to collect, collate and analyse the results from the investigation
4 be able to draw conclusions from the investigation.

Assessment and grading criteria

This table shows you what you must do in order to achieve a pass, merit or distinction grade, and where you can find activities in this book to help you.

To achieve a **pass** grade the evidence must show that you are able to:	To achieve a **merit** grade the evidence must show that, in addition to the pass criteria, you are able to:	To achieve a **distinction** grade the evidence must show that, in addition to the pass and merit criteria, you are able to:
P1 state the objectives and hypothesis relating to the investigation **See Assessment activity 3.1**	**M1** analyse the research information and discuss its relevance to the planned experiment **See Assessment activity 3.1**	**D1** evaluate the different approaches considered for the investigation, justifying the hypothesis chosen **See Assessment activity 3.1**
P2 produce a list of relevant research resources using a recognised protocol for recording them **See Assessment activity 3.1**		
P3 produce a realistic working plan for the experiment including health and safety assessments **See Assessment activity 3.1**		
P4 demonstrate the required manipulative skills to assemble relevant equipment and materials **See Assessment activity 3.2**	**M2** justify the choice of experimental techniques (and their modification if any) as a means of increasing accuracy, reliability and validity **See Assessment activity 3.2**	**D2** evaluate the effectiveness of the investigative procedures, suggesting how these could be improved **See Assessment activity 3.2**
P5 safely carry out the planned investigation **See Assessment activity 3.2**	**M3** justify the statistical techniques used by relating them to the validity of their findings **See Assessment activity 3.3**	
P6 demonstrate the ability to accurately record the results obtained, using scientific protocols **See Assessment activities 3.2 and 3.3**		
P7 analyse the results obtained using appropriate statistical techniques **See Assessment activity 3.3**		
P8 explain the conclusions gained from the investigation **See Assessment activity 3.4**	**M4** justify the conclusions made, drawing on primary and secondary research data **See Assessment activity 3.4**	**D3** using scientific protocols evaluate the outcomes of your investigation **See Assessment activity 3.4**
P9 present the conclusions in a format that uses accepted scientific protocol and language **See Assessment activity 3.4**		

How you will be assessed

Your assessment could be in the form of:

- diagrams, graphs and tables
- presentations
- written reports
- your laboratory notebook.

Ben, 17 years old

Completing this unit helped me to write up my science investigations properly and my work began to look very professional. I was also able to use the skills I learned in science to organise my work better in other subjects.

I was particularly interested in the practical aspects of the unit, and think that I now understand why our science technician takes a lot of time and effort when preparing our chemicals and apparatus.

My laboratory notebook looks very scientific and I used it regularly, especially for diagrams and rough calculations. I was able to complete all the calculations with practice.

The pictures in the book helped to explain many different things and our visit to an industrial chemicals plant was really interesting. We could actually see the technicians using some of the same types of apparatus as we used in the school laboratory.

Catalyst

Science everywhere

When we observe any phenomenon happening in the world around us and ask ourselves what, how or why it is happening, we have unavoidably entered into **scientific investigation**.

The factors which determine the flow of raindrops down a vertical pane of glass, for example, are varied and quite involved. They include: pH, droplet size and shape, particulate matter, observation altitude, gravitational field strength, surface tension, frictional surface…

In pairs or as a group, discuss how you would investigate the factors affecting the speed of raindrops. Work on a simple **hypothesis** and make a short presentation to the class.

3.1 Investigation planning

In this section:    

Key terms

COSHH – Control of Substances Hazardous to Health regulations for use in education and industry, 1999.

CLEAPSS – Consortium of Local Education Authorities for the Provision of Science Services – information cards on chemical hazards.

ASE – Association of Science Education.

Variables – aspects which change during an investigation: independent variables changed by the learner, dependent variables change as a result.

When planning scientific investigations, it is essential to approach the task methodically and in a step-by-step fashion. Very often, the planning stage is not given enough attention, and this lack of preparation can sometimes account for incorrect or false results and conclusions.

The difference between short and long science investigations is generally in the outcome. Shorter investigations tend to have a limited aim while longer ones are more open-ended and sometimes unpredictable. For both, however, you need to consider the following investigation design principles.

1 Choice of investigation and aim or objective.
2 Suitability of producing a preliminary test.
3 Formulating a hypothesis.
4 Risk assessments.
5 Research tools – write down titles of books, websites, journals, etc.
6 The variables – can they be measured and controlled? Which are kept constant?
7 The method (working plan):
   - equipment used
   - how the measurements are to be taken and the observations recorded
   - analysis techniques used for data
   - treatment of possible errors in the investigation.

Choice of investigation

There are countless suitable and interesting investigations in all three scientific disciplines. The final decision is usually made by consideration of the equipment available, the degree of difficulty, the results which can be obtained and the time needed. Here are some examples which are suitable in most educational establishments.

Physics	Chemistry	Biology
factors affecting electrical resistance	factors determining rates of reaction	testing for reducing sugars
resistivity of various metal wires	deterioration of iron tablets	factors affecting the rate of photosynthesis
investigating stress and strain of different materials	determination of concentrations of unknown samples	investigating permeability of membranes
resistance change with temperature for thermistors	vitamin c content in fruit juices	the actions of enzymes in digestion
improving a pinhole camera with lenses	oxidation of ethanol to ethanoic acid	factors affecting plant growth
using a light gate to investigate freefall	electrolysis of aqueous salt solutions	effect of polluted water on seed development
determination of relationship between aileron angle and lifting force for aircraft wing models	determination of chlorine content in public baths	effects of caffeine on water fleas (*Daphnia*)

table continued

determination of terminal velocity of a ball through fluid using light gates	electrical charge calculations from electrolysis of copper(II) sulphate, silver nitrate, lead(II) bromide	plant responses to stimuli
finding the factors which affect the length of a ski jump	enthalpy changes in combustion of alcohols or flammable liquids	factors affecting transpiration in plants
determination of relationship between frequency and length for taut wire	testing water hardness	the effect of penicillin on bacterial growth
		yeast respiration factors

PLTS

Creative thinker

You will practise this skill when designing and setting up your objectives and hypothesis.

Suitability of producing a preliminary test

A preliminary test is just like a practice run but with a difference. In science investigations, the results that we are expecting may fall within a certain range of values and a preliminary test can give us an indication of what values to work with. In titrating an acid and a base, for example, the concentration of the unknown sample may fall well outside the volume capacity of the burette. So, before carrying out a detailed titration, it is good practice to perform a rough titration. Not all scientific investigations need a preliminary test; you will have to decide for yourself.

Formulating a hypothesis

In most cases, a hypothesis is an assumption based on your knowledge, understanding of the topic and observations. When carrying out an investigation, we can hypothesise about the outcome and may change this when other observations are made.

For example, say you want to carry out an investigation into which chemical elements are necessary for plant growth. The hypothesis could be 'Nitrogen is needed for plant growth', but your observations show that other elements are also important; so your hypothesis can be changed during your work and further investigations can be carried out.

Remember, a hypothesis can always be tested and disproven, but not proven completely.

Did you know?

The hypothesis that links animal and plant imprints in rocks to prehistoric organisms could only be supported if time travel was possible.

Risk assessments

Before any practical work is performed, complete a risk assessment thoroughly and have it checked by your tutor. Your place of learning has a responsibility to ensure that this is carried out following the Health and Safety at Work Act of 1974. Remember, a risk assessment is your way of minimising the potential risks involved in your activity.

Investigations in all three science disciplines carry risks but chemistry has more than most. When carrying out a chemical investigation, use every available resource to help you prepare your risk assessment and to fully understand what potential risks are associated with the substances that you are about to use.

Before any investigation:

- identify the equipment and substances
- research the hazards and potential risks
- identify measures to deal with spills and breakages
- have your risk assessment checked.

Activity 3.1A

Where can you find information on using hazardous substances?

Research tools

Correct and useful research is a skill which takes time and the build up of knowledge to be carried out successfully. However, here are some important pointers to help you develop your research abilities.

- **Always use more than one source of information** – there are many sources available on the same subject material, including tutors, textbooks, the Internet, newspapers, magazines, journals and television.
- **Carefully choose your research titles** – when searching for information on acid/base titrations, for example, be specific. You may get a lot of information about titrations in general which will take a lot of time to read and may not all be relevant.
- **Select material from an authoritative source** – the professional experience and qualifications of the author(s) are important, and when more than one source agrees in content, you can develop confidence in the sources you are using.
- **Avoid plagiarism of other people's work** – sometimes you may need to include a direct quotation of a researched note to help with your report. Keep these to a minimum, do not change the quotation at all, and include a suitable reference to indicate where the article came from.

Source listing protocol

Your research sources must be listed at the back of your final report in a recognised manner. Most countries use the *Harvard System* or variations of it.

- **Example of textbook referencing:**
  BTEC Level 3 in Applied Science Student Book (Pearson, 2010) ISBN 9781846906800
- **Example of Internet webpage referencing:**
  Wells, D. (2001) *Harvard referencing* [Online]. Available: http://lisweb.curtin.edu.au/guides/handouts/harvard.html [14 Aug 2001]. (Samantha Dhann, 2001)

Your research sources should be listed in two categories:

- **references** – work quoted or referred to in your report
- **bibliography** – all work used for your report, but not necessarily quoted or referred to.

The variables

Factors which can change or be changed in an investigation are **variables**. Identifying which variables you can reliably change yourself during an investigation is important if it causes another factor to change as a result – this is 'cause and effect'. In electricity, Ohm's law describes the relationship between voltage and current through a conductor. When the voltage is increased, the current increases. Similarly, for springs, Hooke's law describes the relationship between load and extension. When the load is increased, the extension increases.

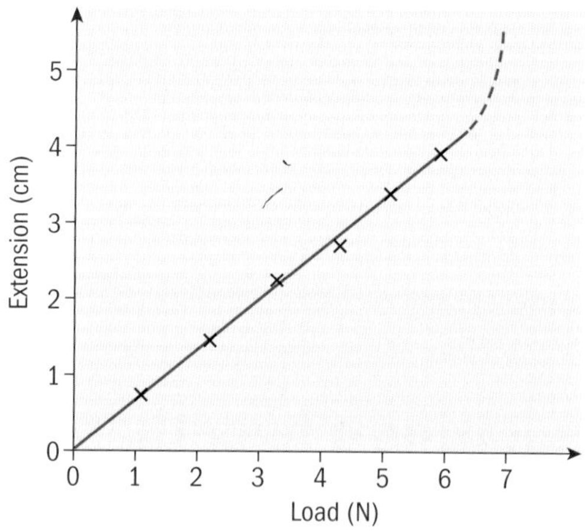

Graph of load v extension showing Hooke's law.

Case study: Risk assessment

David works as a production operative for a large chemical manufacturer. His work involves weighing substances with precision and accuracy and brings him into regular contact with dangerous chemicals which need very careful handling.

David is fully aware of the dangers in using these substances and how to dispose of them safely: 'Preparation is fundamental in preventing accidents and reducing risks. I attend many short courses on risk assessment and how to deal with incidents in and around the laboratory'.

In many cases it is important to include a control experiment within the main investigation to make sure that the 'effect' produced by the 'cause' is real and not due to other factors. For example, when investigating the possible change in oxygen (**dependent variable**) produced by the plant *Elodea* (Canadian pondweed) caused by an increase in light (**independent variable**), it is suitable to have an identical investigation set up in complete darkness to compare to the main investigation (**the experiment**). As all the other factors – amount of water, quality of water, pH and temperature – are kept constant, they are the **control variables**.

In science, there are many instances when a simple cause and effect relationship is not totally obvious. In climate studies, scientists involved with the difficult task of collating and analysing data are confident that there is a direct relationship between carbon dioxide level increases (as a result of industrialisation) and a rise in average global temperatures. Others argue that the Earth's climate is too complex and is affected by many natural variables such as solar energy, volcanic activity, oceanic effects, natural aerosols and Earth's orbital changes.

The method

Producing a working plan (method) for an investigation involves the use of research notes, tutor guidance and preliminary testing (where appropriate). If the working plan is thorough, the investigation should produce useful results and no real surprises. You should try to cover the following points:

- apparatus to be used – with a labelled diagram
- theory – research notes on the topic of your investigation
- notes of preliminary tests (if appropriate)
- step-by-step instructions on how the investigation will be performed
- Health and Safety (risk assessment)
- a prediction – what you think will happen
- a hypothesis (if possible) – suggestions for explaining what may be happening
- variables – which will be kept constant? which will change?
- appropriate method of presenting your results
- analytical technique to be used which is best suited to the results and data
- measures you will take to address possible errors (e.g. repeat readings, calibration).

Your investigation planning stage will take time to develop and needs to be monitored by your tutor and/or a technician regularly before the investigation can properly begin.

Assessment activity 3.1

BTEC

Your work as a junior research technician in the science department of a renowned national university allows you to develop your investigation procedures at every opportunity. You must set up an investigation and demonstrate your activity to visiting sixth formers from a local secondary school.

1 In your chosen investigation, state what you plan to achieve. **P1** Use research notes, and list your notes in rough according to the suggested protocol ready to reproduce them for your final report. **P2** Produce a concise hypothesis from your research. **P1**

2 Draw a table of your research references showing the information from each concerning your chosen topic of investigation. Analyse which ones are relevant to your work and which ones are not and discuss giving reasons. **M1**

3 Write out a method using information from this section. **P3** Outline any other methods you may have considered and give the reasons to justify your final choice. **D1**

Grading tips

To achieve **P1**, **P2** and **M1** make sure you provide good points of reference and background information from many sources. List each source as you work through them.

To achieve **P3** include descriptions of the main aspects covered under 'Method' above.

It is essential that you choose an investigation which has variations of approach so that you can evaluate them to achieve **D1**.

3.2 Carrying out investigations

In this section:

Key terms

Accuracy – closeness of readings to actual value.

Precision – the degree of uncertainty of a measurement; usually the size of the unit of measurement used.

Health and Safety

Before carrying out any practical work, setting up apparatus or even gathering materials for the investigation, it is always wise to revisit your risk assessment and make yourself fully aware of the hazards involved and how best to minimise the risks.

The health and safety of you and others in the laboratory should be constantly in your thoughts as you begin your investigation, and be there continually throughout.

Wear goggles and a laboratory coat.

Using equipment

Notes and diagrams from your method are now ready to be put into practice with the setting up of apparatus. Your tutor and/or technician will provide you with a list of aspects that will form the basis of the assessment of your practical skills. These should include:

- your awareness of Health and Safety issues
- competence in your assembly of equipment
- your ability to physically manipulate the equipment to obtain results
- skills in observation and record keeping
- your adherence to good laboratory practice (GLP)
- **accuracy** and **precision** of measurements.

Correct observational technique when measuring with a burette.

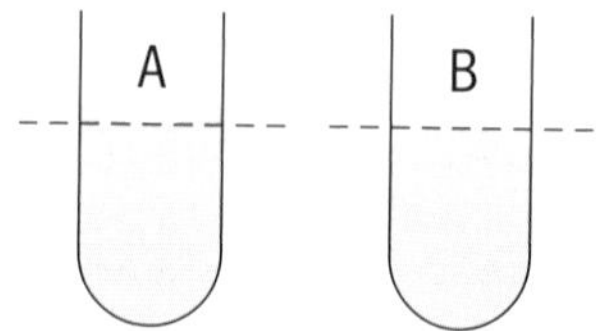

A = meniscus view and reading for most liquids
B = meniscus for mercury

Correct meniscus reading.

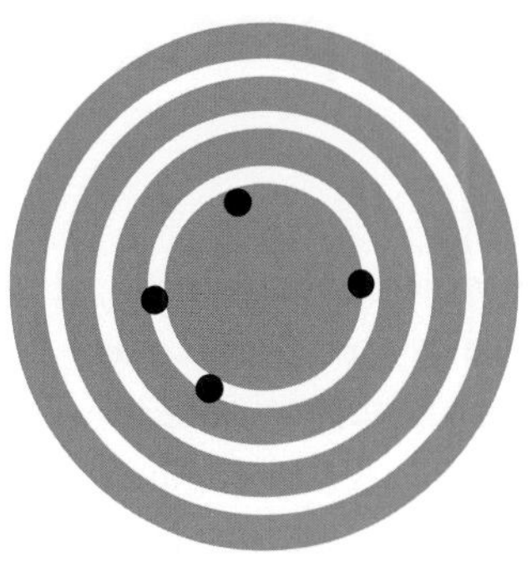

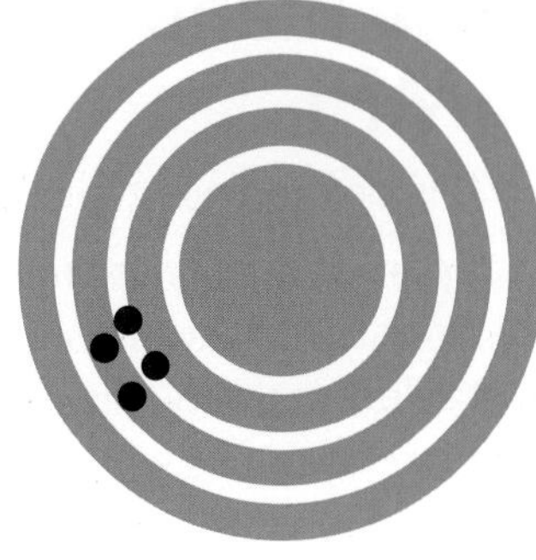

The difference between precision and accuracy.

Activity 3.2A

List six aspects of your practical work which will be assessed by your tutor or technician.

Good laboratory practice (GLP)

The principles of GLP were completed in 1981 by the Organisation for Economic Cooperation and Development (OECD). These were adopted by the European Union and have subsequently been updated.

In the UK, laboratories that wish to promote their technical competence and the overall quality of the systems they operate may want to achieve listing in the United Kingdom Accreditation Service (UKAS) register of companies. This organisation is sponsored by the Department for Business, Innovation & Skills and is committed to maintaining quality in line with strict international laboratory standards.

In the wider context, GLP ensures that tests carried out in non-clinical working laboratories are reliable and well regulated, assessing the hazards and risks to the public and the environment.

Industries using these principles include:

- industrial chemicals
- food and food additives
- agrochemicals
- pharmaceuticals
- veterinary medicine
- cosmetic chemicals.

In the context of your school or college, GLP relates to the general, well established principles of carrying out a practical investigation using safe procedures and suitable scientific techniques.

Safety

In terms of safety your GLP should follow these rules:

- produce a detailed risk assessment
- wear appropriate eye protection and a laboratory coat when the risk assessment indicates they are needed
- tie back long hair
- wear waterproof dressings on wounds
- wash hands before and after the investigation
- do not take food or drink into the laboratory and do not eat or drink in the lab
- do not handle mains sockets and plugs with wet hands
- report breakages and damaged equipment
- know when to use a Bunsen (yellow) 'safety' flame
- handle glassware very carefully
- locate the gas and electrical emergency cut-offs
- make sure you know how to use eye-washing facilities.

Case study: GLP

Nizar works as a soil technician for a regional analytical laboratory in the UK. His duties include the preparation for chemical and biological analysis of soils that may come directly from the Environment Agency.

The information gathered from each analysis is carefully organised into a database (e.g. LIMS, Laboratory Information Management System) and reports are generated once the analysis is complete. Nizar recognises the importance of following GLP, which underpins the reliability of the analysis that he carries out.

Functional skills

English

You will be developing your English skills when carrying out research (safety in the lab).

Did you know?

Many of Sir Isaac Newton's laboratory notes have still not been fully interpreted. Scientists of that time often used unusual jargon in their works and Newton also used his own unique symbols.

Technique

In terms of technique your GLP should follow these rules:

- take liquid measurements with your eye in line with the surface
- repeat readings up to three times whenever possible
- conduct preliminary tests if suitable
- use correct **base** and **derived** units and symbols of measurement consistently (see page 149)
- record results in a laboratory notebook
- work in pairs or small groups.

Good clinical practice (GCP)

Good GCP is defined as:

'a standard for the design, conduct, performance, monitoring, auditing, recording, analyses, and reporting of clinical trials that provides assurance that the data and reported results are credible and accurate, and that the rights, integrity, and confidentiality of trial subjects are protected'

Source: Medical Research Council

In general, GCP is a quality standard for studies involving humans. It became effective from 1997 in the European Union, Japan and USA, setting down the principles of sound ethics, conduct and record keeping. It helps to ensure that clinical trials are reliable, confidential and safe.

Activity 3.2B

What is the difference between Good Laboratory Practice and Good Clinical Practice?

Good manufacturing practice (GMP)

GMP is defined as:

'that part of quality assurance which ensures that medicinal products are consistently produced and controlled to the quality standards appropriate to their intended use and as required by the marketing authorisation (MA) or product specification. GMP is concerned with both production and quality control.'

Source: Medicines and Healthcare Regulatory Agency, www.mhra.gov.uk

The Medicines and Healthcare products Regulatory Agency was set up to regulate quality and safety in the manufacture of medicines and medical devices. The agency is committed to ensuring:

- benefits to the public justify the risks taken
- availability of information
- quicker access to medical products and treatments
- thorough investigation of defective medicines and appliances.

PLTS

Self-manager

You will have the opportunity to practise this skill when organising laboratory time, dealing with the pressure of experimental procedures and time/equipment constraints, and seeking advice from your tutor.

Laboratory notebook

Considered a lost art, but nevertheless a very important document that shows the work carried out during investigations. Many well known scientists have had their work interpreted and analysed later from notebook recordings that they completed during practical investigations. To use a laboratory notebook successfully, follow these guidance points.

1. Use black pen. Do not use pencil – except for graphs and diagrams – and don't use correction fluid or rip pages out.
2. Start each task with a fresh page.
3. Include the title, what the task is and the date.
4. Make notes on safety issues; identify hazards.

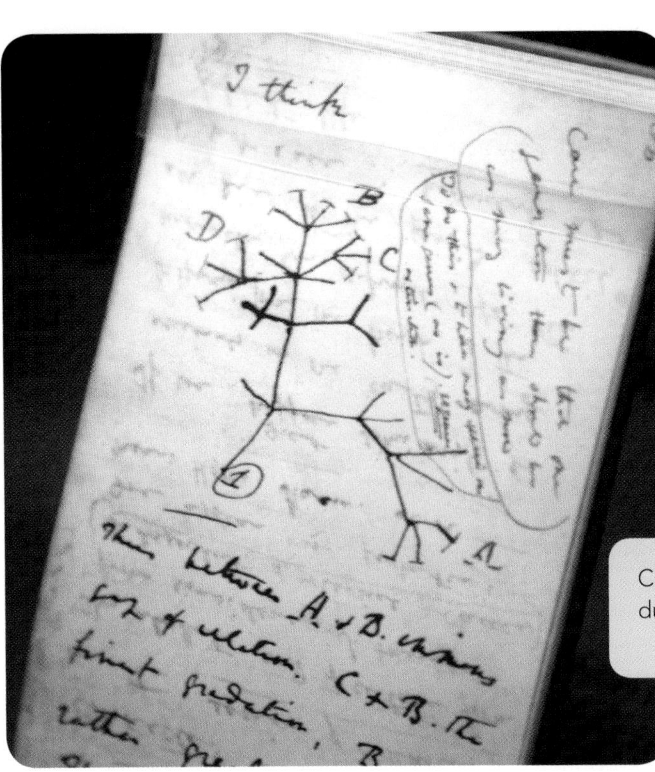

5 Comment using bullet points and highlight those of particular interest.
6 Drawings should be made of specialist equipment only and well labelled.
7 Tables and graphs must be clear and labelled.
8 Make your calculations and present them clearly.
9 Draw general conclusions and link to the hypothesis.
10 Damage will occur, so protect your notebook as best you can.

Charles Darwin's notebook used during his voyage on the 'Beagle'.

Assessment activity 3.2

BTEC

In your capacity as a junior laboratory technician in a research department you are called upon to produce a demonstration of Good Laboratory Practice. You are to carry out a practical demonstration to visiting students which allows you to do this.

1 Arrange and assemble your apparatus from your working plan, following the required health and safety guidelines. **P4**
2 Carry out your practical activity using Good Laboratory Practice, observing safe working practices and demonstrating good manipulative and observational skills. **P5**
3 Make notes of your observations and measurements in a suitable format using your laboratory notebook. **P6**
4 Jusifty in your notebook why the equipment you are using is suitable. Comment on alternative methods to get the same result if this applies to your activity, and include a discussion on the accuracy and precision of methods used. **M2**
5 Evaluate your procedure and provide a detailed list of improvements that could be made to your procedure. **D2**

Grading tips

First identify the apparatus needed and place at your workstation before assembly to achieve **P4**. A practice run may serve to demonstrate your manipulative skills for the actual investigation.

Exaggerate your observations somewhat to make sure that your tutor or technician witnesses your work to achieve **P5**.

Your laboratory notebook should be positioned close to your work without the possibility of it being damaged. Don't worry about the roughness of your notes, although to achieve **P6** your results must be accurately recorded.

To achieve **M2**, make a list of modifications you made to complete the activity, such as suitable thermistor values, concentration of caffeine, or voltage levels for electrolysis.

For **D2** you should break down the procedure and apparatus used into a series of steps. Outline improvements that could be made to each step, explaining how the suggested improvement or change may affect the outcome.

3.3 Organising and analysing results

In this section:   

Mean – the average of all the numbers within a set of results.

Frequency – how often a particular value occurs in a set of values.

Make sure your figures add up.

Organisation of data

See Unit 8 for more information on statistics.

In many scientific investigations, a vast amount of data in the form of numerical figures can be generated. These numbers must be organised in such a way that analysis of the data is made simpler. The following example shows the steps needed to organise data for analysis.

Consider a set of 20 agar plates showing bacterial colony counts under identical conditions after 24 hours:

1, 2, 4, 3, 6, 7, 6, 8, 3, 9, 6, 7, 7, 6, 5, 4, 5, 6, 5, 8

Initial statistical analysis

Firstly construct a frequency table.

Colony count	Tally	Frequency
1	I	1
2	I	1
3	II	2
4	II	2
5	III	3
6	~~IIII~~	5
7	III	3
8	II	2
9	I	1
		20

Mode: looking at the frequency table the colony count which occurs most often is 6, so the mode is 6.

Mean: using the first set of figures for the 20 agar plates of bacterial colonies after 24 hours, the mean is 5.4.

Class intervals and frequency tables

When the data values are much larger, it is useful to group the figures into ranges of values which are called **class intervals**.

The same agar plates were kept under suitable conditions for a further 24 hours. The colonies grew in number:

30, 46, 45, 43, 53, 42, 51, 55, 61, 44, 50, 52, 35, 37, 54, 62, 68, 58, 56, 46

The range is from 68 to 30 (i.e. 38).

A suitable class interval width is 5 for this value, so 38 ÷ 5 = 8 (1 s.f.).

Class interval	Tally	Frequency
30–34	I	1
35–39	II	2
40–44	III	3
45–49	III	3
50–54	~~IIII~~	5
55–59	III	3
60–64	II	2
65–69	I	1
		20

Modal class interval: from this table the class interval that occurs most often is 50–54.

Standard deviation

The standard deviation indicates how closely the data are positioned around the mean. Using the same sample of bacterial colonies after 24 hours, the graph will look like this:

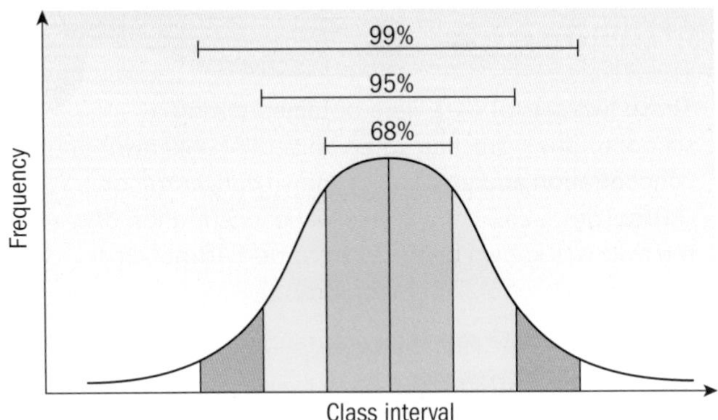

Standard deviation graph of bacterial colonies grown on agar plates.

Note Generally, data like these are normally distributed. This means that about 68% of the data lie within one standard deviation, 95% within two standard deviations and 99% within three standard deviations.

Standard deviation is calculated as follows.

1 Calculate the mean.
2 Subtract the mean from each of your datum values to get the deviations.
3 Square each deviation, and add them all up.
4 Divide this figure by one less than your sample number.
5 The **standard deviation** (s) is the square root of this value.

Unit 8: See page 171 for more information on standard deviation.

The *t*-test

This statistical method is used by many people to compare the means of two samples which may or may not look very different. The final figure is then checked in *t*-tables to determine the percentage probability in terms of how significant the differences in means are. There are two types, for un-matched pairs and matched pairs.

Worked example

Un-matched pairs

From microscopic analysis, the heart rates of two sets of water fleas (*Daphnia*), group A and group B, were recorded in cool river water. Caffeine solution of 0.01% concentration had been added to the water for group A. There were 10 fleas in each sample group (n = number of sample, i.e. 10). Is there a significant difference between the heart rates of the two groups?

Heart rate A	Heart rate B
113	68
111	56
136	62
121	78
108	82
109	64
117	66
122	78
132	77
116	81

Answer

Means ($\bar{x} = \frac{\Sigma x}{n}$): $A = 118.5$, $B = 71.2$

The sum of the squares of each value in each table [Σx^2]:

$A = 141\,225$, $B = 51\,438$

The squares of the totals in each column [$(\Sigma x)^2$]: $A = 1\,404\,225$, $B = 506\,944$

$\left[\frac{(\Sigma x)^2}{n}\right]$: $A = 140422.5$, $B = 50694.5$

Σd^2 using $\left[\Sigma x^2 - \frac{(\Sigma x)^2}{n}\right]$: $A = 802.5$, $B = 743.6$

Standard deviation (s^2) using $\left[\frac{\Sigma d^2}{n-1}\right]$: $A = 89.2$, $B = 82.6$

Variance of difference between means (sd^2) using $\left[\frac{s1}{n1} + \frac{s2}{n2}\right]$: $A = 8.92$, $B = 8.26$

$= 17.18$

Σd using [$\sqrt{sd^2}$]: $\sqrt{17.18} = 4.14$

$t = \left[\frac{\bar{x}1 - \bar{x}2}{\Sigma d}\right]: \frac{47.3}{4.14} = \mathbf{11.43}$

Looking at *t*-tables with 18 degrees of freedom (sample number – 2), the value obtained is much higher than even the 99.9% probability (0.001) which shows as 3.92 in the *t*-tables.

The heart rates of *Daphnia* in group A are significantly higher than those of group B, which can be linked in this case to caffeine.

Functional skills

Mathematics

You will be practising your maths skills when using statistical methods to find out if data collected are viable and you can draw conclusions from them.

Unit 8: See page 185 for more information on the *t*-test.

Units in science

We come into contact with units in all aspects of life. A supermarket with an offer on potatoes, for example, always provides the units. There is a considerable difference between selling potatoes at £1.00 per kilogram (kg) and £1.00 per pound (lb).

Using the appropriate unit of measurement is equally important. The distance between cities is measured in kilometres (km), not centimetres (cm), even though they are both units of distance measurement. Some common units and prefixes are shown below.

Volume: litres (dm^3), millilitres (ml)
Area: m^2, cm^2, mm^2
Resistance: ohms
Prefixes: micro (10^{-6}), milli (10^{-3}), kilo (10^3), mega (10^6)

In all your measurements and calculations in science, you must quickly begin to use the correct and appropriate units associated with the numerical figure. It is common to leave units out when performing calculations or making brief notes, but this is a habit which can cause serious confusion when work is being assessed. The table below gives some SI base units and examples of derived units.

Unit 6: See page 149 for more information on base units, derived units and SI units.

Base units	Derived units
second (s)	force (N)
kilogram (kg)	acceleration ($m\ s^{-2}$)
metre (m)	volt (V)
ampere (A)	speed ($m\ s^{-1}$)
mole (mol)	energy (J)
kelvin (K)	charge (C)
candela (cd)	pascal (Pa)

Assessment of accuracy and precision

Your investigation findings should now be assessed in terms of the accuracy of the results and precision of the readings or measurements. Guidance on understanding the difference between accuracy and precision has been given on page 65.

Activity 3.3A

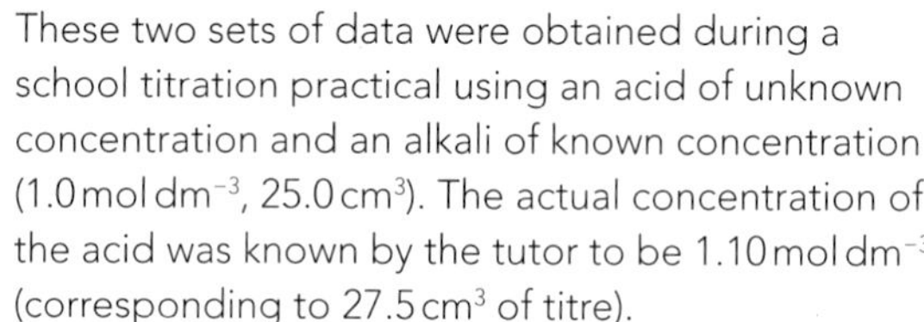

These two sets of data were obtained during a school titration practical using an acid of unknown concentration and an alkali of known concentration ($1.0\ mol\ dm^{-3}$, $25.0\ cm^3$). The actual concentration of the acid was known by the tutor to be $1.10\ mol\ dm^{-3}$ (corresponding to $27.5\ cm^3$ of titre).

Which of the data sets is (a) accurate, (b) precise? Explain your answer.

Data set 1	Data set 2
Volume of titre (cm^3)	Volume of titre (cm^3)
29.5	27
29.5	28
29.0	28

Repeatability and reliability

If your method and results can be reproduced almost exactly by someone else at a later date, then your investigation is **repeatable**. If the data are similar after many repeats, then you can be confident that the results are **reliable**. This is the basis of scientific work. Our theories can then be built upon until they become firmly established.

To be valid in scientific terms, ask the question: Will someone else be able to repeat my investigation to get the same or similar results?

Errors

When measurements are taken it is very important to take into account the possible error value in the reading, particularly when dealing with smaller scales of measurement.

In most cases, the **probable** error in measurement is quoted so that the value obtained can fall within a range of values and still be acceptable in science.

- Measurement of thin constantan wire
  - diameter measured value: 0.38 mm
  - probable error: 0.02 mm
  - range: 0.36 to 0.40 mm
  - label diameter: 0.38 mm ± 0.02 mm
- Resistor components for electronic circuits
  - resistor value: 1000 ohm
  - tolerance: 10%
  - range: 900 to 1100 ohm

It is common practice to estimate the probable error using the **precision** of the scale from which your readings are to be taken. Some useful examples are depicted.

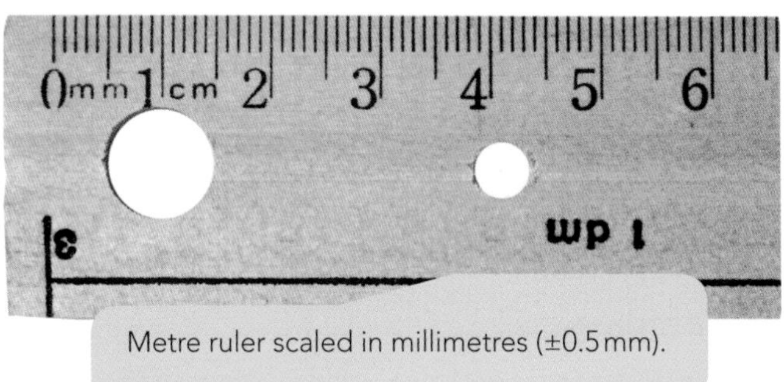

Metre ruler scaled in millimetres (±0.5 mm).

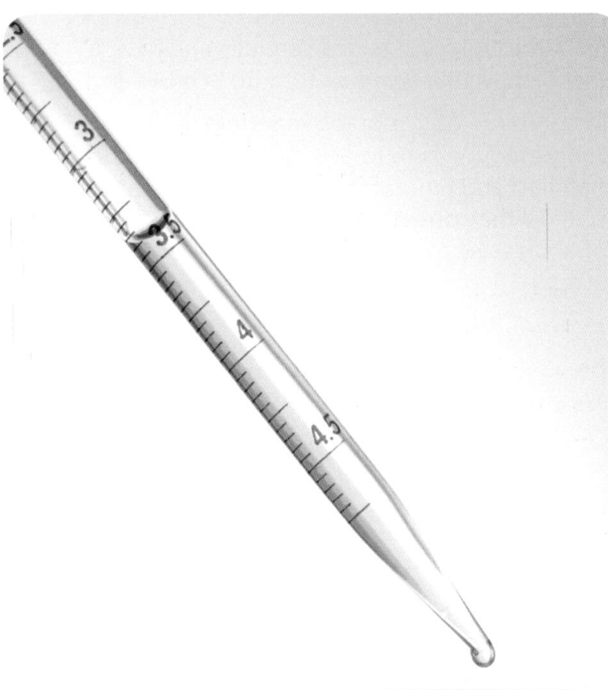

Pipette scaled in 0.05 ml (±0.025 ml).

Analogue DC ammeter scaled in 0.2 A (±0.1 A).

Guidelines for reducing errors

- Repeat readings.
- View meters, gauges and glass measuring devices at 90°.
- Check the calibration of equipment if possible (systematic errors).
- Use a second meter to confirm the reading from the first meter (systematic errors).
- Accept that digital equipment may be no more reliable than analogue.
- Make sure you understand the principles of your investigation.

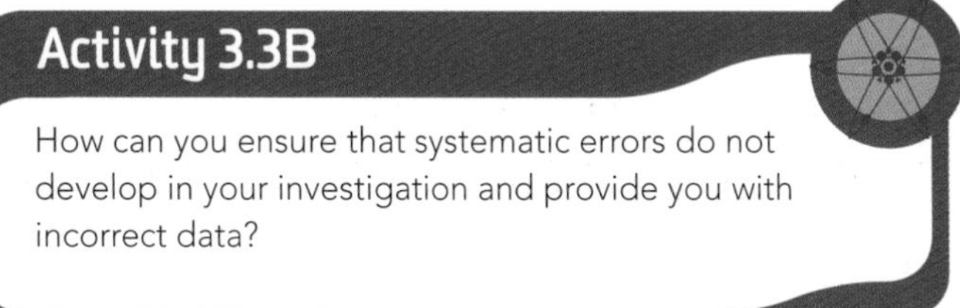

Activity 3.3B

How can you ensure that systematic errors do not develop in your investigation and provide you with incorrect data?

Assessment of information sources

The research information used for your planning stage in the investigation should have been correctly referenced and logged in your notes. If a number of sources were used, the working plan should contain only that information which will have been judged to be common to more than one source. This skill takes time and a build up of knowledge to develop successfully. You should not be too concerned if much of your information is discovered to be irrelevant when the results are completed.

Developing experience in scientific information research is fundamental to expanding the knowledge and skills that you will need to be a competent science learner. It is useful to identify sources of reliable information quickly and to table them in an appropriate manner. The following example may help.

Information source	Brief details of information	Useful	Not useful
Data and Data Handling for AS Level Biology, Bill Indge	suggested use of glass container of water, placed between lamp and *Elodea* to help eliminate effects of increased temp. on rate of photosynthesis	✓	

Assessment activity 3.3

M3 P6 P7 BTEC

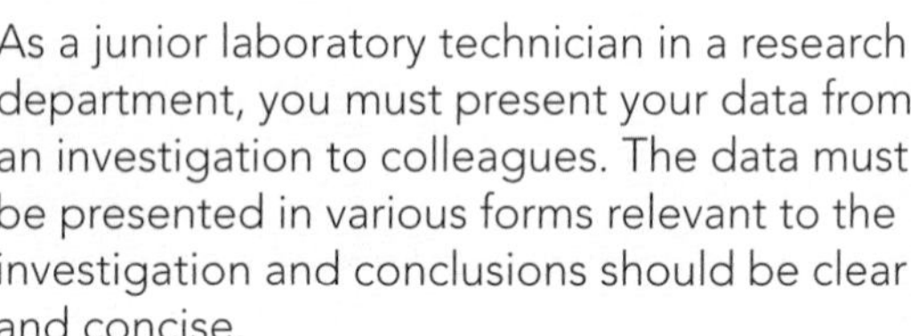

As a junior laboratory technician in a research department, you must present your data from an investigation to colleagues. The data must be presented in various forms relevant to the investigation and conclusions should be clear and concise.

1 Produce a table of results for an investigation which includes correct units, accuracy of recording and precision to acceptable significant figures. P6
2 Conduct statistical tests to analyse your data and draw a conclusion. P7
3 Comment on the method you used to analyse the data. Does the conclusion meet your expectations and does the statistical method provide evidence for this? M3

Grading tips

Your laboratory notebook will provide evidence for you to achieve P6 if it is clear and kept in a reasonable state.

Work through your maths carefully and have it checked by your peers or tutor to achieve P7.

To complete M3, a graph may help. If *t*-tests are performed, be sure to use the correct method for matched and un-matched pairs. Is your statistical method the correct one to use?

3.4 Evaluating investigations

In this section:    

Key terms

Correlation – a link which exists between two related factors.

Proportion – the ratio of relationship between two variables when one of them is varied.

Conclusion – decisions made about the investigation based on analysis of the data or observations produced.

Evaluation – the judgement of quality of the investigation.

Presenting data

Data can be presented in a number of ways:

- tables
- charts (bar, pie and histograms)
- graphs
- photographs and video
- sketches.

If possible, a number of different ways to present data should be used. Graphic presentations are very useful and easily reproduced to good effect on computers.

Graphs

Graphs help us to visualise the comparisons and relationships of data immediately and are not as daunting as looking at tables of numbers. Choosing the correct graph will depend on the data obtained, as shown in the table below.

Data type	Graph type	Description
Discrete (discontinuous) (no. of layers of insulation, no. of masses added….)	pie chart bar chart	labels on axes are for measured value and whole numbers for frequency, count or percentage
Continuous (length, time, temperature, mass, voltage…)	histogram line graph	histogram bars touch for continuation of values horizontal axis scale from low to high and any value
Categoric (compounds, metals, animals, plants…)	bar chart	specific word labels, e.g. metals (copper, tin, iron…) axes labelled as for discrete charts

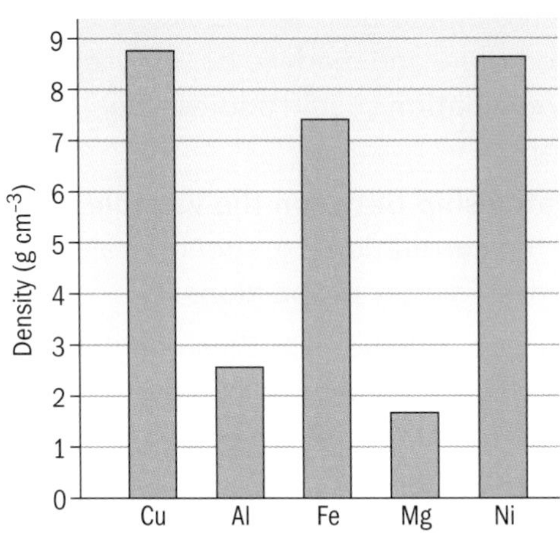

Bar chart showing densities of metals ($g\,cm^{-3}$).

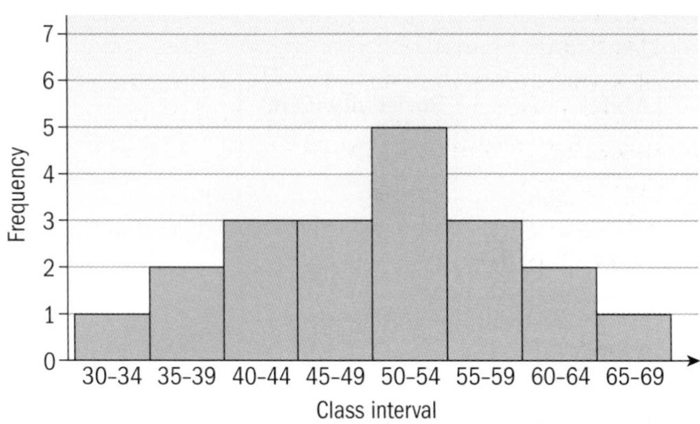

Histogram of bacteria growth in class intervals (see page 68).

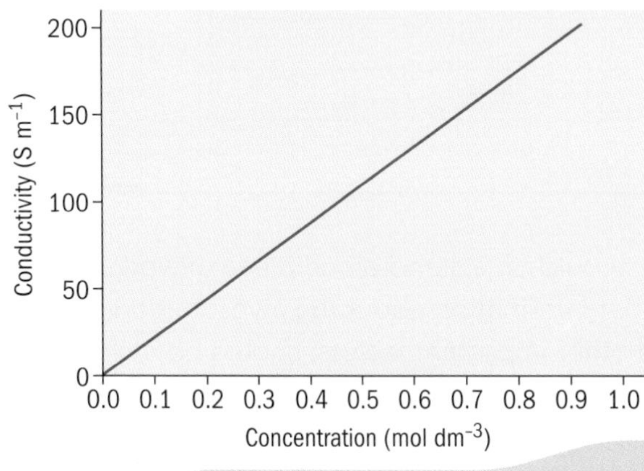

Line graph showing conductivity/concentration for HCl.

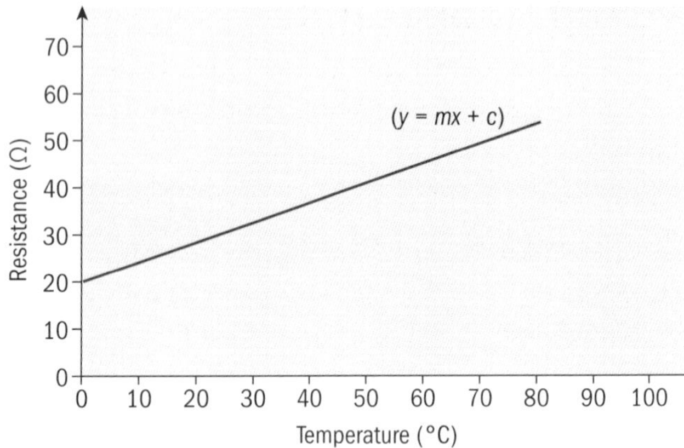

$y = mx + c$ line graph showing that the y-axis is intercepted at a point above zero. In this case it indicates that a wire coil has significant resistance at 0 °C.

Activity 3.4A

Which graph type should be used for the following sets of data?

Data set A:

Metal	Resistivity (Ω m)
copper (Cu)	1.7×10^{-8}
constantan	4.9×10^{-7}
aluminium (Al)	2.8×10^{-8}
silver (Ag)	1.6×10^{-8}

Data set B:

Frequency (Hz)	$\frac{1}{L}$ value for steel wire (d = 0.5 mm)*
155	4.0
122	2.8
100	1.65
82	1.22
60	0.80

*where L = length; d = diameter.

A graph should be a stand-alone piece of work. This means that information about the investigation, e.g. title and other important factors, should be included with the results on the same sheet. If the following points are included, the investigation's main focus can be viewed at a glance.

- **Heading:** title of investigation and variables plotted.
- **Axes labelled:** including correct units and measurement scales.
- **Axis non-zero indicator:** if the scale on an axis does not start from zero, an indication on the axis, usually a zigzag, is placed between the origin and first scale point. If this is not done, the proportions will be viewed incorrectly.

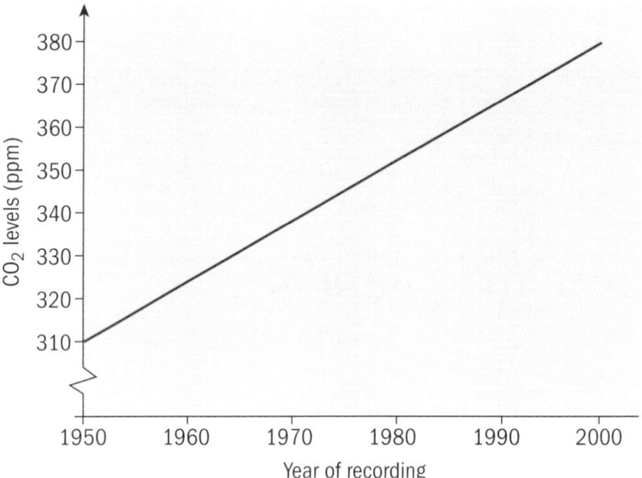

Graph showing the use of zig-zag to eliminate a large y-scale and focus on a narrow data section.

Scientific conclusions and evaluation

This section of a report brings together the experimental planning, technique and recording of data. The results tables and graphs, and observations noted are now complete and ready to be analysed. Your **conclusion/evaluation** should address the following questions.

1 **Is there a relationship between the variables?** Comment on the **correlation**, i.e. strong/weak, positive/negative. Identify **proportionality**.

2 **Do the results support the hypothesis?** If the method was suitable and followed closely, the hypothesis need not always be correct.

3 **Have you identified errors?** Explain where errors may have affected the results and how you incorporated them into your work. Identify anomalous readings and make a valid comment.

4 **Did you experience any significant experimental problems?** Comment on the suitability of equipment and the difficulty of measurement.

5 **Is there room for improvement?** There is no such thing as a perfect investigation. Something can always be improved upon. You can include the applications of your investigation in an industrial context at this point.

Did you know?

Some of the oldest recorded science notes are those from ancient Egypt depicting a unique picture of planetary positions. Luckily, as the science involved is astronomy, it has been possible to use computers to date this event to 1534 BC by using known planetary movements.

Activity 3.4B

List five questions needed to be answered when developing useful scientific conclusions/evaluations.

PLTS

Reflective learner

You will practise this skill when reviewing and evaluating practical work, and preparing work using scientific protocols

Functional skills

English

You will develop your English skills when writing up investigations using scientific protocol and formal, specialised language.

Writing your report

The full report of your investigation can only be completed once all the aspects of the investigation have been covered:

- the plan
- the practical work
- data collection and recording
- data presentation
- evaluation
- list of references and bibliography.

Written work completed on paper and in laboratory notebooks during the investigation is usually unclear, disorganised and not very neat. When all the information is gathered together, the format of the final report must be decided:

- written or word-processed bound report
- computer-based presentation
- video/photographic.

To make your report easier to read, make sure that each main section within the report is clearly headed and started on a new page: Investigation plan, Results, Data analysis, Evaluation, References and Bibliography.

Quality check

When you have written the first draft of your report, make sure that you perform the following quality checks.

- **Spelling**. Ensure that you are using the correct language setting for your spell checker, e.g. English (UK).
- **Grammar**. Keep your sentences short and clear. Remember you are writing a scientific report not a novel.
- **Avoid slang and jargon**. Use the correct scientific terms.
- Explain **acronyms** and **abbreviations**.

Case study: Report writing

Lisa is an assistant chemist for a well known pharmaceutical company. Her job is to prepare solutions for analysis and to collate information for reporting:

'I enjoy my work but report writing was never one of my strengths. I have now written many different pieces of work but always use the same format, which helps enormously. Nowadays, reports are usually computer generated but I may still have to produce a standard report for management.'

- Use **past tense** where appropriate. Your report has been fully completed after the work has been done, so your sentences should begin:

  'The apparatus was set up as shown in the diagram…'

  'Results were recorded in the table shown…'
- Use the **third person** where appropriate. Reports in science can be written in the first person (i.e. using I, We or You), although this should be used sparingly. It is usually more acceptable to write in the third person, but the use of either style seems to be a matter of opinion.

  'From the results obtained, it can be concluded…' (third person)

  or

  'From my results, I can conclude…' (first person)
- Correctly use **active/passive** voice. It is generally considered better to include both these styles of writing in scientific reports than to firmly adhere to either one. In most well written reports, a combination of active and passive writing helps to make the reading fluid and more understandable. It is more common practice to use the active voice when referring to your own opinions and work within the investigation.

Active voice: 'Our percentage yield of 71% was not as we had expected from our background research and preliminary tests. We can assume that some of the loss of yield was caused by the number of transfers which we carried out during the investigation.'

Passive voice: 'The percentage yield of 71% was not as expected given background research and preliminary test results. It can be assumed that some of the loss of yield could have been caused by the number of transfers carried out during the investigation.'

Assessment activity 3.4

BTEC

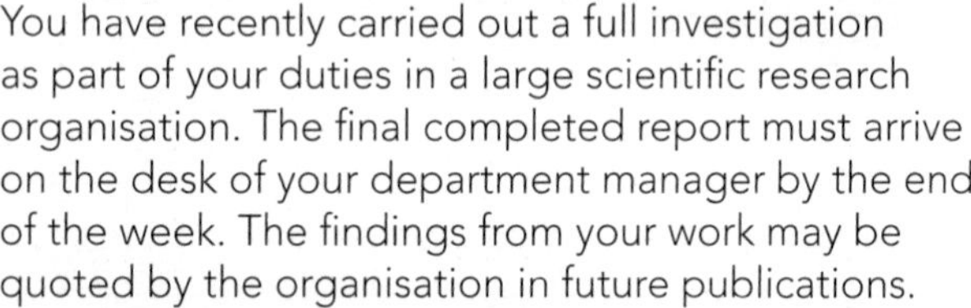

You have recently carried out a full investigation as part of your duties in a large scientific research organisation. The final completed report must arrive on the desk of your department manager by the end of the week. The findings from your work may be quoted by the organisation in future publications.

1. Present your results in tables or graphical forms **P9** and draw conclusions from these, explaining any relationships which may appear in the results, and if the results confirm or negate your hypothesis. **P8**
2. Provide sound reasons for your conclusions and reinforce your points with research information. **M4**
3. Evaluate the reliability of your findings and the investigation technique, including suggested improvements or possible further testing. **D3**

Grading tips

Tables and graphs must be labelled with the correct units. Scatter points should be small and accurately displayed on suitable graph paper to achieve **P9**, and you should then explain in clear words what the graphs/tables are actually showing for **P8**.

Give a clear statement from background research and attempt to link it to your results to achieve **M4**.

Try to demonstrate the quality of your investigation to the reader by considering every stage within your report to achieve **D3**.

WorkSpace

Obianuju Ekeocha

Biomedical Scientist

My main responsibility in the laboratory involves using automated biomedical instruments and manual techniques to analyse blood specimens. I then interpret, validate and report the results to the doctors.

My typical day starts with the daily maintenance of the laboratory equipment, ensuring it is functioning correctly. This will entail equipment calibrations, quality control checks and troubleshooting. Most of the tests are automated and, as such, it is of utmost importance that the equipment functions with reliable precision and accuracy.

Next, I carry out identification checks on the blood samples received from various sources, ensuring that information on the patient's request form matches with the information on the patient's blood specimen.

To maintain the integrity of the laboratory process, I constantly look out for factors that could introduce error and I reject contaminated or sub-standard specimens.

I carry out the requested tests by automated and/or manual techniques.

After the analysis, I carefully go through each result, interpret and validate it.

At the end of the day I generate reports of all the findings and send them out to the doctors in a format that is most accessible to them. These test results will often influence the medical treatment a patient will receive.

I find it very satisfying to know that in the course of my work I provide vital scientific information about a patient that enables the doctors to diagnose diseases properly or treat their patients effectively.

One of the tests I often carry out is the full blood count (FBC) which entails the analysis of blood in order to count and measure its components. The FBC is an automated test. However, if the test result is deranged (i.e. out of the normal range) I might need to carry out some further investigations. At this point, more manual techniques may be necessary, e.g. making a blood film, staining and examining it with a light microscope. These manual techniques can only be performed by trained and competent biomedical scientists.

Think about it!

1 How would you plan an investigation on a patient's blood sample?

2 What data would you need to collect and what questions would you need to ask?

Just checking

1. What is the difference between an independent and dependent variable?
2. What are control variables?
3. Why must Good Laboratory Practice be followed?
4. List at least five points to follow when using a laboratory notebook.
5. What is 'standard deviation'?
6. What condition must be satisfied before you can declare that your experimental results are **reliable**?
7. Which six aspects must your report cover?
8. When is it suitable to use the **active voice** in report writing?

Assignment tips

To get the grade you deserve in your assignments remember to do the following.

- Make clear notes in your laboratory notebook. Diagrams should be well labelled and numerical data tabled. Begin by completing all the required tasks for the **Pass** criteria. For some of these in this unit a suitable list and brief description will be enough, but make sure that the list covers all the key points discussed in your tutor sessions and the assignment tasks.
- To achieve **Merit** and **Distinction** criteria you should include explanations where requested. Use a variety of websites and textbooks for the topic and compare the information to be sure that there are similarities. Make a complete listing of all websites and research material that you have used and store them in your notebook.

Some of the key information you'll need to remember is as follows.

- The processes involved in developing an investigation and the methods used to reach a conclusion from the initial hypothesis. Correct recording of references and bibliography, and the production of well researched, comprehensive risk assessments.
- Good Laboratory Practice, techniques used to ensure that the investigation is carried out under safe conditions. Recording of data and observations is methodical and shows integrity.
- Calculating the mean, standard deviation and assessment of means comparison using *t*-testing. Awareness of the measurements used and the magnitudes of errors when using specific laboratory equipment.
- The structure of a scientific report including headings and sub-headings, suitable data analysis and presentation. Conclusions must be drawn from the data gathered and the assessment of graphical visualisation. Evaluation should be based on investigative planning and technique, and the production of a well recorded appendix section that should highlight the references and bibliography used.

You may find the following websites useful as you work through this unit.

For information on...	Visit...
normal distribution and standard deviation	Normal Distribution and Standard Deviation
writing a science report	Report Writing

Credit value: 10

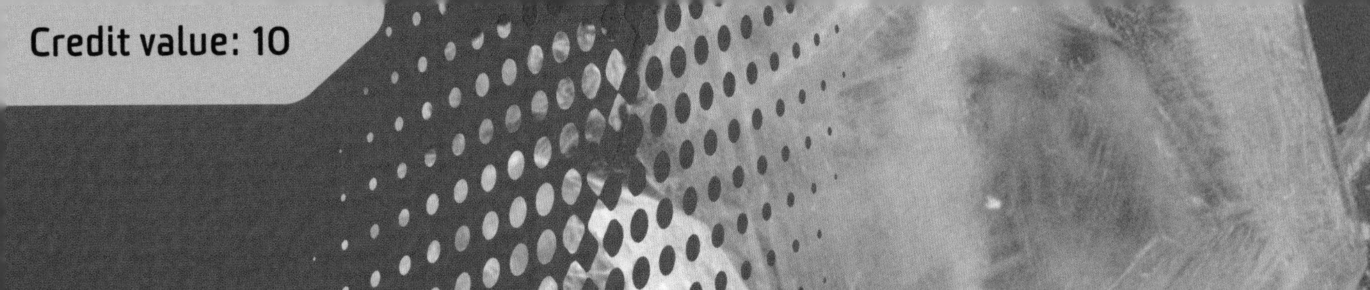

4 Scientific practical techniques

The principles of modern scientific practical techniques used in the laboratory have been developed over many years by brilliant scientists the world over. It is testimony to their achievements that the techniques and equipment used more than 100 years ago still form the basis of analytical chemistry today.

In this unit, you will be able to develop some of these useful skills in the preparation of standard solutions and the titration of substances, relating your work to real-life situations. You will also apply these skills to the physical testing of materials, such as for tensile strength and electrical performance. Emphasis will be placed on practical ability throughout. Health and safety issues will be studied and observed, whilst identification and correct use of equipment terms will need to be consistently practised.

You will have the opportunity to develop skills in many techniques, including quantitative and qualitative analysis, separation techniques, correct sampling methods, and estimation of purity. Finally, the successful completion of this unit will depend on your ability to master use of the range of instruments and sensors available, and to produce suitable measurements from the many and varied chemical, physical and biological analyses carried out.

Learning outcomes

After completing this unit you should:

1 be able to use analytical techniques

2 be able to use scientific techniques to separate and assess purity of substances

3 be able to use instruments/sensors for scientific investigations.

Assessment and grading criteria

This table shows you what you must do in order to achieve a pass, merit or distinction grade, and where you can find activities in this book to help you.

To achieve a **pass** grade the evidence must show that you are able to:	To achieve a **merit** grade the evidence must show that, in addition to the pass criteria, you are able to:	To achieve a **distinction** grade the evidence must show that, in addition to the pass and merit criteria, you are able to:
P1 carry out quantitative and qualitative analytical techniques **See Assessment activity 4.1**	**M1** explain how accuracy may be ensured in the techniques used **See Assessment activity 4.1**	**D1** evaluate the quantitative and qualitative analytical techniques used, suggesting improvements for future investigations **See Assessment activity 4.1**
P2 demonstrate use of scientific techniques to separate substances **See Assessment activity 4.2**	**M2** describe the factors that influence purity **See Assessment activity 4.2**	**D2** evaluate the accuracy of the methods used to estimate the purity of the samples **See Assessment activity 4.2**
P3 estimate the purity of samples using scientific techniques **See Assessment activity 4.2**		
P4 use instruments/sensors to test substances or materials **See Assessment activity 4.3**	**M3** justify the choice of instruments in the practical exercises **See Assessment activity 4.3**	**D3** evaluate the accuracy of the measurements taken **See Assessment activity 4.3**

How you will be assessed

Your assessment could be in the form of:

- presentations
- written scientific reports
- practical laboratory tasks
- demonstrations.

This unit helped me to become confident in using different types of glassware and instruments in the lab.

I particularly enjoyed the many practical activities which we had to complete. I was able to get over my fear of using maths in science because we used formulas regularly until they were no longer a problem. I gained a good understanding of the techniques used in laboratories and realised that they use the same procedures everywhere.

Our class received a guest visit from a university research chemist, who set up a demonstration of separating techniques they use in the lab. I was amazed to find out that the work we were doing in school was very similar.

Most of my time for this unit was spent carrying out science investigations and checking equipment. I was able to complete physics, chemistry and biology tasks and none of them was boring. Hands-on practical laboratory work was the main reason I decided to study science at level 3.

By completing this unit I have increased my interest in working in a laboratory environment, and would like to improve my qualifications further to hopefully find employment using some of the skills I have developed.

Catalyst

'Water, water everywhere...'

In large developed cities around the world such as London, drinking water is recycled. It is estimated that water consumed in London has already 'passed through' dozens of other people. Consider what steps you should take when trying to purify water for drinking.

- How can you separate out solid waste of differing particle sizes?
- Can you eliminate the cloudy appearance of the water afterwards?
- How can you be sure that the water which appears clean doesn't contain harmful bacteria or chemicals?

In groups, discuss these problems, research some aspects of sewage treatment and produce a simple flow chart of your suggestions. You can then present your findings to the class.

4.1 Analytical techniques

In this section:   

Key terms

COSHH – Control of Substances Hazardous to Health

Quantitative analysis – practical experimentation that produces numerical results (how much there is).

Qualitative analysis – practical experimentation that produces observational results, such as colour, odour or transparency.

Anion – negative ion formed when electrons are gained by an atom.

Cation – positive ion formed when electrons are lost from an atom.

Diatomic – two atoms of the same element combining to form the naturally existing molecule.

Stoichiometry – the ratio of the amount of one substance that reacts with another in a chemical reaction.

Safety first

The importance of observing essential safety rules and practices whilst working in a laboratory can never be overstated. A situation can very quickly develop from what may have been a relatively minor incident to a more serious issue if certain basic procedures are not in place.

Risk assessments must be completed and checked by tutors before each series of practical activities is carried out. Remember the following points.

- Research the topic.
- Be familiar with the glassware and instruments used.
- Learn the hazard symbols.
- Be aware of correct field sampling techniques.
- List the chemicals to be used and identify the risks using, for example, CLEAPSS Student Safety Sheets.
- Produce a full risk assessment.

Unit 2: See pages 49 and 50 for more information on Hazcards and COSHH.

Unit 3: See page 61 for more information on risk assessment.

Quantitative analysis

Counting the number of atoms and molecules in a chemical substance is the basis of **quantitative analysis** and is not as difficult as it may first appear. Atoms of different elements have different masses that are clearly displayed in the **periodic table**. Chemists must learn some important definitions before calculations concerning quantities of atoms and molecules can be made.

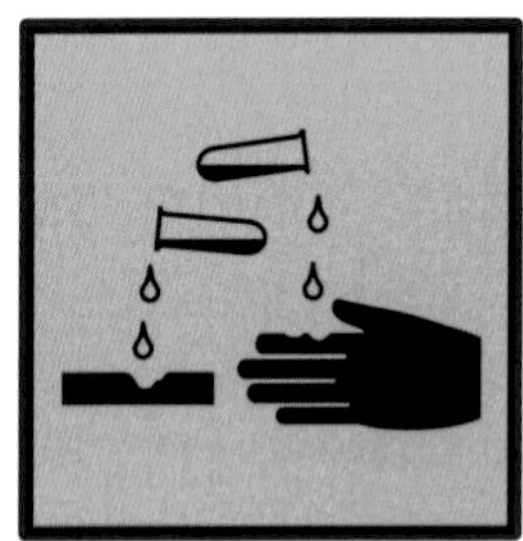

Standard hazard symbols.

Unit 1: See page 4 for more information on the periodic table.

- Relative atomic mass (A_r): the average mass of an atom of an element compared to one twelfth of the mass of an atom of carbon 12.
- Relative isotopic mass: the mass of an atom of that isotope compared to one twelfth of the mass of an atom of carbon 12.
- Relative molecular mass (M_r): the sum of all the relative atomic masses of all the atoms in a molecule.

The average mass of atoms is used because the abundance of naturally occurring isotopes of different elements is taken into account. Chlorine, for example, occurs naturally as Cl-35 and Cl-37. The percentage abundance for each isotope is 75.5% and 24.5% respectively. To find the relative atomic mass of chlorine:

$$A_r = \frac{(75.5 \times 35 + 24.5 \times 37)}{100} = 35.5 \text{ (3 s.f.)}$$

All relative atomic masses shown in the periodic table are calculated from the known abundance of the naturally occurring isotopes of each element.

Moles

When you write chemical formulas, you need to find out how many atoms of each element are present in a given molecule. For example, water (H_2O) contains two hydrogen atoms and one oxygen atom. From looking at the periodic table:

A_r of hydrogen = 1.0

A_r of oxygen = 16.0

M_r of water (H_2O) = 2.0 + 16.0 = **18.0**

The ratio of atoms in a molecule is important. In this case, the ratio of hydrogen to oxygen atoms is 2 : 1. When any mass of water is weighed, the ratio of the atoms will always be the same. So, 18 g of water will have 2 g of hydrogen and 16 g of oxygen.

One mole is the amount of a substance which has the same number of particles as there are atoms in 12 g of carbon 12.

This number is a constant known as **Avogadro's constant**, and has the value of $6.023 \times 10^{23}\ mol^{-1}$.

Worked example

1 How many moles are there in 5 g of magnesium oxide?

2 How many moles are there in 10 g of sodium chloride?

(Hint: use the periodic table to check.)

Answers

M_r of MgO: 24.0 + 16.0 = 40.0 g mol^{-1}

moles in 5 g of MgO = $\frac{5}{40.0}$ = 0.125 moles of MgO

M_r of NaCl: 23.0 + 35.5 = 58.5 g mol^{-1}

moles in 10 g of NaCl = $\frac{10}{58.5}$ = 0.17 moles of NaCl

Activity 4.1A

Using the molar calculations in the Worked example above, calculate how many atoms there are in 0.125 moles of MgO and 0.17 moles of NaCl.

Standard glassware used in laboratory analysis.

Empirical and molecular formulas

Chemists use the empirical formula to show the simplest, whole number, ratio of elements that are present in a compound.

Name of compound	Chemical formula	Simplest ratio of elements in compound
sodium chloride	NaCl	1 : 1
water	H_2O	2 : 1
carbon dioxide	CO_2	1 : 2
methane	CH_4	1 : 4

The difference between the **empirical formula** and the **molecular formula** is that the latter provides the total number of atoms which are present in a molecule of a compound. For many simple compounds, the formulas are the same. The following table illustrates some differences.

Compound	Empirical formula	Molecular formula
butane	C_2H_5	C_4H_{10}
hydrogen peroxide	HO	H_2O_2
hexane	C_3H_7	C_6H_{14}
benzene	CH	C_6H_6

Balancing chemical equations

The law of conservation of mass applies in chemical equations.

Mass is neither created nor destroyed in chemical reactions.

This means that the number of atoms on one side of the equation must exactly match the number of atoms on the other side. Those elements and molecules that react together are called the reactants, and those that are produced from the reaction are called the products.

Step 1 Know the reactants and products.

Natural gas burns in oxygen to produce carbon dioxide gas and water.

Step 2 Write a word equation.

methane + oxygen → carbon dioxide + water

Step 3 Write formulas for elements and compounds.

$CH_4 + O_2 \rightarrow CO_2 + H_2O$

Note that gaseous elements (except those in group 8 of the periodic table) like hydrogen and oxygen are **diatomic** so they must be written as H_2 and O_2.

Step 4 Balance the chemical equation.

$CH_4 + 2O_2 \rightarrow CO_2 + 2H_2O$

The skill of balancing equations becomes easier with practice. Remember that the chemical formulas themselves must not be changed; a number must be placed in front of the associated formula.

Count the number of each element on the left and count the number of each element on the right. A 2 in front of the formula multiplies each atom by 2. So $2H_2O$ means:

$2 \times H_2 = 4$ hydrogen atoms

$2 \times O = 2$ oxygen atoms

Did you know?

The word 'mole' is the English version of a German abbreviation (mol) for molecular weight: *Molekulargewicht*.

Balancing ionic equations

An atom becomes an **ion** when it either loses or gains an electron. The term **redox** refers to the transfer of electrons that occurs during chemical reactions. When atoms of an element lose electrons, it is called **oxidation**. When electrons are gained, it is called **reduction**. As the process of electron transfer allows the reaction, oxidation and reduction occur simultaneously.

To work out the ionic equation for a **redox displacement reaction**:

first write down the word equation

zinc + copper sulfate → zinc sulfate + copper

Then write down the chemical symbols, including the state symbols, which indicate whether the reactants and products are solid (s), liquid (l), gas (g), or in aqueous solution, i.e. dissolved in water (aq).

$Zn(s) + CuSO_4(aq) \rightarrow ZnSO_4(aq) + Cu(s)$

This reaction explains the displacement of copper ions from a solution of copper sulfate by zinc atoms. The result is that zinc atoms become zinc ions in bonding with sulfate ions and copper is removed to become copper atoms.

To focus our attention on the main reaction taking place, the equation can be written as:

$Zn(s) + Cu^{2+}(aq) \rightarrow Zn^{2+}(aq) + Cu(s)$

This is the ionic equation. Notice that the number of ions, and also the charge, is balanced, that is, the same on both sides of the equation.

To work out the ionic equation for a **precipitation reaction**:

calcium chloride + sodium carbonate →
calcium carbonate + sodium chloride

$CaCl_2(aq) + Na_2CO_3(aq) \rightarrow CaCO_3(s)\downarrow + 2NaCl(aq)$

The above reaction explains the chemistry in the formation of sea shells. The important part of the reaction is the joining together of calcium and carbonate ions:

Ca^{2+} and CO_3^{2-}

The Cl^- and Na^+ ions are not involved and are called **spectator** ions.

The ionic equation for the reaction is written as:

$Ca^{2+}(aq) + CO_3^{2-}(aq) \rightarrow CaCO_3(s)$

Standard solutions

In quantitative analysis, standard solutions are prepared for use in titration. These are solutions for which the concentrations are accurately known. Methods can vary according to the chemicals used; for example:

- **non-specified concentrations** – this is a common technique for preparing a standard solution for use in quantitative analysis where the concentration of the solution does not need to be known. This method has very limited use.
- **specified concentration by weighing** – in many cases, the concentrations of standard solutions need to be known, for use in titrations of acids and bases or other chemical reactions, for example.
- **standards from known concentration solutions** – essentially a method of diluting a known concentration standard solution to provide another, less concentrated standard solution. This method is useful when the substance is difficult to obtain as a pure solid, such as hydrochloric or sulfuric acid.

When making up these standard solutions you should consider the following points relating to the accuracy of your technique.

1. Refer to the appropriate CLEAPSS Student Safety Sheets and ask your tutor/technician for details of the actual quantities required to make the standard solution you need.
2. Use measuring cylinders that are appropriate for the standard volumes required, e.g. when making 1000 ml of 5 mol dm^{-3} HCl, use a 500-ml measuring cylinder to measure the concentrated acid and a 1000-ml measuring cylinder to measure the necessary volume of water.
3. Many solutions need to be made up wearing a face-shield, protective gloves and using a fume cupboard. Learners will not be allowed to make up some such solutions because they lack the training to do so.
4. For small quantities, use a graduated glass pipette of appropriate volume instead of a measuring cylinder. Make in a beaker (allow to cool – the **meniscus** position can change if there are temperature differences) and transfer to a volumetric flask, washing out the beaker into the filter funnel and washing out the filter funnel before finally making up to the mark.
5. Shake the mix well.

Worked example

1 Non-specified concentrations

Preparing a solution of sodium carbonate:

Step 1: Mass of beaker and solid sodium carbonate (a) = 60.20 g

[Tip the solid sodium carbonate into the funnel and weigh the beaker again]

Step 2: Mass of beaker and any sodium carbonate remaining (b) = 52.30 g

Step 3: (a) – (b) = 7.90 g

Step 4: Concentration of solution (made up to 250 cm^3 (V) with distilled water in a volumetric flask)

$= \frac{7.90}{250}$ g cm^{-3}

$= 0.0316 \times 1000$ g dm^{-3}

= 31.6 g dm^{-3}

Step 5: Using values from the periodic table, convert the concentration of the solution to mol dm^{-3}.

M_r of $Na_2CO_3 = 2 \times 23.0 + 1 \times 12.0 + 3 \times 16.0 = 106.0$

Step 6: The concentration of the solution = $\frac{31.6}{106.0}$

= 0.298 mol dm^{-3}

continued overleaf

2 Specified concentration by weighing

Preparing a standard solution of 0.1 mol dm^{-3} potassium chloride:

Step 1: From the periodic table

M_r of KCl = 39.1 + 35.5 = 74.6

Step 2:

1.0 mol dm^{-3} = 74.6 g KCl per 1000 cm^3 distilled water

0.1 mol dm^{-3} = $\frac{74.6}{10}$ g KCl per 1000 cm^3 distilled water

= 7.46 g

So, to prepare a standard solution of 0.1 mol dm^{-3} potassium chloride you would need to dissolve 7.46 g of KCl in 200 ml of water, and transfer carefully into a 1000-ml volumetric flask and make up to the mark.

3 Standards from known concentration solution

Preparing a standard solution of 200 cm^3 of 0.1 mol dm^{-3} hydrochloric acid from 1.0 mol dm^{-3} HCl stock:

Step 1: Use the formula $V_S = V_F \times \frac{C_F}{C_I}$

Step 2: Definitions:

V_S is volume of solution needed to be diluted to give the required concentration

V_F is the final volume required = 200 cm^3

C_F is the final concentration required = 0.1 mol dm^{-3}

C_I is the initial concentration = 1.0 mol dm^{-3}

Step 3: Calculate the volume of solution needed to be diluted:

$V_S = 200 \times \frac{0.1}{1.0}$

= **20 cm^3**

So, to prepare a standard solution of 200 cm^3 of 0.1 mol dm^{-3} hydrochloric acid from 1.0 mol dm^{-3} HCl, you would measure out 20 cm^3 of 1.0 mol dm^{-3} HCl stock in a volumetric flask and make up the volume to 200 cm^3.

Titration

This is an experimental technique for finding out the concentration of an unknown solution. It forms an essential part of many industrial processes.

Procedure

1. Set up the apparatus as shown in the diagram. Wear your eye protection. Open the tap.
2. Rinse the burette with distilled water first, then rinse again with a small amount of your prepared standard solution (e.g. 10 cm^3).
3. Close the tap and fill the burette with your standard solution – 'the titre' – to just above the highest graduation mark. Slowly open the tap and drain until the meniscus 'sits' on the mark.
4. There should be no air bubbles in the tip (if there are, repeat the above procedure until clear of bubbles). You are now ready to titrate against the solution to be analysed.
5. Using the procedure outlined on page 109, transfer with a pipette into a conical flask a known volume of solution to be analysed.
6. Add two or three drops of indicator (e.g. methyl red, phenolphthalein) to the transferred solution.
7. Titrate. Your technique should be slow and steady, controlling the tap whilst glimpsing the burette reading.
8. On the first signs of colour change (end point approaching), reduce the flow rate and gently swirl the conical flask, adding the titre drop by drop until a full end point is reached. Touch the end of the burette to the inside of the conical flask. Record the burette reading.

It is good practice to perform a rough titration first so that you have an indication of where to expect the end point.

Remember

- Remove the funnel before the final zeroing of the burette.
- Insert a thin piece of wire or hot water to remove any blockages in the burette tip.
- Wherever possible, use the burette to dispense the acid into the alkali to ensure that there is no possibility of blockage.
- Use a pipette filler for all solution transfer stages.
- You should avoid skin contact with the liquids you are using and wash your hands when the activity is over.

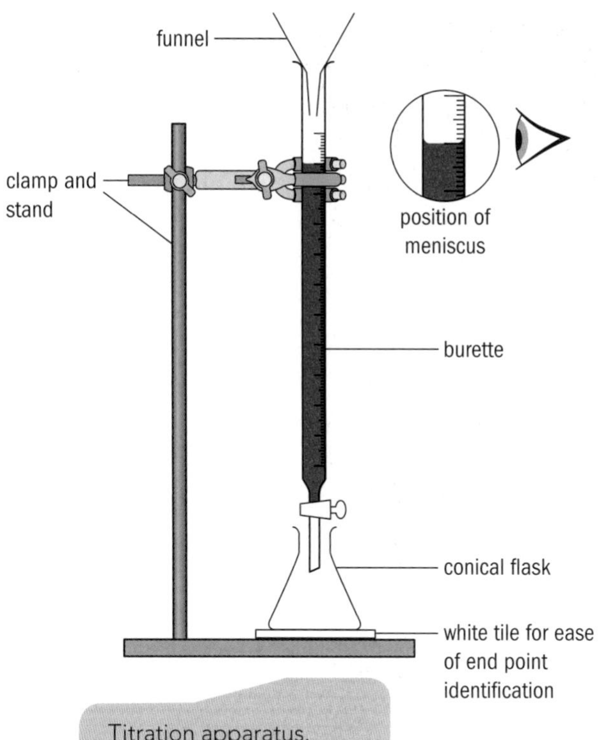

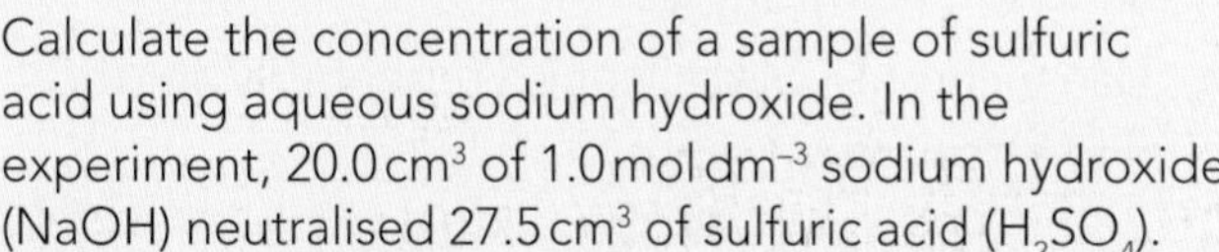
Titration apparatus.

At each stage of the titration carefully consider where errors may occur and how your technique may have contributed to them. Preventing **parallax** errors is easily achieved by ensuring that the burette is looked at from a perpendicular point of view, for example to reduce quantitative errors. Spills or losses of solution in the transfer process are common. Rinsing the burette and glass pipette for every activity will help to reduce contamination. The meniscus is a source of error if not fully taken into account (see page 64). This possible error may be eliminated if the volume to each graduation is checked by pipetting water into a beaker and weighing on a sensitive balance. An estimate of the error can then be made empirically.

Worked example

Calculate the concentration of a sample of sulfuric acid using aqueous sodium hydroxide. In the experiment, 20.0 cm^3 of 1.0 $mol\,dm^{-3}$ sodium hydroxide (NaOH) neutralised 27.5 cm^3 of sulfuric acid (H_2SO_4).

Step 1: Work out the balanced chemical equation:

$2NaOH(aq) + H_2SO_4(aq) \rightarrow Na_2SO_4(aq) + 2H_2O(l)$

2 mol of sodium hydroxide are needed to neutralise 1 mol of sulfuric acid so the **stoichiometry** is 2 : 1.

Step 2: Work out the amount of sodium hydroxide in moles.

Change the volume into dm^3:

$20.0\,cm^3 = \frac{20}{1000} = 2.00 \times 10^{-2}\,dm^3$

Then work out the number of moles, remembering that the concentration is 1.0 $mol\,dm^{-3}$:

$= 2.00 \times 10^{-2}\,dm^3 \times 1.0\,mol\,dm^{-3}$

= **0.02 mol**

Step 3: Work out how much sulfuric acid is neutralised, remembering that the stoichiometry is 2 : 1:

= ½ × amount of NaOH

= ½ × 0.02 mol

= **0.01 mol**

Step 4: Change the volume of sulfuric acid to dm^3:

$= 27.5\,cm^3$

$= \frac{27.5}{1000}$

$= \mathbf{2.75 \times 10^{-2}\,dm^3}$

Step 5: Work out the concentration of sulfuric acid:

$= \frac{0.01}{2.75 \times 10^{-2}}\,mol\,dm^{-3}$

$= \mathbf{0.364\,mol\,dm^{-3}}$

Functional skills

Mathematics

You will develop your calculation skills when rearranging equations involving concentration, volume and number of moles.

PLTS

Effective participator

You will develop this skill when carrying out quantitative and qualitative analytical techniques.

Qualitative analysis in inorganic chemistry

Identifying chemical substances from a particular characteristic in a reaction is **qualitative analysis**. Testing for **cations** and **anions** is essential in industry, and the techniques used should be performed often enough to be able to provide clear distinctions between the various ions present. It is important that you carry out qualitative tests for chemical reactions regularly to help you to develop skills in the practical applications of chemistry.

In water sampling, for example, it is important to identify the presence of specific salts by their anions or their cations.

- Chloride anions in large quantities can be corrosive to metal pipe work and machinery.
- Fluoride anions in small quantities help to prevent dental cavities but are fatal in larger quantities.
- Metal cations which leach into rivers and lakes can have serious effects on aquatic organisms.

In general, the techniques used for qualitative analysis involve separating out impurities followed by instrumental analysis.

Testing for cations

Safety note: safety goggles must be worn and skin contact with the acids and alkalis avoided. Any splashes on the skin should be washed off at once with cold running water.

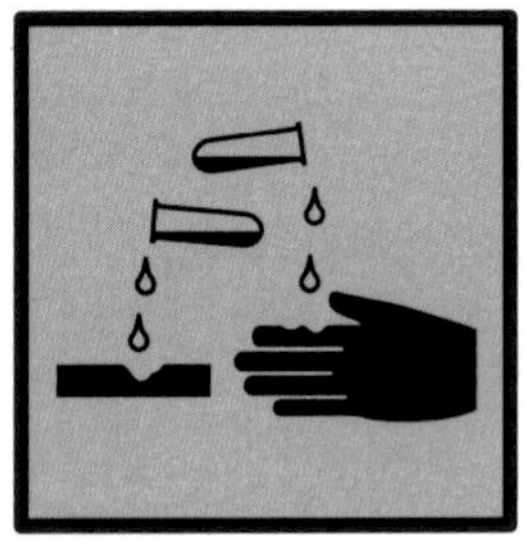

Chemicals required

- sodium hydroxide (2 M) (CORROSIVE)
- sulfuric acid (2 M) (CORROSIVE)
- ammonia solution (2 M)
- hydrochloric acid (conc.) to clean the wire for flame testing (CORROSIVE)
- indicator papers.

Apparatus

- safety goggles
- test-tube rack
- Bunsen burner
- heatproof mat
- spatula
- test tubes
- dropper
- wire for flame test
- tile.

Procedure

- Carry out the tests one at a time to avoid confusion of your results.
- Test the lime water for a cloudy precipitate reaction with CO_2 by blowing through it with a straw.
- Make up the ammonia solution fresh because it deteriorates with time.
- Use nichrome wire, cleaned repeatedly, for the flame testing.
- Note safety aspects – concentrated acids should be kept separate; silver nitrate should be kept in a brown bottle because it decomposes in light; it must be kept off the skin.

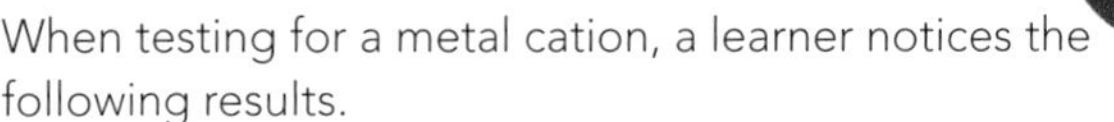

Activity 4.1B

When testing for a metal cation, a learner notices the following results.

- A precipitate is formed when dilute NaOH is added.
- The precipitate is white in colour and does not dissolve when excess NaOH is added.
- A flame test produces a red colour.

Use the flow chart to identify the metal cation.

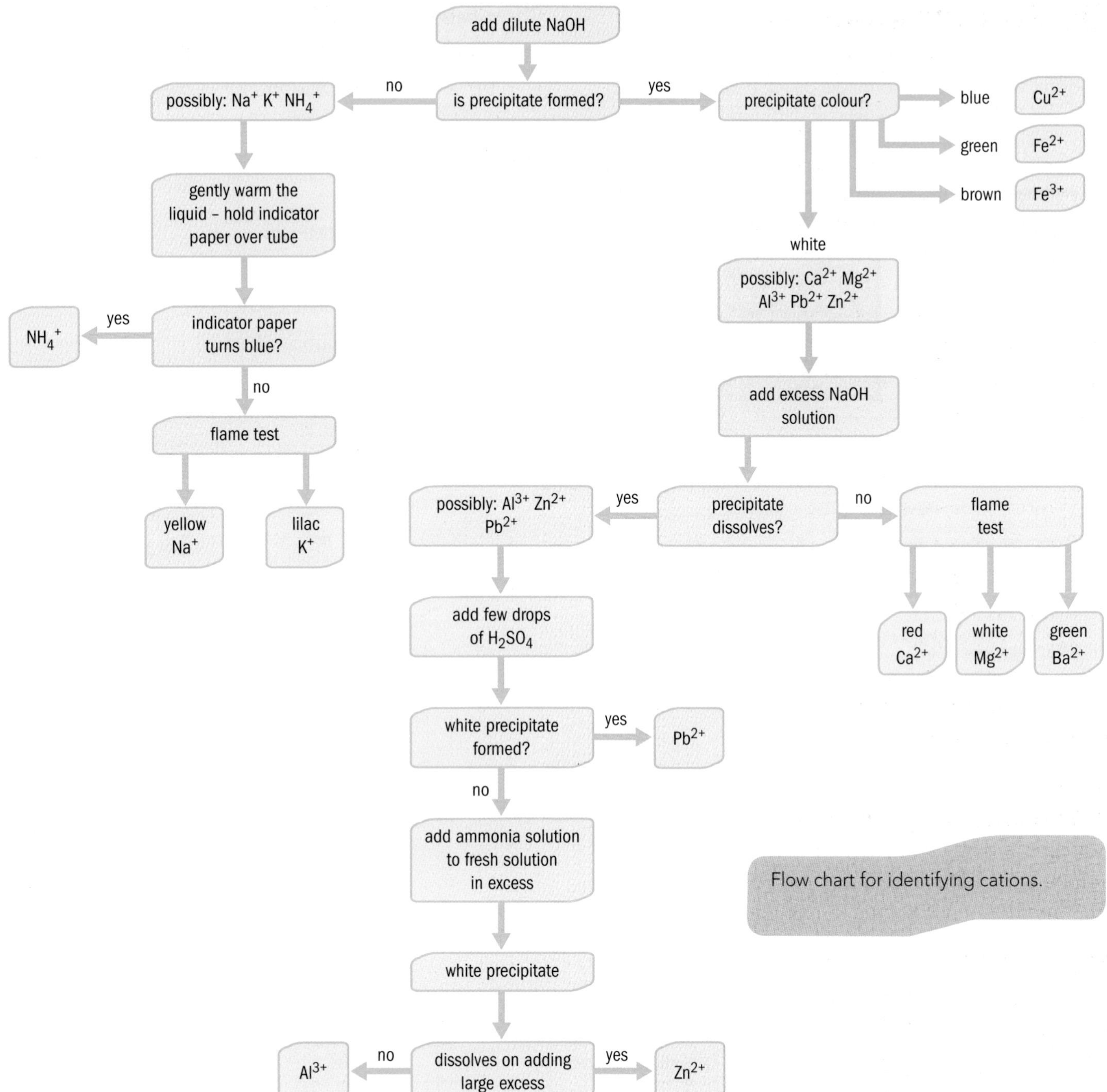

Flow chart for identifying cations.

Testing for anions

Safety note: safety goggles must be worn; avoid skin contact with all the chemicals used. If any come into contact with your skin wash off with lots of running water. Wash your hands when the activity is over.

Chemicals required

- hydrochloric acid (2 M) (IRRITANT)
- nitric acid (2 M) (CORROSIVE)
- ammonia solution (2 M)
- sulfuric acid (conc.) (CORROSIVE)
- lime water
- iron(II) sulfate (HARMFUL)
- potassium dichromate(VI) solution (VERY TOXIC)
- silver nitrate solution (CORROSIVE)
- barium chloride solution (TOXIC).

add dilute HCl to the solid and warm gently

gas given off?

yes → possibly: SO_2 CO_2 H_2S

- SO_2 potassium dichromate orange → green → SO_3^{2-} sulfite
- CO_2 lime water turns milky → CO_3^{2-} carbonate
- has rotten egg smell → S^{2-} sulfide

no → add 3 drops of barium chloride to solution

white precipitate?

yes → SO_4^{2-} sulfate

no → take fresh solution → add equal volume of dilute nitric acid → add 3 drops of silver nitrate solution

precipitate?

yes → add ammonia solution until excess

- white precipitate which dissolved → Cl^- chloride
- slightly soluble cream precipitate → Br^- bromide
- insoluble yellow precipitate → I^- iodide

no → add few crystals of iron(II) sulfate to fresh solution. Heat gently to dissolve. Slowly drip conc. H_2SO_4 into tube

brown ring formed?

yes → NO_3^- nitrate

no → unknown anion

Flow chart for identifying anions.

Apparatus

- safety goggles
- test-tube rack
- Bunsen burner
- heatproof mat
- spatula
- test tubes
- dropper.

Qualitative analysis in organic chemistry

Organic chemistry is the branch of chemistry that deals with carbon-based compounds. These organic compounds form the basis of all life on earth. They contain other elements in addition to carbon; for example, hydrogen and oxygen (as in carbohydrates). Other elements, such as nitrogen and sulfur, also form part of some of these complex compounds. These can be used to provide many products including: paints, plastics, explosives and food.

Amino acids are the chemical 'building blocks' which make up **proteins**. The importance of proteins becomes clear when we consider that they form muscles, organs, tendons, glands, hair and nails. They are also fundamental in the body's ability to grow, repair itself and maintain cells. **Starch** is a natural glucose polymer which stores energy in the form of **reducing sugars** such as glucose, fructose and lactose.

Activity 4.1C

What is the essential difference between qualitative and quantitative chemistry?

Suggested foods for testing

- bread
- milk
- baby food
- butter or margarine
- cereal
- egg white
- banana
- orange juice.

Test for proteins

Wear your eye protection.

1 Mash up the food with a pestle and mortar. Add a little water to form a suspension.

2 Put approximately 2–3 cm^3 of the suspension into a test tube.

3 Add a few drops of potassium/sodium hydroxide (CORROSIVE) (biuret A) until the suspension clears.

4 Add a few drops of copper sulfate (biuret B) and shake. A purple colour indicates that protein is present.

The amino acids liberated by the hydroxide in biuret A form a complex ion with the Cu^{2+} ion which is soluble and coloured purple.

Test for starch

Wear your eye protection.

1 Put a small potassium iodide crystal (IRRITANT) onto the food to be tested from a suitable range of foods.

2 Put one drop of sodium chlorate(I) solution onto the food allowing it to cover the potassium iodide crystal.

3 A blue-black colour indicates the presence of starch.

Note that chlorine from the sodium chlorate(I) solution reacts with potassium iodide to form potassium chloride and iodine. It is the iodine, which is an orange-brown colour, that turns blue-black in the presence of starch.

Test for reducing sugars

Wear your eye protection.

1 Mash up the food with a pestle and mortar. Add a little water to form a suspension.

2 Put approximately 2–3 cm^3 of the suspension into a test tube.

3 Add approximately 2–3 cm^3 of Benedict's solution to the test tube and shake.

4 Boil water in a beaker over a Bunsen burner and place the test tube in the beaker of boiling water for 2 minutes. The temperature should be at least 80 °C.

5 A green or brown precipitate forming indicates that reducing sugar is present. Benedict's solution contains copper sulfate that is reduced by sugar on heating. Sugars that have this effect are called reducing sugars, and include glucose, lactose and fructose.

Case study: Kathy Sadler, senior science technician

I work with two other junior technicians in a very busy comprehensive school science department. My work is demanding and can be very varied. My responsibilities include:

- organising equipment, sensors and chemical solutions (such as standards) for practical activities
- organising apparatus cleaning and storage schedules
- supervising the work of junior technicians
- updating equipment lists and ordering new stock
- revising work using COSHH (including Hazcards)
- equipment maintenance.

I have been a technician for many years and thoroughly enjoy my work. Every day I encounter something that is unusual or different and this encourages me to use my skills to help in all three science disciplines.

Assessment activity 4.1

As a newly appointed technical assistant within a large chemical plant, you must demonstrate that you have retained good laboratory practices in the preparation and use of equipment after achieving your qualifications. A senior technician will be assessing your practical techniques.

1. Wearing your eye protection, prepare a suitable volume of a standard solution of 1.0 mol dm^{-3} sodium carbonate and use it to determine the concentration of an unknown sample of hydrochloric acid. **P1**
2. Wearing your eye protection, carry out flame tests to determine the presence of cations in a suitable sample of road grit by comparing the colours you obtain with known research. **P1**
3. With reference to Unit 3 and other research material, explain how you ensured accuracy in your techniques for tasks 1 and 2 above. **M1**
4. Outline improvements which you could make in your practical tasks 1 and 2. **D1**

Grading tips

To achieve **P1** for both tasks 1 and 2, you must include safety considerations and a risk assessment. Your notes must be clear and well organised.

There should be discussion about the equipment used and calibration units to achieve **M1** together with comparison techniques for the researched flame colours.

To achieve **D1**, a full evaluation of the procedures must be carried out. Use multiple appropriate reference sources to compare techniques. Do you have a specialist spectrometer to determine the actual colour of the light from flame testing?

4.2 Scientific techniques

In this section:

Key terms

Insoluble – does not readily dissolve in the solvent used.

Purity – freedom of a substance from other matter of different chemical composition; in chemistry, elements and compounds are pure, mixtures are not.

Melting point – the temperature at which a solid turns to a liquid.

Miscible – two liquids that mix together completely, e.g. water and alcohol.

Immiscible – two liquids that do not mix completely, e.g. oil and water

R_f – in chromatography, the distance travelled by the compound divided by the distance travelled by the solvent; known as the retention factor.

Homogeneous – samples which are uniform throughout.

Separation techniques

In chemistry, it is necessary to isolate one substance from another to be able to analyse the substance further or to prepare it for use in other ways. When carrying out the separation of substances, chemists must consider:

- how complete the separation is
- how efficient the separation is
- whether the substance can be separated from other compounds.

The type of separation technique used will depend on the properties of the substances that are to be separated. If their properties are similar, the separation may take many difficult steps.

The most common properties of substances that can be used to separate them are: volatility, solubility, electrical, magnetic and adsorption (which is the ability of a material to attract molecules of a gas or liquid to its surface).

Precipitation

Precipitation is a useful method to separate and prepare salts that are **insoluble** and separate out during the reaction. The yellow colour in lines on the sides of roads is due to 'chrome yellow' made from the precipitation reaction between potassium (or sodium) chromate(VI) and lead nitrate. The procedure illustrated on page 93 can be used with many nitrates and soluble metal salts, e.g. silver nitrate and sodium chloride. The following point should be taken into account if a pure sample is to be achieved.

The appropriate choice of spectator ions will avoid problems of low solubility that would result in unwanted precipitation of a contaminant into the precipitate required. To obtain a pure sample it is necessary to wash the precipitate at least three times with warmed distilled water to ensure soluble materials are washed out. Correct drying preparation prior to weighing will provide a more accurate yield figure.

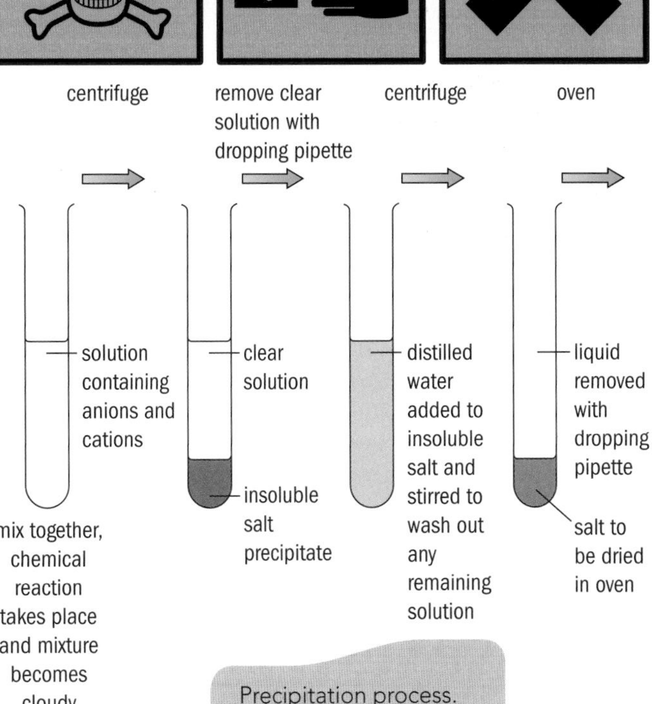

Precipitation process.

Crystallisation (re-crystallisation)

Crystallisation is a simple method to separate a soluble salt dissolved in aqueous solution. The method relies on the slow evaporation of the solution over a heat source (hot pan). The slower the evaporation is, the larger the size of crystals produced. Once crystals begin to form, heating should stop to allow further crystal formation. The process of crystal formation helps to purify the substance. This is because the structure of the crystal is dependent on the atoms required for the lattice pattern. Unsuitable atoms (impurities) do not take up positions in the lattice.

The presence of more than one soluble substance will result in impurities in the recovered solute (dissolved solid). Subsequent processing, correctly termed re-crystallisation, relies on the difference in solubility of the substances, so the least soluble substance crystallises out first, and can then be re-crystallised again to purify the product.

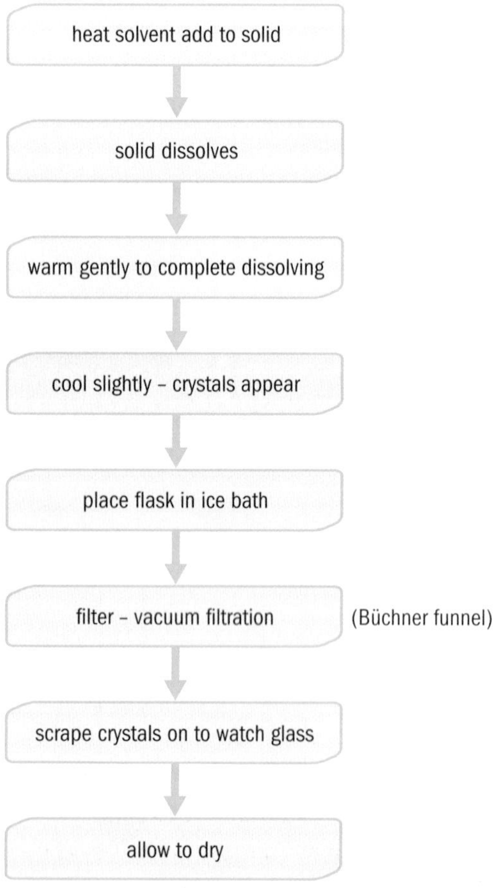

Re-crystallisation process.

Functional skills

ICT, Mathematics and English

You will develop your ICT skills using the centre's ICT systems to find information for the unit, obtaining experimental data, and collecting health and safety information from books and websites.

Drawing conclusions about final calculated results and their reliability will help you practise your maths skills.

Discussing risk assessment and representative sampling gives you the opportunity to develop your English skills.

PLTS

Effective participator, self-manager and team worker

You will practise your effective participator skills by demonstrating the use of scientific techniques to carry out separation of substances, estimating the purity of substances by using scientific techniques, and using instruments/sensors to test substances or materials.

Organising time and resources when carrying out practical tasks will help to develop self-management skills.

Sharing data/comparing analysis data gives you the opportunity to practise your skills as a team worker.

Did you know?

Crystallisation takes place naturally in salt lakes. Salts in solution crystallise out when the lake evaporates due to climatic conditions. Some salt deposits can be hundreds of metres thick.

Filtration

When solids are mixed physically with liquids, passing the mixture through filter paper placed in a funnel allows the liquid to flow through and leaves the solid residue behind. Results are improved if:

- the filter paper is fluted, giving a bigger surface area for the liquid to flow through
- any solid left in the beaker is rinsed and added to the filter paper.

On an industrial scale, filtration is an important step in the purification of water for consumption. The water containing solid matter is passed through filter beds of various grades of sand and rock where aerobic bacteria act to break down organic substances.

Filtration removes particulate solids from gas or liquid environments. The **porosity** of the filter should be chosen to remove solids of the desired size; for example, very fine particulates from vehicle exhausts, which contribute towards smog and are hazardous to health, can be monitored by filtration techniques. A complex filter can be made from several filters of different porosities graded coarse to fine. Filtration methods are specifically designed for the task.

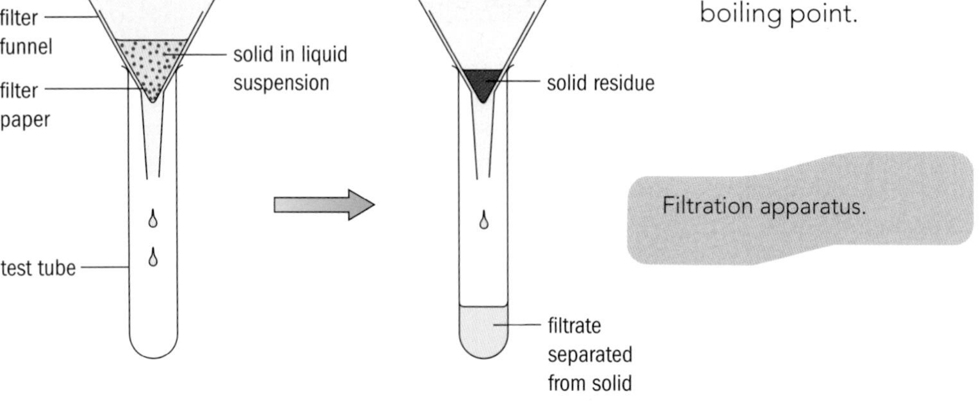

Filtration apparatus.

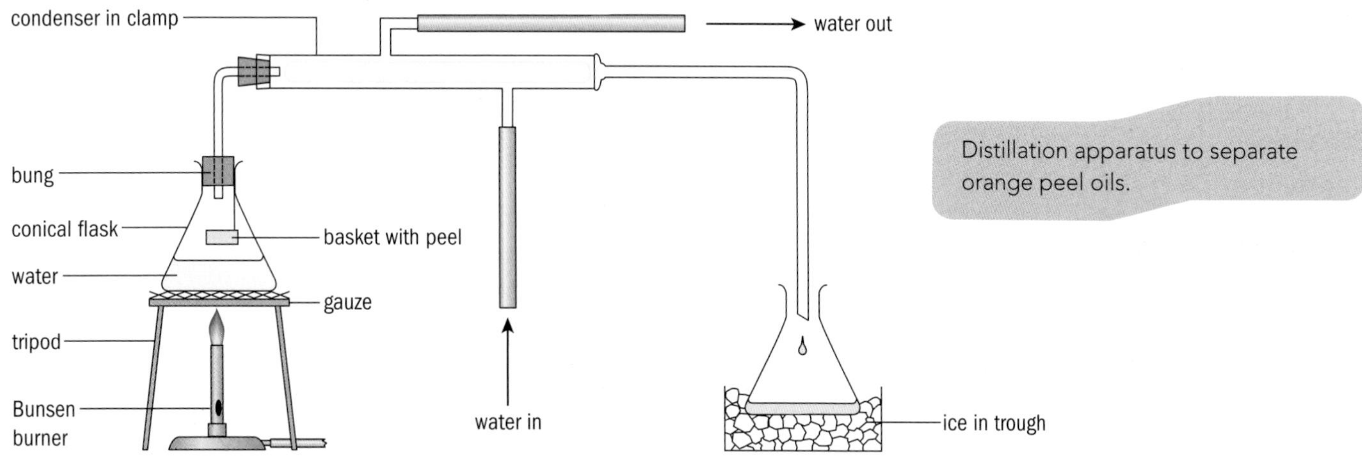

Distillation apparatus to separate orange peel oils.

Drying

In chemistry terms, drying involves trying to remove all the water from a chemical compound. In organic compounds, water may be locked up because some organic substances are partly **miscible** in water. To remove water from these organic compounds, drying agents (salts) are added to take up the water. The salts become hydrated as they do this. Examples of these salts include magnesium sulfate ($MgSO_4$), calcium sulfate ($CaSO_4$) and calcium chloride ($CaCl_2$).

When dry, some samples can be lost because they can easily be blown away in a draught. Suitable containers should have loose fitting lids (not airtight) to prevent this.

Distillation

Distillation is the process of separating out compounds within a liquid state because of the differences in their boiling points – a *physical* property. It relies on the principle that all liquid compounds have a characteristic vapour pressure that causes boiling to occur when it matches the atmospheric pressure. If the vapour pressure is low, the compound will have a high boiling point.

This method has been used for thousands of years in the separation of perfumes. It also provides the basis for crude oil processing, industrial dry cleaning and the production of alcoholic drinks like whisky.

Solvent extraction

Solvent extraction provides a method of separating compounds based on their differing solubility in two

immiscible liquids; that is, liquids that do not mix, like water and petrol. For example, suppose a compound is currently dissolved in water but will dissolve more readily in petrol. When petrol is added to the solution and the mixture is shaken, the compound will move out of the water and into the petrol. Because the two liquids are immiscible, they will separate out when left to stand.

Iodine can be found in naturally occurring materials such as seaweed. It has many uses, including medical disinfectant and chemical staining. Iodine is soluble in water but is more soluble in other solvents such as cyclohexane. In the presence of both, it will dissolve in cyclohexane. The following solvent extraction process makes use of these properties.

Extracting iodine from seaweed (tutor demo)

Safety notes: Eye protection required. Both iodine and cyclohexane are harmful by inhalation and skin contact; cyclohexane is also highly flammable. Ensure the cyclohexane is not poured out while heating with a naked flame is taking place. Hydrogen peroxide is an irritant, especially to the eyes. Note additional precautions in step 5 which, if not followed, may result in the tap funnel shattering explosively.

1 In a fume cupboard, burn a quantity of *Laminaria* seaweed (kelp) to ash. Gently heat a quantity of the ash in a beaker of water. The iodine will dissolve in the water.

2 Filter this into a conical flask.

3 Add a few drops of dilute sulfuric acid and approximately 20 ml of hydrogen peroxide (1.5 M) (filtrate turns yellow). These help to fully release the iodine from the seaweed.

4 Pour the filtrate into a tap funnel which contains a suitable quantity, e.g. 20 ml, of cyclohexane.

5 **Caution:** there will be pressure in the tap funnel after you have shaken it because cyclohexane is volatile. You need to release pressure safely: hold the funnel upright and 'unscrew' the stopper slowly to release the pressure. Do this between shaking sessions and finally before removing stopper.

So proceed as follows: put a stopper in the tap funnel and shake the contents, remembering to release the pressure several times. The water will become colourless and the cyclohexane layer will turn red.

6 Put the tap funnel back in the clamp stand and let the two liquid layers settle out. Release the stopper carefully. Firstly, run off the water layer and discard it and then run off the iodine-rich layer of cyclohexane into a second flask.

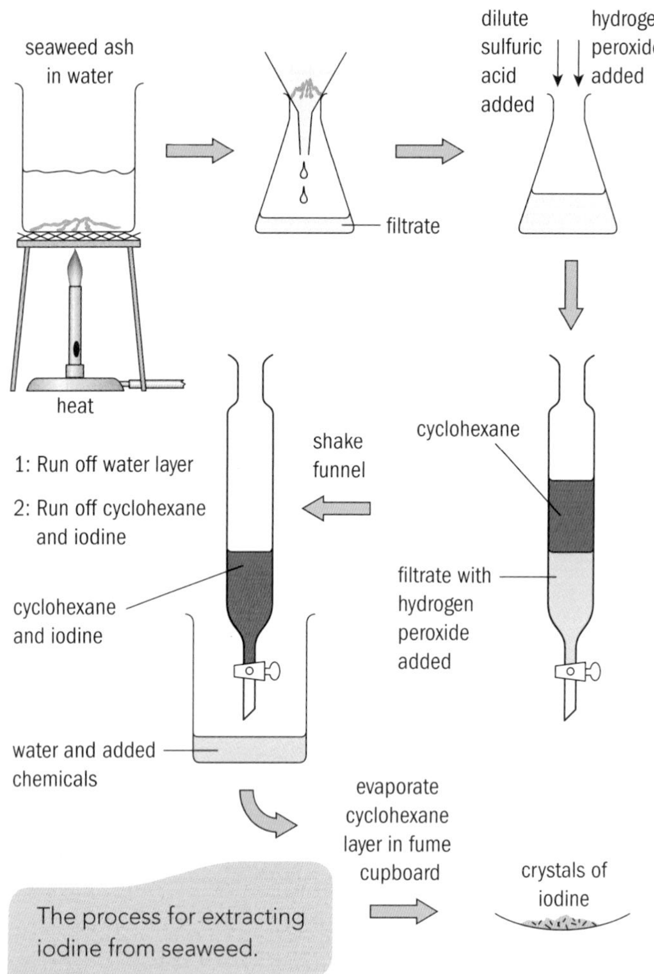

The process for extracting iodine from seaweed.

7 Leave the cyclohexane to evaporate off in a fume cupboard to obtain crystals of iodine. **Do not use a naked flame**; stand the flask in a bowl of very hot water if you wish to speed up the evaporation.

Paper chromatography or thin-layer chromatography

Chromatography is a method of separating mixtures and identifying substances so that **purity** can be established. Substances identified may be organic or inorganic and the principle has many applications; for example, in forensic investigation of a crime scene, DNA and RNA sequencing, quick identification of substances in the drugs industry. The basic principle relies on a liquid or gas moving over a stationary paper or powder. Different compounds in the mixture will move different distances and can therefore be identified.

There are many forms of chromatography in use in industry. These include high-performance liquid chromatography (HPLC) and gas–liquid chromatography (see pages 102 and 103). The basic versions of paper chromatography and thin-layer chromatography (TLC) can be carried out using simple apparatus in the laboratory.

Typical applications would be to separate the components in a mixture for comparison or identification purposes (fraud involving alteration of a cheque or document could be detected, for example). Also, pure pigments can be identified as only one component is present.

Materials

- chromatography paper or TLC plate
- test tubes
- soft pencil and ruler
- suitable glass vessel with lid or glass plate
- appropriate solvent; if flammable make sure no one is using a Bunsen or other naked flame
- a selection of mixtures, e.g. water-soluble coloured inks.

Procedure

1 Dissolve each sample in the solvent, using a separate test tube for each.
2 Draw a reference line in pencil on the chromatography paper, or TLC plate, at a point 2 cm from the bottom. Draw another line 1 cm from the top (pencil is insoluble and therefore will not interfere with the test).
3 Draw small crosses equally spaced on the bottom line in pencil, one for each sample of the mixtures you have dissolved in the solvent.
4 Place a drop of a sample onto a cross and allow to dry. Then, add another drop. Repeat once more. Do this for each sample on each cross. This provides sufficient sample to obtain satisfactory results
5 Pour the solvent into the glass vessel to a depth of 1 cm.
6 If a paper chromatogram is used, roll it into a tube and stand it upright in the vessel. If a TLC plate is used, stand it upright against the side of the vessel.
7 Put the lid or glass plate on the vessel and wait for the solvent to reach the top line you have drawn.
8 Then take the paper or TLC plate out to dry. Use your pencil to mark the centre of each spot on the paper. There will be more than one spot per sample if they are mixtures. A singe spot identifies a pure compound.
9 Record the distance of x (bottom to top pencil lines) and y (bottom pencil line to each spot).

Calculation

Work out the **retention factor** (R_f) for each substance that has separated out using the equation:

$$R_f = \frac{y}{x}$$

Note that the retention factor is always the same for a particular chemical substance if the chromatography procedure is kept constant. This is a useful method for identifying specific chemical substances.

Electrophoresis

When an electrical field is applied to a solution, charged particles (ions) or molecules in the solution will begin to move (migrate) towards the electrical plate (electrodes) of opposite charge. In this way, particles are separated from the solution. This process can have many applications including:

- determining the molecular weights of proteins
- investigating the physical characteristics of macromolecules.

Pathogens (microorganisms causing disease) are known to adhere to host cells by using proteins. Separation of their micro-components using this method allows a comparison between good and bad strains and helps scientists to analyse their differences.

There are many types of electrophoresis including:

- gel electrophoresis
- capillary electrophoresis
- isoelectric focusing
- laser
- alternating field
- pulsed field.

Centrifugation

A centrifuge is used to separate fine solids that may prevent the use of a filtration method. It can also be used to separate a solid suspended in a liquid. A centrifuge spins at high speed causing the particles to separate according to their density. The denser the particle the more inertia it has, and therefore the further it moves from the axis of the centrifuge. In this way, a centrifuge separates the particles in a

heterogeneous mixture (see 'Sampling', page 98) because of differences in density.

When using this method, it is important to ensure that the system is balanced by putting a similarly filled tube opposite the investigation tube.

Activity 4.2A

Suggest the separation techniques you would need to use when obtaining a sample of halides (compounds of halogens such as fluoride, chloride, bromide, etc.) from road grit, used to grit roads during icy conditions.

Five-step correct operation of a centrifuge

1 Carefully place the test tube containing the heterogeneous mixture in the centrifuge holder.
2 Balance this tube with another directly opposite filled with an equal amount of liquid.
3 Close the cover of the centrifuge and switch it on.
4 Centrifuge for 3 minutes. Switch off and wait for the device to stop before removing the tube.
5 Repeat if necessary to ensure that the precipitate has collected at the bottom of the test tube, identified when there is a clear distinction between the precipitate and solution.

A longer period of centrifugation is more suitable when yield of product is important.

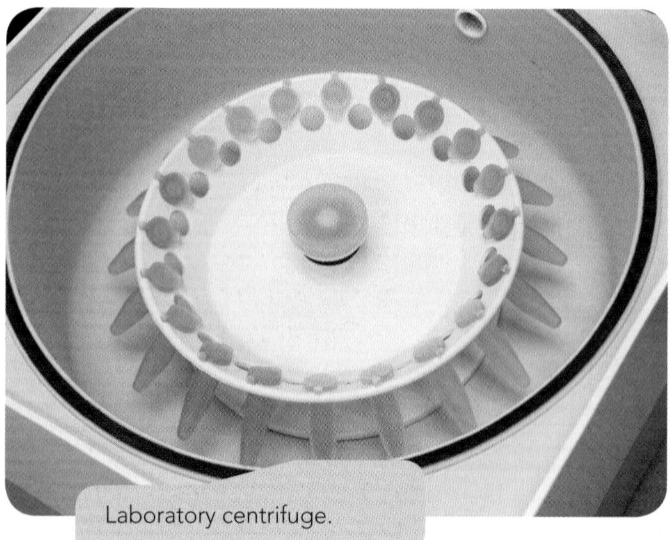

Laboratory centrifuge.

Did you know?

NASA uses virtually all of the chemical separation techniques discussed in this section to help them identify the specific chemicals that are contained within soil samples on the surface of other planets and moons in our Solar System. This will help to indicate the presence of essential chemicals, necessary for life to exist.

Sampling

Biological, chemical and physical testing and analysis of substances depend significantly on the methods used to sample the substances and to prepare them for the analysis. If the sampling is not carefully planned, the sample itself may not be representative. This means that the results would be inconclusive or even invalid, and would produce a serious source of statistical error.

In simple terms, the unusual small black flake in a box of cornflakes does not really provide a representative sample of the whole box.

Sampling solids

The sampling process must take account of the variations within the substance to be sampled. For example, powders tend to show a distribution in particle size because of the way the particles pack together. If you watch a load of mixed sand and chippings being tipped from a tipper truck onto the ground, you will see that the fine sand filters through the pile leaving the larger chippings on the top. For this reason, in chemistry, spatulas and other scoops should be flat so as not to favour larger or smaller particles (gravity plays an important part in the settling of particles of different sizes in a pile, smaller particles settling to the bottom in a process termed a 'kinetic sieve'). In addition, spatulas are manufactured from a metal resistant to acid corrosion, to a given volumetric amount and of a shape which can be used to deliver substances into test tubes.

It is good practice to take a number of samples of the substance and to homogenise the sample as explained below. Smaller particles tend to pile up easily on a spatula and larger ones can fall off. It is vital for getting a representative sample that there is a combination of both.

The sampler must also be fully conversant with correct sampling techniques to reduce bias in their method and ensure that the samples are clearly random and suitable for the investigation, noting the procedure and reasons for its use. If at all possible, other sampling devices should be used such as a **chute** or **spinning riffler**, both of which separate larger quantities of material through holes in their main parts.

- **Homogeneous** samples are those that are the same throughout. A packet of flour can be said to be homogeneous. A sample will appear the same irrespective of where in the bag you sample from. Sometimes, a substance or product may need to undergo homogenisation – a process that helps to distribute the particles within the substance equally. You homogenise some things on a regular basis at home. When opening a fresh yoghurt pot or tin of paint, it is advisable to stir the product to redistribute the ingredients throughout the container. The product and its viscosity become more uniform and viscous.
- **Heterogeneous** samples are those made up of different parts within the substance. Soil is composed of many different minerals, vegetation and rock fragments. Each time you take a small sample from the main sample it will have a different make up, and so scientists separate the sample into category piles, e.g. wet, rocky, vegetative areas, then sample the pile at the top, middle and bottom, with at least half of the samples in the bottom third. The samples taken are then homogenised.

Planning your sampling (solids/liquids/gases)

A sampling plan should include the following.

- **Labelling:** dates and times written on sampling containers. Names of the sampler and substance collected.
- **Type of container:** glass or plastic depending on substance and conditions. Generally, if samples are environmental, plastic is the best choice for practicalities of transportation and safety.
- **Cleanliness of container:** if contaminants are present before sampling then spurious results will arise.
- **Selection of appropriate sampling points:** e.g. deciding on the best point on a river for environmental water sampling. This depends largely on the purpose of the sampling. If biodiversity is being investigated, stream disturbance should allow organisms to flow into a catch net downstream and the varying habitats must also be sampled. Water analysis will require samples from the cross-section of the river, at varying depths and at tributary points, and at locations where there is a significant change in width and flow rate.
- **Storage:** the storage location of the sample for identification and retrieval is important. You should be aware of the correct storage conditions (dry, warm, light, dark, etc.) and procedures to ensure that your samples do not deteriorate.

Biological sampling

Field studies in biology are, of course, concerned with living plants and organisms. Methods of determining the appropriate sample size must therefore take this into account.

- When sampling a particular plant or slow-moving insect species, it is usual to use a **quadrat** of 1 m^2. This enables you to find the average density per square metre of the species in a given location. The quadrat is placed at random in 10 different places in the location being studied. The number of the particular species under investigation within the quadrat is counted. Then the average number of species per square metre is calculated for that location using the data from all 10 places. Results for other locations can then be compared using a suitable statistical analysis technique.

Unit 8: See page 179 for more information on quadrats.

- Mobile insects and small animals are more difficult to sample and require a number of different capture techniques. These include: catching moths by using a night light; baiting with food in an open jar overnight; using a net for fish or larger flying insects; digging in soil for worms; or looking under rocks for seashore creatures.

In food sampling, industrial processes involve the mass use of agar plates. A food sample is liquefied and then diluted to varying degrees. (In contaminated food samples, the number of bacterial cells is too high to produce a countable result and so dilution is needed.) The diluted sample is spread onto an agar plate containing the nutrients required for bacterial growth. A particular bacterium under investigation is then

Case study: Isabelle, technical support worker

I work for the Environment Agency (EA), which has many different departments dealing with all aspects of the environment. I am based in the water quality section, which monitors the quality of surface and ground waters (streams, rivers, lakes, sea and drainage conditions).

My job involves some desk work but also a lot of field tasks that I really enjoy. It is my responsibility to take samples of soil, water and flora/fauna for analysis in one of the many laboratories operated by the EA.

In a typical field task, I will have to arrange many things:

- sampling equipment – test tubes, bottles, plastic containers, scoops, labels, meters (flow rate for example), tape measures, maps...
- suitable clothing and footwear – waterproofs and wellingtons if conditions are difficult, good walking boots, rucksacks, plastic bags.
- sampling points – I usually check the area from maps and decide which are the most suitable sampling points depending on geology, industrial activity, agricultural activity, etc. My supervisor confirms my decisions.

identified by colony sizes within the **selective growth medium** used to identify the bacterium (e.g. *Escherichia coli*, *Listeria* or *Salmonella*). In the sampling process, the agar plates and dilution tubes are clearly labelled: time, bacterium type, agar broth and number of dilutions.

Sampling liquids and gases

If the sample required is of a flowing liquid, the apparatus used must allow the liquid to flow through it, and should consist of an inlet, a flow chamber and a non-return valve in its simplest form. To provide a representative sample of flowing liquid or gas, consideration must be given once again to the homogeneity of the sample, because even liquids and gases may be different in composition throughout. This can be due to chemical differences, temperature variations, viscosity fluctuations or the presence of particulate matter.

To solve this sampling problem the following guidelines are used and form a sound basis for all laboratory gas/liquid sampling.

- Sample at multiple points within the flow channel of gas or liquid.
- Record the flow rate of the gas or liquid at points along the same plane.
- Sample gases **isokinetically** when particulate matter and aerosols are involved.

In industrial outlet ducts such as gas chimneys or waste-water pipes, sampling meters are situated at various points along the duct length, and numerous samples are taken along the same plane at any given point.

In laboratory work, correct pipette sampling of solutions is crucial if both random and systematic errors are to be greatly reduced. The following guidance points will help to maintain precision and accuracy when sampling solutions for further analysis.

- Connect the pipette to the pipette filler and check for leaks when filling.
- Pre-rinse the pipette tip before any sampling is carried out.
- Stir the solution with a glass rod to homogenise the liquid.
- Try to arrange it so that the pipette and solutions are at the same temperature if possible. Place the pipette down on a suitable surface after each pipetting to prevent your body heat from warming it up.
- Ensure that the tip of the pipette is fully immersed in the solution, to exclude the possibility of including air bubbles, and that the pipette is held vertically.
- If the solution is more viscous (as is the case when laboratories carry out blood sampling) best results are achieved when a small amount of the liquid is put into the tip.
- When pipetting solvents, wet the tip with the solvent prior to use.
- Depress the plunger with a constant motion, which reduces interrupted flow.

See page 109 for more information on using a glass pipette and safety filler.

Isokinetic gas sampling

Ensuring the correct speed of gas movement into the sampling device is important to ensure that a representative sample of gas is provided. The desired sampling point is reached when the speed of the gas entering the sampler matches the speed of the gas stream itself. This method is used when monitoring, for example, industrial chimney stack output, exhaust fumes or environmental air pollution. These samples contain various sizes and compositions of particulate matter.

Estimation of purity

You would probably think twice about drinking a glass of clear water if you knew that there were even a small number of microscopic organisms in it.

Using the melting point or boiling point

Estimating how pure a substance is in chemical terms is very important in many industrial applications, for example, pharmaceuticals. **Melting points** and **boiling points** are known very accurately for most elements and compounds and are listed in data books. A substance can be identified as impure by comparing the experimental values of melting or boiling point with those from a data book.

- Water, for example, boils at 100 °C at 1 bar atmospheric pressure, but the boiling point is increased if salt (NaCl) is present in the water.
- In the manufacture of sugar, the liquid that is boiling is a solution of sugar or syrup. If the concentration of sugar rises, the boiling point rises (at constant pressure).

In general, any increase in concentration of a solution also increases the boiling point.

The melting point is the temperature at which a substance exists in both liquid and solid state, somewhere from the start of the first signs of liquid to the total disappearance of solid. Observations of melting provide chemists with an indication of the purity of organic substances. Pure solids melt at a definite temperature (usually within 1 or 2 degrees Celsius) but impure solids will become soft and melt over a range of temperature. Generally, the addition of another element lowers the melting point of the substance. Inorganic solids melt at very high temperatures and are unsuitable for testing in school laboratories.

Procedure for finding the melting point of synthesised aspirin

Safety note: wear your eye protection; take care not to burn yourself on the melting point apparatus.

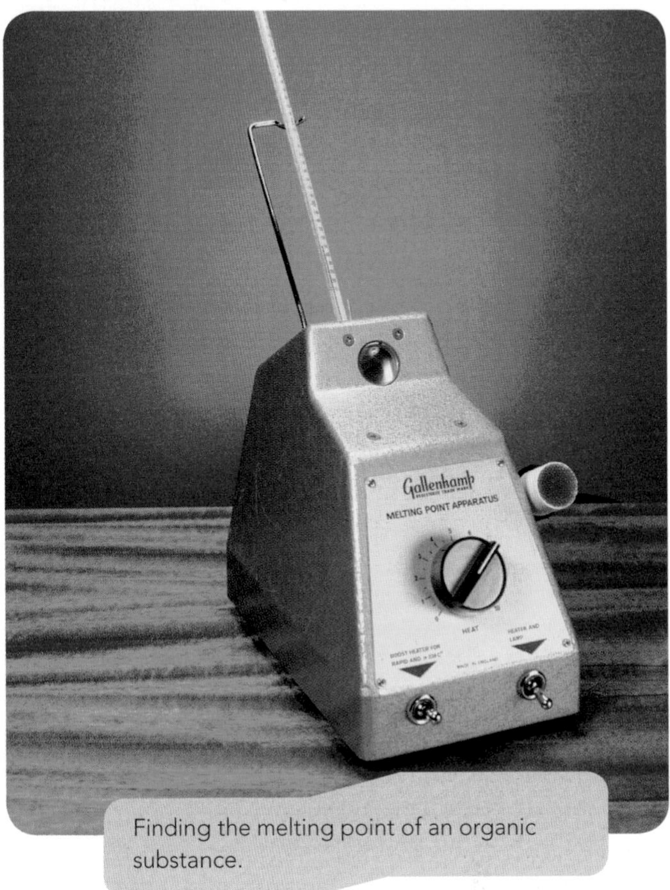

Finding the melting point of an organic substance.

Apparatus

- eye protection
- aspirin samples
- pestle and mortar
- capillary tubes
- Bunsen burner
- electronic melting point equipment
- thermometer (0 °C to 200 °C)
- magnifying glass.

Method

1 Grind the aspirin into a fine powder using the pestle and mortar.

2 Prepare a suitable number of capillary tubes by breaking gently and sealing one end by heating in a hot Bunsen burner flame for a few seconds.

3 Switch on the melting point equipment. Refer to the data tables for approximate temperature values for melting of the compound. Set the equipment to a point below the expected melting point. *Note that the rate of heating will influence the reading of melting point.* Read the thermometer.

4 Dip the capillary tube into the powder until sufficient depth (a few millimetres) of sample is contained in the tube – gently tap or scrape the glass to allow the solid to fall to the bottom. Place the sample tube into the hole next to the thermometer in the melting point equipment.

5 Observe the melting of the aspirin sample through the magnifying glass. At the point when the compound begins to melt, record the temperature using the thermometer. When the melting is complete, record the final thermometer reading. This is the full temperature range of the melting point.

6 Compare your results with the data tables.

Activity 4.2B

Mandelic acid is used in skin care and as an antibacterial. It starts to become liquid at 120 °C and is completely liquid at 122 °C. What is its actual melting point? If the melting point range was 118 °C to 122 °C, what would this tell you about its purity?

Gas–liquid chromatography

This technique has many applications; for example, analysis of banned substances used in sports competitions, animal fat contamination in vegetable oils, alcohol concentrations in a motorist's blood sample. When used with a mass spectrometer (which determines the molecular weight of the substances), gas–liquid chromatography can be used to detect components in explosives, trace elements in forensic analysis, substances in food and plants, in environmental analysis, and many other applications.

- The liquid sample being analysed is injected into the column. This comprises a coiled steel tube packed with porous rock on which is adsorbed a liquid solvent.
- The heat from the thermostatically controlled oven boils the sample to produce a vapour.

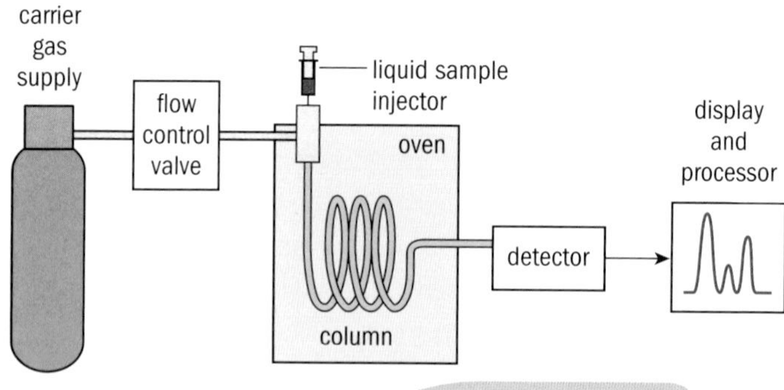

Gas–liquid chromatography system.

- An inert gas (such as helium) is also pumped into the column.
- As the components of the sample move through the column, some will be carried with the inert gas (mobile phase) while others will dissolve into the liquid solvent (stationary phase).
- The molecules travelling in the mobile phase will take less time to travel through the column than those travelling in the stationary phase.
- The time taken for each component of the sample to pass through the column to the detector is the **retention time** and depends on the solubility of each component of the inert gas or liquid solvent.
- The display on the processor shows a series of peaks. Each peak corresponds to the retention time of a different component of the sample. This enables you to identify each component. The area under each peak is a measure of the amount of each component present in the sample. This gives you a measure of the purity of the sample.
- Note that if the sample is pure, i.e. it only contains one chemical, there will be a single peak on the recorder.

High-performance liquid chromatography (HPLC)

This method also has numerous applications, including the analysis of proteins, water quality, additives and contaminants in food; quality control – assessing the purity of raw materials, monitoring how substances degrade over time, etc. HPLC used together with mass-spectrometry can analyse the metabolism of biological substances used in medical treatments, for example, and is generally able to detect traces of chemicals within a complex chemical mixture.

- The liquid solvent used is forced through the column at high pressure.

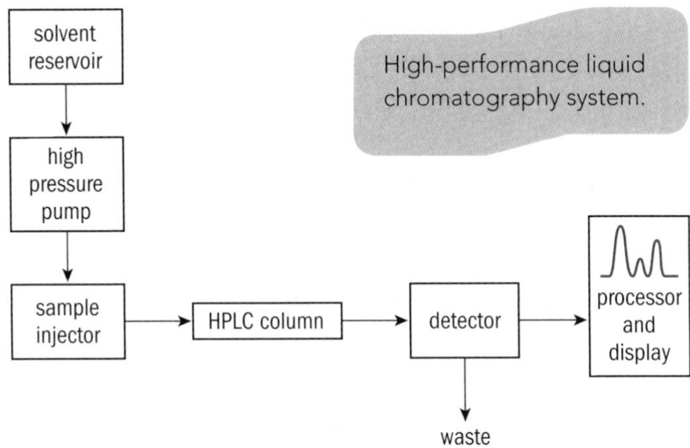

High-performance liquid chromatography system.

- The particle sizes of the packing material in the column are very small and provide a very large surface area for the sample to interact with. This enhances the separation of the components of the sample.
- Detection is automated and very sensitive; ultraviolet (UV) absorption is often used.
- Many organic compounds can be identified by the amount of UV radiation they absorb at particular wavelengths. The components of the sample coming out of the column are passed through UV radiation. Allowing for the absorption of UV by the solvent itself, the display on the processor will indicate which organic compounds are present.
- This will indicate the purity of the sample under test.

Spectroscopy

Scientists use spectroscopy to determine the purity of a chemical substance because it is very reliable and extremely accurate. Spectroscopy uses the principle that substances *absorb*, *emit* or *scatter* electromagnetic radiation. Each chemical substance interacts with the electromagnetic radiation to a different extent, dependent on the wavelength of the radiation. Note that shorter wavelengths have greater energies. The spectrum that results indicates which substances are present in the sample.

Absorption: electromagnetic radiation is absorbed by an atom in the chemical substance when the amount of energy is equal to the difference between electron energy levels. As a result, there are dark lines in the spectrum corresponding to the wavelength of the absorbed energy. Unpredicted dark lines will give an indication of impurities present in the sample.

Absorption spectrum.

Emission: when an electron in an atom has moved to a higher energy level and returns to a lower energy level, energy equal to the difference between the energy levels is emitted as an electromagnetic wave. This will correspond to a bright line for that wavelength in the resulting spectrum. Unpredicted bright lines will give an indication of impurities present in the sample.

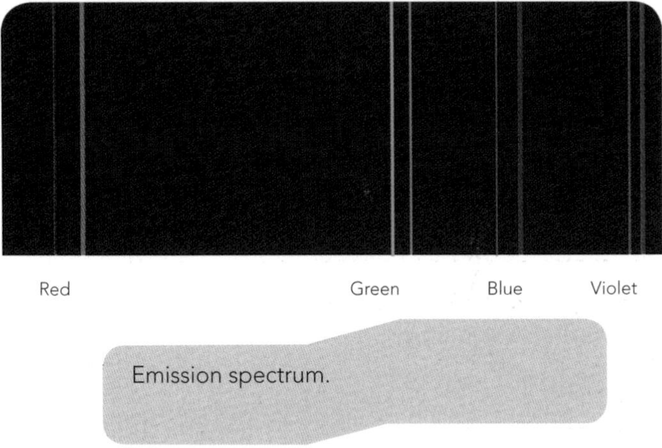

Emission spectrum.

Scattering: information about a chemical substance is determined from the way it deflects electromagnetic waves in various directions. The pattern obtained is specific to that chemical substance. Energy is not always changed when this happens.

Spectroscopy is used in a vast number of applications for physics, chemistry and biology. The various types of spectroscopic methods have specific names related to the part of the electromagnetic spectrum used; for example, ultraviolet spectroscopy and infrared spectroscopy.

Activity 4.2C

In spectroscopic absorption analysis of a solvent, how would impurities be identified?

Ultraviolet spectroscopy

Ultraviolet (UV) spectroscopy involves the determination of the purity of a substance using the absorption of UV wavelength electromagnetic energy. As different molecules absorb radiation of different wavelengths, the resulting absorption spectrum provides a method of identification for particular molecules. The energy absorbed is directly related to the energy between electron levels in the atoms that make up the molecules.

Beer's law provides the relationship between the absorbance of a solution and its concentration:

$$A = \varepsilon bc$$

where A is the absorbance, ε is the constant of proportionality, b is the path length and c is the concentration.

This spectroscopic method is used in the quantitative analysis of protein and DNA samples. The ratio of the protein building blocks, amino acids, to the DNA can be found by comparing the absorbance of UV waves in each. DNA absorbs in the 260-nm range and if the amino acid absorbs in the 280-nm wavelength the relative concentration of each can be calculated.

Infrared spectroscopy

In infrared (IR) spectroscopy the energy absorbed corresponds to the increased vibration in the bonds that join the atoms in each molecule. The wavelengths used in IR spectroscopy are in three distinct groups: near infrared (NIR), mid infrared (MIR) and far infrared (FIR). These descriptions correspond to the position of the wavelengths relative to the red end of the visible wavelengths in the electromagnetic spectrum.

Infrared radiation of a frequency (proportional to 1 / wavelength) that matches the natural structure of the molecular bond is absorbed. Each bond type, e.g. carbon–carbon or carbon–oxygen, has a distinct frequency. As a result, this characteristic can be used to determine the types of bond present in the substance being analysed.

The method is used extensively in research and industry, in areas such as forensic science and the manufacture of polymers. It also provides a quick way of analysing exhaust gases from cars. This is because the carbon–oxygen bond in carbon monoxide and the bonds in nitrogen oxides and unburnt fuel all have distinct absorption characteristics.

Reference data

The term 'reference data' refers to a comprehensive listing of components and particulars associated with them. In science, this usually means: names and formulas of chemical elements and compounds, crystal structure, mathematical formulas, physical and chemical properties, definitions of units, tables of constants, definitions of terms...

In research and education, work or social activities, and many more areas, the need to refer to other literature on the subject in question is essential. In science, in particular, the need to find information in the form of explanatory documents or data tables is very important. It is vital that many different sources are used for you to be sure that there is at least a general consensus of opinion.

You may find the following sources of information useful for reference.

- Chemistry Resources
- 100 Best Reference Sites for Science Students Online
- Lide, D. R. *Handbook of Chemistry and Physics.* CRC Press, 2004. ISBN13: 978 08493 05979.

Other techniques for estimating purity

Refractive index

When light passes from one medium to another (e.g. from air to water, or from air to glass) it is bent. This is called **refraction**. It is a result of the interaction of light waves with the change in the density of the medium. The ratio of the velocity of light in a vacuum to its velocity in another medium is called the **refractive index**. This varies significantly between solids, liquids and gases. The refractive index is also affected by the temperature of the compound and the wavelength of light used, so both have to be recorded.

A liquid can vary in its composition with the addition of other substances. This means that its refractive index will change slightly when different substances are in solution. A **refractometer** is a device that can measure these differences. These can then be cross-referenced to known data values to identify the compound. Typical refractive index values for most compounds lie between 1.30 and 1.70. The refractive index for ice is 1.31 and for water at 20 °C is 1.33.

Polarimetry

Light waves which vibrate in a single plane are said to be **polarised**. If this light passes through a solution, it can be rotated slightly to the left or right depending on the chemical composition. The observed amount of rotation can be matched up with known values to provide an identification of the compound that is present. The equipment used to measure how much the light is rotated is called a **polarimeter**.

Assessment activity 4.2

Competency in carrying out chemical experiments develops over time with practice. As a junior technician with a large chemical supplier, you must demonstrate that your skills are improving by carrying out a significant number of practical activities using separation techniques.

1. Wear your eye protection. Carry out separations using the following techniques: chromatography, distillation, precipitation, centrifugation, filtration, drying, crystallisation, solvent extraction and electrophoresis. P2
2. Wear your eye protection. Estimate the purity of chemical compounds using: chromatography, titration and melting point techniques. P3
3. Benzoic acid is very useful in industry. Its compounds are used in food preservatives and flavourings; insect repellent; medicines such as fungal treatments; and many other applications. Wear your eye protection. Carry out a re-crystallisation of benzoic acid and describe what factors may affect the purity of your final product . M2
4. Evaluate the accuracy of your results for each method in task 2. D2

Grading tips

To achieve P2, you must carry out all the techniques listed in task 1, including the use of an electrophoresis kit.

To achieve P3, you must be familiar with using the melting point apparatus. You can use your evidence for titrations from Assessment activity 4.1 if your tutor is satisfied with your technique.

For M2 in task 3, calculate the percentage yield recovered from your sample. Also determine the melting point and compare this to reference data values.

Discuss the difference between reference and experimental values. Which aspects of your procedure may have affected the purity of your product?

To achieve D2, compare your results from task 2 to those supplied by the laboratory technician. Are they accurate? Can you account for any differences because of your technique? Are there any other explanations?

4.3 Instruments and sensors

In this section:   

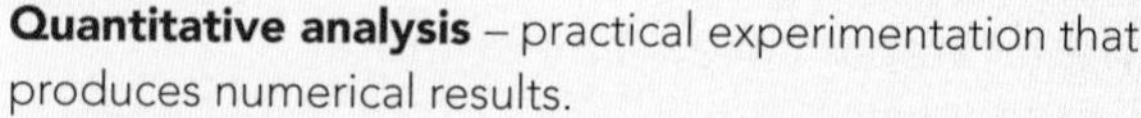

Quantitative analysis – practical experimentation that produces numerical results.

Scientific parameter – a limiting factor or boundary.

Coverslip – thin piece of glass placed on top of a specimen for observation under a microscope.

Cuvette – a specially designed clear container for use in spectroscopic analysis.

Ocular lens – eyepiece of a microscope.

Meniscus – the curved surface of a liquid in a container.

Linearity – a feature of measuring instruments where the displayed reading is proportional to the measured quantity.

Limit of proportionality – the point beyond which the stress divided by strain is no longer a constant for a material.

Siemen (S) – the SI unit of electrical conductance (1/Ω).

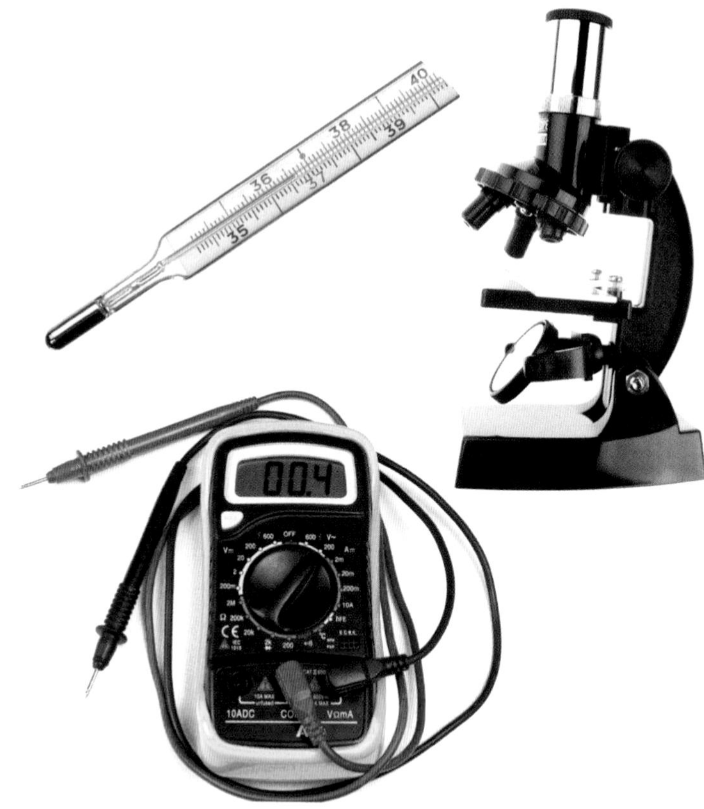

Some essential equipment and sensors.

Basic instruments and sensors

Using laboratory instruments correctly is paramount if the investigative work of a scientist is to have purpose and outcome. All equipment must be handled with care before, during and after use. It is fair to say that more complex equipment is generally more expensive (although there are some very expensive pieces of glassware in use) and so there are serious financial considerations when carrying out scientific practical work.

You may be familiar with the proverb 'using a sledge hammer to crack a nut'. The correct choice of instrument or sensor in the science laboratory usually depends on the specific result(s) required. For example, if the purpose of the activity is to provide a simple idea of the **scientific parameters** of a main investigation, then the choice of equipment may depend less on precision and accuracy and more on speed and availability.

Linearity of instruments and sensors refers to the scale representation of measurement. A linear scale voltmeter, for example, will be divided into equal intervals that will represent equal changes in voltage measurements.

Essential instruments that you need to know how to use are the:

- microscope
- pH meter
- pipette
- colorimeter
- top-pan balance.

Precision and measurement errors

The **precision** of an instrument is the smallest change in quantity that the instrument can measure. For example, the smallest division on a 30-cm ruler is 1 mm so you can make measurements with a precision of 1 mm. The **maximum absolute uncertainty** or **error** of an instrument is half of the precision, in this case 0.5 mm. We can write this error as ±0.5 mm. The percentage error in a measurement is defined as:

$$\text{percentage error} = \frac{\text{maximum error}}{\text{measurement}} \times 100\%$$

The sensitivity of many instruments decreases with use and time. The instrument may consistently provide incorrect readings. These are **systematic errors**. Recalibration for each practical experiment, checking zero points, using additional meters, etc., are ways of reducing systematic errors.

When repeat measurements of the same quantity made under the same experimental conditions produce different readings, they are **random errors**. An example is timing the swings of a pendulum. In these cases, the mean value should be calculated from a range of repeats to reduce the error.

When you are conducting an experiment you need to consider carefully the precision to which you want to make your measurements. If you are going to measure length, what precision do you need to know the value to?

- 1.0 mm, so a ruler with millimetre divisions will be fine.
- 0.01 mm, so you need to choose a micrometer or a vernier calliper depending on the length to be measured.

Microscope

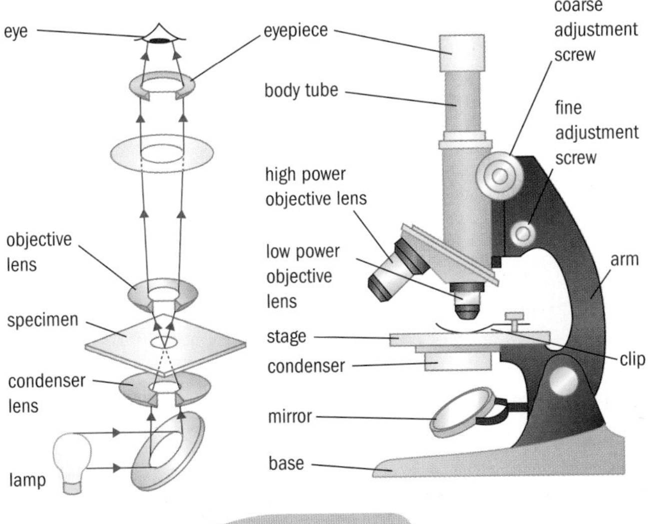

Laboratory light microscope.

Unit 11: See page 196 for more about the light microscope.

Carry the microscope with both hands; take care with glass slides and **coverslips**; beware of the bright light; and turn focus knob very slowly to prevent damaging the glass slides/coverslips and lenses.

Standard operation

Safety note: If you have to use a microscope with a mirror that uses daylight illumination, never use it where the Sun's rays can hit the mirror directly as the magnification will cause the light to permanently damage your retina. Choose a spot near a window that is not in direct sunlight.

1 Put the specimen on a glass slide and cover it with a coverslip.

2 Place the slide over the central hole of the *stage* and secure it with the *clips*.

3 Rotate the nosepiece to ensure that the lowest magnification *objective lens* is over the slide.

4 Position the *mirror* to reflect the maximum amount of lamp light or natural light up through the microscope.

5 Look down the *eyepiece*, also called the **ocular lens**. Adjust the iris diaphragm to limit the dazzling effect of the light.

6 Carefully rotate the *adjustment screw* until the lens is close to the slide. Look from the side of the microscope as well as down the *eyepiece* when you do this. Adjust the *coarse focus* first and then the *fine focus*.

7 When using higher magnification, adjust the focusing carefully.

Calibration

The total magnification of the lens system in use is simply a case of multiplying the magnification of the ocular lens (eyepiece) by the magnification of the objective lens, e.g.

ocular lens = ×10 and objective lens = ×20, so the total magnification = ×200

Before you can make any measurements using a microscope, you need to calibrate the ocular micrometer that is in the eyepiece. This is done using a stage micrometer which has a precise scale.

You should read these instructions in conjunction with the diagrams shown.

1 Choose an objective lens and position the stage micrometer onto a glass slide.

2 Focus the ocular micrometer and then align both scales so that the left-hand lines on both scales exactly overlap.

3 Look along the scales and find a point where there is another pair of measurement lines overlapping.

4 Count the ocular micrometer units to this point and count the stage micrometer units to the same point. Note the *units* of the stage scale because these are exact (you will find this etched on the stage micrometer, e.g. 0.01 mm).

5 Now calculate the conversion factor for the ocular units:

$$1 \text{ ocular unit} = \frac{\text{(number of stage units)}}{\text{(number of ocular units)}} \times \text{(length of 1 stage unit in mm)}$$

Note: measurements that you make using a microscope are often very small. As a result, it makes sense to use an equally small unit of measurement. Instead of using millimetres, you should use microns. A micron is 10^{-6} m, or 10^{-3} mm. So, to change from millimetres to microns (μm), you multiply by 1000.

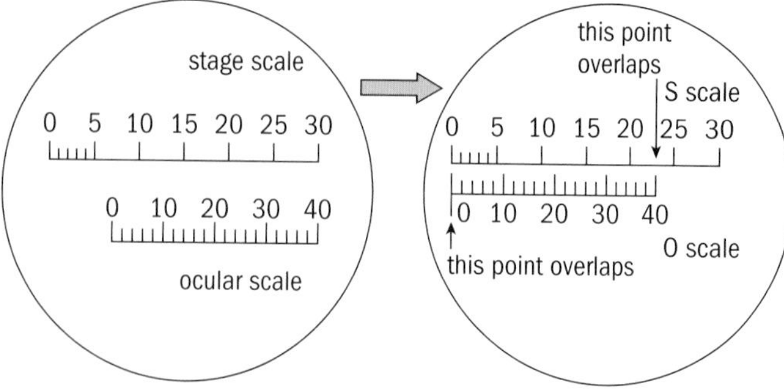

40 units on the ocular scale correspond to 23 units on the stage scale
1 unit on the stage scale is 0.01 mm
So 1 unit on the ocular scale is equivalent to $\frac{23}{40} \times 0.01$ mm = 0.00575 mm
This is equivalent to 5.75 μm.

Ocular and stage micrometer calibration of a light microscope.

PLTS

Self-managers and independent enquirers

You will practise your self-management skills by handing in assignment work to meet deadlines.

Evaluating the accuracy of the methods used to estimate and to measure the purity of the compounds prepared gives you the opportunity to practise your skills as an independent enquirer.

Accuracy

To determine the overall accuracy of your measurements, you must relate the measured values to the scales involved. The precision of a microscope provides values to microns (μm) and so the accuracy of your measurements should reflect this. In the example shown in the diagram, the precision of the ocular scale is 0.006 mm (to 1 s.f.). This corresponds to a maximum absolute error of ±0.003 mm (to 1 s.f.). Note that this is the same as ±3 μm (to 1 s.f.).

Worked example

The parenchyma cell of a plant is measured as 92 μm using the ocular scale shown in the diagram. What is the percentage error of the measurement?

Answer

Each division on the ocular scale represents 0.006 mm so the maximum absolute error is 0.003 mm, which is the same as 3 μm.

$$\text{percentage error} = \frac{\text{maximum error}}{\text{measurement}} \times 100\%$$

$$= \frac{3}{92} \times 100\% = 3.3\%$$

This gives a range of accepted values from 89 μm to 95 μm.

Did you know?

Around the year 1590, two Dutch spectacle makers, Hans and Zacharias Janssen, experimented by putting two lenses in a tube separated by a small distance. They had invented the compound microscope.

pH meter

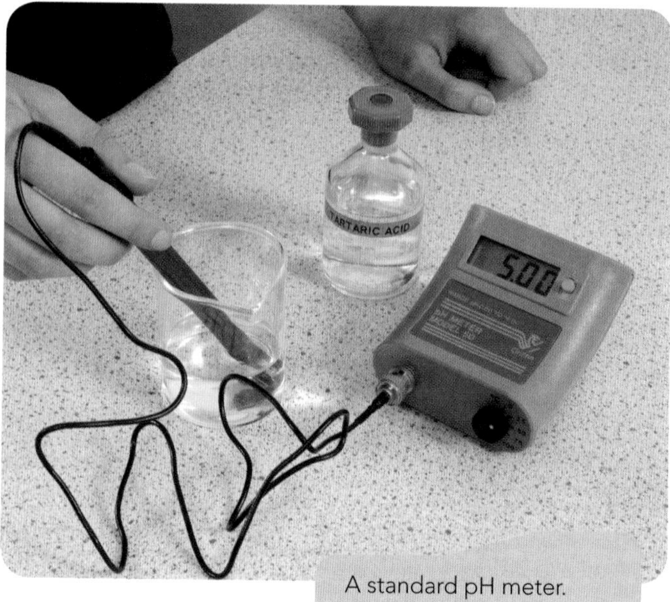

A standard pH meter.

The acidity of a chemical substance can be defined as its ability to donate a proton to another molecule or atom in an ionised state. When hydrogen chloride is dissolved in water it ionises completely and produces $H^+(aq)$ and $Cl^-(aq)$ ions.

As the hydrogen ion is essentially a proton, acids are associated with transfer of the hydrogen ion. A pH meter is a very precise voltmeter. It registers the ion activity of a solution by measuring the potential difference between two electrodes placed in that solution. The potential difference is amplified and converted to a pH reading that is displayed on the scale.

Handle the pH meter carefully; avoid the possibility of building up a static electrical charge that may affect the meter.

Standard operation and calibration

The method of calibration refers directly to the electrodes that are placed in the solution. Standard exact solutions of known pH can be bought from chemical supplies as **pH calibration buffers**. Examples of standard calibration buffers at 25 °C are given in the table below.

1 Obtain suitable calibration buffer solutions of pH 4.01, 7.01 and 10.01.
2 Depending on the model of pH meter you have, set the temperature control to account for the temperature of the buffer solution.
3 Rinse the electrodes with distilled water and immerse in the 7.01 buffer solution.
4 Ensure the meter reads 7.01 and adjust if necessary.
5 In the same way, calibrate the meter using the 4.01 buffer solution for acids and the 10.01 buffer for alkalis. Remember to rinse the electrodes with distilled water before you immerse them in each buffer solution.

Solution	Chemical formula	Concentration	pH value
hydrochloric acid	HCl	0.100 M	1.094
sodium hydrogen carbonate	$NaHCO_3$	0.025 M	10.01

Accuracy

The difference between the electronic measure of pH and the litmus paper/universal indicator paper versions is clearly in the precision to which the measurement is made. The reading on the pH meter shown gives values to 0.01 and so the maximum absolute error is quoted as ±0.005, whereas litmus/universal indicator papers will only indicate whether a solution is acidic or alkaline.

Activity 4.3A

What would be the possible result of not cleaning the electrodes of an electronic pH meter when recording the pH of numerous solutions?

Glass pipette

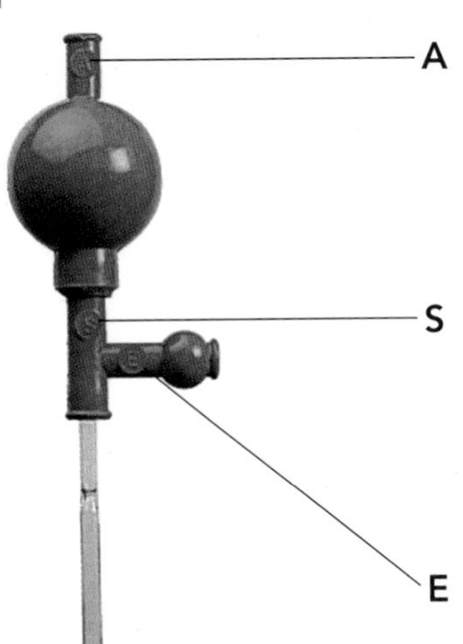

A laboratory glass pipette and rubber safety filler.

Safety notes: wear your eye protection; use and dispose of glassware carefully; report any breakages immediately; take care when inserting the glass volumetric pipette into the rubber safety pipette filler.

Standard operation (for glass pipette)

This procedure is for use with the rubber safety pipette filler shown in the picture. If you are using a pipette with a manual release piston you should adjust these instructions accordingly.

1 Choose a glass volumetric pipette of the required total volume or a pipette with a linear scale.
2 Carefully insert the volumetric pipette into the rubber safety pipette filler.
3 Squeeze A to expel the air and provide suction.
4 Insert the pipette into the solution.
5 Squeeze S to draw up the solution to a point well above the volume measuring line.

6 Remove the pipette from the solution ensuring there are no air bubbles in the liquid column. Touch the tip of the pipette on the inside of the beaker to help draw out excess air that may sit in the lower end of the pipette.

7 Squeeze E to release the solution slowly until the **meniscus** 'sits' exactly on the volume measuring line. Again, make sure there is no air in the end of the pipette.

8 Transfer the solution into another container by squeezing E. When all the solution is transferred, touch the side of the beaker to allow capillary action to draw out any liquid left in the pipette. This is part of your measured volume and should not be left behind.

Calibration

The volumetric pipette has been precisely engineered using sophisticated glass manufacturing techniques. The measurement lines on the glass surface are generally calibrated to read in millilitres (ml) or cubic centimetres (cm^3) for those with a linear scale, and a printed volumetric quantity for those with a single measurement line.

Accuracy

Accuracy of the volumetric pipette depends on the technique used. If the measurements are calibrated to 1.0 ml, for example, the accuracy for measuring purposes lies in the region of ±0.5 ml. This level of accuracy is dependent on the user. To develop accuracy in your technique, you should:

- keep the volumetric pipette vertical, reading measurements at 90° against a white background
- ensure that the meniscus 'sits' precisely on the measuring line
- expel air or liquid at the various points explained in the procedure
- repeat the procedure if in any doubt.

With practice, you will find you can use a pipette to measure out volumes of liquid very accurately.

Colorimeter

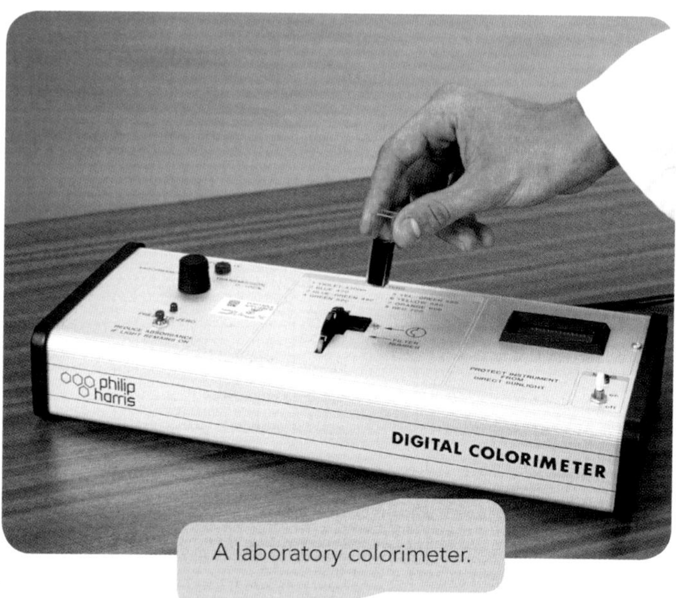

A laboratory colorimeter.

The colorimeter is used to measure the concentration of dilute solutions. It provides an easy method that is accurate and very sensitive. The physical principle used is absorption spectrophotometry whereby the concentration of the solution is related to the amount of light energy absorbed by a **chemical species** at a particular wavelength.

Handle the colorimeter carefully to prevent damage to delicate parts.

Standard operation and calibration

1 Switch on. Allow some time for the light source and detector to stabilise.

2 Choose the appropriate wavelength setting related to the solution to be tested from tables.

3 Set the display to 'Transmittance' by pressing the mode key.

Case study: Reegan, quality control technician

I work in one of the world's largest manufacturers of chemical laboratory supplies. My job is to ensure that each of the products made is of the required standard.

The company supplies: chemical glassware, consumables, chemical reagents and specialist sensitive equipment such as balances and centrifuges.

If any of our products is found to be faulty when used in a laboratory, there could be some serious accidents, and so I think my work is very important.

My job is very enjoyable and allows me to travel to different parts of the world to see how the manufacture of some pieces of equipment is changing.

4 Adjust the display to 0.0%T. The sample compartment must be empty and the cover closed when you do this.

5 Carefully fill a **cuvette** to the mark with reagent blank solution (usually distilled water or other transparent solvent). Clean fingerprints or spills from the side of the cuvette using a lens cleaner.

6 Following the manufacturer's instructions, position the cuvette in the colorimeter and close the cover.

7 Adjust the display to read 100.0% using the Transmittance/Absorbance control.

8 Press the mode key and select the status indicator. The status indicator must now be switched to 'Absorbance' and the display should show 0.0. If it does not show this, adjust the Transmittance/ Absorbance control again. The Mode key should now be changed to 'Transmittance'.

9 Remove the blank solution cuvette and insert your sample solution cuvette according to the manufacturer's instructions.

10 Record the %T value. Change the Mode key to 'Absorbance' and record the %A value.

Accuracy

The calibration in the colorimeter is set at 0 to 100%. Any value obtained for absorbance is provided as a percentage and so probable error or accuracy should be written as ± 1%.

The measured values of percentage absorbance are directly proportional to the concentration of the chemical species at the given wavelength and so provide a linear measurement (see Beer's Law, page 104).

Top-pan balance

The standard top-pan balance (also referred to as top-loading balance) used in chemical laboratories is a precision-made device. As a result, it incorporates very sensitive mechanical components that are extremely susceptible to vibrations or damage. In all applications for the purpose of this course, a balance of sensitivity to 1 mg (0.001 g) is easily sufficient. However, in industrial applications analytical balances with sensitivities to 1/10000 of a gram (0.0001 g) are necessary.

Handle the top-pan balance very carefully to prevent costly damage to the component parts.

Laboratory top-pan balance.

Standard operation

1 Turn on the balance.

2 Ensure the reading is 0.000 g. If not, press 'tare' to zero the balance.

3 Check that the weighing pan of the balance is dry and free from other materials left over from any previous investigations.

4 Place a suitable lightweight container on the weighing pan. In some instances you could use a clean, dry piece of paper.

5 Record the mass of the container/paper in your laboratory notebook.

6 Remove the container/paper carefully and place on a clean part of the table.

7 Use a spatula to transfer a quantity of the substance to be weighed into the container.

8 Place the container and substance on the weighing pan. If there is not enough of the substance, carefully add more using a spatula and a gentle tapping motion of the finger over the container. Do not knock the balance. If there is too much, it is best to take the container off the balance and remove some of the substance using a spatula.

9 Record the final mass and calculate the mass of the substance by deducting the mass of the container.

Accuracy

With time and excessive use, the balance can become less sensitive and as a result its accuracy will become suspect. The method for calibrating the balance is dependent on the instrument's manufacture. A general calibration can be carried out using reference masses supplied with the instrument.

Measurement of material properties

The essential purpose of carrying out scientific investigation is to provide some form of observable or measurable result which can be used to help understanding, confirm certain expectations or give absolute values of quantities for further use.

Measurements in the laboratory are usually confined to aspects of material properties (such as melting point or boiling point, see pages 100 and 101) that involve mastering techniques of fundamental importance in developing good scientific practice. The following determinations will enable you to experience a broad range of measurement techniques:

- resistance
- resistivity and conductivity
- tensile strength
- compressive strength
- elasticity
- refractive index
- turbidity
- viscosity.

Resistance

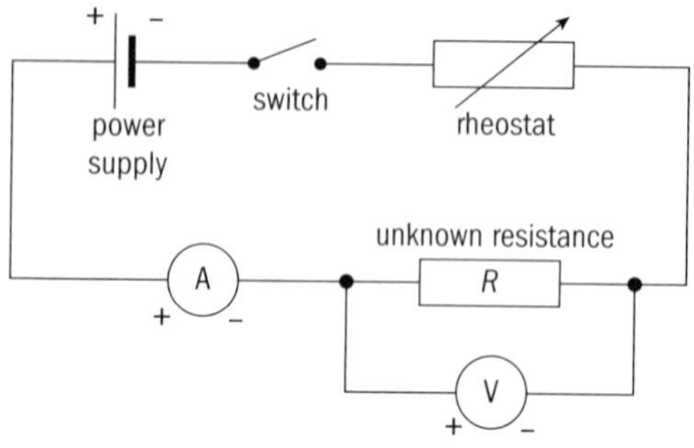

Circuit to measure the resistance of a conductive material.

The ability of an object to oppose the flow of electrical current (or the flow of heat when referring to thermal resistance) is called its resistance. Electrical resistance is measured in ohms, symbol Ω. Finding the resistance is simply a case of measuring the voltage applied and the current through a conductive material, and applying Ohm's law to calculate the value.

Apparatus

- cell or low-voltage power supply
- rheostat (variable resistor)
- ammeter and voltmeter
- choice of known resistors and conductive materials of unknown resistance.

Procedure

To calibrate the experimental set-up, first take a known resistor. Connect the components in the circuit as shown in the diagram, making sure that the meters are connected the correct way around. The voltmeter should be the last item to be connected. Adjust the rheostat so that it is at maximum resistance. Close the switch. Note the readings on the ammeter and voltmeter. Repeat this as the resistance of the rheostat is decreased in stages.

Put all your results in a table, like the one below, and use the formula for Ohm's law to find the resistance values. Calculate the average of these resistance values to find the resistance value of the resistor. Compare this to the known value of the resistor. The closer the two values are, the more accurate your measurements have been.

Now repeat the procedure for each conductive material of unknown resistance; you will need a results table for each material you investigate. You won't need a column for the resistance in these tables because you are going to plot your measurements on a graph and calculate the resistance from the gradient of the graph.

Rearranging the formula that represents Ohm's law gives:

$$\text{resistance } (\Omega) = \frac{\text{potential difference (V)}}{\text{current (A)}}$$

Potential difference (V)	Current (A)	Resistance (Ω)

Analysis

For each conductive material, draw an *x–y* scatter graph with current on the *y*-axis and voltage on the *x*-axis. A straight line will confirm that the material is an **Ohmic conductor**.

Resistivity and conductivity

The measure of a solid material's ability to allow electrical current to flow through it is called its conductivity (σ, pronounced 'sigma') and is measured in **Siemens** per metre ($\mathrm{S\,m^{-1}}$). The value is obtained by calculating the resistivity (ρ, pronounced 'rho') of the material by experimentation and using the formula:

$$\sigma = \frac{1}{\rho}$$

Resistivity (and, therefore, conductivity) is a property of the material and can be used as a means of identification.

Activity 4.3B

A student carried out an electrical investigation of various thin wires. In his conclusion he included the phrase 'wires with a high value of resistivity also have a high value of conductivity'. Is the student correct? Explain your answer.

Did you know?

Thin strands of copper wire of only 50 microns (μm) are currently used in some electronic speaker systems. This is the thickness of an average human hair.

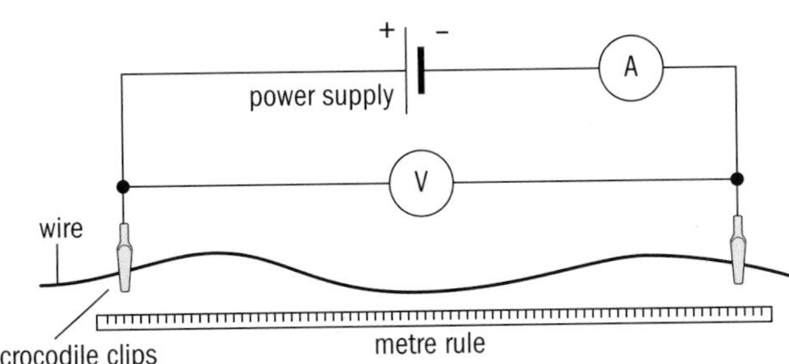

Circuit set-up for measuring the resistivity of a metal wire.

Apparatus

- 12V DC power supply
- ammeter and voltmeter (analogue, digital or both)
- choice of metal wires and thicknesses (e.g. copper: 0.22 mm and 0.52 mm, constantan: 0.27 mm and 0.40 mm, nichrome: 0.26 mm and 0.57 mm)
- crocodile clips
- micrometer screw gauge.

Procedure

Set up the apparatus as shown in the diagram and measure a 1-metre length of each wire. Take one of the wires and measure its diameter in three places; calculate the average and record the diameter.

Set the power supply to a low voltage to reduce the current and therefore the heating effect on the wire. This is because heating the wire would affect the readings. Move the position of the crocodile clip to reduce the length of the measured wire by 10 cm each time (from 100 cm to 30 cm). Record the values of current and voltage for each length.

Repeat this process for the other thicknesses of wire of the same type and then for the other types of wire. You will need separate tables for each wire and each diameter. Check the thicknesses with the micrometer screw gauge.

Length (m)	Current (A)	Voltage (V)	Resistance (Ω)

Analysis

1 **Using the formula to find resistivity and conductivity**

Example: choosing any given length and diameter of copper wire and its corresponding resistance

0.8 m length, 0.22 mm diameter and resistance of 0.5 Ω

$$\rho = \frac{RA}{l}$$

where: ρ is the resistivity (Ω m), *R* is the resistance (Ω), *A* is the cross-sectional area ($\mathrm{m^2}$), *l* is the length of wire (m).

The diameter of the wire = 0.22 mm. So, the cross-sectional area

$(\pi \times r^2) = \pi \times 0.11^2 = 0.038\,\text{mm}^2$

Converting this figure to square metres

$= \dfrac{0.038}{1 \times 10^6} = 3.8 \times 10^{-8}\,\text{m}^2$

So

$\rho = \dfrac{0.5 \times 3.8 \times 10^{-8}}{0.8} = 2.38 \times 10^{-8}\,\Omega\,\text{m}$

Conductivity (σ) is the reciprocal of the resistivity value. So

$\sigma = \dfrac{1}{2.38 \times 10^{-8}}$

$= 4.2 \times 10^7\,\text{S}\,\text{m}^{-1}$

2 Using a graph to find the resistivity and conductivity

For the values obtained for each wire, plot an *x*–*y* scatter graph with resistance values on the *y*-axis and lengths on the *x*-axis.

The gradient (slope) of the graph is equivalent to $\dfrac{\rho}{A}$

Therefore, the resistivity, ρ = gradient × A

Copper is such a good conductor that the resistance of the wire is low and errors due to the resistance of poor connections will be very significant.

Graphical methods of processing data are usually more accurate and should be used whenever possible because they depend on taking many readings. Drawing a line of best fit is similar to calculating an average.

You can compare the results you get from this investigation with the values given in data books. This will enable you to think about possible sources of error. Note that it is also useful to use both analogue and digital meters for this activity. This will provide you with an insight into the differences that may occur.

Conductivity of solutions

The electrical conductivity (EC) of a solution is much lower than that of the metal wires just described. For this reason, it has the unit microSiemens per centimetre ($\mu\text{S}\,\text{cm}^{-1}$) and is measured using a standard EC meter.

The electrical conductivity of a solution provides valuable information about the concentration of dissolved salts. When salts dissolve they separate into their constituent ions. These ions become the *carriers* of the electrical current. The greater the concentration of dissolved salts the higher the electrical conductivity is.

Electrical conductivity is used in environmental analysis when monitoring and assessing the possible action needed to maintain water quality in reservoirs. Dissolved salts can enter the water in several ways:

- directly from the rocks and soil in the surrounding area
- from waste water and sewage
- run-off from nearby roads
- atmospheric input like acid rain
- from agricultural activity.

Water entering our homes has an electrical conductivity of approximately $55\,\mu\text{S}\,\text{cm}^{-1}$, pure water has a value of approximately $0.06\,\mu\text{S}\,\text{cm}^{-1}$ and values for seawater can be in excess of $60\,\text{mS}\,\text{cm}^{-1}$. (Note that the unit for the electrical conductivity of seawater is milliSiemens not microSiemens.)

Tensile strength

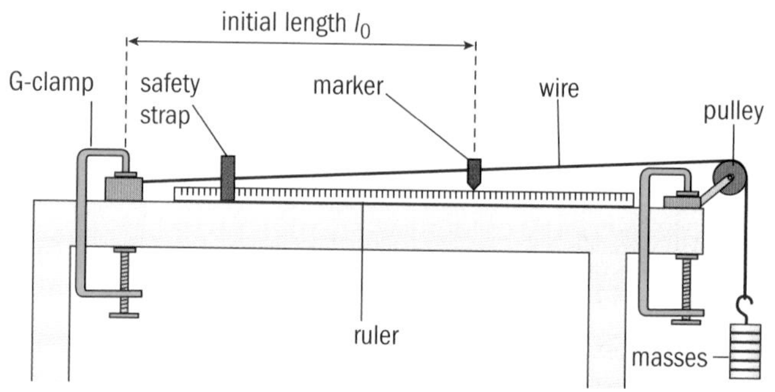

Simple method of measuring tensile strength. Note that the safety strap stops the wire whiplashing upwards should it break.

When a material is stretched by increasing the load on it, it will eventually break. Its tensile strength is defined by:

$$\text{tensile strength} = \frac{\text{the maximum load that the material can withstand before breaking}}{\text{the original cross-sectional area of the material}}$$

The original cross-sectional area is used here because the cross-sectional area decreases as the material is stretched. Tensile strength is measured in units of force per given unit area and so can be written as $\text{N}\,\text{mm}^{-2}$, $\text{N}\,\text{m}^{-2}$ and so on.

Up to the **limit of proportionality** (point A on the graph), the strain (extension/original length) is proportional to the stress applied (the load/cross-sectional area). The $\dfrac{\text{stress}}{\text{strain}}$ value is a constant within this limit and is known as the Young modulus (E).

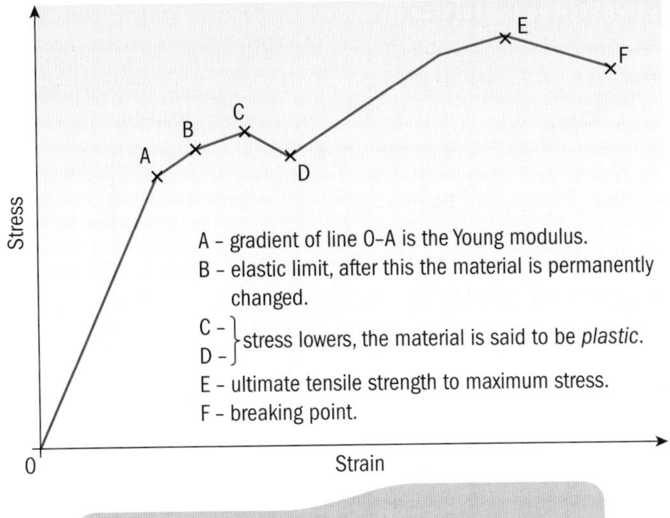

Graph of stress versus strain for a metal wire.

Apparatus

- eye protection
- suitable laboratory bench
- two G-clamps
- 1–2 metre lengths of wire
- safety fixings
- marker tape shaped to give a pointer
- micrometer screw gauge
- pulley
- metre rule (with millimetre divisions)
- masses.

Procedure

Safety note: eye protection must be worn because the wire may break and flick back into your eye if excessive loads are applied. Keep feet and hands away from the 'drop zone' where the masses will fall if the wire snaps.

Set up the apparatus as shown in the diagram with a small mass for initial tension in the wire. Make a note of the initial marker position. Use a micrometer screw gauge or vernier scale to measure the diameter of the wire in several places. Record the initial length of the wire from the fixed point to the marker; this is l_0. Add masses to the end of the wire and note the extension. Check the wire diameter carefully for each load before stretching, because this is the figure to be used in the calculation of cross-sectional area (the diameter decreases when the wire is stretched).

Load (or tension) (N)	Extension (mm)

Analysis

Plot a graph of load (T) on the y-axis against extension (e) on the x-axis and work out the value for the gradient.

$$\text{Young modulus } (E) = \frac{\text{stress}}{\text{strain}} = \frac{\text{load} \div \text{area}}{\text{extension} \div \text{initial length}} = \frac{T/A}{e/l_0}$$

Re-arranging this equation gives:

$$T = \frac{AE_e}{l_0}$$

So the gradient of your graph of load against extension will equal $\frac{AE}{l_0}$

Because you have measured the initial length (l_0) and the initial cross-sectional area (A), you can work out the Young modulus (E) from the gradient of your graph. The final step is to compare your value for E with the values given in data books.

Compressive strength

This is the amount of stress (stress = force/area) required to produce a minor but permanent change in shape of a material when it is subject to compression (being crushed). It is calculated in using:

$$\text{compressive strength} = \frac{\text{maximum load (N)}}{\text{original cross-sectional area (m}^2\text{)}}$$

The original cross-sectional area must be used because when the material is distorted the cross-sectional area can change significantly. The unit of measurement is the pascal (Pa) which has units of force/area ($N\,m^{-2}$). Compressive strength is still sometimes quoted using the imperial units of tons per square inch (1 ton per square inch is approximately 13 MPa). Some typical values for compressive strength are given in the table.

Material	Compressive strength (MPa)
steel	505
concrete	41
cast iron	up to 827
granite	138

Elasticity

Elasticity is a property that allows a material to return to its original shape once stretched.

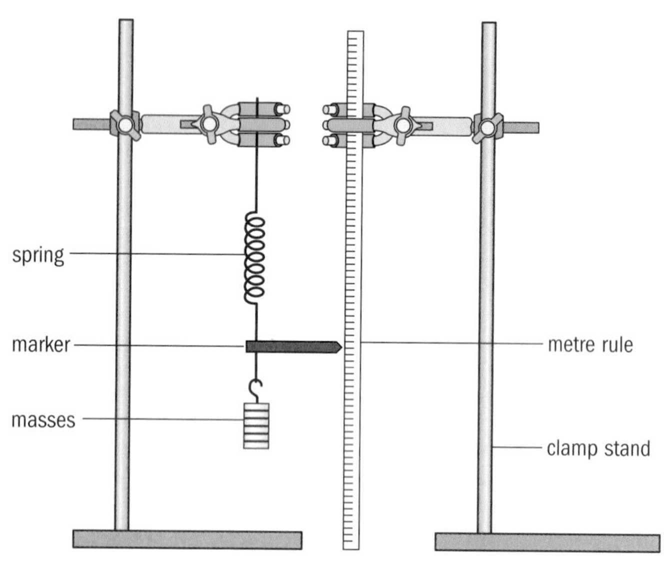

Determining the stiffness of a spring.

Apparatus

- two clamp stands, clamps and bosses
- G-clamps to hold stands securely on the bench
- metre rule
- choice of springs
- masses.

Procedure

Safety notes: ensure that masses cannot fall and cause injury. Conduct a preliminary test to assess the safe maximum mass that can be supported by the apparatus and the parameters of measurement of the spring without overextending it.

Set up the apparatus as shown in the diagram. Add a single mass at a time noting the extension of the spring by viewing the marker against the rule. Record results in a table.

Load (N)	Extension (mm)

Analysis

Provided the elastic limit (the point beyond which the material can no longer return to its original shape) is not exceeded, then a graph plotted of load (tension) on the *y*-axis against extension on the *x*-axis should produce a straight line. The extension of a spring is proportional to the tension (Hooke's law) and the gradient calculated is the stiffness constant, *k*. Its unit is $N\,m^{-1}$.

Unit 3: See page 62 for more information about Hooke's law.

Refractive index

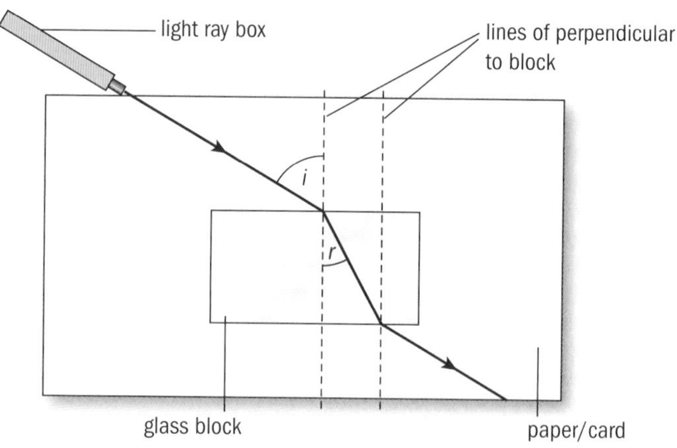

Determining the refractive index of glass.

Apparatus

- blocks in a range of transparent materials for testing
- sheets of white paper
- protractor, rule and pencil
- light ray box and power supply.

Procedure

Safety note: Do not tap glass blocks together; small shards of glass may break off and fly about!

Place the transparent block (glass, perspex or other suitable material) on a sheet of paper and draw round it. Draw in a line perpendicular to the straight edge of the block. Position the ray box at an angle to the block so that the light ray enters the block where you have drawn your perpendicular. You will notice that this incident ray is refracted (bent) as it enters the block and then bent in the opposite direction where it exits the block.

Mark a point at both ends of the incident ray. Then mark where the light ray leaves the block and mark another point on this ray. Remove the light box and the block. Then draw in the light ray. Measure the angle to the normal (perpendicular) of the incident ray and the refracted ray. Repeat this several times for accuracy.

Analysis

Using Snell's law:

$$\text{refractive index} = \frac{\sin i}{\sin r}$$

The refractive index is a property of a material in a particular medium. It is related to the change in the speed of light as it passes from that medium into the material, e.g. from air into water or from glass into perspex. You should note that in data-book tables the refractive index is always quoted for light passing from a vacuum into the material (see the table below for some examples). What you are calculating here is the refractive index for each of your materials in air.

Measurements of refractive index can also be used to determine the concentration of a solution. This is because light entering the solution will be refracted to different degrees dependent on the concentration of the solution.

Material	Refractive index
air (at standard temperature and pressure, STP)	1.0003
carbon dioxide (CO_2)	1.00045
ethanol	1.361
water	1.333
Pyrex®	1.47
impure glass	1.49–1.80
sodium chloride	1.5
diamond	2.42

Turbidity

In general terms, turbidity is a property of a liquid that is related to the loss of transparency associated with the amount of particulate matter suspended in the liquid. The amount of light that passes through the liquid is, therefore, dependent on the clarity of the liquid.

The large-scale measurement and treatment of turbidity is very important for the delivery of clean water for industrial and domestic use. Silt, mud, bacteria and chemical precipitates are all responsible for increases in turbidity. This increase can affect machinery and also lower the effectiveness of chlorine in destroying harmful microorganisms.

Measurement is carried out by using an electronic meter or a scaled tube that is calibrated to provide a measure of the turbidity depending on how far down the tube the observer can see clearly. The unit of measurement is the turbidity unit (TU). Drinking water should have a turbidity value ≤5 TU.

Viscosity

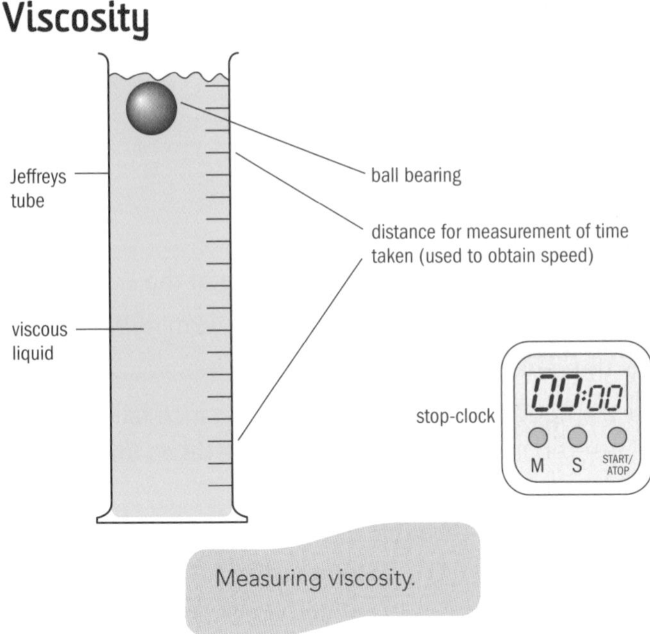

Measuring viscosity.

When a liquid like treacle is compared to water, there is an obvious difference in the ease with which each of them flows. The amount of force required to move the treacle is greater than is required to move the water. It is more viscous. Viscosity (η) of a liquid is a measure of its resistance to flow and is related to the structure of the molecules and their arrangement. It is measured in $N\,s\,m^{-2}$.

Apparatus

- Jeffreys tube
- viscous liquids
- ball bearing
- micrometer screw gauge
- stop clock.

Procedure

Safety notes: If engine oil or similar is used avoid skin contact and wash hands when finished. Make sure the long glass tube is stable before you start work.

Set up the apparatus as shown in the diagram and record the diameter and mass of the ball bearing (before you put it in the viscous liquid). Also mark the distance over which the ball will be timed. The terminal velocity (maximum falling rate) of the ball bearing should be achieved very quickly so the 'start timing' position can be set about two ball-bearing diameters from the top of the tube. The 'stop timing' position should be a similar distance from the base of the tube.

Carefully release the ball bearing into the tube. Record the time taken for it to fall from your upper mark to your lower mark. Repeat at least three times and average the results.

Analysis

Using Stokes' law:

$$F = 6\pi\eta R v$$

where:

F is the force due to the viscous drag of the liquid;

η is the viscosity of the fluid to be determined;

R is the radius of the ball bearing;

v is the speed of the ball bearing: distance fallen (measured in metres) divided by time taken (measured in seconds).

When the ball bearing is moving at constant speed, the upward force due to the viscous drag of the liquid will equal the downward force due to gravity. The force due to gravity is equal to the mass of the ball bearing (m) multiplied by the acceleration due to gravity ($g = 9.81\,\text{m s}^{-2}$):

$F = m g$

Substituting for F in the Stokes' law formula and re-arranging gives:

$$\eta = \frac{m g}{6\pi R v}$$

This experiment can provide valuable information about the flow of liquids and gases through pipe work. The influence of dust and dirt particles, temperature differences, water content or gas bubbles can affect the viscosity of liquids and gases. Some of these factors can be investigated in this practical activity. A suitable light gate/computer interface can replace the physical method described.

Activity 4.3C

Two volcanic eruptions A and B produce two very different kinds of lava. The viscosity of A is greater than B. What difference would this make to the rate at which it flows down the mountain? How is it possible for the rate of flow of volcano A to be quicker than volcano B?

Assessment activity 4.3

P4 M3 D3 BTEC

Making the appropriate choice with regard to instruments and being able to use them with care and precision are the hallmark of a competent laboratory technician. As a junior technician, you must demonstrate that your skills and instrument knowledge are improving by carrying out a significant number of practical activities using biological, physical and chemical techniques.

1 Carry out suitable practical activities that demonstrate the use of instruments and sensors used in each of the three science disciplines. P4
2 Explain the function of each of the types of equipment/sensors used to achieve P4 and justify why each was a suitable choice. M3
3 For at least one of the practical activities, provide details of the evaluation of accuracy of the procedure used, comparing it to an alternative method of measurement. D3

Grading tips

Using a microscope, top-pan balance and colorimeter would provide the evidence needed to achieve P4, although another chosen from within the text from physical science would also be appropriate.

Some additional research is needed in task 2 to supplement the information provided in this section on the functions of instruments and concepts involved. Is a multimeter really more suitable than an analogue voltmeter/ammeter, for example? To achieve M3, you should provide an explanation for all three experimental set-ups you have used in Task 1.

To achieve D3 you must include a calculation or full explanation of the difference in accuracy between two pieces of equipment used for the same measurement. In the example given for M3 (digital multimeter and analogue voltmeter), attempt to assess the error attributed to the analogue meter and account for possible differences in both meter readings.

WorkSpace

Tony Bristow

Associate Principal Scientist, Pharmaceutical Development, AstraZeneca

AstraZeneca is one of the world's leading pharmaceutical companies.

I am a specialist mass spectrometrist. Mass spectrometry is an analytical science tool which is used to determine the structure of molecules – an essential tool for pharmaceutical development.

I work with a wide range of colleagues across not just Pharmaceutical Development but across AZ as a whole. I will be approached by colleagues from analytical or chemical sciences to assist with a problem. I then select the most appropriate mass spectrometry technique to apply. I have access to a wide range of state-of-the-art mass spectrometers, which are capable of producing different types of data.

I am also involved in coaching, mentoring and training colleagues across different AstraZeneca sites. In 2009 this included a trip to Bangalore in India to train colleagues there.

I am the Vice-Chair of the British Mass Spectrometry Society whose aims are to encourage participation and promote knowledge and advancement in the field.

I genuinely enjoy my job as there is no routine. Every day is different and I am extremely keen on the daily problem solving which is an integral part of my job.

I solve problems for a range of colleagues all with the same goal of developing new pharmaceuticals. The time taken to solve a problem can be variable. On occasion it may be just 10 minutes assisting someone with data interpretation. At the other end of the scale, depending on the nature of the problem, I could be involved in an issue which could last for several weeks, where I am attempting to characterise every impurity in a sample in order to understand a chemical process.

Think about it!

1 Would you prefer to be the person who is performing a scientific experiment, or the person who is responsible for understanding and processing the actual results?

Just checking

1. What is the definition of a mole of a substance?
2. What is the purpose of preparing standard solutions?
3. For each method of separation, give a simple sentence to explain the principle involved.
4. What is meant by 'representative sample'?
5. Explain, briefly, the principle of spectroscopy.
6. Why is it important to correctly calibrate equipment before use?
7. Give the basic principles of determining a measure of four properties of materials (e.g. conductivity, tensile strength, melting point and conductivity of solutions).

Assignment tips

The key to successful laboratory practice is the manner in which you plan, research, carry out and evaluate.

- Each stage of the practical activity should be recorded in your laboratory notebook for reference and clarity. If you have a clear direction as to where the activity is going and what your expected results are, you will be able to identify errors more easily.
- Make sure you become fully conversant with the operation of the equipment and learn about the limitations. Laboratory safety is not simply confined to the possibility of injury but should also include the damage that can occur to delicate and expensive equipment.
- Carry out all the techniques outlined in this unit to help you develop an understanding of the chemical separation techniques and how identifying substances can be achieved through a number of step-by-step procedures.
- Remember that, in chemistry, rough observations and approximations are of very little use and play no valid part in the industrial context, apart from assessing the possible parameters of a thorough investigation. When taking measurements, ask for a second opinion and repeat tests where appropriate.
- Some calculations provided in this unit appear quite complicated at first glance, but are simply formulas from which results can be produced by inputting your experimental figures. If results appear incorrect, it is of more value to understand where the errors could have been made than to produce a textbook answer.

You may find the following websites useful as you work through this unit.

For information on...	Visit...
filtration	ThinkQuest
sampling and extraction, analysis and monitoring	Pollution Issues
sample preparation	Encyclopædia Britannica Online
safety rules	Laboratory Safety for Schools

Credit value: 10

5 Perceptions of science

The days of the scientist caricature as portrayed in many television and newspaper reports may have disappeared to some extent, but the perceived typical scientist and the characteristic laboratory still remain a powerful image in the minds of the general public.

Understanding the way in which science is reported and the interaction between the media and society, and an appreciation of the influence of the reporter on public perception are vital to the science community in attempting to make difficult concepts more accessible and gaining public support.

In this unit you will gain an understanding of the developments in scientific theories and theory testing. You will appreciate the difficulties faced by many theorists and the need for acceptance by the scientific community.

You will learn about the influential role of the media and how reporting can affect public perception, sometimes to the detriment of the population as a whole. You will have the opportunity to develop your own research technique and discover that many advances in science have a direct association with new and difficult moral or ethical issues.

Finally, you will learn how the need for funding, from government, the public sector and private concerns, affects the development of science and its uses, and you will begin to understand the links between passive scientific advancement and the misuse of science for financial and political gain.

Learning outcomes

After completing this unit you should:

1 know how scientific ideas develop
2 understand the public perception of science, as influenced by the media
3 be able to investigate the ethical and moral issues associated with scientific advances
4 know the relationship between science, commerce and politics.

Assessment and grading criteria

This table shows you what you must do in order to achieve a **pass, merit** or **distinction** grade, and where you can find activities in this book to help you.

To achieve a **pass** grade the evidence must show that you are able to:	To achieve a **merit** grade the evidence must show that, in addition to the pass criteria, you are able to:	To achieve a **distinction** grade the evidence must show that, in addition to the pass and merit criteria, you are able to:
P1 describe the development of a scientific theory, highlighting the processes involved **See Assessment activity 5.1**	**M1** differentiate between those questions that science is currently addressing, those that science cannot yet answer and those that science will never be able to answer **See Assessment activity 5.1**	**D1** explain why sometimes there is resistance to new scientific theories **See Assessment activity 5.1**
P2 identify public perception about science **See Assessment activity 5.2**	**M2** explain whether concerns raised about science in the media are justified **See Assessment activity 5.2**	**D2** analyse whether the media makes a positive contribution to the public's perception of science **See Assessment activity 5.2**
P3 explain how the media has influenced public perception of science **See Assessment activity 5.2**		
P4 report on the ethical and moral issues related to scientific developments **See Assessment activity 5.3**	**M3** discuss how the ethical and moral issues related to scientific developments will affect society **See Assessment activity 5.3**	**D3** evaluate whether the ethical and moral issues are important enough to stop scientific developments **See Assessment activity 5.3**
P5 identify how different groups and organisations have an influence on science **See Assessment activity 5.4**	**M4** describe how different groups and organisations have an influence on science **See Assessment activity 5.4**	**D4** compare and contrast how different campaigns, by groups and organisations, influence science **See Assessment activity 5.4**

How you will be assessed

Your assessment could be in the form of:

- diagrams and posters
- PowerPoint presentations
- written reports
- a reference log.

Danni, 18 years old

I really enjoyed the 'Perceptions of science' unit because it made me realise the importance of science in almost every part of life. I was amazed to find out how different styles of media reports on the same science topic could influence different groups of the public.

I was able to concentrate my PowerPoint presentation on a subject that I find personally interesting – the theory of plate tectonics.

The section on scientific advances helped me to understand why there are so many demonstrations from pressure groups and the public concerning the ethical and moral issues involved. I didn't know that information from war crimes has helped to advance science and I never imagined that people would sell their organs whilst they are still alive, for example.

Completing this unit has made me more aware of the media reports that I read and has shown me the importance of finance and politics in developing science.

Catalyst

Media hype?

Can we believe all that we read in the newspapers and all that we see and hear on television? We are given a daily diet of information through the media on scientific topics, but is this information accurate and how much is media sensationalism?

Research a particular recent event by looking at newspaper headings and news bulletins. Decide within your group whether the media have produced a factual and unbiased account of the story, or have tended to misrepresent information. Present your findings to the class in a clear PowerPoint presentation.

5.1 Scientific theories

In this section:

Key terms

Placebo – a false or fake drug used in testing groups of people for the effects of a newly developed medicine, for example.

Control – chosen sample used to compare with actual drug-testing sample.

Hypothesis – an explanation of what is happening based on observation.

Concordant – in science terms, agreeing sets of data.

Theory – a developed understanding of how something may work or happen produced by many people using many observations.

Charles Darwin, who developed the theory of evolution.

Development of theories

At many times throughout our lives we will encounter a situation which produces a question needing an answer. You can probably recall the last time something unusual happened which caused you to try to explain it. This simple process is the basis for scientific theories.

Our understanding of the world around us has developed from asking questions related to our observations and our attempts to explain what and how a particular event has happened. To help us, we need to experiment with ideas, produce models which allow us to visualise what may be happening and try to recreate an event to support (or refute) our **theory**.

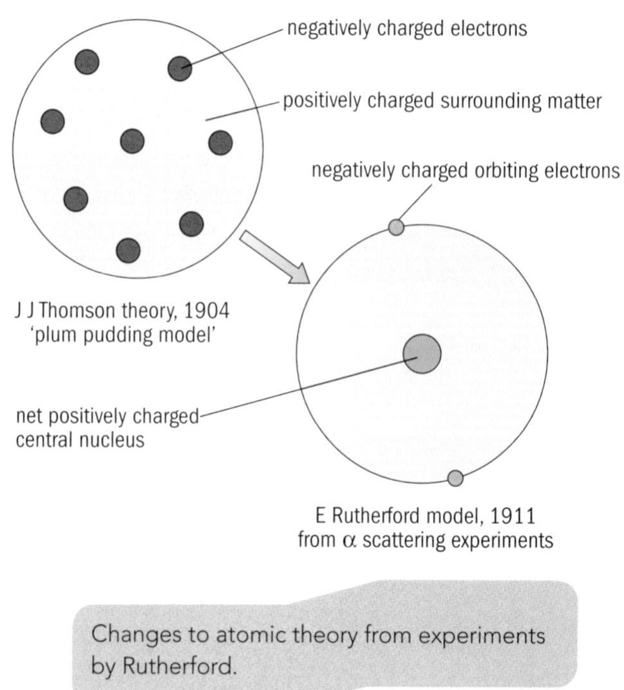

Changes to atomic theory from experiments by Rutherford.

Developing your own theory, step by step

1 Observe and question – bread becomes mouldy when left for a few days.

2 **Hypothesis** – I suspect that an organism from the air could be growing on the surface of the bread.

3 Carry out an experiment – place one piece of sliced bread on a window sill, one in a plastic bag on a window sill and one inside a plastic bag in a fridge for 1 week.

4 Share your findings with others – make general and definite statements (predication) supported by your experiment. For example, mould grew well in the plastic bag on a window sill, less well outside a plastic bag on a window sill and least in a bag inside the fridge.

5 Others repeat your experiment and have the same results (verification).

6 Produce your theory – microorganisms from the air grow well with heat and moisture.

Your theory is now ready for more experimental work

Testing theories is the job of pure and applied scientists. Measurements in science have to be **reliable**, in other words a consistent value obtained for each measurement. If colonies of mould were of different sizes for exactly the same conditions the results would be unreliable and the findings invalid.

When trying to gather together data from large-scale testing, it is essential that detailed planning is carried out prior to the investigation.

In medicine, for example, testing the effects of a newly developed drug on a selected group of people has to take place with clearly defined objectives. Those taking part will have been given full details of the testing and possible side effects before agreeing to take part. This is similar for patients undergoing medical treatment and is called informed consent.

Within the selected group, some are chosen unknowingly as the **control** group. Everyone is given something which resembles the new drug, but the control group is given the **placebo**. The experimental conditions are kept the same for everyone to ensure fair testing. This way scientists can determine, with repeated results, whether or not the effects observed can be attributed to the drug itself.

Placebos are not used much nowadays. It is more ethical to give one group of patients the new drug and another group are given the best of the previous drugs used for the condition.

Activity 5.1A

What is the difference between a hypothesis and a theory?

Collection and analysis of data

Repeating experiments is crucial in science and ensures that reliability of the results can be reported with some degree of confidence. This is essentially what makes scientific investigations different from many reported cases which tend to make headline news in the media, such as UFO sightings or alleged ghostly apparitions. If the occurrence can be repeated, the findings are reliable, and if the results from different experiments agree closely with each other, they can be said to be **concordant**.

When experiments are carried out there can be a tendency on the part of the experimenters to affect the outcome of the experiment either consciously or sub-consciously. This is bias. To help reduce this, conditions can be employed to introduce **blinds** or **double blinds**:

1 Blinds – the experimenter has full knowledge of the conditions of the experiment but does not know the outcome. The experimental results produced can be influenced by the experimenter if a prediction based on this knowledge is made. When butter manufacturers want to advertise their product, they sometimes offer a number of competitor products to a blindfolded taster who chooses which butter tastes better. The person offering the test knows which products belong to which company.

2 Double blinds – the experimenter does not know what the experimental outcome is likely to be and does not have enough knowledge to be able to predict the outcome. In this case, the possibility of influencing the outcome is greatly reduced. In the example of the butter manufacturer, neither the taster nor the person offering the test knows which product belongs to which company.

Well-known scientific theories

- atomic theory
- Big Bang theory
- germ theory of disease
- cell theory
- theory of evolution
- plate tectonic theory
- theory of gravity
- theory of general relativity
- string theory
- quantum mechanical theory.

Case study: The theory of evolution

Evolution is the process by which species change over a long period of time. As the species evolve, their characteristics, which help them survive and reproduce, are passed down. This is called 'natural selection'.

Observation and questioning stage:

1831: Darwin joins a 5-year scientific expedition on the survey ship *HMS Beagle* and begins developing ideas as a result of his experiences. He is already aware of beliefs in adaptation of life for different environments.

1835: *HMS Beagle* reaches the Galapagos islands.

1836: Darwin returns to England. His collections from the voyage are praised by the scientific community.

Hypothesis and experiment stages:

1837: The hypothesis that the environment evokes adaptive changes in species is tested. Darwin starts his first notebook on evolution. He finds evidence from his study of the fossil record: he observes that fossils of similar relative ages are more closely related than those of widely different relative ages.

Sharing findings stage:

1858: Alfred Russell Wallace concurs with Darwin's findings.

1859: Darwin publishes *On The Origin of Species by Means of Natural Selection.*

1871: Darwin's book *The Descent of Man* is published, applying his theories of evolution to humans.

1872: *The Expression of Emotions in Man and Animals* is published.

1873: Darwin publishes a book about human evolution, sexual selection and the descent of man. Professional and public anger at his findings is decreasing.

1876: Darwin publishes *The Effects of Cross and Self Fertilisation in the Vegetable Kingdom.*

1881: Darwin publishes *The Action of Worms*, highlighting the significant part that these organisms play in soil movement.

Repeat by others stage:

Scientific advances helped to provide more testing of Darwin's work after his death in 1882. Mendel and De Vries continued the work on gene mutation. Biochemistry and molecular biology advances provided supporting evidence for Darwin's hypothesis.

Theory acceptance stage:

Twentieth century: The evolution theory had enormous difficulty winning acceptance from the scientific community, but by the middle of this century its updated version – called the 'modern synthesis' – was accepted as a valid theory by many biologists.

Bibliography

More information about Darwin can be found in the **ThinkQuest online library**, **Wikipedia online encyclopedia** and the **Darwin website**.

More information about evolution can be found in the **BioWeb educational website**.

What significant developments have helped to firmly establish the theory of evolution?

Ethics committees

These are groups of people whose aim is to consider and advise on ethical issues that arise within education and research. They are usually scientists, philosophers and religious clerics who will all make valid contributions when discussing the moral implications of research studies.

Uncertainties

In science, the acceptance of a theory is usually arrived at through a general consensus of opinion from scientists around the world. It can take many years for this to happen. The *facts* which can bring about an almost undisputed acceptance of a theory can take a lot longer, and generally rely on advances in technology for experimentation.

The following table gives a list of known theories and the evidence provided which has supported them. It is important to remember that theories can never be proven because new evidence at some point in the future may show them to be incomplete, inaccurate or even totally wrong.

Theory	Supporting evidence
the Big Bang	observed red shift of galaxies, cosmic microwave background measurements, distribution of the galaxies, relative abundance of light elements
evolution	adaptation characteristics in species, transitional fossil records, development of feathers, teeth and skeletons in vertebrates, resistance to toxins in some animals
atomic theory	law of conservation of mass, alpha particle deflection experiments, emission spectra of hydrogen observations, discovery of the predicted neutron, successful use of Schrödinger equation and Pauli exclusion principle

Scientific questions

Asking the right questions in science is vital if we are to search for the right answer. Look at the following question:

'Are humans responsible for global warming?'

It is easy for most to answer simply *yes*, *no* or *I don't know* based on stories in the news, articles and headlines in the newspapers or from brief discussions with friends and colleagues. But, to answer this question meaningfully, a scientific mind needs to ask:

1 What exactly is global warming?

2 Is global warming actually happening?

3 Can it be caused by natural activity?

4 What is the scientific evidence to support it?

A scientist may answer the question in the following way:

'If we define global warming as an accelerated rate of lower atmospheric temperature rise, then the evidence over the last 50 years seems to suggest that global warming is taking place. It is known that CO_2 is a gas that can prevent the transfer of heat (a greenhouse gas) from the Earth's surface into space. We can, therefore, suggest a link between the recorded atmospheric temperature rise and the recorded rise in CO_2 levels over the same period.

It is unlikely that natural phenomena alone are responsible for such a sharp increase in CO_2 over such a short time period, although there is evidence for very quick temperature changes in ancient ice core samples.

We know that fossil fuels produce large amounts of CO_2 on combustion, and since humans have become increasingly dependent on this energy source during the last 100 years, we can conclude that our use of fossil fuels is a significant factor.'

In this example it is clear that a direct yes or no is not given. Instead, the scientist draws on known scientific principles and confirmed measurements to make a valid conclusion based on the evidence.

When asking questions in science:

- be specific
- think about how the question can be tested or answered
- research the topic to gain a clearer understanding
- be prepared to change your original question in light of new ideas.

Science – an answer for everything?

Historically, science has been able to find acceptable answers for many theories and ideas, our understanding resulting in advances in technology which in turn help us to develop as a civilisation.

There are many theories and questions which are the subject of research to find answers at the moment and many which may never be answered.

Did you know?

Scientists and mathematicians around the world are currently working on a 'theory of everything' which would be able to link all the known forces and their associated particles together to explain how the Universe works.

1. Questions science has answered

Is the Earth flat?

Can mankind adapt to living in the oceans and in space?

Can the speed of sound be achieved?

2. Questions science is currently addressing

Is there a cure for cancer?

Can we stop the damage to our environment?

Is there accessible water on other planets?

3. Questions science cannot yet answer

Is there a completely environmentally friendly alternative to fossil fuels?

Is there intelligent life elsewhere in the Universe?

Can humans be frozen and brought back to life many years in the future?

4. Questions science may never answer

How did the Universe begin and how will it end?

Is there a God?

Do ghosts really exist?

Is it ever right to use nuclear weapons?

Is it right to clone humans?

When new theories are suggested, especially when attempting to replace existing and well-established theories, there is always resistance to accept them. In addition, many important theories usually have 'sub-theories' running alongside which attempt to explain particular failings in the main theory. An example of this is the Big Bang theory for the origin of the Universe. To help explain how heat could have been transmitted throughout the Universe in the time given, an inflation theory has been introduced.

Peer review

A scientific paper can only be published when it has been scrutinised by qualified experts researching in the same field of science. These are peers and they will assess the work based on its originality and scientific competence. This process is called **peer review** and is a form of quality check. In all the examples above, there have been many peers working within the same scientific field who have had a significant influence on the acceptance of the proposed theories. Their work sometimes adds a further dimension to the theory.

Assessment activity 5.1

P1 M1 D1 BTEC

Your work as a TV researcher involves gathering together as much information on subject material as possible. It is your job to use or discard the material depending on its relevance for your programme.

1 Choose one scientific theory and show its main points of development. P1

2 Make a simple list into three categories (questions that science is addressing, questions that science cannot yet answer and questions that science may never answer) and describe the issues in one question from each category. Explain why the questions are different. M1

3 Choose one scientific theory and present the arguments for and against it. D1

Grading tips

You should follow the general pattern for theory development shown in the case study for evolution to help you achieve P1. To achieve M1 you should include some explanation as to why science may never be able to answer a question, and possible examples of questions that were once thought impossible to answer but which have subsequently been answered. The arguments for and against your theory for D1 should be detailed enough to show clearly the problems which make the theory difficult to accept.

PLTS

Team working

You will develop your team working skills when carrying out group research and producing a group document or presentation.

Functional skills

English

You will develop your English skills when presenting and listening to work from research.

5.2 Public perception and the media

In this section:

Key terms

Media – a means of communication to the general public, e.g. TV, newspaper, radio.

Target audience – a particular group of people (e.g. social, professional, intellectual, religious, cultural, age) for which an article is intended.

Caricature of a scientist popularised through television and other forms of media.

Meat eating to be banned!

Pet cloning a reality

GM food causes obesity

There is life on Mars

LHC could destroy planet

Would you believe headlines like these?

Media

The power of the **media** to influence public opinion is well documented throughout history and much use has been made by many historical figures to further their political ideals or to drive advances in business and commerce.

When reporting on science issues, the information presented depends largely on the type of media being used. Three important forms of media – television, the Internet and newspapers – are highlighted below:

Media	Details
television	Very large **target audience** from all backgrounds in society; virtually every home in the UK has a TV. Instantly accessible source of media information; some tight controls in place from regulatory bodies.
newspaper	Cheap and popular source of news and information. Regulatory bodies take a much more relaxed approach to style of reporting; varied methods used to promote newspaper publications; political biasing and open support of political parties within many publications.
Internet	Very large target audience. As popular as, if not more popular than, TV. Used as a resource for information about past, present and future events. Replacing written forms of research and entertainment. Developing social and cultural networks, e.g. blogs, online magazines, Facebook, etc.

Reporting science

The way in which science is reported varies to fit the forms of media in which it is presented. Journals specialising in specific areas of science will contain reports that are more detailed than newspaper articles, for example.

The language used by the media is crucial in portraying the message. The overall content of a science topic report will be tailored to the target audience and so the amount of technical information, complicated words and focus of the report will vary greatly. The following comments are taken from introductory paragraphs of reports published just before the Large Hadron Collider (LHC) was due to be activated in September, 2008:

1 'This Large Hadron Collider (LHC) is a powerful and complicated machine, which will smash together protons at super-fast speeds in a bid to unlock the secrets of the Universe.' *BBC News* © bbc.co.uk/news

2 'The world's biggest machine, the Large Hadron Collider (LHC) is being switched on tomorrow. Most scientists are very excited, but a few fear the LHC could bring about the end of the world – and even if it doesn't, how much is it costing anyway?' *MoneyWeek* magazine

3 'We could be welcoming our first time-travellers from the future this year after a giant scientific experiment is switched on in May. Or that is what a couple of Russian scientists would have us believe.

(...) It will accelerate them close to the speed of light and smash them together in an attempt to recreate conditions that existed in the first billionth of a second after the Big Bang that created the cosmos. But the forces unleashed could tear the fabric of the universe, causing a rift in space and time like that in Doctor Who spin-off Torchwood....

(...) That would turn the experiment, called the Large Hadron Collider, into the world's first time machine, say Irina Aref'eva and Igor Volovich, of the Steklov Mathematical Institute in Moscow.' *Skymania* (**Will time travel begin this year?**)

4 'Come 10 September, when the warren of caverns and tunnels has been checked out and sealed off – using high-security, iris-scanning locks – the first protons will be whipped up to nearly the speed of light through a chain of smaller accelerators on the CERN site. Then they will be injected into one of the LHC's two adjacent beam-pipes at an energy of 0.45 trillion electron volts (TeV).' *New Scientist* (**World's most powerful accelerator set to switch on**)

Example 1: Concise paragraph, easy to understand language explaining the concept to a large, varied audience.

Example 2: Emphasises concerns and costs to an audience of probable mature non-scientists with interests in world financial matters.

Example 3: Begins with a concise explanation using well-known scientific terms and drifts into the realms of science fiction. Targets a younger audience.

Example 4: Very descriptive paragraph with no explanation of scientific terms aimed at intelligent readers with a knowledge of the concept of particle acceleration.

Media giants

Organisation name	Estimated annual revenues (US$ billion)
Time Warner	27
General Electric	100
Viacom	20
The Walt Disney Co.	23
Bertelsmann	15
News Corporation	13

The table above shows the considerable wealth of some of the largest media groups. Whilst helping to promote positive messages to the people of the world, the business emphasis of groups like these appears to be firmly in support of commercial interests. This global cartel is quite uncompetitive and actually encourages joint equity ventures among its members. It is highly likely that this amount of money will sometimes influence media representation. For example, company bosses may be susceptible to promoting a particular direction of news because of some hidden agenda.

These powerful organisations deliver science news directly to large populations and, as a result, have a great responsibility to ensure that the information is well researched and balanced. The effects of the reporting style will influence different groups of people in different ways, for example:

- Religious groups – science reporting needs to be understanding of religious points of view so that science can be effectively incorporated into religious beliefs. There are many religious clerics who have carried out very important scientific work and contributed greatly to our understanding of scientific principles. The two disciplines can be studied together in most aspects.
- Age groups – older and younger people will need to be given science information which will consider their different levels of understanding. Young people need to build on knowledge whilst older people may not fully understand the new technologies and vocabulary used and may feel alienated from modern life.
- Pressure groups – highly motivated people can sometimes use science news to bring about changes in government policy and public perception. Careful reporting considerations can ensure that these causes are more justifiable.

Science, films and television

Science has always provided excellent subject material as the focus for movie makers. Film directors tend to take a concept or idea with a basis in science and develop a series of events that appears to demonstrate the positive and negative aspects of the topic in graphic and sensational ways.

Movie title	Science topic and description
Jurassic Park	DNA taken from fossilised amber to reproduce an extinct animal
Deep Impact	a large comet on a collision course with the Earth which may cause extinction of all life
Outbreak	transmission of a deadly human virus and the attempts to cope with its spread
Gattaca	set in the not too distant future this shows what the world could be like when humans without a perfect genetic profile are discriminated against in all areas of life

It is very difficult for documentary-style film makers to present a totally balanced programme, free from all bias. Researchers for these programmes have to look very closely at the information obtained from a wide variety of sources and enlist the help of expert scientists. Very often, a documentary that sets out to provide a balanced viewpoint eventually delivers a programme that can be perceived as biased.

Did you know?

Friday 13th may be regarded as an unlucky day but on this date in April 2029 an asteroid – 2004 MN4 (Apophis) – the length of three football pitches will pass very close to the Earth (possibly 30 000 km) and be visible to the naked eye. It will head back out into space and return on exactly the same date in 2036. Be prepared for some sensational headlines.

Media influence

The influence of media reports can be demonstrated in the following examples:

- Measles, mumps and rubella vaccination (MMR) – some popular newspapers in the UK published articles claiming there was a link between the MMR vaccination and autism.

Following the publication of articles like this, MMR vaccinations dropped by almost 14% due to the concerns of worried parents. The findings in the report which prompted the media headlines have subsequently been found to be false.

- Swine 'flu outbreak

Some newspaper reports in South America predicted that the virus would affect many millions of people in a very short time when the virus was first noted in Mexico. They caused widespread panic and were regarded as totally irresponsible journalism.

Activity 5.2A

When news needs to reach the general public informing them of a deadly new virus, which form of media would be most suitable and why?

Attitudes to science

It is in the interests of most democratic governments and societies as a whole to gauge the level of public understanding and general attitude to science, not least to help guide future government policies relating to science education and technological development.

Regular surveys for reports on public attitudes to science are commissioned by the *Research Councils UK* and *Department for Innovation, Universities and Skills*. The findings of the most recent report – 2008 UK – are highlighted below:

- 64% of the population believes that the government is in control of science innovation, and funding.
- Science is generally viewed in a positive light.
- 46% of the population believes that the benefits from science outweigh the potential harm, even when animal testing is taken into account.
- Media representation adds to confusion and distrust concerning science issues such as the conflicting accounts of climate change.
- Since public money is also used for science research, most feel that they would prefer better media reporting and prior consultation for new developments, especially the implications of ethical and moral issues.
- Most people have a limited knowledge of the regulation of science and engineering.
- Broadly, younger age groups agree with those over 25 years.
- 84% of the population agree that scientists should be less connected to business.

Generally, the public are in support of science and innovation but would like to be better informed, have more input into scientific debate and prefer scientific research to have less connection to private business to limit the tendency to put commercial interests ahead of scientific advancement for the good of the population.

Governments have a very difficult job to maintain and control scientific research whilst ensuring that public confidence, safety and ethical or moral issues are considered.

It is widely accepted that the UK is the world leader in particular areas of science such as biotechnology pharmaceuticals. The **allocation of public money** to scientific research and development needs to be firmly controlled and accurate financial records must be kept. The public will expect those in charge of spending their money to be held to account in the event of any of it going astray.

Funding for research comes from government, industry or charitable concerns.

Approval for funding for any line of research is given serious consideration before any financial commitment is made. In commercially funded ventures, the emphasis on approving research proposals is generally on close-term profitable developments, but there are good examples where longer-term research has been approved such as the production of the computer mouse or the continuing research into **quantum computing**.

Safeguards for scientific research

It is in everyone's interest to make sure that basic principles of honesty within science research are upheld. A lack of honesty will certainly help to undermine public confidence in a subject that, historically, has suffered from widespread mistrust. Safeguarding scientific research is the responsibility of all of us, particularly members of the science community itself. This applies to universities, scientific journals, research institutes, funding bodies and science societies. Maintaining good scientific practice is at the heart of scientific development.

Unit 3: See page 65 for more information on good scientific practice.

Even the ideas and work of scientists are not exempt from theft. In modern-day science, the **intellectual property** of many scientists needs protection from unauthorised use and plagiarism from within the scientific community. Theft of intellectual property, apparently rife in science, is thought to be a direct consequence of issues such as poor quality mentoring, political pressure, insufficient funding and ill-advised high regard for the mass publication of material.

The system of safeguarding scientific research is largely self-monitored within the scientific community. In general, hospitals, universities and other research institutes use review boards that approve research studies. Reported findings are then subjected to scrutiny by experts in the field.

Assessment activity 5.2

BTEC

You are a science journalist working for a well-known newspaper group taking part in a recorded public debate on the way in which science is reported in the media. You are asked to comment on some specific science reports which have made headlines recently.

1 Develop a questionnaire and use it on a wide audience from different backgrounds which will enable you to gather information on how the public perceive science. P2 Choose a well-known scientific article, film or documentary, dating back no more than 10 years, highlighting how it was perceived by the public. P3

2 Gather together a number of headline reports for one science topic from a variety of sources and list their concerns. Attempt to find scientific explanations for the topic and give your opinion on whether or not the reports are justified. M2

3 Analyse information in this unit and other extensive research to help you decide if the media reports on scientific issues are constructive or destructive. D2

Grading tips

Be careful with your questions in your questionnaire and limit to 10 for P2. The task for criterion P3 is a general TRUE or FALSE issue, and can be linked to the questionnaire by including articles about science and asking the question 'Do you believe the article or not – and why?'

For M2 and D2, a PowerPoint presentation can be given that highlights some important headline reports. You can research each topic for scientific explanations and sum up your opinion on media reporting.

5.3 Scientific advances, ethical and moral issues

In this section:   

Key terms

Moral – what is perceived to be right or wrong in human behaviour.

Ethical – the philosophy of questioning right or wrong issues.

Dolly, the world's first cloned adult animal (born 5 July 1996; died 14 February 2003).

Animal cloning is a form of genetic manipulation, taking a single cell from a parent organism to produce a genetically identical animal with exactly the same DNA. Although there are some complications which can develop, it is hoped that further development of the method will help in such areas as protection of endangered species.

Scientific advances

There has been an explosion in technological advances over the last 50 years. In general, thanks to developments in microprocessor and materials technology, the world has witnessed a greater rate of scientific development during this time than at any other time in history. However, with our new-found abilities to enhance the quality of life or search for answers to difficult questions, there are many **moral** and **ethical** questions to be raised and debated.

The benefits of science advancement are seen in all areas of life: healthcare, space exploration, transport, business, communication, food, clothing, etc., but sometimes the risks which are inbuilt in many areas of scientific research can produce significant drawbacks to the use of the new technology and delay its progression. There are examples where the risks have not been properly addressed or understood and have had terrible consequences or have delayed the use of scientific advances.

- **Thalidomide** – a drug used in the late 1950s and early 1960s as a sedative and to relieve the effects of morning sickness in pregnant women. Many children suffered serious birth defects.
- **Nuclear power** – there is still limited worldwide development of this kind of energy source because of a lack of communication between scientists and the public relating to the risks involved.
- **DDT** – was used as an agricultural pesticide for decades until a scientific report drew attention to the possible environmental effects which could develop. As an effective pesticide, DDT can eliminate significant disease-carrying insects such as the malaria-transmitting *Anopheles* mosquito. It is virtually non-toxic to birds but is highly dangerous for aquatic life.

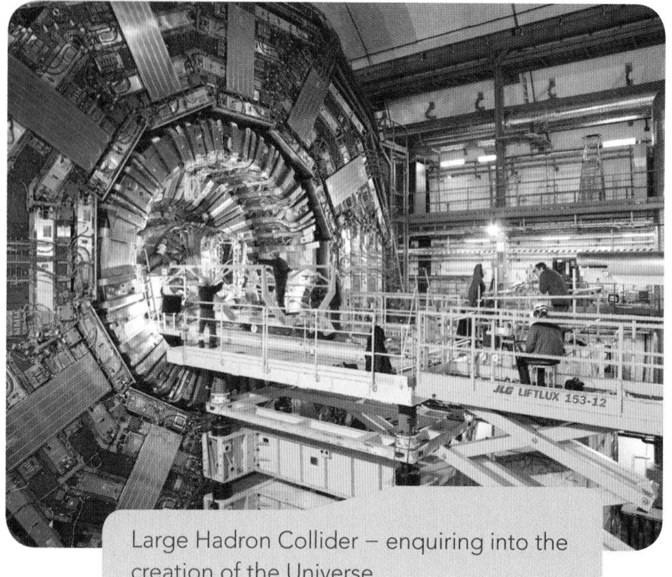

Large Hadron Collider – enquiring into the creation of the Universe.

Scientific advancements, such as the development of the Large Hadron Collider, can raise moral issues which centre on the religious argument that we are attempting to search for an answer to the Creation, which would disregard the actions of God. However, many scientists would argue that if we as a species are able to develop technology to this extent, if God does exist, it would be exactly what was expected of us.

Artificial intelligence – face recognition technology.

Artificial intelligence is described as the science of making apparently intelligent machines (machines which 'learn' rather than being programmed to perform tasks), but it is perhaps better to refer to artificial intelligence as artificial human-like abilities. Since intelligence is not clearly defined and links cognitive mechanisms with physiological and biochemical operations, the performance of even the most advanced computers must be limited to the understanding of intelligence by its designers.

Recently, a team of scientists in the USA led by Dr Craig Venter, claim to have produced the world's first form of artificial life. Although not artificial intelligence, the breakthrough could herald a global change in the way we tackle disease, food shortage or climatic change, for example. The principle involves developing a synthetic genetic software from pieces of DNA and implanting them into a host bacterial cell with its DNA removed. The cell then adopts the new information supplied and continues to multiply.

Activity 5.3A

What are the important benefits to the consumer from genetically modified crops?

Ethical and moral issues

As science makes progress, questions relating to the simple rights and wrongs of our behaviour are voiced.

Advances in science are made through years of study and experimentation, usually with the intention of satisfying our natural curiosity of the processes in our Universe, but also in many cases driven by a need to protect our species from disease, disaster and even from ourselves.

Scientists accept that it is essential to question the morality of our experimentation at all stages but that it is society that must decide whether or not the benefits outweigh the concerns. Scientists in all branches of science should work by the accepted principles of scientific investigation sometimes referred to as the three Rs – **R**igour, **R**espect and **R**esponsibility.

Did you know?

The musician and entertainer Roy Castle died in 1994 of smoking-related lung cancer but never actually smoked. It is thought that he developed the disease by performing in smoke-filled bars and clubs. His death added extra weight to a campaign which helped to bring about a UK-wide smoking ban in public places.

Area of science	Main points of concern
animal/ human organ donors	More than 30 people die while waiting for a donated organ every day worldwide. Animal organs have a virtually zero success rate when transplanted into humans. Not enough research has been carried out into the possibility of new viruses developing from animal to human organ transplantation. Although readily available, animal organs are dismissed on religious or animal rights grounds. Using animal organs rather than human organs is more easily justified because of the importance placed on human life.
self-inflicted illness	Smoking has been an addictive pastime for many thousands of years. The recent banning of smoking in public places in the UK was seen as an infringement of a person's right to smoke even though it is intended to prevent serious illness or death to smokers and non-smokers. Obesity, measured using the **body mass index (BMI)** calculation, increases the risk of diabetes, some cancers, liver and heart disease.
biological screening	Testing biological samples from whole communities or families who show no signs of an illness or disease must be carried out with good pre-planning. This method can help to avoid inherited diseases, for example, or inform the medical community early and provide useful data, but the information given to people may be a misdiagnosis or develop a false sense of security. There is also concern that life insurance providers might gain access to this information and use it to deny some members of society access to life insurance cover.
animal/ human drug testing	Drugs tested on animals for human benefit do not always work in humans and sometimes it cannot be avoided that the animals suffer a lot of pain. Usually the animals are killed as part of the test or when the series of tests is over. Human voluntary testing offers payment in exchange for the subject allowing themselves to be tested. Low earners, students, etc. are typical volunteers and sometimes serious problems can arise.
sources of scientific data	Some data can be obtained from illegal methods. In many parts of the world, regulations are not as strict on experimental practices and some unethical scientists are given more freedom to use human 'guinea pigs'. During large-scale wars, such as World War II, people were experimented on in horrendous ways by the Nazi regime.
genetic manipulation	It is argued that the transfer of genes from one type of cell to another (animal or plant) can create unknown health hazards which will continue through generations of the species. There is also concern about poor regulation of the use of this technology.

Nazi experiments – a case in point

Testing and experimenting on human subjects is a practice that continues today, both legally and illegally. It has been widely accepted that an atomic bomb detonated in Bikini Atoll, Marshall Islands, in 1954, was part of an investigation into the effects of nuclear fallout on a human population by the United States of America.

During war time, terrible crimes against humanity become more prevalent as the natural balance of law and order is distorted.

During World War II, 1939–1945, experiments carried out by the Nazi regime in Germany involved the mass injury and deaths of many thousands of people. This was said to be an attempt to understand the limits and adaptability of the human body.

The man chiefly responsible for these outrageous practices was Dr Josef Mengele. His work and that of others was well documented. The experiments included: radiation tests, injections with viruses, pressure tests, surgical organ removals, artificial insemination, sterilisation, extremes of temperature and investigations on 1500 sets of twins.

The data and records recovered from these atrocities have been examined and used for scientific advancement. As an example, Professor of Biology at the Victoria University in Canada – Dr John Hayward – used the Nazi war experiments to research into cold-water survival clothing at sea. He claimed that, although he did not want to, using the data was justified because at least it was being put to a constructive use.

Assessment activity 5.3

You are a researcher for a local journal publication assigned to gather information needed for a meeting of the representatives of an ethics committee.

1 Make a list of at least five areas of scientific advancement and describe, briefly, the ethical and moral issues associated with each. P4
2 Choose one prominent area of scientific advancement and list the positive and negative attitudes which were shown. Summarise how either point made may have an effect on society. M3
3 Choose one prominent area of scientific advancement, research from a wide variety of sources and present a report looking at the issues which surround this subject. Decide if the issues are important enough to stop the research. D3

Grading tips

A three-column table of science advances, benefits and ethical/moral issues should be sufficient to achieve P4. You will need to concentrate research into a topic which was well publicised, list the positive and negative comments and provide a brief explanation as to where the debate stands at the moment for M3. Your personal viewpoint is required for D3 based on extensive research of one chosen area and your evaluation of both sides, concerns and scientific advantages.

PLTS

Independent enquirer

You will be working as an independent enquirer when you produce information about public opinions on scientific advancement.

Functional skills

English

You will use your English skills comparing and selecting arguments and opinions from a wide selection of sources.

5.4 Science, commerce and politics

In this section:

Key terms

Vested interest – to claim support for a cause that can benefit the interests of the supporter.

Pressure group – organisations funded publicly and privately whose aims are to influence the public and politicians to have an effect on decisions by government on a variety of issues.

Civil rights – the entitlement of people to certain accepted standards of life including: personal safety, healthcare, protection from discrimination, etc.

Demonstration banner from AFAR (Alliance For Animal Rights), a pressure group.

Society and politics

Political parties and political groups

Political parties are organisations responsible for maintaining and developing the governing structure within a country. In democratic societies, they are publicly elected parties entrusted to provide leadership and make decisions which represent the interests of the majority of its people. In the UK the main political parties have developed over hundreds of years but the familiar modern party ideology which defines each of them is relatively new. Conservative and Liberal party politics have grown since the early nineteenth century, whilst Labour has only really developed since the early part of the twentieth century.

Political groups are interest groups who seek to bring about change in government policy without putting up any of their members for election. In this respect, they can also be called **pressure groups** because they can influence the support of political parties by voting for those parties which adopt a specific area within their electoral campaign. This may include: environmental issues, anti-abortion and nuclear disarmament, for example.

Whichever political party is elected, it is government responsibility to make decisions regarding finances and organisation of research and development in science. Political groups will monitor these decisions and continue to influence policies on specific issues through political ties and public support.

Activity 5.4A

Who provides a lot of money for research and development and where does this money come from?

Important global priorities facing all governments, for which science will be expected to provide solutions, are:

- climate change
- security (terrorism, Internet)
- human diseases
- population increase.

Pressure groups (also called lobby or protest groups)

These are organisations funded publicly and privately. Their aims are to influence the public and politicians to bring about decisions made by government on a variety of issues. Voluntary pressure groups, as they are also known, are independent of any political system and free to voice their opinions without restraint or constraint for the good of the world's populations.

Did you know?

Whilst docked in New Zealand in July 1985, the Greenpeace research ship *Rainbow Warrior* was sunk with explosives and a photographer was killed by the French secret service with support from the French government.

Group name	Fundamental aims, campaigns and funding
Campaign for Nuclear Disarmament (CND)	Set up to attempt to draw attention to the problem of nuclear war and to eventually bring about the complete disarmament of nuclear weapons, from all nations. It is funded by supporters and CND members. See the CND website for more information.
Greenpeace	Non-profit-making organisation on a global scale. Addresses issues through direct intervention concerning the planet, its environment and sustainability. Has helped to bring about abolishment of nuclear weapons testing, radioactive dumping at sea and highlighted the issue of whaling. Funded by members and foundation grants. See the Greenpeace website for more information.
Voluntary Euthanasia Society (VES)	Organisation which campaigns for a change in the law to allow the right of persons with a painful terminal illness to have medical help in ending their own lives. Set up by lawyers, doctors and religious clerics. Funded by members and public donations. See the Dignity in Dying website for more information.
Amnesty International	Campaigning for the recognition of human **civil rights** internationally. Work is supported in more than 140 countries with 1 million members. Funding is from members' donations and subscriptions. See the Amnesty International website for more information.
Friends of the Earth	Largest international environmental pressure group network in the world. See the Friends of the Earth website for more information.
Action on Smoking and Health (ASH)	A public health charity which has as its focus tobacco use and the eventual elimination of tobacco-related illnesses. Set up by the Royal College of Physicians in 1971, funded by subscriptions, supporters and donations. See the ASH website for more information.
Global Internet Liberty Campaign (GILC)	Developed from a meeting of the Internet Society in Montreal. Campaigns for freedom of speech and the privacy of information for Internet users. Funding is from donations, subscribers and company members. See the GILC website for more information.
Privacy International (PI)	A watchdog-style organisation set up in 1990 to monitor government and corporations, campaigning on issues related to freedom of information, video surveillance, ID cards, police information, medical data (such as DNA databases), etc. They are a non-profit, private limited company. See the PI website for more information.
Environmental Campaigns (ENCAMS)	Charity organisation set up to highlight the issues required in a sustainable, litter-free environment. They work with communities, businesses, local authorities and others. Their slogan is 'Keep Britain Tidy'. See the Keep Britain Tidy website for more information.

Concerns profiled

The importance of pressure groups and other organisations in highlighting scientific developments and their effects on society should not be understated. By raising the profile of some issues, society in general is forced to take notice of science and debate the concerns intelligently. Example concerns are:

- DNA database – the sampling and storing of DNA information from the public as a matter of routine under police investigation is widely regarded as unethical. It is argued that, in many cases, DNA analysis is still limited to predictions from statistics and that DNA should be taken only from those convicted of a crime, not all those arrested.
- Artificial reproduction – a biomedical technique which provides the means for infertile couples to conceive and have a baby by manipulating gametes/embryos in laboratory conditions and transferring them into the human body. There are two basic forms of this artificial insemination (AI): AIH (using sperm from the husband) and AID (using sperm from a donor). Concerns arise because in AID a third-party anonymous donor is used. This is not accepted by many religious groups for a number of reasons. They argue: it introduces different genetic material and changes the definition of 'family'; additional embryos produced in the laboratory are discarded and die; embryos are treated as a commodity and part of a commercial venture.

International pressures

Where serious concerns arise from the unpopular and dangerous policies of some governments, other countries may want to bring about a change by adopting strong international pressure tactics such as limiting trade links, or the exact opposite by offering trade and economic incentives.

The regular summits held by most of the world's most powerful nations, the G20, are well-known platforms where international discussion can help to change policies of governments. Essentially developed to bring about world economic stability, climate change and other science-related issues have yet to be part of the agenda.

In December 2009, the UNFCC (United Nations Framework Convention on Climate Change) Climate Change Conference took place in Copenhagen, Denmark. It was accepted that the meeting of the G20 would not result in any real discussion of the environmental issues, but it is certainly likely that continued international pressure may help to put this matter on the agenda in the future.

Social groups and popular science

For the purposes of political voting, advertising, entertainment, education and for many other reasons, the population in a country is categorised into social groups. They will share similar opinions, have similar aspirations, do similar things and like similar objects. In general, they will all have similar interests. Because the categories are so varied, we all belong to more than one social group and, as a result, we can both influence and be influenced by others.

Social groups are generally categorised by:

- age
- personal wealth or income
- education
- employment or profession
- health
- interests and motivations
- relationships.

The teaching of science across social groups has to take into account a lot of factors and must therefore be flexible in order to be able to bring the concepts and issues of science to a mass audience.

Popular science is seen as a means of delivering difficult but important topics to all social groups by making the language used more accessible whilst containing enough science material to maintain accuracy. This allows a continuation of debate at all levels and at least ensures that the public, in general, can make informed decisions.

Scientific funding

Scientists and science researchers need to be paid, and materials need to be paid for. Money to fund ongoing research in science is provided by a number of bodies:

- government – through universities, agencies and research councils (such as the Medical Research Council and Natural Environment Research Council)
- large corporations – by research and development departments
- foundations for science
- charity organisations – including cancer, AIDS and malaria.

The availability of funds depends on the research undertaken, and is generally awarded as Fellowships, International Grants, Science Bursaries, Scholarships or Studentships. When private funds are awarded, there is always a great consideration given to the cost-effectiveness of the research and many concerns are raised about the influence of the funding contributor on the direction of the science. Where a cure for a disease is being developed in drug research, for example, organisations may not allocate significant funding if most people affected by the disease are too poor to pay for the drug, the costs to develop it are not earned back or it is not profitable.

In 1990, the Human Genome Project (HGP) was started and funded by governments and private finances. Now complete, the project has identified all genes in human DNA and the information gathered will help in such areas as medical diagnosis and treatment, agriculture, healthcare, environmental considerations and forensics.

Space exploration costs

Edwin (Buzz) Aldrin on the surface of the Moon, 1969. Will we ever return?

When humans first landed on the Moon in 1969, it was the culmination of a unique drive to prove the technological superiority of one superpower over another. It is ironic that the manned exploration of space today seems to be less advanced than it was more than 40 years ago. This is mainly due to the extraordinary costs involved.

Each time the Space Shuttle sets off on a voyage to the International Space Station (ISS), the cost approaches US$700 million. There is significant work being carried out to develop manned space flight further including a permanent Moon base and expeditions to Mars, but the progress of these projects will undoubtedly depend on the willingness of governments to spend the vast sums of money necessary.

Payments for organs

The waiting list for organ transplants is increasing. Some patients turn to family and friends and the advances made in **immunosuppression** now mean that there does not need to be a genetic match. Organs such as kidneys and lungs are offered for use by a donor on death or for financial gain whilst still alive. The latter practice is not encouraged in the UK but is becoming more common throughout the world.

In the UK, donor cards are carried by potential donors in the event of death but it is also allowable in UK law for a person to donate an organ to help save the lives of others whilst they themselves wish to carry on living.

Where there is a continued shortage of organs, there will always be organ 'trafficking', selling organs for money, within and between countries.

Funding problems

Recent studies and media reports (e.g. BBC *Panorama: The NHS Postcode Lottery – It Could be You)* have highlighted the variation in drug availability throughout the UK. Their findings have hinted at the apparent differences in availability of medicines depending on the area of the UK in which people live. A government body (National Institute for Health and Clinical Excellence, NICE) was set up in 1999 to help eliminate this kind of problem, but the issue is still present all over the UK.

On 1st April 2007 NHS prescription charges were abolished in Wales, but are still paid by patients in the rest of the UK. This situation may change, of course, in response to economic conditions prevalent at any given time.

The cost-effectiveness of prescription drugs is an issue which is reviewed by NICE in the UK. The threshold cost has not changed since the organisation was set up and it is argued that the limit is too low and does not take into account the rising costs of equipment, materials and expertise in drug manufacture.

The availability of prescription medicines for particular illnesses also depends largely on a principle known as quality-adjusted life years (QALY). To be accepted as a treatment, the drug has to be able to improve the quality of life for patients measured on a calculated scale. If, for example, the cost to produce the treatment is extremely high and the improvement for a patient's quality of life is low, then the treatment is not usually offered.

Vested interests

Unfortunately, not all scientific studies will be carried out for the sole benefit of the general population and not all conclusions will be reported in the same way. A report in *Environmental Health Perspectives*, published in 2005, highlighted the point that from a large number of studies relating to the effects of bisphenol A (an organic compound used in the manufacture of plastics) on animals the reported outcomes were very different. Most studies funded publicly found harmful effects on animals, whilst none of the studies funded by the chemical industry found evidence of harmful effects.

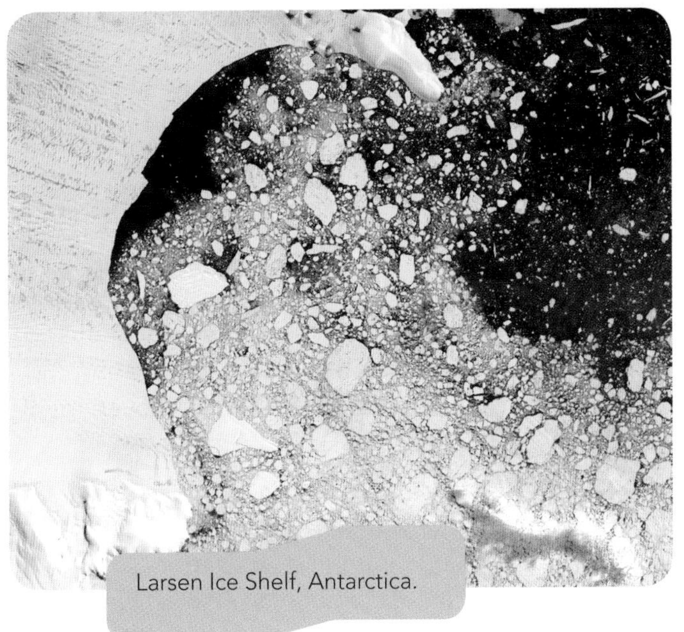

Larsen Ice Shelf, Antarctica.

Atom bomb 'Little Boy' explosion in Hiroshima, Japan, 6th August 1945.

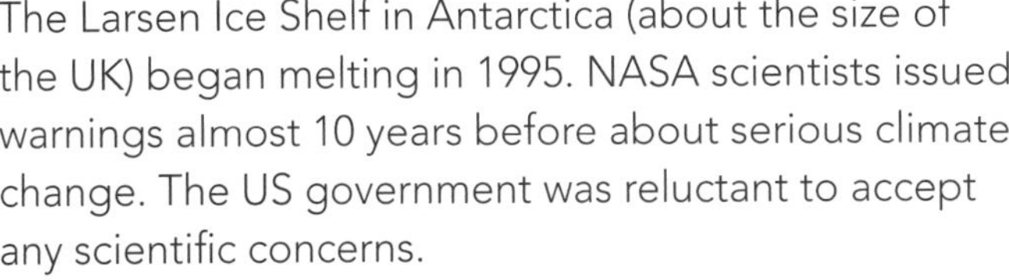

The Larsen Ice Shelf in Antarctica (about the size of the UK) began melting in 1995. NASA scientists issued warnings almost 10 years before about serious climate change. The US government was reluctant to accept any scientific concerns.

The science research for the nuclear fission process (the Manhattan Project) was well funded by the US government. Advances in this area of physics have produced both destructive nuclear weapons and useful nuclear power.

Assessment activity 5.4

M4 D4 P5 BTEC

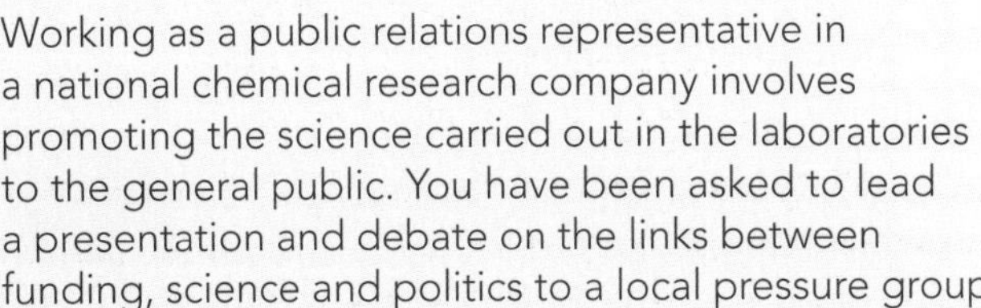

Working as a public relations representative in a national chemical research company involves promoting the science carried out in the laboratories to the general public. You have been asked to lead a presentation and debate on the links between funding, science and politics to a local pressure group.

1. Produce a list of pressure groups and identify their contributions to achieving scientific advancement and recognition. P5
2. Describe what strategies are used by political, social and pressure groups to affect science research. M4
3. Highlight specific campaigns from different groups and organisations that have affected scientific research. Compare each campaign style. D4

Grading tips

Describe at least five well-known pressure groups, giving a brief description of a successful campaign each has used to bring about a change of government policy for P5. To achieve M4, choose at least one type of group and give a description of a case where each group has brought about science advancement to meet its own interest. Make a list of some prominent campaigns to the government on scientific matters and compare the methods used to effect change for D4.

PLTS

Creative thinker

You will develop your creative thinking skills questioning the motives of pressure groups, industry and politicians for scientific research.

Functional skills

ICT

You will use and develop your ICT skills when using a variety of websites for information regarding pressure groups, politics and industry and examples of influence.

WorkSpace

Lucy Goodchild

Press Officer, Faculty of Natural Sciences, Imperial College London

I have a background in genetics and microbiology, and completed an MSc in the History of Science. I now work as part of the Research Communications Group at Imperial College London – I help reporters to cover science stories and I promote research done by scientists in the Faculty of Natural Sciences.

Every day is different but, typically, my mornings will be spent reading the newspapers scanning for any media coverage on Imperial. I will also take phone calls from journalists who may want to know more details from a scientific point of view about topical events such as the volcanic eruption in Iceland or maritime oil spills.

I will have numerous requests for radio and print interviews and it is my responsibility to know the correct member of staff to talk to about an issue.

I will often meet academics to talk to them about their research and to see if it is newsworthy. I also do a great deal of public speaking, working with scientists to help them communicate the ideas within their work.

I write press releases and web stories, make videos, hold audio interviews, create photo features and write stories for the College's internal magazine.

It is an incredible honour to meet and work with so many talented and interesting academics. To be able to encourage people's interest in difficult subject areas is equally satisfying.

People are often sceptical about the authenticity of media reports and, unfortunately, science stories can occasionally be misreported, thereby compounding this scepticism. In one recent case, some facts from two different stories were confused leading to one fictitious article!

Think about it!

1 What would you do if you saw something misreported?

2 How would you convince a scientist that talking to the media could be beneficial?

Just checking

1 Provide an example of informed consent.
2 List the main pieces of supporting evidence for the theory of evolution.
3 What are the most popular forms of media?
4 Provide an example of how media reporting has influenced public behaviour.
5 What does 'ethical and moral' mean for science?
6 What are the main points of concern regarding human and animal organ donors?
7 List four areas of government priority related to science.
8 How can the outcomes of science research be influenced by private funding?

edexcel

Assignment tips

To get the grade you deserve in your assignments remember to do the following:

- Make point by point notes on the definitions of terms used throughout this unit and refer to them frequently to help you complete the set activities at the end of each section. View many aspects and reports with an open mind and read all references carefully. Attempt to produce work which shows both sides of the argument, especially for the section on ethical and moral issues within science, and give your personal viewpoint only when directed.
- Begin by completing all the required tasks for the Pass criteria. For some of these in this unit a suitable list and brief description will be enough, but make sure that the list covers all the key points discussed in your tutor sessions and the assignment tasks.
- To achieve Merit and Distinction criteria you should include detailed explanations where asked. Research is fundamental and a variety of websites and text books for the topic is essential to provide further information. Make a note as a reference for websites you have used. Many sites provide valuable links which should be explored. Specific journals are important, but beware of some bias in reports and accounts.

Some of the key information you'll need to remember includes the following:

- There is a standard procedure in the development of scientific theories and no theory can ever be completely proven.
- Understanding questions is an essential aspect of developing investigations and must be given serious consideration.
- Media representation of science-related matters can be both constructive and destructive. The power of the media in reporting science to the general public is evident and wording of science reports varies according to the target audience.
- Advances in science usually include an associated ethical and moral issue. Scientists should work to strict ethical codes of conduct when investigating science of a sensitive nature. History is fraught with examples of inhumanity in the name of scientific advancement.

Suggested research topics for project work:

Use of pesticides
Battery farming
Nuclear energy
Human cloning
Use of illegal drugs for medical purposes
MMR vaccine
Donating organs
Effect of our lifestyle on climate change
Animal and human testing of pharmaceuticals
Storage of smallpox and other 'eradicated' viruses
Intensive care treatment
Assisted suicide
Radio/mobile phone mast siting
Maintaining nuclear weapons
IVF treatment
DNA database
Screening for inherited diseases

You may find the following websites useful as you work through this unit.

For more information on ...	Visit:
scientific theories and investigation	Worsley School website and Kosmoi photographs
media reports and scientific journalism	websites of the Australian Press Council and the BBC
scientific advancement, ethical and moral issues	technology and human development
science, commerce and politics	history learning site and Scientists for Global Responsibility

Credit value: 5

6 Using mathematical tools for science

Mathematics touches all our lives. Whether you work as a health visitor monitoring the mass of newborn babies or as a NASA technician assisting with the design of new engines, you will work with numbers and very likely you will want to display data. Being able to leave your calculations to the appropriate significant figures and in standard form will be vital. This unit will provide you with plenty of examples to allow you to master this skill.

Your data may be discrete or continuous; for example, when analysing fingerprint types in a forensic lab or the rate of bacterial growth in a hospital microbiology laboratory. Your collected data may be linear or non-linear. Understanding what data are and how to interpret and handle data is an important skill that you will develop by studying this unit.

Measurements are only as accurate as the measuring instrument you use, as well as the skill of the experimenter. In this unit you will look at various instruments that are used in science to measure quantities. You will learn about the types of errors (called uncertainties) that may result when making measurements.

This is an important unit, not only because it provides you with invaluable Level 2 revision of some interesting and challenging mathematical concepts, but also because you will develop numeracy skills that will help you with whatever career you follow.

Learning outcomes

After completing this unit you should:

1 be able to use mathematical tools in science
2 be able to collect and record scientific data
3 be able to display and interpret scientific data.

Assessment and grading criteria

This table shows you what you must do in order to achieve a **pass, merit** or **distinction** grade, and where you can find activities in this book to help you.

To achieve a **pass** grade the evidence must show that you are able to:	To achieve a **merit** grade the evidence must show that, in addition to the pass criteria, you are able to:	To achieve a **distinction** grade the evidence must show that, in addition to the pass and merit criteria, you are able to:
P1 carry out mathematical calculations using suitable mathematical tools **See Assessment activity 6.1**	**M1** use standard form to solve science problems **See Assessment activity 6.1**	**D1** use ratios to solve scientific problems **See Assessment activity 6.1**
P2 carry out mathematical calculations using algebra **See Assessment activity 6.1**	**M2** use mensuration to solve scientific problems **See Assessment activity 6.1**	**D2** use algebra to solve scientific problems **See Assessment activity 6.1**
P3 collect and record scientific data **See Assessment activity 6.2**	**M3** describe the process involved in accurately collecting and recording scientific data **See Assessment activity 6.2**	**D3** compare methods of data collection **See Assessment activity 6.2**
P4 identify errors associated with collecting data in an experiment **See Assessment activity 6.2**	**M4** calculate any errors associated with scientific data collected in an experiment **See Assessment activity 6.2**	**D4** explain how errors can be minimised in data collected in the experiment **See Assessment activity 6.2**
P5 select the appropriate formats for displaying the scientific data that have been collected **See Assessment activity 6.3**	**M5** interpret the trend in the scientific data collected in an experiment **See Assessment activity 6.3**	**D5** calculate scientific quantities from linear and non-linear graphs **See Assessment activity 6.3**
P6 interpret scientific data **See Assessment activity 6.3**		

How you will be assessed

The criteria in this maths unit can be assessed through various methods. Worksheets could be one way, where you have to solve scientific problems using algebra and ratios. You can be assessed on how you collected scientific data. For example, you may be asked to conduct an experiment and then display your data in the form of graphs and charts.

Gul, 16 years old

This unit is useful regardless of what area of study you go into. This is because I learned maths skills that were related to the real world; it wasn't just maths without a purpose. The algebra used to be dull but using it in a science context helped make it more interesting and easier. The ratio section was particularly useful and it is helping me with the mole calculations in chemistry. Linking the maths tools to experiments I found interesting, as I could see how the maths was being used and it made sense to me. The other aspects that I enjoyed about this unit are the use of ICT, as I could use Excel to plot and analyse my data. The measurement section was also useful as we are taught how to improve our measurement techniques. This was particularly useful when we carried out experiments, both in the physics and chemistry modules. The 'Using your scientific calculator' section was also useful, as I could never understand how to make good use of the calculator before this course.

Catalyst

Maths for science

In groups of three or four, discuss what kind of maths may be used by the following professionals:

- health visitor
- pharmacy technician
- forensic scientist
- manufacturing calibration control technician.

For example, do they use fractions, algebra, charts or graphs? What do they use the maths to do in their work? What maths tools are common to all and which tools might only apply to each profession?

Present your results on a clip chart or a poster.

6.1 Using mathematical tools in science

In this section:      

Key terms

Standard form – a way of writing down small and large numbers easily using powers of ten.

Unit – a symbol that follows a number; it tells us about the quantity.

SI units – units that have been agreed internationally.

Imperial units – units that were previously used mainly in the UK and US.

Metric units – a decimal system of units used by the SI system.

Prefix – letter added to the start of the name of a unit that represents powers of ten (both positive and negative).

Denominator – the bottom number of a fraction.

Numerator – the top number of a fraction.

Ratio – a comparison of the sizes of different numbers.

Mensuration – calculation of areas and volumes.

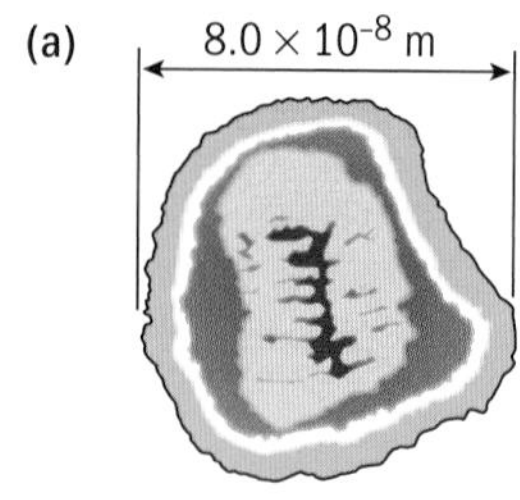

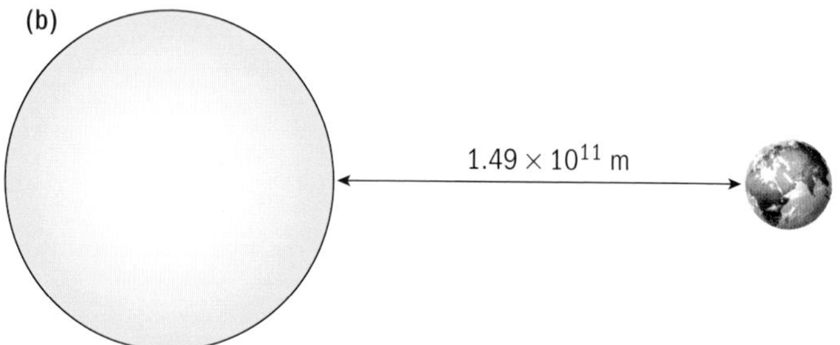

(a) Influenza virus particle. **(b)** The distance between the Sun and the Earth.

Standard form

Scientists have to handle very small and very large numbers; for example measuring the size of cells or calculating the memory capacity of a memory stick. They use **standard form** to make these numbers easier to handle.

Light takes 8 minutes to travel from the Sun, where it is emitted, to the Earth. The speed of light is approximately 300 000 000 metres per second. The distance between the Sun and the Earth is about 149 000 000 000 metres; that's 149 million kilometres. These numbers are difficult to use written like this, so scientists change them into standard form.

Numbers in standard form are written as numbers between 1 and 10 multiplied by a power of 10. So the speed of light would be written as 3.0×10^8 m s^{-1} (count the zeros) and the distance between the Sun and the Earth as 1.49×10^{11} m.

6 000 000 000 000 000 000 000 000 kilograms

6.0×10^{24} kg

number between 1 and 10

the power of 10 is increased by moving the decimal point to the left (e.g. 0.60×10^{25}) or decreased by moving the decimal point to the right (e.g. 60.0×10^{23})

24 is the power of 10, which means 10 × 10 × 10 × 10 × 10 × 10 × 10 × 10 × 10 × 10 × 10 × 10 × 10 × 10 × 10 × 10 × 10 × 10 × 10 × 10 × 10 × 10 × 10 × 10: this is 10 multiplied by itself 23 times

Writing large numbers in standard form.

Activity 6.1A

The mass of the Sun is estimated to be 1 900 000 000 000 000 000 000 000 000 000 kg. Write this number in standard form.

Very small numbers, such as the mass of a proton (0.000 000 000 000 000 000 000 000 00167 kilograms) or the width of a DNA molecule (about 0.000 000 0025 metres) are just as difficult to use as very large numbers. We write them in standard form too. So the mass of the proton is written as 1.67×10^{-27} kg, and the diameter of an average DNA molecule is written as 2.5×10^{-9} m.

0.000 000 000 000 000 000 000 000 000 000 981 kilograms

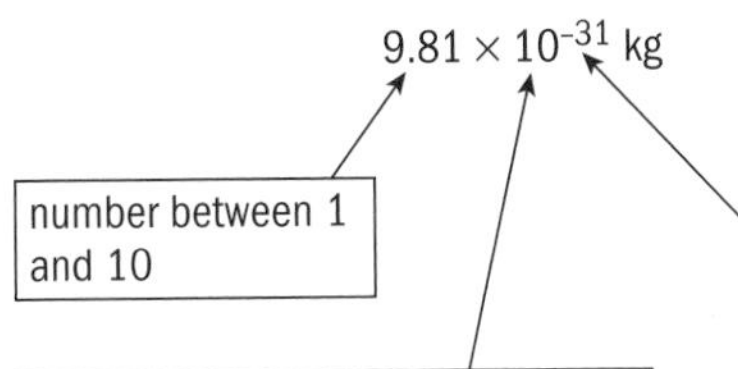

Writing small numbers in standard form.

Did you know?

When solving problems in standard form, a negative power means a small number and a positive power means a large number.

PLTS

Independent enquirer

This type of question will allow you to develop your skills as an independent enquirer as you identify a mathematical problem and display the data in an appropriate format.

SI (metric) units

In the UK, there are two systems of measurements in use: **imperial** and **metric**. For example, there are still signposts which show the distance in miles and milk is still sold in pints; these are examples of imperial **units**.

At the same time, potatoes are sold in kilograms and petrol in litres; these are examples of **SI units**. There are seven base units which have been agreed upon internationally. These are shown in the table below. All other quantities that are used in science are based on these base units.

Physical quantity	Name of unit	Symbol for unit
length	metre	m
time	second	s
mass	kilogram	kg
temperature	kelvin	K
amount of substance	mole	mol
electric current	ampere	A
luminous intensity	candela	cd

Converting between SI and imperial units

Quantity	Conversion factor	
mass	1 kg	~ 2.2 lb
	1 ton	~ 1000 kg
	1 stone	~ 6.9 kg
length	1 inch	~ 2.5 cm
	1 foot	~ 30 cm
	5 miles	~ 8 km
capacity	1 litre	~ 1.75 pints
	1 gallon	~ 4.5 litres

Worked example

John converted 35 lb of biofuel ingredients to 6 litres of diesel.

1 What mass of biofuel ingredients, in kilograms, did John use?

2 How many gallons of diesel did John produce?

Answer

1 Step 1: 1 kg = 2.2 lb so 1 lb = 1 ÷ 2.2 = 0.456 kg

Step 2: 35 lb = 0.456 × 35 = 15.9 kg

So, 35 lb is about 16 kg.

2 Step 1: 1 gallon = 4.5 litres so 1 litre = 1 ÷ 4.5 = 0.222 gallons

Step 2: 6 litres = 0.022 × 6 = 1.33 gallons

So, 6 litres is about 1.33 gallons.

Prefixes

As well as standard form, there is another way to make large and small numbers neater. Scientists often use a **prefix** in front of the base unit to represent the power of ten. The table below shows the prefixes that you need to know.

Prefix	Symbol	Factor
giga	G	10^9 (1 000 000 000)
mega	M	10^6 (1 000 000)
kilo	k	10^3 (1000)
deci	d	10^{-1} (0.1)
centi	c	10^{-2} (0.01)
milli	m	10^{-3} (0.001)
micro	μ	10^{-6} (0.000 001)
nano	n	10^{-9} (0.000 000 001)
pico	p	10^{-12} (0.000 000 000 001)

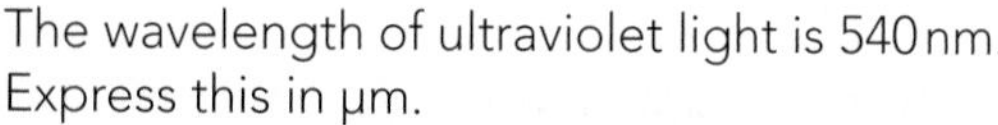

Worked example

The wavelength of ultraviolet light is 540 nm. Express this in μm.

Answer

1 nm = 10^{-9} m so 1 m = 10^9 nm

1 μm = 10^{-6} m so 1 μm= $10^{-6} \times 10^9$ nm = 10^3 nm so 1 nm = 10^{-3} μm so 540 nm = 540×10^{-3} μm = 0.54 μm

Ultraviolet light has a wavelength of 0.54 μm.

Converting mass, area and volume

In science, converting metric units in mass, area and volume is particularly important. The table below shows some common conversions that are encountered in science.

Length	Mass	Volume
1 km = 1000 m	1 tonne = 1000 kg	1 litre = 1000 ml
1 m = 10 dm	1 kg = 1000 g	1 dm^3 = 1 litre
1 m = 100 cm	1 g = 1000 mg	1 m^3 = 1000 litre
1 m = 1000 mm		1 cl = 10 ml
1 cm = 10 mm		1 cm^3 = 1 ml

Note A tonne is the unit for a metric 'ton' as opposed to a ton which is an imperial unit.

Worked example

1. A chemical container has a volume of 1.2 dm^3. Convert this into millilitres.
2. A granite block has dimensions of 12 cm × 6.4 cm × 0.2 cm. What is its volume in SI units?

Answer

1. From the table, 1 dm^3 = 1 litre so 1.2 dm^3 = 1.2 litres = 1200 ml
2. Convert the dimensions into metres:

   12 cm = 0.12 m (dividing by 100),

   6.4 cm = 0.064 m and 0.2 cm = 0.002 m

   The volume will be:

   $0.12 \times 0.064 \times 0.002 = 1.54 \times 10^{-5}$ m^3

Activity 6.1B

A sheet of metal was estimated to have an area of 0.934 m^2. Convert this value into cm^2. Leave your answer in standard form.

Hint: Remember, when you convert from a large metric unit to a small one, e.g. metres to centimetres, you multiply by the conversion factor (in this case, 100); when you convert from a small metric unit to a large one, you divide by the conversion factor.

Using your scientific calculator

Scientific calculators vary in both data capability and processing power. You should read your calculator's manual to benefit fully from the calculator capability. When performing fractions using a calculator, remember to use the 'Bracket' function so the **denominator** and **numerator** are calculated separately. Remember that the 'EXP' function means that you are multiplying by powers of 10 (e.g. 2EXP3 means 2×10^3 which is 2000). Note that some calculators may have a $\times 10^x$ key in place of an EXP key.

Did you know?

When you use a calculator you must make sure that you do each part of the calculation in the correct order, otherwise you'll get the wrong answer. Remember, to get the right order use BIDMAS: brackets, indices, divide, multiply, add, subtract.

Worked example

Use a calculator to solve this calculation. Give your answer in standard form.

$$\frac{5 \times 10^6 - 6 \times 3.2 \times 10^4}{2.4 \times 10^3 + 66.6} \times 5$$

Answer

Remember BIDMAS: use the Bracket key of the calculator:

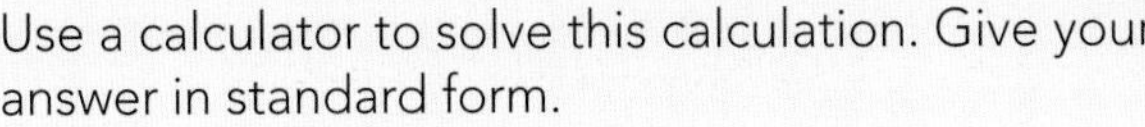

and then multiply answer by 5.

ANS × 5 = 9.7×10^3

Activity 6.1C

Use your calculator for the following calculation. Give your answer in standard form.

$(1.3 \times 10^{-15}) \div (7.2 \times 10^{-22}) + 3.4 \times 10^7$

PLTS

Independent enquirers

By using the calculator correctly and displaying the answer in standard form you will have developed independent enquiring skills.

Rounding off

Imagine you work in a plant science lab. You have to find the average mass of grains from a new breed of wheat. If you collect 500 grains and find their combined mass is 34.23 g, you can calculate that one grain has an average (mean) mass of 0.06846 g. But if your scales can only measure the combined mass to two decimal places, your answer for one grain can't be accurate to five decimal places. You must round off the answers to a specific number of decimal places or significant figures. In this case the answer will be 0.07 g.

Decimal places (d.p.)

When rounding up or down, find the last digit you want and look at the next digit to the right: if it is 5 or more, then round up the last digit, if it is less than 5, then leave the last digit as it is.

Worked example

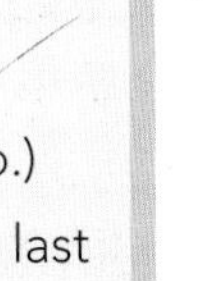

The mass of a neutron is 1.6752×10^{-27} kg. What is this to two decimal places?

Answer

1.6752×10^{-27} kg rounds to 1.68×10^{-27} kg (2 d.p.)

(The last digit you want is '7'. The digit after the last digit you want is '5', so round up to 1.68.)

Significant figures (s.f.)

When you do scientific calculations you need to be aware of the precision to which you should give your answer. The significant figures of a number are those digits which have meaning relevant to your calculation. Answers to calculations often produce more digits than the accuracy of the original data.

Worked example

1 Write 6.025301 to three significant figures.

Answer

6.025301 = 6.03 (3 s.f.)

2 Write 0.000120543 to four significant figures.

Answer

In this case the first significant figure is the first digit that isn't a zero. The second, third, etc. significant figures follow from the first, whether they are zero or not.

0.000120543 is written as 0.0001205 (4 s.f.)

Using fractions, percentages and ratios in science

Scientists often need to compare numbers. For example, an ecologist may look at the percentage of an area covered by a particular plant species.

An ecologist counting plant species in a field.

Fractions

In science, you will need to be able to simplify fractions and to use them. A fraction is described by a **numerator** (top number) and **denominator** (bottom number). For example, $\frac{3}{5}$ is a fraction with a numerator of 3 and denominator of 5.

Percentages (%)

Percentages can be turned into fractions by dividing by 100. So, 30% is the same as $\frac{30}{100}$. The percentage can be read as '30 out of a 100'.

Worked example

1 An engine powered by biofuel was found to be 52% efficient. Express this as a fraction.

2 In the first year of an environmental investigation, the number of bees in a park in Sheffield was estimated to be 2000. In the second year they found that the count decreased to 1500. What was the percentage decrease over the period of investigation?

Answers

1 $52\% = \frac{52}{100} = \frac{13}{25}$ (simplifying)

2 The decrease in the number of bees was $2000 - 1500 = 500$ so the decrease is $\frac{500}{2000} = \frac{1}{4}$ (as a fraction) or 25% (as a percentage).

We can say that the percentage change is 25%.

Activity 6.1D

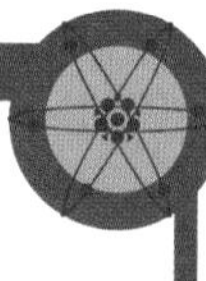

A compound consists of 22% nickel. How much nickel will there be in 500 g of the compound?

Functional skills

Mathematics

By carrying out this activity you will have identified a problem and the mathematical methods needed to tackle it.

Ratios

Ratios tell us how big or small one quantity is compared to another and so have no units.

Worked example

1 The mass of the Earth is estimated as 6.0×10^{24} kg and the mass of the moon is estimated as 7.3×10^{22} kg. What is the ratio of the mass of the Moon to that of the Earth in its simplest form?

2 A metallurgist determines that a metal comprises 30% copper, 5% nickel and 65% tin by mass. Express this composition as a ratio.

Answers

1 Step 1: The ratio of the mass of the Moon to that of the Earth can be written as $7.3 \times 10^{22} : 6.0 \times 10^{24}$.

Step 2: To express this ratio in its simplest form, divide both sides by the mass of the Moon:

1:82

(This means that the mass of the Earth is about 82 times bigger than the mass of the Moon.)

2 Ratio of the masses is 30:5:65 which we can simplify to 6:1:13 (dividing all the numbers by the smallest number; 5 in this case).

The ratio of the mass of the Moon to the mass of the Earth is 1:82.

Using algebra in science

Scientists try to understand the world by finding the mathematical rules that describe it. We can write these rules down without using actual numbers. Instead we use symbols, usually letters, to represent quantities and how they link with each other.

Did you know?

Algebra was invented over a thousand years ago by the Arab mathematician Muhammad ibn Musa Al-Khwarizmi.

We can write an equation in words or in symbols. You need to remember the following points.

- There is an equals sign between the left-hand side (LHS) and the right-hand side (RHS) of the equation. What is on the LHS should always be equal to what is on the RHS.
- We can transpose, or rearrange equations, or we can substitute numbers into them.
- In science, most answers will have a unit as well as a number.

Substitution

This just means putting numbers in place of the symbols. The worked example will help you understand this.

Worked example

Calculate the speed required for a molecule of air to escape the Earth's atmosphere. This speed is called the escape velocity and is given by the equation:

$$v_{esc} = \sqrt{\frac{2GM}{R}}$$

where G is a constant = 6.67×10^{-11} N m^2 kg^{-2}

M is the mass of the Earth = 5.97×10^{24} kg

and R is the radius of the Earth = 6.38×10^6 m.

Answer

Step 1: Substitute (put in) the values G, M and R into the equation:

$$v_{esc} = \sqrt{\frac{2 \times 6.67 \times 10^{-11} \times 5.97 \times 10^{24}}{6.38 \times 10^6}}$$

Step 2: Arrange the powers of ten together:

$$v_{esc} = \sqrt{\frac{2 \times 6.67 \times 5.97 \times 10^{-11} \times 10^{24}}{6.38 \times 10^6}}$$

Step 3: Use a scientific calculator to calculate the numerical value in the square root:

$$v_{esc} = \sqrt{1.2483 \times 10^8}$$

Step 4: Taking the square root:

$$v_{esc} = 1.12 \times 10^4 \text{ m s}^{-1}$$

(Leave the answer to 3 s.f. because the values of G, M and R were given to 3 s.f.)

Activity 6.1E

The time for one swing of a pendulum, used in a grandfather clock, can be approximated using the equation:

$$T = \sqrt{\frac{4\pi^2 L}{g}}$$

Calculate T, if the length L of the pendulum is 50 cm and $g = 9.81\ \text{m s}^{-2}$.

Hint: Make sure that you convert the lengths into SI units!

PLTS

Creative thinker

Obtaining the correct answer for T will have developed your creative thinking skills.

Transposition of equations

Transposing means rearranging the equation to get a symbol as the subject, the symbol representing a quantity. The rule is that whatever is done to the left-hand side (LHS) needs to be done to the right-hand side (RHS). The following worked examples show the procedure of transposing equations.

Mensuration

In science you will have to calculate areas and volumes of different shapes; this is called **mensuration**. Engineering structures, such as solar panels and loudspeakers, use cone shapes in their design. Blood vessels in the human body, however, are cylindrical.

Worked example

Example 1

The relationship between voltage (V) and current (I) for a resistor (R) is given by:

$V = I \times R$

Transpose this equation to make the current, I, the subject.

Answer

Divide both sides by R then rearrange the equation with I on the left-hand side:

$$\frac{V}{R} = \frac{I \times R}{R} \rightarrow \frac{V}{R} = I \text{ so } I = \frac{V}{R}$$

Example 2

The number of moles is given by the equation:

$n = \frac{m}{M_r}$ where m is the mass of the chemical and M_r is the relative molecular mass. Transpose this equation to make the mass the subject.

Answer

Multiply both sides by M_r and then rearrange the equation so that m is on the left-hand side:

$$n \times M_r = \frac{m}{M_r} \times M_r \rightarrow nM_r = m \text{ so } m = nM_r$$

Example 3

The distance, s, travelled by a car moving with a constant acceleration, a, for time t is give by the equation:

$s = ut + \frac{1}{2}at^2$ where u is the initial speed.

Answer

Transpose this equation to make a the subject.

First, to leave the term including a on its own on the RHS, subtract ut from both sides:

$$s - ut = ut + \frac{1}{2}at^2 - ut \rightarrow s - ut = \frac{1}{2}at^2$$

Now multiply both sides by 2:

$$2 \times (s - ut) = 2 \times \frac{1}{2}at^2 \rightarrow 2 \times (s - ut) = at^2$$

Now divide both sides by t^2 and rearrange the equation to get a on its own on the left-hand side.

$$\frac{2 \times (s - ut)}{t^2} = \frac{at^2}{t^2} \rightarrow \frac{2 \times (s - ut)}{t^2} = a$$

$$\text{so } a = \frac{2 \times (s - ut)}{t^2}$$

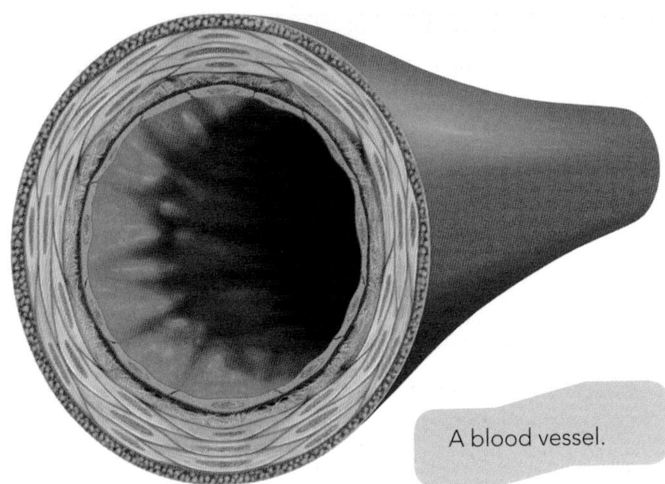

A blood vessel.

Volume of some objects

The volume of a cone is given by the formula:

$$\text{volume} = \frac{1}{3}\pi \times r^2 \times h$$

where r is the radius of the cone and h is the vertical height.

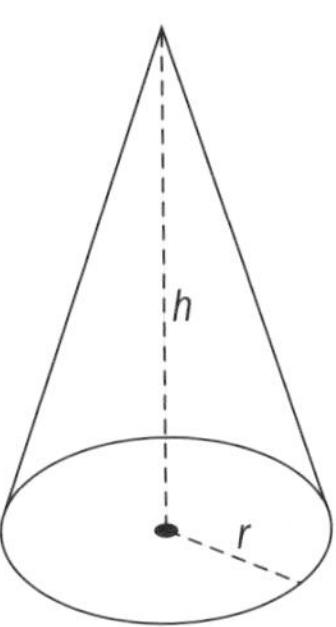

The volume of a sphere is given by the formula:

$$\text{volume} = \frac{4}{3}\pi \times r^3$$

where r is the radius of the sphere. Its surface area is given by:

$$\text{surface area} = 4\pi \times r^2$$

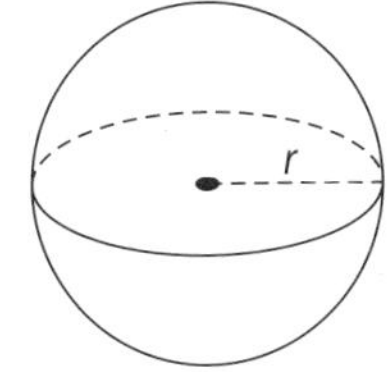

The volume of a cylinder is given by the formula:

$$\text{volume} = \pi \times r^2 \times h$$

where r is the radius of the circular base and h is the height (or length of the cylinder).

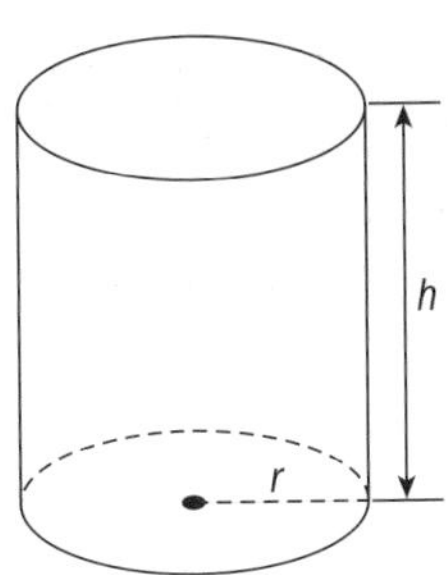

Worked example

The density of the Earth is estimated to be 5530 kg m^{-3} and the radius of the Earth approximately 6.4×10^6 m. The following equation allows the mass of the Earth to be calculated:

$$\rho = \frac{m}{V}$$

where m is the mass, V is the volume and ρ is the density.

Calculate the volume of the Earth, assuming it is perfectly spherical. Then use the density equation to estimate the mass of the Earth. Give your answer in standard form and to two significant figures.

Answer

$$\text{volume of the Earth} = \frac{4}{3} \times \pi \times (6.4 \times 10^6)^3$$

$$= 1.1 \times 10^{21}\ \text{m}^3$$

Note that the units must be included. Rearranging the density formula and substituting for the volume, V, gives:

$$\text{mass of the Earth} = \rho \times V = 5530 \times 1.1 \times 10^{21}$$

$$= 6.1 \times 10^{24}\ \text{kg}$$

Activity 6.1F

A loudspeaker has a radius of 11 cm and a height of 7.5 cm. Calculate the volume of the cone.

Surface area of some objects

The formula for total surface area of a cylinder has two terms: one for the two circular faces and one for the curved surface:

$$\text{total surface area} = 2 \times \pi \times r^2 + 2 \times \pi \times r \times h$$

where r is the radius of the cylinder and h is the height (or length) of the cylinder.

The surface area of a sphere, radius r, is given by:

$$\text{total surface area} = 4 \times \pi \times r^2$$

Assessment activity 6.1

P1 M1 D1 P2 M2 D2 BTEC

1 A small mirror designed for a European telescope has dimensions of 7 inches × 8 inches. Convert into SI units. P1

2 It is estimated that light takes 100000 years to travel the full distance across our galaxy, the Milky Way. If light travels 3.0×10^8 m in 1 second, how long is the Milky Way? Give your answer in standard form. M1

3 A block cut from asteroid rock was measured to have dimensions: 13.3 mm × 12.4 mm × 15.2 mm.

   a Convert the dimensions into SI units. P1

   b Find the volume of the block. Give your answer in standard form and to the appropriate number of significant figures. P1, M1

4 You are working as a clinical chemist in charge of the testing of blood samples. You are required to mix a sodium chloride solution with water in the ratio of 2:5.

   a Starting with 780 ml of the sodium chloride solution, how much water needs to be added? D1

   b Give your answer in standard form to an appropriate number of significant figures. M1

5 You are a geologist assigned to estimate the density of the Earth. You know that density is given by: $\rho = \frac{m}{V}$. You need to first find the volume (V) of the Earth, which can be found by the equation: $V = \frac{4}{3}\pi \times R^3$. The radius of the Earth (R) can be found by using the equation: $g = \frac{GM}{R^2}$. You are given the following values:

   $G = 6.67 \times 10^{-11}$ N m^2 kg^{-2}, $M = 5.97 \times 10^{24}$ kg and $g = 9.81$ m s^{-2}.

   a Calculate the radius of the Earth and then the volume of the Earth. P2, M2, D2

   b Use your answer to part **a** to determine the density of the Earth. Give your answer in standard form to the appropriate number of significant figures. D2

6 You are working at the National Blood Service Clinical Studies Unit. Recently you have been investigating blood flow in arteries. A 4-cm-long section of an artery of cylindrical shape is found to have a diameter of about 0.44 mm.

   If the average mass of blood in this artery section is estimated to be 6.9×10^{-3} g, determine the density of the blood. P2, M2, D2 Leave your answer in standard form and to the correct number of significant figures. P1, M1

Density (ρ) is given by the formula: $\rho = \frac{m}{V}$

Grading tips

To achieve P1 in question 1, remember that in science most quantities have a unit; without the unit you will not know what the quantity is. To achieve M1 in question 2, remember to use standard form. Hint: before you start, change the years into seconds.

When attempting P1 in question 3**a**, the number of significant figures in the final answer should not be any more than those given in the question. When working towards M1 in question 3**b**, don't get confused with the notation on your calculator; this will not be in standard form.

Learners often make the mistake of trying to solve ratio questions in one step. To achieve D1 in question 4**a**, make sure you first find the total volume of the solution required and then subtract the solution of sodium chloride to get the volume of water required.

For P2 in question 5**a**, remember if you divide by a number on one side of an equation you will need to do the same thing to the other side. For D2 in question 5, be careful when you transpose the equations.

The key to answering problems like question 6 is first to identify the shape M2, write out the equation and then use the basic rules of algebra to rearrange the equation to get the subject P2. Make sure that you have converted all dimensions to SI units. Substituting the value correctly and leaving your answer to the appropriate number of significant figures with the correct unit will go towards meeting P1, M1 and D2.

6.2 Collecting and recording scientific data

In this section: 

Key terms

Precision – the smallest change in quantity an instrument can measure.

Data logger – automated system used to make and record measurements.

Random errors – chance occurrence in any experiment.

Systematic error – errors in the experimental technique or instrument.

Collecting data

Science is all about collecting and interpreting data. These data can be primary or secondary. Data that are collected specifically to answer your research question are called **primary data**. If you were working as a microbiologist, investigating processes within a cell, you might use a **data logger** like the one shown below to collect your data. A data-logging system allows a computer to collect the data even when the operator is not there.

Data collection in a laboratory; a data logger is attached to a pH probe.

Data that have previously been collected, often for another purpose, are called **secondary data**. The map below is a satellite picture taken by the NASA *QuikScat* satellite. The map reveals the ocean areas where winds could produce the most energy. These data were collected continuously by microwave radar, which tracks the speed and power of the winds, controlled by a computer system. The original (or primary) use of the data is to predict the formation of tropical cyclones and other storm systems. However, you could use it as secondary data to help you if you were involved in selecting a location for a wind farm that could be used for producing electricity.

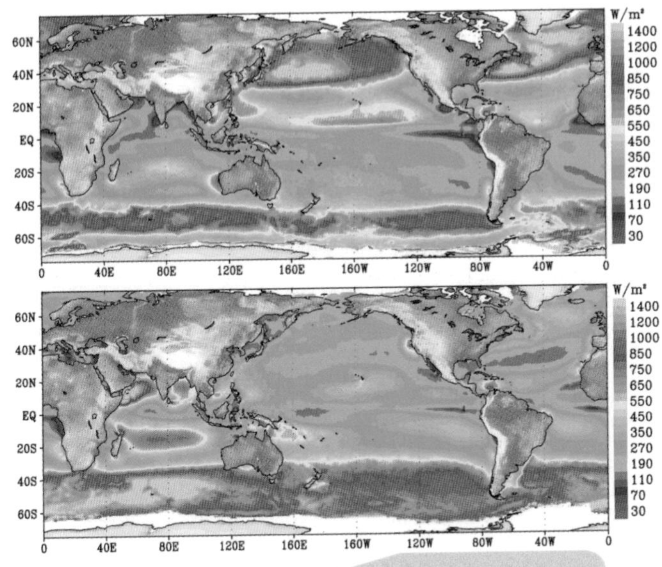

Data collected by a satellite measuring the power generated by the wind in a northern winter. Red and white areas indicate where high energy is available; blue areas show lower energy.

Activity 6.2A

Give one other example of (a) primary data and (b) secondary data that you have used.

Did you know?

The Large Hadron Collider experiment will collect enough data to fill about 100 000 dual-layer DVDs every year.

Errors

The accuracy of any measurement depends on both the skill of the experimenter and the equipment used. The **precision** of an instrument is the smallest change in quantity that it can measure. For example, a string used in a pendulum could be 12.3 cm long according to a centimetre rule with a precision of 0.1 cm (1 mm). Even if the string is in fact 12.28 cm or 12.34 cm long, you will still read about 12.3 cm with the 30-cm rule. We say that there is a maximum absolute uncertainty or error of half of the precision, in this case 0.05 cm (or 0.5 mm). We can write this error as ±0.05 cm. The string has a maximum length of 12.35 cm and a minimum length of 12.25 cm, so the measurement is written as (12.30 ± 0.05) cm.

Did you know?

In science, 'error' does not mean 'mistake'. Error tells us how far a result might be from the true value. The difference is caused by the limitations of the equipment or the technique used.

Instrument	Precision	Maximum absolute error
centimetre rule 	1 mm	0.5 mm
measuring cylinder 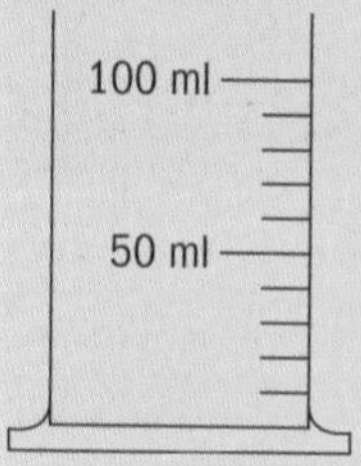	10 ml	5 ml
balance 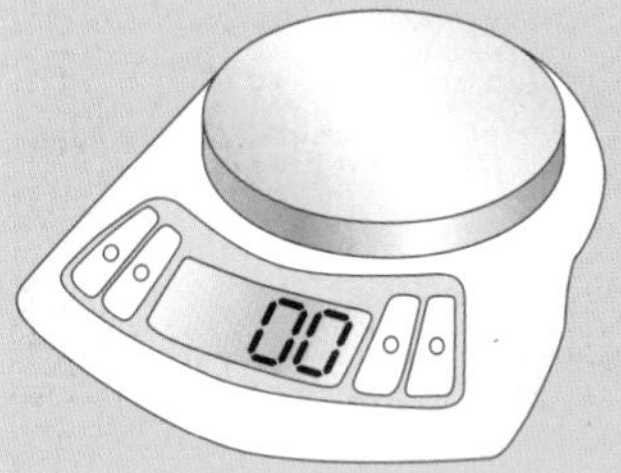	0.1 g	0.05 g
micrometer 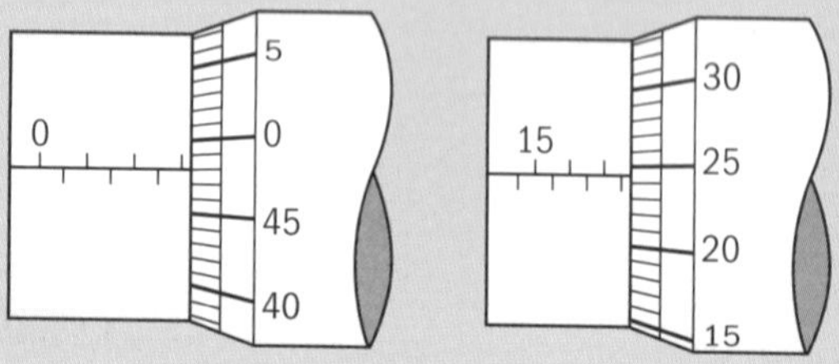	0.01 mm	0.005 mm

Sometimes, in the case of a timing experiment, human reaction time will be important and can be longer than the precision of the timing instrument.

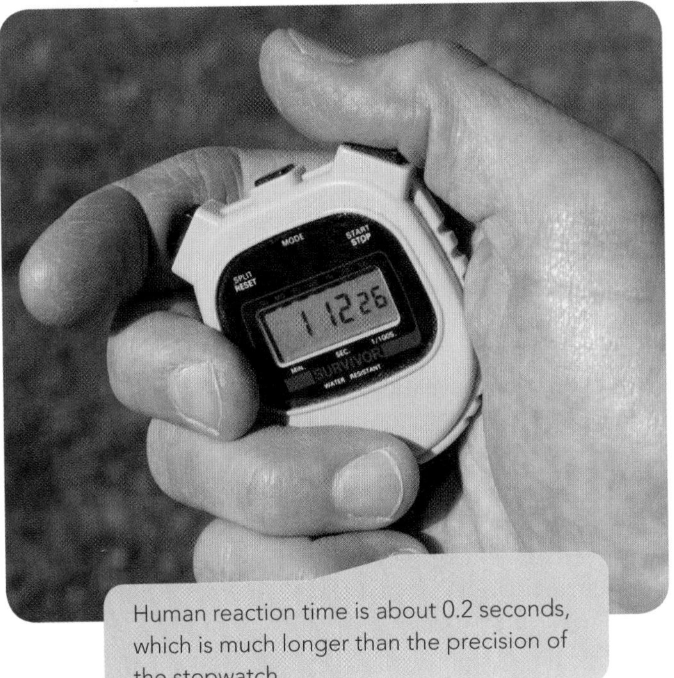

Human reaction time is about 0.2 seconds, which is much longer than the precision of the stopwatch.

There will be other situations when the spread of your data may be greater than the precision. In this case, the error will be related to the spread of the data as shown in the worked example.

Worked example

A micrometer was used to measure the diameter of a metal wire. The diameter of the wire was measured at different points so as to improve the reliability of the results. Calculate the error.

Measuring point	1	2	3	4
Diameter of wire (mm)	2.22	2.26	2.30	2.34

Answer

As you can see from the table of results, the diameter varies between 2.22 mm and 2.34 mm. So, the range is 2.34 – 2.22 = 0.12 mm.

The average thickness is 2.28 and so we can say that the error in the mean will be the range ÷ 2 = 0.06 mm. This value is greater than the precision of 0.005 mm.

Therefore, the error (or the uncertainty) is ±0.06 mm.

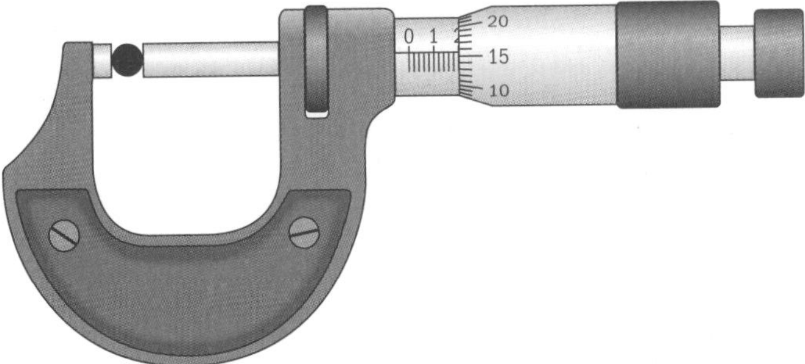

A micrometer being used to measure wire diameter.

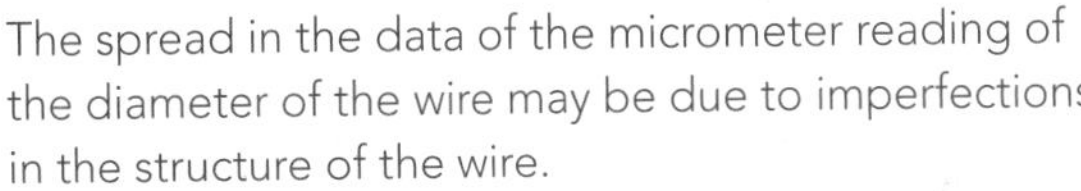

Did you know?

The spread in the data of the micrometer reading of the diameter of the wire may be due to imperfections in the structure of the wire.

Sometimes it is useful to compare measurement errors as a maximum percentage error.

$$\text{Maximum percentage error} = \frac{\text{maximum error}}{\text{measurement}} \times 100\%$$

While carrying out experiments, there may be errors that relate to the apparatus (e.g. a mass balance that is not zeroed); these are called **systematic errors**. Such errors can be eliminated by correctly calibrating the equipment.

Sometimes errors are caused by the experimenter not being careful (e.g. an experimenter daydreaming while doing an experiment and noting the wrong value down); these are called **random errors**. Random errors can be minimised by taking averages.

Activity 6.2B

A calcium carbonate solution was prepared. Using a measuring cylinder a technician reported that there was 45 ml of solution.

1 What is the precision of the instrument?

2 Write down the maximum and minimum values of the solution.

3 What is the maximum percentage error?

Assessment activity 6.2

BTEC

As a trainee engineer, you are working to develop a state-of-the-art greenhouse. You have been assigned to investigate how the temperature varies in a greenhouse between 11 am and 3 pm during spring and summer. So far, you have collected the data for Monday as shown in the table. You used a watch to note the time and a digital thermometer to measure the temperature.

		Temperature (°C) at time			
		11:00	12:00	13:00	14:00
Day	Monday	18.03	19.67	24.45	26.45
	Tuesday				
	Wednesday				
	Thursday				
	Friday				

1 Carry out a similar experiment and display your data in a table. P3 Describe the stages you followed in collecting the data. M3 How would the method of collecting the data be different if you used a data logger? D3

2 Write down some of the errors that could occur in the experiment. P4 Calculate the maximum percentage error in the average temperature. M4 Explain how the errors you have listed can be minimised? D4

Grading tips

To achieve P3 in question 1, don't forget to include both primary and secondary data. For M3 and D3 you need to consider methods of collecting the data and refer to how you collected your data (primary) and any secondary data.

In question 2 to meet P4 you must include both random and systematic errors. For M4, show your working. A common mistake when attempting D4 is to generalise and not refer to the experiment.

6.3 Displaying and interpreting scientific data

In this section:   

Continuous data – data that can take any value.

Discrete data – data that can only take certain values.

Tangent – a line that passes through a point on a non-linear graph and has the same gradient as the graph at that point.

Types of data

Data may be **continuous**, such as the thickness of a material, or **discrete**, such as the number of earthworms in a soil sample. Other data can be a mixture of numbers and names; for example, the types of fingerprints found on a glass of water at the scene of a crime (loops, whorls and arches).

Did you know?

Scientists handling experimental data need to know the definitions of mean, median and mode. For example, for the nine values:

1.28, 1.24, 1.23, 1.27, 1.23, 1.24, 1.29, 1.23, 1.21

Mean: average value =

(1.28 + 1.24 + 1.23 + 1.27 + 1.23 + 1.24 + 1.29 + 1.23 + 1.21) ÷ 9 = 1.25 (2 d.p.)

Median: middle number = 1.24 (the fifth number in order)

Mode: number that occurs most often = 1.23

Unit 8: See page 170 for more information on mean, median and mode.

Displaying data

A forensic scientist analysing fingerprints needs to show her results to others. She can do this by plotting the data as a bar chart or pie chart, as shown opposite. For some data, such as the amount of gas given off in a chemical reaction or the heights of different plants (that is, continuous data), it is more appropriate to use histograms. The difference between the bar chart and the histogram is that in the bar chart the frequency is shown by the height of the bar and there are normally gaps between the bars; in a histogram the frequency is represented by the area of the bar.

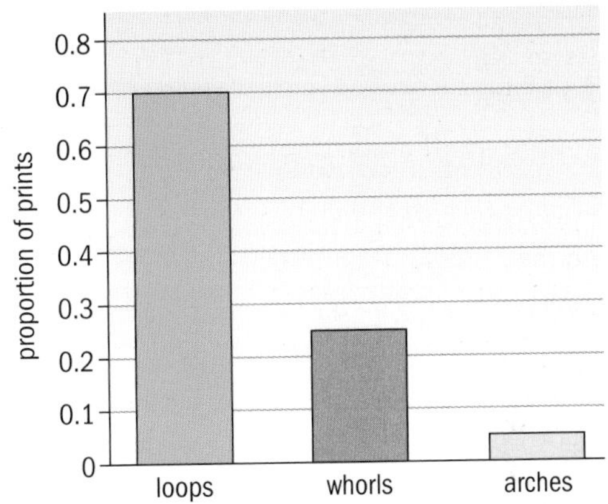

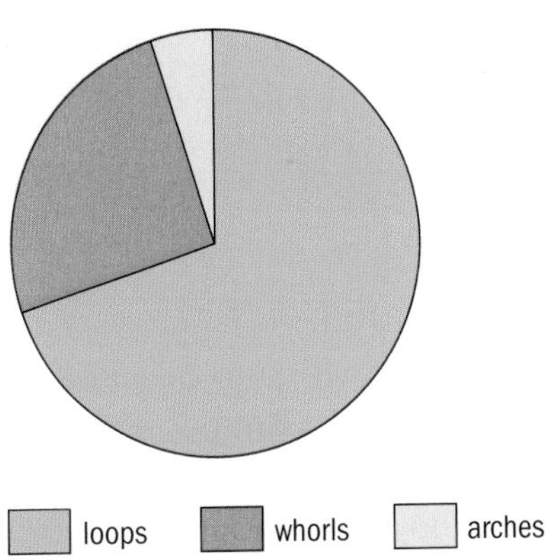

Two ways of showing results of fingerprinting tests (discrete data). In the bar chart, height represents the proportions of different types of fingerprints found at the crime scene. In the pie chart, the sectors of the 'pie' represent the proportions of fingerprint types.

Activity 6.3A

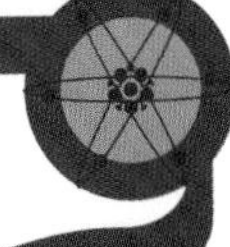

Give examples of continuous and discrete data.

Case study: Equal groupings

Jill is a geneticist studying the lifespan of fruit flies, as this may give information on human lifespans; 50 fruit flies were grown and their lifespan measured, in days. The data are shown below.

12	10	34	36	32	3	45	67	41
26	33	19	50	45	37	58	34	56
8	27	44	61	66	40	22	36	45
37	21	31	35	48	57	54	32	44
67	26	35	39	24	17	11	14	9
38	13	64	63	9				

Display the data in a grouped frequency table.

Draw a histogram of the data.

Answer

Step 1: We need to divide the data into groups and decide the width interval. Group the data with an equal interval of 10 (seven groups, with first group 0–9 and seventh group 60–69).

Step 2: Determine the frequency (the number of data points that fall within the interval).

Step 3: Draw the table:

Lifespan (days)	0–9	10–19	20–29	30–39	40–49	50–59	60–69
Frequency	4	7	6	14	8	5	6

Step 4: Use a spreadsheet to construct a histogram of the data.

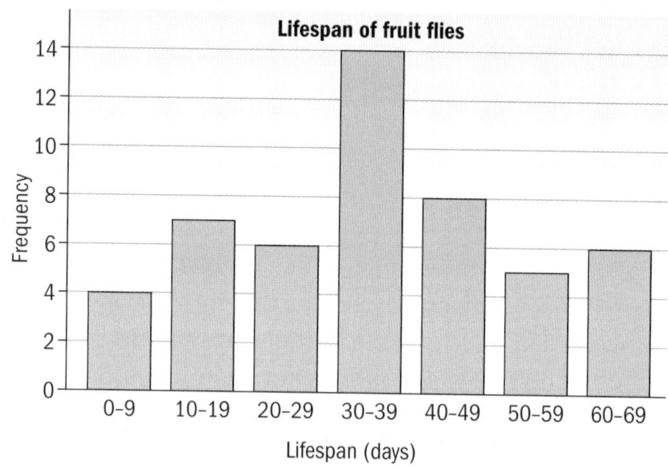

Histogram of the fruit fly data.

Linear graphs

Sometimes, an experiment may be carried out to investigate whether two quantities are related. For example, when investigating the electrical relationship in a wire with unknown characteristics, the current and voltage may behave as shown in the diagram.

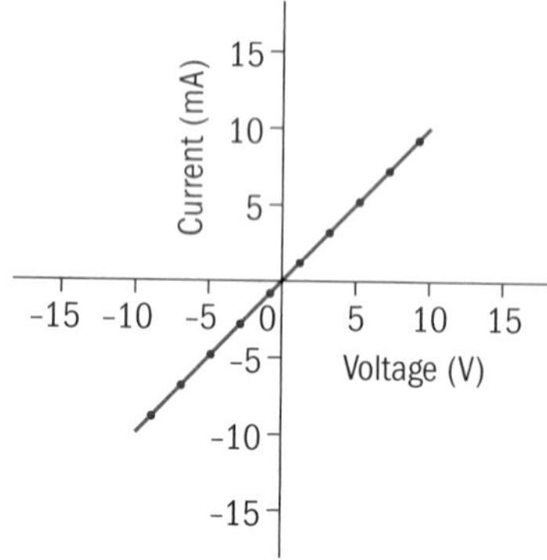

A graph showing how current varies with voltage for a wire.

The graph indicates that the current is proportional to the voltage (the current and voltage increase by the same rate). We can get lots of information from this graph; for example, the gradient (the slope of the line) will give us the conductance of the wire (the inverse of the resistance). Reversing the voltage makes no difference to the gradient of the line. The resistance is the same whichever direction the current flows. There are many examples in science where plotting two quantities can give lots of information.

Non-linear graphs

When two quantities are plotted there are many occasions when they don't give a straight line but a curve. This is called a non-linear graph. An example of a non-linear graph is given opposite for a chemistry experiment involving the release of hydrogen gas.

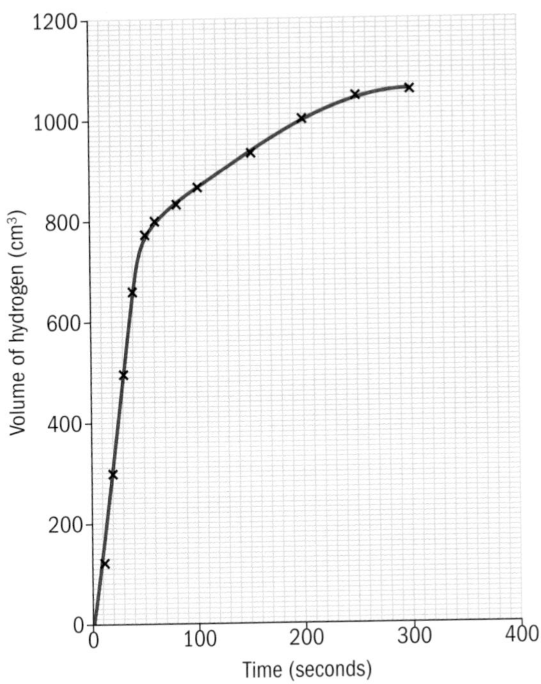

Plot of hydrogen gas versus time during a chemical reaction.

In this graph the amount of hydrogen gas increases rapidly at the start, but after 200 seconds the gas is only increasing gradually. We find out how fast the chemical reaction gives off the gas at first by drawing a tangent at 0 seconds and measuring the gradient.

Did you know?

The trajectory of a cricket ball is a parabola, which is an important type of non-linear graph.

Assessment activity 6.3

BTEC

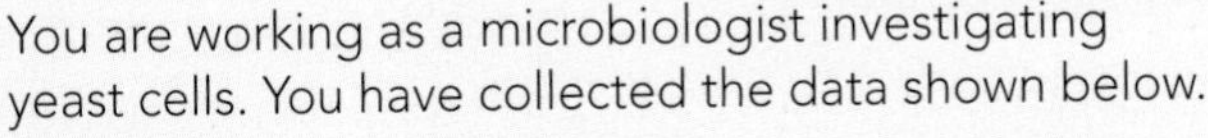

You are working as a microbiologist investigating yeast cells. You have collected the data shown below.

Time (hours)	Number of cells
0	2
1	5
2	15
3	40
4	110
5	300
6	800

1 Display your data by plotting an appropriate graph. P5
2 Explain the trend in the graph you have plotted. M5
3 Estimate the rate of growth by taking tangents at three different points on the curve. D5
4 Explain what is happening to the rate of growth of the yeast cells. P6

Grading tips

When displaying data make sure that you label your graphs or charts, otherwise you will not meet P5.

To achieve M5 you will need to cover the full timescale of your experiment, not just the start or the end, and explain possible reasons for the trend in your graph.

For D5 make sure that you include the units and give your answer to the correct significant figures.

For P6 make sure you use appropriate mathematical terminology.

WorkSpace

Doctor Chris Tofts

Chief Mathematics Officer (concinnitas ltd)

Concinnitas is a leader in the application of science and mathematics to the analysis, design, development and management of complex systems.

My responsibilities include gathering customer modelling requirements, developing models to illustrate the issues the customer faces, analysing and visualising model results so that customers can understand them, presenting results and implications to customers and developing potential alternative solutions if necessary, and extending software support as required.

I meet with customers to understand their problems and to discuss any data they have and the range of solutions available to them. I then build models of the customer problem with tools such as DEMOS2K, Savant, Excel, Mathematica or directly in C or Perl. I look at the results of the model and follow up in interesting areas. Finally, I write up explanations as to how the models represent the customer's issues and what the data they generate implies. Presenting these results in a form the customer can understand and exploit is critical. Some days I work in the development of the wider subject we call Service Science at the University of Swansea.

One of the most satisfying aspects of my job is modelling a very wide range of problems. I have modelled and analysed systems ranging from ants, shrimps, asynchronous hardware, printers, knowledge transfer in social groups, to effectiveness and measurability of policy interventions, and availability of huge supercomputers.

Think about it!

The service (queuing) time of a system is derived by a simple application of Little's law to be given by the formula $W = S/(1 - U)$, where S is the time to serve a single customer and U is the utilisation of the server.

1 If $S = 10$ seconds, how big can U be before the customer has a total time of 20 seconds before they complete service?

2 Thinking about the till at a supermarket how could you measure S and U? What happens to W if you underestimate U by 25%?

Just checking

1 What are the SI units for: (a) length; (b) mass?
2 Convert (a) 2 litres into pints; (b) 4 kilometres into miles.
3 Use a calculator to solve the calculation: $1.2 \times 10^{-5} \times 1.2 \times 10^{11}$. Leave your answer in standard form.
4 A superconducting wire has a diameter of 2 mm. Calculate the cross-sectional area (hint: the shape of the end of the wire is circular) of the wire. Leave your answer in SI units of area and in standard form.
5 Give an example in science when a graph is (a) linear and (b) non-linear.
6 Give an example of a random error and a systematic error.
7 A chemical was measured to have a mass of (22 ± 2) mg. (a) What is the precision of the instrument? (b) What is the percentage error of the mass?

Assignment tips

- Remember: most quantities in science will have a unit. Without the unit, or the wrong unit, the quantity will not make sense and none of the grading criteria will be met.
- Make sure when using a calculator that you make use of the brackets, as the order will matter. Don't use calculator notation in your answer as that will not be standard form and M1 will not be met.
- Where you have used algebra, mensuration or ratios to solve a scientific problem, you have to demonstrate a step-by-step method in order to satisfy the grading criteria P1, P2, M2, D1 and D2; don't just write the answer down!
- When carrying out experiments, make a note of the precision of the equipment and any possible errors that may occur in an experiment; these factors are important in meeting criteria P4, M4 and D4.
- When plotting graphs, make sure that your data points cover at least 50% of the graph paper and that all axes are labelled correctly, with the correct units.

You may find the following websites useful as you work through this unit.

For information on...	Visit...
plotting line graphs	BBC GCSE Bitesize Geography
algebra and other maths concepts	AAA Math

Credit value: 5

8 Using statistics for science

Statistical techniques are useful in many areas of work, not only in science. They are used in opinion polls during elections, as well as for forming opinions about shopping habits. In science, biologists use statistics to help in their understanding of genetics. Psychologists use statistics to understand human behaviour. Radiographers use statistics to understand how a radioactive material may behave. In all cases, the use of statistics depends on collecting data, displaying the data collected and then finally interpreting the data.

This unit will allow you to develop statistical tools covering various applications in science and beyond, such as working out the mean, mode, median and standard deviation, both manually and using scientific calculators and ICT. Probability is increasingly important in genetics and so you will learn the probability rules. Advanced statistical tools such as the chi-squared test, the *t*-test and correlations will be brought to life using many examples. You will learn how to plot linear and non-linear graphs accurately and also how to correctly interpret these graphs.

Learning outcomes

After completing this unit you should:

1. be able to use statistical techniques to investigate scientific problems
2. be able to perform statistical tests to investigate scientific problems

Assessment and grading criteria

This table shows you what you must do in order to achieve a **pass, merit** or **distinction** grade, and where you can find activities in this book to help you.

To achieve a **pass** grade the evidence must show that you are able to:	To achieve a **merit** grade the evidence must show that, in addition to the pass criteria, you are able to:	To achieve a **distinction** grade the evidence must show that, in addition to the pass and merit criteria, you are able to:
P1 carry out statistical calculations to investigate a scientific problem **See Assessment activities 8.1 and 8.2**	**M1** perform a calculation using probability to investigate a scientific problem **See Assessment activity 8.1**	**D1** interpret shapes of distributions in scientific data **See Assessment activity 8.1**
P2 perform a chi-squared test to support a scientific hypothesis **See Assessment activity 8.2**	**M2** interpret the results of the chi-squared test **See Assessment activity 8.2**	**D2** evaluate the validity of the interpretation of the results of the chi-squared test **See Assessment activity 8.2**
P3 perform a *t*-test on data collected from a laboratory experiment **See Assessment activity 8.2**	**M3** interpret the results of the *t*-test **See Assessment activity 8.2**	**D3** evaluate the validity of the interpretation of the results of the *t*-test **See Assessment activity 8.2**
P4 carry out an appropriate correlation method to investigate data collected from a laboratory experiment **See Assessment activity 8.2**	**M4** interpret the results of the correlation **See Assessment activity 8.2**	**D4** evaluate the validity of the interpretation of the results of the correlation **See Assessment activity 8.2**

How you will be assessed

The criteria in this maths unit can be assessed using various methods. Worksheets could be one way, where you have to solve scientific problems using standard deviation or other statistical parameters. You may be assessed on how you collected scientific data. For example, you may be asked to conduct a biological experiment and then display your data in the form of graphs and charts. This could then be extended to deciding if there is a relationship between the data you have collected.

Danlgal, 16 years old

This unit enabled me to understand statistical tools that I had not studied before in school – for example the chi-squared test, *t*-test and correlation testing. I have learned how to plot non-linear graphs which have always been difficult for me. What made the subject interesting is that it was related to the real world, not only in science, but also to how data are collected in shopping centres. What makes this course very different, and much better than maths lessons in school, is that we used our experimental data to learn about the maths. We realised the purpose of doing the maths, as it linked to the science experiments that we were carrying out. It even linked to a college trip we went on! I also became more confident in handling data, and the unit made me more aware of questioning conclusions that are deduced from statistical analysis. As I am going to study at university I have no doubt that this course has given me vital skills that I will use in my future career.

Catalyst

Statistics for life

In groups of three or four, discuss what kind of statistical tools you can use in your everyday life. For example, this could be when going to the cinema, going on holiday or buying a video/DVD. Compare your findings with other groups. Discuss what information you can obtain from using these tools. Discuss further how reliable you believe the conclusions would be. Then think about how these tools can be used in science. Present your results on a flipchart or a poster.

8.1 Statistical techniques

In this section:   

Key terms

Mode – the data value that occurs the most often.

Median – the middle data value when arranged in order.

Mean – sum of all the data values divided by the total number of values.

Standard deviation – gives a measure of how far data values are from the mean value.

Venn diagram – diagram arranged in circles that is used to calculate probability of an event.

Normal distribution – bell-shaped distribution of data; the area under the curve gives the probability of something happening.

Standard error of the mean – this is used in sampling and is the standard deviation of the sample mean.

Mean, mode and median

If you want to draw conclusions from a set of data, it is useful to be able to calculate an average for the data. An average is a typical value. There are three main types of average: the **mean**, **mode** and **median**.

Mean

The mean of a set of data values is the sum of all the values divided by the total number of values. It can be expressed as:

$$\bar{x} = \frac{\Sigma x_i}{n}$$

where $\bar{x}$ is the mean, Σ is the Greek upper case letter 'sigma' which means the 'sum of', so

$$\Sigma x_i = x_1 + x_2 + \ldots x_n$$

where x_i is a particular value and n is the number of values.

Mode

The mode is the value that occurs most often. In some sets of data, where there are several values that occur the same number of times, there will be more than one mode.

Median

The median is the middle value when the data are arranged in order. If there is an even number of values, the median is the midpoint between the middle two values.

Worked example

John has built a mini greenhouse for his biology project. He has recorded the temperature over the last 10 days as shown below:

25 °C, 28 °C, 24 °C, 22 °C, 25 °C, 29 °C, 27 °C, 27 °C, 26 °C, 23 °C

What is the mean temperature recorded in his greenhouse? What is the mode? What is the median?

Answer

There were 10 measurements taken, so $n = 10$. The mean, n, is given by:

$$\bar{x} = \frac{\Sigma x}{n} = \frac{25+28+24+22+25+29+27+27+26+23}{10}$$

$$= 25.6$$

So the mean temperature recorded is 26 °C [to 2 significant figures (s.f.)].

Mode: 25 °C and 27 °C occur the most often so the modal temperatures are 25 °C and 27 °C.

Median: arranging the order gives:

22, 23, 24, 25, 25, 26, 27, 27, 28, 29

There are 10 numbers so we take the mean of the 5th and 6th, which in this case will be

$\frac{(25+26)}{2} = 25.5$ so the median is also 26 °C (to 2 s.f.).

Unit 6: See page 151 for more information on significant figures.

Activity 8.1A

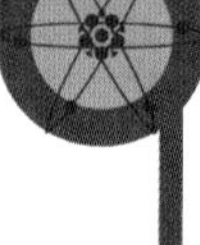

A researcher checked the shoe size of seven people as they walked into a shoe shop. The data collected are:

5, 6, 7, 7, 7, 8, 9

Calculate the mean, mode and median of the size of shoes.

Standard deviation

A way of measuring how far a set of data values is spread about the mean value, $\bar{x}$, is by using the **standard deviation**, s. For a sample of n data values, it is given by:

$$s = \sqrt{\frac{\Sigma(x - \bar{x})^2}{n - 1}}$$

Worked example

You are working with a team of biologists who have been investigating a type of herb that could be used to cure a disease. You have just measured the heights (in cm) of different specimens of the same herb. The data you have collected are:

10.2, 10.3, 10.4, 10.4, 10.5, 10.6, 10.6, 10.6, 10.8, 10.9

Calculate the standard deviation.

Answer

Step 1: Calculate the mean, $\bar{x}$:

$\bar{x} = (10.2 + 10.3 + 10.4 + 10.4 + 10.5 + 10.6 + 10.6 + 10.6 + 10.8 + 10.9)/10$

$\bar{x} = 10.53$

Step 2: Construct a table, as shown below. Column A shows the data. Column B is the difference between the data value and the mean. Column C is the square of the value in column B.

A	B	C
height (cm)	$x - \bar{x}$	$(x - \bar{x})^2$
10.2	−0.33	0.1089
10.3	−0.23	0.0529
10.4	−0.13	0.0169
10.4	−0.13	0.0169
10.5	−0.03	0.0009
10.6	0.07	0.0049
10.6	0.07	0.0049
10.6	0.07	0.0049
10.8	0.27	0.0729
10.9	0.37	0.1369

Adding column C gives: $\Sigma (x - \bar{x}) = 0.421$

$$s = \sqrt{\frac{\Sigma(x - \bar{x})^2}{n - 1}} = \frac{0.421}{9} = 0.22$$

So, the standard deviation of the height of the herbs is 0.2 cm (1 d.p.). (Note that the standard deviation is given to the same number of decimal places as the original data.)

If you know all the possible data values, and are no longer just working with a sample, then this equation can be simplified to:

$$s = \sqrt{\frac{\Sigma(x - \bar{x})^2}{n}}$$

Using calculators and ICT to calculate statistical parameters

You will have realised that when you calculate the standard deviation using the method outlined in the previous example, there are several time consuming steps involved. Scientific calculators that work in 'statistical mode' can calculate the standard deviation very quickly and will save you a lot of time; time you could use to interpret the results. Spreadsheet programs such as Microsoft Excel can also be used.

Worked example

The diameter of wire was measured at different points, using a micrometer. These results (in mm) were obtained:

2.34, 2.34, 2.35, 2.37, 2.38

Use a scientific calculator to determine the standard deviation.

Answer

Step 1: Using the Casio (fx-83 MS). Press [MODE] and then select [2]. 'SD' will be displayed showing that the calculator is now using statistical mode.

Step 2: Input data value [2.34] and press [M+]. The calculator will display $n = 1$ (this tells you it is storing the first data value). Repeat for other values: $n = 5$ should now be displayed.

Step 3: Retrieving the data. To get the mean, $\bar{x}$, select [SHIFT] and [2]. You will get the options $\bar{x}$, $x\delta n$ and $x\delta n - 1$. Select [1] $\bar{x}$ for the mean and press [=]; you should get 2.356.

Step 4: To get the standard deviation, s, press [SHIFT] and then press [2]. Now press [3] [xδn–1] (this is for the standard deviation). Press [=]; you should get 0.018. Press [MODE] and [1] to clear the data.

Thus, the mean diameter is 2.36 mm with a standard deviation from the mean of 0.02 mm. (Note that both values are given to the same number of decimal places as the original data.)

If you have a different calculator model then follow the instruction manual that relates to your model. The procedure will be similar.

Activity 8.1B

A chemist produces a chemical powder using a new technique. He tries the experiment eight times. The mass (g) of powder produced in each experiment is:

2.23, 2.13, 2.32, 2.28, 2.24, 2.26, 2.31, 2.27

Use the statistical mode of a scientific calculator to determine the mean and standard deviation of the mass.

Functional skills

Mathematics

This activity gives you the opportunity to develop your maths skills, by identifying a problem (the need to calculate the mean and standard deviation) and using the method needed to tackle it (using the mathematical facilities of a scientific calculator).

Finding the standard deviation using a spreadsheet program

To find the mean (average), median, mode and standard deviation using Microsoft Excel, for example, the formulas to use are:

=AVERAGE(cell range) for average

=MODE(cell range) for mode

=MEDIAN(cell range) for median and

=STDEV(cell range) for the standard deviation.

(a)

Wire testing.xls

	A	B
1	Measuring point	Diameter of wire (mm)
2	1	2.34
3	2	2.34
4	3	2.35
5	4	2.37
6	5	2.38
7		
8		
9	mean (average) =	2.356
10	mode =	2.34
11	median =	2.35
12	standard deviation =	0.018165902
13		

Sheet1 / Sheet2

(b)

Wire testing.xls

	A	B
1	Measuring point	Diameter of wire (mm)
2	1	2.34
3	2	2.34
4	3	2.35
5	4	2.37
6	5	2.38
7		
8		
9	mean (average) =	=AVERAGE(B2:B6)
10	mode =	=MODE(B2:B6)
11	median =	=MEDIAN(B2:B6)
12	standard deviation =	=STDEV(B2:B6)
13		

A spreadsheet showing **(a)** the calculation of mean, mode, median and standard deviation for a set of data and **(b)** the formulas used. Note that the calculated values should be displayed to two decimal places to match the precision of the original data.

The significance of standard deviation and its value in processing data is covered later in the chapter.

Using probability in science

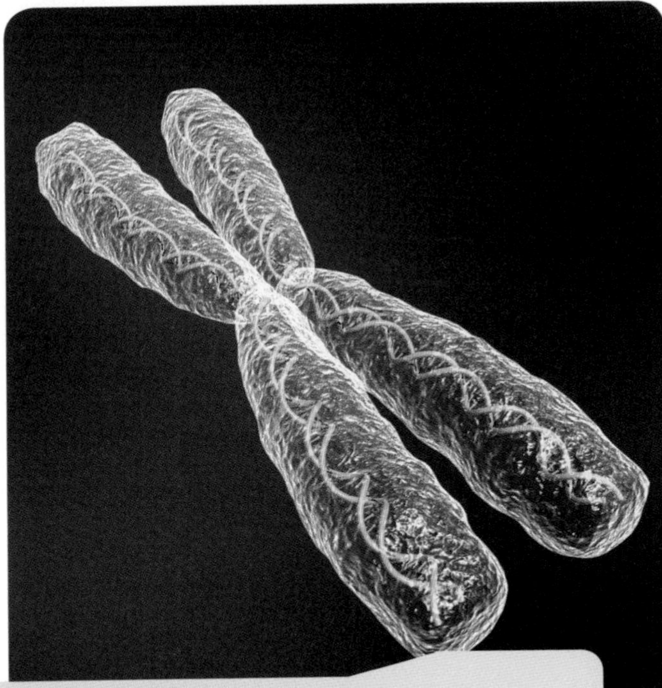

Probability is used to calculate the risks of genetic diseases in humans.

Probability is a method used by scientists to assess how likely something is to happen. For example, the nucleus of a radioactive material randomly disintegrates to form other nuclei. The rate of chemical reaction is related to the probability of collisions between atoms that have sufficient energy for a reaction to occur. In genetics, probability is used to work out the chance of offspring being born with certain hereditary diseases.

Understanding the basics of probability is key to understanding how science works.

Did you know?

That 8% of men and only 0.5% of women are colour blind. This is due to a genetic disease that is inherited by children from their parents. The chance or probability is related to the type of colour blindness.

How do we describe probability?

We can measure probability by using numbers. A probability (P) of 0 means a particular event will not happen, while a probability of 1 means an event will be certain to happen. Any probability between 0 and 1 measures the chance of it happening. The nearer the probability is to 1, the more chance there is of the event taking place.

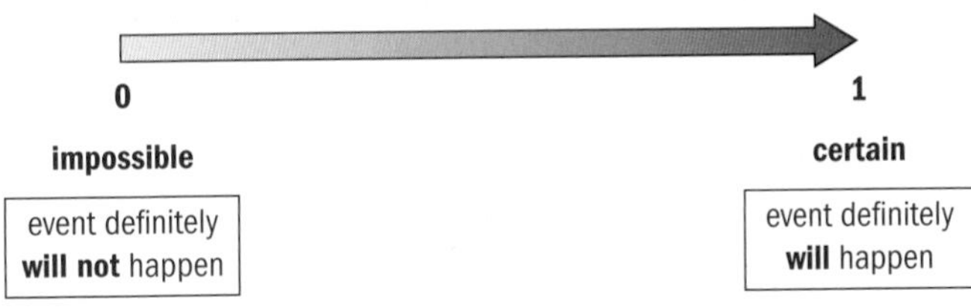

Probability scale.

Worked example

A die has six faces. The number 4 is printed on one of the faces. The die is thrown. What is the probability of the die landing with the number 4 face up?

Answer

Step 1: The die has six faces. The faces are numbered 1, 2, 3, 4, 5 and 6. So there are six outcomes.

Step 2: There will be one outcome with the number 4 face up.

Step 3: The probability of the face being a 4 will be:

$\frac{1}{6}$

This assumes that the die is fair so that all outcomes are equally likely. (If the die had been tampered with so that one side was heavier than the rest, each outcome would not be equally likely.)

So, $P(4) = \frac{1}{6}$

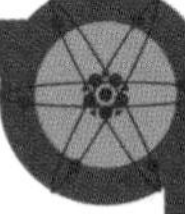

Activity 8.1C

Find the probability of throwing an odd number with a fair die.

In statistics language, the six possible outcomes when a die is thrown are called the **sample space**, S, and one particular outcome is referred to as an **event**. If there are two events, e.g. throwing a die and tossing a coin, a table can be used to represent the sample space.

Case study: Genetics

Amy is a biologist working on a genetics experiment involving plants (Mendel's experiment). The types of gene (or alleles) present are:

T for a tall plant (the dominant gene)

S for a short plant (the recessive gene).

Amy has been asked to determine the probability that the plants will have offspring that are tall plants.

Answer

This is similar to having two coins (with heads and tails for each coin).

Step 1: Draw the table showing all possible outcomes.

	T	**S**
T	T, T	S, T
S	T, S	S, S

Step 2: Work out how many outcomes produce tall plants. With T being the dominant gene, these are (T, T), (T, S) and (S, T).

Step 3: Work out the probability of these outcomes.

P(TT, TS or ST) = ¾ = 75%

So, there is a 75% probability that the offspring will be tall plants. What is the probability that the offspring will be short plants?

Worked example

A fair die and fair coin are thrown. Calculate the probability of getting a 4 and a 'tail'.

Answer

Step 1: Draw a table of the sample space (all possible outcomes).

	1	**2**	**3**	**4**	**5**	**6**
Head	1, H	2, H	3, H	4, H	5, H	6, H
Tail	1, T	2, T	3, T	4, T	5, T	6, T

Step 2: Note that there are 12 possible outcomes and only one outcome that gives a 4 and a tail. So the probability of getting a 4 and a tail is:

$$P(4, T) = \frac{1}{12}$$

You can use the procedure in this worked example to assess the outcome of a genetics experiment.

PLTS

Independent enquirer

You will develop this skill when you identify the set of data.

Venn diagrams

The probabilities of events can be calculated using **Venn diagrams**. Each region of a Venn diagram represents data that satisfy certain conditions. In the diagram below, each set of data, X and Y, is represented by a circle. Region A represents the data that are in set X but not in set Y. Region B represents the data that are in set Y but not in set X. Region T represents the data that are in both set X and set Y. Q represents the data that are not in either set X or set Y.

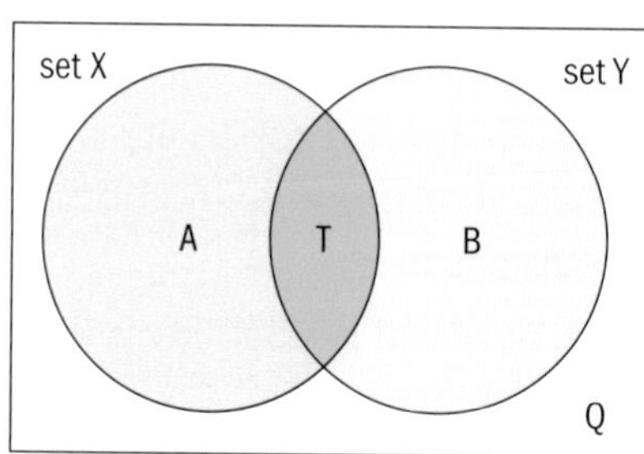

Venn diagrams can be used to calculate probabilities.

The probability of data belonging to X or Y is written as: *P*(X U Y) where U means the **union** between X and Y. You say this as 'X union Y'.

The probability of data belonging to X and Y is written as: *P*(X ∩ Y), where ∩ means the **intersection** between X and Y. You say this as 'X intersection Y'.

Worked example

Jamie is working for the Institute of Cancer Research. He is looking at the effects of two new drugs. His study involves 100 patients: 40 received drug A, 30 received drug B and 16 received both drug A and drug B. His first task is to calculate the probability that a patient chosen at random will receive both drug A and drug B.

Step 1: To draw the Venn diagram, you need to find out how many patients only took drug A (call this x), how many only took drug B (call this y) and how many didn't take either of the drugs (call this z).

Step 2: There are 40 who received drug A so:

$x + 16 = 40$, which means $x = 24$

Step 3: There are 30 who received drug B so:

$y + 16 = 30$, which means $y = 14$

The total number of patients is 100 so the number of patients who did not take drug A or drug B will be

$100 - 24 - 14 - 16 = 46$

So, $z = 46$

Step 4: Knowing the number of people who took both drugs, as well as x, y and z, you can draw the Venn diagram.

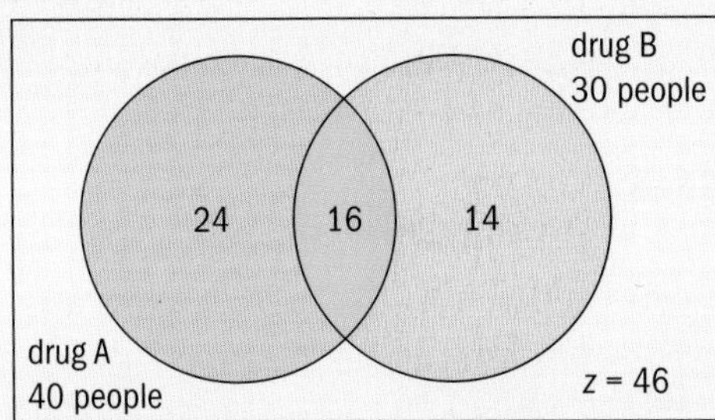

Step 5: The probability that a patient will take both drug A and drug B is

$P(A \cap B) = \frac{16}{100} = 16\%$

The addition rule of probability

If you need to work out the probability of a patient taking either drug A or drug B, you will need to use the **addition rule** of probability. This is represented as $P(A \cup B)$ and can be obtained using the following rule:

$$P(A \cup B) = P(A) + P(B) - P(A \cap B)$$

Where $P(A)$ and $P(B)$ are the probability of selecting at random a patient who has taken drug A and B, respectively. So, $P(A)$ will be $\frac{40}{100}$ and P(B) will be $\frac{30}{100}$. $P(A \cup B)$ has already been calculated as $\frac{16}{100}$.

So,

$$P(A \cup B) = \frac{40}{100} + \frac{30}{100} - \frac{16}{100} = \frac{(70 - 16)}{100} = \frac{54}{100}$$

The probability of a patient taking either drug A or drug B is 54%.

Mutually exclusive events

When two outcomes cannot occur at the same time, they are said to be **mutually exclusive**. This means that $P(A \cap B) = 0$.

The multiplication rule for independent events

Consider two events; for example, tossing a fair coin (call that C) and rolling a fair die (D). These events are said to be **independent** because rolling a 6 on a die does not influence the outcome of the coin toss.

The rule for working out the probability of independent events occurring together is called the **multiplication rule**:

$$P(C \cap D) = P(C) \times P(D)$$

Worked example

A building company buys two solar panels to be used in an energy-efficient bungalow. The probability that one solar panel will malfunction in the first year is 0.01. What is the probability that both solar panels will malfunction in the first year?

Answer

Step 1: Let event A be that the first solar panel malfunctions in the first year. Let event B be that the second solar panel malfunctions in the first year.

Step 2: $P(A) = 0.01$ and $P(B) = 0.01$ so

$P(A \cap B) = 0.01 \times 0.01 = 0.0001 = 0.01\%$

Conditional probability

Refer back to the worked example on the drug trial and ask this question about the patients who are taking drug B: what is the probability that they are also taking drug A?

Altogether, 30 took drug B, of whom 16 also took drug A. The probability that a patient takes drug A given that they also take drug B can be written as $P(A \mid B)$. This is called **conditional probability**. This can be written as:

$$P(A \mid B) = \frac{P(A \cap B)}{P(B)} = \frac{16/100}{30/100} = \frac{16}{30}$$

Frequency distributions

Data can be **discrete**, such as the blood types of patients which can only be from a specific set of types, or **continuous**, such as the heights of those patients which can take any value from a continuum.

Unit 6: See page 161 for more information on types of data and displaying data.

Continuous data can be grouped or ungrouped. Grouped data are data that have been organised into groups (frequency distribution); here is an example:

mass (g)	number of fish
0–5	8
6–10	10
11–15	24
16–20	12

We refer to the data for mass as the 'class' and the number of fish as the 'frequency'.

Ungrouped data appear as a list of numbers that haven't been organised into groups; for example, the diameter of eight copper wires were measured to be: 33 mm, 23 mm, 18 mm, 29 mm, 28 mm, 10 mm, 45 mm and 30 mm.

Frequency distributions of data, including their shape, allow scientists to observe possible patterns in data and display their data clearly. The shape of distributions tends to be **unimodal** with one single peak. The **normal distribution** of data is common in science. This is a symmetrical, bell-shaped distribution with one peak. It can occur, for example, when measuring people's heights or the breathing rate of insects. Skewed distributions, with a single peak and a tail to the positive or negative end can occur, for example, in chemical reaction times or the distribution of wages in a large organisation.

Sometimes, the shape of a distribution can show two distinct peaks; this is called a **bimodal** distribution. The two peaks may arise because the sample data comprise two groups. For example, if you plotted the height distribution of a group of 1000 people from the same ethnic group, there would be two peaks: one corresponding to the mean height of the men in the sample and the other corresponding to the mean height of the women in the sample. This is because, on average, men are taller than women.

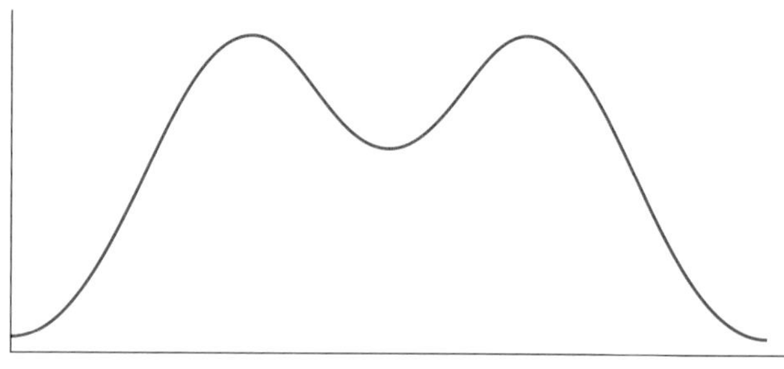

Bimodal distributions have two distinct peaks.

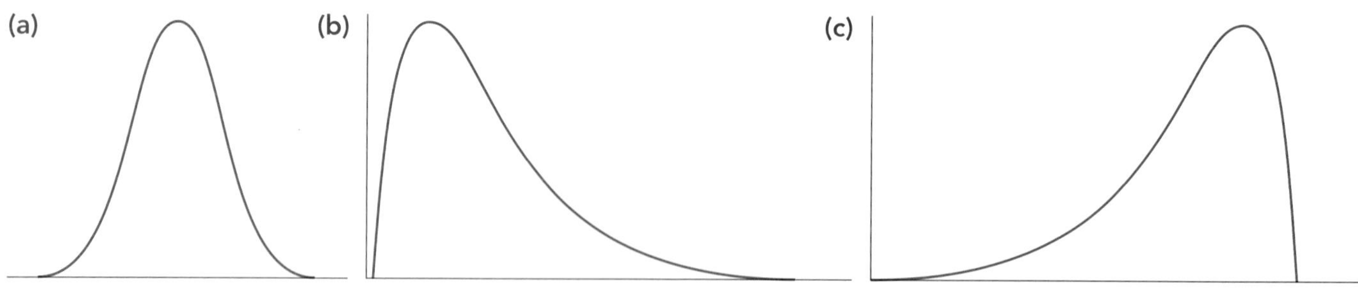

Unimodal distributions have a single peak: **(a)** normal distribution, **(b)** positively skewed distribution, **(c)** negatively skewed distribution.

PLTS

Creative thinkers

This activity allows you to select the appropriate scientific data in the appropriate format.

Did you know?

The observation of the colour of galaxies gives a bimodal distribution, showing two distinct peaks corresponding to blue and red.

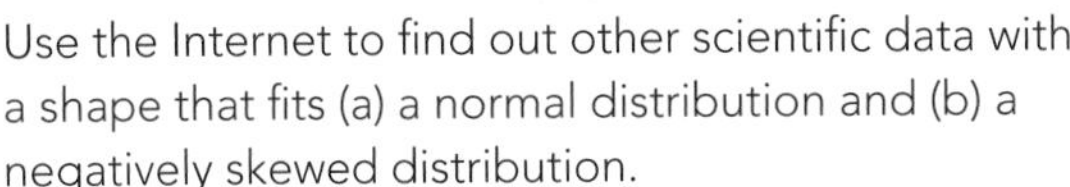

Activity 8.1D

Use the Internet to find out other scientific data with a shape that fits (a) a normal distribution and (b) a negatively skewed distribution.

The normal distribution

The normal distribution is one of the most important distributions in science. It resembles a bell shape. This is because many measured quantities that occur in science tend to follow normal distributions; for example, if you measure the length of 100 sheep or the mass of 100 leaves, the distribution of the data would tend towards a normal distribution. Knowing that quantities follow the normal distribution allows scientists to calculate the probability of how a particular quantity may behave. This is represented by the area under the curve. For example, if a biologist studying a tropical insect finds that the lifespan follows a normal distribution, then she may be able to calculate their probable lifespan.

In a normal distribution, 68% of the data are within one standard deviation of the mean; 95% of the data are within two standard deviations of the mean; and 99.9% of the data are within three standard deviations of the mean.

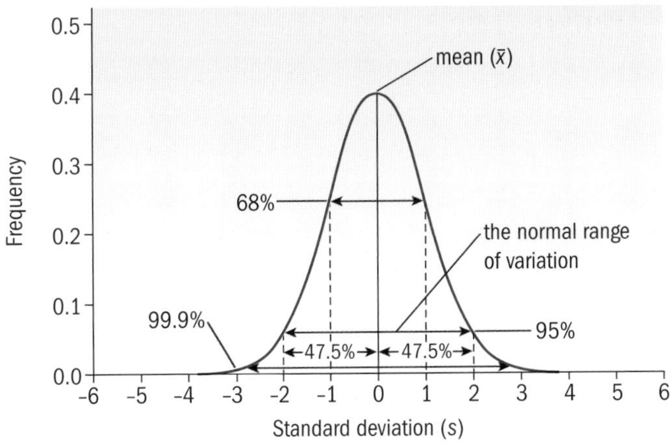

As the normal distribution is symmetrical, half of the area will lie on each side of the mean, i.e. 47.5% will lie between $\bar{x} + 2s$ and 47.5% will lie between $\bar{x} - 2s$.

The normal distribution can be written as:

$$X \sim N(\bar{x}, s^2)$$

where X is the normal variable, ~ means distribution, N stands for normal, $\bar{x}$ is the mean and s^2 is the variance (the square of the standard deviation).

The area under the curve of a normal distribution, taken between two values, will give the probability of an outcome occurring between these two values. Between values a and b this can be written as:

$$P(a < X < b)$$

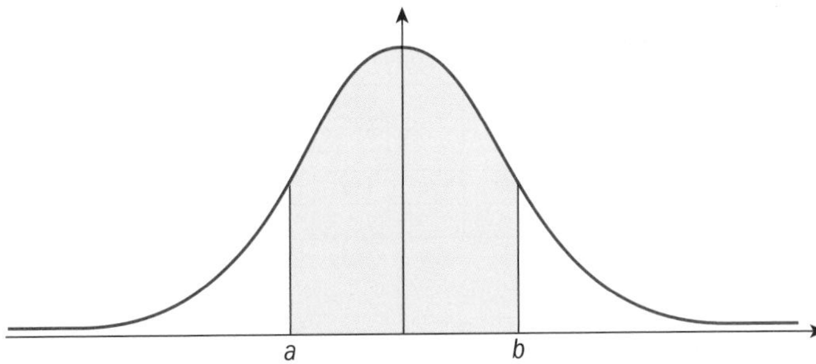

Sketch of normal distribution. The shaded area shows the probability of an event occurring between *a* and *b*.

The area between two values can be found by using standard normal distribution tables.

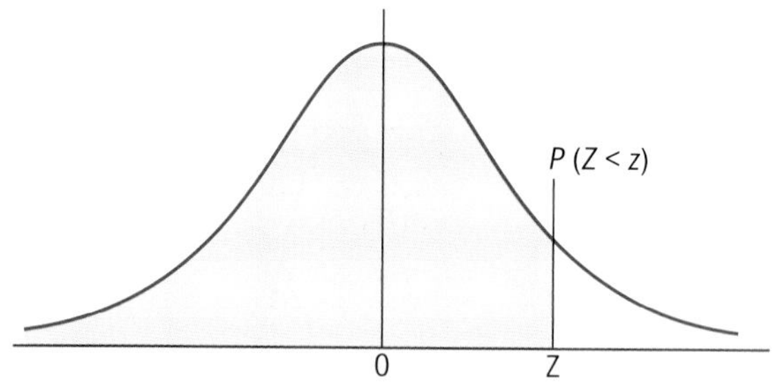

z	0.00	0.01	0.02	0.03	0.04	0.05	0.06	0.07	0.08	0.09
0.0	0.5000	0.5040	0.5080	0.5120	0.5159	0.5199	0.5239	0.5279	0.5319	0.5359
0.1	0.5398	0.5438	0.5478	0.5517	0.5557	0.5596	0.6536	0.5675	0.5714	0.5753
0.2	0.5793	0.5832	0.5871	0.5910	0.5948	0.5987	0.6026	0.6064	0.6103	0.6141
0.3	0.6179	0.6217	0.6255	0.6293	0.6331	0.6368	0.6406	0.6443	0.6480	0.6517
0.4	0.6554	0.6591	0.6628	0.6664	0.6700	0.6736	0.6772	0.6808	0.6844	0.6879
0.5	0.6915	0.6950	0.6985	0.7019	0.7054	0.7088	0.7123	0.7157	0.7190	0.7224
0.6	0.7257	0.7291	0.7324	0.7357	0.7389	0.7422	0.7454	0.7486	0.7517	0.7549
0.7	0.7580	0.7611	0.7642	0.7673	0.7704	0.7734	0.7764	0.7794	0.7823	0.7854
0.8	0.7881	0.7910	0.7939	0.7967	0.7995	0.8023	0.8051	0.8078	0.8106	0.8133
0.9	0.8159	0.8186	0.8212	0.8238	0.8264	0.8289	0.8315	0.8340	0.8365	0.8389
1.0	0.8413	0.8438	0.8461	0.8485	0.8508	0.8531	0.8554	0.8577	0.8599	0.8621
1.1	0.8643	0.8665	0.8686	0.8708	0.8729	0.8749	0.8770	0.8790	0.8804	0.8830
1.2	0.8849	0.8869	0.8888	0.8907	0.9825	0.8944	0.8962	0.8980	0.8997	0.9015
1.3	0.9032	0.9049	0.9066	0.9082	0.9099	0.9115	0.9131	0.9147	0.9162	0.9177
1.4	0.9192	0.9207	0.9222	0.9236	0.9251	0.9265	0.9279	0.9292	0.9306	0.9319
1.5	0.9332	0.9345	0.9357	0.9370	0.9382	0.9394	0.9406	0.9418	0.9429	0.9441
1.6	0.9452	0.9463	0.9474	0.9484	0.9495	0.9505	0.9515	0.9525	0.9535	0.9545
1.7	0.9554	0.9564	0.9573	0.9582	0.9591	0.9599	0.9608	0.9616	0.9625	0.9633
1.8	0.9641	0.9649	0.9656	0.9664	0.9671	0.9678	0.9686	0.9693	0.9699	0.9706
1.9	0.9713	0.9719	0.9726	0.9732	0.9738	0.9744	0.9750	0.9756	0.9761	0.9767
2.0	0.9773	0.9778	0.9783	0.9788	0.9793	0.9798	0.9803	0.9808	0.9812	0.9817
2.1	0.9821	0.9826	0.9830	0.9834	0.9838	0.9842	0.9846	0.9850	0.9854	0.9857
2.2	0.9861	0.9865	0.9868	0.9871	0.9874	0.9878	0.9881	0.9884	0.9887	0.9890
2.3	0.9893	0.9896	0.9898	0.9901	0.9904	0.9906	0.9909	0.9911	0.9913	0.9936
2.4	0.9918	0.9920	0.9922	0.9924	0.9927	0.9929	0.9931	0.9932	0.9934	0.9936
2.5	0.9938	0.9940	0.9941	0.9943	0.9945	0.9946	0.9948	0.9949	0.9951	0.9952
2.6	0.9953	0.9955	0.9956	0.9957	0.9959	0.9960	0.9961	0.9962	0.9963	0.9964
2.7	0.9965	0.9966	0.9967	0.9968	0.9969	0.9970	0.9971	0.9972	0.9973	0.9974
2.8	0.9974	0.9975	0.9976	0.9977	0.9977	0.9978	0.9979	0.9980	0.9980	0.9981
2.9	0.9981	0.9982	0.9982	0.9983	0.9984	0.9984	0.9985	0.9985	0.9986	0.9986

z	3.00	3.10	3.20	3.30	3.40	3.50	3.60	3.70	3.80	3.90
P	0.9986	0.9990	0.9993	0.9995	0.9997	0.9998	0.9998	0.9999	0.9999	1.0000

Table of values for the area under the standard normal distribution curve.

Using standard normal tables

In order to use the same set of tables for all possible values of the mean and variance, the variable X is standardised so that the mean is 0 and the standard deviation is 1. The standardised normal variable is called Z so the expression for the distribution becomes:

$$Z \sim N(0,1)$$

Worked example

In this example you will use the standard normal distribution.

1 Find the area under the standard normal curve between 0.5 and 2.0 standard deviations from the mean.

2 Use standard normal tables to find $P(Z < 1.25)$.

Answers

1

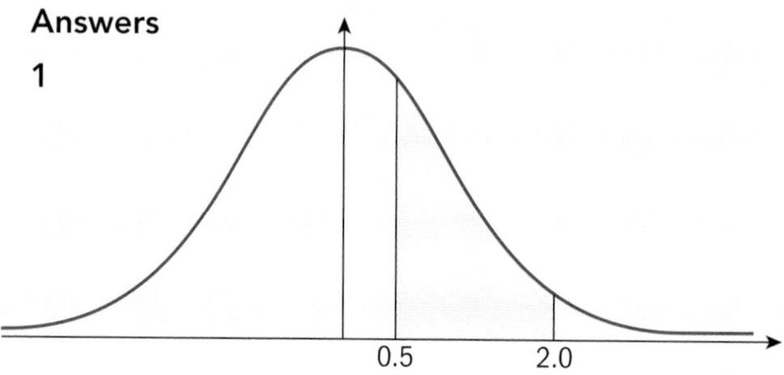

From the standard normal table given above, the area to the left of $z = 2.0$ is 0.9773 and the area to the left of $z = 0.5$ is 0.6915. Therefore, the area between $z = 0.5$ and $z = 2.0$ is:

$$\Phi(2.0) - \Phi(0.5) = 0.9773 - 0.6915 = 0.2858$$

2

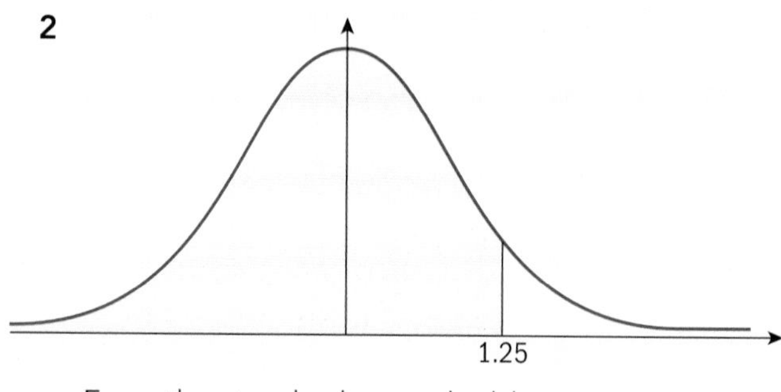

From the standard normal table:

$$P(Z < 1.25) = 0.8944$$

In order to use the standard normal tables, the normal variable, X, needs to be converted to a standard normal distribution. This is done by calculating the value Z using:

$$Z = \frac{X - \bar{x}}{s}$$

The standard normal tables give the area under the curve for a particular value, z, where z is the number of standard deviation units from the mean. This value is what is obtained from the normal tables.

This area under the curve as far as a particular value, z, can be written as $\Phi(z)$ (pronounced 'phi z').

$$\Phi(z) = P(Z < z)$$

Case study: Manufacturing

Ahmed works in the quality control department of a company producing steel parts for a major car manufacturer.

A machine produces ball bearings with diameters normally distributed with a mean of 15.0 mm and a standard deviation of 1.0 mm. Ahmed needs to find the probability that the diameter of a randomly selected ball bearing will be less than 16.5 mm.

Answer

X is the diameter of the ball bearing, $\bar{x} = 15.0$ mm and $s = 1.0$ mm:

$X \sim N(15.0, 1^2)$

Ahmed needs to find the probability that the diameter is less than 16.5 mm, i.e. $P(X < 16.5)$. To be able to use the standard normal tables, X needs to be standardised.

X is used in the formula: $Z = \frac{X - \bar{x}}{s} = \frac{(16.5 - 15.0)}{1.0} = 1.5$

So, $P(X < 16.5)$ becomes $P(Z < 1.5)$

So, $P(Z < 1.5) = \Phi(1.5) = 0.9332$

So, the probability that the diameter of a randomly selected ball bearing will be less than 16.5 mm is 0.93 or 93%.

What is the probability that it will be greater than 16.5 mm? What is the probability that it will be greater than 16.7 mm?

Sampling

In statistics, the word **population** is used to describe the entire set of individuals, items or data values from which a statistical **sample** can be taken. The population could be large or small, and could be any object under investigation; for example, blood cells taken from patients, a culture of bacteria, or the people who support a particular political party.

Some investigations require you to consider the entire population; other investigations require you to consider a sample of that population. For example, the population could be all the patients who were admitted to Accident and Emergency over a particular weekend, whereas a sample could comprise 50 patients randomly selected from the population of A&E patients.

Similarly, political polls, e.g. Ipsos MORI and Gallup, which canvas the views of the voting public, use samples. Asking the views of all members of the voting public would be prohibitively expensive and would take too long.

Note that the larger the sample is, the more representative of the whole population it will be, and the more reliable any conclusions drawn from the statistical analysis will be.

Did you know?

Polls done by Ipsos MORI and Gallup tend to have a sample size that is greater than 1000.

Sampling can be random or non-random. In ecological field investigations, quadrat sampling is used. A quadrat is normally a square area under which the investigation can take place. For example, if sampling the types of birds that live in a forest, it would be impossible to determine exactly the number of birds that exist in the entire forest. However, areas are randomly picked out (for example 10 quadrat, 1 km^2) and the birds identified. This would be a sample of the forest. Other examples are shown in the table.

Activity 8.1E

What is the difference between cluster sampling and stratified sampling?

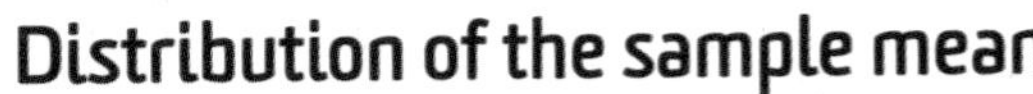

Distribution of the sample mean

When you have analysed a random sample you often want to know how the sample is distributed and how representative it is of the population from which it was taken. For example, how does the mean of the sample compare to the population mean? How confident are you with the data provided by your sample?

Assuming the population follows a normal distribution, the mean of the sample ($\bar{X}$) can be worked out from the following formulas:

$\bar{X} \sim N\left(\bar{x}, \frac{s^2}{n}\right)$ remembering that $X \sim N(\bar{x}, s^2)$

where n is the number in the sample.

	Sampling	
	random sampling	non-random sampling
Type 1	drawing lots	cluster select groups having similar characteristics and take sample of each group (cluster)
Example	select, via lots, 20 insects out of a population of 100 and look at their mass distribution	investigate the growth rate of bacteria in similar ponds within a district
Type 2	stratified population divided into layers	quota population divided into categories depending on criteria such as age, income level
Example	investigate the effect of a new drug on males and females	used in market research on the high street or opinion polls regarding an election

Worked example

A group of ecologists is studying the availability of food for a population of frogs. The masses of a sample of male frogs taken from a pond can be modelled by a normal distribution, with a mean mass of 70 g and standard deviation 5 g. Four male frogs are chosen at random. Find the probability that their mean mass is less than 65 g.

Answer

Step 1: X is the mass of a male frog and $X \sim N(\bar{x}, s^2)$ where $\bar{x} = 70$ g and $s = 5$ g. As the distribution is normal, the distribution of $\bar{X}$ is also a normal distribution:

$$\bar{X} \sim N\left(\bar{x}, \frac{s^2}{n}\right)$$

continued

Step 2: So,

$\bar{X} \sim N(70, \frac{25}{4})$

$\bar{X} \sim N(70, 6.25)$

Step 3: Standardise $\bar{X}$: $P(Z < (\frac{65-70}{\sqrt{6.25}})$

So,

$P(\bar{X} < 65) = P(Z < (\frac{65-70}{\sqrt{6.25}})$

$P(\bar{X} < 65) = (Z < -2)$

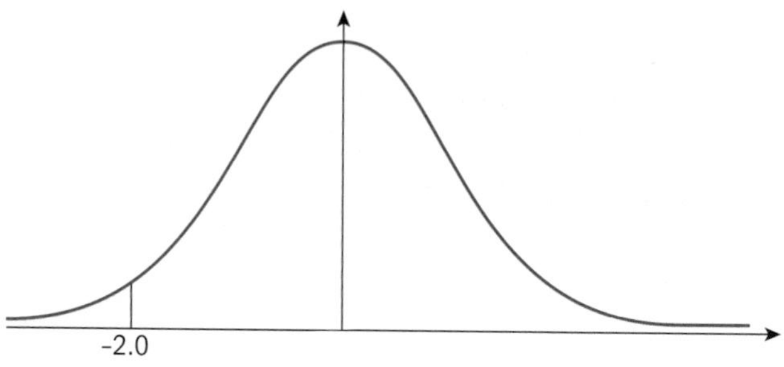

Step 4: Because the curve is symmetrical, you can use the value from the standard table for the normal distribution for $z = 2.0$ and take that value from 1:

$P(\bar{X} < 65) = \Phi(-2) = 1 - \Phi(2)$

$= 1 - 0.9773$

$= 0.0228$

So, the probability that the mean mass of the sample frogs is less than 65 g is 0.023 or 2.3%.

Functional skills

Mathematics

This example allows you to develop the maths skills involved in drawing and justifying conclusions.

Standard error of the mean and confidence limits

As the mean and standard deviation are estimates of a population, there will be a sampling error. The standard error of the mean is given by:

$$SE = \frac{s}{\sqrt{n}}$$

You need to give a figure of confidence to your data; this is described by **confidence limits**, which are related to the standard error. For normally distributed data, 95% of data fall within two standard deviations of the mean. The 95% confidence level is, therefore, adequate for most scientific investigations. Based on standard normal tables, the 95% confidence limit can be calculated from:

95% confidence limit = mean value ± 1.96 × SE

This tells us that we are 95% confident that the mean value lies between these two limits and there is a probability of 5% that the value falls outside these limits. Sometimes 95% is not sufficient and something like a 99.9% confidence limit is required (e.g. in clinical trials).

Worked example

The mean of five data values is 2.36 mm and the standard deviation is 0.018 mm (see the Worked example on page 177).

1 What is the standard error of the mean?

2 What are the 95% confidence limits, assuming the data are normally distributed?

Answers

1 $SE = \frac{s}{\sqrt{n}} = \frac{0.018}{\sqrt{5}} = 0.008\,\text{mm}$

2 The 95% confidence limits are given by:

mean value ± 1.96 × SE = 2.36 ± 1.96 × 0.008

= 2.36 + 0.02 and 2.36 − 0.02

That is, we are 95% confident that the diameter of the wire will lie between 2.34 mm and 2.38 mm.

Assessment activity 8.1

1 You are working as an applied biologist in the biology department of a local university. You have been investigating tissue cells, and you have taken a sample of 9 cells and measured the diameters to be: 123 μm, 126 μm, 129 μm, 122 μm, 125 μm, 128 μm, 125 μm, 124 μm and 125 μm.

   a Determine the mean, mode and median diameter of the cells.

   b Determine the standard deviation, using both manual and ICT methods.

   c Calculate the standard error of the mean. What is the 95% confidence limit? P1

2 You are working as a microbiologist in a local hospital. You have been growing a number of microorganisms. The probability of growing type A is 0.3 and that of type B is 0.5 and $P(A \mid B) = 0.25$. Find:

   a $P(A \cap B)$

   b $P(A \cup B)$ 

3 You are an ecologist studying a tropical insect. You have just determined that the lifespan is normally distributed. The mean lifespan of this insect is 144 days and the standard deviation of its lifespan is 16 days. Find the probability that the next insect studied lives:

   a less than 140 days

   b more than 156 days. D1

Grading tips

When attempting question 1, remember to convert micrometres to metres. To reach P1 you will have to show that you have used a calculator, so all calculator steps used need to be identified clearly.

When calculating probabilities for M1, make sure that you draw and correctly label the Venn diagrams.

For D1, make sure you understand how to use standard normalised tables; look at the worked examples for more assistance in interpreting distribution shapes.

8.2 Using statistical tests to investigate scientific problems

In this section:

Key terms

Null hypothesis – this states that there is no significant difference between two sets of data (e.g. because the data occurred by chance).

Degrees of freedom – number of variables that are used to make a calculation.

Independent samples – quantities that are not related scientifically.

Significance level – also called confidence level, is used in hypothesis testing. It is a figure used to reject or accept the null hypothesis. Experimentalists use figures ranging from 1% (0.01) to 5% (0.05) significant level.

Correlation testing – method used to see if there is a relationship between quantities.

The chi-squared test

Statistical tests, such as the chi-squared test (χ^2 test), may be used to test or support a scientific hypothesis, or to see if there is a relationship between two quantities or factors. For example, you may want to compare two sets of experimental data to see whether there is any difference between them. This can be done using the χ^2 test. The **null hypothesis** will state that there is no significant difference between the two sets of data, and this will be accepted or rejected after the data have been subjected to statistical tests.

The χ^2 test is particularly useful in biology, geography and psychology.

Did you know?

The chi-squared test was invented by the statistician Karl Pearson in 1900. The test was first used to examine Mendel's work on genetics.

Step-by-step method of carrying out the χ^2 test

Step 1 Tabulate your results, recording the **observed value** (*O*) and **expected value** (*E*). You may need to calculate the expected values based on a particular model.

Step 2 For each pair of values, calculate $(O - E)$, square this to find $(O - E)^2$ and then divide by *E* to find:

$$\frac{(O - E)^2}{E}$$

Step 3 Calculate χ^2 by adding all the values of $\frac{(O - E)^2}{E}$:

$$\chi^2 = \Sigma \frac{(O - E)^2}{E}$$

Step 4 Determine the **degrees of freedom** in the data. The degrees of freedom, *n*, for one column (or row) is given by:

$n = 1$ – (number of items of data in that column or row)

Step 5 The critical values of the χ^2 distribution are tabulated at different confidence levels (*p*) for a range of degrees of freedom, as shown in the following table.

		p	
		0.05	0.01
Degrees of freedom	1	3.841	6.635
	2	5.991	9.210
	3	7.815	11.345
	4	9.488	13.277
	5	11.070	15.086
	6	12.592	16.812
	7	14.067	18.475
	8	15.507	20.090
	9	16.919	21.666
	10	18.307	23.209
	11	19.675	24.725
	12	21.026	26.217
	13	22.362	27.688
	14	23.685	29.141
	15	24.996	30.578

In biology, the critical value used tends to be at the $p = 0.05$ level. This is called the 5% level, which means that there is a 5% probability that the observed data occurred by chance, when χ^2 equals this critical value.

- If the calculated value of χ^2 is greater than or equal to the critical value, the observed data differ significantly from the expected data. The larger the value of χ^2 the more confident we can be that there is a difference between the observed and expected data (that is, the null hypothesis is incorrect).
- If the calculated value of χ^2 is less than the critical value, the null hypothesis has to be accepted, which means that there is no significant difference between the observed and expected data.

Activity 8.2A

Use the Internet to find one set of scientific data (which could include psychology) that the χ^2 test could be used for.

Worked example

A genetic model suggests that if a red tropical flower self-pollinates, the expected outcome is that of red, pink and yellow flowers in the ratio 1 : 3 : 2.

A botanist grew 300 plants from the self-pollinated seeds of this tropical plant. The observations that he made are shown in the table.

	Red flowers	Pink flowers	Yellow flowers
observed	45	160	95

Use the χ^2 test to determine whether these results are consistent with those predicted by the genetic model.

Answer

Step 1: Calculate the expected result (E) for each flower colour.

The number of flowers in the sample is 300. Since the model predicts a ratio of 1 : 3 : 2, the expected results would be $\frac{1}{6} \times 300$ for red, $\frac{3}{6} \times 300$ for pink and $\frac{2}{6} \times 300$ for yellow flowers. The expected results are entered in a table.

continued

	Red flowers	Pink flowers	Yellow flowers
observed	45	160	95
expected	50	150	100

Step 2: Determine $\frac{(O - E)^2}{E}$ for each flower colour.

	Red flowers	Pink flowers	Yellow flowers
observed	45	160	95
expected	50	150	100
$\frac{(O - E)^2}{E}$	$\frac{(45 - 50)^2}{50}$ = 0.5	$\frac{(160 - 150)^2}{150}$ = 0.67	$\frac{(95 - 100)^2}{100}$ = 0.25

Step 3: Add these up to get χ^2:

$\chi^2 = 0.5 + 0.67 + 0.25 = 1.42$

Step 4: Determine the number of degrees of freedom, n. There are three columns of observed data, so

$n = 3 - 1 = 2$ degrees of freedom

Step 5: Determine the critical value. From the χ^2 table, for $n = 2$ at the 5% level, this is 5.991.

Since $1.42 < 5.991$, the null hypothesis is valid. This means that there is no significant difference between the observed and expected data, so the botanist can accept the genetic model. As the value of χ^2 is more than three times the critical value we can have confidence in the result.

Functional skills

Mathematics

In this example you will develop your skills in dealing with routine and non-routine maths problems in an unfamiliar context, in this case the work of a botanist.

Chi-squared test using contingency tables

This is used when the relationship between quantities is being tested; only the test using a 2 × 2 contingency table will be considered here. In this case, the data table will be comprised of R rows and C columns.

χ^2 is calculated as before, but the degrees of freedom will be given by:

$n = (R - 1) \times (C - 1)$

The worked example demonstrates the χ^2 test using contingency tables, as used in science.

Worked example

Using contingency tables

Consider an investigation into colour blindness: the table below shows observed values for males and females. Determine if there is a link between male and female colour blindness.

	Males	Females
colour blind	56	14
not colour blind	754	536

Answer

Step 1: The null hypothesis will state that there is no difference between the occurrence of colour blindness in males and females. This means that if χ^2 is less than the critical value the null hypothesis will be accepted, and if χ^2 is greater than the critical value then the null hypothesis will be incorrect. The greater the value the more confident we can be.

Step 2: Calculating the **expected values** using a 2 × 2 contingency table: if there is no difference between males and females then we should observe the same fraction of colour-blind individuals. The fraction of colour-blind people will be:

$$\frac{56 + 14}{56 + 14 + 754 + 536} = 0.05147$$

The expected number of colour-blind males will be:

0.05147 × no. of males = 0.05147 × 810 = **42**

The expected number of colour-blind females will be:

0.05147 × 550 = **28**

The fraction of people that are not colour blind will be:

$$\frac{754 + 536}{56 + 14 + 754 + 536} = 0.9485$$

The expected number of males that won't be colour blind will be:

0.9485 × 810 = **768**

The expected number of females that won't be colour blind will be:

0.9485 × 550 = **522**

continued

Step 3: A contingency table can then be constructed for the observed and expected values.

Observed	**Males**	**Females**
colour blind	56	14
not colour blind	754	536
Expected	**Males**	**Females**
colour blind	42	28
not colour blind	768	522

Step 4: Calculation of $\frac{(O-E)^2}{E}$

Colour blind:

$\frac{(O-E)^2}{E}$	$\frac{(56-42)^2}{42} = 4.7$	$\frac{(14-28)^2}{28} = 7$

Not colour blind:

$\frac{(O-E)^2}{E}$	$\frac{(754-768)^2}{768}$ = 0.255	$\frac{(536-522)^2}{522}$ = 0.374

Step 5: Calculate χ^2:

$$\Sigma\frac{(O-E)^2}{2} = 4.7 + 7.0 + 0.26 + 0.37 = \mathbf{12.33}$$

Step 6: Calculate the degrees of freedom: the table comprises $R = 2$ and $C = 2$, so the degree of freedom will be = (2 – 1) × (2 – 1) = 1.

Step 7: From the table on page 188, for $p = 0.05$ the critical value is **3.84**.

Step 8: Conclusion: χ^2 is much greater than the critical value and so the evidence suggests that the null hypothesis has to be rejected. That is, the data suggest that the fraction of males with colour blindness is greater than that of females, and the difference cannot be attributed to chance; therefore, there may be another reason. Indeed, even using a confidence level of 0.01, the value of χ^2 is still twice as high, giving us more confidence that the null hypothesis can be rejected.

Hint: To calculate the expected values for each cell in the table:

$$\frac{\text{(row total)} \times \text{(column total)}}{\text{grand total}}$$

Activity 8.2B

State how you would use the two χ^2 tests.

The *t*-test

The *t*-test is often used when you are testing unrelated **independent samples** of data, for example from two separate experiments, and the two sets of data have a normal distribution.

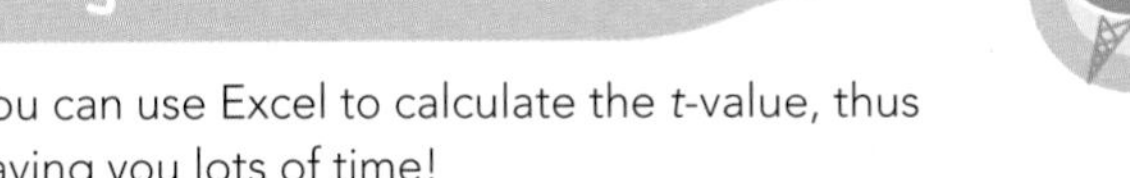

Did you know?

You can use Excel to calculate the *t*-value, thus saving you lots of time!

Steps in carrying out the *t*-test (independent samples)

When comparing two sets of data, the null hypothesis states that there is no significant difference between the two sets of data.

Step 1 Calculate the mean for each set of data, $\bar{x}_1$ and $\bar{x}_2$. Remember that the mean is given by

$$\bar{x} = \frac{\Sigma x}{n}$$

Step 2 Calculate the magnitude of the difference between the two means. Note the magnitude means you are only interested in the value and not its sign (positive or negative), so you take the **modulus**. (This is indicated by drawing two vertical lines around the expression.)

difference between the means = $|\bar{x}_1 - \bar{x}_2|$

Step 3 Calculate the standard error in the difference. Remember the standard error is given by $\frac{s}{\sqrt{n}}$, where s is the standard deviation and *n* is the number of measurements in the sample. So for two samples, the standard error in the difference will be:

$$\sqrt{\left(\frac{s_1^2}{n_1} + \frac{s_2^2}{n_2}\right)}$$

Step 4 Calculate the value of *t* using the formula:

$$t = \frac{\text{difference between the means}}{\text{standard error in the difference}}$$

Step 5 Calculate the number of degrees of freedom:

degrees of freedom = $(n_1 + n_2 - 2)$

where $n_1 + n_2$ represent the total population of the two samples.

Step 6 In the table below, find the critical value that corresponds to the number of degrees of freedom, for the **significance level** you are working to.

degrees of freedom	significance level					
	20%	10%	5%	2%	1%	0.1%
1	3.078	6.314	12.706	31.821	63.657	636.619
2	1.886	2.920	4.303	6.965	9.925	31.598
3	1.638	2.353	3.182	4.541	5.841	12.941
4	1.533	2.132	2.776	3.747	4.604	8.610
5	1.476	2.015	2.571	3.365	4.032	6.859
6	1.440	1.943	2.447	3.143	3.707	5.959
7	1.415	1.895	2.365	2.998	3.499	5.405
8	1.397	1.860	2.306	2.896	3.355	5.041
9	1.383	1.833	2.262	2.821	3.250	4.781
10	1.372	1.812	2.228	2.764	3.169	4.587
11	1.363	1.796	2.201	2.718	3.106	4.437
12	1.356	1.782	2.179	2.681	3.055	4.318
13	1.350	1.771	2.160	2.650	3.012	4.221
14	1.345	1.761	2.145	2.624	2.977	4.140
15	1.341	1.753	2.131	2.602	2.947	4.073
16	1.337	1.746	2.120	2.583	2.921	4.015
17	1.333	1.740	2.110	2.567	2.898	3.965
18	1.330	1.734	2.101	2.552	2.878	3.922
19	1.328	1.729	2.093	2.539	2.861	3.883
20	1.325	1.725	2.086	2.528	2.845	3.850
21	1.323	1.721	2.080	2.518	2.831	3.819
22	1.321	1.717	2.074	2.508	2.819	3.792
23	1.319	1.714	2.069	2.500	2.807	3.767
24	1.318	1.711	2.064	2.492	2.797	3.745
25	1.316	1.708	2.060	2.485	2.787	3.725
26	1.315	1.706	2.056	2.479	2.779	3.707
27	1.314	1.703	2.052	2.473	2.771	3.690
28	1.313	1.701	2.048	2.467	2.763	3.674
29	1.311	1.699	2.043	2.462	2.756	3.659
30	1.310	1.697	2.042	2.457	2.750	3.646
40	1.303	1.684	2.021	2.423	2.704	3.551
60	1.296	1.671	2.000	2.390	2.660	3.460
120	1.289	1.658	1.980	2.158	2.617	3.373
∞	1.282	1.645	1.960	2.326	2.576	3.291

Step 7

- If the calculated value of *t* is less than the critical value, there is no significant difference between the two sets of data and the null hypothesis is accepted.

- If the calculated value of t is equal to or greater than the critical value, then the null hypothesis is rejected. This means that the two sets of data differ significantly.

Activity 8.2C

What is the null hypothesis?

Worked example

Two types of fertiliser (A and B) were to be tested. Fertiliser A was added to eight plots of land and fertiliser B was added to eight different plots of land. From then on, each plot was managed in the same way, with the same number of potato plants being planted in each plot. The yield of potatoes from each plot was recorded, as shown below.

Considering the 5% level, is there a significant difference between the yields due to the fertiliser used?

		Yield/kg	
		Fertiliser A	Fertiliser B
Plot	1	17	18
	2	10	9
	3	6	8
	4	8	11
	5	12	14
	6	9	10
	7	13	15
	8	11	17

Answer

Step 1: Calculate the mean and standard deviation.

$\bar{x}_A = \mathbf{10.75}$ and $s_A = \mathbf{3.37}$ and $\bar{x}_B = \mathbf{12.75}$ and $s_B = \mathbf{3.77}$

continued

Step 2: Difference between the means:

$\bar{x}_A - \bar{x}_B = \mathbf{2.00}$

Step 3: Standard error in the difference:

$$\sqrt{\left(\frac{s_1^2}{n_1} + \frac{s_2^2}{n_2}\right)} = \sqrt{\frac{3.37^2}{8} + \frac{3.77^2}{8}} = \mathbf{1.79}$$

Step 4: Calculate the value of t:

$$t = \frac{\text{difference between the means}}{\text{standard error in the difference}} = \frac{2.00}{1.79} = 1.12$$

Step 5: Calculate the degrees of freedom:

$(n_1 + n_2 - 2) = 8 + 8 - 2 = 14$

Step 6: Determine the critical value (at the significance level of 5%). From the table of t values above, this is **2.145**.

Because $t < 2.145$, there is no significant difference between the yields.

Although we are 95% confident that there is no significant difference between the effects of the two fertilisers, the test can be made more reliable by increasing the number of data points (only eight were used in this experiment).

PLTS

Reflective learner

By reflecting on what the results of this worked example mean, you will develop this skill.

The *t*-test for matched pairs

This test is used when two samples of data are linked in some way, and each set of data is normally distributed.

Suppose you wanted to investigate whether the numbers of greenfly on wild roses are related to the height of the roses: the table shows the results for 12 plants at two different heights.

	1	2	3	4	5	6	7	8	9	10	11	12
2.5 m	20	18	17	22	19	16	18	21	16	25	23	24
0.8 m	16	11	16	19	14	13	15	17	11	22	21	13

Calculate the difference, D, between the pairs in the sample:

4, 7, 1, 3, 5, 3, 3, 4, 5, 3, 2, 9

Calculate the mean difference:

$$\bar{D} = \frac{\Sigma D}{n} = 4.1$$

Calculate the standard deviation of the difference:

$$s = \sqrt{\frac{\Sigma(D - \bar{D})^2}{n - 1}} = 2.2$$

Calculate the standard error:

$$SE = \frac{2.2}{12} = 0.18$$

Calculate the value of t:

$$t = \frac{\bar{D}}{SE} = \frac{4.1}{0.18} = 22.8$$

Calculate the number of degrees of freedom:

no. of pairs of data – 1 = $n - 1 = 12 - 1 =$ **11**

Now look up the critical value (p) that corresponds to the degrees of freedom in the table on page 191. In this case, the critical value = 3.106 at the significance level of 1%. As the value of t is much greater, there is a significant difference between the two, and we can be more than 99% confident that the number of greenfly increases with increasing height.

Correlation methods in science

When undertaking experiments and investigations you are likely to come up with one measurement influencing another. Sometimes, quantities change but don't necessarily affect each other. Looking at patterns and possible relationships between quantities is therefore an important part of a scientist's work. **Correlation testing** helps us look for any relationships that may exist between two quantities.

Types of correlation

- When there is no clear pattern between the data, there is **no correlation** (see diagram a).
- When there is an indication that as y increases, x increases, there is a **positive correlation**. When the line of best fit is drawn, it has a positive slope (see diagrams b and d).
- When there is an indication that as y increases, x decreases, there is a **negative correlation**. When the line of best fit is drawn, it has a negative slope (see diagram c).

Evaluating the validity of the interpretation of the results

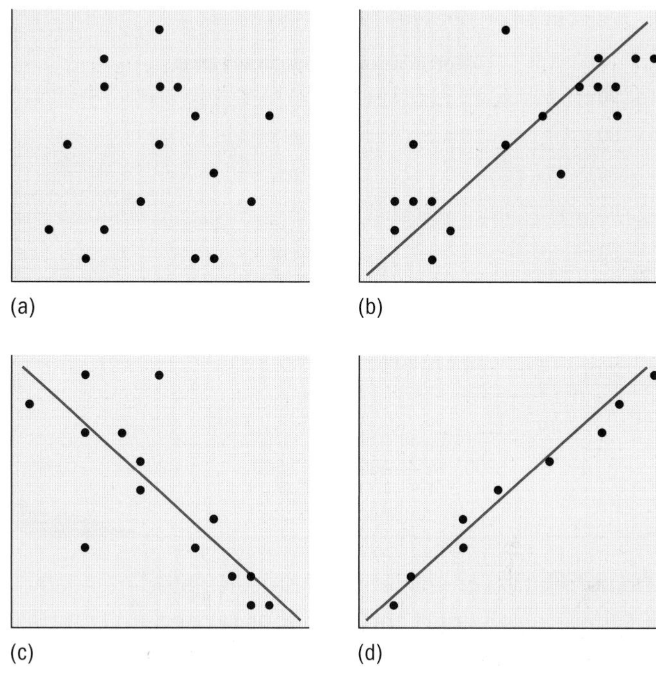

Different types of correlation showing lines of best fit for the data: **(a)** no correlation; **(b)** weak positive correlation; **(c)** weak negative correlation; **(d)** strong positive correlation.

- When most of the data points lie close to the line of best fit, there is **strong** correlation (see diagram d).
- When the data points are more widely scattered around the line of best fit, there is **weak** correlation (see diagrams b and c).

Correlation coefficient

Drawing a straight line through the data points of a graph is not precise enough to determine the strength of correlation. A correlation coefficient, R^2, is often used to do this. It is a measure of how far the plotted points lie from the line of best fit.

R^2 can lie between 0 and 1.

- When $R^2 = 1$, there is perfect correlation between the variables (all plotted points will be exactly on the line of best fit).
- The closer R^2 is to 0, the weaker the correlation will be.

Activity 8.2D

Look at the graph below and decide whether there is a correlation between length and mass.

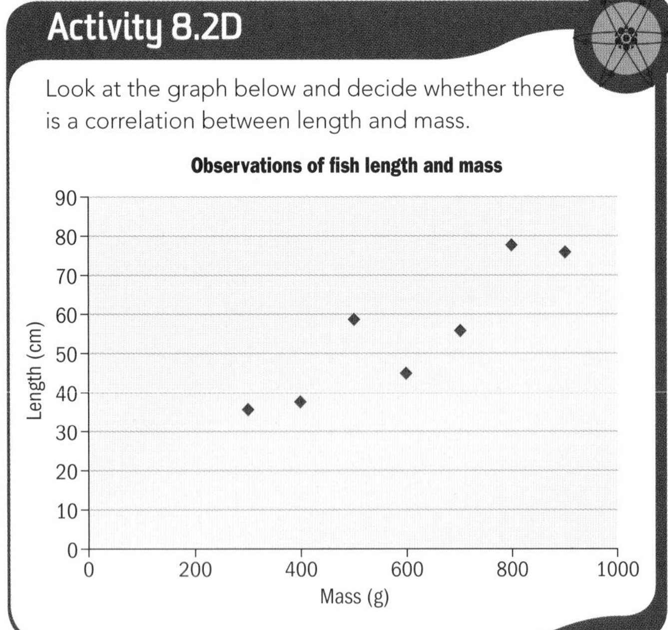

The correlation coefficient can be calculated manually, but this is a very long and laborious procedure. ICT methods enable correlation coefficients to be calculated much more easily.

Regression

When it is not easy to draw a line that best fits the data, but you know that one variable causes a change in the other, then the equation of the straight line or the equation of the curve needs to be determined; this is called **regression**.

The equation of a straight line is given by:

$$y = mx + c$$

where m is the slope (gradient) and c is the intercept on the y-axis (where $x = 0$). Note that the slope (m) can be positive or negative. Determining the equation of the line from the data is extremely important in science. The slope and intercept will correspond to physical quantities that may need to be determined from your experimental data.

You can use a calculator or a spreadsheet program to do this. Manual methods take a lot of time which could be better spent interpreting the data.

The linear regression of data set 2 (in the worked example) is shown on the graph. The equation of the straight line was obtained by right-clicking any data point on the graph, selecting **Add Trendline** and then clicking on **Linear**, followed by clicking on **Options** and then selecting **Display equation on chart**. The equation that fits the line is:

$y = 0.575x - 0.37$.

This states that the slope is 0.575 and the intercept on the y-axis is -0.37.

Worked example

Determine the correlation coefficient for each set of data displayed below.

Data set 1:

Temperature (°C)	Mass (g)
29	22
39	23
34	26
34	20
34	11
40	18
37	27
37	22
17	29
47	15

Data set 2:

Voltage (V)	Current (A)
2	0.8
4	1.8
6	3.1
8	4.5
10	5.2

Answer

Step 1: Plot a scatter graph of each data set using a spreadsheet program, e.g. in Microsoft Excel: choose **XY** (**Scatter**) in the list of chart options.

Remember that when plotting data, where possible you should first determine which variable is the independent variable and which variable is the dependent variable. The independent variable is always plotted on the x-axis and the dependent variable is plotted on the y-axis. For data set 1, temperature is the independent variable; for data set 2, voltage is the independent variable.

Step 2: To add a line of best fit and calculate R^2, right-click any data point on the graph and select **Add Trendline**. Make sure that **Linear** is selected then display the **Options** tab and select **Display R-squared value on chart**.

Step 3: A line of best fit to the graph is added to the graph and the value of R^2 is displayed.

continued

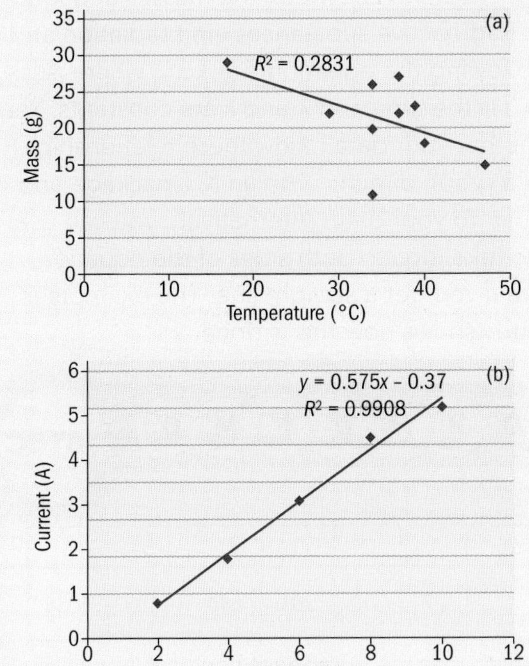

Graph of **(a)** data set 1 and **(b)** data set 2 showing the lines of best fit and the values of R^2. The graph for data set 2 also has its equation displayed (see 'Regression' page 188).

For data set 1, R^2= 0.2831 and for data set 2, R^2 = 0.9908.

This indicates there is no correlation between temperature and mass for data set 1, but good correlation between voltage and current for data set 2. In considering the validity of your data, consider the range of your data. For example, in data set 2, the range was only between 2 and 10 volts. Do you expect this relationship to continue at higher voltages such as 100 V? What about negative voltage values: would the same relationship exist? In data set 1, look at the range of mass values in the experiment: would you expect that there is no relationship for higher masses?

Case study: Radiography

Paul is a medical physicist working in the radiography department of a local hospital. He has carried out an experiment over the past 10 days to investigate the activity of a radioactive source. His data are shown in the table.

Time (days)	Activity (Bq)
0	90.0
1	33.1
2	12.2
3	4.48
4	1.65
5	0.606
6	0.223
7	0.0821
8	0.0302
9	0.0111
10	0.00409

Method 1: Use a spreadsheet program to plot the data, e.g. using Microsoft Excel select **XY** (**Scatter**) from the chart options. Follow the earlier procedure for adding a trendline and its equation but select **Exponential** this time. A fit will be drawn on the data and you should find that the equation: $y = 90.02e^{-1x}$ will be displayed. This is an exponential decay equation which appears in various fields in science.

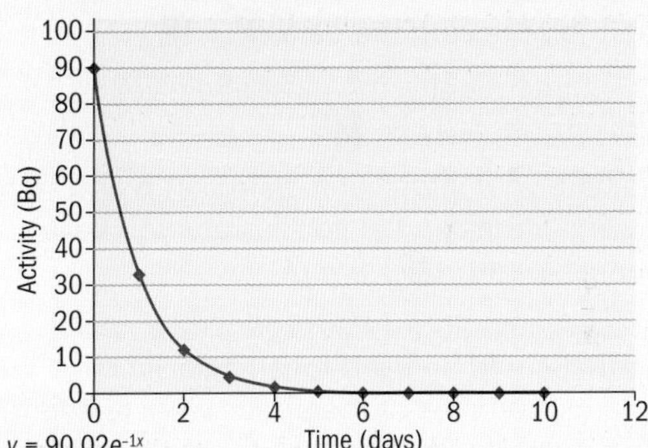

If you try a linear regression, what value of the correlation coefficient, R^2, will you get? (You may need to right-click and choose **Clear** to remove the previous trendline before you try this.)

Method 2: The activity is believed to follow the relationship:

$A = ke^{-\lambda t}$

Taking logs to base e on both sides gives:

$\log_e A = -\lambda t + \log_e k$

Comparing this to the equation of a straight line, $y = mx + C$, where m is the gradient and C is the intercept on the y-axis (when $x = 0$), we see that

$\log_e A = y$, $x = t$, $m = \lambda$ and $C = \log_e k$

So a plot of $\log_e A$ against t should give a straight line, with a negative slope.

What about non-linear graphs?

The data collected from science investigations can often be represented by a non-linear equation. Therefore, it is an important skill to be able to determine the equation of the non-linear graph that fits the data. These equations could be in the form of $y = kx^2$, $y = ke^x$ or $y = ke^{-x}$ where k is a numeric constant. For example, investigating the gravitational forces between masses, one finds that the force follows an inverse square law and can be written as: $F = kR^{-2}$, where R is the distance between the masses and k is a constant. Radioactive substances emit radiation and decay by following the equation $A = ke^{-\lambda t}$, where A is the activity, t is the time and λ and k are constants. You can test whether your data follow these relationships by plotting a graph directly onto an ICT package and applying a 'fit' to your data. Alternatively, you can use algebra and take logs to both sides of the equation, and observe if you get a straight line. The example in the case study shows how this is done.

Assessment activity 8.2

BTEC

1 You are working for a pharmaceutical lab that has developed a new drug that may treat people with a particular skin disease. In a pilot study of 500 patients, you are required to determine if there is any evidence to suggest that the new drug is more effective than the old drug.

	New drug	Old drug
cured	94	139
not cured	155	111

a Use the χ^2 test on the data. P1, P2

b Which treatment is the most effective? Explain your answer. M2

c Evaluate your interpretation in part **b** and state any limitations on the conclusions you have made in part **b**. D2

2 A medical technologist has produced a new plastic that could be used for artificial limb transplants. Samples of the plastic were tested for their breaking points, and then the samples were treated with a special coating (GPa) and retested.

breaking point before GPa	7.1	6.4	7	7.9	7.4	7.6
breaking point after GPa	9.3	8	6.5	9.8	8.6	8.3

a Use the appropriate *t*-test for the experiments. P3

b On the evidence of the results, does the coating strengthen the plastic? M3

c Evaluate your interpretation in part **b** and explain how reliable it is. D3

3 You are working for a biochemical company looking at growth of bacteria. You have recorded the number of bacteria on different days of the week.

Time/days	1	2	3	4	5	6
No. of bacteria	27	74	201	546	1485	4034

It is believed that the number of bacteria, N, varies exponentially (that is $N = ke^{\lambda t}$).

a Using ICT, prove that the number of bacteria does increase exponentially. P4

b Determine the value of the constants k and λ. What is the number of bacteria present at the start? M4

c Evaluate your interpretation of the results in part **b** and comment on their reliability. D4

Grading tips

To help you meet P2, M2 and D2, consider whether you are comparing two sets of data from the same experiment, or are you testing whether there is a relationship between two sets of data; are you testing for a contingency table?

For P3, think about what type of *t*-test you are going to use; do you have independent or paired samples? In considering the validity of your conclusions for D2, think about the assumptions: do you have a normal distribution? How big is your sample and will this affect the reliability of your results?

For P4, remember the independent variable is always plotted on the *x*-axis (in this case it is 'time'). Use logs to verify the relationship. In the interpretation of the data for M4, what does the intercept physically represent? The reliability of your conclusions for D4 will normally relate to your sample size (data points) and how close to a straight line you got on the graph (think about regression).

WorkSpace

Paul Hewson

Lecturer in Statistics, University of Plymouth

My responsibilities include lecturing (preparing material, teaching, marking) as well as research on statistical methods. However, the ability to interpret statistics is so 'in-demand' that my responsibilities also include consultancy and giving support to many other people, including researchers, businesses, local councils and hospital doctors.

I like the fact that there is no routine in my job. Some days I work alone writing complex programmes to run on modern multiprocessor computers. These programmes might solve problems such as predicting the number of people in each of 8,000 GP practices in England who might have a heart attack next year.

Most days I spend a lot of time meeting other people – ranging from students who want help with coursework, to hospital consultants who need guidance on data analysis for their research. I was recently looking to see if there were spatial patterns in allergy referrals for different types of allergy.

In my line of work you get the chance to look into everyone else's business which is fascinating! I also enjoy meeting a wide variety of people. It is very rewarding having a core set of skills which can contribute to solving so many different types of problems.

I am currently working with two Year 12 Nuffield placement students. We are looking at how health organisations determine whether an area has a problem with 'acute malnutrition'. Some of the methods that are used to do the data analysis may underestimate the true extent of malnutrition. The Nuffield students are also thinking about other ways in which the analysis could be inaccurate – one student suggested larger families might be more likely to suffer from malnutrition and this would bias the results. We are examining how misleading such a bias could be.

Think about it!

1 Why do you think that being able to analyse data is so important?

2 What data do you think Google, Facebook, BeBo and others might hold on you? Who do you think would be interested in that data and why?

Just checking

1 State the equation that is used to calculate: (a) the mean and (b) the standard deviation.
2 In probability, what does *P*(*XUY*) mean?
3 What is the additional rule of probability?
4 What is the difference between continuous and discrete data? Give an example of each.
5 Write down an expression for the standard normal distribution.
6 Write down the equation for the χ^2 test. Give an example in science that uses the *t*-test.
7 Write down the equation of a straight line.
8 What is meant by 'non-linear relationship'?
9 Give an example in science that uses a power relationship.
10 Write an expression for this power relationship.

Assignment tips

- Remember that using a calculator and/or ICT to calculate standard deviation is part of meeting the grading criteria. Practise using a scientific calculator before completing the assignment.
- When calculating mean, mode and median, ask yourself: does the answer make sense? This is a useful way of checking your answer. Answers only tend to make sense if they are correct!
- When using the statistical tests (χ^2 and *t*-test) don't forget to show all your workings out.
- When plotting graphs, make sure that the axes are fully labelled and that the correct units are shown. Use standard symbols to represent physical quantities; for example, *t* is used for time, not *T* (which is period), and *v* is used for velocity, not *V* (which is volume).
- When determining the correlation coefficient, make sure you show all your steps; it may be a good idea to 'print screen' the worksheet if you are using Excel.
- When attempting the merit and distinction criteria, you need to show that you have interpreted your data correctly and thought about whether your interpretation is reliable.

You may find the following websites useful as you work through this unit.

For information on...	Visit...
a useful calculator for using the χ^2 test	The Free Statistics Calculators website
tutorials for various statistical tests	Statistics Tutorial: Hypothesis Tests

Credit value: 10

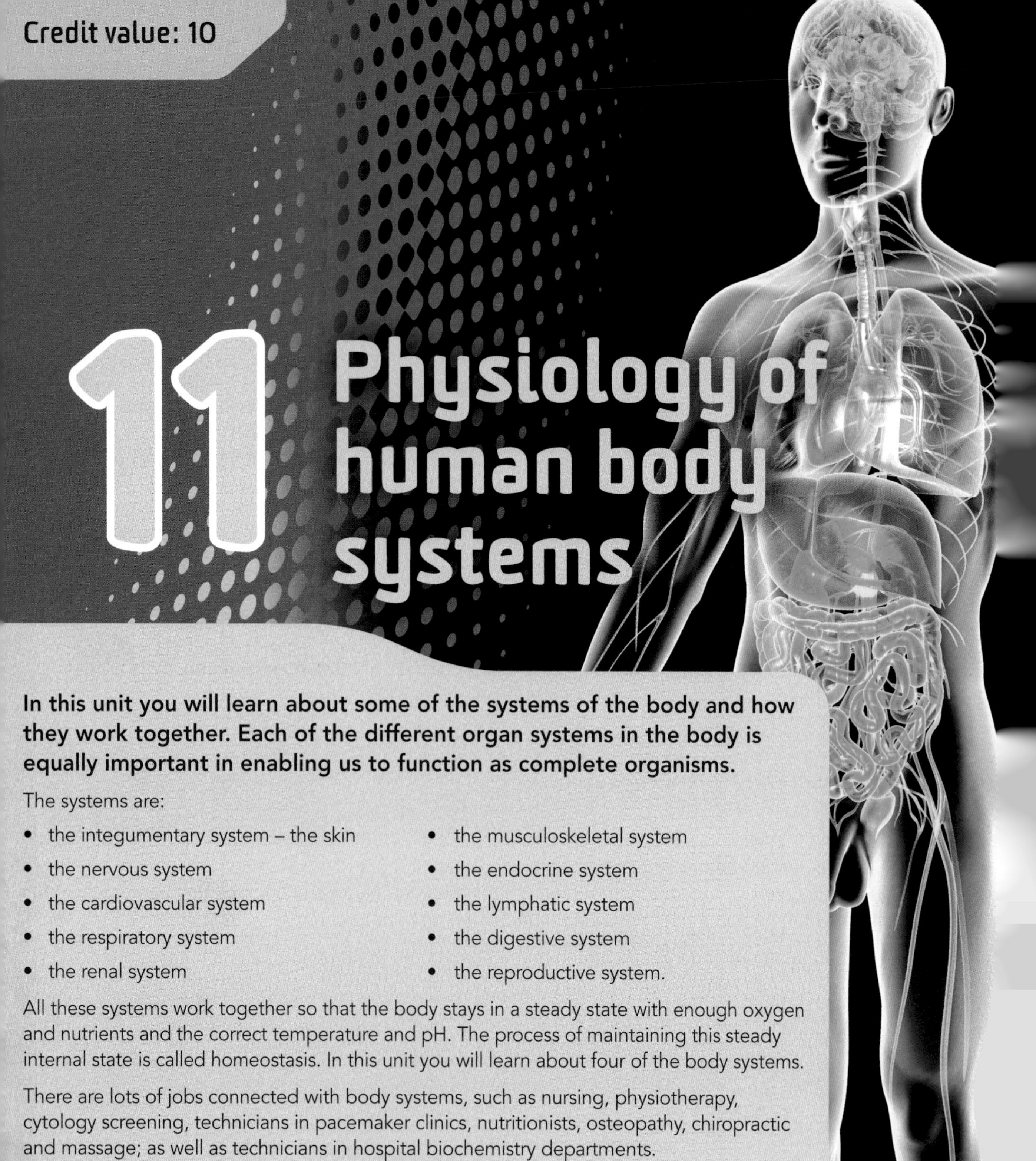

11 Physiology of human body systems

In this unit you will learn about some of the systems of the body and how they work together. Each of the different organ systems in the body is equally important in enabling us to function as complete organisms.

The systems are:

- the integumentary system – the skin
- the nervous system
- the cardiovascular system
- the respiratory system
- the renal system
- the musculoskeletal system
- the endocrine system
- the lymphatic system
- the digestive system
- the reproductive system.

All these systems work together so that the body stays in a steady state with enough oxygen and nutrients and the correct temperature and pH. The process of maintaining this steady internal state is called homeostasis. In this unit you will learn about four of the body systems.

There are lots of jobs connected with body systems, such as nursing, physiotherapy, cytology screening, technicians in pacemaker clinics, nutritionists, osteopathy, chiropractic and massage; as well as technicians in hospital biochemistry departments.

Learning outcomes

After completing this unit you should:

1 know the levels of organisation within the human body

2 be able to relate the structure of the circulatory system to its function in a multi-cellular organism

3 be able to relate the structure of the respiratory system to its function

4 be able to relate the structure of the digestive system to its function

5 understand the immunological function of the lymphatic system.

Assessment and grading criteria

This table shows you what you must do in order to achieve a pass, merit or distinction grade, and where you can find activities in this book to help you.

To achieve a **pass** grade the evidence must show that you are able to:	To achieve a **merit** grade the evidence must show that, in addition to the pass criteria, you are able to:	To achieve a **distinction** grade the evidence must show that, in addition to the pass and merit criteria, you are able to:
P1 describe the organisation of the eukaryotic cell in terms of the functions of the organelles **See Assessment activity 11.1**	**M1** use diagrams or micrographs to compare and contrast the four tissue types **See Assessment activity 11.1**	**D1** explain the relationship between cells, tissues, organs and organ systems in the organisation of the human body **See Assessment activity 11.1**
P2 describe the four different tissue types **See Assessment activity 11.1**		
P3 take measurements related to the cardiovascular system, relating the results to the function of the cardiovascular system **See Assessment activity 11.2**	**M2** explain the need for transport systems in a multi-cellular organism **See Assessment activity 11.2**	**D2** explain how the digestive, cardiovascular and respiratory systems are interrelated **See Assessment activity 11.4**
P4 take measurements related to the respiratory system, relating the results to the function of the respiratory system **See Assessment activity 11.3**	**M3** explain the need for ventilation systems in a multi-cellular organism **See Assessment activity 11.3**	
P5 use appropriate chemical tests to identify different dietary nutrients **See Assessment activity 11.4**	**M4** use chemical equations to show how the main food groups are dealt with in the digestive system **See Assessment activity 11.4**	
P6 explain how these dietary nutrients are processed through the digestive system **See Assessment activity 11.4**		
P7 describe the structure and purpose of the lymphatic system **See Assessment activity 11.5**	**M5** explain how the lymphatic system protects the body **See Assessment activity 11.5**	**D3** explain the difference in lymphatic function in health and disease state **See Assessment activity 11.5**

How you will be assessed

Your assessment could be in the form of:

- a poster, e.g. explaining the need for a transport system in a multi-cellular organism
- a laboratory practical, e.g. investigating the nutritional content of foods
- a model, e.g. to show how the lymphatic system protects the body
- a leaflet, e.g. to illustrate different tissue types and their features.

Jason, 18 years old

My name is Jason. I want to be a children's nurse. If I do well enough on the BTEC National course I will be able to study for a degree in nursing. I have enjoyed Unit 11 because I have learnt about how the body works. I especially enjoyed the section where we took measurements about blood pressure and pulse rate and saw how the readings change when we exercise. I have learnt about the different body systems and how they all work together to keep the body functioning properly. The assignments have helped me because I have done research and found things out for myself. Giving presentations to the rest of the class has helped build my confidence. We have had guest speakers who have told us about the jobs they do in the health service. I am hoping to get some work experience in a hospital where I can find out more about being a children's nurse.

Catalyst

All systems go

Match each system with the correct function

System	Function
1 Integumentary	A Links outside air to the blood. In lungs oxygen diffuses into blood and carbon dioxide diffuses from the blood.
2 Musculoskeletal	B For carrying oxygen and nutrients to cells so they can carry out chemical reactions; transports toxic waste away from cells. The blood is also involved in protecting us from infectious diseases.
3 Nervous	C Food is broken down into smaller molecules that can pass across the gut wall and enter the blood to be carried to all cells.
4 Endocrine	D Enables us to make appropriate responses to stimuli and to modify behaviour with experience – to learn.
5 Cardiovascular	E This protects us from mechanical damage, damage from chemicals, entry of pathogens, is waterproof and stops us from drying out, and helps keep the body temperature stable.
6 Lymphatic	F Supports us, gives anchorage to all the soft organs inside the body, enables us to move and protects vital organs. Bone marrow stores fat and also makes blood cells. Bones also store calcium and other minerals, which can be released into the blood to go to parts of the body as needed.
7 Respiratory	G Regulates growth, development and metabolism (all the chemical reactions that go on in cells and keep us alive).
8 Digestive	H Removes toxic waste such as urea and helps regulate the salt and water content and pH of the body.
9 Reproductive	I Consists of vessels that take escaped fluids back to the blood and of lymphoid organs that play a crucial role in defence against infectious diseases.
10 Renal	J Enables us to produce children and pass on our genes to the next generation.

11.1 Organisation within the human body

In this section:    

Key terms

Magnification – the number of times larger an image appears, compared with the object's actual size.

Resolution – the ability to distinguish two separate points as distinct from each other (to see them clearly as two objects).

Ultrastructure – the fine detailed structure of the inside of cells as seen with electron microscopes.

Organelles – the small, distinct structures inside cells. Each has a specific function.

Tissue – a collection of cells that work together to perform a particular function.

Organ – a collection of tissues that work together to perform a particular function.

Organ system – a group of organs that work together to accomplish a common purpose – an overall life function.

Surface area to volume ratio – the surface area divided by the volume. Small objects have a large surface area to volume ratio and the larger the object the smaller the surface area to volume ratio.

Gland – one or more cells that make a particular product, called a secretion.

Connective tissue – the most diverse, abundant and widely distributed type of tissue in the body. Its functions include binding and support, protection, insulation, movement, and transportation of substances within the body.

Matrix – substance between cells (plural: matrices)

Tendon – tough fibrous tissue that joins muscle to bone.

Light and electron microscopes

Humans have used lenses to magnify objects for about 800 years, firstly as spectacles and telescopes. The first light microscope was made about 400 years ago. From the late seventeenth century scientists started to use them to look at living organisms or parts of them.

Did you know?

In the mid-1600s Anton van Leeuwenhoek (pronounced *lay van hook*) used magnifying glasses to count threads in cloth. He ground and polished lenses to give a magnification of ×270. He then built a simple light microscope, with one lens, and looked at very small organisms living in water.

Light microscopes

Compound light microscopes have the features shown in the diagram.

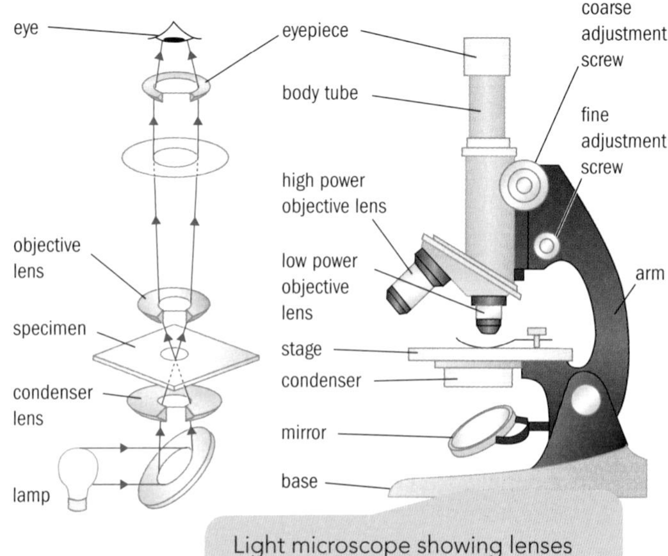

Light microscope showing lenses and light paths.

Some have a light source underneath the stage and some use a lamp directed onto a mirror. The specimen is mounted on a small glass slide with a glass coverslip and placed on the stage so it is over a central hole. Light passes through the condenser lens, through the specimen and is focused by the objective lens and the eyepiece lens.

Magnification

There are three or four objective lenses and each magnifies the object by a different amount, e.g. ×4; ×10; ×40; ×100, and produces a magnified image.

The eyepiece lens magnifies the image again, so if a ×10 eyepiece is in place then the total **magnification** with a ×4 objective lens is 4 × 10 = ×40.

This means you see an image of the object which is 40 times larger than the actual size of the object. A camera can take photographs of the image and this is called a photomicrograph.

Activity 11.1A

If you look at a specimen under the microscope with a ×10 eyepiece and a ×40 objective lens, what is the total magnification of the object?

If you now use a ×15 eyepiece but the same objective, what is the total magnification?

Light microscopes are relatively cheap and easy to use. However they can only magnify objects clearly up to about ×1500 (higher if blue light is used).

Resolution

Resolution refers to the clarity of the image you see. Because light travels in waves and has a wavelength of about 500 nm (0.0005 mm), then any object smaller than 200 nm is seen as a blur. Also, if two small objects are closer together than 200 nm then you would see them as one blurred object. The resolution of light microscopes is 200 nm.

So although light microscopes enabled scientists to see tissues and cells and some of the structures inside cells, they had limited magnifying ability due to their low resolution. The finer details of cell structure have only been seen since the 1930s when electron microscopes were developed.

Electron microscopes

These microscopes use a beam of electrons instead of a beam of light. High voltage makes the electrons travel fast so they have a wavelength of 0.005 nm – about 100 000 times smaller than that of light. This enables a much higher resolution. For biological specimens the resolution is not as high as theoretically possible because the electromagnetic lenses are not perfect, but it is about 1 nm. This means you can see objects 1 nm in diameter clearly when images are magnified up to 200 000 times.

Activity 11.1B

How many times greater is the resolving power of an electron microscope, compared with a light microscope?

Transmission electron microscopes

The electrons pass through a very thin prepared sample. They pass through less dense parts more easily than through dense parts. The electron beam is focused by magnets onto a screen, photographic paper, or a type of digital camera to give an electron micrograph. In either case the 2D image is in shades of dark to light grey.

Scanning electron microscopes

A fine beam of electrons is scanned over the surface and bounced off to make an image on a screen. This has a lower resolution but a great depth of focus so the image is in 3D. Technicians can add colour to the image.

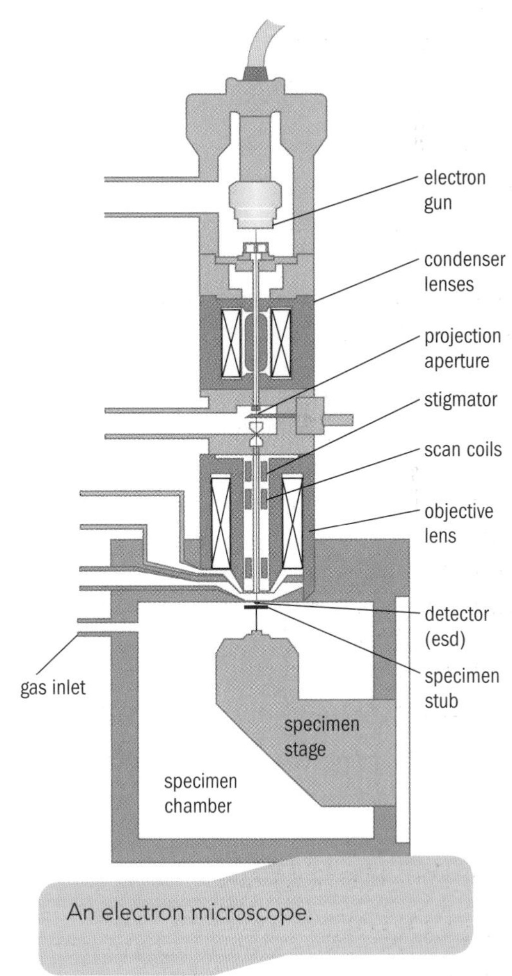

An electron microscope.

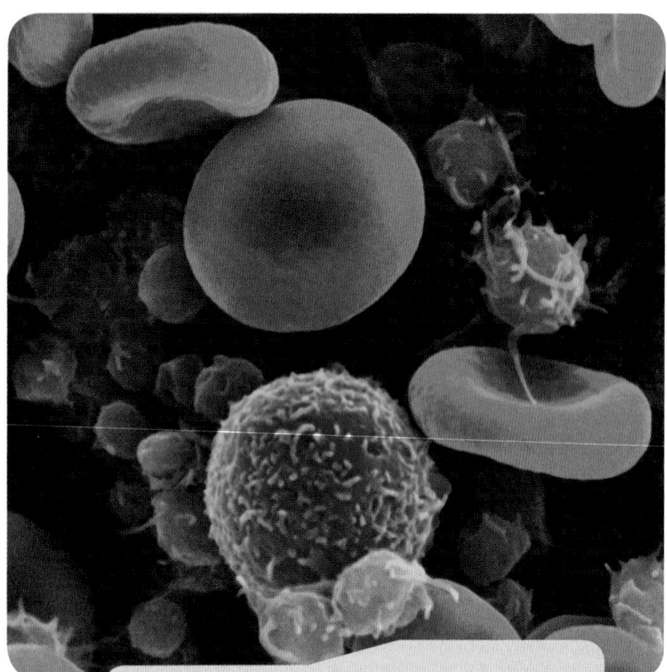

False-colour scanning electron micrograph of blood cells. Note the red blood cells (red), the white blood cell (blue) and the thrombocytes (yellow).

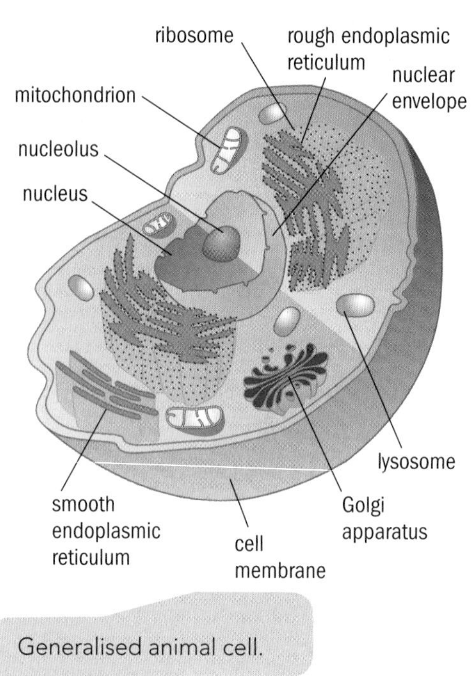

Generalised animal cell.

Molecules in air interfere with electron beams so specimens have to be placed in a vacuum. Electron microscopes are very expensive and a lot of skill and training is needed in order to be able to use one. Specimens have to be stained with heavy metal salts, mounted on a copper grid and placed in a vacuum.

The ultrastructure of an animal cell

We all begin life as one cell. It divides into a ball of cells and then some become blood, bone, nerves, muscle, skin and internal organs. Each type of cell in these tissues is specialised to carry out particular functions. However they all have certain features in common. When biologists describe these features they talk about a generalised animal cell. This is just a way of describing the **organelles** that can be found in animal cells.

Cells have a 3D structure

When you look at cells through a microscope you see them cut into thin sections or flattened on the slide and they appear two-dimensional. However, they have depth as well as length and breadth.

Each cell is surrounded by a plasma membrane. Animal cells are eukaryotic, which means they have a nucleus bound by a nuclear envelope (double membrane). Outside the nucleus and within the plasma membrane is granular cytoplasm and within the cytoplasm are the other organelles.

Division of labour

Each organelle has a specific function. Nearly all of the organelles have membranes around them which separates them from other parts of the cell. Each has enzymes so it can carry out its specific biochemical reactions. All of these reactions keep us alive.

The table below briefly describes the functions of the main organelles in an animal cell.

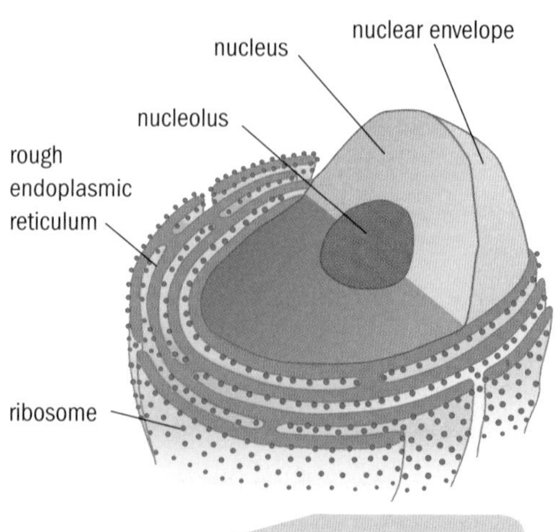

Nucleus and endoplasmic reticulum.

Organelle	Function
plasma membrane	Made of phospholipids and protein. It keeps each cell separate from the outside environment. It regulates transport of substances into and out of the cell. It has antigens (special proteins) on the surface so the immune system can recognise it as 'self'.
cytoplasm	The matrix of the cell in which organelles are embedded. It contains water (about 70%) and proteins, some of which form a supporting network of microtubules. Some proteins are contractile; these proteins help to move organelles within the cell.
nucleus	Houses the genetic material, DNA, of the cell, in the form of chromatin. It has a double membrane or envelope. There are pores in the envelope so that some substances can pass into or out of the nucleus.
nucleolus	A dense area in the nucleus, without a membrane around it, where RNA and ribosomes, both needed for protein synthesis, are made.
endoplasmic reticulum (ER)	A series of flattened membrane-bound sacs called cisternae. Rough ER is covered with ribosomes and is where proteins are assembled. The proteins pass into the cisternae and travel to the Golgi apparatus. Smooth ER does not have ribosomes. Here, steroids and lipids are synthesised.
Golgi apparatus	Here, proteins made at ribosomes are modified and packaged into vesicles or lysosomes.
vesicles	Membrane-bound sacs derived from Golgi apparatus. They may hold particular chemicals or may carry proteins to the plasma membrane to be released outside the cell.
lysosomes	Large spherical organelles derived from Golgi apparatus and containing digestive enzymes, some of which may break down invading microorganisms. Keeping these enzymes inside the lysosome prevents them from breaking down the cell.
mitochondria	Spherical or sausage-shaped organelles with an inner and outer membrane. Here the stages of aerobic respiration occur.
ribosomes	Tiny organelles, made of two subunits and having no membrane around them. They are made of RNA and protein. Here, proteins are assembled.
centrioles	Small tubes of protein fibres. There is no membrane around them but they are in an area of the cell called a centrosome. They form the spindle when the nuclei of cells divide in mitosis and meiosis. Some centrioles form cilia on the surface of certain cells.

Levels of organisation

Some organisms consist of only one cell. They are small and therefore have a large **surface area to volume ratio** and can obtain oxygen and nutrients through their plasma membranes. Inside the cell the organelles carry out all the functions that keep the cell alive. All living things carry out life processes. They move, excrete waste, respond to stimuli, obtain nutrients, respire, grow and reproduce.

Differentiation and specialisation

Humans are multi-cellular organisms. They consist of many cells. They are larger and have a smaller surface area to volume ratio. This means that many cells are not in direct contact with the outside environment so cells become specialised to perform specific functions such as:

- carrying oxygen
- removing toxic waste
- dealing with invading pathogens
- communication
- movement
- covering the body surface and lining its cavities.

We all start as one cell – a zygote (fertilised egg). This cell has the ability to become any kind of cell. It is a stem cell. Its nucleus contains 46 chromosomes, containing altogether about 20000 genes. As it divides again and again, a ball of cells forms. These cells pass over a special area, within the developing embryo, and certain genes get switched on or off. As a result the cells become different from each other. They are each specialised to carry out a certain function. They also become organised into **tissues**. A tissue is a group of similar cells that works together to perform a specific function.

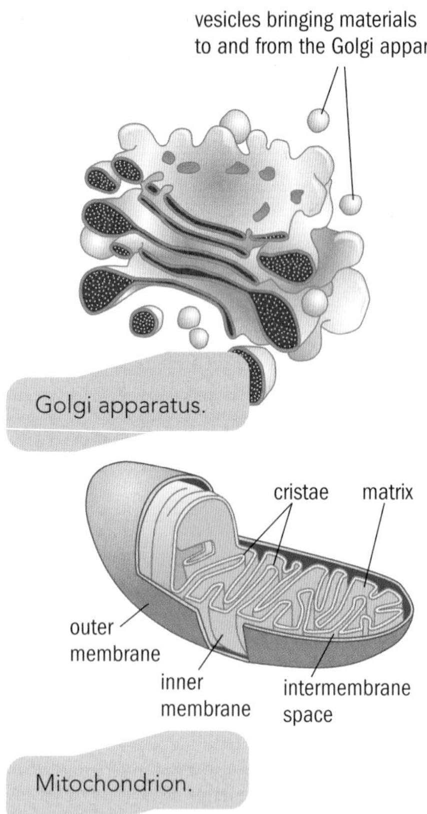

Golgi apparatus.

Mitochondrion.

Organ systems

Your **organs**, such as brain, eyes, spinal cord, heart, lungs, liver, stomach, intestines, pancreas, kidneys, ovaries and bladder, are inside body cavities. Your muscles and bones are also individual organs. Each organ may contain more than one tissue type. The study of tissues is called histology. The shape and structure of each organ is its anatomy. Knowing about histology and anatomy of an organ helps scientists understand how it works – its physiology.

Organs are organised into systems. Each system carries out a life process.

It is impossible to say which organ or system in the body has most importance. Certainly we can live without the appendix and without a spleen. You might think the liver is most important as it carries out a great many functions, but all systems are interrelated.

Did you know?

Animal, plant, fungi and protoctist cells are eukaryotic. Bacterial cells are prokaryotic because they do not have a nucleus or any membrane-bound organelles or centrioles. They have smaller ribosomes than eukaryotic cells. You may learn more about these in Unit 15. Viruses don't have cells and are described as akaryotic.

Your blood carries oxygen to cells for respiration, but if your lungs aren't working, there won't be enough oxygen in the blood. If your heart stops working, cells will not receive nutrients and oxygen and will die. If your kidneys fail, the build up of toxins in the blood will cause other organs to fail. If you lose a lot of skin, for example by burning, then your body will lose fluid and become infected, so other systems will fail.

Activity 11.1C

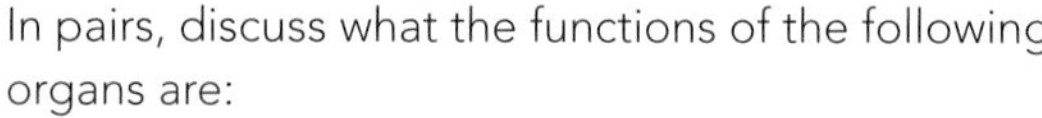

In pairs, discuss what the functions of the following organs are:

heart, stomach, small intestine, pancreas, kidney, bladder, biceps muscle, eye.

Did you know?

Your skin is your largest organ. It weighs about 4.5 kg.

Tissue types

There are four basic tissue types in the human body. They are:

- epithelium
- muscle
- connective tissue
- nervous tissue.

The four tissue types interweave to form the fabric of the body.

Case study: Cytology screener

Jane works in a hospital as a cytology screener. (Cytology is the study of cells.) She looks down a microscope at slides of cervical smears to see if any of the cells look abnormal. Cells that have larger nuclei and a different shape from usual may be becoming cancerous.

Cytology screeners spend many hours peering at cells down a microscope. How do you think the chances of making a mistake and missing a cancerous cell could be reduced?

Epithelial tissue

There are two types:

- Covering and lining epithelium, found on all surfaces such as skin, covering the walls of body cavities, covering all organs and lining the blood vessels (capillaries, arteries and veins) and heart. The walls of the alveoli in lungs are made of one layer of flattened epithelial cells.
- Glandular epithelium makes up the **glands** such as sweat glands in skin, salivary glands, goblet cells in the lining of the respiratory tract, mammary glands and intestinal glands.

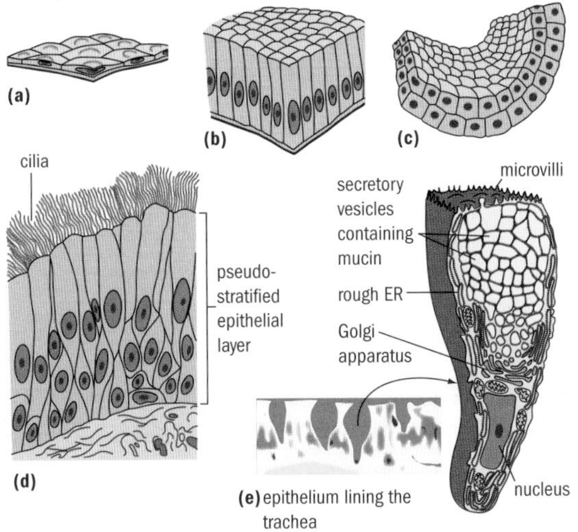

Some types of epithelial tissue. (a) Squamous epithelium: forms walls of alveoli and capillaries and linings of other blood vessels. (b) Simple cuboidal (cube-shaped) epithelium – this lines most of the digestive tract. (c) Stratified (layered) cuboidal epithelium – this forms the ducts of the salivary glands. (d) Pseudostratified (false-layered) ciliated (having cilia) columnar (column-shaped) epithelium lining the trachea. There is only one layer of cells but, because the nuclei are at different positions, it appears that there is more than one layer. (e) One goblet cell from the epithelium of the trachea. This is a glandular cell. Note the large amounts of rough ER and ribosomes. This is where the mucin proteins are assembled. The Golgi apparatus then packages the mucin molecules into secretory vesicles.

Muscle tissue

There are three types of muscle tissue:

- Cardiac muscle is found only in the heart. It contracts to make the heart beat but is *not* under voluntary control. It contracts at a steady rate set by the heart's pacemaker but certain nerves can speed or slow the pacemaker when necessary.
- Skeletal muscle is found in the muscles attached to bones. We can voluntarily control their contraction and relaxation, although they sometimes contract in response to reflexes. This type looks striped (striated) when seen under the microscope.
- Smooth muscle is found in the walls of hollow organs such as the stomach and other organs of the gastrointestinal tract, and bladder, arteries, veins and lymph ducts, and in the respiratory passages. It does not have any striations and is not under voluntary control – it has involuntary innervation.

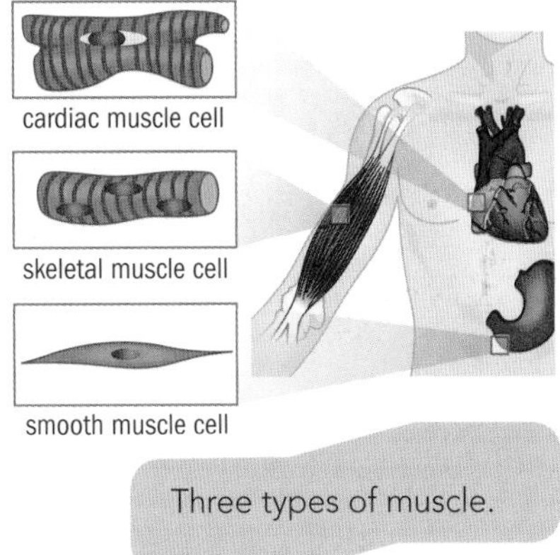

Three types of muscle.

Muscle tissue has four main functions:

- maintaining posture
- stabilising joints
- movement
- producing heat.

Muscle tissue has a good blood supply.

Did you know?

Your muscles make up nearly half of your body mass.

Nervous tissue

There are two major cell types:

- Neurones – conduct nerve impulses.
- Supporting cells, such as Schwann cells, found in the peripheral nervous system. Here they form the myelin sheaths that insulate the neurones and neuroglia or glial cells that are in the brain and spinal cord. Glial cells can divide.

Activity 11.1D

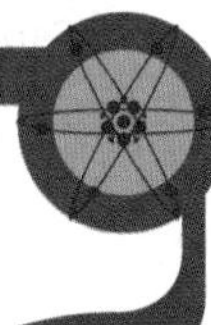

Most brain tumours are gliomas (consisting of glial cells). Why do you think this is the case?

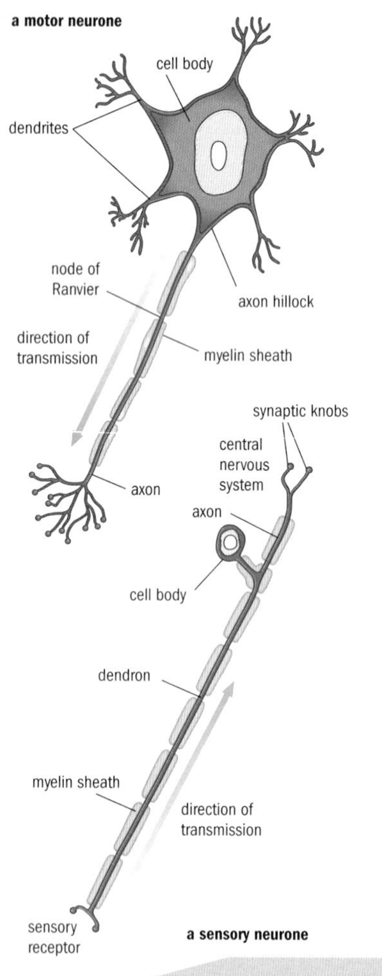

A sensory neurone and a motor neurone.

Connective tissue

This is found everywhere in the body. There are four types, **cartilage, bone, blood** and **connective tissue proper**. All are derived from the same embryonic tissue and are very diverse. All have three parts

- **matrix** (unstructured material that fills spaces between cells)
- fibres (collagen, elastic and reticular fibres) in the matrix
- cells that make the matrix and the fibres in it.

There are different cell types which make different matrices and fibres and hence make different tissues:

- Fibroblasts make connective tissue proper – areolar, adipose, reticular, collagenous and elastic tissues.
- Chondroblasts make the matrix for cartilage.
- Osteoblasts make the matrix for bone.
- Haemocytoblasts (haematopoietic stem cells) make blood.

Once these cells have secreted the matrix they become less active and maintain the matrix.

Most of an embryo's skeleton is cartilage. Adults have cartilage covering the ends of our bones, between ribs and sternum, covering the discs between vertebrae, at the end of our nose and in our ear lobes. Cartilage matrix contains collagen fibres.

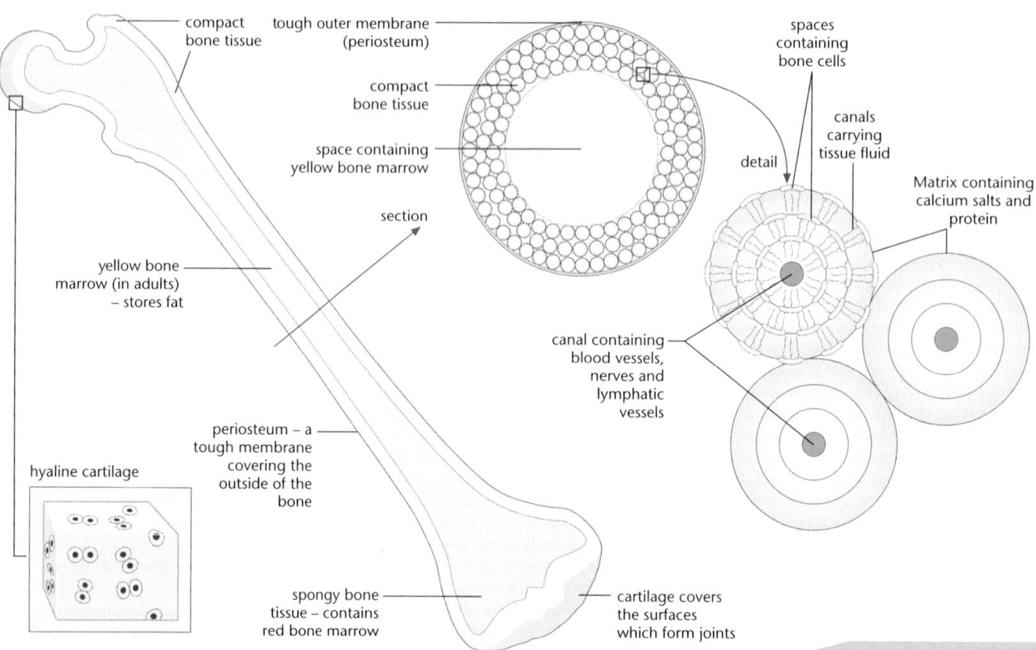

Section through a long bone showing the distribution and structures of cartilage and compact bone.

The embryonic skeleton becomes bone as the matrix becomes hardened with calcium salts. It also contains collagen fibres. Our skeleton protects organs, supports us and the joints allow us to move.

Blood consists of red and white blood cells and thrombocytes, in a fluid matrix (the plasma). It is a transport system (see page 205) and helps protect us from infections.

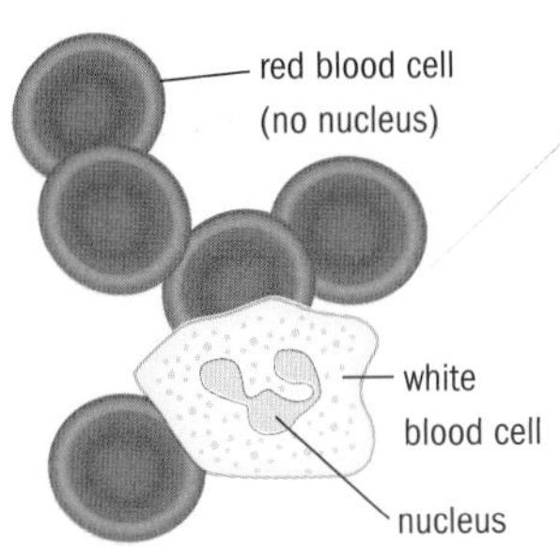

Blood is a type of connective tissue.

There are two sub-groups of connective tissue proper – loose (areolar, adipose and reticular) and dense (collagenous and elastic). Dense regular connective tissue forms **tendons** that join muscle to bone.

The main cell type is fibroblasts, which secrete the matrix. There are also fibres, which may be:

- collagen – a fibrous (string-like) protein
- elastin – a coiled protein that can stretch and recoil
- reticular – a different type of collagen that forms a network.

Activity 11.1E

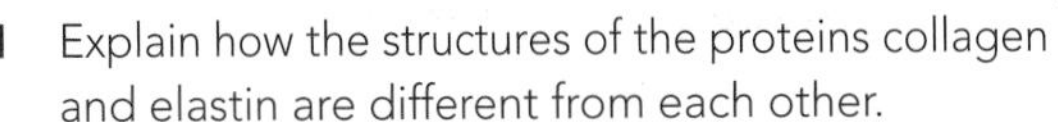

1 Explain how the structures of the proteins collagen and elastin are different from each other.

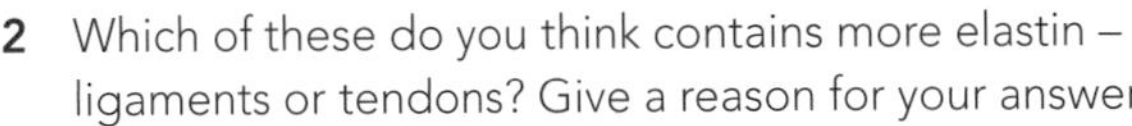

2 Which of these do you think contains more elastin – ligaments or tendons? Give a reason for your answer.

Did you know?

You have 206 bones in your body. More than half of these (106) are in your hands and feet. When you were born you had over 300 bones but some fuse together as you grow. A giraffe has seven bones in its neck, the same number that you have.

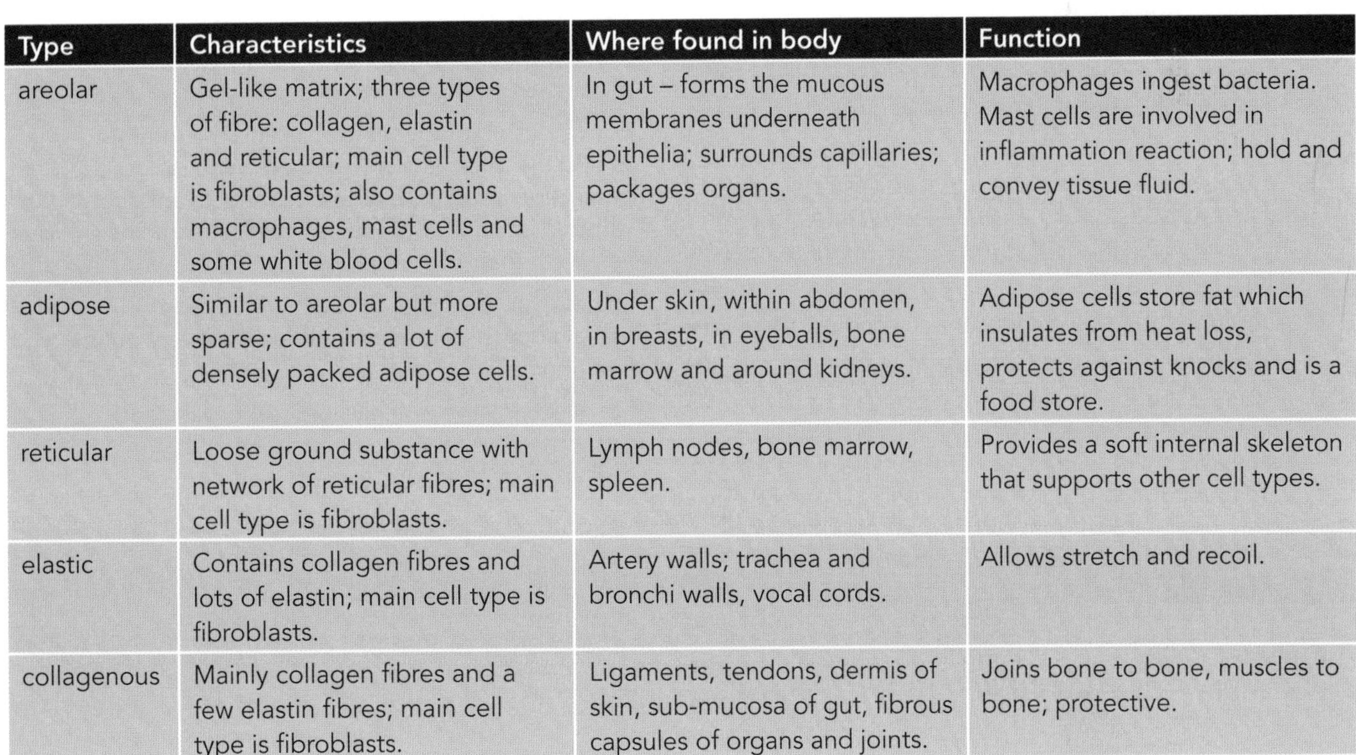

Type	Characteristics	Where found in body	Function
areolar	Gel-like matrix; three types of fibre: collagen, elastin and reticular; main cell type is fibroblasts; also contains macrophages, mast cells and some white blood cells.	In gut – forms the mucous membranes underneath epithelia; surrounds capillaries; packages organs.	Macrophages ingest bacteria. Mast cells are involved in inflammation reaction; hold and convey tissue fluid.
adipose	Similar to areolar but more sparse; contains a lot of densely packed adipose cells.	Under skin, within abdomen, in breasts, in eyeballs, bone marrow and around kidneys.	Adipose cells store fat which insulates from heat loss, protects against knocks and is a food store.
reticular	Loose ground substance with network of reticular fibres; main cell type is fibroblasts.	Lymph nodes, bone marrow, spleen.	Provides a soft internal skeleton that supports other cell types.
elastic	Contains collagen fibres and lots of elastin; main cell type is fibroblasts.	Artery walls; trachea and bronchi walls, vocal cords.	Allows stretch and recoil.
collagenous	Mainly collagen fibres and a few elastin fibres; main cell type is fibroblasts.	Ligaments, tendons, dermis of skin, sub-mucosa of gut, fibrous capsules of organs and joints.	Joins bone to bone, muscles to bone; protective.

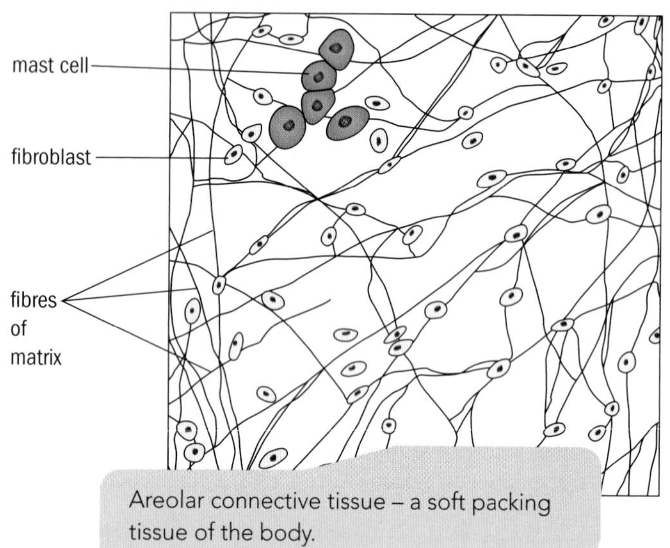

Areolar connective tissue – a soft packing tissue of the body.

PLTS

Independent enquirer and Reflective learner

You will develop your skills as an independent enquirer when researching and as a reflective learner when selecting relevant information.

Assessment activity 11.1

BTEC

You are a public awareness officer working in a tissue engineering department.

You need to convey information to visitors about levels of organisation in the body.

1 Draw a large annotated diagram of a typical (generalised) animal cell to show its ultrastructure. Label each organelle and indicate its function. Indicate which level of organisation this describes. P1

   You also need to produce a leaflet illustrating the different types of tissues and their features. You have already started this leaflet. You need to now add more.

2 Describe the features of connective tissue. P2

3 Use diagrams and micrographs to compare and contrast the four tissue types (epithelial, muscle, nervous and connective). Make a large poster-sized table to show the features of each type, where they are found in the body, their function and how their structure enables them to carry out their function. Include labelled diagrams of examples of each tissue. M1

4 Draw a flow diagram to explain the relationship between cells, tissues, organs and organ systems in the organisation of the human body. D1

Grading tips

For P1 think of: organelles → cells → tissues → organs → organ systems → organism.

For P2 use biology textbooks and the Internet to find the information you need. Write a paragraph about one type of connective tissue: where it is found, what type of cells it contains and what their functions are.

To help you gain M1, for each tissue type explain how its structure enables it to carry out its function.

For D1 explain how cells, tissues, organs and organ systems relate to each other so that the human body functions as a whole.

11.2 The circulatory system

In this section: 

Key terms

Diffusion – movement of substances from a region of high to low concentration. Diffusion may be through a membrane. It does not use energy from ATP.

Gaseous exchange – exchange of oxygen and carbon dioxide at a respiratory surface. For example, at the lungs oxygen enters the bloodstream and carbon dioxide enters alveoli.

Thoracic cavity – chest cavity.

Myocardium – main layer of heart wall; made of cardiac muscle.

Atria – top two chambers of the heart. Blood enters these from the body or lungs.

Ventricles – bottom two chambers of the heart. Blood leaves these to go to the lungs or body.

Valves – allow one-way movement. In the heart there are four valves: atrioventricular between atria and ventricles; semilunar at the base of each artery leaving the heart.

Systemic circulation – the route blood takes, in vessels, from heart to body tissues and back to heart.

Pulmonary circulation – the route blood takes, in vessels, from heart to lungs and back to heart.

Lumen – the space inside a blood vessel.

Living organisms need a supply of nutrients and many also need oxygen. They also have to remove their toxic waste products.

Why do we need a transport system?

Single-celled organisms are small, which means that their surface area is large compared with their volume; they have a large surface area to volume ratio. Hence, they can obtain substances by **diffusion** through their relatively large plasma membrane. These substances have to diffuse only short distances so can diffuse at a faster rate and meet the organism's needs.

Multi-cellular organisms have a much smaller surface area to volume ratio. Many of their cells are not in contact with their surroundings so they cannot solely rely on diffusion to supply all their organs with oxygen and nutrients, as the distance from their surface to all cells is too far. We are multi-cellular and have special surfaces for **gaseous exchange** and for obtaining nutrients.

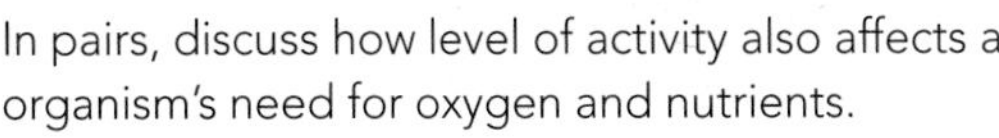

Activity 11.2A

In pairs, discuss how level of activity also affects an organism's need for oxygen and nutrients.

There is also a special transport system, the blood.

Structure and functions of blood

Blood is a tissue. The plasma is the fluid matrix and the cells float in it.

There are also plasma proteins in the blood. These are part of the blood; they are not nutrients being carried.

Our blood is always in vessels (arteries, arterioles, capillaries, venules and veins) where it circulates due to the pumping action of the heart. The heart, blood vessels and blood make up the cardiovascular system.

Blood carries:

- oxygen from the lungs to all cells of the body so they can respire aerobically
- carbon dioxide from respiring cells to the lungs to be breathed out
- nutrients from the intestines to all cells of the body
- urea from excess protein in the diet, from the liver to the kidneys to be removed
- hormones from endocrine glands to their target tissues
- heat from respiring tissues to the organs to keep them warm, or to skin to be lost.

Blood helps to regulate:

- the body temperature – by absorbing and distributing heat
- the pH in body tissues – some blood proteins act as buffers and keep the pH stable
- the volume of fluid in circulation – salts and blood proteins prevent loss of excess fluid into the spaces between cells.

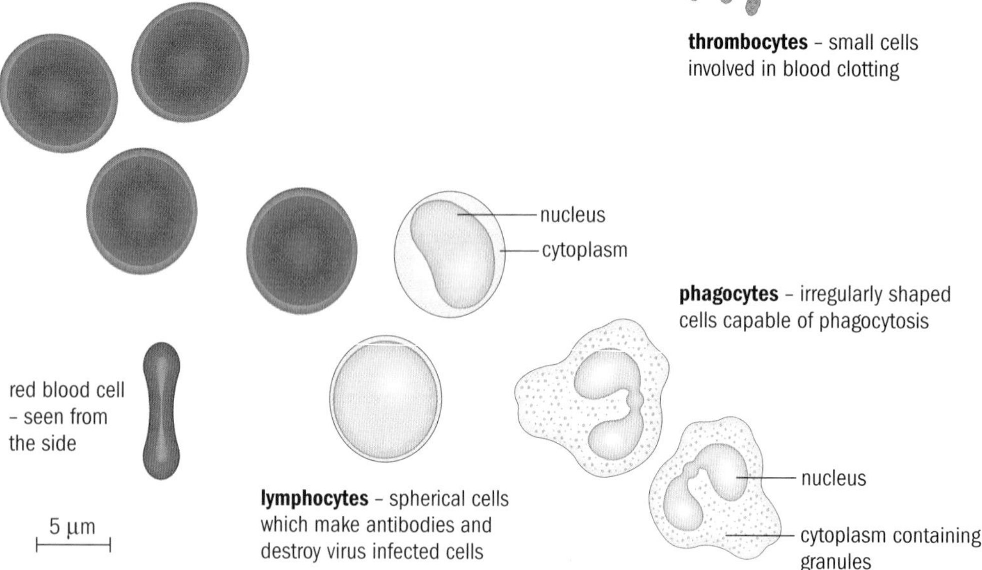

Blood tissue contains different types of cells.

Blood protects us by:

- clotting to prevent bleeding and entry of pathogens
- defending us against infections.

All blood cells come from blood stem cells in bone marrow.

Erythrocytes

Erythrocytes (also known as red blood cells) are small biconcave discs. They have no nucleus, mitochondria, Golgi, ribosomes or endoplasmic reticulum. They are packed full of haemoglobin, which contains iron and carries oxygen from the lungs to respiring cells. Erythrocytes also carry some of the waste carbon dioxide back to lungs. They live for about 3 months and are then broken down in the liver, spleen or bone marrow.

Did you know?

You have about 5 million red blood cells in each mm^3 of blood. This means you have about 25 million million red blood cells in your body. This is a mind-boggling number. If you counted them all, at the rate of 2 per second, without stopping, it would take you 400 000 years!

Leucocytes

Leucocytes are white blood cells. There are different types:

- **Lymphocytes** are small, about the same size as erythrocytes. Most of them are in the lymphoid tissue, rather than in the blood (see page 229). B lymphocytes make antibodies. T lymphocytes attack virus-infected cells and cancerous cells.
- **Neutrophils** go to sites of inflammation and ingest bacteria and fungi.
- **Monocytes** go into tissues and become macrophages, ingesting pathogens and activating lymphocytes to make antibodies.
- **Eosinophils** attack parasitic worms and are involved in allergic reactions.
- **Basophils** (and mast cells in connective tissue) release histamine when pathogens invade.

Thrombocytes

Thrombocytes are small cells that activate a cascade of events which leads to blood clotting when you cut yourself.

Structure and functions of the heart

The heart is a transport system pump. It is a muscular bag divided into four chambers. It has a mass of about 300 g and is about the size of your clenched fist. It lies in the **thoracic cavity**, flanked by the lungs and enclosed in a fibrous bag of connective tissue, called the pericardium.

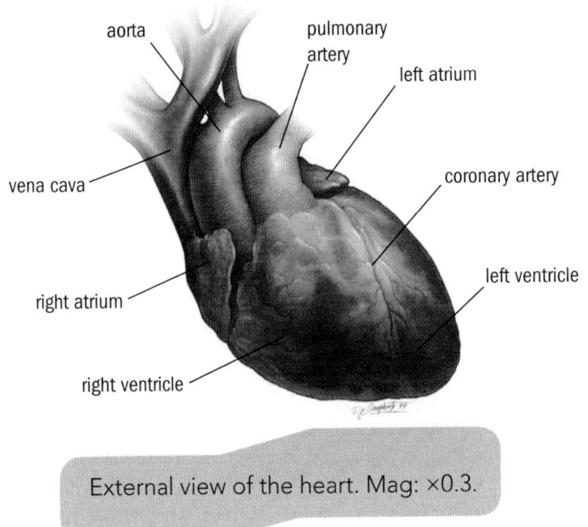

External view of the heart. Mag: ×0.3.

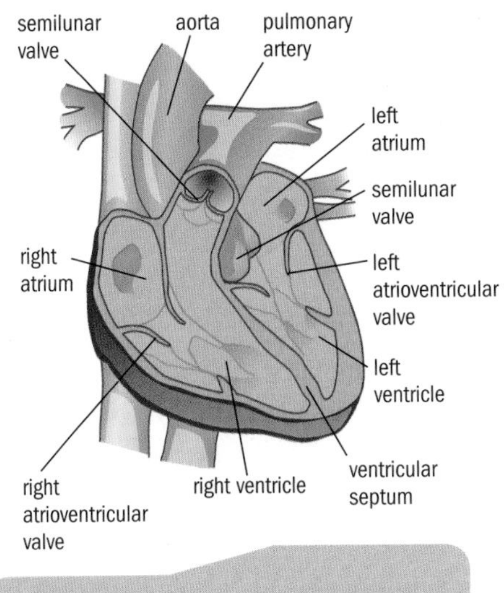

The internal structure of the heart.

Notice the coronary arteries that supply oxygen and nutrients to the heart muscle.

Activity 11.2B

Discuss with a partner why the heart muscle needs a good supply of oxygen and nutrients.

The heart wall is mainly **myocardium**, made of cardiac muscle. It is this that contracts to make the heart beat and pump blood into the arteries.

Worked example

There are about 800 times more erythrocytes than leucocytes in the body. How many leucocytes do you have in your body?

We have 25 000 000 000 000 erythrocytes. If there are 800 times more erythrocytes than leucocytes, then the number of leucocytes is 25 000 000 000 000 divided by 800.

You can also say this is 25×10^{12} divided by 8×10^{2}. This is 3.125×10^{10} or 31 250 000 000.

The cardiac cycle

1 Both **atria** relax and fill with blood from the veins.

2 Atria contract and force the AV **valves** open. Blood enters the ventricles, which fill.

3 The currents of blood entering the **ventricles** make the AV valves close so blood cannot flow back into the atria.

4 The ventricle walls contract and increase pressure in the ventricles.

5 When the pressure in the ventricles is higher than the pressure in the arteries it forces the semilunar valves open and blood enters the arteries.

6 The semilunar valves close to prevent backflow into the ventricles.

One whole cycle takes about 0.8 seconds.

The regularity of the heartbeat is set by a patch of tissue in the wall of the right atrium, called the sinoatrial node or the pacemaker. Purkyne fibres, special muscle fibres that don't contract but conduct electrical impulses, carry the impulses from atria to ventricles.

Case study: Haematology

Leander works as a technologist in a hospital laboratory in the haematology department. She carries out lots of tests on blood samples, such as blood cell counts, blood clotting time and tests for leukaemia and other blood disorders. Her boss is a haematologist – a qualified doctor who specialises in blood. Their work is important and they will be part of a multidisciplinary team dealing with diseases such as haemophilia, lymphoma, vitamin B_{12} deficiency and bone marrow transplants.

For which of the diseases mentioned above would the team include an oncologist?

The ventricles contract from the apex (tip) upwards and force blood out of the ventricles into the arteries.

The atria stop contracting just before the ventricles start to contract. ECG traces can be used to detect abnormalities such as slow or fast resting heart rate, an enlarged heart, heart block, atrial arrhythmia, infection, mechanical damage or heart attack, premature contractions and ventricular fibrillation.

Doctors also measure blood pressure. There are two readings: the higher value is the systolic pressure and shows the pressure generated when the left ventricle of the heart contracts to push blood out into the aorta. The lower value is the diastolic pressure when the left ventricle is relaxing and filling. It is important for people to know their resting blood pressure because if it is high (hypertension) it can increase their risk of heart attack or stroke. When we exercise the systolic blood pressure increases temporarily to help meet the muscles' increased demand for oxygen.

Activity 11.2C

Find out what the normal range of resting blood pressure readings is for children, adults aged 20–40, 40–60 and adults aged over 60.

The volume of blood pumped out of the heart at each beat is called the stroke volume. The cardiac output is the volume of blood pumped out of the heart in 1 minute.

So, cardiac output (CO) = stroke volume (SV) × heart beat rate (HBR).

Worked example

1 A person has a resting heart rate of 70 beats per minute. His cardiac output is 4.2 dm^3/min. What is his resting stroke volume?

CO = SV × HBR

So SV = CO/HBR

SV = 4200 cm^3/70 = 60 cm^3.

2 What would you expect to happen to his heart rate, stroke volume and cardiac output when he is running?

Heart rate would increase. Stroke volume would increase therefore cardiac output would increase. All these ensure more blood is pumped each minute to supply the contracting muscles with more oxygen for the increased respiration needed to make more ATP for muscle contraction.

Did you know?

The ancient Greeks thought the heart was the seat of intelligence. People have also thought the heart was the seat of the emotions. We now know it is just a pump but our emotions can affect the heart rate.

Taking measurements of heart function

Doctors listen to heart sounds using a stethoscope. They may tell if a patient has a faulty valve. A microphone placed over the heart can also detect damaged or stiffened valves.

Electrocardiograms (ECGs) measure the electrical activity of the heart.

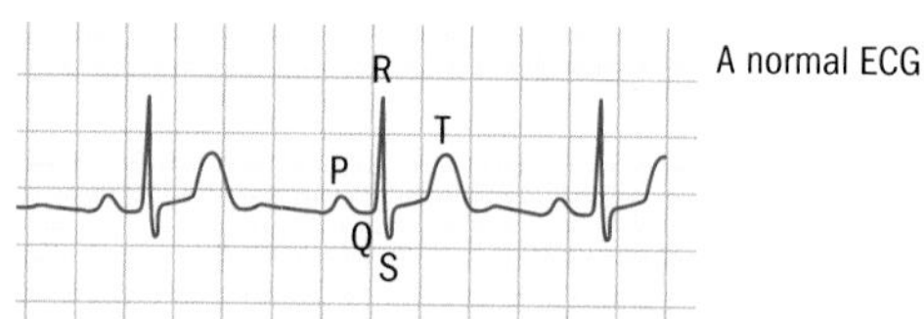

A normal ECG

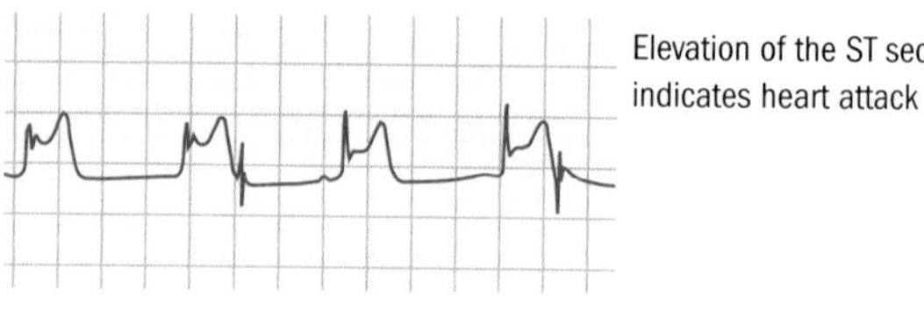

Elevation of the ST section indicates heart attack

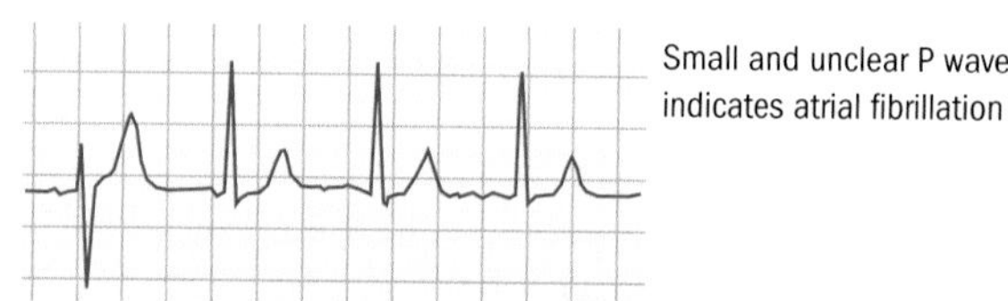

Small and unclear P wave indicates atrial fibrillation

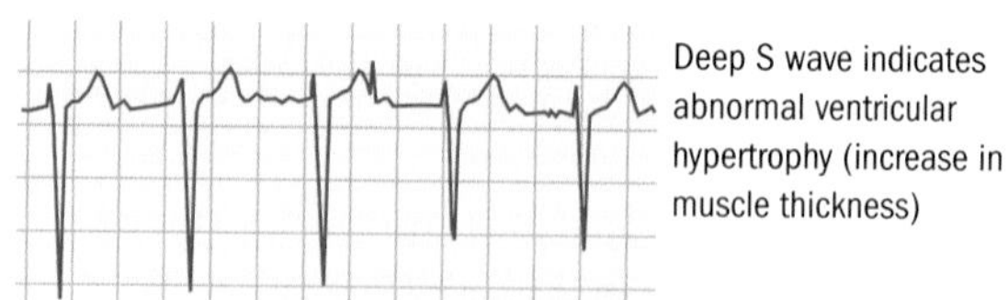

Deep S wave indicates abnormal ventricular hypertrophy (increase in muscle thickness)

A normal ECG trace (top) compared with others that indicate heart problems.

The P wave shows when the atrium begins to contract. The QRS wave shows when the ventricle begins to contract. The T wave shows when the ventricles stop contracting.

Structure and functions of blood vessels

A closed circulation

Your blood is always in vessels – either arteries, arterioles, capillaries, venules or veins. The vessels form a closed transport system that begins and ends at the heart.

A double circulation

In humans, as in all mammals, the blood flows from the heart to body tissues and back, then to the lungs to get rid of carbon dioxide and take in oxygen, then back to the heart to be pumped around the body. So there are two separate circulations – **systemic** (body) and **pulmonary** (lungs). This is very efficient and supplies all the body tissues with enough oxygen for their needs.

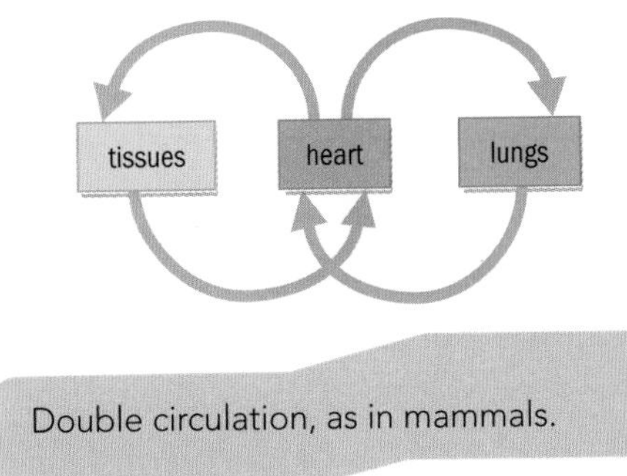

Double circulation, as in mammals.

Did you know?

Ancient Greek doctors thought that blood moved, like an ocean, through the body. In 1620 William Harvey showed that blood circulated in vessels.

Arteries and arterioles

Blood leaves the ventricles of the heart in **elastic arteries**. These have thick walls to withstand the high pressure, with many elastin fibres as well as smooth muscle. They have small **lumens**. The walls expand and recoil and this keeps the blood under continuous pressure. This expansion and recoil is the pulse, transmitted through all arteries at each heartbeat. You can feel a pulse where an artery passes near the surface of the body. The large elastic arteries divide into smaller **muscular arteries**. These carry blood to the body organs. The middle layer of their wall contains less elastic tissue and more smooth muscle than the elastic arteries. These arteries divide into smaller **arterioles**. These are small vessels with smooth muscle cells wrapped around the endothelium. If they dilate, more blood flows into the capillaries.

Capillaries

These are very small vessels with thin walls, made of one layer of endothelial cells (a type of epithelium). The diameter of each capillary is just big enough to allow one erythrocyte through at a time. Capillaries form networks or capillary beds in tissues and plasma is forced out through the small gaps in their walls; it bathes the cells, delivering oxygen and nutrients and carrying away wastes.

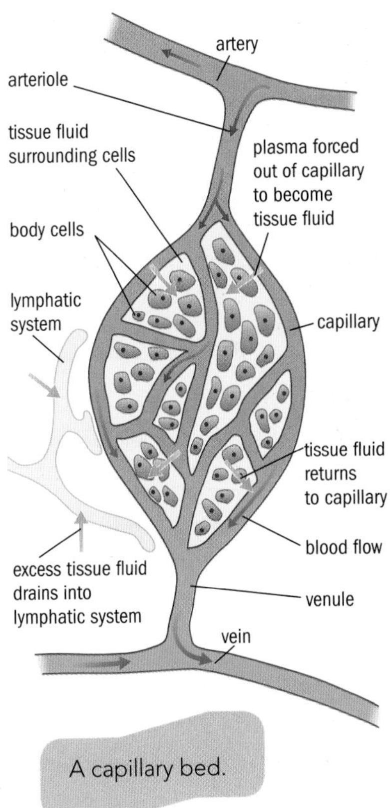

A capillary bed.

Venules and veins

Capillaries join up to form **venules**. They have small diameters, very thin walls and white blood cells can move through their walls into tissues. Venules join to form **veins**. Vein walls have three layers but are thinner than artery walls and have little smooth muscle or elastin. The blood pressure in veins is low so there is no danger of them bursting. Veins have valves to prevent backflow.

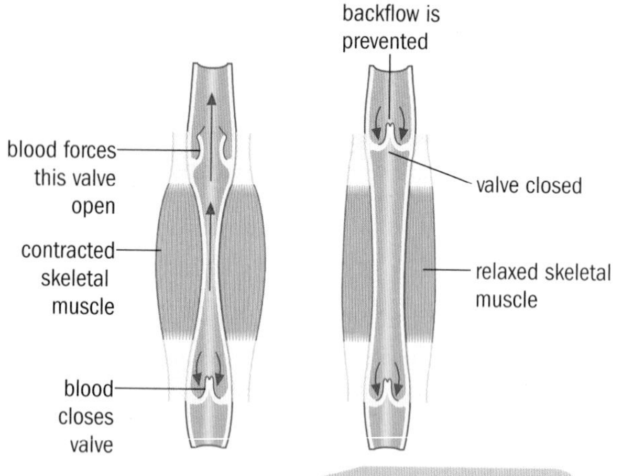

LS (longitudinal section) veins showing how valves prevent backflow of blood.

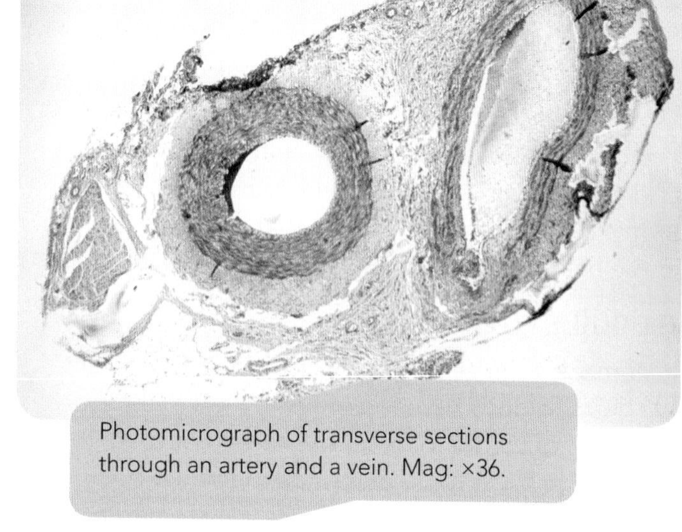

Photomicrograph of transverse sections through an artery and a vein. Mag: ×36.

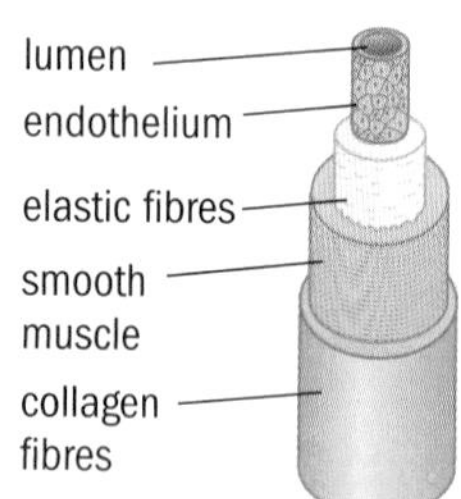

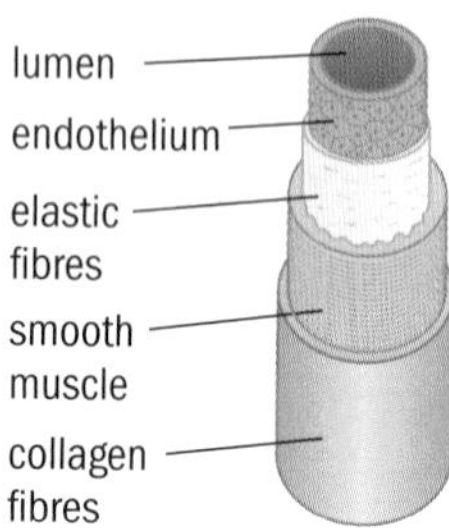

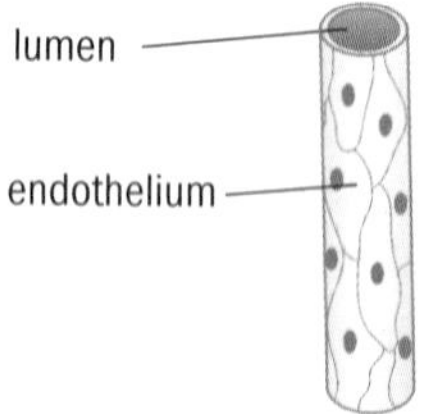

Walls of an artery, a vein and a capillary.

Activity 11.2D

Find out how skeletal muscles and breathing help blood return to the heart in veins. Share your ideas with the rest of the class.

Taking measurements

You can measure your own pulse rate, which tells you your heart beat rate. You are actually measuring the pulse in an artery but it tells you your heart rate.

Doctors and nurses measure blood pressure in a large artery in the arm but it indicates the pressure in the left ventricle when it is contracting (systole) and relaxing (diastole). Measurements are given in millimetres of mercury, e.g. 120/80 mm Hg.

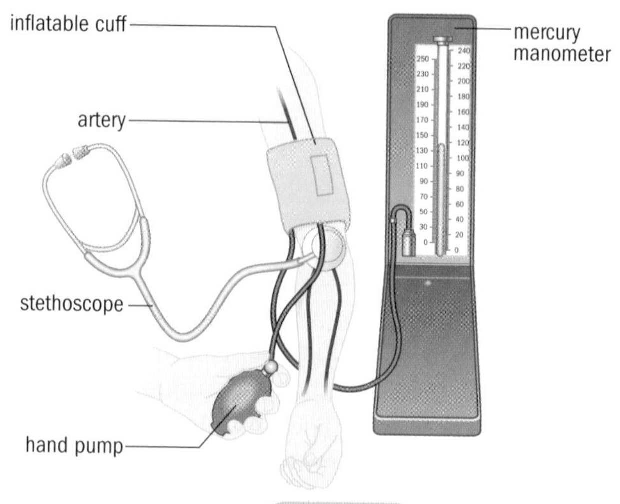

Measuring blood pressure.

Examining blood components

Doctors often take samples of blood from patients and send them to a laboratory. In the laboratory various tests may be carried out to check or find out, for example:

- if the patient has abnormal blood glucose or insulin levels and may be diabetic
- if the patient has an underactive thyroid gland
- if the patient has an infection of the blood
- the patient's liver function by looking for abnormal levels of enzymes that may have leaked from damaged liver cells into the blood
- kidney function
- levels of leucocytes, as high levels may indicate infection or leukaemia and low levels may indicate a problem of the immune system
- a low erythrocyte count, which may indicate anaemia
- if the patient has an infection such as malaria.

Blood tests may be carried out regularly on some patients, such as those who are HIV positive to check their T cell count. Some cancer tests, such as those for prostate cancer, also involve measuring the levels of certain antigens.

Activity 11.2E

Explain why it is not correct to say that arteries always carry oxygenated blood and veins always carry deoxygenated blood.

Activity 11.2F

Make a table to compare the structures and functions of arteries, arterioles, capillaries, venules and veins.

PLTS

Self-manager, Independent enquirer and Effective participator

You will develop your self-management and independent enquiry skills to plan and carry out the investigations and your effective participator skills when presenting information to the rest of the class.

Assessment activity 11.2 M2 P3

BTEC

You are a technician in a cardiology department.

You need to produce a poster for the patients' waiting room.

1 Explain the need for a transport system in a multi-cellular organism. Use annotated diagrams on the poster. Use biology textbooks and the Internet to help you. M2

2 As part of your job you sometimes need to take measurements of heart function under different circumstances.

   Under the supervision of your teacher, you can take measurements of heart rate and blood pressure under varying circumstances. Measure the resting heart rates and resting blood pressure for everyone in your class. Find the range and average. Compare these with published norms. Then, investigate the effect of exercise, excitement/fear or caffeine on heart rate and blood pressure. You will need to plan your activity and indicate what you expect to happen. Record your measurements in a table. Explain your observations in terms of the function of the cardiovascular system. Present your findings to the rest of the class. P3

Grading tips

For M2 remember to refer to the number of cells, surface area to volume ratio, diffusion rates and distances, and levels of activity. You will also need to mention the substances that are needed by the organism and why they are needed, and how they are carried by the blood.

For P3 measure heart rates and resting blood pressure three times for everyone and find the means (averages). Follow the rules for tabulating data. When presenting your findings remember to explain why heart rate and blood pressure vary when we are afraid or exercising. Explain how the changes help the body to function under these circumstances.

11.3 The respiratory system

In this section:  

Key terms

Breathing – movements of ribcage and diaphragm that cause air to enter and leave the lungs.

Gaseous exchange – exchange of oxygen and carbon dioxide at a respiratory surface. For example, at the lungs, oxygen enters the bloodstream and carbon dioxide enters the alveoli.

Diffusion gradient – the difference in concentrations of a substance, for example oxygen, on either side of a surface. The steeper the gradient (the greater the difference) the more diffusion will occur.

Respiration – the release of energy from digested food. This occurs in cells. Aerobic respiration requires oxygen.

Endocytosis – method of transporting large particles into a cell. The plasma membrane invaginates and forms a vesicle containing the particle.

Exocytosis – method of transporting large particles or molecules out of a cell. A vesicle containing the particle moves to and fuses with the plasma membrane and releases the particle.

Tidal volume (TV) – volume of air breathed in and out in one normal breath.

Inspiratory reserve volume (IRV) – maximum extra volume of air that you can breathe in with one deep breath in.

Expiratory reserve volume (ERV) – maximum extra volume of air you can breathe out with one big breath out.

Vital capacity – tidal volume + inspiratory reserve volume + expiratory reserve volume.

Residual volume – air that stays in the lungs after a big forced breath out.

Dead space – air in the trachea and bronchi that is not involved in gaseous exchange.

Respiration

Respiration is the release of energy from organic molecules such as those found in food.

All the cells in our bodies respire to obtain energy. Some of the energy is used to make ATP and some is released as heat. The heat helps to keep our body temperature stable. ATP is used:

- to actively transport sodium ions out of cells so that the water potential does not become too low and cause water to enter cells by osmosis, which would make cells burst
- to provide the energy needed for synthesis of large molecules such as proteins and lipids
- to move large molecules into and out of cells by **endocytosis** or **exocytosis**
- to replicate DNA so cells can divide.

Some of our cells, e.g. erythrocytes and chondrocytes, respire anaerobically but most cell types respire aerobically as well. For this they need oxygen.

Activity 11.3A

1 The food we eat contains potential energy that is released during respiration. Where has the energy in food come from?

2 Why is it advantageous for erythrocytes to be unable to carry out aerobic respiration?

Did you know?

ATP was discovered in 1929. Since then, scientists have found that, in addition to its role as energy currency inside cells, it has many other functions. It carries critical messages between cells. Some nerves release ATP as a neurotransmitter. It also plays a part in blood clotting, vasodilation, vasoconstriction and retinal function. Its release in the embryo during development signals eye and inner ear development.

Breathing (pulmonary ventilation)

This is how air, containing oxygen, is brought into the lungs.

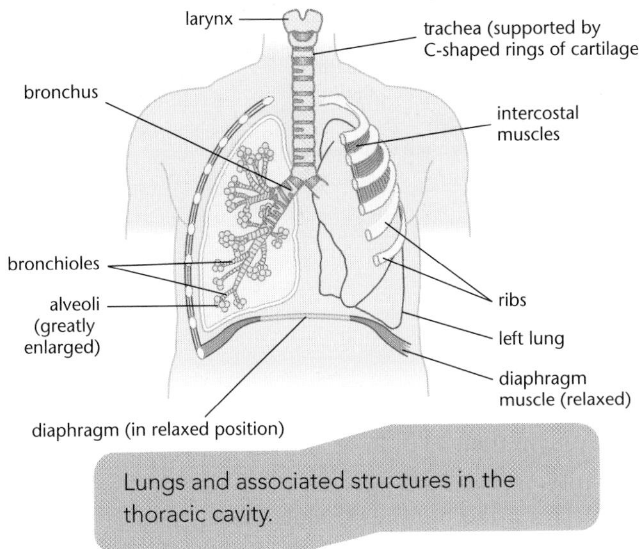

Lungs and associated structures in the thoracic cavity.

When we **breathe** in (inspiration):

1 Muscles between the ribs contract and raise the ribcage up and out.
2 The diaphragm muscles contract and flatten the diaphragm.
3 Together, these two actions increase the volume of the thoracic cavity.
4 This reduces the air pressure inside the thorax.
5 Air moves, down the pressure gradient, from the atmosphere, into the alveoli of the lungs. Here, oxygen and carbon dioxide are exchanged.

When we breathe out (expiration)

1 The ribcage drops down and in.
2 The diaphragm muscle relaxes and the diaphragm domes.
3 These two actions reduce the volume of the thorax and increase air pressure inside it.
4 The elastic lungs snap back to their original size and air is pushed out of the lungs.

The passage of air

Air can enter through the mouth, but if we breathe through the nose, hairs filter out particles and the air is warmed before it reaches the lungs. It passes through the pharynx (back of the throat), larynx (voice box) and along the trachea, bronchi and bronchioles, to the alveoli. The walls of the trachea and bronchi contain cartilage to keep them open, and their epithelial tissue contains goblet cells that produce mucus, which traps small particles and pathogens. The epithelial tissue also has cells with cilia; the cilia beat and waft the mucus up to the larynx, where it can be swallowed so that stomach acid kills the pathogens.

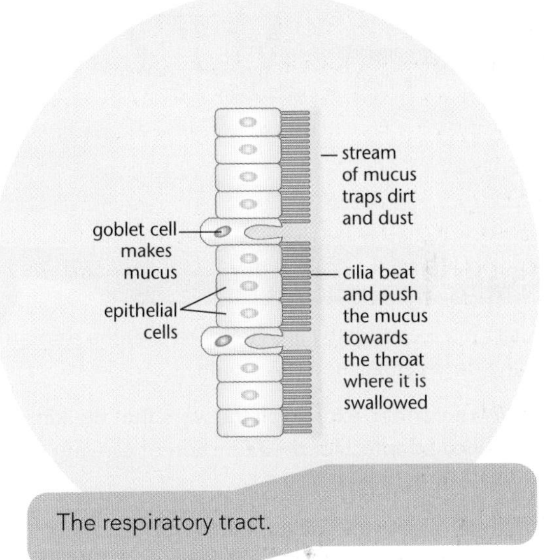

The respiratory tract.

Gaseous exchange

The lungs are well adapted for **gaseous exchange**.

- They contain millions of alveoli (air sacs) which give a very large surface area for diffusion of gases.
- Alveoli have walls made of squamous (flattened) epithelial cells just one layer thick, giving a short diffusion pathway.
- Each alveolus is very close to a capillary. The oxygen from air in the alveoli diffuses the short distance across these cells and then across the thin walls of blood capillaries, into the blood.
- The flow of blood in the capillaries maintains a steep **diffusion gradient**.
- Oxygen from alveolar air diffuses down its gradient into the blood and joins with haemoglobin in erythrocytes.
- Carbon dioxide, carried back from respiring cells in the blood, diffuses down its gradient into the alveoli and is breathed out. If it stayed in the blood it would be toxic as it would lower the blood pH.

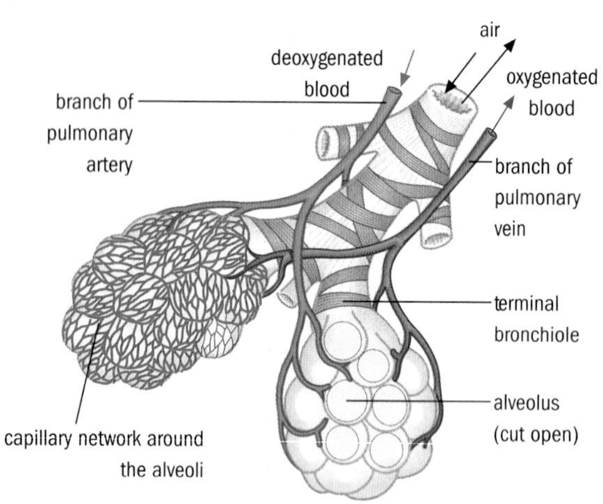

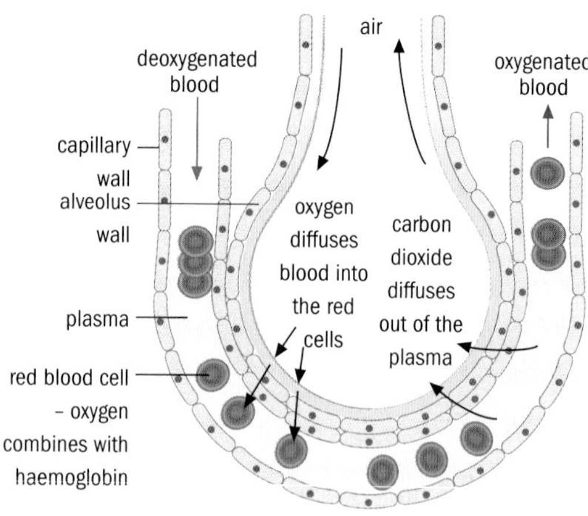

Gaseous exchange at the alveoli.

Activity 11.3B

1. List the reasons why it is better to breathe through the nose rather than the mouth.
2. With a partner, explain all the ways that the lungs are well adapted for their function of carrying out gaseous exchange.
3. Explain clearly the difference between breathing, gaseous exchange and respiration.

Lung function tests and measurements

Spirometry

A simple way to investigate pulmonary ventilation (breathing) is to use a spirometer to measure volumes of air moving into and out of the lungs. This can be done with the subject at rest and during or after exercise.

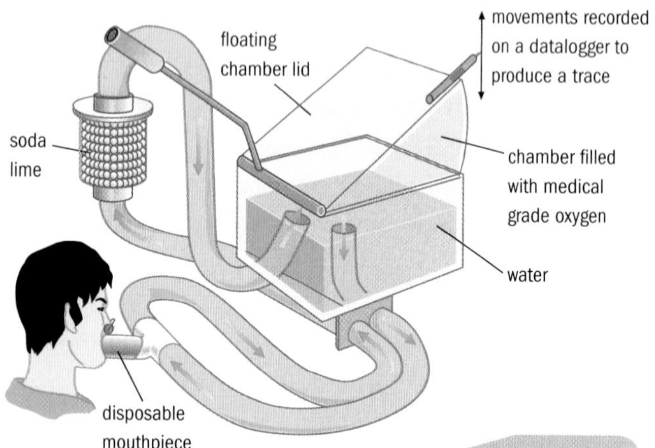

Person using a spirometer.

The chamber of the spirometer is filled with medical grade oxygen and floats on a tank of water. As the person breathes in through a mouthpiece, oxygen is taken from the chamber, which sinks down. If the chamber is linked to a datalogger then a trace on the computer shows the volume of oxygen taken in at each breath. This is the **tidal volume (TV)**. Tidal volume at rest is about 0.5 dm^3.

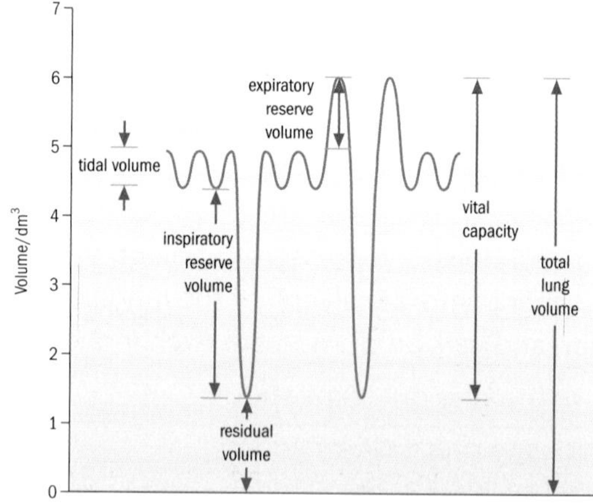

Spirometer trace showing tidal volume (TV), residual volume, inspiratory reserve volume (IRV), expiratory reserve volume (ERV), vital capacity (VC) and total lung volume.

Activity 11.3C

From the spirometer trace above, calculate the person's TV, IRV, ERV and VC.

If the person takes a deep breath in the trace dips and this volume is called the **inspiratory reserve volume (IRV)**. This is about 2.5–3.0 dm^3.

If he then breathes out as much air as possible the trace shows the **expiratory reserve volume (ERV)**. This is about 1 dm^3.

The maximum volume of air breathed out after taking a deep breath in shows the person's **vital capacity**. Vital capacity can also be worked out using the equation:

VC = TV + IRV + ERV

Vital capacity varies according to a person's age, gender and size, and fitness. It may be anything between 3 and 6 dm^3 but is usually around 4.0–4.5 dm^3.

The lungs can never be totally emptied, otherwise the alveoli would collapse and not open again to let in air. The air left in the lungs is called the **residual volume**. This is normally about 1.2 dm^3.

Total lung volume is vital capacity + residual volume.

Measuring oxygen uptake

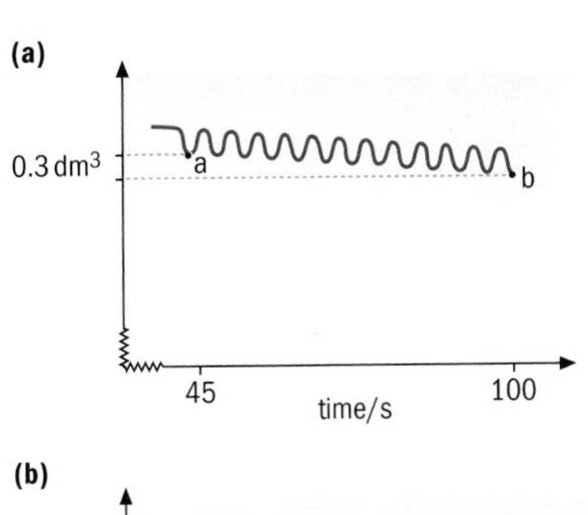

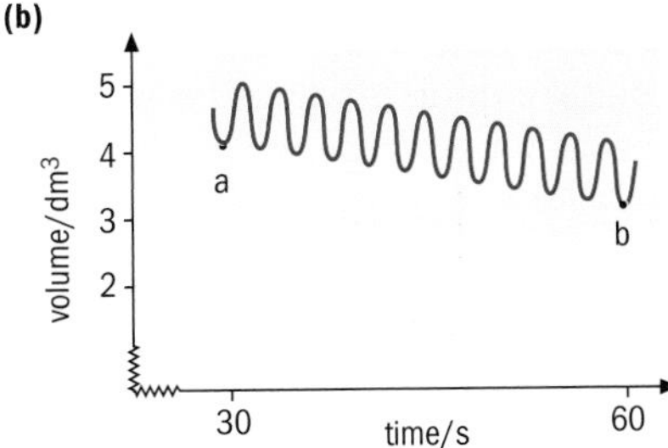

Spirometer traces of a subject **(a)** at rest and **(b)** during exercise.

The soda lime absorbs the carbon dioxide the person breathes out so gradually the volume of oxygen in the chamber reduces. The trace slopes downwards. This slope can be used to measure the volume of oxygen being used in a specific time period, such as when at rest or when exercising.

Activity 11.3D

1 Use the spirometer trace above to calculate the rate of use of oxygen by this person (a) per minute when at rest and (b) when exercising.

2 Use the trace above to calculate the total volume of air entering the lungs per minute (breaths per minute × tidal volume).

The air in the nose, trachea and bronchi is not involved with gaseous exchange, so this volume is known as **dead space**.

Peak flow

When you breathe out as much and as fast as you can through a peak flow meter, it measures speed of air flow and indicates how well the lungs are working. It is particularly useful for asthmatic people, to monitor their lung function. You do the test three times and take the best reading. The graph shows peak expiratory flow rates for females of average height.

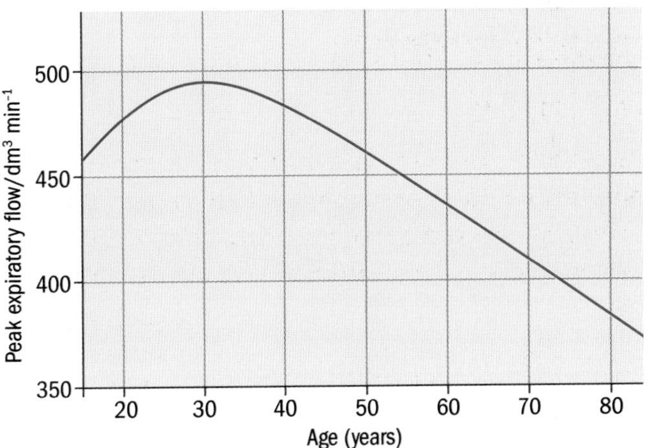

Peak expiratory flow rates for females of average height.

Case study: Personal trainer

David is a personal trainer in a gym. He needs to assess the fitness levels of his clients as they begin their exercise programmes and during them. He will take various measurements, such as height, weight, body fat percentage, body mass index, as well as heart rate and blood pressure. He will also take lung function measurements, such as peak flow, vital capacity and tidal volume.

1 Name two other lung function tests that David might carry out.

2 Why is it essential that David carries out these tests before the clients begin their exercise programmes?

Assessment activity 11.3

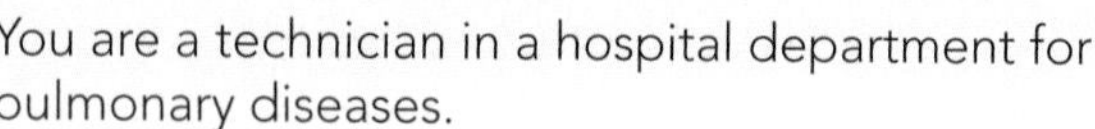

You are a technician in a hospital department for pulmonary diseases.

1 Produce a poster for the waiting room, explaining why we need a ventilation system. M3

2 You also need to carry out lung function tests on a variety of patients.

Under the supervision of your tutor, use the spirometer and measure your tidal volume, ERV, IRV and vital capacity. Collect data for your whole class and tabulate them. Calculate the range and average values for (a) males and (b) females in your class. How do these compare to published values for normal range? Measure your peak expiratory flow using the peak flow meters. P4

Grading tips

For M3, on the poster: explain why multicellular organisms need a ventilation system. A ventilation system is the system that brings oxygen in contact with the gaseous exchange surface. In humans this is breathing. You need to make reference to the fact that they have a blood transport system to carry oxygen to all cells of the body, and explain how oxygen gets from the atmospheric air into the blood.

For P4 make sure that you follow the instructions for using the equipment and that your tutor is there to help you. Take at least three measurements for each lung function test for each person and then find the means. For peak flow, however, you take three readings for each person and then use their best (highest) reading. Make sure you put all the data into a table and follow the rules for tables – no units in the body of the table, only in column headings. Make sure you find out the published values for the normal ranges.

PLTS

Self-manager, Independent enquirer, Effective participator

You will develop your skills of self-management when planning and carrying out these tests; independent enquiry skills when finding the published normal range values; effective participator skills when carrying out the practical investigations.

11.4 The digestive system

In this section: 

Key terms

Balanced diet – a diet containing the right proportion of nutrients and the correct amount of energy to maintain good health.

Prostaglandins – hormone-like substances made in the body from certain fatty acids. They have a wide range of physiological functions such as regulating blood pressure and inflammation.

Hormones – chemicals (many are proteins) made in endocrine glands and carried in the blood to target tissues where they help control growth and development.

Steroid hormones – hormones made from cholesterol.

Antibodies – proteins involved in the immune response.

Coenzymes – chemicals that aid enzymes.

Polymer – large molecule made of repeating sub-units called monomers.

Monosaccharide – single sugar, such as glucose, fructose, galactose.

Disaccharide – double sugar made from a condensation reaction between two monosaccharides, e.g. sucrose, maltose, lactose.

Polysaccharide – large carbohydrate molecule made from many glucose residues joined together.

Condensation reaction – a reaction where two small molecules are joined, with elimination of a water molecule, to make a larger molecule.

Hydrolysis reaction – a reaction where a large molecule is split into two smaller ones with the addition of water.

Digestion – breakdown of large molecules to smaller ones, by hydrolysis.

Gut – the alimentary canal, also known as the gastrointestinal tract or digestive tract. It runs from the mouth to the anus.

Polypeptide – polymer of amino acids. Large polypeptides are also called proteins.

Egestion – the expulsion of faeces from the anus.

Absorption – the process by which small, soluble molecules produced by digestion are taken up from the gut. This involves them passing across the plasma membranes of epithelial cells lining the gut wall, passing through these cells and into the bloodstream or lymphatic fluid.

Assimilation – the products of digestion are taken up and used or stored in cells of the body.

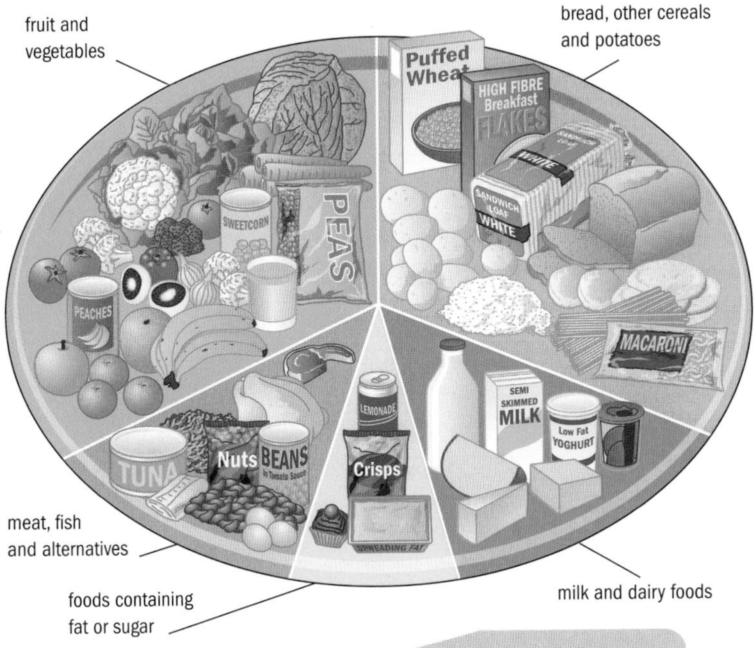

fruit and vegetables

bread, other cereals and potatoes

meat, fish and alternatives

foods containing fat or sugar

milk and dairy foods

This plate indicates how much of each food group should be included in a balanced diet.

Why do we need to eat a balanced diet?

Food contains energy and the materials that we need for growth, development, repair and reproduction. However, we don't just need any old food. We need to eat a **balanced diet** that gives us the right amount of the necessary nutrients to maintain good health and organ functions. The energy consumed should be sufficient for energy used, so that we:

- maintain a healthy weight
- have strong immune systems
- are strong and can learn and work productively
- are happy.

There are seven components to a balanced diet. They may also be described as nutrients. The table shows why each is needed in the diet and gives some examples of each.

Nutrient	Role in the body	Examples of foods
carbohydrates	the main source of energy	potatoes, rice, cassava, millet, bread
lipids	Used to make cell membranes; can be stored and insulate organs from heat loss; protect organs from knocks; can be used as a source of energy; contain fat-soluble vitamins (A, D and E); some are used to make **prostaglandins**, and **steroid hormones** such as testosterone and oestrogen; long-chain omega 3 fatty acids in fish oils protect us from heart disease.	butter, cheese, oils (olive, palm, sunflower), oily fish such as mackerel, kippers, tuna and salmon, chips, avocado, fatty meat
proteins	Used to make enzymes, **hormones, antibodies**, haemoglobin, muscles, cell membranes, collagen and elastin in connective tissue, bone, and for growth and repair of tissues.	soya, milk, meat, eggs, fish, nuts, cheese
vitamins	B group vitamins are **coenzymes** and aid processes such as respiration.	cereals, wholemeal bread, green leafy vegetables
	Vitamin A is needed for dim-light vision, for healthy epithelial tissue and for the immune system.	liver, green vegetables, orange vegetables and fruits
	Vitamin C helps make connective tissue and helps us absorb iron from food.	citrus fruits, blackcurrants, kiwi fruits, potatoes, peppers
	Vitamin D helps us absorb calcium from food and deposit it in bones; it also protects us from cancer and heart disease.	oily fish, egg yolk, liver; made in skin when exposed to sunlight
	Vitamins A, C and E are also anti-oxidants and protect our DNA from damage by free radicals and hence protect us from cancer and heart disease.	E: nuts, wholemeal bread
minerals	iron (Fe) needed for making haemoglobin that carries oxygen in erythrocytes	red meat, spinach, dried apricots
	calcium (Ca) hardens bones and teeth, helps blood clot, helps nerve conduction, helps muscles contract	milk, yoghurt, cheese, broccoli
	sodium (Na) and potassium (K) for nerve conduction and heart function	Na: table salt; K: bananas
	phosphorus (P) to make cell membranes and ATP and for bones	cheese, pork, tuna, nuts
	magnesium (Mg) for bones	green vegetables
	iodine (I) for thyroxine	table salt, seafood
	zinc (Zn) for insulin	meat, cereals, nuts, milk
fibre	Indigestible plant matter, such as cell walls – it adds bulk to food and takes up water when in the gut. It helps us pass faeces easily. Soluble fibre helps to lower blood cholesterol levels.	insoluble fibre in leaves, fruit, wholemeal bread, bran, bananas, apples and pears with skin, corn, potatoes with skin, soluble fibre in oatmeal, baked beans
water	Makes up 70% of us – blood plasma is 90% water, so water is needed for transport in the body. Needed for sweat, tears and urine.	water, milk, drinks such as tea and coffee, juice, fruit and vegetables

The foods we eat also contain DNA and we use the parts of digested DNA to make more of our own DNA. Fruit and vegetables also contain other chemicals, such as polyphenols, which improve our health.

Activity 11.4A

1 Write down everything you ate yesterday. Now write down the nutrients you think were in those foods. Are you eating a balanced diet?

2 Discuss with a partner why it is better to eat fruit and vegetables rather than take vitamin supplements.

Did you know?

For every kilogram of extra fat you store in adipose tissue, your body makes 2 miles of new capillaries to serve the fat cells.

You can see from the table above that some nutrients, notably protein, have a structural role – they are needed to make the structures in our bodies, such as connective tissue, muscle, bone and skin. Some, such as enzymes, antibodies and haemoglobin, have specific functions. Other nutrients, particularly vitamins, have a regulatory role – they allow metabolic reactions to take place so that our bodies function properly. However, the distinction can be a bit blurred, for instance vitamin A is used to make a chemical that is part of the structure of rod cells in the retina of eyes and without it we cannot see in dim light.

Vitamin D is a hormone

Vitamin D has important regulatory roles. Although it can be obtained in the diet, the main source of vitamin D is by the action of ultraviolet light on the skin.

1 Cholesterol in the skin is converted to an inert form of vitamin D.
2 This is then changed, in the liver and kidneys, to the active form.
3 This active form, calcitrol, is a hormone and is extremely important for aiding absorption of calcium from food and depositing it in bones.
4 It is released into the blood and carried by a protein to its target organs.
5 It is fat soluble and passes through the plasma membranes of target cells and to their nuclei where it binds to a special receptor.
6 The activated receptor then triggers specific genes to make proteins that aid calcium absorption.
7 If this activation occurs in cells of the gut, then more calcium is absorbed from food.
8 If it happens in cells of bone, more calcium enters bone.
9 This receptor is also involved in regulating cell division and can protect us from cancer.

Vitamin D also aids the immune system and helps protect us from infections.

It may also protect us from heart disease as it helps regulate blood cholesterol level and blood pressure.

However, having too much vitamin D can increase the risk of heart disease. This illustrates a general principle – we need optimum amounts of vitamins and minerals (and other nutrients) – too little or too much can be harmful.

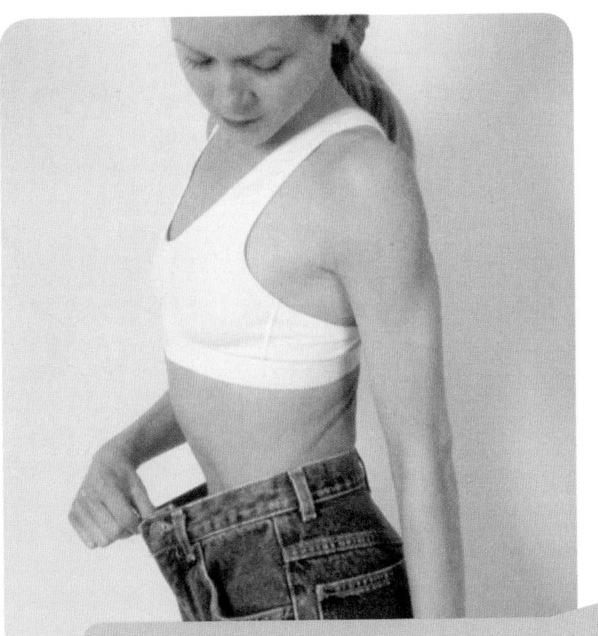

Eating too little or too much are both examples of malnutrition and are harmful to health.

Chemical stucture of nutrients

Carbohydrates

Carbohydrates all contain the elements carbon, hydrogen and oxygen in their molecules. In this nutrient group there are:

- simple sugars (**monosaccharides**) such as glucose, fructose and galactose
- **disaccharides** such as sucrose, maltose and lactose
- **polysaccharides** – e.g. starch found in plants and glycogen in liver and muscles.

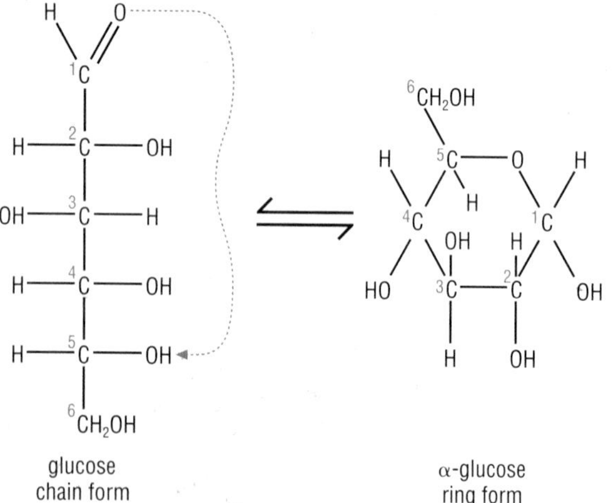

Carbons are numbered 1–6 to show how the ring structure is made. The numbering also allows us to note where further bonds are formed

Chain and ring forms of glucose (a monosaccharide).

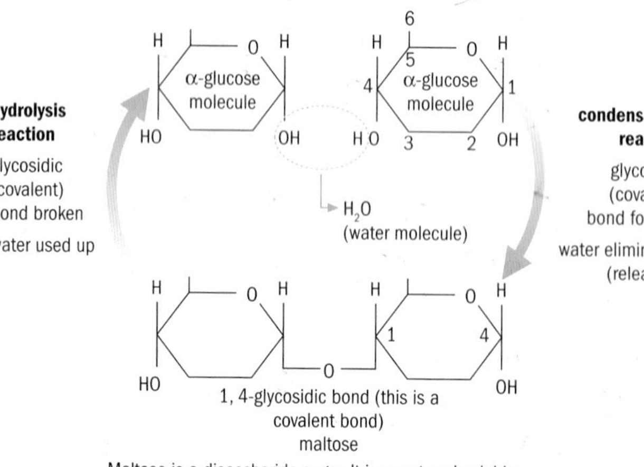

Maltose is a disaccharide sugar. It is sweet and soluble.

Making and breaking bonds.

Sugars we eat are respired to release energy or converted to glycogen for storage.

Activity 11.4B

Find out the structural formula of cellulose.

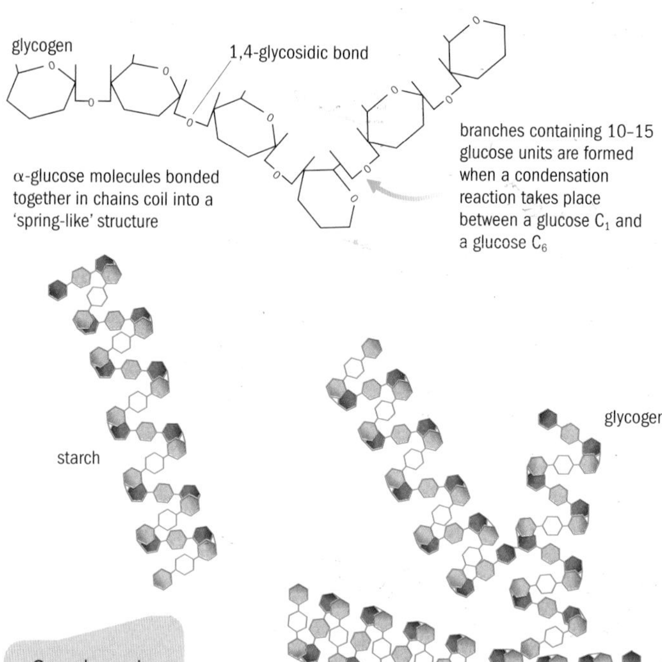

Starch and glycogen.

Lipids or fats

Lipids or fats are eaten mainly in the form of triglycerides. These are made up of glycerol joined to three fatty acid chains. Their molecules also all contain carbon, hydrogen and oxygen.

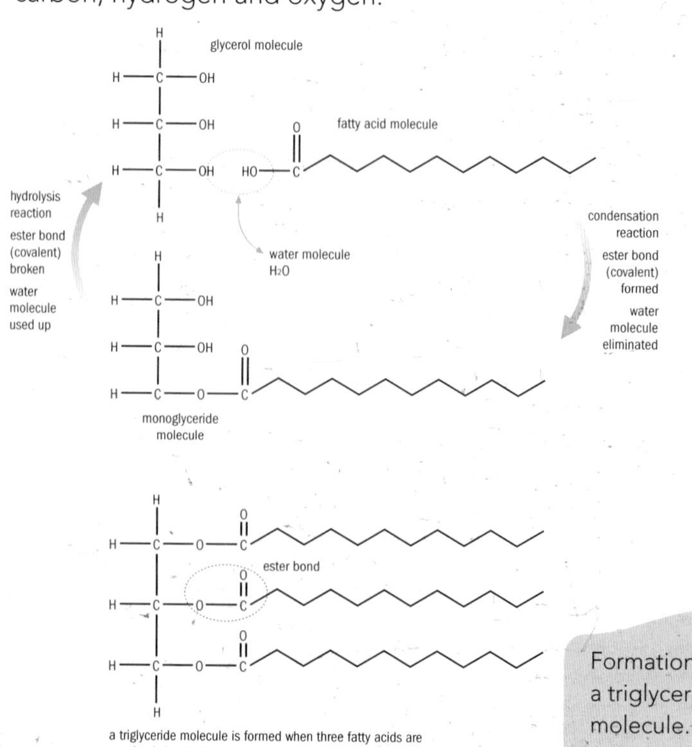

Formation of a triglyceride molecule.

Proteins

Proteins are **polymers** made from amino acids. They may also be called **polypeptides**. There are 20 different amino acids and all contain carbon, hydrogen, oxygen and nitrogen. Two of them also contain sulfur. Each amino acid has an amine group, which is basic, and a carboxyl group, which is acidic. Each one has a different R group. Two amino acids join in a condensation reaction (where a molecule of water is eliminated) by a peptide bond.

The sequence of amino acids in a protein is the **primary structure** of the protein. The chain may then pleat or coil. This is called the **secondary structure**.

H

H_2N — C — COOH

amine group

carboxyl group

R

side chain

The general structure of amino acids.

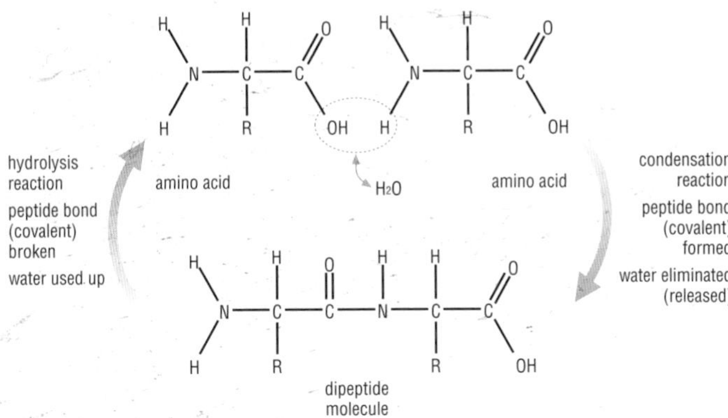

Two amino acids can be joined to make a dipeptide. The dipeptide can be split back into two amino acids.

This coiled and pleated chain may then fold into a very specific three-dimensional shape. This is called the **tertiary structure**. Each protein has a different 3D shape and this is what enables it to perform its particular function.

Some proteins, such as insulin and haemoglobin, are made of more than one polypeptide chain. Such proteins have a **quaternary structure**.

Activity 11.4C

1 Find out the names of all 20 amino acids. Some of them are called 'essential amino acids' because they have to be eaten as we cannot make them in our bodies. Which ones are essential amino acids?

2 Find out which fatty acids are 'essential in the diet' and why they are essential.

Biochemical tests for nutrients

Because these nutrients have molecules with different chemical structures they have different chemical properties. This means there are specific biochemical tests that can be used to identify which nutrients are present in different foods.

Did you know?

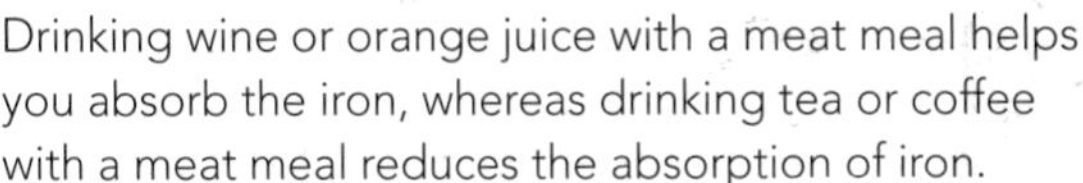

Drinking wine or orange juice with a meat meal helps you absorb the iron, whereas drinking tea or coffee with a meat meal reduces the absorption of iron.

Food in your gut is still outside your body

When you eat food, it passes down your oesophagus into your stomach and then into your intestines. Whilst in the **gut**, the food is still outside of your body proper. If this seems a strange concept, just think of your body as a very elongated doughnut. The hole through the middle is like the space inside your gut; it is continuous with the outside world. However, at each end of our digestive tract there is a circular sphincter muscle. When these muscles contract they close off the digestive tract. Whilst the food is in this space, it is safe – no one else can steal it and you can digest it at leisure.

Nutrient	Biochemical test	Positive result
starch	Add iodine/KI solution.	colour change from brown to blue/ black
sugars – all except sucrose	Add Benedict's solution and heat in a water bath at 80 °C.	colour change from blue to green/ yellow/red
sucrose sugar	Heat with dilute hydrochloric acid to hydrolyse the sucrose to glucose and fructose. Cool and neutralise by adding sodium hydrogencarbonate solution. Add Benedict's reagent and heat in a water bath at 80 °C.	colour change from blue to green/ yellow/red after hydrolysis with acid. Before hydrolysis the Benedict's test was negative
lipids	Add ethanol to the food and shake to dissolve any fat. Carefully pour the ethanol into a test tube of water.	a white emulsion (droplets of fat that come out of solution in the water and are suspended in the water) near the top of the water
protein	Add Biuret reagents (dilute sodium hydroxide followed by dilute copper sulphate).	colour change from blue to mauve/ lilac/purple
vitamin C	Add a solution of the food being tested to some DCPIP solution.	the blue colour of DCPIP disappears
minerals	Clean some nichrome wire by dipping in nitric acid and placing it in the hot part of a Bunsen flame, until there is no colour in the flame. Now dip the wire in some food to be tested and place it in the hot part of the flame. Observe any colours.	orange/yellow flame indicates sodium lilac flame indicates potassium white flame indicates magnesium red flame indicates calcium

What is digestion?

Digestion means breaking down. For food to be absorbed across the cells of the gut wall into your bloodstream to be carried to all the cells of your body, the large molecules have to be broken down into smaller ones. This is digestion.

Digestion begins in the mouth

1 **Ingestion** is when you take food into the mouth. You usually bite off pieces using your incisors (front) teeth.

2 When you chew food, the premolar and molar (back) teeth break large pieces into smaller pieces. This is **mechanical** digestion.

3 At the same time, saliva, from salivary glands and containing enzymes, pours into the mouth and begins to **chemically** digest food. The enzyme amylase, in saliva, breaks down starch to maltose sugar. Lingual lipase, made in the mouth, begins to digest lipids.

4 The tongue rolls the chewed food into a ball (a bolus) and pushes it to the pharynx (back of the mouth). The swallowing reflex happens and the epiglottis closes over the glottis to prevent food entering the trachea.

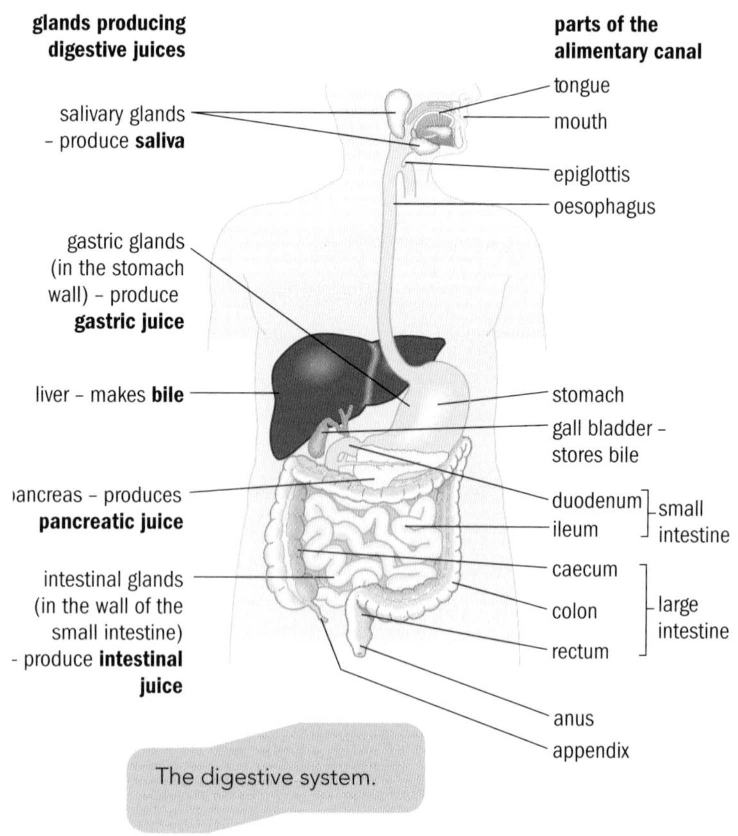

The digestive system.

Case study: Dietitian

Arwyn is a dietitian. He specialises in advising managers who are responsible for providing food to people in settings such as hospitals, schools and the armed forces. He advises the managers about the nutritional content of foods and energy requirements of the different groups being catered for, and helps design appropriate menus. He has a professional qualification at degree level.

1. **In what ways would you expect the recommended diets for soldiers, school children and hospital patients to differ?**
2. **What other factors, besides dietary factors, would need to be considered by the managers who provide food for such groups?**

How does food pass along the digestive system?

Once you have swallowed the bolus, it is pushed and squeezed through the oesophagus and the rest of the gastrointestinal tract by **peristalsis**. There are two layers of smooth muscle in the gut wall and when the circular layer contracts behind the bolus it pushes food along, like pushing a marble inside a rubber tube by squeezing behind the marble.

There is no digestion in the oesophagus but its walls secrete mucus to help food move easily.

Digestion in the stomach

Your stomach is a muscular bag about the size of a cola can when empty. Special cells, called oxyntic cells, in gastric pits in the stomach lining, secrete hydrochloric acid. This makes the pH low and kills many of the bacteria on the food.

Chief cells secrete pepsinogen, an inactive form of the enzyme pepsin. The hydrochloric acid cuts off some amino acids from pepsinogen, changing its shape to expose the active site so it becomes the active pepsin. Pepsin works best at pH 2.0–3.0. It starts to digest proteins by breaking each long protein into two shorter chains. Pepsin is particularly good at digesting the protein collagen, which is found in connective tissues in meat.

Salivary amylase stops working after about an hour as the pH becomes too low.

Lingual lipase continues to work.

As the stomach churns, fats melt, proteins begin to be digested and the whole acid mix of semi-digested food is called chyme (pronounced *kyme*). After about 4 hours, the sphincter muscle between the stomach and duodenum relaxes and food leaves the stomach.

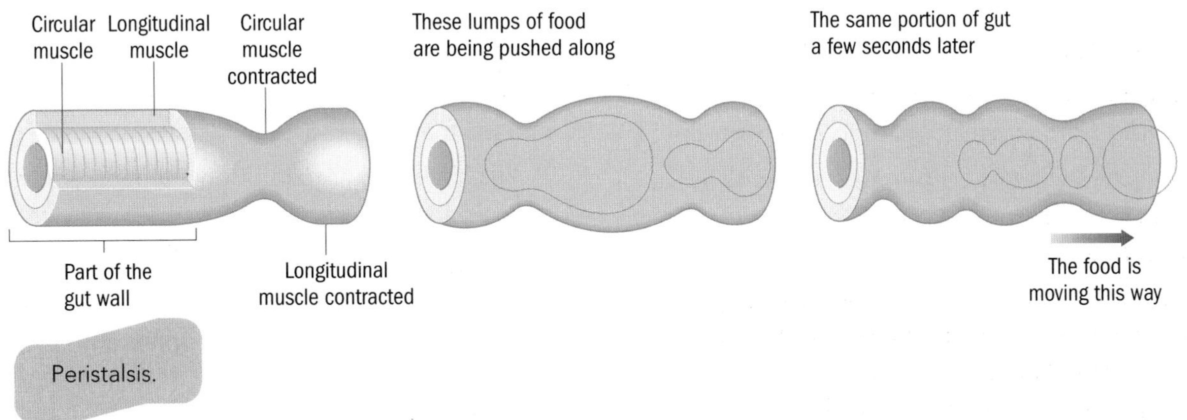

Peristalsis.

Digestion in the small intestine

When chyme enters the duodenum, hormones travel in the blood and trigger the pancreas to release some **pancreatic juice** and the gall bladder to release **bile**.

Bile is made in the liver and stored in the gall bladder. It does not contain enzymes but it has:

- salts that emulsify fats, increasing their surface area for lipase enzymes to work on
- hydrogencarbonate ions that neutralise the acidic chyme
- waste products from the breakdown, in the liver, of old erythrocytes.

Pancreatic juice contains several enzymes:

- trypsinogen, activated by enterokinase in the small intestine to trypsin, digests the partly digested proteins into smaller polypeptides
- carboxypeptidase breaks amino acids away from the ends of the polypeptides
- elastase digests elastin fibres in connective tissue in meat
- amylase digests starch to maltose
- lipase digests lipids to fatty acids and monogylcerides
- nuclease digests nucleic acids.

The epithelial cells (enterocytes) in the **ileum** contain enzymes embedded in their cell membranes. These enzymes protrude into the intestine so food comes into contact with them. In the ileum:

- maltase digests maltose to glucose
- lactase digests lactose to glucose and galactose
- sucrase digests sucrose to glucose and fructose
- enterokinase activates trypsinogen to trypsin
- peptidases break down polypeptides to amino acids.

Activity 11.4D

1 The enzymes made in the pancreas are secreted in an inactive form, and are then activated in the small intestine where they work. Why do you think they are secreted in an inactive form?

2 Pepsin is described as an endopeptidase because it breaks protein molecules in the middle. Explain why carboxypeptidase is described as an exopeptidase.

The products of digestion (mainly glucose, amino acids, fatty acids and monoglycerides) are absorbed into the blood from the ileum. This is dealt with below. What is not absorbed passes into the large intestine.

The large intestine

Bacteria that live in the large intestine ferment some of the cellulose in the food residues and produce gas. These bacteria also make some B vitamins and vitamin K that we use.

In the colon (part of the large intestine) water and minerals are absorbed into the blood.

The main function of the large intestine is to push its contents along to the rectum, where these **faeces** can be **egested** from the anus (**egestion**).

Faeces contain:

- undigested food residues
- mucus
- sloughed-off epithelial cells from the digestive tract lining
- millions of bacteria
- water.

The products of digestion

Small molecules such as alcohol, aspirin and glucose (from high-energy drinks) are absorbed from the stomach into the bloodstream.

By the time food has been in the ileum for 6–12 hours, all the useful nutrients have been digested to smaller molecules:

- Carbohydrates have been broken down to simple sugars, mainly glucose.
- Lipids have been broken down to fatty acids and monogylcerides.
- Proteins have been broken down to amino acids.
- Nucleic acids have been broken down to their constituent molecules (sugars, phosphates and nitrogenous bases.

Unit 18: See page 322 for more information on nucleic acids.

These products, along with water and some minerals, are absorbed from the ileum into the bloodstream or lymphatic fluid.

How the structure of the ileum wall enables it to absorb the products of digestion

- The wall of the ileum is highly folded, giving a large surface area for **absorption**.
- There are also numerous finger-like projections, called villi, that further increase the surface area for absorption.
- The epithelial cells lining each villus have microvilli on their plasma membranes, increasing the surface area for absorption of the products of digestion even more.
- Many of the products of digestion are produced right next to the epithelial cells of the villi as the enzymes concerned with the final stages of digestion are embedded in the plasma membranes of these cells.

Molecules can pass across the plasma membranes of the epithelial cells lining the villi, and then into blood or lymph, by:

- diffusion
- facilitated diffusion
- active transport
- and, in the case of water, by osmosis.

Glucose and amino acids

Glucose and amino acids are absorbed by a type of active transport involving co-transport of sodium ions.

1. ATP is used to actively transport sodium ions out of the epithelial cells.
2. This lowers the concentration of sodium ions in the cell.
3. Sodium ions enter from the ileum, by facilitated diffusion down their gradient, through a co-transporter protein channel in the plasma membrane.
4. They bring with them either glucose or amino acid molecules.
5. The resulting high concentration of amino acids and glucose in the cells leads to them passing out of the other side of the cell by facilitated diffusion.
6. They can then diffuse into the bloodstream.

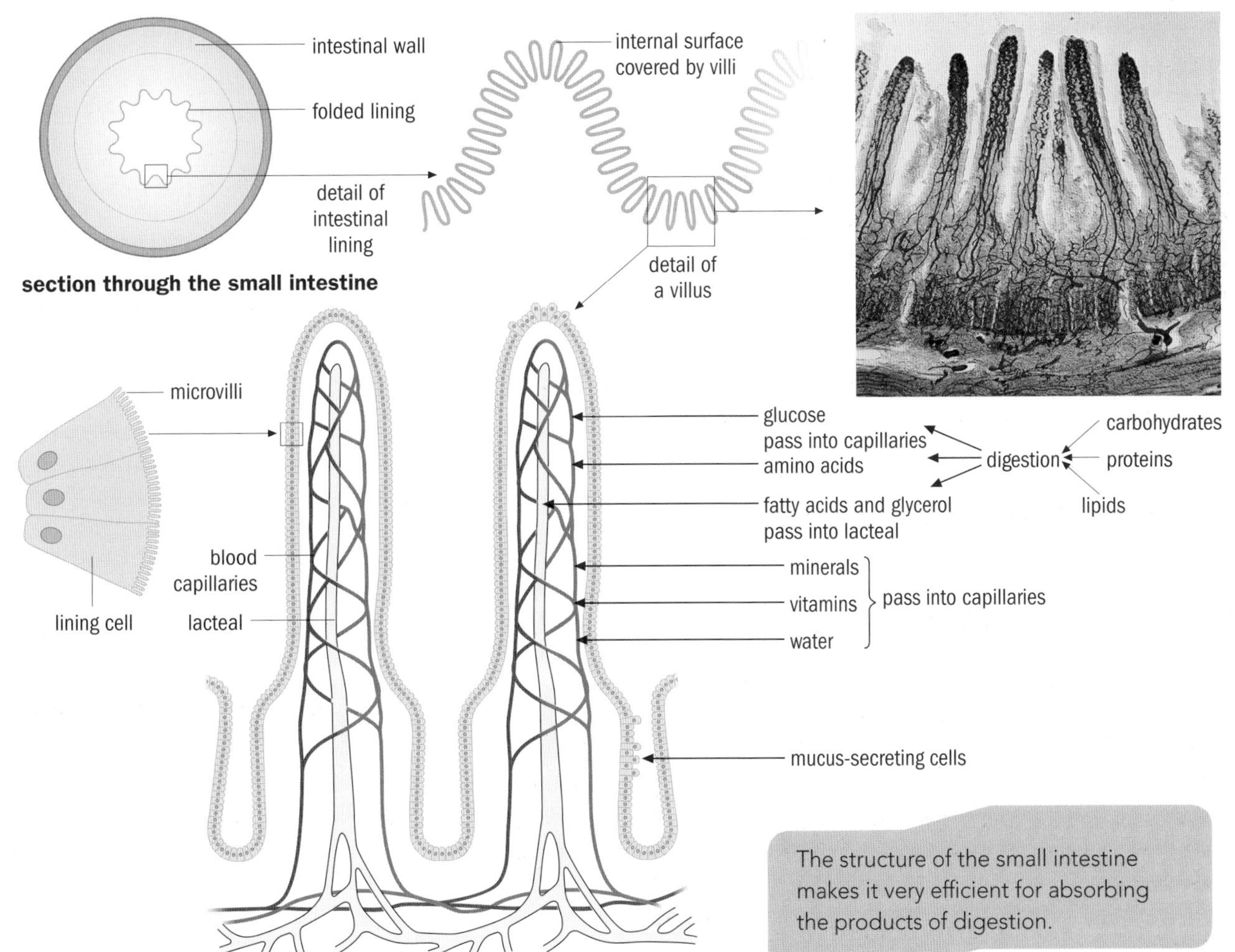

The structure of the small intestine makes it very efficient for absorbing the products of digestion.

Fatty acids and monoglycerides

Fatty acids and monoglycerides are fat soluble and can diffuse through the plasma membranes into the epithelial cells. Here, they pass to the smooth endoplasmic reticulum and are converted back to triglycerides. They pass to the Golgi apparatus and are wrapped in a protein coat to become chylomicrons. These chylomicrons pass into the lacteal in the centre of the villus and then pass through the lymph fluid and eventually enter the bloodstream when the thoracic lymph duct drains into the subclavian vein (see page 228 for more information on lymph).

Some short-chain fatty acids diffuse into the bloodstream in the villi.

Water

Water enters the bloodstream by osmosis.

Other ions

Other ions are absorbed from the small intestine, mainly by active transport through special channels in the plasma membranes of the epithelial cells. Vitamin D is needed to help the absorption of calcium.

Assimilation

When the nutrients have been taken to cells, they are used or stored for later use.

Glucose

The blood glucose levels must be maintained at about 90 mg glucose per 100 cm^3 blood, so that it can be delivered to all cells for respiration, to make the ATP needed to power biochemical reactions in cells. Any extra glucose is taken into liver or muscle cells to be stored as glycogen. Some may also be changed into fat to be stored.

Lipids

Some are used for respiration (heart muscle cells respire fatty acids), some are used to make cell membranes and some are stored in adipose tissue.

Amino acids

These are used to synthesise new proteins for growth and repair. Some may be respired to make ATP.

Vitamins and minerals

Vitamins and minerals are carried to the tissues that need them and used for their functions as outlined on page 218.

The liver and pancreas

The nutrient molecules that pass into the blood in the ileum are carried to the liver in the hepatic portal vein.

In the liver:

- excess amino acids are deaminated, which means the amine group is removed and changed to ammonia, which is very soluble and toxic. The ammonia is then quickly changed to urea, which is less toxic and soluble enough to be carried, in the blood, to the kidneys to be removed
- excess glucose is converted to glycogen (a polysaccharide) to be stored. Insulin, secreted from the pancreas, is needed for this to happen.

As well as secreting digestive enzymes, the pancreas is an endocrine organ and secretes two hormones.

- When blood passing through the pancreas has a high glucose content this triggers the pancreas to produce insulin.
- Insulin particularly targets liver and muscle cells. It makes their plasma membranes more permeable to glucose so these cells take up the extra glucose from the blood.
- It then activates enzymes in the cells to turn glucose into a polysaccharide, glycogen.
- This restores the blood glucose level to normal and provides an energy store.
- Muscle cells can use this glycogen. They break it down to glucose and respire it to make ATP for contraction. However, they do not release glucose into the bloodstream, only the liver does this.
- If blood glucose levels drop then the pancreas secretes another hormone, glucagon, into the blood. Glucagon is carried around the body and targets liver cells.
- Glucagon causes liver cells to break down glycogen to glucose and release it into the bloodstream to restore the normal blood glucose level.

- Neither hormone enters its target cells. They both fit onto special receptors on the target cells' plasma membranes and this activates a series of biochemical reactions inside the cells.

Activity 11.4E

Where do you think the products of nucleic acid digestion are stored in cells?

What are they used for?

PLTS

Self-manager, Effective participator, Independent enquirer, Creative thinker and Reflective learner

You will develop your self-management, effective participator and independent enquiry skills as you carry out the practicals, record the results and share your information with the others in your class. You will further develop your skills of self-management, creative thinking and reflective learning as you synthesise relevant information you have learnt about the three systems and show, in an interesting and understandable way, how these systems are interlinked. You will also develop creative thinking skills to design your poster.

Assessment activity 11.4

1 You are a forensic laboratory technician investigating the nutritional content of foods.

Under the supervision of your teacher, use the appropriate biochemical tests to identify different dietary nutrients in a range of foods and in a mystery solution. Work in pairs and record your results to share with the rest of the class, so that as a class you have tested a wide range of foods. Carry out a risk assessment before you start. P5

2 You are a patient information officer for the gastrointestinal department of a hospital. Produce a large poster showing the structure of the digestive tract and indicate, by annotations, where digestion of carbohydrates, lipids and proteins takes place. In each case show the enzymes that are produced and explain their action. Show clearly the products of digestion. Ensure you add to your poster explanations of the roles of the liver and pancreas in processing digested nutrients. P6

3 When indicating the food that is being broken down by enzymes and the subsequent products of digestion, use chemical equations. M4

4 You are a journalist writing an article for a health magazine. In your article you need to explain to people how the cardiovascular, respiratory and digestive systems are interrelated. You need to illustrate your article with diagrams. D2

Grading tips

For P5 make sure that you test each food and the mystery solution for sugar, starch, lipids and protein. Put your results into a big table. Show how each test was carried out, what the result was and what this shows.

For P6, also add to your poster some information to show how the products of digestion become assimilated into body tissues.

For M4 use information in the 'Chemical structure of nutrients' section (page 220) to help you with chemical equations.

For D2, when writing your article think about the following: the digestive system allows us to break down food molecules so they are small enough to pass across the gut wall into the blood. The blood (part of the cardiovascular system) then carries the nutrients to cells to be stored or used (think of all its uses). Some of the nutrients will be respired in cells to make ATP (Why do we need ATP?) and this also needs oxygen. The respiratory system enables oxygen from air to enter the blood and blood carries the oxygen to cells. What part do breathing and gaseous exchange play? How does removal of carbon dioxide fit in here? How does blood circulate in the body – what parts do vessels and the heart play?

11.5 The lymphatic system

In this section: D3 M5 P7

Antigen – a molecule, usually a protein, with a specific 3D shape, on the plasma membrane of cells.

B lymphocyte –a type of leucocyte that can divide to form plasma cells that secrete antibodies, and memory cells that provide us with an immunological memory.

T lymphocyte –a type of leucocyte that can divide to give helper T and killer T lymphocytes. Helper T lymphocytes stimulate B lymphocytes to divide; killer T lymphocytes destroy infected 'self' cells

Macrophage – a type of leucocyte derived from a monocyte. They can squeeze out though capillary walls and into tissues where they ingest bacteria, viruses or other debris.

The lymph system consists of two parts, each of which performs important functions.

The two parts are:

- a network of lymphatic vessels that collect tissue fluid and return it to the bloodstream
- lymphoid organs and tissues that house phagocytes and lymphocytes.

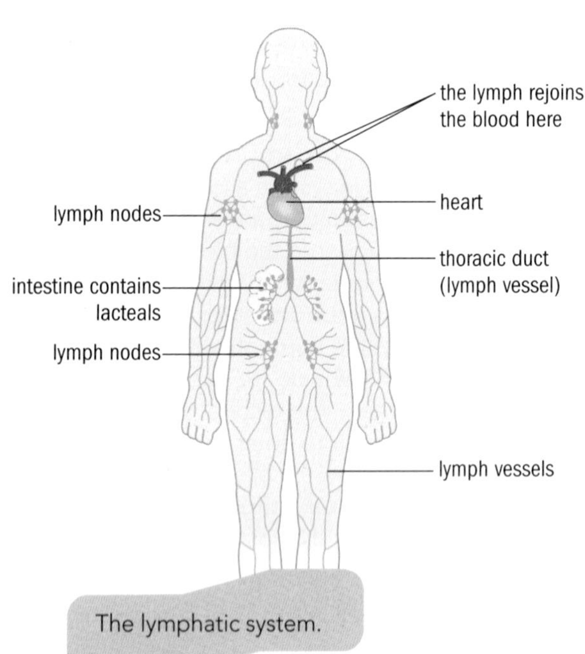

The lymphatic system.

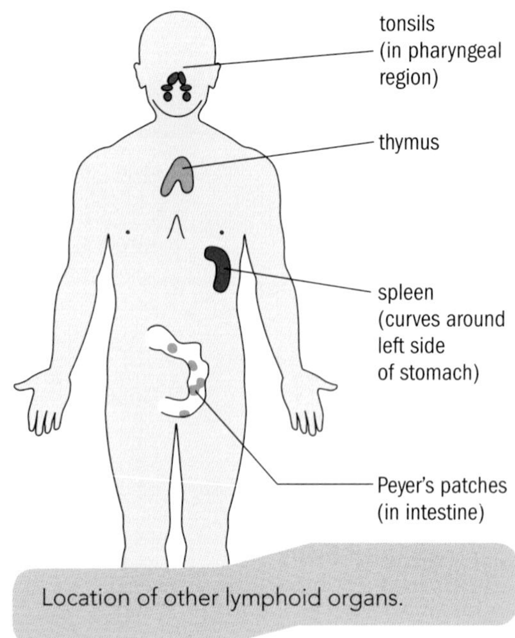

Location of other lymphoid organs.

Lymphatic vessels

These form a one-way system and the lymph fluid inside them flows towards the heart. There are three types of lymph vessel – capillaries, trunks and ducts.

Lymph capillaries

The smallest lymph vessels are called lymph capillaries. These thin-walled, blind-ending vessels in between cells of body tissues collect tissue fluid that has been forced out of blood capillaries but not returned to them. They are drainage vessels. Their walls are made of a single layer of overlapping endothelial cells (a type of epithelium) and this makes them very permeable. Fine filaments anchor the endothelial cells to surrounding structures so, when tissues swell with fluid, the endothelial cells of the lymph capillary wall are pulled apart slightly. As a result tissue fluid between cells, and any proteins in it, can enter the lymph capillaries. The cells making up the wall of these capillaries act like one-way swing doors. When the lymph capillaries are full of fluid they push on the endothelial cells of the walls and shut any gaps between them so fluid cannot leak out again.

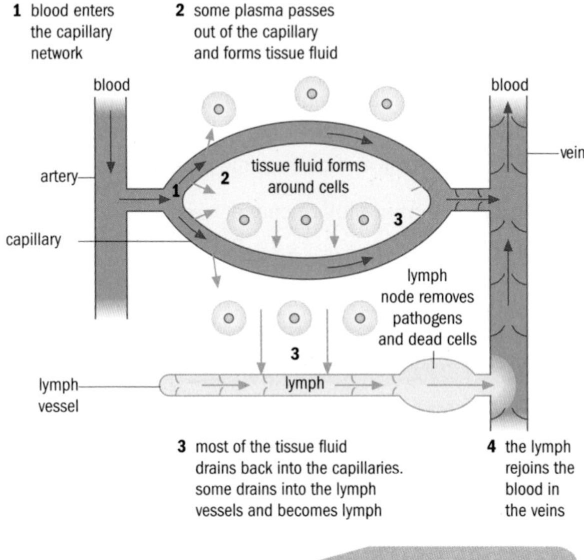

How tissue fluid and lymph are formed.

The lacteals in villi of the ileum are specialised lymph capillaries. The fluid in them is milky white as it also contains absorbed lipids. This fluid is called chyle (pronounced *kyal)*.

Did you know?

All tissues in the body have lymph vessels in them, *except* bone, teeth and the central nervous system.

Lymphatic ducts

The lymph capillaries join to make larger and thicker-walled lymphatic collecting vessels, which join to form lymphatic trunks and then lymphatic ducts. The walls of lymphatic ducts have three layers and are similar to, but thinner than, the walls of veins.

Two large ducts in the thoracic region deliver their lymph fluid into the subclavian veins.

By collecting all the tissue fluid that didn't return to blood capillaries, and delivering it back to the bloodstream, the lymph system enables the cardiovascular system to keep working. The lymph system also collects digested lipids from the ileum and delivers them to the blood.

How does lymph fluid flow?

There is no pumping mechanism within the lymphatic system and the fluid in them is at low pressure.

- The contraction of skeletal muscles near the vessels helps propel the lymph fluid in them.
- There are valves in lymph vessels that prevent backflow of lymphatic fluid.
- In addition, when you breathe in, the pressure in the thoracic cavity reduces and this helps lymph fluid move along the thoracic ducts.
- Lymph trunks and ducts have smooth muscle in their walls and when this contracts it helps propel the fluid.
- These vessels are wrapped in connective tissue near to arteries so the pulsating action of arteries also helps lymph fluid to move.

However, the flow of lymph fluid is very slow but this means that the rate of return of lymph fluid to the blood is equal to the rate of loss of tissue fluid from the bloodstream. If you are physically active, the lymph fluid flow rate speeds up, as does production of tissue fluid from capillaries.

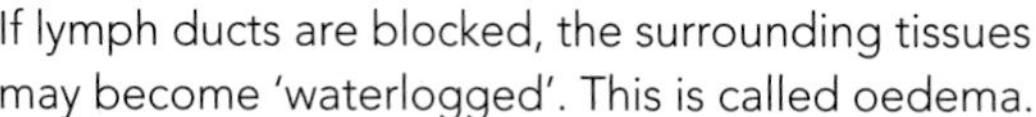

Did you know?

If lymph ducts are blocked, the surrounding tissues may become 'waterlogged'. This is called oedema.

If lymph vessels are surgically removed, new ones eventually grow to replace them.

Lymphoid organs and tissues

Lymph nodes

There are hundreds of small lymph nodes clustered along the lymphatic vessels. They are bean-shaped and less than 2.5 cm long. Each is surrounded by a fibrous capsule. Strands of connective tissue from this capsule extend inwards and divide the node into compartments. Many reticulate fibres support the internal structure and the lymphocytes. Lymph nodes contain many **B lymphocytes**, which may be dividing to produce plasma cells. They also contain **T lymphocytes** that are in transit as these cells circulate between the blood, lymph nodes and lymphatic vessels.

As lymphatic fluid flows through these nodes macrophages engulf bacteria, cancer cells and other particles, such as bits of debris from dead cells. This cleanses the lymph fluid before it enters the blood.

The lymphocytes, which began life in the bone marrow but have migrated to lymph nodes and other lymph organs, can mount an immune response against any pathogens in the lymphatic fluid.

Activity 11.5A

Discuss the following with a partner.

1 Why is it beneficial for lymph fluid to be filtered through lymph nodes before it enters the bloodstream?

2 Why do you think it is particularly dangerous to someone with breast cancer if the underarm lymph nodes become swollen and contain cancer cells?

3 When a patient has an infection, they often go to bed and rest. Why might this hinder the recovery process?

Other lymphoid organs

All the lymphoid organs shown above are composed of reticular connective tissue (see page 203) and contain reticular fibres and free cells – mainly lymphocytes.

- The **spleen** extracts old erythrocytes, thrombocytes (platelets), bacteria and viruses from blood. It stores iron from old erythrocytes and stores thrombocytes.
- The **thymus gland** 'educates' T lymphocytes. T lymphocytes are made in the bone marrow and go to the thymus gland where they are sorted. Any that have receptors that would make them mount an immune response against the body's own (self) tissues are destroyed. Any that are not capable of mounting an immune response are also destroyed. Thus, only immunocompetent T cells remain.
- The **tonsils** form a ring of lymphatic tissue around the back of the throat and around the base of the tubes that connect the pharynx to the middle ear. The tonsils filter out pathogens that are inhaled or taken in with food.
- **Peyer's patches** are similar in structure to the tonsils but are in the wall of the ileum. They contain macrophages that can ingest bacteria and prevent them from crossing the intestinal wall.

Did you know?

If you have swollen glands, infected with bacteria or viruses, these are painful. They are called buboes. This was one of the symptoms of bubonic plague.

Swollen glands that are full of cancer cells are not usually painful.

Activity 11.5B

What do you think would happen if bacteria from the gut managed to pass through the gut wall?

How macrophages, B lymphocytes and T lymphocytes defend us from infections

Macrophages

Macrophages are attracted to invading pathogens by chemicals. They ingest the pathogen and break it down into smaller pieces. They then put pieces of the pathogen's outer membrane (or coat in the case of viruses) that contains the pathogen's **antigens** onto its surface membrane. It is now called an **antigen presenting cell**.

T lymphocytes

- Each antigen has a specific shape. Somewhere in your lymphatic system there is one T lymphocyte that has receptors that fit this antigen.
- The macrophage 'searches' for this T lymphocyte.
- When it is found, the T lymphocyte docks with the antigen on the surface of the macrophage.
- This stimulates the T lymphocyte to multiply and produce helper T cells and killer T cells. Helper T cells stimulate B cells to divide.
- Killer T cells destroy our cells that are infected with viruses and also destroy cancer cells.

B lymphocytes

There is also a B lymphocyte with receptors to fit these antigens.

- When that has been found it multiplies, in the lymph node, and produces lots of identical B lymphocytes.
- Some become plasma cells. These produce antibodies which enter the bloodstream and combat the pathogens. Antibodies may clump viruses together to stop them entering cells or they may coat bacteria so macrophages and neutrophils can ingest them.
- Some form memory cells which stay in the body, ready for a quicker and greater response if that pathogen invades again.

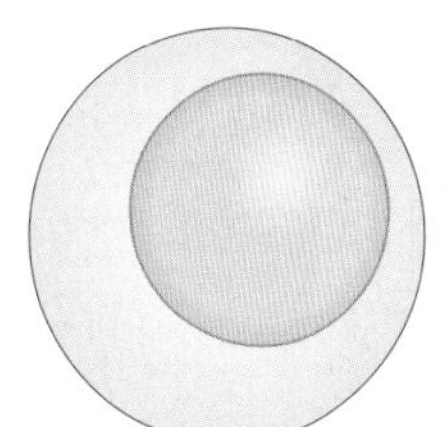

Plasma cells – derived from B lymphocytes. Plasma cells make antibodies

Lymphocytes.

Activity 11.5C

1. Find out about the structure of antibodies and make a 3D model of one. Explain to the rest of the class how antibodies work.
2. Draw an annotated flow diagram to show the stages of the immune response, as described above.

Assessment activity 11.5

D3 M5 P7 BTEC

You are an information officer for a hospital immunology department. You need to produce a poster to display when students visit your department.

Make a poster to show the structure and functions of the lymphatic system. You will need to show the positions of vessels, nodes and other lymphoid organs in the body. Add annotations that describe the functions of these parts of the lymphatic system. P7

Make models or 3D posters or an annotated flow diagram to explain how the lymphatic system protects the body. You need to include reference to the filtering at nodes and to the roles of macrophages and B and T lymphocytes. M5

Prepare a presentation to explain how the lymphatic functions during good health and infection may differ. Remember that every day our bodies are exposed to pathogens and the lymph nodes filter out bacteria but we do not suffer any symptoms and are not aware of this. If pathogenic bacteria invade in suitable numbers then we need to mount an immune response. D3

Grading tips

For P7 make your poster large and clear. Use biology textbooks and the Internet for information about the lymph vessels, lymph nodes and other lymphoid organs.

For M5 you need to show *how* the lymphatic system protects the body so explain how filtering at the nodes helps and how macrophages and B and T lymphocytes help.

For D3 clearly show how filtering by lymph nodes works all the time – when we are healthy. Explain how the immune response occurs when we are infected.

PLTS

Independent enquirer, Reflective learner and Effective participator

You will develop your skill as an independent enquirer when researching the information you need; reflective learner when using information sources to extract relevant information and synthesise it; effective participator when giving your presentation.

WorkSpace

Tom Warrender

Founder and Director, 'Classroom Medics'

I design and run workshops for educational establishments. These workshops provide hands-on use of medical equipment which test the parameters of the human body. I also create videos which show how this equipment works. My background is in human physiology and the subject inspired me to want to show students how studying physiology can open doors to many careers.

On a day when I am running a workshop, I will travel to a school or college and set up the different pieces of equipment such as an electrocardiogram (for measuring heart rate), a video otoscope (for looking in the ear), and a patient simulator (a lifesize mannequin which simulates human reactions and diseases).

Stan, our patient simulator, is an important member of our team! He has a heartbeat, he sweats and he blinks. We use him to simulate, for example, how the body reacts to drugs, what treatments are needed for certain diseases, and the effect of various injuries. He is attached to a monitor that studies his vital signs.

What I enjoy the most about my job is using all the equipment and designing practical experiments which enhance students' understanding of the human body. It is also extremely satisfying to help students make a more informed decision about their health or career choice.

We very occasionally encounter the problem of a student fainting when a fake blood sample is taken, but in general students find the hands-on aspect of the workshop extremely interesting because the equipment gives them a greater understanding of physiological concepts.

One of the areas we concentrate on both in the workshops and the videos is sports physiology. We use our equipment to measure reaction times, power output and strength.

Think about it!

1 Having studied human physiology, which careers could be open to you?
2 What type of equipment would you use in the field of sports physiology?

Just checking

Working in groups, discuss what you know about body systems and answer these questions.

1. Draw a eukaryote cell and label the plasma membrane, nucleus, mitochondria, endoplasmic reticulum, Golgi, ribosomes.
2. Make a flow diagram to indicate the levels of organisation within a multicellular organism.
3. Name six systems in the body.
4. State the functions of the systems you have named.
5. Draw a mind map/spider diagram to show how the cardiovascular, respiratory, digestive and lymphatic systems are interrelated.
6. Draw a mind map to show how food affects our health (show positive and negative effects).

Assignment tips

To get the grade you deserve in your assignments remember to do the following:

- If your assignment asks you to prepare a scientific report, then remember that you need to include a plan, equipment, risk assessment, results, and a conclusion of your findings.
- If you are presenting information in a poster, it needs to be visually stimulating and not too crowded. Make the text big enough so that it is easy to read.
- If you are writing a report for a magazine, remember who is going to be reading it and present the information so it is easy for them to understand. Make sure you include only the necessary information. Make sure you understand what you are writing as you may be asked questions about it. Make sure that you can pronounce all the words correctly so that you can read aloud the article to your class if asked to.

Some of the key information you'll need to remember includes the following:

- cell structure of eukaryotic cells
- the different tissue types – epithelial, nervous, muscle and connective tissues
- levels of organisation: cells are grouped into tissues; tissues are grouped into organs; organs are grouped into systems which make up the whole organism
- the difference between magnification (how many times enlarged) and resolution (clarity of image) and the advantages and disadvantages of light microscopes and electron microscopes
- the need for a transport system and the structure and functions of the cardiovascular system (blood, blood vessels and heart)
- the structure and functions of the respiratory system (nose, mouth, trachea, lungs, ribcage and diaphragm)
- the structure and functions of the digestive system (gastrointestinal tract plus liver and pancreas)
- how the cardiovascular, respiratory and digestive systems are interrelated
- the immunological function of the lymphatic system.

You may find the following websites useful as you work through this unit.

For information on...	Visit...
The Wellcome Trust	The **Wellcome Trust** website
cells	**CELLS Alive!**
The Society of Biology	**Society of Biology** website
The lymphatic system	**Lymphoma information** website

Other useful sources of information are journals and magazines such as *New Scientist, Scientific American, National Geographic, Focus* and *The Economist.*

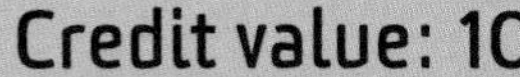

13 Biochemistry and biochemical techniques

Over the last 40 years, biochemistry has successfully explained so many living processes that now in almost all areas of science that involve living things, such as botany and medicine, scientists are carrying out biochemical research in order to increase our understanding of natural processes. Today, the main focus of biochemistry is to understand how biological molecules are involved in the processes that occur within all living cells. This knowledge should greatly improve future studies of entire organisms.

This unit provides you with an insight into the chemical reactions that allow all living things on the planet to survive. Biochemistry is the study of these chemical processes. The chemicals involved need to have structures that allow them to do certain jobs in any living organism. One part of a biochemist's job is to identify these chemicals using laboratory techniques and be able to examine their complex structures. Your body has the ability to change the speed of biochemical reactions, so it is important that a biochemist knows how such reactions can be controlled. One such reaction, respiration, is essential to your survival as it provides you with the energy you need to exist. Understanding such a vital reaction can show biochemists how the human body actually works. Biochemistry overlaps with many other areas of study such as pharmacology, physiology, microbiology and clinical chemistry.

Learning outcomes

After completing this unit you should:

1. be able to investigate properties of water and biological molecules in living organisms
2. understand the structure of proteins
3. be able to investigate the factors that affect the activities of enzymes in biological systems
4. know the difference between aerobic and anaerobic respiration.

Assessment and grading criteria

This table shows you what you must do in order to achieve a pass, merit or distinction grade, and where you can find activities in this book to help you.

To achieve a **pass** grade the evidence must show that you are able to:	To achieve a **merit** grade the evidence must show that, in addition to the pass criteria, you are able to:	To achieve a **distinction** grade the evidence must show that, in addition to the pass and merit criteria, you are able to:
P1 identify the structure and function of biological molecules **See Assessment activity 13.1**	**M1** interpret data obtained from experiments designed to separate biological molecules **See Assessment activity 13.1**	**D1** discuss the relationship between structure and function for carbohydrates and lipids **See Assessment activity 13.1**
P2 use laboratory separative techniques for the characterisation of biological molecules **See Assessment activity 13.1**		
P3 describe the primary and secondary structures of proteins **See Assessment activity 13.2**	**M2** explain the tertiary and quaternary structures of proteins **See Assessment activity 13.2**	**D2** discuss the relationship between the structure and function of proteins **See Assessment activity 13.2**
P4 describe the structural features of an enzyme **See Assessment activity 13.3**	**M3** explain how the rates of enzyme reactions are affected by changes in temperature **See Assessment activity 13.3**	**D3** explain why enzyme activity is affected by pH **See Assessment activity 13.3**
P5 use laboratory methods to investigate factors that affect the enzyme rate of reaction **See Assessment activity 13.3**		
P6 identify the main differences between anaerobic and aerobic glucose degradation **See Assessment activity 13.4**	**M4** compare the sites of ATP production and consumption during aerobic and anaerobic breakdown of glucose in cells **See Assessment activity 13.4**	**D4** evaluate the regulation of glycolysis in terms of energy requirements in cells **See Assessment activity 13.4**

How you will be assessed

Your assessment could be in the form of:

- presentations
- practical laboratory tasks
- written scientific reports.

James, 17 years old

This unit helped me see how relatively simple molecules are so important to the human body. Without these chemicals my body wouldn't operate in the correct way. It was interesting to see what the biochemicals look like and how this decides what role they carry out in the body. I enjoyed seeing what kind of work biochemists do, as I am considering starting a career in the area of biochemical laboratory work after I've finished my studies. I was surprised at the number of different uses enzymes can have in various industries. I liked carrying out the practical activities that investigated enzymes. I think I have improved my laboratory technique through practice, as my results provided enough evidence to form a convincing conclusion. I also spent time carrying out extensive research on the Internet about what some enzymes do in my body, and making my own model enzyme helped me visualise how these biological catalysts work.

I enjoyed looking at how the body can use the energy it gets from food. I had never heard of the ATP molecule that is our source of energy. On first sight, some of the pathways looked very complicated, but I could see what was happening after I followed a glucose molecule through the diagrams.

Catalyst

Chemicals in your body

Think of three chemicals that you know are in your diet. Discuss in small groups what you think your body needs them for and where in the body they are found. What might happen to your body if it does not take in these chemicals?

13.1 The properties of water and biological molecules in living organisms

In this section:   

Key terms

Polarity – the property of molecules having an uneven distribution of electrons, so that one part is positive and the other part is negative.

Functional group – the group of atoms in a compound that are responsible for its reactions.

Monomer – a molecule that joins with others to form a polymer.

Polymer – a large molecule consisting of repeating units or monomers.

Covalent bond – a type of chemical bond between atoms that share pairs of electrons.

Solvent – normally a liquid that dissolves another chemical to form a solution.

Hydrogen bond – an attractive force between a hydrogen atom and either a nitrogen, oxygen or fluorine atom.

pH – a measure of the acidity or alkalinity of a solution.

Condensation reaction – a reaction where a small molecule, usually water, is expelled from the reactants.

Hydrophilic – describes a molecule, or part of a molecule, that can form hydrogen bonds with water or other polar solvents, allowing them to dissolve.

Hydrophobic – describes a molecule, or part of a molecule, that is non-polar and is repelled by water, but may dissolve in non-polar solvents.

Complementary base pairs – two nucleotides on opposite nucleic acid strands that form hydrogen bonds between each other.

Water properties

Water is a common but very important chemical. Your body weight is made up of about 70% water. Water is involved in food digestion, transporting chemicals and controlling body temperature. Your body needs to replenish the water it has, so many people think it is a good idea to drink about two litres of water every day. Without water you could become tired, dizzy, and eventually so dehydrated that your body could not function properly.

Polarity

A water molecule is made from one oxygen atom and two hydrogen atoms, bonded together by two **covalent bonds**. Each bond consists of a shared pair of electrons: one electron from the oxygen atom and one electron from the hydrogen atom.

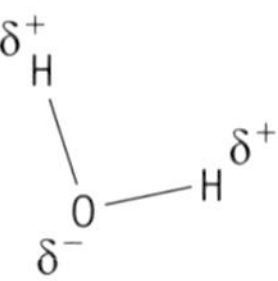

A water molecule showing its polarity.

The oxygen atom has the ability to attract the negative electrons away from the two hydrogen atoms. This means that the oxygen–hydrogen bond has a negative end on the oxygen and a positive end on the hydrogen. This gives the molecule **polarity**. This type of covalent bond is called a polar bond. You should note that these charges are very small. The oxygen atom in water has two pairs of electrons that are not involved in bonding. These are called lone pairs of electrons.

Water as a solvent

Water is often referred to as the universal **solvent** because it can dissolve so many chemicals. The polarity of the water molecule enables water to dissolve other chemicals that have charges on them. The most common example is salt, or sodium chloride, which is found in sea water. Sodium chloride contains positive sodium ions and negative chloride ions. The slightly negative oxygen atom of a water molecule can attract the positive charge on a sodium ion. Many water molecules surround the sodium ions and keep them separated from the chloride ions, so that they are dissolved.

Hydrogen bonding

Because of the polarity of water, one water molecule can form an attraction to another water molecule. The lone pair of electrons on the oxygen atom can attract the slightly positive hydrogen of another water molecule. This attraction, although not a bond as such, is called a **hydrogen bond** and is a type of intermolecular force.

Boiling point

In a glass of water, the water molecules are in motion, continually making and breaking the hydrogen bonds between them. In order for water to boil, all of the hydrogen bonds must be overcome, so that there are no longer attractions between the molecules and the liquid water can become a gas. This is done by providing the water with energy, usually in the form of heat, which makes the water molecules move even faster. The hydrogen bonds are weak, however, because there are so many of them it takes a lot of energy to make all of the water molecules move around fast enough to break all of the hydrogen bonds. If they have enough energy, the water molecules can separate from each other and form bubbles of vapour within the liquid which then rise to the surface and escape. This explains why water has a boiling point of 100°C, which is a high boiling point for such a small molecule. For example, oxygen, which has similar-sized molecules to water, has a boiling point of −183°C; carbon dioxide, which has slightly larger molecules than water, has a boiling point of −57°C.

Melting point

If liquid water is cooled to below 0°C, then the water molecules have very little energy and don't move around very much. So the hydrogen bonds that form between molecules of water don't tend to break again. This results in solid ice being formed. To turn the ice into liquid water again, more energy is needed to break the hydrogen bonds, so the temperature needs to be raised above its melting point of 0°C. The hydrogen bonds give water a relatively high melting point for such a small molecule. Oxygen melts at −218°C and carbon dioxide melts at −78°C.

Water's involvement as a buffer solution

Water is involved in maintaining solutions in the human body at the correct **pH**. This is a measure of the acidity or alkalinity of a solution. Buffer solutions, such as that found in blood, stop the pH changing if small amounts of hydrogen ions or hydroxide ions are added. Without water these solutions would not be able to keep the pH at the required value.

Activity 13.1A

Working with a partner, discuss three properties of water that are directly the result of the polarity of the hydrogen–oxygen bond.

Biological molecules and their structural characteristics

Living organisms are made up of many different types of chemicals. Each one has a different structure and a different role to play.

Carbohydrates

Carbohydrates have many functions in organisms. These include acting as an energy source, energy storage and structural support. Carbohydrates are made from carbon, hydrogen and oxygen atoms. The general formula for a carbohydrate is $C_n(H_2O)_n$ where n is the number of carbon atoms.

Carbohydrates that are energy sources

The most common energy source in the body is a sugar called glucose. Glucose releases energy from its bonds when respiration occurs.

glucose + oxygen → carbon dioxide + water + energy

The energy is stored in a chemical called adenosine triphosphate (ATP), which can be used by cells as an energy supply.

Glucose molecules can exist as a straight chain or as a ring, and there are two different ring structures.

- Alpha glucose has the –OH **functional group** on carbon 1 below the plane of the ring.
- Beta glucose has the –OH functional group on carbon 1 above the plane of the ring.

These two forms have different functions in living things. For example, only alpha glucose can undergo respiration in plants and animals. This is because the enzyme used to speed up respiration can only fit with the shape of alpha glucose.

Glucose is described as a monosaccharide. This is a simple carbohydrate that has the ability to form long chains. Two monosaccharide units (or **monomers**) can be joined together in a reaction that removes one water molecule from the two units. Because water is made, this is called a **condensation reaction**.

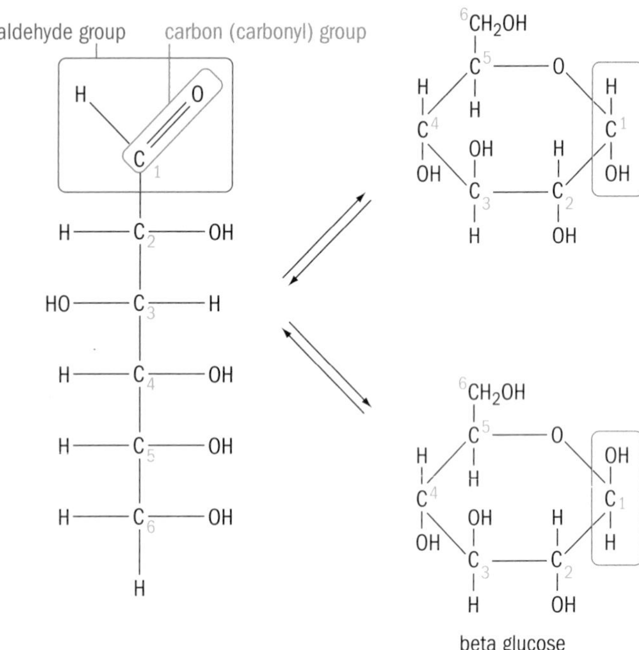

The structure of a glucose molecule as a chain and a ring.

glycosidic bond

Two monosaccharides form a disaccharide in a condensation reaction.

A disaccharide is formed from the two monosaccharide units. These join together by forming a bond between carbon 1 of one glucose molecule and carbon 4 of another glucose molecule. This bond is called a 1,4 glycosidic bond. The bond can be broken using water in a hydrolysis reaction, forming two monosaccharides again.

A polysaccharide is produced when a large number of monosaccharide units form glycosidic bonds. Polysaccharides are called **polymers** because they are made up of many identical monomers in a chain. Polysaccharides have functions that include energy storage and structural support.

Activity 13.1B

Working in pairs, one person should explain to the other what a condensation reaction is, then the other person explains what a hydrolysis reaction is. Try and explain the difference together.

Carbohydrates that are used for energy storage

When thousands of alpha glucose monomers are joined together by glycosidic bonds, a polysaccharide called amylose is formed. Because of the shape of the glucose units and the position of the 1,4 glycosidic bond, amylose has the shape of a coiled spring.

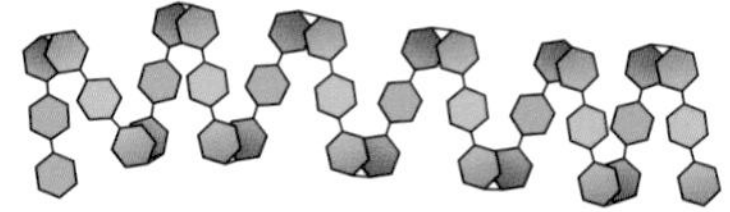

Amylose has the shape of a coiled spring resulting from 1,4 glycosidic bonds.

Plants store their food in the form of starch. This is a mixture of amylose and another polysaccharide called amylopectin. Plants can use enzymes to break down starch into glucose monomers. These can then be used in respiration to release energy for the plant to grow.

Animals store energy in the form of glycogen. This is a polysaccharide that contains alpha glucose monomers arranged in a branched chain. Glycogen can be broken down by enzymes into alpha glucose monomers. These can then be used in respiration to release energy for the body to use.

Both starch and glycogen contain alpha glucose monomers bonded together in long chains. When energy is needed, a monomer can be broken from the end of the polysaccharide and used in respiration.

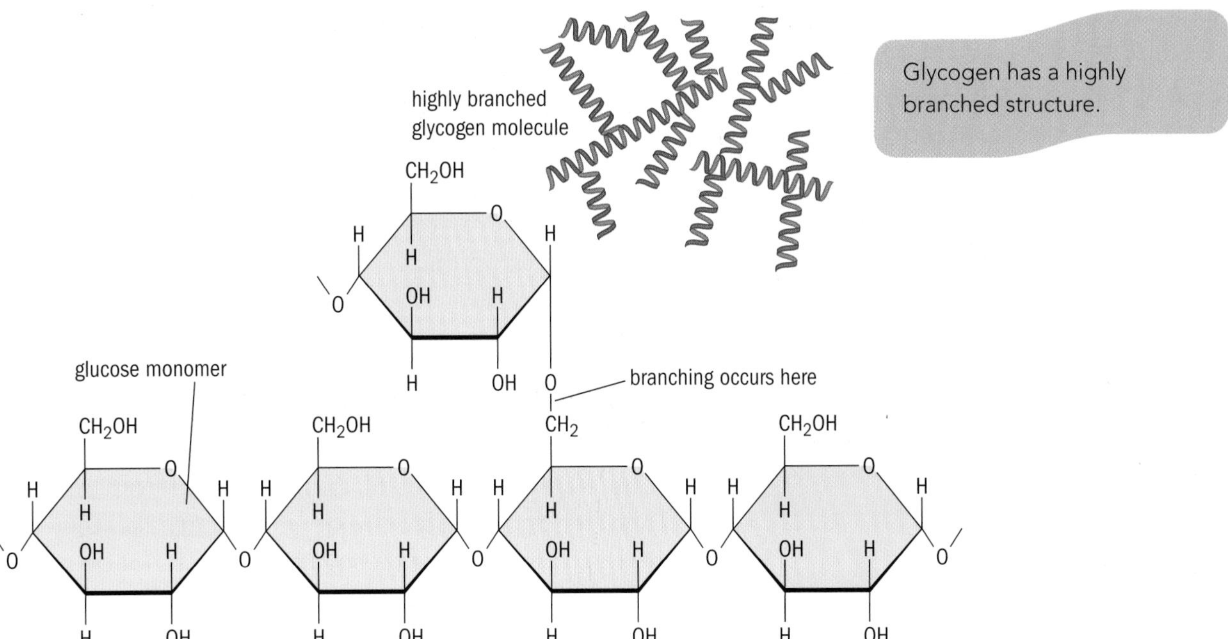

Glycogen has a highly branched structure.

Carbohydrates that provide structural support

When beta glucose monomers are joined together by condensation reactions, the polysaccharide that is produced is not coiled like amylose. A beta glucose chain is straight. This is because a beta glucose molecule has a slightly different shape.

When thousands of beta glucose monomers are joined together, cellulose is produced. Cellulose is only found in plants, where it forms cell walls that are very strong. This function of cellulose is a result of its structure. Cellulose chains are straight, so the chains line up against each other and form hydrogen bonds between adjacent hydrogen atoms and oxygen atoms.

Beta glucose monomers can form a straight-chained polymer.

hydrogen bonding is possible between these OH groups and an adjacent cellulose chain.

Adjacent cellulose chains are linked by hydrogen bonds.

The combined strength of these hydrogen bonds gives a very strong structure called a microfibril. Many microfibrils can be held together by hydrogen bonds forming a macrofibril, which is almost as strong as steel.

Because of the structure of cellulose, plant cells are very strong and give the whole plant a rigid structure.

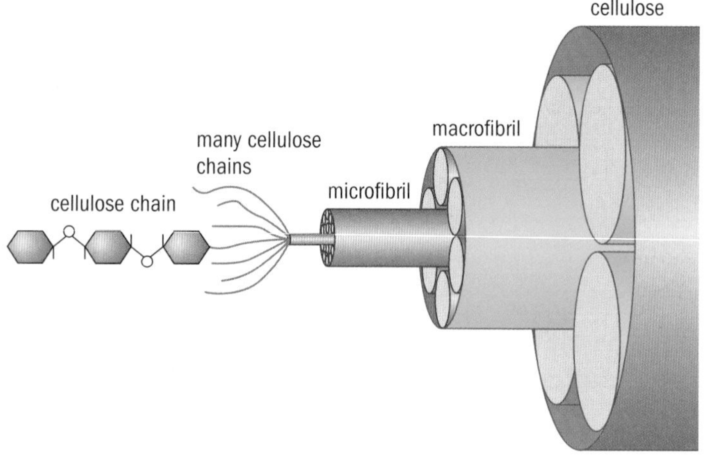

The microfibrils and macrofibrils that make up cellulose fibres.

Activity 13.1C

How does the difference between the positions of the –OH group on the first carbon of alpha glucose and that of beta glucose affect the shape of the polymers they form?

Lipids

Lipids are molecules that can be used as an energy source. However, they are mainly used as an energy *store*. They are used as a component of cell membranes, insulating materials in the body, protective layers and hormones. Lipids are made from carbon, hydrogen and oxygen atoms. They are commonly called fats when solid, and oils when liquid.

Lipids are made from glycerol and fatty acids. Glycerol contains three –OH functional groups which can react with three fatty acids to form a lipid.

The structure of glycerol.

There are many types of lipid; they differ from one another depending on what type of fatty acid they are made from. Many of these fatty acids cannot be made by your body, so they need to be ingested as part of your diet.

The structure of a fatty acid.

All fatty acids have a carboxyl functional group (COOH) on one end and a long hydrocarbon chain on the other end. The hydrocarbon chain can be either:

- saturated, containing just carbon–carbon single bonds (C–C), or
- unsaturated, containing carbon–carbon double bonds (C=C).

Did you know?

Unsaturated fats (or lipids made from unsaturated fatty acids) in your diet tend to reduce the amount of cholesterol in your body when compared to saturated fats.

Saturated fatty acids

Saturated fatty acids form straight chains. This means that many fatty acid molecules can line up against each other easily, forming attractions between all of the molecules. These attractions require energy to

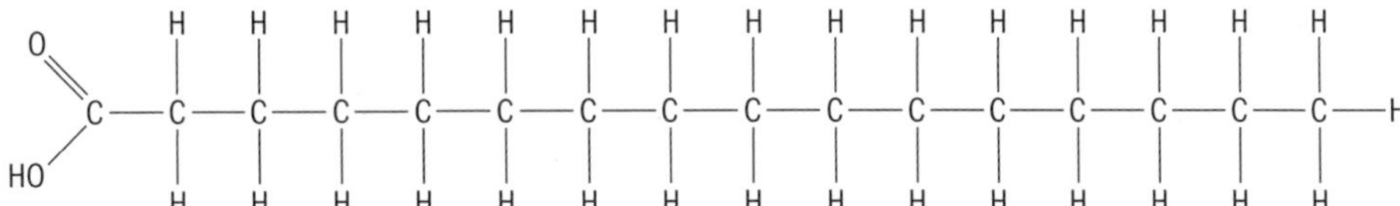

The structure of palmitic acid.

overcome them. As a result, saturated fatty acids have relatively high melting points and tend to be solids at room temperature, and lipids containing saturated fatty acids also have high melting points.

Palmitic acid is an example of a saturated fatty acid. This is because it has the maximum number of hydrogens bonded to the carbons in the hydrocarbon chain; it has no C=C bonds. It has a melting point of 63°C so is solid at room temperature.

Unsaturated fatty acids

Unsaturated fatty acids form bent chains. Unlike saturated fatty acids, their shape means that they push away from each other and attractions between molecules are not formed as strongly. They have relatively low melting points and tend to be liquids at room temperature. Lipids containing unsaturated fatty acids also have low melting points because of their bent shape.

Oleic acid is an unsaturated fatty acid because it has a double bond between a pair of carbons in the hydrocarbon chain. It has a melting point of around 13°C so is liquid at room temperature.

Activity 13.1D

Explain the difference in structure between saturated fats and unsaturated fats. How does this affect the melting points of these two types of fat?

Triglycerides

Triglycerides are a type of lipid used for energy storage, insulation and protection. They are found in fatty tissue under your skin and surrounding your organs. Triglycerides are insoluble in water.

A triglyceride molecule is made when one molecule of glycerol joins together with three fatty acid molecules. The three –OH functional groups of the glycerol molecule form bonds with the acid functional groups of three fatty acids. The three new bonds that are formed are called ester bonds. The reaction is a condensation reaction because water is made.

Triglycerides can be used to generate more energy than carbohydrates based on their mass. However, carbohydrates can generate energy more quickly. The ester bonds in triglycerides must be broken using water in a reaction called hydrolysis. This is the condensation reaction in reverse. The fatty acids can then undergo respiration, releasing energy used to form ATP molecules.

The structure of oleic acid.

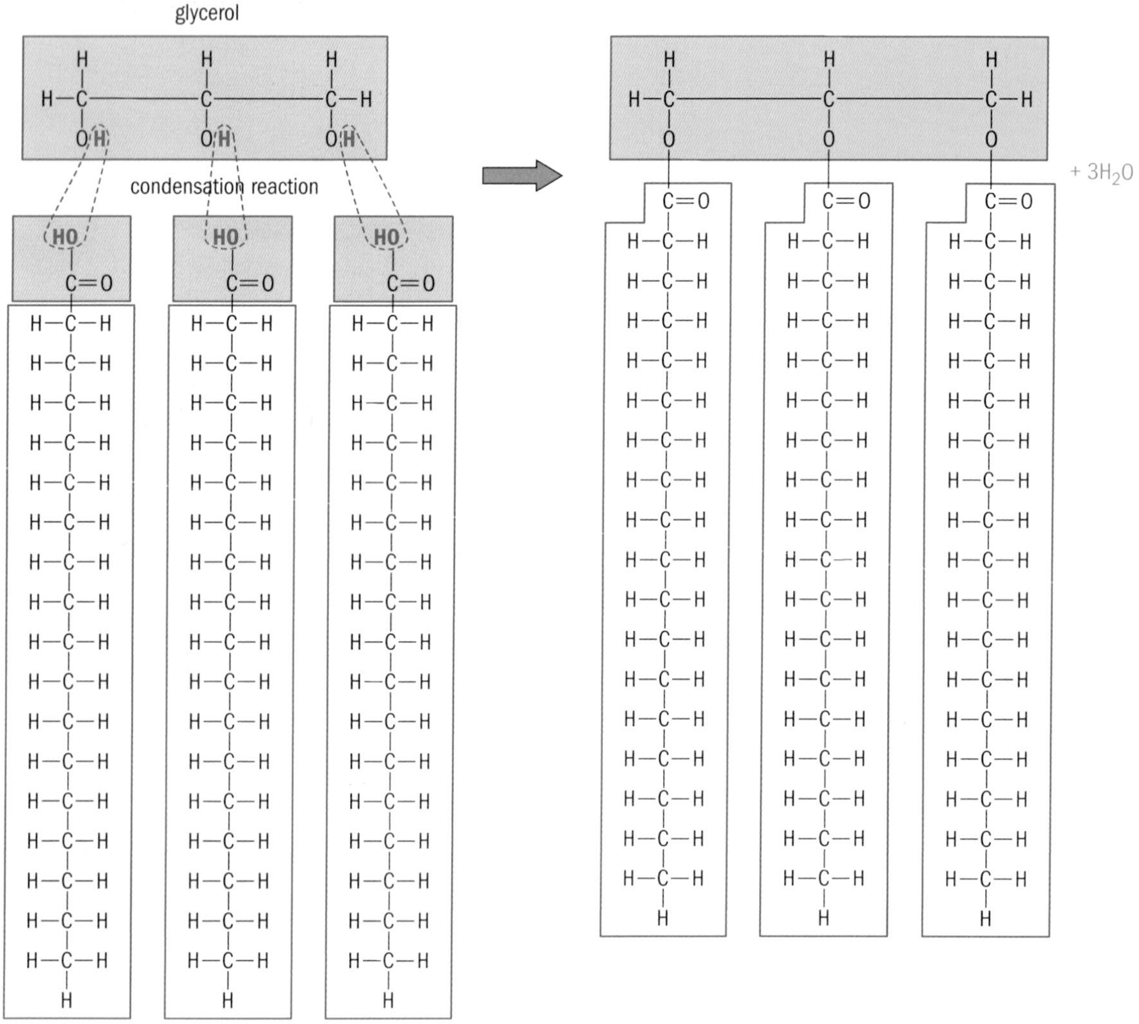

glycerol

condensation reaction

$+ 3H_2O$

three fatty acid molecules

triglyceride

A triglyceride is formed from a condensation reaction.

Phospholipids

Phospholipids are a type of lipid found in cell membranes. They are made when one molecule of glycerol bonds with two fatty acids and one phosphate group via a condensation reaction. Phospholipids have a negatively charged phosphate group which is soluble in water, or **hydrophilic**. They also have an uncharged hydrocarbon chain which is insoluble in water, or **hydrophobic**. This dual solubility allows phospholipids to form cell membranes.

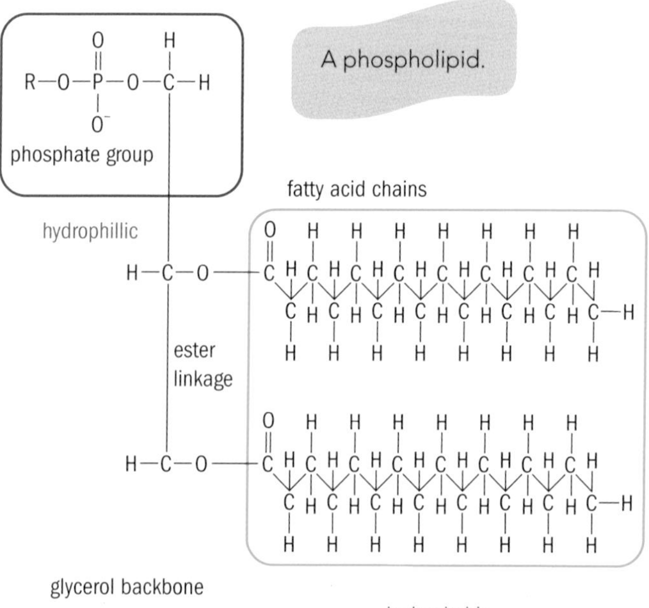

A phospholipid.

Other lipids

There are other lipids which don't have the glycerol-based structure of triglycerides and phospholipids. Cholesterol is one example. Cholesterol is used in cells to provide rigidity. It is an important lipid because it is also used to form steroid hormones such as testosterone and oestrogen. Hormones play roles in communication within and between cells.

The structure of cholesterol.

> **Did you know?**
>
> If your body has very high cholesterol levels you could suffer from cardiovascular disease.

Proteins

The cells in your body are 50% protein. Proteins have many functions, including forming structures (such as muscle), enzymes, hormones and antibodies. They are important for an organism's growth and repair.

Proteins are studied later in this unit on page 251.

Nucleic acids

Nucleic acids contain the genetic information that produces every living cell on the planet. They include DNA (deoxyribonucleic acid) and RNA (ribonucleic acid). They are called nucleic acids because they were first found in the nuclei of cells and they are weak acids. These acids are polymers made from monomers called nucleotides.

Unit 18: See page 322 for more information on nucleic acids.

Nucleotides are made up of three parts:

- a phosphate group
- a five-carbon sugar molecule, which is deoxyribose in DNA or ribose in RNA
- an organic nitrogenous base.

There are five different organic nitrogenous bases, which all contain the elements carbon, hydrogen, oxygen and nitrogen. These are adenine, thymine, guanine, cytosine and uracil. Thymine is found only in DNA and uracil is found only in RNA.

A phosphate group.

Deoxyribose and ribose.

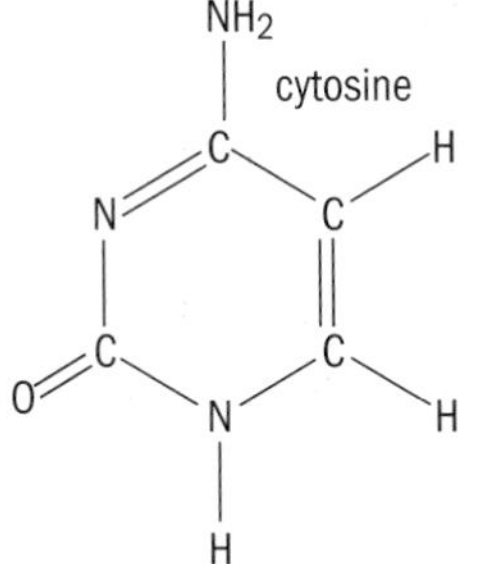

The five nucleotide bases. A, adenine; G, guanine; T, thymine; U, uracil; C, cytosine.

A nucleotide is formed when the hydroxyl functional group (–OH) of the sugar reacts with the hydrogen atom of an –OH from the phosphate. This reaction expels water, so it is a condensation reaction. The bases also undergo a condensation reaction between an –OH functional group on the sugar and a hydrogen atom from the base.

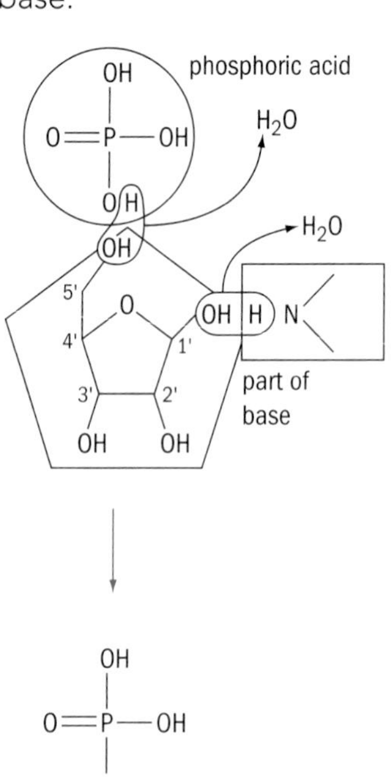

The formation of a nucleotide.

Activity 13.1E

Use labelled diagrams to explain how nucleotides are formed.

Nucleotides are the monomers that bond together to form the nucleic acid polymer. Nucleotides join together by a condensation reaction. The reaction occurs between the –OH functional group from the phosphate group of one nucleotide and the hydrogen from the –OH group on carbon 3 of the sugar of another nucleotide. The bonds that link the nucleotides together are strong phosphodiester bonds. Note that the bases do not take part in this polymerisation. When many nucleotides bond together, a sugar–phosphate backbone is formed, with bases extending from it.

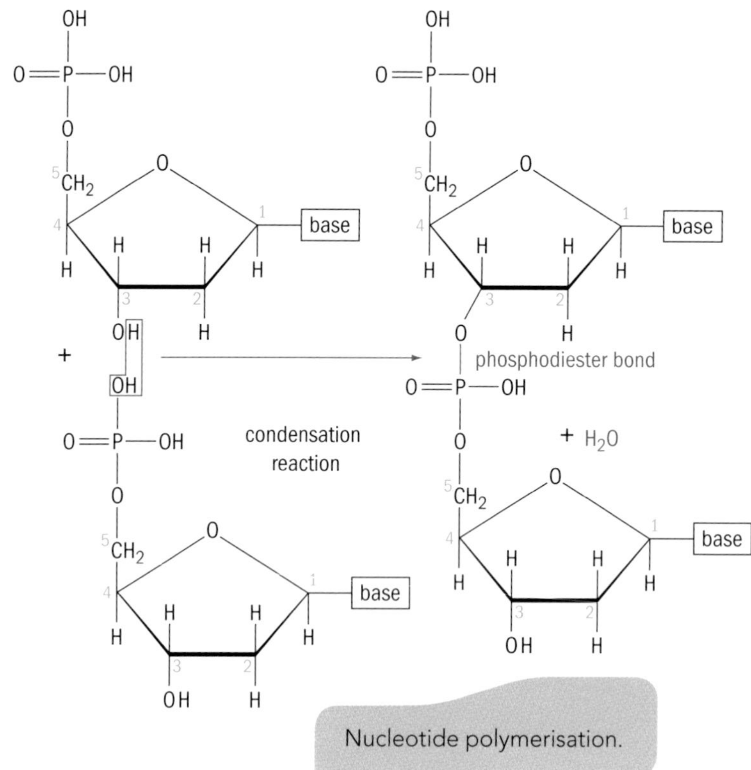

Nucleotide polymerisation.

DNA is a nucleic acid, or polynucleotide, that contains a deoxyribose sugar unit. DNA stores genetic information that builds organisms. DNA can do its job in the nucleus of a cell because of its structure.

DNA is made of two polynucleotide chains that lie alongside each other. The sugars of the polynucleotides point in opposite directions, so the two chains are antiparallel. The two strands twist around each other to form a double helix.

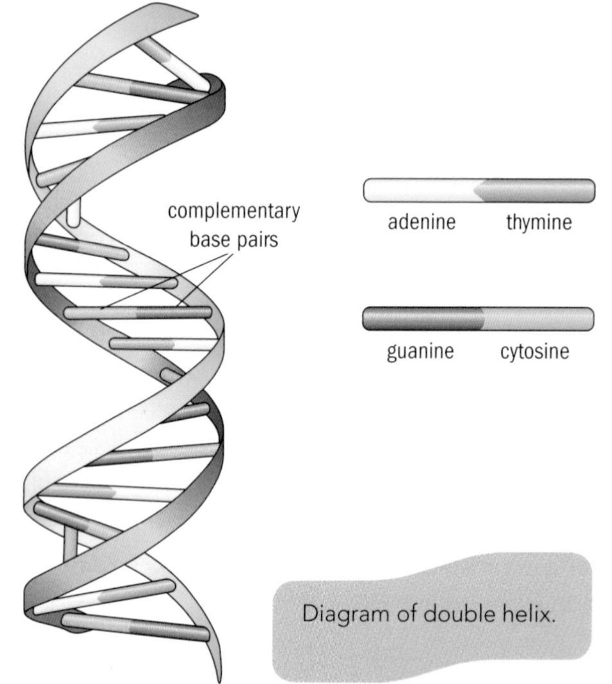

Diagram of double helix.

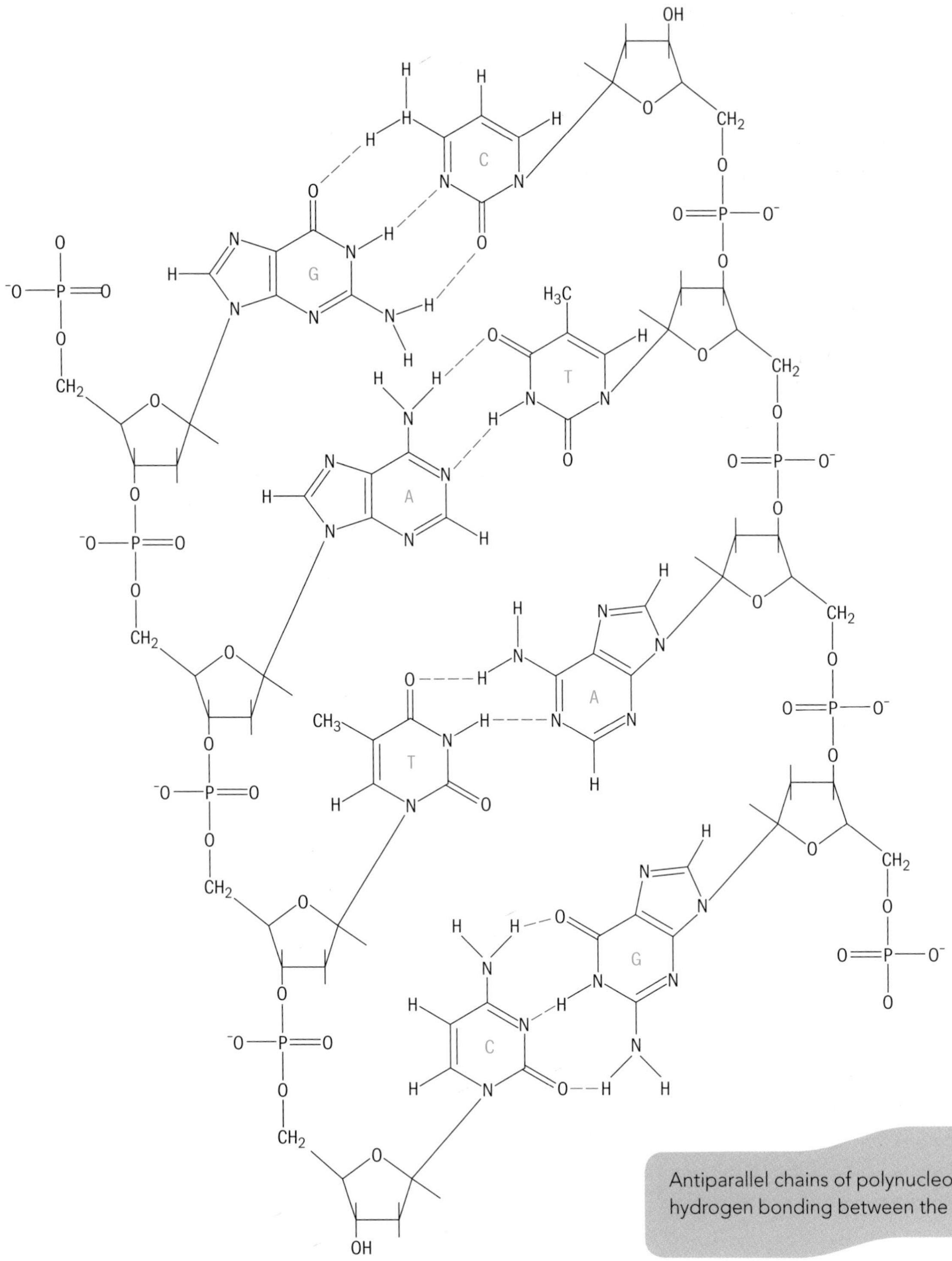

Antiparallel chains of polynucleotides with hydrogen bonding between the bases.

The two strands are joined together by hydrogen bonds between the organic nitrogenous bases. This makes DNA very stable. The bases form base pairs. A only binds to T; C only binds to G. These are called **complementary base pairs**.

DNA stores the information that is used to build organisms in the order of the base pairs that are bonded together in the double helix. This sequence gives the genetic code. Because DNA is a very long polymer, large amounts of genetic code can be stored. This code is very important for the organism to function properly, and the stable double helix helps protect the information. The weak hydrogen bonds which provide this stability can be broken, in order for the genetic code to be used. However, the hydrogen bonds can easily be formed again, so the double helix has the ability to be unzipped and zipped up again.

Unit 18: See page 324 for more information on how the genetic code is used and the unzipping of the DNA double helix.

RNA is a nucleic acid like DNA; however, there are some important differences.

- RNA contains the sugar unit ribose, not deoxyribose.
- RNA can contain the base uracil instead of thymine.
- RNA is usually a single polynucleotide.

RNA is used to read the genetic code from DNA after it has been 'unzipped'. RNA carries the genetic code from the DNA in the nucleus to the ribosomes, where the code is read and amino acids are assembled into proteins.

Unit 18: See page 324 for more information on RNA and protein synthesis.

Laboratory techniques

The biological molecules described above can all be analysed in the laboratory using a variety of methods.

Electrophoresis

Electrophoresis can be used to separate DNA fragments so that they can be identified. The technique uses the fact that DNA chains have a negative charge because they contain phosphate groups. This means that they can be made to move using an electric current. Shorter fragments of DNA will move faster than longer lengths, so the different fragments can be separated.

Electrophoresis uses a gel plate made from a type of sugar covered in a buffer solution. The buffer solution prevents any large changes in pH. In order to pass a current through the gel plate, a power supply is connected to each end of it using electrodes.

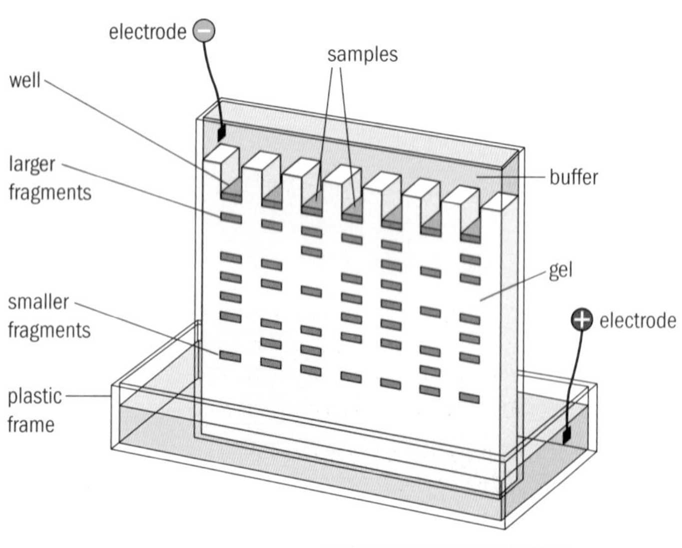

Experimental set-up for electrophoresis.

Before the DNA can be separated, a restriction enzyme is used to split it into fragments. The DNA samples made can then be placed into small wells cut out of one end of the gel plate. The plate is then submerged in a tank containing a buffer solution, and a current is passed through the plate for about 2 hours. The DNA is negatively charged, so it will be attracted towards the positive end of the plate. The DNA fragments can then be viewed by staining them with a dye.

After separating the DNA fragments they can be analysed further in order to find the base-pair sequences. This is done by comparing the distances travelled along the plate by the DNA fragments with known values.

Safety note: Disposable gloves must be worn and skin contact avoided when using the dyes used for DNA staining.

Battery operated electrophoresis systems are preferred and must limit the voltage to less than 40 V dc. Any electrical system delivering greater than 40 V dc should have an interlock to protect the user from access to connections or electrolytes when the power is on.

Did you know?

Every person has their own unique sequence of base pairs within their DNA. This allows scientists to identify an individual's DNA in a process known as DNA fingerprinting.

Gas–liquid chromatography

Gas–liquid chromatography (GLC) is used to analyse compounds that can be vaporised. Uses of GLC include testing the purity of a substance, and separating the different components of a mixture and determining the relative amounts of each component.

Unit 4: See page 102 for more information on GLC.

Gel-permeation chromatography

This form of chromatography is normally used to analyse polymers. It uses a column which contains a cross-linked polymer having small pores of a certain size. The polymer to be analysed can be passed through the pores by dissolving it in a liquid solvent. The size of the polymer being analysed determines how quickly it passes through the pores. The smaller the polymer the greater amount of time it spends in the pores, the larger the polymer the less time is spent in the pores. The amount of the polymer passing through the column can be detected using an ultraviolet detector, which can provide data about the molecular weight of the polymer.

Thin-layer chromatography

Thin-layer chromatography (TLC) can be used to separate compounds, such as amino acids from a protein, and then identify them. It can also be used to compare known substances with a mixture of unknown substances. This type of analysis uses the fact that some compounds are more soluble in solvents than others. The more soluble a compound is in a solvent then the faster it can move through the solvent.

If you were going to try to identify a mixture of unknown amino acids, a good way to proceed would be to choose two or three known amino acids to compare with your mixture.

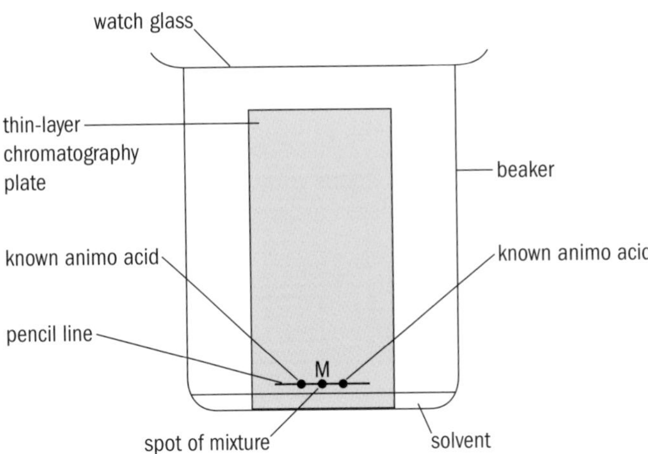

Experimental set-up for thin-layer chromatography.

Unit 4: See page 96 for detailed instructions on how to carry out a TLC procedure.

Once the procedure is complete, spots of the separated compounds may be visible. If not, a dye or even ultraviolet light can be used to locate the compounds. In the case of amino acids, once the chromatogram is dry it can be sprayed with ninhydrin which turns amino acids a brown colour.

Safety note: Eye protection must be worn. UV light is hazardous; it can damage the eyes. Disposable gloves should be worn and skin contact avoided when using dyes.

Ninhydrin is harmful if swallowed and irritating to the eyes, skin and respiratory system. (Ninhydrin spray is extremely hazardous to the eyes.)

For the mixture of amino acids, you would see several spots at different distances from the baseline on the plate. If you were comparing known amino acids with your mixture of amino acids, substances that were identical would have moved the same distance up the plate.

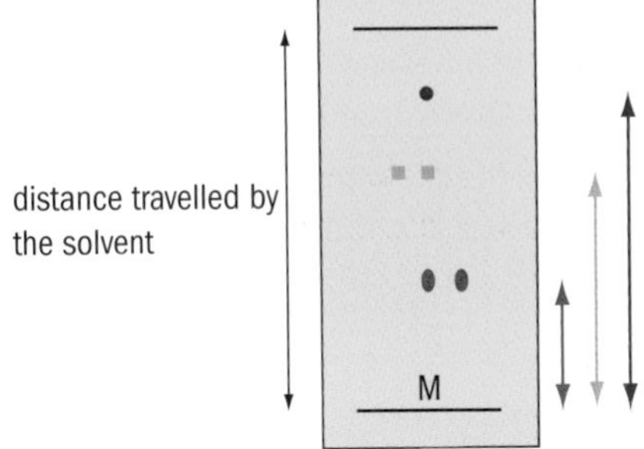

A chromatogram showing how different compounds in a mixture will move different distances.

Case study: analysis in the food industry

Rebecca is a trainee laboratory technician working in the food industry. She is developing her laboratory skills by using different analytical techniques. Today, she is analysing a mixture of amino acids using TLC in order to improve her laboratory technique. This technique can be used to identify compounds in substances, such as food, thus determining whether they contain any compounds that may be harmful or illegal. Many additives and colourants used in foodstuffs are harmful, but can be recognised quickly and easily using TLC. Other compounds may have been added to the food inadvertently, such as pesticides and insecticides, and these may also be identified using this technique.

1 **What measurements does Rebecca need to take in order to work out which amino acids she is analysing?**

2 **Why is TLC a good technique to use in order to identify unknown substances?**

The compounds can also be identified using their retention factors, R_f. These values can be worked out by measuring the distance the solvent has moved up the sheet and the distance each spot has moved up the sheet.

$$R_f = \frac{\text{distance travelled by compound}}{\text{distance travelled by solvent}}$$

These R_f values can be compared with known values for an identical solvent in order to identify the compounds.

Assessment activity 13.1

BTEC

You are a research scientist working in the biochemistry department of a university. In order to show potential students what type of work your department specialises in, an open day has been organised. To prepare for the open day the following tasks need to be completed.

1 Produce a poster for your laboratory wall that describes one of the biological molecules you have studied. Use diagrams to describe its structure. Use bullet points to show how the molecule is used in living organisms. **P1** Label your structure diagram to show which parts of the structure let the molecule fulfil its function. Produce two such posters covering carbohydrates and lipids. **D1**

2 Carry out TLC of two known amino acids and a mixture of two unknown amino acids. **P2**

Safety note: Eye protection must be worn. Avoid skin contact with the gels and amino acids.

3 Calculate the R_f values for the two known amino acids and the mixture. Produce a crib sheet explaining how these R_f values can be calculated using the chromatograms from task 2. **M1**

Grading tips

For the **P1** task, make sure that you describe how the structure of the molecule relates to the role it carries out.

The **D1** task will require more discussion, so paragraphs of text should be used in addition to any labels.

For the **P2** task, make sure you can explain why each step of the process is required.

PLTS

Independent enquirer

You will develop this skill when you research information to include on your poster.

13.2 The structure of proteins

In this section: 

Key terms

Amino acid – a compound that contains a carboxyl group (–COOH) and an amino group (–NH_2) attached to the same carbon atom.

Intramolecular bond – a force or bond that holds together different parts of the same molecule.

Biochemistry involves studying living things. Many biochemists work in medicine, researching the causes of disease to find a way of treating the illness. This may involve finding out about protein structures. If you have ever taken an antibiotic, then a biochemist will have found out how the antibiotic acts on proteins within your body and will have established that it will treat your particular illness.

Proteins are very important molecules in our cells. They are involved in nearly all cell functions. Each protein within the body has a specific job or function. Some examples are given in the table below.

Protein	Function
Collagen	Used to form structures in the human body such as bone, muscle and hair
Amylase	Acts as an enzyme to break down starch in the mouth
Haemoglobin	Transports oxygen around the body via the blood stream
Rhodopsin	Detects light in the retina of the eye

Did you know?

Proteins account for 50% of the organic material in your body and 17% of your body weight.

Proteins are made of small molecules called **amino acids**. Amino acids are made of the five elements carbon, hydrogen, oxygen, nitrogen and sulfur. They are called amino acids because they contain an amino group and an acid group. There is a central carbon atom (called the 'alpha carbon'), with four chemical groups attached to it:

- a hydrogen atom
- an amino group
- a carboxyl group
- a variable R group (or side chain).

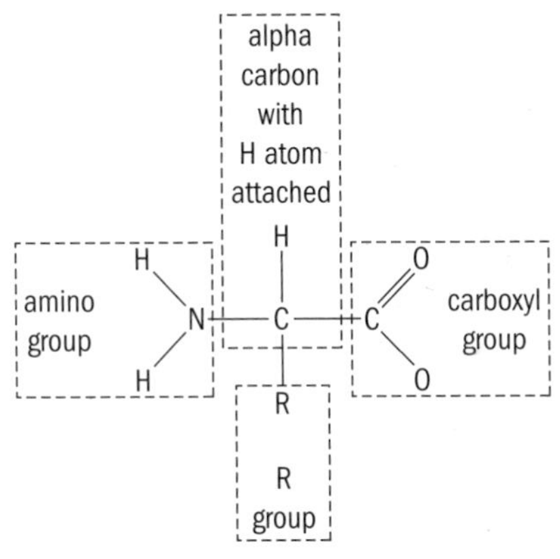

The structure of an amino acid molecule.

There are 20 different R groups which means that there are 20 different amino acids. These are shown in the figure overleaf. There are three letter and one letter abbreviations for each amino acid. The proteins in the human body are all made from different combinations of amino acids, so proteins can have lots of different properties.

To make protein structure easier to understand, it is broken down into four levels:

- primary
- secondary
- tertiary
- quaternary.

We will look at each in turn.

Primary structure

The simplest level of protein structure is called the primary structure. This tells you which amino acids are in the protein and in what order they are joined together. The amino acids are joined together by peptide bonds.

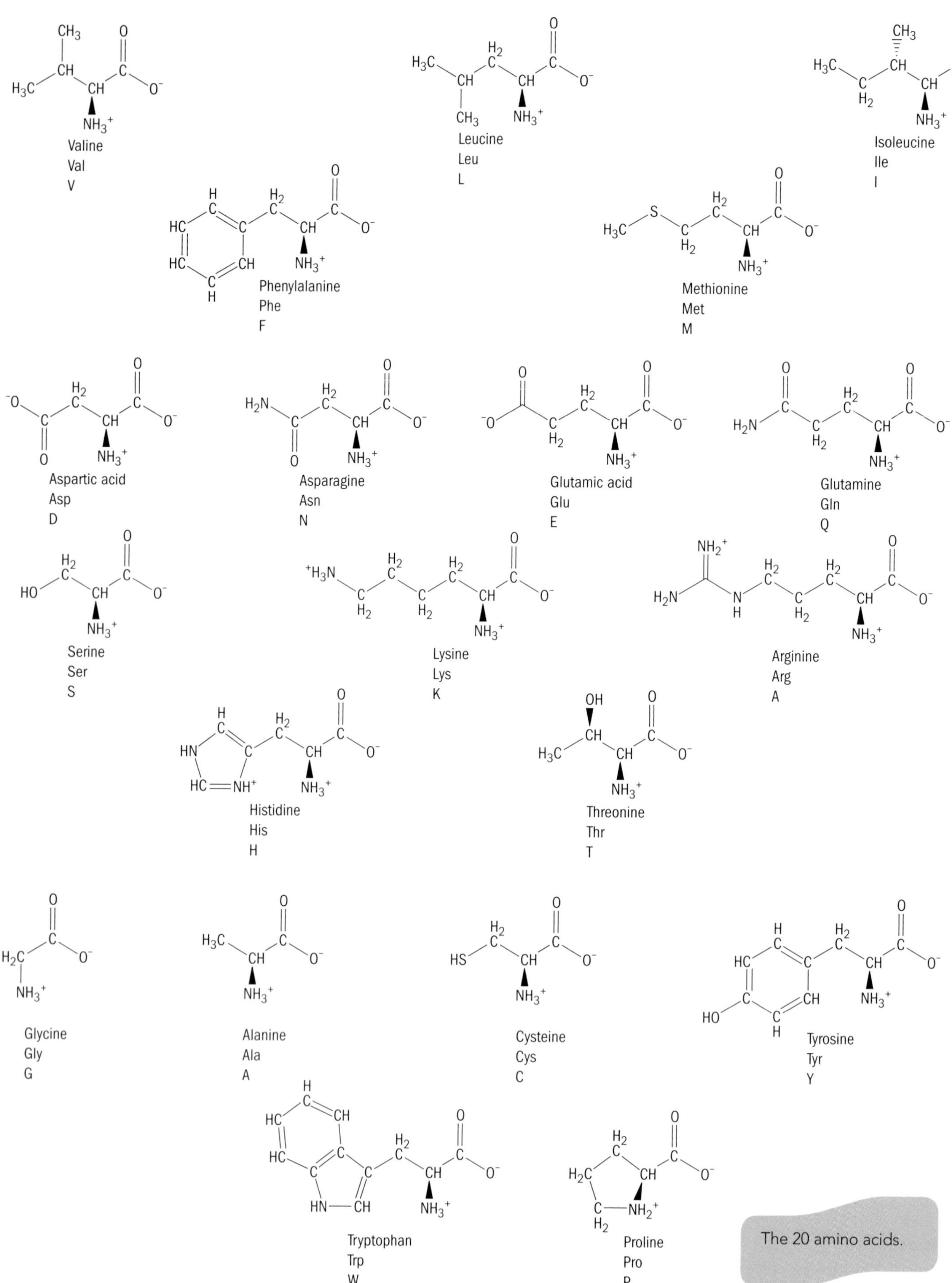

The 20 amino acids.

The reaction involves making a molecule of water, so it is called a condensation reaction. When two amino acids join together like this, a dipeptide is formed. The diagram below shows a peptide made from the amino acids glycine and alanine.

glycine + alanine

H_2O removed via condensation

peptide bond

Two amino acids undergo a condensation reaction to form a dipeptide.

In a peptide the amino acids may be referred to as **residues**. Note that the R group for glycine is H and the R group for alanine is CH_3.

Three amino acids form a tripeptide. The tripeptide shown below is made of glutamine, cysteine and glycine.

Glu Cys Gly

A tripeptide.

Many amino acids can be joined together to form a polypeptide.

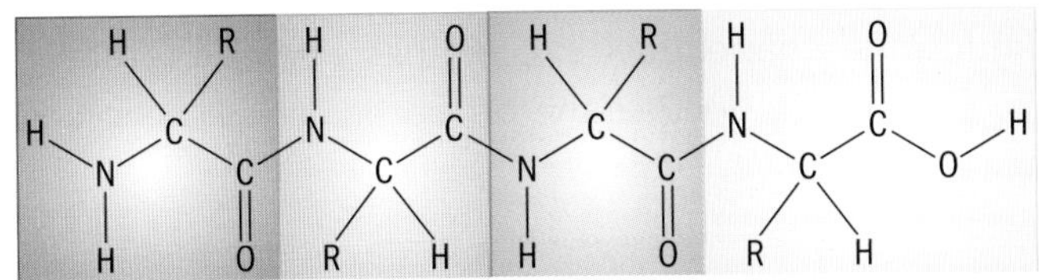

A polypeptide.

When you draw a peptide chain, the amino group ($-NH_2$) which hasn't been used in a peptide bond is written at the left-hand end. The unreacted carboxyl group (–COOH) is written at the right-hand end. The $-NH_2$ end of the peptide chain is known as the N terminal, and the –COOH end is the C terminal.

When the polypeptide is very long, more than about 50 amino acids, then it is called a protein. A protein chain (with the N terminal on the left) is shown in the diagram below.

NH_2—CH—C—N—CH—C—N—CH—C—N—CH .... —C—N—CH—C—N—CH—COOH

A protein chain.

The 'R' groups come from the 20 amino acids that make up the protein. The peptide chain is called the backbone, and the 'R' groups are called the side chains.

This primary structure can be shown using the abbreviations for the amino acids. Using the three-letter abbreviations given in page 252, a part of a protein chain can be represented as shown below.

....Tyr.Gly.Lys.Pro.Val.Gly.Lys.Lys.Arg.Arg.Pro.Val.Lys.Val.Tyr.Pro.Ala.Gly.Glu....

A protein chain using abbreviations.

In a protein the polypeptide chain can be very long, and may contain hundreds of amino acids. The primary structure is just the order of amino acids in the polypeptide chain.

Activity 13.2A

Draw a peptide containing three amino acids in the order: glycine, alanine and cysteine.

Secondary structure

The chain of amino acids tends to form orderly shapes, which we call the secondary structure. This is the next level of protein structure. These secondary structures are held in shape by hydrogen bonds. Each bond is a weak force of attraction between a lone pair of electrons on an oxygen atom and a hydrogen atom attached to a nitrogen atom.

δ- δ+
C=O-------H—N
H bond

Hydrogen bond between an oxygen atom and a hydrogen atom.

In this secondary structure, the hydrogen bonds are a type of **intramolecular force**. This means that they are forces of attraction which exist between different parts of the *same* molecule. Hydrogen bonds may also exist as intermolecular forces between individual molecules. In a protein, these intramolecular forces are between different parts of the backbone. The two most common secondary structures are the alpha-helix (written α-helix) and the beta-pleated sheet (written β-pleated sheet).

α-Helix

The α-helix is formed when the polypeptide chain is wound round to form a loosely coiled spring shape. It is held together by hydrogen bonds. Although each hydrogen bond is only a weak force, there are so many hydrogen bonds that their combined effect results in a very strong structure.

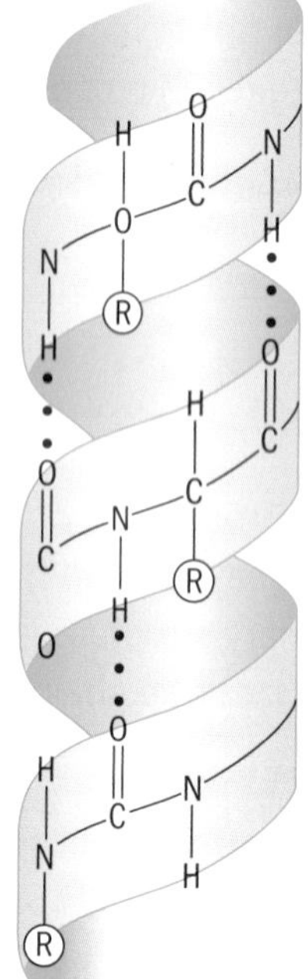

An α-helix.

• • • hydrogen bond
(R) = amino acid side chain

β-Pleated sheet

In a β-pleated sheet the chains are folded so that they lie alongside each other. The protein chain is also held in position by hydrogen bonds.

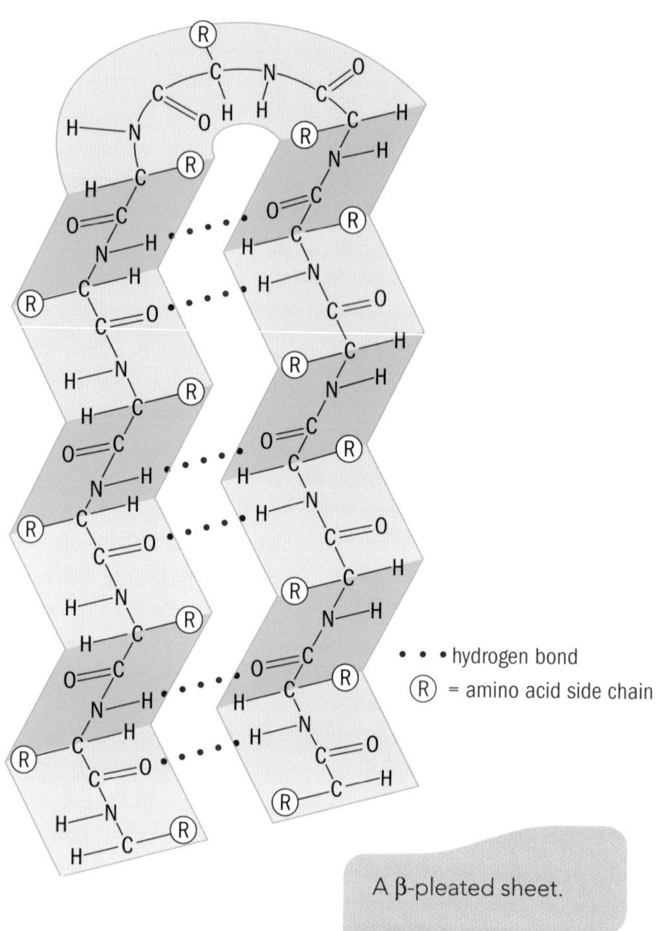

A β-pleated sheet.

Activity 13.2B

Describe what is meant by the secondary structure of proteins and give two examples.

Tertiary structure

The tertiary structure of a protein shows how the secondary structures fold themselves into a three-dimensional shape. Each protein has a unique tertiary structure, which decides the properties and function of the protein.

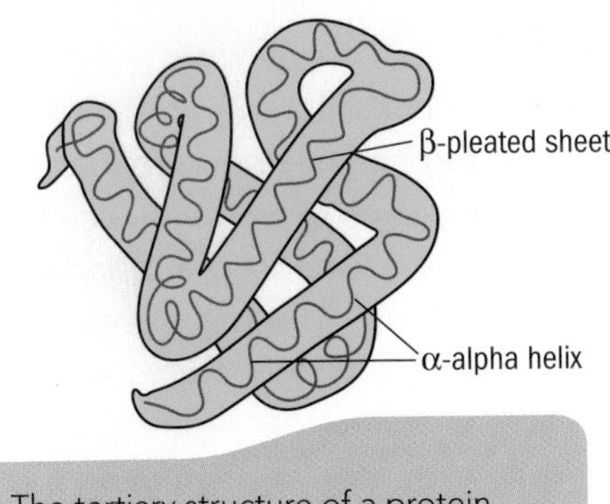

The tertiary structure of a protein.

The tertiary structure of a protein is held together by interactions between the amino acid side chains or the 'R' groups. There are means by which this can happen:

- hydrogen bonds
- ionic bonds (hydrophobic bond)
- sulfur bridges (or S–S links).

A protein can contain all of these bonds, holding the tertiary structure in its three-dimensional shape. The final three-dimensional shape of a protein can be classified as globular or fibrous.

Hydrogen bonds

Many amino acids contain groups in the side chains that have a hydrogen atom attached to either an oxygen atom or a nitrogen atom. Hydrogen bonding can occur between such groups, if there is a lone pair of electrons on either the oxygen atom or the nitrogen atom.

An example could be where a protein contains two serine residues. Serine contains an –OH group in its side chain. If the protein chain was folded so that the two serine residues were close together, then a hydrogen bond could form between the lone pair on the oxygen of one serine residue and the hydrogen of the other serine residue.

residue from serine
peptide chain
$CH_2–O^{\delta-}–H$
hydrogen bond
$CH_2–O–H^{\delta+}$
residue from serine

Tertiary structure showing a hydrogen bond.

Ionic bonds

An ionic bond can form between amino acids which contain a carboxylic acid group (–COOH) and an amino group ($–NH_2$). A hydrogen ion can be donated from the –COOH to the $–NH_2$ group, producing $–COO^-$ and $–NH_3^+$. So if a protein contains a residue such as aspartic acid close to a residue such as lysine, then an ionic bond can form between the two parts of the protein. These bonds are stronger than hydrogen bonds.

residue from lysine
peptide chain
$CH_2CH_2CH_2CH_2NH_3^+$
ionic bond
CH_2COO^-
residue from aspartic acid

Tertiary structure showing an ionic bond.

Sulfur bridges

Another way in which the tertiary structure can be held in place involves the amino acid cysteine. A sulfur bridge is a type of bond that can form between two cysteine residues that are close together when the protein chain is folded. Each cysteine residue loses a hydrogen in the form of H^+. This is a covalent bond between the two sulfur atoms (S–S), which is stronger than a hydrogen bond.

peptide chain
sulfur bridge

Tertiary structure showing a sulfur bridge.

Globular proteins

The vast majority of proteins are globular (or roughly spherical), including haemoglobin, the oxygen-carrying protein in red blood cells. Haemoglobin consists of four globular subunits arranged in a roughly spherical structure. The oxygen binds to a prosthetic (helper) group which is bonded within the quaternary structure (see page 257 for explanation of quaternary structure). Prosthetic groups are organic groups that are bonded to proteins and allow the proteins to carry out their biological role.

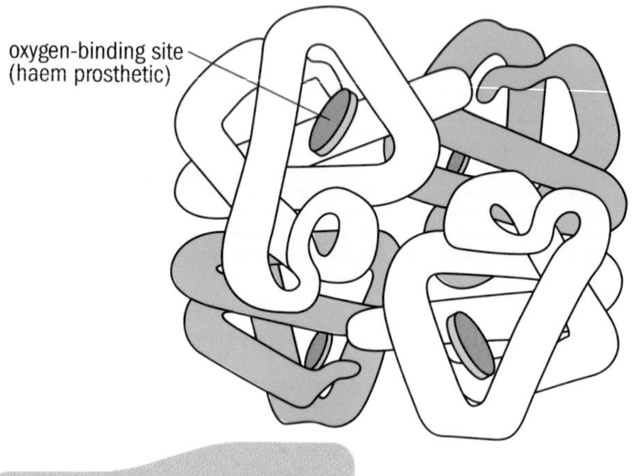

Haemoglobin showing the haem prosthetic.

Fibrous proteins

Fibrous proteins consist of different protein strands coiled around one another. They normally have a structural function. One such example is collagen which is found in the body's tendons, supporting its internal organs, and in bone. This has three α-helix polypeptides held in a very strong structure.

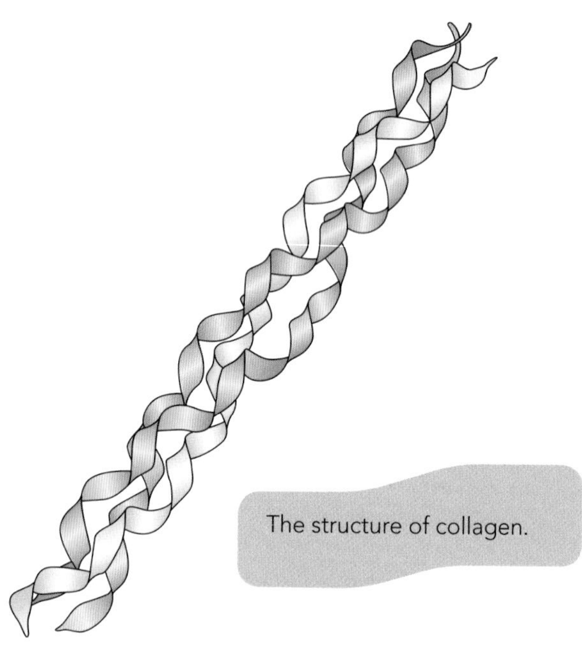

The structure of collagen.

Case study: The Protein Data Bank

Helen is a biochemist carrying out research at a university. She is working with other researchers trying to find out the molecular structure of a protein. This group is also researching how changes in the protein's structure affect its function within cells.

One part of Helen's job involves using gel electrophoresis to find the amino acid sequence of one part of the protein. The sequence can then be combined with other sequences found by her team in order to describe the protein's structure. This job is important because macromolecules carry out most of the functions of cells. The structure of proteins allows them to carry out functions within cells. If a protein didn't have a specific three-dimensional shape then it couldn't do its job in the cell. The tertiary structure of molecules depends on the sequence of amino acids in the protein backbone (the primary structure).

Once the structure of a protein has been discovered, it can be placed in a database called the Protein Data Bank (PDB). This database contains 3D structures of large biological molecules, such as proteins. The data, obtained by biochemists from around the world, can be accessed on the Internet. The PDB is a key resource, as scientists can see the structure of proteins, which can help in the development of new pharmaceuticals. These interact with the protein structure and help treat a patient's illness.

1 **What information does Helen need in order to describe the structure of the protein?**

2 **Why is gel electrophoresis the best method to use for determining protein structure? Would any other methods work as well?**

Activity 13.2C

Select appropriate pairs of amino acids that could form the following types of intramolecular bond between them, as part of a protein's tertiary structure:

(a) hydrogen bond

(b) hydrophobic bond

(c) sulfur bridge.

Quaternary structure

A quaternary structure is only found in proteins containing more than one polypeptide chain. This structure shows how the different polypeptide chains are arranged together. If a protein contains more than one polypeptide chain, then each chain is called a sub-unit. Hydrogen bonds, ionic bonds and sulfur bridges can hold the various chains in their particular shapes.

Did you know?

The collagen strands that give your skin its strength and structure are stronger than steel wires of the same diameter.

Assessment activity 13.2

BTEC

You work within the sales department of a company that specialises in providing proteins for use in washing powders. You have been given the task of introducing your company's expertise to a potential customer. Your task is to prepare materials to use for a meeting with the client's senior executive.

1 Draw a flow chart to show how amino acids can react to form dipeptides, polypeptides and proteins. P3

2 Prepare a presentation by researching an enzyme of your choice. Include a picture of the enzyme's structure. Identify the secondary structures that are within the picture and describe how they maintain their shape. P3 Describe and explain the tertiary and quaternary structure of the enzyme. M2

3 Proteins with different structures carry out different biological roles. Prepare a short essay that you can refer to in the meeting, which discusses how the structure of a protein can affect its function. D2

Grading tips

In task 1, you should use diagrams and abbreviations to show the primary structure and the bonds that form between the amino acids, to help you achieve P3 .

To meet M2 in task 2, you will need to use detailed explanations and labelled diagrams.

To meet D2 in task 3, you may find it useful to refer to several examples of different proteins, e.g. proteins that provide structural support, enzymes, etc.

Functional skills

ICT

You will develop your skills in ICT when you bring together information for these tasks and then present it in an appropriate format.

13.3 The factors that affect the activities of enzymes in biological systems

In this section:   

Key terms

Catalytic activity – the increase in the rate of a reaction caused by a catalyst or an enzyme.

Substrate – the molecule that is affected by the action of an enzyme.

Specificity – a property of an enzyme; enzymes can only increase the rate of reaction of a certain type of chemical reaction.

Active site – the area of an enzyme where the substrate binds.

Induced fit mechanism – the process involving a substrate interacting with an enzyme which causes the active site to change shape so that the substrate can bind.

Denatured – when the tertiary structure of a protein is changed, leading to a loss in function.

Equilibration – the process of making the temperatures of an enzyme and its substrate identical.

Enzymes are biological catalysts. A catalyst is a chemical involved in (but not changed by) a chemical reaction. Enzymes speed up the rate of reactions in the human body. The amount by which the reaction rate is increased is called the **catalytic activity**. This allows the chemical reactions that make life possible to take place at the relatively low temperature of the body. Because enzymes are not changed by the chemical reactions they affect, they can do their job over and over again.

There are thousands of different enzymes in human cells, each controlling a different chemical reaction. For example, the enzyme called peptidase, which increases the rate of protein breakdown in your food, will not work on starch. Starch breakdown is catalysed by the enzyme amylase found in saliva. The **specificity** of each enzyme ensures that they won't catalyse the breakdown of both starch and protein.

Many pharmaceuticals (or medicines) prevent or treat illnesses by stopping certain enzymes from working. These drugs are called enzyme inhibitors. Here are some examples.

- The inhibitor of the enzyme HIV protease can control AIDS.
- The inhibitor of angiotensin converting enzyme (ACE) can treat high blood pressure and heart failure.
- The inhibitor of the enzyme HMG-CoA reductase can reduce your level of cholesterol.

On the other hand, most enzymes are essential for our good health. Malfunctioning of an important enzyme in the body may lead to abnormalities within the body that can cause life-threatening illnesses.

One example is that of mutations in DNA repair enzymes. Defects in these enzymes cause cancer, as the reactions which repair mutations in the DNA cannot take place. This causes a build up of mutations and results in many types of cancer in the sufferer.

Enzymes are essential for the human body to work properly, and so scientists need to understand what they are, how they work and how they can be controlled. This knowledge lets us make important use of enzymes, in medicine and industry.

Did you know?

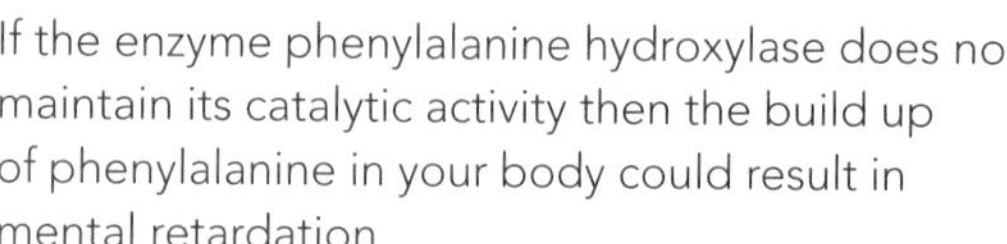

If the enzyme phenylalanine hydroxylase does not maintain its catalytic activity then the build up of phenylalanine in your body could result in mental retardation.

Enzyme function

Enzymes are proteins, and the job they do is determined by their structure. The specific reaction they speed up takes place in a small part of the enzyme called the **active site**, while the rest of the protein holds the active site in shape.

In a chemical reaction a **substrate** (S) is converted into a product (P):

$S \rightleftharpoons P$

In a reaction which can be catalysed by an enzyme, the substrate binds to the active site of the enzyme to form an enzyme–substrate (ES) complex. The enzyme will only bind that particular substrate because other molecules cannot fit into the active site. The substrate then reacts and turns into the product, which is still attached to the enzyme. Finally, the product is released. This is called a 'lock and key mechanism'. The enzyme is then left with an empty active site so it can start again.

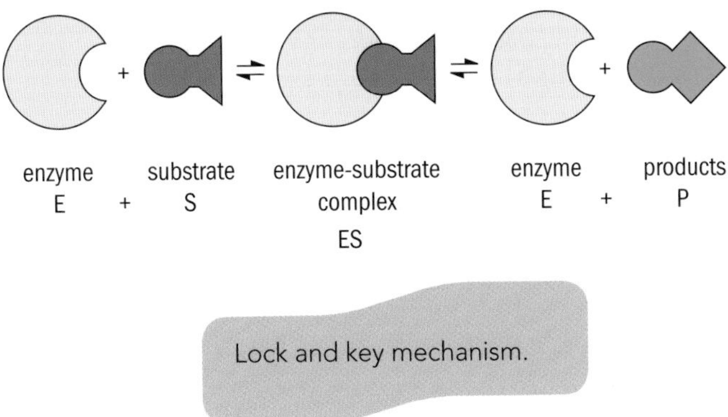

Lock and key mechanism.

The active site actually changes shape when the substrate molecule fits into it. This distorts the substrate in the active site, turning the molecule halfway into the product. This distorted substrate is called the transition state. The transition state is more likely to turn into the product than the substrate. This is called the **induced fit mechanism**.

The reaction is made more likely for various reasons.

- If a bond in the substrate needs to be broken, then the enzyme might stretch and weaken it.
- Alternatively, the enzyme can change the pH, water concentration, or charge in the active site.

Co-factors

Many enzymes need co-factors in order to work. These can be metal ion co-factors (such as Fe^{2+}) or organic co-factors (such as haem). Many of these co-factors are taken from vitamins in the human diet. The enzyme with its co-factor is called a holoenzyme and is active. Just the protein part of the enzyme without its co-factor is called the apoenzyme, which is inactive.

There are two types of organic co-factor – prosthetic groups and co-enzymes. Prosthetic groups are attached to the enzyme. Co-enzymes, such as NADH, detach from the enzyme. A co-enzyme's role involves moving chemical groups from one enzyme to another, but in the process they are chemically changed. So co-enzymes are used up by enzymes, and need to be regenerated by your cells.

Activity 13.3A

Describe the stages that occur when an enzyme catalyses the breakdown of a dipeptide into two amino acids.

Factors that affect the rate of enzyme reactions

Temperature

Each enzyme has a temperature at which it works fastest. For some human enzymes this is about 40°C. (Actually, many human enzymes work best at higher temperatures; however, it would require too much energy to maintain a body temperature of, say, 50°C and also many other complex molecules in the body would be denatured at such high temperatures.) If the temperature is increased up to this optimum value, then the rate of the reaction it controls also increases. The temperature increase means that the substrate and enzyme molecules have more energy, so are more likely to collide with each other. More molecules also have the minimum amount of energy that is needed for a reaction to occur upon collision. This is called the activation energy.

Below the optimum temperature, enzyme and substrate molecules have less energy, so fewer collisions result and the rate of reaction is slower. Above the optimum temperature the rate also decreases. The heat energy breaks the hydrogen bonds which form the secondary and tertiary structure of the enzyme, so the enzyme (and the active site) loses its shape. The substrate cannot fit into the active site so the reaction cannot be catalysed. At very high temperatures this change is permanent. When this change in shape happens, the enzyme is said to be **denatured**.

pH

Enzymes also have an optimum pH at which they work fastest. This is around pH 7–8 (the pH of your cells). Some enzymes can work at more extreme pH values, such as protease enzymes in your stomach. These need to work in acidic conditions, so they have an optimum pH of 1.

The pH affects the charge on the amino acids that make the active site on the enzyme. If the pH changes, then the charges on the active site change and the substrate may no longer be attracted to the active site. For example, an amino acid containing a carboxyl group will be uncharged at low pH (COOH), but charged at high pH (COO^-). So a substrate with a positive charge cannot be made to react quickly at a low pH. At very acidic and very alkaline pH values an enzyme can be denatured.

Substrate concentration

If the substrate concentration is increased then the number of substrate molecules that can collide with enzyme molecules increases. This results in more reactions happening and so the rate increases. At higher concentrations, there are so many substrate molecules binding to the enzymes that there will be very few free enzyme molecules left. This means that adding more substrate will not affect the rate further.

Inhibitors

Inhibitors reduce the activity of enzymes, reducing the rate of their reactions. Inhibitors exist in nature, but they can be used as drugs and pesticides. There are two kinds of inhibitors.

- A competitive inhibitor has a similar shape to the enzyme's substrate molecule, and it fits into the active site. It competes with the substrate for the active site, so the reaction is slower. Antibacterial drugs are competitive inhibitors.

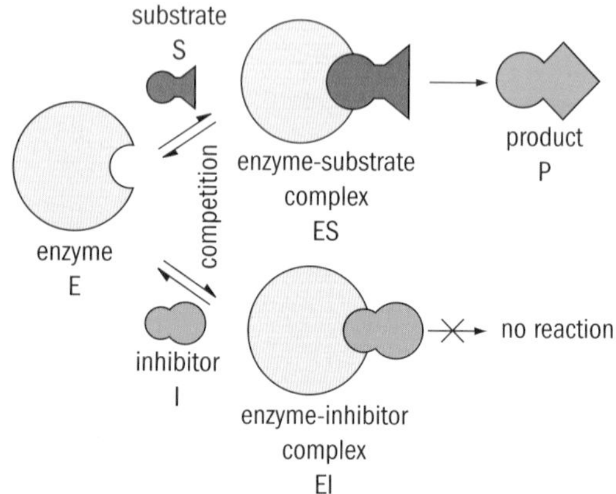

How a competitive inhibitor works.

- A non-competitive inhibitor molecule is different in structure from the substrate and does not fit into the active site. It attaches itself to a different part of the enzyme. This changes the shape of the enzyme, especially the shape of the active site. As a result, the normal substrate for the enzyme cannot bind to the active site. Non-competitive inhibitors reduce the number of active enzymes, so they slow down the overall reaction. Poisons like cyanide, lead ions and insecticides are non-competitive inhibitors.

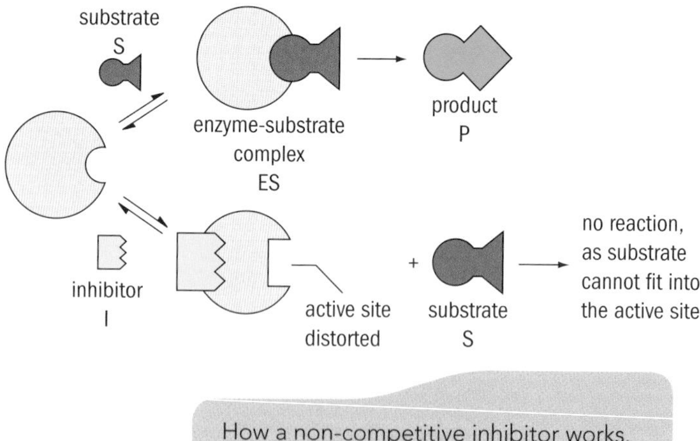

How a non-competitive inhibitor works.

Activity 13.3B

How does the activity of an enzyme change if the temperature is increased to a very high value? Explain what happens to the enzyme when this occurs.

Laboratory techniques

The effects on enzyme activity of pH, temperature, substrate concentration and inhibitors can be measured in the laboratory.

An investigation into enzyme activity needs to have a variable that is changed (the independent variable) and a variable that is measured (the dependent variable). All other variables should be kept constant. These are called controlled variables.

For example, if you wanted to see how changing pH affected an enzyme's activity then you would do several experiments where you changed the pH and measured the rate of reaction. You would keep all other variables the same.

Independent variable	pH
Dependent variable	rate of reaction
Controlled variables	temperature enzyme concentration substrate concentration

When using enzymes in an experiment, it is important that the enzyme and the substrate are at the same temperature as otherwise the results could be incorrect. This process of warming the enzyme and substrate to identical temperatures is called **equilibration**.

The rate of the catalysed reaction can be measured using two types of experiment.

- After the reaction has started, the amount of product made (or the amount of substrate used up) should be measured after a certain amount of time. Once a fixed amount of product has been made (or a fixed amount of substrate used up) then the rate of the reaction can be calculated by 1/time. This experiment should be carried out several times using different values of the independent variable. Plotting a graph of reaction rate (y-axis) against independent variable (x-axis) will show you which value of the independent variable provides the highest rate of reaction.
- After the reaction has started, the amount of product made (or the amount of substrate used up) should be measured at several different times. Then a graph can be plotted of time (x-axis) against product made, or substrate used up (y-axis). This experiment should be carried out several times using different values of the independent variable. This will give you a graph with several plotted lines on it. By drawing a tangent to the steepest part of each line, the gradient at that point can be calculated. This gradient gives you the maximum rate of each catalysed reaction.

Safety note: Eye protection and disposable gloves must be worn. All enzymes are proteins and may produce sensitisation or allergic reactions. Skin contact and the risk of inhalation must be minimised.

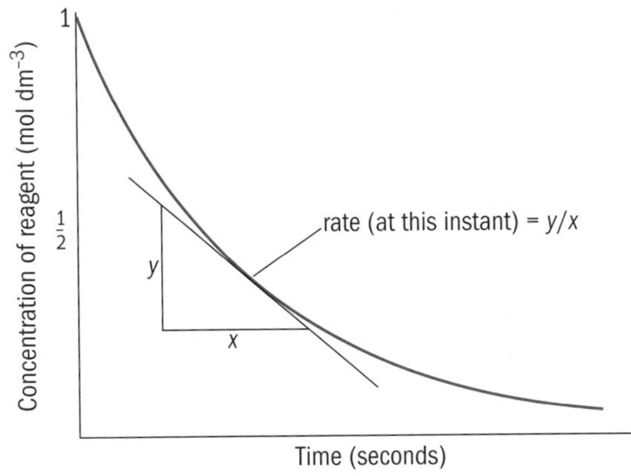

Graph showing how to calculate the gradient of a tangent to a curve.

Whatever method you choose to use and whichever variables you wish to investigate, it is important that you carry out the experiments at least three times. This lets you spot any anomalous results, which may have been caused by faulty apparatus or incorrect technique. It also allows you to determine how precise your experiment has been and whether any of the controlled variables have been changing due to poor experimental technique.

Examples of enzyme investigations

Investigating how temperature affects the activity of the enzyme lipase

Lipase is an enzyme that catalyses the breakdown of fat in milk to produce fatty acids. If you start with a solution of milk, lipase and sodium carbonate, it is alkaline with a pH value of about 10. After the lipase catalyses the reaction, the pH drops to about 8.2 because acid has been made.

This enzyme-catalysed reaction can be investigated using an indicator called phenolphthalein. This is pink in alkaline solutions of about pH 10 but goes colourless when the pH drops below pH 8.2.

In this experiment, the independent variable is the temperature and the dependent variable is the amount of acidic product. Safety precautions for this practical include eye protection.

Safety note: Eye protection must be worn. Phenolphthalein is low hazard, but it is likely to be in a highly flammable solution, so ensure there are no naked flames.

This experiment can be done as follows.

1. Prepare a lipase solution (this may have already been done for you).
2. Set up water baths at a range of temperatures (up to around 60°C) and put a beaker of lipase, containing a 2-cm^3 syringe, into each bath.
3. Label a test tube with the temperature you are currently investigating.
4. Add four drops of phenolphthalein indicator to the test tube.
5. Use a measuring cylinder to place 5 cm^3 of milk in the test tube.
6. Measure out 8 cm^3 of sodium carbonate solution and add it to the test tube. The solution should go pink because it will be alkaline.

7 Put a thermometer in the test tube.
8 Place the test tube in the correct water bath and let it reach the same temperature as the water bath.
9 Remove the thermometer from the test tube.
10 Use the 2-cm^3 syringe to measure out 1 cm^3 of lipase from the water bath for the temperature you are investigating.
11 Add the lipase to the test tube and start the stop clock.
12 Stir the contents with a glass rod until the solution is no longer pink.
13 Stop the clock and record the time.
14 Repeat steps 3 to 13 for each temperature. If you have time, you may wish to repeat the experiment for each temperature three times, and then take an average value.

You should now have a table of results showing the amount of time taken for the pH of the solution to reach about 8.2 at different temperatures. The rate of each reaction is 1/time. Once you have calculated the rate for each temperature you can write a conclusion.

Investigating how changes in pH can change amylase activity

Amylase is an enzyme that catalyses the breakdown of starch. This reaction can be investigated using iodine, which turns a dark-blue colour in the presence of starch but remains orange in the absence of starch.

Safety note: Eye protection and disposable gloves must be worn. Avoid skin contact with all enzyme solutions and iodine.

If the amylase is obtained from saliva, students should collect and use their own and put any contaminated glassware into freshly prepared disinfected solution after use. Glassware can then be washed up in the normal way.

In this experiment, the independent variable is the pH and the dependent variable is the amount of reagent. Buffer solutions should be used to control the pH. This experiment can be done as follows.

1 Put a single drop of iodine solution into each well of a dropping tile.
2 Label a test tube with the pH you are investigating.
3 Place 2 cm^3 of amylase into the test tube using a syringe.
4 Using a different syringe, add 1 cm^3 of buffer solution of known pH to the test tube.
5 Add 2 cm^3 of starch to the amylase solution. Start the stop clock. Stir the test tube with a plastic pipette.
6 After 10 seconds, use the plastic pipette to place one drop of the mixture in the first well of the dropping tile. Do not stop the clock. The iodine solution will turn dark blue. Put the remaining solution from the pipette back into the test tube and stir.
7 Wait 10 seconds. Remove another drop of the mixture and add it to the next well containing iodine.
8 Repeat step 7 until the iodine solution stays orange when you drop on the amylase solution.
9 How many iodine drops did you need? Each drop tested represents 10 seconds of reaction time.
10 Repeat this method for a different pH, provided by another buffer solution.
11 Plot a graph of reaction time (*y*-axis) and pH (*x*-axis).

Activity 13.3C

List all the variables that should be kept constant if you wanted to investigate how pH affects the activity of an enzyme.

Commercial applications of enzymes

Enzymes are used in the chemical industry and for other industrial applications when specific catalysts are required. Most naturally occurring enzymes only catalyse a small number of reactions and can be denatured by industrial conditions, such as high temperatures. Because of this, protein engineering is an active area of research. Biochemists who work in this area attempt to create new enzymes with properties that will be useful in industry. Enzymes have now been designed to catalyse reactions that don't happen in nature.

Enzymes can also be used in diagnostic testing. This involves measuring the amount of enzymes in a sample of blood from a patient. High levels of enzymes in blood serum would suggest that cells have been damaged, causing the enzymes to be released from within the cells into the blood.

Application	Enzymes used	Uses
diagnostic testing	alanine transaminase	to diagnose liver damage
	aspartate aminotransferase	to diagnose heart disease
molecular biology	restriction enzymes, DNA ligase and polymerases	to manipulate DNA in genetic engineering, pharmacology, agriculture, medicine and forensic science
food industry	amylases from fungi and plants	to make sugars from starch; in baking to catalyse the breakdown of starch in flour to sugar; yeast fermentation of sugar generates the carbon dioxide that makes bread rise
	glucose isomerase	to catalyse the conversion of glucose into fructose in production of high fructose syrups from starchy substances; fructose syrups have enhanced sweetening properties and lower amounts of calories than sucrose
baby foods	proteases, such as trypsin	to predigest baby foods because young babies do not have the enzymes required to break down many proteins in their diet
brewing industry	enzymes from barley	to catalyse the breakdown of starch and proteins to produce sugar, amino acids and peptides; these products are used by yeast for fermentation
	industrially produced barley enzymes	in the brewing process as a man-made substitute for the naturally occurring enzymes in barley
	proteases	remove the cloudiness in beer that can be present when stored
dairy industry	rennin	in the manufacture of cheese
	lipases	to enhance the ripening of blue moulds in some cheeses
meat tenderiser	papain	to soften meat prior to cooking
biofuels	cellulases	to catalyse breakdown of cellulose in plants to form sugars for use in fermentation to produce biofuels
biological detergents	proteases	to help remove protein stains from clothes
	amylases	as detergents for dishwashers to remove starch residues
	lipases	to assist in the removal of fatty stains
contact lens cleaners	proteases	to clean proteins from the lens, avoiding infection

Assessment activity 13.3

You are working as a laboratory technician within a company that has started to look at enzymes as a possible new area of research. You and your team must familiarise yourselves with the theoretical and practical details regarding enzymes by carrying out the following tasks.

1 Working in small groups, each person uses mouldable putty to build their own enzyme. Include an active site and a co-factor. Demonstrate how your enzyme works using a plastic bead as a substrate. As a group, write a series of bullet points describing how the structure of the enzyme allows it to catalyse the reaction of the substrate. P4

2 Carry out an investigation into how changes in pH can change the activity of the enzyme amylase. Produce a table of results, an appropriate graph and a conclusion. Base your method on the one provided in the text. P5 Write a report explaining why the changes in pH affect the enzyme activity. D3. See safety notes on pages 261 and 262 before carrying out this investigation.

3 Research how temperature affects enzyme activity. Produce a series of bullet points which explain how temperature changes can alter the effectiveness of an enzyme. M3

Grading tips

To attain P4, you should use the correct terminology to describe the reaction and the structure of the enzyme itself. The primary structure and the types of amino acid side chain should be mentioned.

For the P5 task, two investigations should be carried out covering pH and temperature. Your report should include a description of your method.

To reach D3, you will need to make detailed references to the enzyme structure.

For M3 in task 3, you may need to refer to examples of enzyme-catalysed reactions.

PLTS

Independent enquirer

You will develop this skill when you plan and carry out the investigation into amylase activity in task 2.

13.4 Aerobic and anaerobic respiration

In this section: 

Key terms

Anaerobic respiration – a reaction in which foodstuffs release energy in the absence of oxygen.

Aerobic respiration – a reaction in which foodstuffs (usually carbohydrates) release energy. This reaction requires oxygen.

Pasteur effect – The effect that oxygen has on yeast during fermentation (oxygen decreases the fermentation rate).

Metabolism refers to the thousands of chemical reactions that take place in living cells. These reactions can be linked together into metabolic pathways.

- Reactions that release energy are called *catabolic* reactions. They are reactions that involve the breakdown of a molecule, e.g. respiration.
- Reactions that use up energy are called *anabolic* reactions. They are reactions that produce a molecule, e.g. photosynthesis.

The structure of adenosine triphosphate (ATP).

Respiration is an important reaction. This is because it can use the energy stored in the bonds of carbohydrates (normally glucose), lipids and proteins to produce a chemical called adenosine triphosphate (ATP).

Each molecule of ATP contains adenine, ribose and three phosphate groups. The bond between the two phosphate groups on the right-hand side can be broken by an enzyme, immediately releasing small amounts of energy. By this process, ATP is broken down into adenosine diphosphate and inorganic phosphate.

This energy is used in your cells for all of the bodily processes, such as muscle contraction and enzyme-catalysed reactions. In one day, you will use about your own body weight of ATP; however, your body only ever contains about 50 g at any one time. This means that ATP is being continually used and produced.

This is the word equation for cellular respiration:

glucose + oxygen → carbon dioxide + water (+ energy)

In reality, respiration is a complex metabolic pathway of 30 separate steps. To understand respiration in detail we can break it up into four stages. Stage 1 is anaerobic which means that energy generation does not need oxygen. Stages 2, 3 and 4 are the aerobic stages meaning that energy generation requires oxygen.

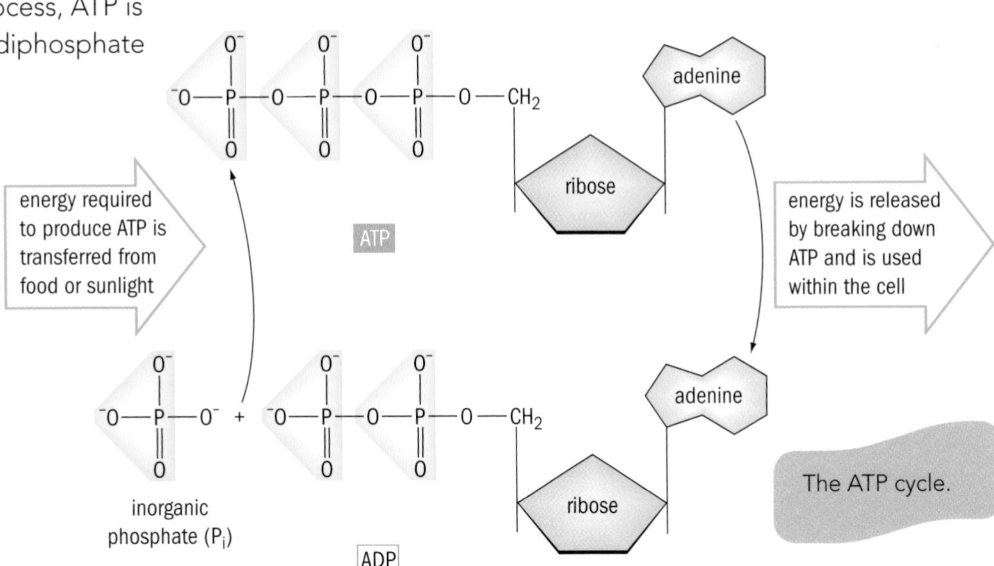

The ATP cycle.

Stage 1: glycolysis

This occurs in the cytoplasm of cells.

Glycolysis involves the breakdown of glucose into a compound called pyruvate. Only two ATP molecules are formed, but the pyruvate can then be used to form more ATP in stages 2 and 3. This pathway consists of 10 reactions, each catalysed by a different enzyme. None requires oxygen, so this stage is called **anaerobic respiration**.

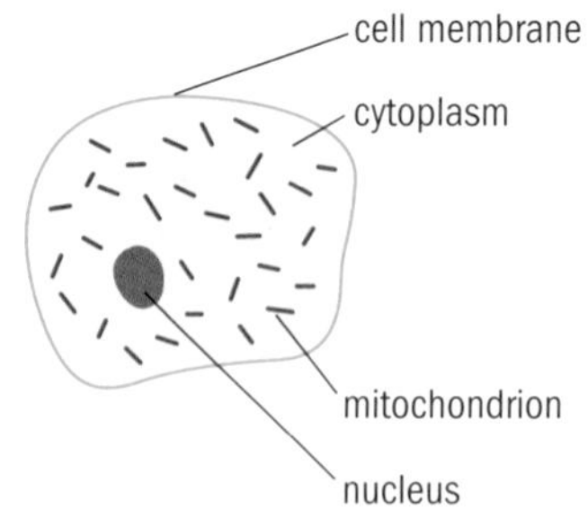

The main structures of a cell, important in cellular respiration.

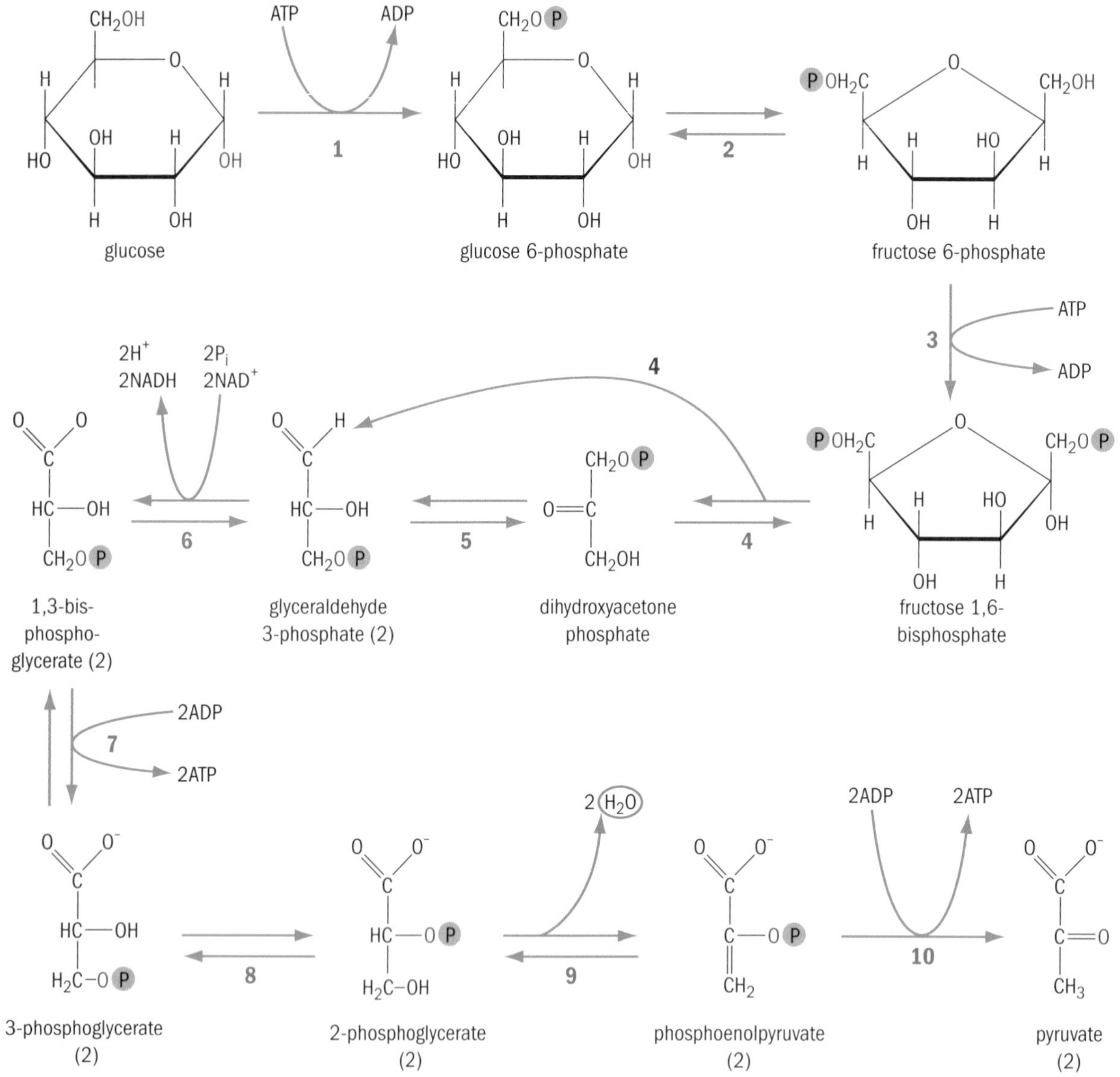

Enzymes:

1 hexokinase
2 glucose phosphate isomerase
3 phosphofructokinase
4 fructose diphosphate aldolase
5 triose phosphate isomerase
6 glyceraldehyde phosphate dehydrogenase
7 phosphoglycerate kinase
8 phosphoglyceromutase
9 enolase
10 pyruvate kinase

Glycolysis.

The first three reactions involve activating a glucose molecule by attaching phosphate groups to it so that it can be split up. This is called phosphorylation.

1. An ATP molecule is hydrolysed (split by water) and the phosphate group produced is attached to carbon 6 of the glucose molecule.
2. Glucose 6-phosphate is turned into fructose 6-phosphate.
3. Another ATP molecule is hydrolysed and the phosphate group bonds to carbon 1 of the fructose 6-phosphate, producing fructose 1,6-bisphosphate.

The next reactions involve splitting the fructose 1,6-bisphosphate so that the products can be used to form the energy source of ATP molecules.

4. Fructose 1,6-bisphosphate is split into two molecules.
5. An enzyme catalyses the production of two molecules of glyceraldehyde 3-phosphate.

The next reaction is called the oxidation (meaning loss of electrons) of glyceraldehyde 3-phosphate.

6. Two hydrogen atoms are removed from each glyceraldehyde 3-phosphate. This reaction is catalysed by an enzyme that needs another molecule for it to work (a co-enzyme). This molecule is nicotinamide adenine dinucleotide, or NAD. Each NAD molecule can remove one hydrogen atom, becoming reduced NAD, or NADH. This hydrogen atom stores energy, which is used to make more ATP later. In this reaction, two NAD molecules are reduced.

The next four reactions produce pyruvate, which is used later to form more ATP.

7–10

Four different enzymes convert each 1,3-bisphosphoglycerate into a molecule of pyruvate. Over these four reactions, two more ATP molecules are formed per 1,3-bisphosphoglycerate.

Overall, glycolysis uses one molecule of glucose to form four molecules of ATP (although two ATP molecules have been used up), two molecules of reduced NAD and two molecules of pyruvate.

Activity 13.4A

What is the overall reaction carried out by glycolysis? How many different enzymes are required to carry out the whole process?

Stage 2: the link reaction

The next stage of respiration takes place within a cell organelle called a mitochondrion (plural, mitochondria). Mitochondria have an inner and outer membrane. Between these membranes is the intermembrane space. Inside the inner membrane is the matrix, which is a mixture of DNA, enzymes and lipids.

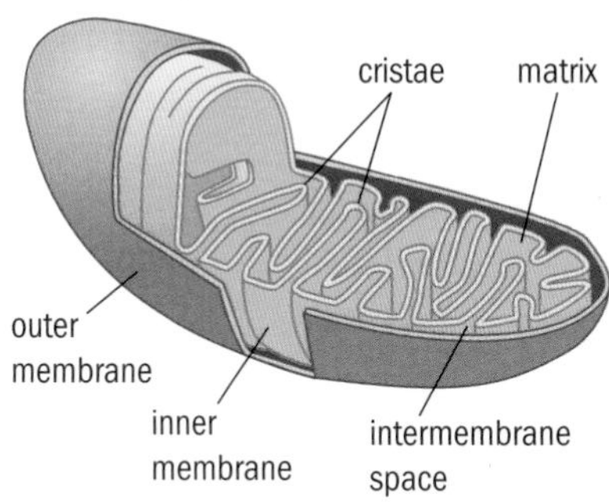

The inner structure of a mitochondrion.

The link reaction takes place in the matrix. The pyruvate made during glycolysis is changed into acetate during this reaction. Pyruvate is a three-carbon molecule. Enzymes remove one carbon dioxide and two hydrogen atoms from each pyruvate molecule. A two-carbon compound called acetate is formed. This attaches to a co-enzyme called co-enzyme A (coA), forming acetyl co-enzyme A.

The carbon dioxide diffuses through the mitochondrial and cell membranes and into the bloodstream, where it is carried to the lungs to be breathed out. Carbon dioxide is a product of respiration. The hydrogen atoms are taken up by NAD, forming reduced NAD (or NADH).

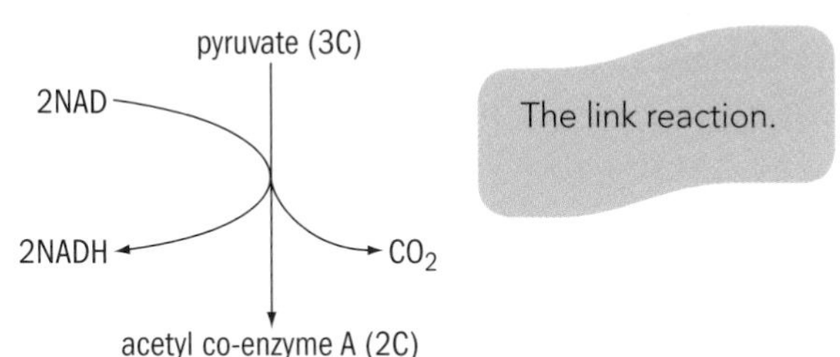

The link reaction.

So, from the two molecules of pyruvate produced in stage 1, two molecules of reduced NAD are made. No ATP is made by the link reaction; however, acetyl co-enzyme A is used to generate energy in stage 3 and reduced NAD is used in another metabolic pathway to generate ATP.

Stage 3: the Krebs cycle

The next stage is the Krebs cycle, which also takes place in the mitochondrial matrix. It is classed as **aerobic respiration**. The cycle consists of five enzyme-catalysed reactions.

1 The acetate compound is removed from the acetyl co-enzyme A produced in the link reaction. The acetate then reacts with oxaloacetic acid to form citric acid.

2 Carbon dioxide and two hydrogen atoms are then removed from the citric acid to form a five-carbon compound. Reduced NAD, or NADH, and an H^+ is made from the two hydrogen atoms.

3 There are two reactions within this step; two four-carbon compounds are made in turn. Carbon dioxide and two hydrogen atoms are removed from the five-carbon compound forming the first four-carbon compound. Another reduced NAD, or NADH, and an H^+ is made from the two hydrogen atoms. The first four-carbon compound is changed into a different four-carbon compound, and one ATP is produced.

4 This four-carbon compound is changed into another four-carbon compound. During this reaction a co-enzyme called flavin adenine dinucleotide (FAD) accepts two hydrogen atoms forming reduced FAD (or $FADH_2$).

5 Two hydrogen atoms are removed from the four-carbon compound, forming reduced NAD again. The oxaloacetic acid used in reaction 1 is also reformed.

The Krebs cycle produces one ATP molecule from one acetate molecule. However, one glucose molecule produced two pyruvates, which in turn produced two acetates. This means that one glucose molecule has produced two ATP molecules. More ATP can be made later from the other products: six reduced NAD molecules (three from each acetate molecule) and two molecules of reduced FAD (one from each acetate molecule).

Activity 13.4B

Why do cells only contain relatively small amounts of oxaloacetic acid?

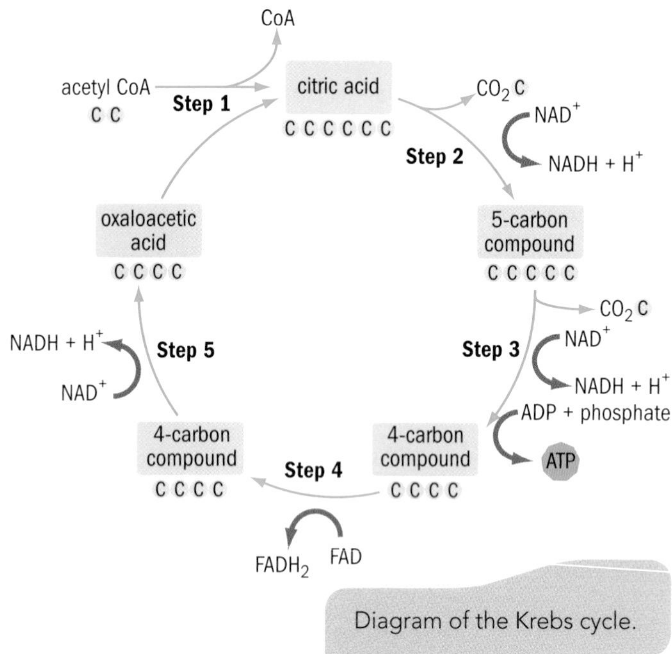

Diagram of the Krebs cycle.

Stage 4: oxidative phosphorylation

Oxidative phosphorylation takes place inside the inner mitochondrial membrane, using large protein complexes. There are four complexes: I, II, III and IV. In oxidative phosphorylation the hydrogen atoms from reduced NAD (made in the previous stages) release all their energy to form ATP.

Reduced NAD molecules bind to complex I and release their hydrogen atoms as protons (H^+) and electrons (e^-) into the matrix. The NAD molecules can then be used again in the Krebs cycle to collect more hydrogen. Reduced FAD molecules (made in stage 4) bind to complex II to release their hydrogen atoms as protons and electrons into the matrix.

The electrons are passed along through all of the complexes. This is the electron transport chain (ETC). In complexes I, II and IV the electrons give up some of their energy, which is then used to pump protons across the inner mitochondrial membrane. The protons are pumped from the matrix into the cytoplasm by complexes I, III and IV.

In complex IV the electrons are combined with protons and oxygen molecules to form water, another product of respiration. This is the only stage which uses oxygen, but the whole respiration process cannot happen without it.

$$4H^+ + 4e^- + O_2 \rightarrow 2H_2O$$

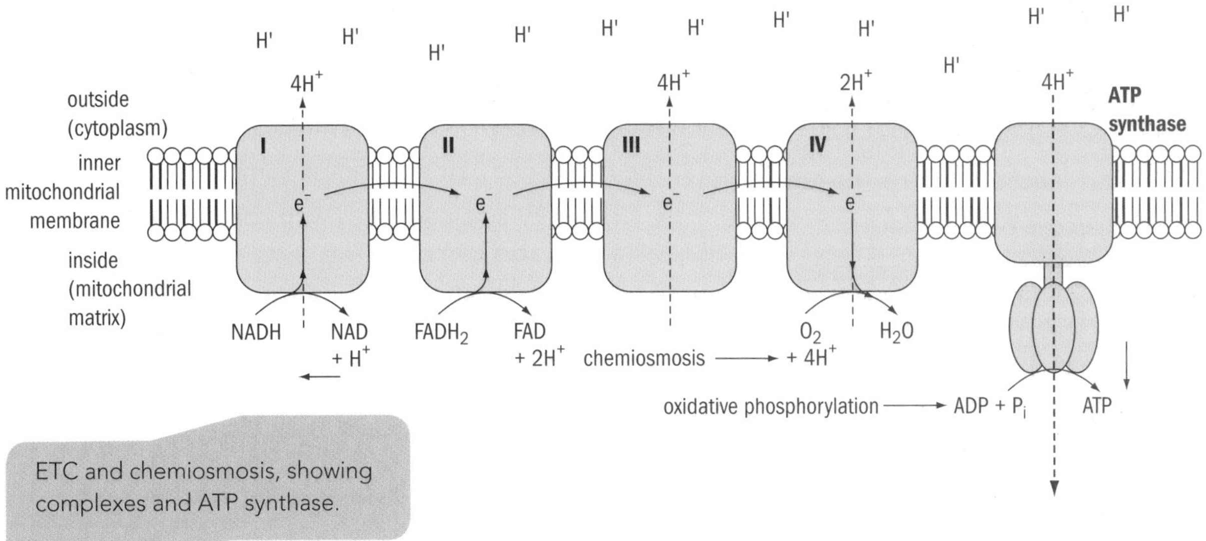

ETC and chemiosmosis, showing complexes and ATP synthase.

Because the energy of the electrons was used to pump protons out of the matrix, there are more protons on one side of the inner mitochondrial membrane compared to the other. This is called a proton gradient, and is a store of potential energy. This potential energy is used to generate ATP using the ATP synthase enzyme. The protons cannot pass through the inner mitochondrial membrane without being pumped through. However, the ATP synthase enzyme has a channel through which the protons can pass. This flow of protons across a membrane because of the proton gradient is called chemiosmosis. When this happens the energy from the protons is used to make ATP.

The energy physically spins part of the enzyme around, which joins adenosine diphosphate (ADP) with inorganic phosphate (P_i) producing ATP. This reaction is called phosphorylation. As oxygen is needed for this stage to operate, the process is called oxidative phosphorylation.

In summary, stage 4 uses up the reduced NAD and reduced FAD produced in the earlier stages. The energy that was stored in the glucose molecule at the start of stage 1 has now been used to produce ATP molecules. An estimate of the number of ATP molecules made overall is given in the following table.

Respiration stage	Molecules used/produced	Total ATP produced after oxidative phosphorylation (stage 4) (numbers are estimates)
Stage 1, glycolysis	– 2 ATP	2
	+ 4 ATP (2 per 1,3-bisphosphoglycerate	
	+ 2 reduced NAD (1 per 1,3-bisphosphoglycerate)	5
Stage 2, link reaction	+ 2 reduced NAD (1 per pyruvate)	5
Stage 3, Krebs cycle	+ 2 ATP (1 per acetate)	2
	+ 6 reduced NAD (3 per acetate)	15
	+ 2 reduced FAD (1 per acetate)	3
One molecule of glucose undergoes four stages of respiration to generate approximately 30 molecules of ATP, with the products being carbon dioxide and water.		

Glycolysis regulation

If cells don't need as much energy from ATP or they require more energy, then the respiration process can be controlled. This regulation occurs during stage 1. In glycolysis there are three reactions that can be regulated:

1 the formation of glucose 6-phosphate

3 the formation of fructose 1,6-bisphosphate

10 the formation of pyruvate.

The enzymes that catalyse these reactions can be slowed down by inhibitors. Your body can control the amount of inhibitors in a cell. So the amount of respiration can be increased if a cell requires more energy or decreased if there is a shortage of glucose.

Reaction 1 is catalysed by an enzyme called hexokinase. One ATP molecule is used by the reaction which makes glucose 6-phosphate. This product can be converted into glycogen as an energy store. When glycogen is broken down, the glucose 6-phosphate can then undergo glycolysis after reaction 1. Hexokinase is inhibited by the product of the reaction that it catalyses (glucose 6-phosphate).

Reaction 3 uses one ATP molecule and is catalysed by phosphofructokinase. The product of reaction 3 is then converted into dihydroxyacetone phosphate, which can be converted into a triglyceride. Triglycerides can be converted into dihydroxyacetone phosphate, which can then undergo glycolysis after reaction 3. High levels of ATP inhibit phosphofructokinase.

Reaction 10 produces one molecule of ATP and is catalysed by pyruvate kinase. This enzyme is inhibited by high levels of ATP.

Fatty acid metabolism

Glucose is not the only source of energy we have. The biggest energy store in your body is in the form of fatty acids, which are stored as triglycerides (fats). Triglycerides store more energy per gram than glucose; however, unlike glucose they are not soluble in water. This means that triglycerides need to be broken down by enzymes into fatty acids and glycerol before the fatty acids can undergo respiration. This takes time, so fats are used as your energy source during long periods of exercise.

Fatty acids are very long molecules, containing many carbon and hydrogen atoms. They can generate many protons from these hydrogen atoms, so lots of ATP molecules can be made during the oxidative phosphorylation stage (using the proton gradient as an energy source).

The metabolic pathway can be summarised as follows.

1. The fatty acid combines with co-enzyme A to form a fatty acid complex. This reaction requires energy from an ATP molecule.
2. The fatty acid complex enters the mitochondrial matrix and is broken down into acetyl groups attached to co-enzyme A. During this reaction reduced NAD and reduced FAD are formed.
3. The acetyl groups detach from the co-enzyme A and are used in the Krebs cycle. Each acetate group forms three reduced NAD molecules, one reduced FAD molecule and one ATP molecule.
4. The reduced NAD and reduced FAD undergo oxidative phosphorylation, producing ATP.

Respiratory quotient

It is useful to compare respiration when glucose is metabolised with respiration when fatty acids are metabolised. This is difficult to do; however, the respiratory quotient (RQ) can be used to help us understand respiration. The RQ is used in calculations to measure your basal metabolic rate (BMR).

After measuring the amount of oxygen you breathe in and the amount of carbon dioxide you breathe out, the RQ value can be calculated:

$$RQ = \frac{CO_{2\ \text{eliminated}}}{O_{2\ \text{consumed}}}$$

RQ values range from 1.0 (the value expected for glucose respiration) to 0.7 (the value expected for fatty acid respiration). If your diet is a mix of fat and glucose then an average value of 0.85 is expected.

Anaerobic respiration

During vigorous exercise, your body may not be able to take in enough oxygen for aerobic respiration to occur. Only stage 1, glycolysis, can form ATP when there is no oxygen present because it is an anaerobic process. The reduced NAD formed by glycolysis needs to lose hydrogen atoms and turn back into NAD, otherwise glycolysis cannot occur and no ATP will be made.

Lactate fermentation

In human cells, NAD is made by lactate fermentation. Pyruvate accepts the hydrogen atoms from reduced NAD, forming NAD. The NAD can be used in glycolysis, forming more ATP. Pyruvate is turned into lactate in the process.

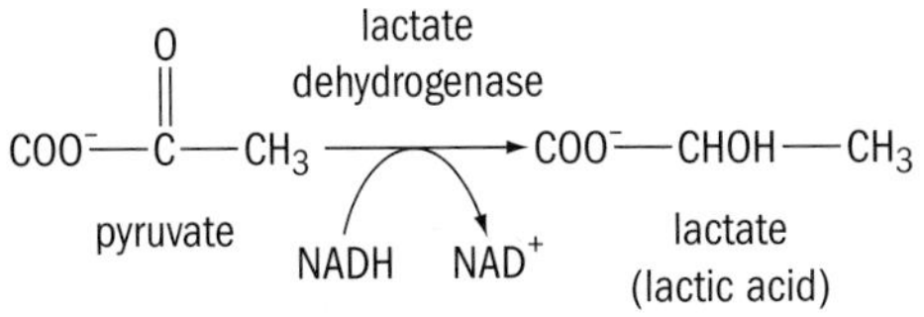

$$COO^- - \overset{O}{\overset{\|}{C}} - CH_3 \xrightarrow[NADH \rightarrow NAD^+]{\text{lactate dehydrogenase}} COO^- - CHOH - CH_3$$

pyruvate → lactate (lactic acid)

Lactate fermentation.

Lactate can be turned back into pyruvate when more oxygen is available. If lactate levels build up because anaerobic respiration continues to be the only way your cells can make ATP, then your body muscles will start to ache (we know this as 'cramp'). This is because lactate affects how effectively the muscles' enzymes work.

Alcoholic fermentation

In fungi, such as yeast, NAD is made by alcoholic fermentation. Pyruvate is converted into ethanal, producing carbon dioxide. Ethanal accepts the hydrogen atoms from reduced NAD, forming NAD. The NAD can be used to accept more hydrogen atoms from glucose in glycolysis. Ethanal is turned into the alcohol called ethanol.

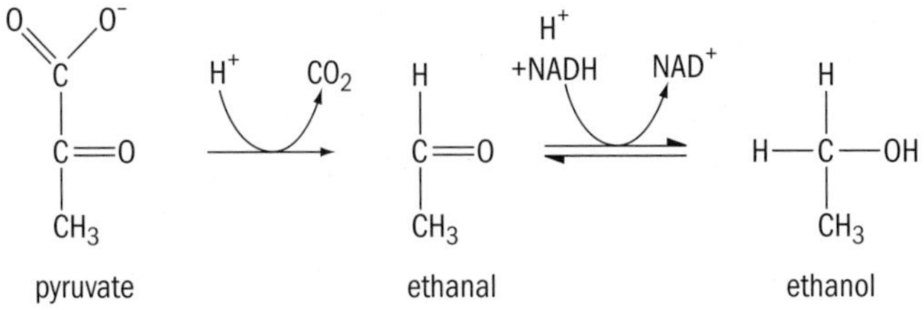

$$\text{pyruvate} \xrightarrow[H^+ \;\; CO_2]{} \text{ethanal} \xrightarrow[H^+ + NADH \;\; NAD^+]{} \text{ethanol}$$

pyruvate: $COO^- - C(=O) - CH_3$; ethanal: $H - C(=O) - CH_3$; ethanol: $H - CH(OH) - CH_3$

Alcoholic fermentation.

This process is used on an industrial scale to form ethanol that can be used to make beers and wines. The yeast does not need oxygen to survive; however, it grows faster when it has a source of oxygen. In the brewing process, yeast is grown in aerobic conditions to speed up the process. When there is enough yeast, anaerobic conditions are used to make the yeast produce ethanol. This is because of the **Pasteur effect:** oxygen slows the rate of fermentation. If yeast undergoes aerobic fermentation then ATP production increases. However, ATP inhibits phosphofructokinase, the enzyme that catalyses reaction 3 of glycolysis. Therefore, respiration slows down.

This table compares the sites of aerobic and anaerobic respiration.

Type of respiration	Site
glycolysis (anaerobic)	cytoplasm
link reaction (aerobic)	mitochondrial matrix
Krebs cycle (aerobic)	mitochondrial matrix
oxidative phosphorylation (aerobic)	inner mitochondrial membrane
lactate fermentation (anaerobic)	muscle cells

Case study: Healthcare technician

Keira works as a technician in a lab for a company within the healthcare industry. She works with dieticians who advise patients regarding the food they eat. One part of her job involves measuring RQ values for patients. A respirometer is used. Keira is responsible for the respirometer being clean before asking a patient to breathe into it. She needs to ensure that the patient is sitting down and is relaxed before taking the RQ measurement.

Keira carries out the measurement a minimum of three times. The RQ values that are produced can show if the patient's diet contains large amounts of carbohydrate or large amounts of protein. Keira compares her patient's results with reference RQ values for known substances that are common in the diet, and then shares her findings with the patient.

1 **Why should the patient be relaxed before taking the RQ measurement?**

2 **Why does Keira carry out the measurement at least three times?**

Assessment activity 13.4

You are a laboratory technician working within a research laboratory in a university. You have been asked to find out about aerobic and anaerobic respiration and report your findings to your team. In order to do this, complete the following tasks.

1 Aerobic and anaerobic exercise can affect the body in different ways. This is because the metabolic pathways for the two types of respiration are different. Using the text and other sources of information, produce a list of differences between the two types of glucose respiration. **P6**

2 Produce a labelled diagram showing the aerobic and anaerobic stages of glucose respiration. Identify the reactions which use and produce ATP molecules. Produce a table comparing the amounts of ATP molecules used and produced by the aerobic stages and the anaerobic stages. **M4**

3 Describe how glycolysis is regulated, and how changing concentrations of ATP may affect this regulation. Evaluate how the inhibition of reactions may affect the amount of energy within the cell. Include a description of how lactate fermentation during anaerobic glycolysis affects muscle cells. **D4**

Grading tips

To help you gain **P6**, include the effects on the body, the products and the amount of energy formed.

Task 2 requires a detailed diagram in order to describe and compare the relevant sites, for **M4**.

To achieve **D4**, a description is needed of how the energy derived from glucose may be stored if the cell doesn't require ATP.

PLTS

Reflective learner

You will demonstrate this skill when you use several sources to reach a conclusion.

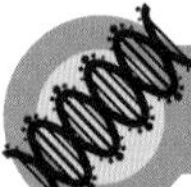

Functional skills

Creative thinker

You will develop this skill when you identify the differences between the two types of glucose respiration.

WorkSpace

Madeleine Strickland

PhD student in chemistry at the University of Bristol

I work in a team which specialises in clarifying protein structures using nuclear magnetic resonance (NMR) spectroscopy. I've worked on an HPV virus protein, E2, and the insulin-like growth factor II receptor (IGF2R), both of which malfunction in certain tumours.

Synthesising a protein typically occurs in five stages – transformation (inserting the DNA of the protein you want to grow into *E. coli* cells), cell growth, induction (using a chemical such as IPTG to force the cells to make that protein), refolding (breaking the disulfide bonds in the protein and rejoining them in the correct place) and purification.

This typically takes up to two weeks in total, which is followed by a couple of weeks on the NMR machine to obtain all of the data you need. Then there is the fun bit – solving the structure of the protein!

Protein structure elucidation isn't for the faint-hearted! A lot of work goes into learning all of the techniques needed for this very specialised job, but it is widely used in drug discovery so you never know which disease you'll be working on next. This year I'll be travelling to Italy and Sweden to learn the latest techniques.

We had trouble growing a new receptor in the *E. coli* cells recently. We could have truncated the protein (which means shortening it at either end), added His tags (six histidine amino acids added to either the beginning or the end of the protein to aid its solubility) or tried various media and conditions for the growth itself. Eventually we discovered that the natural DNA sequence could not be used with *E. coli* as it did not recognise some of the amino acid coding used in humans. We inserted the sequence into yeast instead (*P. pastoris*), which recognises more codes, but this is much more expensive. This time the protein was produced.

Think about it!

1 Which base pairs in DNA code for the following amino acids – alanine, phenylalanine and cysteine?

2 What types of secondary structure can form in proteins?

Just checking

1 Which two amino acids form the dipeptide Gly-Ala?
2 What is the name of the reaction that formed the dipeptide?
3 What is the difference between the two amino acids?
4 What is the role of coenzyme A in respiration?
5 Where in the mitochondrion do the Krebs cycle and oxidative phosphorylation occur?
6 List three factors that can affect the catalytic activity of an enzyme.
7 Why will an enzyme only catalyse a specific reaction?
8 Why can the word monomer be used to describe glucose?
9 Which compounds can react together to form a triglyceride?
10 What shape is a DNA molecule?
11 What is meant by the tertiary structure of a protein?

Assignment tips

To achieve the highest grading criteria possible you must:

- follow instructions for practical work carefully
- work safely
- work as a team
- record your data clearly
- only include research material if it is directly relevant and you can understand it.

The key points you need to remember from this unit are as follows.

- Biological molecules have different structures which decide their function.
- Different analytical techniques are available for each type of molecule.
- The primary structure of a protein defines its other levels of structure.
- Enzymes have a structure which is specific to one type of substrate.
- The rate of reaction that an enzyme catalyses can be affected by pH, temperature and substrate concentration.
- Enzyme activity can be investigated practically by different methods.
- Anaerobic and aerobic respiration generate different amounts of ATP via different routes.

You may find the following website useful as you work through this unit.

For information on...	Visit...
general biochemistry	The Biochemical Society

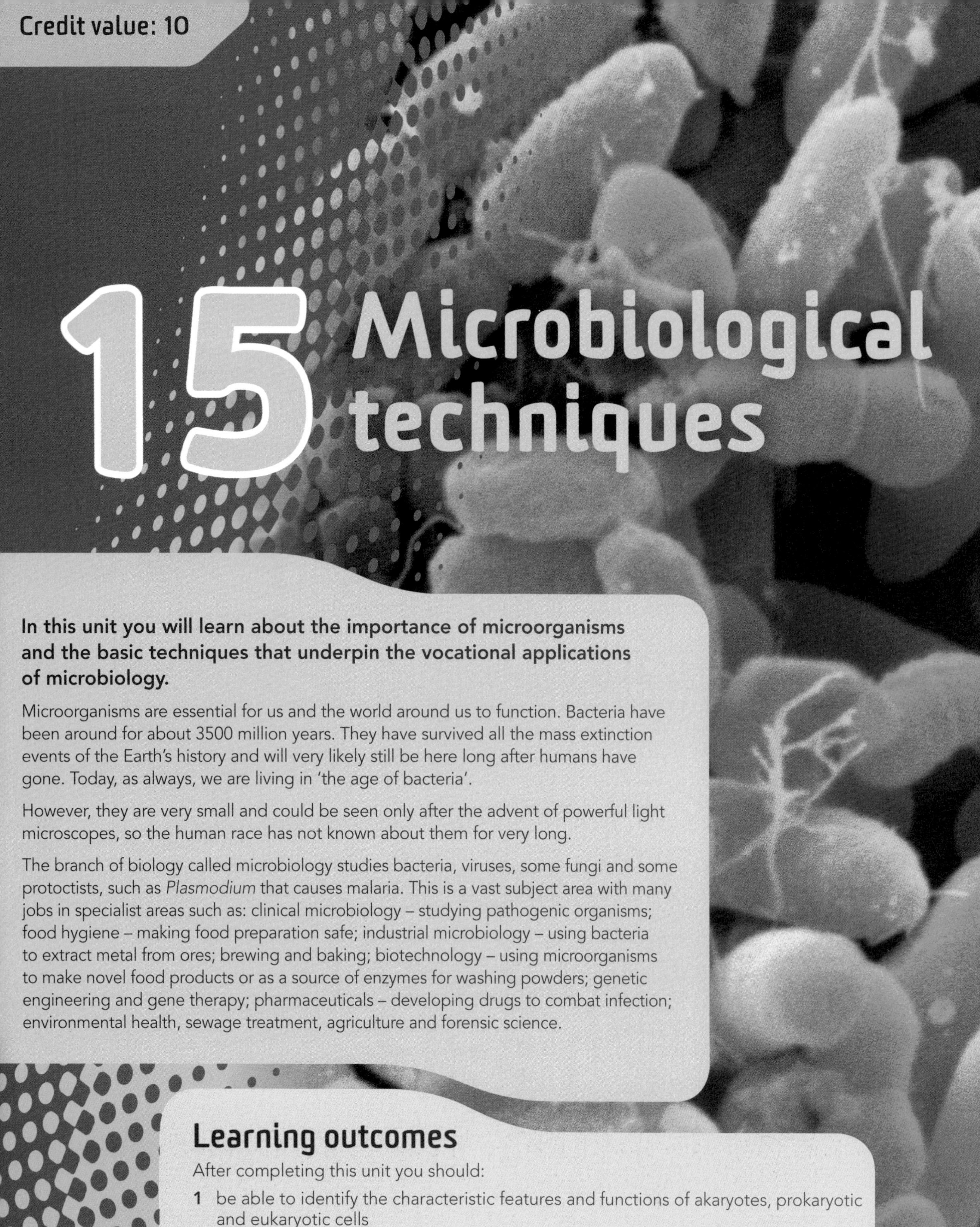

Credit value: 10

15 Microbiological techniques

In this unit you will learn about the importance of microorganisms and the basic techniques that underpin the vocational applications of microbiology.

Microorganisms are essential for us and the world around us to function. Bacteria have been around for about 3500 million years. They have survived all the mass extinction events of the Earth's history and will very likely still be here long after humans have gone. Today, as always, we are living in 'the age of bacteria'.

However, they are very small and could be seen only after the advent of powerful light microscopes, so the human race has not known about them for very long.

The branch of biology called microbiology studies bacteria, viruses, some fungi and some protoctists, such as *Plasmodium* that causes malaria. This is a vast subject area with many jobs in specialist areas such as: clinical microbiology – studying pathogenic organisms; food hygiene – making food preparation safe; industrial microbiology – using bacteria to extract metal from ores; brewing and baking; biotechnology – using microorganisms to make novel food products or as a source of enzymes for washing powders; genetic engineering and gene therapy; pharmaceuticals – developing drugs to combat infection; environmental health, sewage treatment, agriculture and forensic science.

Learning outcomes

After completing this unit you should:

1 be able to identify the characteristic features and functions of akaryotes, prokaryotic and eukaryotic cells
2 be able to use aseptic techniques to culture microorganisms
3 be able to determine the factors that influence the growth of microorganisms
4 know how to identify microorganisms.

Assessment and grading criteria

This table shows you what you must do in order to achieve a pass, merit or distinction grade, and where you can find activities in this book to help you.

To achieve a **pass** grade the evidence must show that you are able to:	To achieve a **merit** grade the evidence must show that, in addition to the pass criteria, you are able to:	To achieve a **distinction** grade the evidence must show that, in addition to the pass and merit criteria, you are able to:
P1 use light microscopy techniques to identify the characteristic features and functions of prokaryotic and eukaryotic cells **See Assessment activity 15.1**	**M1** describe the function of prokaryotic and eukaryotic cell components **See Assessment activity 15.1**	**D1** relate the characteristic features of prokaryotic and eukaryotic cell components to their function **See Assessment activity 15.1**
P2 use data from electron microscopy to identify the characteristic features and functions of akaryotes, prokaryotic and eukaryotic cells **See Assessment activity 15.1**		
P3 carry out practical activities to cultivate microorganisms using aseptic techniques **See Assessment activity 15.2**	**M2** explain the principles underlying the cultivation and aseptic techniques used **See Assessment activity 15.2**	**D2** evaluate the growth conditions in terms of the cultivation techniques used compared with large-scale industrial growth **See Assessment activity 15.3**
P4 carry out practical investigations of factors that influence the safe growth of microorganisms **See Assessment activity 15.3**	**M3** compare the calculated growth rates of microorganisms grown under varying conditions **See Assessment activity 15.3**	**D3** draw valid conclusions from the growth rate calculations, suggesting how this knowledge may be applied in a biotechnological or biomedical context **See Assessment activity 15.3**
P5 describe the main groups of microorganisms by their principal taxonomic characteristics **See Assessment activity 15.4**	**M4** outline how the techniques used to identify microorganisms relate to their structure **See Assessment activity 15.4**	**D4** outline the potential usefulness of a variety of identification techniques in a specific application **See Assessment activity 15.4**

How you will be assessed

- Many of your assessment activities will involve carrying out the practical work safely and competently.
- You may also be required to present what you have learnt during the practical and share it with others in your class.
- You may also give presentations or make posters on various aspects of microbiology.
- You will need to put together a portfolio of evidence of all your assignments.

Sonal, 17 years old

I have really enjoyed this microbiology unit. Before I studied this topic I thought that all microorganisms caused diseases. However, I now realise that only a few cause diseases and without bacteria there would be no life on Earth, as dead bodies would not be decayed and their molecules recycled. Without the bacteria in our guts we would not live. Forensic scientists are now working on using the bacteria left from people's fingerprints to identify them, as we each have a different range of bacteria living on our skin. This unit has a lot of practical work in it and I have developed my laboratory practical skills. After a while, using aseptic technique becomes automatic. Growing bacteria and staining them to look at under the microscope is fascinating. We have visited a food processing factory, a sewage works and a medical microbiology laboratory. I am hoping to study microbiology at degree level and would like to go on to get a job in the food industry or in medical microbiology.

Catalyst

Small is beautiful

Match the organism with its use or disease caused.

Use a combination of inspired guesswork, elimination and looking things up.

Activity	Organism
A brewing	1 *Mycobacterium tuberculosis*
B yoghurt making	2 *Plasmodium vivax*
C causes athlete's foot	3 *Vaccinia* virus
D causes TB	4 *Saccharomyces carlsbergensis*
E causes malaria	5 human immunodeficiency virus
F causes tetanus (lockjaw)	6 *Lactobacillus*
G causes cowpox	7 *Clostridium tetani*
H causes AIDS	8 *Tinea pedis*

15.1 Akaryotes, prokaryotic and eukaryotic cells

In this section:

Key terms

Magnification – number of times larger the image is compared with the original object.

Resolution – the ability to distinguish two objects, which are close together, as separate objects.

Refraction – bending of light rays.

Organelle – structure inside a cell. Each organelle has a specific function.

Metabolic activities – chemical reactions inside living cells that sustain life (also known as metabolism).

Akaryote – viruses; they have no cell structure or organelles.

Nanometre – a unit of length. There are 1000 nm in 1 μm, which means there are 1 000 000 nm in 1 mm.

Endocytosis – method of bulk transport into a cell. The plasma membrane invaginates around a particle and encloses it inside a vesicle within the cell.

Host – organism infected by a parasite.

Light microscopes

In 1673 Anton van Leeuwenhoek used a simple microscope of one biconvex lens enclosed in two metal plates and examined drops of pond water. His microscope magnified about 300 times and he showed that living creatures existed that were too small to be seen with the naked eye.

Since then, light microscopes have become more complex. Light:

- is focused by a condensing lens
- passes through the specimen mounted on a slide and placed on the stage
- is focused by an objective lens to make a real magnified image that is magnified again by an eyepiece lens.

The total **magnification** is calculated by multiplying the objective lens magnification by the eyepiece lens magnification.

Most microscopes have three objective lenses: ×4, ×10 and ×40; and may use a ×10 or a ×15 eyepiece. These give a range of magnifications from ×40 to ×600.

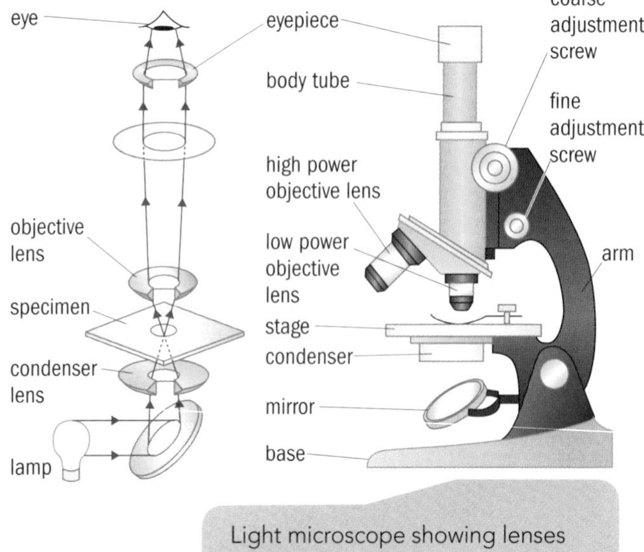

Light microscope showing lenses and light paths.

The microscope most commonly used in schools and colleges is the compound brightfield microscope. It is relatively cheap and easy to use. It can produce clear images with magnifications up to ×600 and can be used to examine living specimens of some microorganisms.

Photographs of images seen with light microscopes can be made. These are called photomicrographs and give a permanent record of what has been examined. They also allow accurate measurements of the specimens to be made.

Activity 15.1A

Explain to a partner how a light microscope produces magnification from ×40 to ×600.

Oil immersion

Oil immersion is a technique used to increase the **resolution** of a light microscope. This enables scientists to examine small specimens, such as bacteria, with a higher magnification, whilst still seeing a clear image.

Unit 11: See page 196 for more information on resolution in light microscopes.

- A special objective lens, usually with a magnifying power of ×100, is used.
- The bacteria are heat-fixed onto a glass slide to make a smear.

- The bacterial smear is stained, excess stain washed off and the slide is dried.
- A drop of immersion oil, which has optical properties (bending power) very like those of glass, is placed on the slide and the slide is placed on the stage.
- The immersion lens is lowered carefully so it is in the oil just touching the slide.
- Now the fine focus knob is slowly turned to move the lens slightly away from the slide, but still in the oil, to bring the bacteria into focus.
- Depending on the eyepiece lens used, the bacteria can be magnified by ×1000 or ×1500.
- At the end of the activity, the objective lens has to be carefully wiped clean, using a special fine lens tissue.

Activity 15.1B

Discuss with a partner how the bacteria are magnified by ×1500.

Phase contrast

Some light microscopes have special phase contrast objective lenses and condenser lenses.

1. As light rays are transmitted through a specimen, the rays are bent (**refracted**) due to variations in density and thickness of the cell components of the specimen.
2. The special lenses in the microscope convert the differences between transmitted light and refracted light into a variation in the intensity of light.
3. This gives a clear image of transparent cells, without having to stain them. The image appears light against a dark background.
4. This means it can be used to observe living cells and has been used to observe the behaviour of chromosomes during (eukaryotic) cell division.

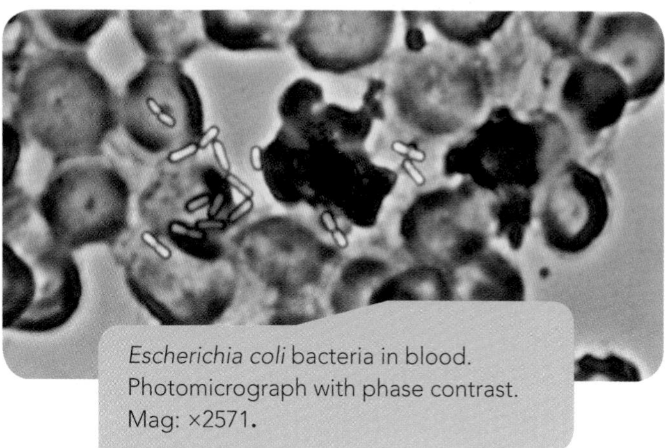

Escherichia coli bacteria in blood. Photomicrograph with phase contrast. Mag: ×2571.

Electron microscopes

The development of electron microscopes since the 1930s has enabled scientists to see very small structures with high magnification and high resolution. They have been able to study the structures of sub-cellular components and the structures of viruses.

Instead of a beam of light, a beam of electrons, having a much shorter wavelength than that of light, is used.

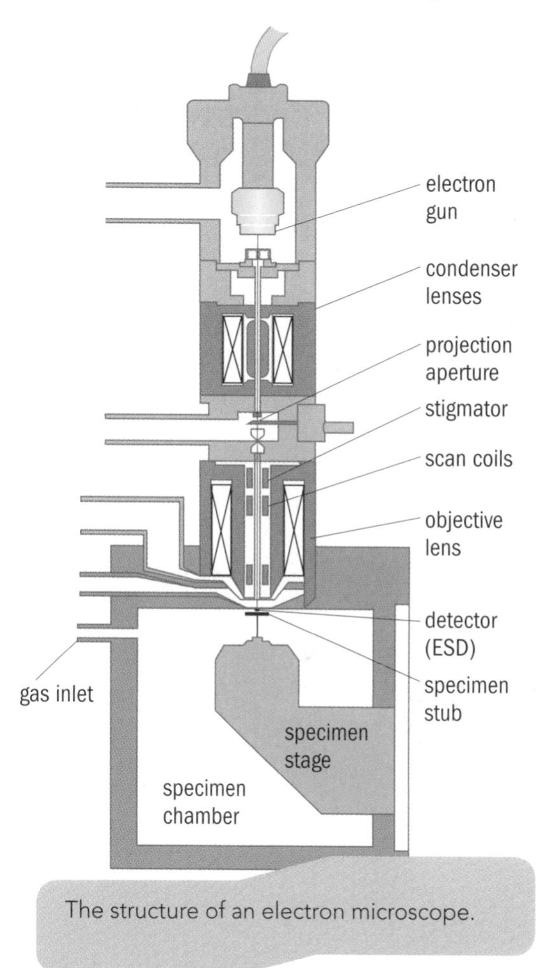

The structure of an electron microscope.

Transmission electron microscope

- A very thin section of the specimen is dried and stained with heavy metal salts.
- It is mounted on a copper grid.
- The copper grid and specimen are placed, in a vacuum, in the microscope.
- The beam of electrons passes through a very thin section of the specimen.
- As the electrons pass through the specimen parts, some of which are denser as they have absorbed the heavy metal stains, a contrasting image is formed on an X-ray film plate or nowadays some type of digital camera.

- This gives an image in shades of grey that can be printed as an electron micrograph.
- Sometimes colour can be added to give a false colour electron micrograph.
- Total magnification may be up to ×1 million.

Scanning electron microscope

- The electron beam is directed onto and bounces off the specimen. Electrons don't pass through the specimen.
- The final image is a 3D view of the surface of the specimen. False colour can be added to the image.
- The total magnification is up to ×100 000.

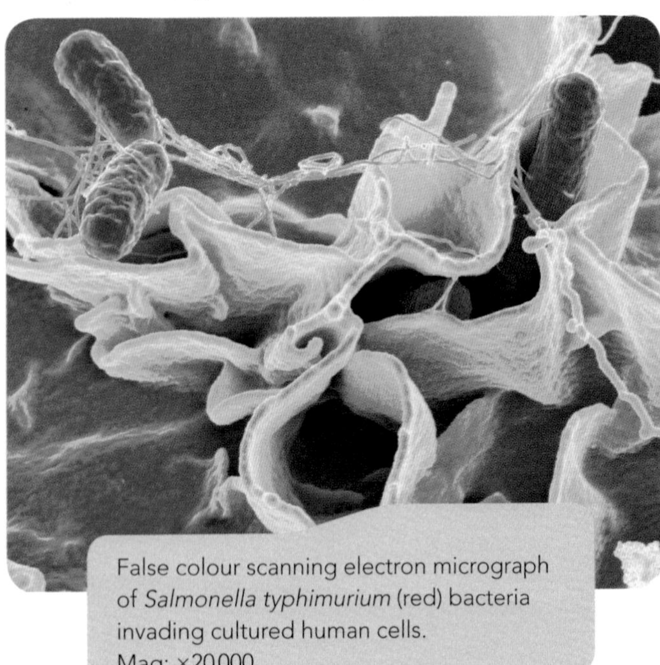

False colour scanning electron micrograph of *Salmonella typhimurium* (red) bacteria invading cultured human cells. Mag: ×20000.

Electron microscopes enable specimens to be seen with much higher magnification and high resolution. However, they are very expensive and to use them requires a great deal of skill and training. Specimens have to be dead so we cannot see any activities in living cells. The images produced are in shades of grey, although false colour can be added.

Did you know?

The photographer Lennart Nilsson developed a way of taking lots of electron micrographs of cells at each stage of invasion by a virus and duplication of the viruses inside the cell. He added false colour and ran the images to make a 'movie'.

Characteristics of living organisms

All living organisms grow, move, need nutrients, obtain energy, can reproduce, excrete toxic waste they have made and can respond to stimuli. This applies to single-celled as well as multi-cellular living organisms. Cells exhibit these characteristics.

Biologists classify living organisms by putting them into groups. Members within each group have similar characteristics, biochemistry and genetic make-up. This classification can also indicate something about the evolutionary relationships between organisms.

At present we classify living organisms into five kingdoms. These are:

- prokaryotes – the bacteria
- fungi – e.g. moulds, yeasts and mushrooms
- protoctists – e.g. algae, protozoa
- plants – e.g. mosses, liverworts, grasses, roses, fir trees, oak trees
- animals – e.g. jellyfish, worms, insects, spiders, fish, amphibians, reptiles, birds, mammals.

Fungi, protoctists, plants and animals are all eukaryotes as they all have eukaryotic cells.

The cell structure of prokaryotes is different from that of eukaryotes – they have prokaryotic cells (see page 283).

Eukaryotic cells

All eukaryotic cells have certain features in common. They have cytoplasm, a plasma membrane, and **organelles** such as mitochondria, rough endoplasmic reticulum, smooth endoplasmic reticulum, ribosomes, secretory vesicles, Golgi apparatus and a nucleus surrounded by a double membrane, the nuclear envelope. Inside the nucleus is a denser area called the nucleolus.

- Some eukaryotic cells (animals and some protoctists) also have centrioles, cilia and undulipodia (formerly called flagella).
- Most plant cells and some protoctist cells also have chloroplasts.
- Some eukaryotic cells (plants, some protoctists and fungi) also have a wall around the plasma membrane.

- Plant cells and algal cells have walls made from cellulose (a polymer of a type of glucose).
- Fungi have walls made from chitin – also a polymer but made of *N*-acetylglucosamine.

Division of labour

Each type of organelle has a specific function but they work together to ensure the survival of the cell. The plasma membrane and cytoplasm also have functions.

Functions of structures found in all eukaryotic cells:

Structure	Function
plasma membrane	Made of phospholipids and protein; keeps each cell separate from the outside environment; regulates transport of substances into and out of the cell, as it has special transport proteins or channels; has antigens (special proteins) on the surface so the immune system can recognise it as 'self'.
cytoplasm	The matrix of the cell in which organelles are embedded; contains water (about 70%) and proteins, some of which form a supporting network of microtubules; some of these proteins are contractile and help to move organelles within the cell.
nucleus	Houses the genetic material, DNA, of the cell, in the form of chromatin; has a double membrane or envelope with pores so that some substances can pass into or out of the nucleus.
nucleolus	A dense area in the nucleus, without a membrane around it, where RNA and ribosomes, both needed for protein synthesis, are made.
endoplasmic reticulum (ER)	A series of flattened membrane-bound sacs called cisternae. Rough ER is covered with ribosomes and is where proteins are assembled; the proteins pass into the cisternae and travel to the Golgi apparatus. Smooth ER does not have ribosomes; steroids and lipids are synthesised here.
Golgi apparatus	Here, proteins made at ribosomes are modified and packaged into vesicles or lysosomes.

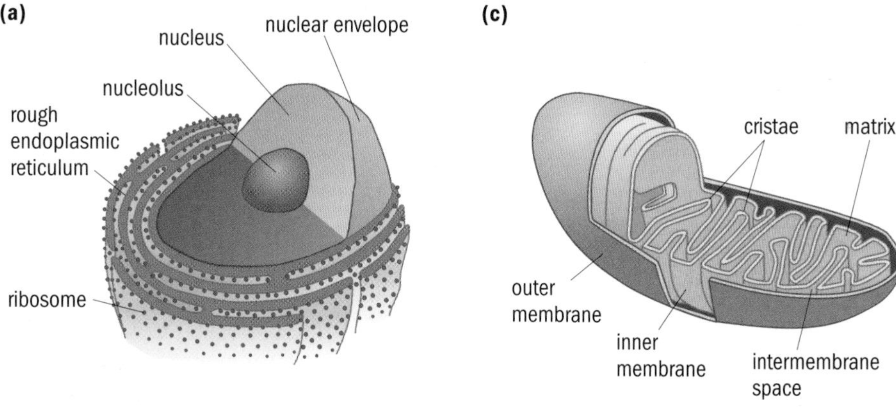

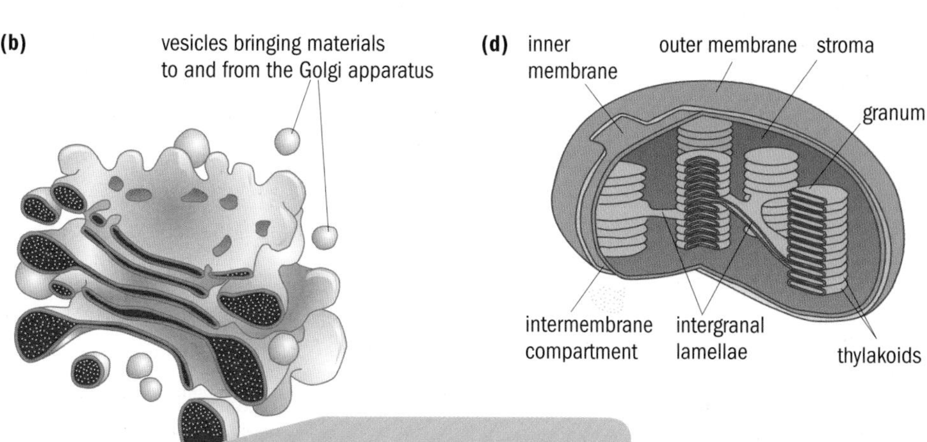

(a) Nucleus and endoplasmic reticulum. **(b)** Golgi apparatus. **(c)** Mitochondrion (2–5 μm long). **(d)** Chloroplast (4–10 μm long).

Structure	Function
vesicles	Membrane-bound sacs derived from Golgi apparatus; may hold particular chemicals or may carry proteins to the plasma membrane to be released outside the cell.
lysosomes	Large spherical organelles derived from Golgi apparatus and containing digestive enzymes, some of which may break down invading microorganisms. Keeping these enzymes inside the lysosome prevents them from breaking down the cell.
mitochondria	Spherical or sausage-shaped organelles with an inner and outer membrane; the stages of aerobic respiration occur here.
ribosomes	Tiny organelles, made of two subunits and having no membrane around them; made of RNA and protein; proteins are assembled here.

Function of other organelles:

Organelle	Function
centriole	Small tubes of protein fibres having no membrane around them but within an area of the cell called a centrosome. They form the spindle when nuclei of cells divide in mitosis and meiosis. Some centrioles form cilia on the surface of certain cells.
cilia	Hair-like extensions on the surface of some animal and protoctist cells. They beat in a synchronised rhythm to enable the cell to move, or, as in the case of ciliated epithelial cells in our respiratory tract, to move mucus with trapped pathogens.
undulipodia (eukaryotic flagellum)	Large cilia – the tail of each spermatozoan is an undulipodium.
chloroplast	Structure with an outer and inner membrane. Inside are stacks of thylakoids (flattened membrane sacs) called grana. The thylakoid membranes contain chlorophyll and the first stage of photosynthesis happens here. The second stage occurs in the fluid-filled stroma.

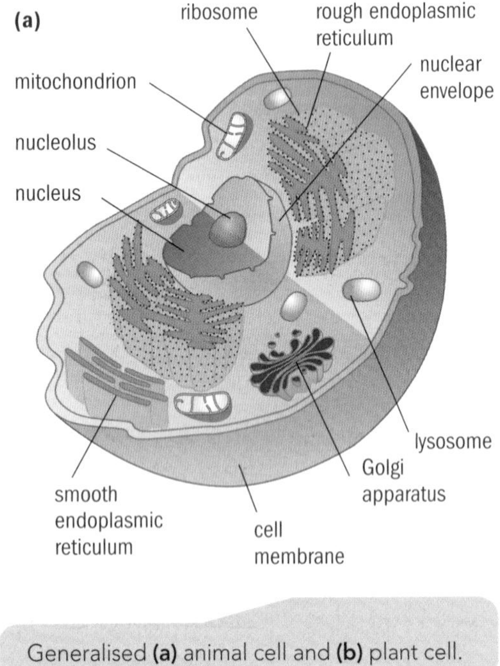

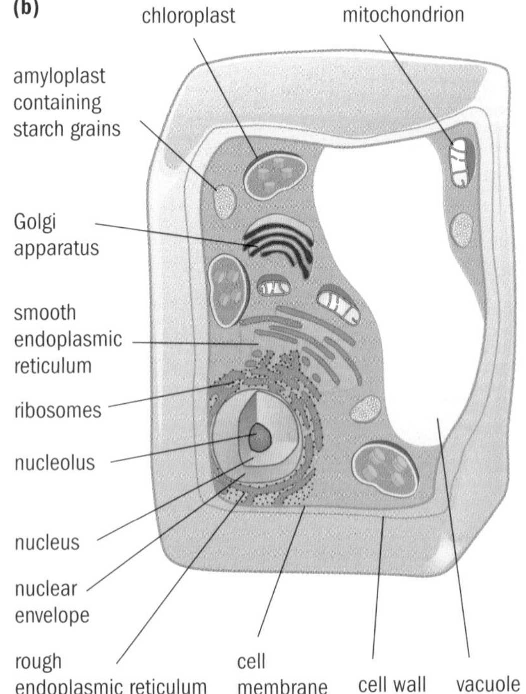

Generalised **(a)** animal cell and **(b)** plant cell.

Eukaryotic microorganisms

Some protoctists, such as *Plasmodium* that causes malaria, and *Entamoeba* that causes amoebic dysentery, are eukaryotes but are small pathogenic organisms and so are studied by microbiologists.

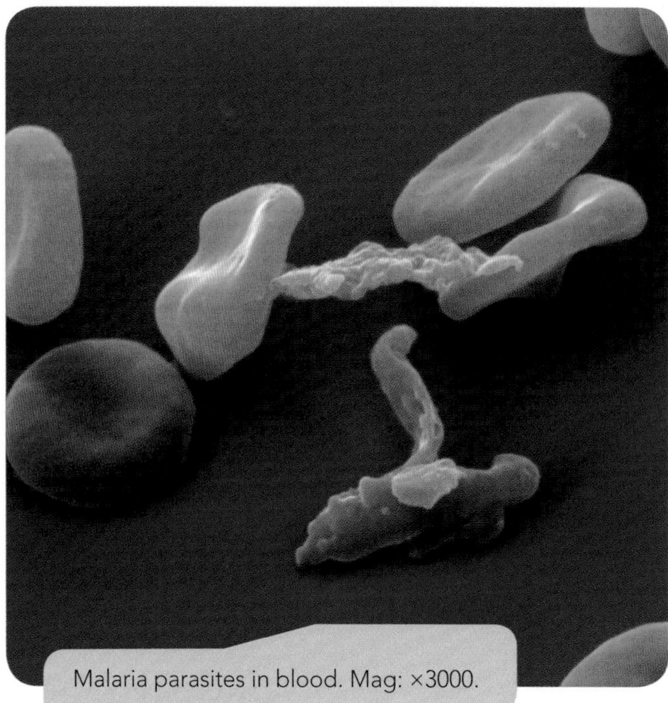

Malaria parasites in blood. Mag: ×3000.

Malaria used to be endemic (always present) in Britain. In Shakespeare's time it was called 'the ague'. The Romans gave it its name – 'malaria' means 'bad air'. They knew that it was associated with swampy areas, which always smelt bad. We now know that the vectors, female mosquitoes, lay their eggs in stagnant water such as swampy regions. If global warming continues, malaria will become endemic in Britain again.

Fungi are also eukaryotic. Some fungi are pathogenic. Thrush, athlete's foot, farmer's lung and ringworm are all caused by fungi.

Fungi have long thread-like hyphae surrounded by cell walls made of chitin. They reproduce by spores.

Prokaryotic cells

The five kingdom classification system puts all bacteria into one kingdom – the prokaryotes. Prokaryotes make up a very large group of single-celled microorganisms that are found in every habitat on the planet Earth.

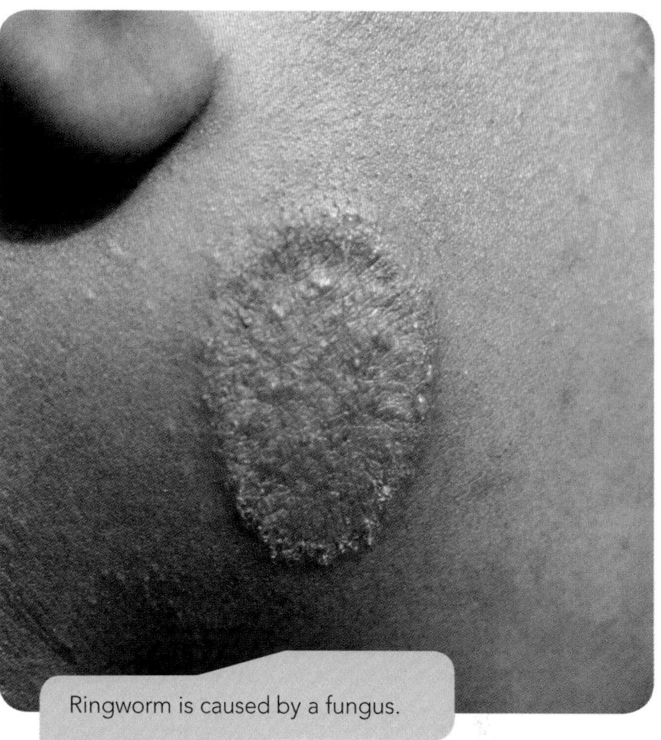

Ringworm is caused by a fungus.

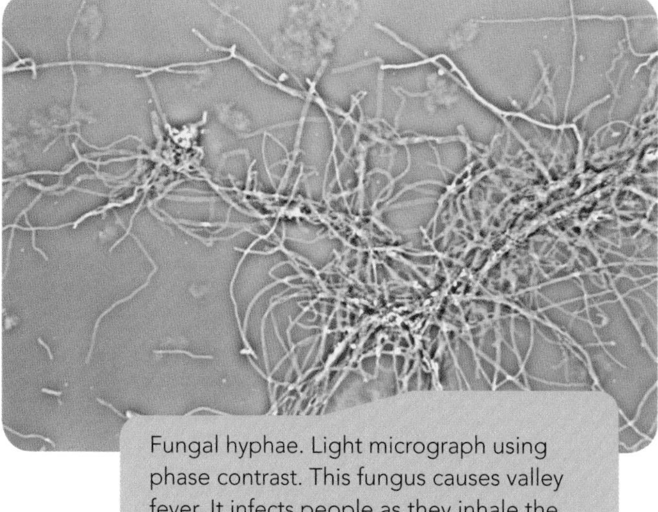

Fungal hyphae. Light micrograph using phase contrast. This fungus causes valley fever. It infects people as they inhale the spores. Mag: ×205.

There are about 5×10^{30} bacteria on Earth and they make up the bulk of the planet's biomass.

Did you know?

You have 10 times as many bacterial cells in your gut as you do cells in your body.

There are nearly as many bacterial cells in a healthy human mouth than there are people on Earth.

There are about 10^8 bacteria in one gram of soil.

Most bacteria are extremely useful as:

- they help recycle nutrients
- they are crucial in the process of breaking down sewage waste
- some fix atmospheric nitrogen and turn it into a form that plants can use to make amino acids and proteins
- those in our gut help us digest food and produce some vitamins for us
- we have special populations of bacteria on our skin that help prevent harmful bacteria from entering the body
- many are used in the biotechnology industry, for example to make yoghurt and cheese and as a source of enzymes for various processes, such as the PCR
- some are used in the biomedical industry to make antibiotics.

Unit 18: See page 355 for more information on PCR.

However, a few species are pathogenic and cause infectious diseases, such as typhoid, cholera, TB, meningococcal meningitis and syphilis.

Almost 20 years ago scientists discovered that prokaryotes consist of two very different groups, or domains, of organisms that evolved independently of each other from a common ancestral form. The two domains are called Archaea (ancient types) and Eubacteria. Here, we are referring to the structure of eubacteria.

Features of prokaryotic cells

Prokaryotic cells are living and carry out **metabolic activities**. They respire and therefore also need nutrients. Some have chlorophyll and can photosynthesise. All grow, reproduce and move; although some may only move at the same speed as continental drift, others move at such a pace that it is equivalent to a human swimming through treacle at 10 miles an hour. They all produce toxic waste that is removed and many can respond to stimuli such as chemicals, light and magnetism; they may move towards or away from the stimulus.

- They are all between 1 and 5 μm in length and may be spherical, rod-shaped or spiral.
- They have a cell wall made of peptidoglycan (previously called murein) consisting of chains of carbohydrate polymers linked by peptides that consist of unusual amino acids.
- The plasma membrane surrounds the cytoplasm and regulates entry and exit of substances. Parts of it are folded inwards to increase surface area, and have enzymes that catalyse various stages of metabolic reactions embedded in it. Some bacteria have chlorophyll in these folded membranes and can carry out photosynthesis. Many have an inwardly folded membrane called a mesosome that has enzymes for making ATP (adenosine triphosphate) during respiration.
- There are no membrane-bound organelles, such as mitochondria, ER, Golgi apparatus or chloroplasts.
- There are ribosomes but they are smaller than the ones in eukaryotic cells.
- The DNA is not associated with histone proteins and is described as 'naked'. It forms one circular chromosome that is in a region of the cytoplasm called the nucleoid.

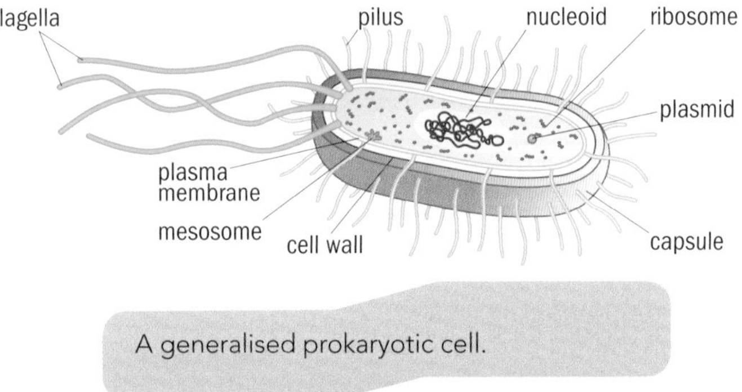

A generalised prokaryotic cell.

Many bacteria also have smaller circular loops of DNA called plasmids. Some have a protective capsule around the cell wall. This may protect pathogens from the host's immune system.

Unit 18: See page 351 for more information on plasmids.

Some have one or more flagella (singular: flagellum) – whip-like structures that help them to move. The internal structure of the prokaryotic flagella is different from the structure of the eukaryotic undulipodia. Flagella rotate, using energy from a hydrogen ion (proton) gradient across the plasma membrane to power the rotation. The flagellum is like an axle and the bacterial cell is moved, as a wheel is, around this axle.

Some photosynthetic bacteria can produce vesicles full of gas. This allows them to alter their buoyancy and move up or down in water with different light intensities and nutrients.

Some have hair-like pili on their surface and these enable them to stick to host cells, to other surfaces or to other bacteria when they are transferring DNA.

Bacteria divide by a process called binary fission. They do *not* undergo mitosis.

Activity 15.1C

1 Which gas do you think collects in the vesicles of photosynthetic bacteria, when there is light and they are carrying out photosynthesis?

2 Explain how keeping this gas inside their cell in vesicles alters their buoyancy.

3 In which direction in the water (up or down) will they move when they are more buoyant?

4 How will this help them to carry out more photosynthesis?

Did you know?

Mitochondria and chloroplasts are probably both derived from bacteria that lived inside host cells. Both of these organelles have plasmids of DNA and small ribosomes, the same as the ribosomes found in bacteria, inside them. They can replicate, which they do in interphase, and are the same size as large bacteria.

Akaryotes – viruses

You have seen that eukaryotic cells have a true nucleus, bound by a nuclear envelope, whereas prokaryotic cells do not have a proper nucleus. Prokaryotic cells also do not have membrane-bound organelles.

Microbiologists study important microorganisms – those that are useful or those that cause diseases. Microorganisms include bacteria (prokaryotic cells), some fungi and protoctists (eukaryotic cells) and another important group – the viruses.

Viruses are described as **akaryotes** because they do not have any kind of nucleus and they do not have any cellular structure. They do not have:

- cytoplasm
- cell membranes
- organelles of any sort
- any metabolism.

Activity 15.1D

Think of at least six diseases caused by viruses. Share your ideas with the rest of the class and produce a large list of viral diseases.

Viruses are extremely small and most cannot be seen with even the most powerful light microscope. They are described as sub-microscopic. The word 'virus' is Latin for 'poison' and was used by Edward Jenner, in the late eighteenth century, although he had no real idea of the nature of the agent that caused smallpox.

Viruses consist of:

- some genes, made of *either* DNA *or* RNA (unlike eukaryotic and prokaryotic organisms they do not contain both types of nucleic acid)
- a protein coat, the capsid, made of protein subunits called capsomeres. This coat protects the genetic information. Some of the proteins on the viral surface are antigens.

Viruses also contain some enzymes. In addition, some viruses, such as *Herpes simplex,* also have a lipid envelope, derived from the membranes of **host** cells they infect.

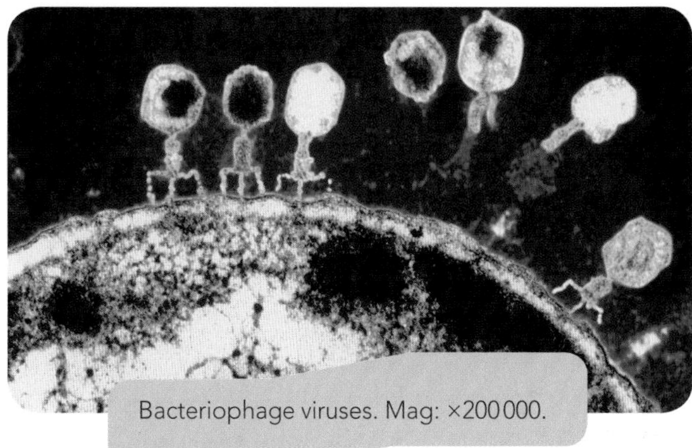

Bacteriophage viruses. Mag: ×200 000.

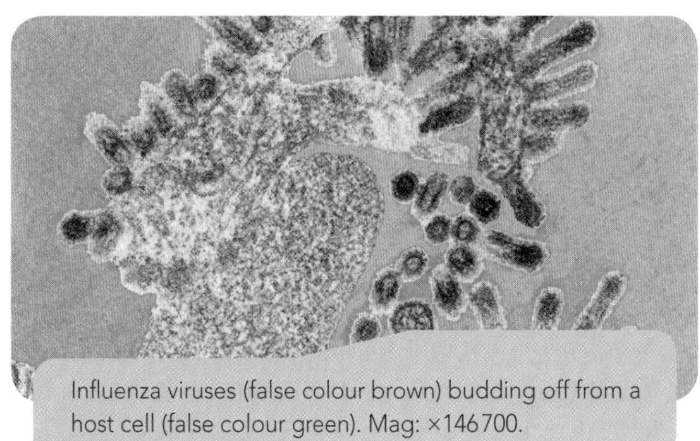

Influenza viruses (false colour brown) budding off from a host cell (false colour green). Mag: ×146 700.

Most viruses are between 10 and 300 **nanometres** in diameter, although some larger ones have been identified. There are different-shaped viruses.

- Some, such as tobacco mosaic virus, are helical – made of a single type of capsomere stacked around a central axis to form a coil or helix.
- Some, such as adenoviruses, are icosahedral, with 12 (or multiples of 12) capsomeres.
- Bacteriophages, viruses that infect bacteria such as T4 phage, have a complex structure with an icosahedral head, a helical tail and a hexagonal base plate with protruding tail fibres.

Some viruses infect Archaea (see page 284). These are quite different from other forms of viruses, with a range of shapes including spindle-shaped, hooked rods and teardrop-shaped.

Are viruses alive?

All viruses are obligate parasites as they can only be replicated when they are inside a host cell. They use the host cell machinery to make many copies of virus particles, which can then come out of the host cell and infect many other host cells. By the time the host has mounted an immune response and has recovered, virus particles will have been spread to other hosts, for example by droplets, touch or body fluids.

Viruses do not have any cell structure or metabolism and do not feed, respire, move or show sensitivity. They do not increase in size. However, they do contain nucleic acid, as do all living organisms on Earth. They also undergo mutation to their genetic material and this changes their antigens. This is how new strains of viruses, such as swine flu, arise.

Akaryotes are not classified along with the other five kingdoms of organisms and have been considered as not truly living. However, there are two groups of bacteria called rickettsias and chlamydias which can reproduce only when inside a host cell, and these are classified as living.

Where did viruses come from?

There are three main theories:

- Viruses may once have been small cells that lived as parasites inside larger cells. They may have lost the genes that controlled their metabolism and allowed them to lead independent lives.
- Viruses may have evolved from bits of DNA or RNA that 'escaped' from larger cells. They could have come from plasmids or 'jumping genes' (transposons) that are molecules of DNA that can replicate and move around within the DNA of a cell.
- Viruses may have evolved from complex protein and nucleic acid molecules at the same time as their host cells.

How do viruses replicate?

Phage viruses attach to bacterial cells and inject their nucleic acid into the bacterial cell, leaving their protein coat on the outside of the bacterium. The viral DNA is copied and read by bacterial ribosomes which make new virus coats. The bacterium bursts and many new virus particles are released to infect more cells.

DNA viruses that infect eukaryotic cells are taken into the cell by **endocytosis**. They pass into the nucleus and the coat is removed, leaving the DNA, which can be transcribed into mRNA (messenger RNA) and then translated into the proteins of the coat.

RNA viruses replicate in the host cell's cytoplasm. Retroviruses, such as HIV, contain RNA and have an enzyme called reverse transcriptase, which copies RNA into DNA. The DNA can be incorporated into the host genome and may stay dormant for quite a long time before making new copies of the virus.

Viruses and cancer

Some viruses cause cancer in their host. The virus in question may insert its genes into a host gene that regulates cell division. It alters the gene to make it become an oncogene, which fails to regulate cell division. Unregulated mitosis leads to a tumour.

- The human papilloma virus causes cancer of the cervix, penis and anus.
- Hepatitis B and C viruses can cause liver cancer.
- The Epstein–Barr virus causes Burkitt's lymphoma and Hodgkin's lymphoma – both cancers of the lymph system.

Why are viruses important?

Viruses can infect humans and livestock (e.g. foot and mouth disease) and crop plants. So they can make us ill and reduce our food production. Because viruses do not have any metabolism, antibiotics are ineffective against them. There are antiviral drugs that prevent them from replicating but one method to prevent viral diseases is to vaccinate the population. This stops the viruses being transmitted to new hosts and so the viruses cannot replicate.

Assessment activity 15.1

BTEC

You are a microbiologist preparing educational material to inform the general public about the differences between bacterial and other types of cells.

1 Examine some photomicrographs and electron micrographs of animal and plant cells. Identify the structures that you can see, such as cell wall, plasma membrane, ribosomes, endoplasmic reticulum, Golgi body, chloroplasts, mitochondria. Examine some prepared microscope slides of eukaryotic cells, such as from plants, protoctists, fungi and animals. Make a poster to show the features of eukaryotic cells. On the poster you should have a large labelled and annotated diagram to show the structures present in an animal cell; a plant cell; a protoctist cell; a fungal cell. P1, P2

2 Produce another poster that shows the features and functions of prokaryotic cells. Use electron micrographs of bacterial cells to help you. Label and annotate the diagrams of bacteria to show the general structures. P1

3 On both posters describe the functions of the components in the annotations. M1

4 Explain how the features/structure of each organelle in a eukaryotic cell enable(s) it to perform its function. For prokaryotic cells, explain how the structures enable metabolic functions to be carried out. D1

5 Under the supervision of your teacher make and examine slides of bacteria, using oil immersion microscopy. P1

6 Indicate the sizes of bacterial cells and the range of shapes. Indicate the usefulness of bacteria and the importance of understanding the structures and functions of pathogens in order to deal with them. P2

7 Use data from electron micrographs, biology textbooks and the Internet to add some information or make another poster about akaryotes. P2

Grading tips

For P1 and P2 make sure your diagrams are large and clear. Do not shade anything, use clear unbroken lines. 'Annotate' means label each structure and then add some notes about it, by the label.

Also for P1 carry out the practical activity. It will enable you to appreciate how small bacteria are and you will also see the different shapes of bacteria. On your poster you may have a large labelled diagram showing the internal structure of a prokaryote cell and its features. You could show two diagrams, one that shows features possessed by all bacteria and another that shows features possessed by only some bacteria.

Also for P2, around your central diagram of a prokaryotic cell you could show all the different shapes of bacteria, with some named examples of each type. Indicate their actual sizes or show how much bigger your drawings are (an idea of scale). Indicate which features pathogens have that can make them difficult for our immune system to deal with.

P2: When making a poster or adding information on akaryotes, make a large labelled and annotated diagram to show the structure and features of a virus. Indicate with satellite diagrams the range of shapes and sizes of different types of viruses. Include, somewhere, information about how they replicate, their possible origins, their importance and whether or not they can be considered to be living.

For M1 include annotations about the functions of each component.

For D1 explain how the structure of each component relates to the function it carries out.

PLTS

Self-manager, Independent enquirer and Effective participator

You will develop your self-management and independent enquiry skills to plan and carry out the investigations and your effective participator skills when presenting information to the rest of the class.

15.2 Culturing microorganisms

In this section:  

Key terms

cm³ – unit of volume equal to 1 ml. There are 1000 cm^3 or 1000 ml in 1 dm^3 (1 litre).

Biocontainment – protocol in microbiology laboratories to prevent contamination of workers and to prevent escape of microorganisms from the lab to the outside environment.

Nutrient broth – liquid medium with nutrients in which microorganisms can be grown.

Nutrient agar – solid medium, containing nutrients, for growing microorganisms.

Inoculation – adding microorganisms to a medium.

Turbidity – cloudiness of a liquid.

Colony – visible group of many bacteria on a solid medium. Each bacterium in one colony has arisen from a single bacterium.

Total count – measurement of number of microorganism cells in a volume of culture medium that includes both dead and living cells.

Bacteriophage – a virus that infects bacteria.

Microbiologists may have to grow, or culture, microorganisms for a variety of reasons:

- to develop a vaccine against them
- to find out how effective a particular chemical is at killing the microorganism or preventing it from multiplying
- to ascertain the optimum conditions for growing the microorganisms so they can then culture it on a large scale to use it in a biotechnological process
- to test if surfaces in food preparation areas or in hospital operating theatres are free of microorganisms
- to test products to make sure they are free from bacterial contamination.

Safety aspects

Biocontainment

This term refers to laboratory safety procedures in microbiology laboratories where pathogenic microorganisms, including viruses, are being cultured. The organisms have to be contained so the workers in the laboratories are not accidentally infected, and to prevent any organisms being released into the outside environment.

Negative containment is primarily the protection of personnel and the laboratory environment from exposure to the pathogens. Personnel have to follow a code of conduct and use aseptic technique (see page 289), and safety equipment such as laminar flow or clean air cabinets.

On a secondary level it is the prevention of escape of pathogens. Personnel may have to remove their normal clothing and put on special clothing before entering the lab. They will wear special footwear (rather like Wellington boots), cover their hair, remove jewellery, and not wear make-up or nail varnish. Before going home they will place their lab clothing in a bag to be sterilised and will shower with antiseptic soap before dressing. There will be a room between a microbiology lab and the outside corridor. It is not possible to have both doors to this room open at the same time. The air pressure outside the lab may also be higher than inside the lab, so air does not flow out of the lab.

For culturing some extremely dangerous (class IV) pathogens there are maximum containment facilities.

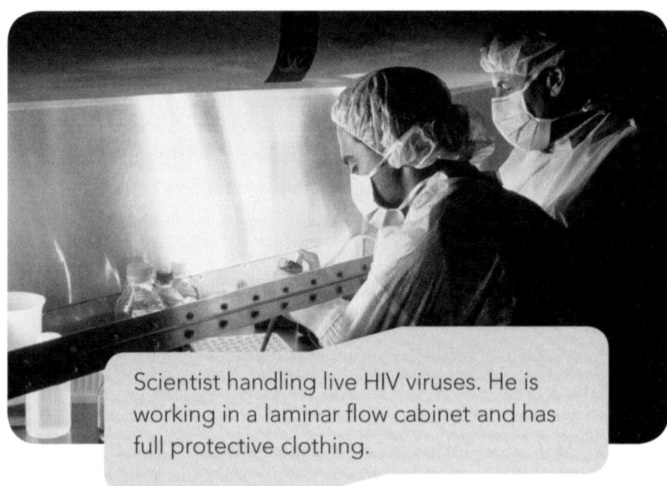

Scientist handling live HIV viruses. He is working in a laminar flow cabinet and has full protective clothing.

Operators manipulate cultures of these pathogens by putting their hands into special 'gloves' inside sealed cabinets.

Laminar flow cabinets are enclosed spaces made of stainless steel with no gaps or joints where microorganisms could collect. They are operated at a negative pressure so that a smooth (laminar) flow of air is constantly being drawn into the cabinet. The air from the cabinet is then filtered so no microorganisms can escape as they would be stopped by the filter.

They may also have an ultraviolet lamp so that, when not in use, this light will kill any microorganisms inside the area. This lamp has to be switched off when the cabinet is in use.

Activity 15.2A

Think of the reasons why the UV lamp should be switched off when the laminar flow cabinet is being used.

Positive containment refers to the use of good aseptic technique to prevent contamination of the microbiological sample.

Did you know?

Alexander Fleming discovered penicillin by accident, because his aseptic technique was not very good! He was about to throw away contaminated agar plates when he noticed that the bacteria he was trying to grow on the plates were not growing near the mould fungus that had contaminated them. He deduced that the mould must be making a chemical that stopped the bacteria growing. The mould was *Penicillium*.

Laboratory safety and aseptic technique

- When you carry out laboratory microbiology practicals you will use only harmless bacteria, fungi or phage viruses, provided by your teacher, but you still need to observe the same safety practices that you would use if they were potentially dangerous.

1 If you have a heavy cold or a stomach upset check with your tutor before you begin a microbiology practical activity.

2 Cover any cuts or grazes with a clean waterproof dressing.

3 Do not eat or drink anything in the laboratory. Also avoid any other hand to mouth operations and don't chew pen tops.

4 Always wear a clean, disposable plastic apron. This protects your clothes and reduces risk of contamination of the microorganisms you are growing. At the end of the session, place this apron in the bag provided. It will be properly disposed of by the technicians.

5 Close windows and doors to prevent airborne contamination.

6 Wash your hands with antiseptic soap before you start. Dry them on a paper towel and place it in the bag for later disposal.

7 Benches should have smooth surfaces so that bacteria cannot collect in crevices.

8 Spray your bench area with disinfectant, leave it to work for 10–15 minutes, wipe with a paper towel and place the towel in the disposal bag.

9 Work near a lit Bunsen burner as you will need to flame loops and also the updraft will prevent bacteria from the air falling onto the culture plates you are inoculating. Place the Bunsen flame on yellow when not in use as this is more visible.

10 Sterilise your instruments before use. Hold wire loops in the hottest part of a blue Bunsen flame until they are red hot. Allow them to cool in the air, dip them in 70% methanol and then flame the loops again to burn off the methanol. Your tutor will show you how to avoid 'spluttering' when heating wet loops.

11 Sterilise glass instruments, such as spreaders, by dipping them in 70% methanol and then flaming them lightly.

12 Keep the methanol in a small beaker and away from the Bunsen flame.

13 Never place bungs or caps from flasks of bacterial cultures on the bench; hold them.

14 Pass the necks of culture tubes through the Bunsen flame before and after using a loop or pipette to take a sample out.

15 Before you inoculate plates or tubes, label them with your name, date and the bacterial culture. Label Petri dishes on the bottom. Do not lick labels!

16 Always tape Petri dish lids on, using two pieces of Sellotape. Never seal the lids completely.

17 Incubate plates upside down so that any condensation runs onto the lid and not onto the bacteria.

18 Report any spillages immediately.

19 Incubate Petri dishes at 30 °C so that if there are any pathogenic contaminants they will be unlikely to grow.

20 When examining Petri dishes, do not remove the lids.

21 Place all used instruments in a pot of bleach or disinfectant after use.

22 Technicians will autoclave used Petri dishes, aprons and towels, before sealing them into a bag for disposal. An autoclave is like a pressure cooker – it will heat things to 121 °C for 20 minutes.

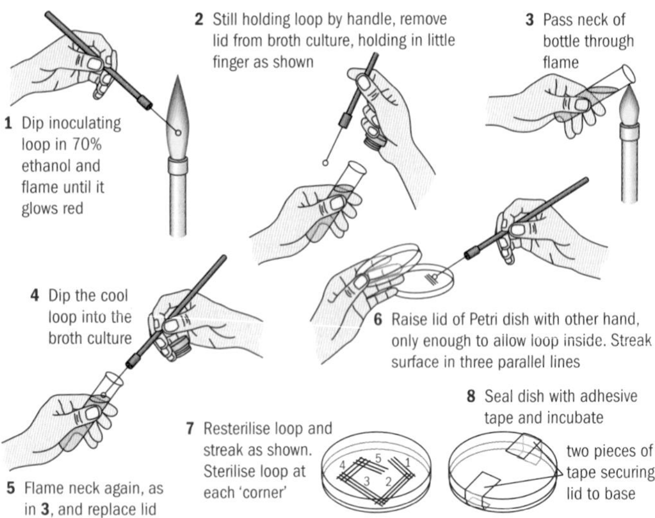

Using aseptic technique to prepare a streak plate.

Membrane filtration

Even though aseptic technique is used during commercial food preparation or for making medicines, the products need to be tested to make sure they are not contaminated by bacteria, mould fungi or yeasts. A sample from each batch of product is passed through a special micro-porous membrane, made from nitrocellulose, with very small pores of 0.45 μm. The residue left behind after filtration is washed to remove any chemicals that might inhibit growth of microorganisms. It is used to inoculate liquid nutrient broth or solid nutrient agar which are then incubated at a suitable temperature. If there is growth of fungi, yeasts or bacteria, the batch is contaminated and can be destroyed.

Drinking water and dialysis fluid can also be screened in this way.

Activity 15.2B

Explain why membrane filtration cannot be used to test if a batch of medicine, food or water is contaminated with viruses.

Preparing liquid and solid media

You can grow microorganisms in liquid media or on solid media. Most bacteria can be grown in labs as long as the media contain a source of the major nutrients: carbon, nitrogen, sulfur and phosphorus. They may also need to have other nutrients: potassium, magnesium, calcium and iron; plus some traces of manganese, cobalt, copper, zinc and molybdenum.

These nutrients are made into a broth and the pH and salinity (salt content) can be adjusted. The **nutrient broth** is placed in test tubes, which are plugged with cotton wool, capped with foil and then sterilised in an autoclave, at 121 °C for 20 minutes. These tubes are then cooled before they are inoculated.

Activity 15.2C

Why do you think the tubes have to be cooled before they are inoculated with microorganisms?

If agar powder is added to the broth, this can be used to make a solid medium. The **nutrient agar** solution is placed in a plugged and capped flask and sterilised in an autoclave at 121 °C for 20 minutes. It is allowed to cool to about 50 °C and may be kept in a water bath at this temperature before being poured into a Petri dish.

Using aseptic technique (see page 289), place the Petri dishes to be poured on a bench. Lift the lid of one dish, just enough to pour some liquid nutrient agar into the dish. Replace the lid and gently swirl the dish so the nutrient agar is evenly distributed. Pour the rest of the plates in the same way.

After a while, the nutrient agar jelly will set.

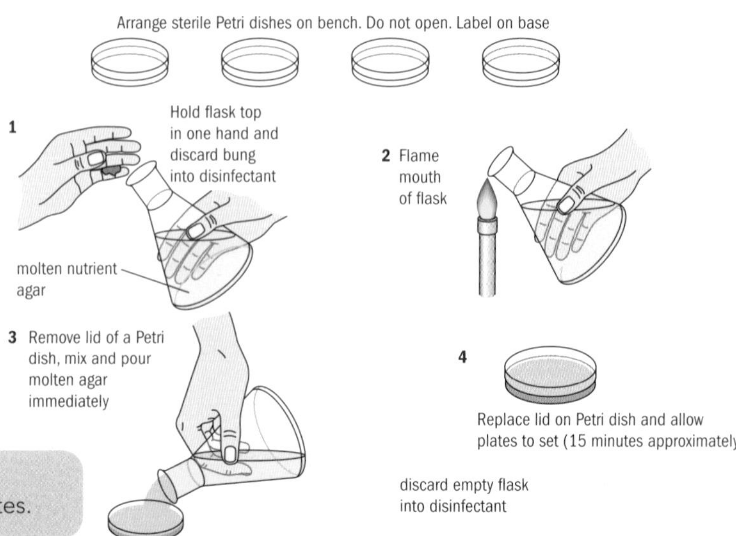

Pouring sterile agar plates.

Nutrient agar and nutrient broth are general purpose media that will allow many bacteria to grow.

Selective media

In some hospital microbiology labs special media are used to either suppress the growth of some types of bacteria or to promote growth of other types of bacteria. This can help to identify specific bacteria associated with a certain infection and make an accurate diagnosis.

- Mannitol salt agar has 7.5% sodium chloride and inhibits the growth of most bacteria except streptococci.
- Blood agar is also useful for promoting the growth of streptococci.
- MacConkey agar can be used to identify gut bacteria, which could indicate possible sewage contamination.

Inoculating liquid media

At all stages, aseptic technique should be followed.

1 Flame the loop and allow it to cool.

2 Take the culture tube and shake it to distribute bacteria. Loosen the cap and then remove the cap from the culture tube, but hold it by your little finger, whilst still holding the culture tube.

3 Flame the neck of the culture tube.

4 Insert the sterile loop into the culture tube and obtain a loopful of culture. Hold this loop whilst reflaming the neck of the tube and replacing the cap. Take care, the tube will be hot!

5 Pick up the tube of liquid to be **inoculated**. Remove the cap (hold onto it) and flame the neck of the tube.

6 Insert the loopful of bacterial culture into this tube and then withdraw it.

7 Flame the neck of the newly inoculated tube again and replace the cap. Place this tube in a rack.

8 Sterilise your loop by flaming it.

The inoculated tubes can be incubated at a suitable temperature and the amount of growth can be measured by colorimetry: the more growth, the more **turbid** (cloudy) the solution.

Inoculating solid media

When one bacterium lands on a suitable solid medium, at a suitable pH and temperature, it will divide by binary fission. Each new cell will grow and divide again, so that, after about 24–48 hours, where there was one bacterium there will be a visible **colony** of several thousand million bacteria. As always, aseptic technique should be used throughout. Follow the instructions above to pour a plate.

Streak plates

The diagram on page 290 shows how to prepare a streak plate. This method spreads out one loopful of bacterial culture so that the bacteria on the last streak are very spread out. The colonies they form are distinct and easy to see.

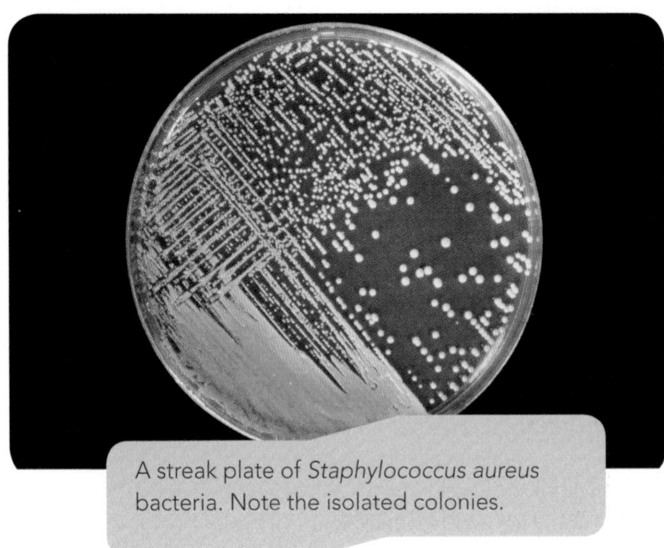

A streak plate of *Staphylococcus aureus* bacteria. Note the isolated colonies.

Lawn plates

Using aseptic technique throughout:

1 Label the base of a nutrient agar plate with your name, date and bacterium to be added.

2 Pipette 1 ml (1 **cm³**) of bacterial culture onto the middle of a solid nutrient agar plate. Remember, you must not use a mouth pipette when doing microbiology work.

3 Use a sterilised bent glass rod to spread this evenly over the agar jelly.

When incubated, this type of inoculation produces a lawn of bacteria rather than clear individual colonies. It is particularly useful if you want to test how effective antibiotics or other antibacterial substances are.

4 You can now use sterilised forceps to place an antibiotic multidisc onto the inoculated agar plate.

5 Tape the lid and incubate the plate at 30 °C for 24–48 hours.

6 Where the antibiotics have diffused into the nutrient agar jelly and prevented the growth of bacteria, there will be clear zones of inhibition around the antibiotic discs.

7 You can examine your plates, without removing the lids, and measure the diameters of the zones of inhibition.

8 The antibiotic that gives the largest zone of inhibition is the most effective against that bacterium.

- Another way is to make a lawn plate, then use a sterilised cork borer to make a well in the centre of the agar jelly.
- Some garlic extract or lemon juice can be placed in the well and the plate incubated, the right way up.
- After 24–48 hours, a clear zone of inhibition around the well shows that the substance prevents the growth of bacteria.

Did you know?

Agar jelly has a peculiar property. Its melting point and setting point are different. It sets at 32–40 °C but will not melt until it reaches 85 °C. This property of having a different melting and setting point is called hysteresis. You need to leave it in a water bath at 50 °C until you are ready to pour it, otherwise it will set in the tube and you will have to heat it up to 85 °C again. Agar is obtained from seaweed. It is also used to thicken ice cream, soups, custard and emulsion paint. It can be used in breweries to clarify beers. This is more acceptable to vegetarians than using isinglass from fish swim bladders, which used to be used. It is also used to grow small plants by tissue culture. It is a polymer of the sugar galactose. We cannot digest it and it swells up in the stomach, making people feel full, so it is also being used to help obese people lose weight.

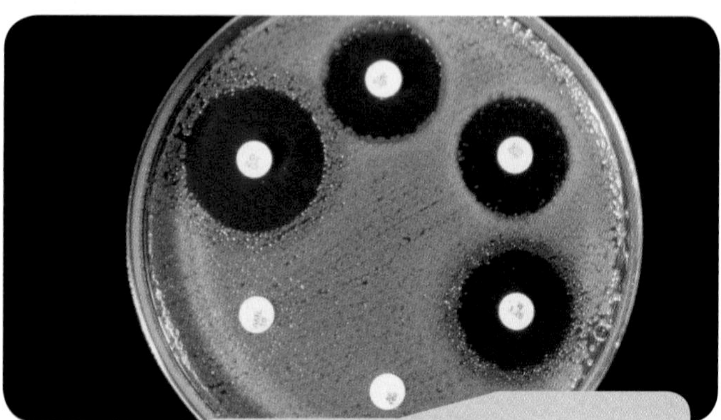

E. coli bacteria are sensitive to four antibiotics here and resistant to two.

What else do microorganisms need to grow?

The ones you will grow also need oxygen, as well as nutrients, a suitable temperature, suitable pH and suitable salinity (saltiness). Some microorganisms are obligate anaerobes – they do not grow if oxygen is present. In medical labs there are special procedures for growing such bacteria.

Growing fungi

Single-celled fungi, such as yeast, grow in a similar way to bacteria. Yeast can be cultured in a liquid medium which is inoculated with a loopful of yeast culture, using aseptic technique. The tube can be incubated for 24–48 hours. Before it is sampled you need to shake the tube as yeast cells may settle on the bottom.

Did you know?

The bloom you see on the surface of grapes is yeast that grows on them. When people first made wine they did not know about the yeast, so wine making was an accidental discovery. Such useful accidental discoveries (like Fleming and penicillin) are called serendipity.

Using colorimetry to measure growth

The amount of growth can be measured using colorimetry to measure turbidity. A colorimeter shines a beam of light through a sample which is placed in a special plastic container called a cuvette. A photoelectric cell picks up the light that has passed through the sample and tells you how much has been absorbed.

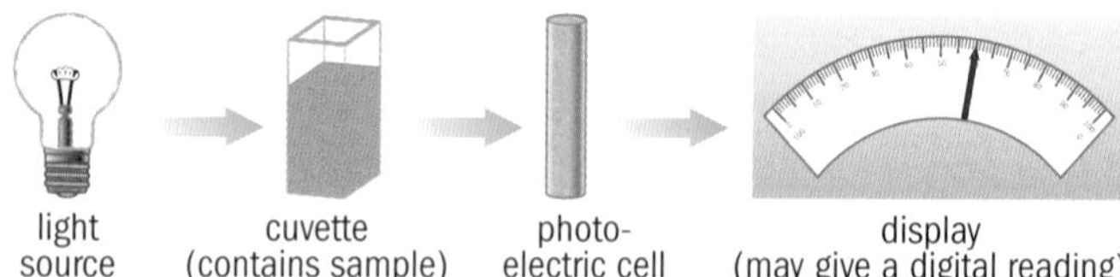

- When using a colorimeter, the device is usually zeroed between each reading by placing an appropriate 'blank' sample to reset the 100% transmission/0% absorption. In this case, the blank used would be uninoculated liquid medium.
- Colour filters are often used for greater accuracy. In this case, a green filter would be used.

Using a colorimeter.

1 Fill a cuvette with uninoculated liquid medium and place it in the special chamber. Use a blue or green filter. As the light shines through the sample set the reading to zero, as this sample is clear so there is no absorption.

2 Now fill a cuvette with some cloudy medium that has yeast cells growing in it.

3 Place it in the colorimeter, use the same filter as before and measure the absorption. The greater the absorption, the greater the growth of yeast.

The colorimeter cannot distinguish between dead and living cells and it cannot distinguish between particles and cells in the culture medium. However, it is quick and easy. This method can also be used to measure the growth of bacteria in a liquid culture.

Using a haemocytometer

Yeast cells are large enough to be seen with a light microscope so they can be counted using a haemocytometer. This is a special slide that has a grid etched onto a middle section that is slightly lower than the rest of the slide.

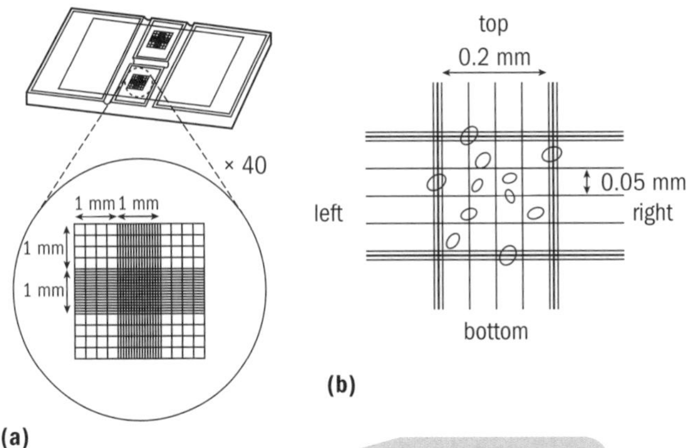

(a) A haemocytometer slide. **(b)** Using a haemocytometer to make a total cell count.

When a special coverslip is placed firmly on the slide it forms a chamber of known depth, 0.1 mm, so we can calculate the volume of liquid over each etched square.

1 Shake the tube of liquid medium in which yeast cells have been growing.

2 Take 1 cm^3 of this liquid and add it to 9 cm^3 of sterile water in another test tube.

3 This dilutes the yeast medium by ×10.

4 Mix the diluted yeast solution well and, using a pipette, allow some of it to trickle into the grooves under the haemocytometer coverslip.

5 Observe the haemocytometer grid under the microscope.

6 Focus on the central grid area, where there are 25 squares each divided into 16 smaller squares.

7 Count the yeast cells in five of the 25 squares (80 small squares in total). Count the central square and the four corner squares. The volume of liquid over 80 small squares is 0.02 mm^3.

8 Now you know how many yeast cells (*n*) are in 0.02 mm^2 you can calculate how many were in 1 cm^3 of the undiluted liquid medium.

9 The number is *n*/0.02 in 1 mm^3 (μl), which is *n*/0.02 × 1000 in 1 cm^3 diluted solution and *n*/0.02 × 1000 × 10 in 1 cm^3 of undiluted solution.

This method is a **total count** as it counts all yeast cells. However, some of them may be dead. It is also very time consuming.

Activity 15.2D

In an undiluted sample of yeast, a total of 120 cells was counted in 80 small squares (0.02 mm^2).

Calculate how many yeast cells were in each cm^3 of this culture.

Mycelial discs

Many fungi consist of thread-like hyphae. As they grow the hyphae produce a sort of mat, called a mycelium.

1 Using aseptic technique you can cut a disc from a fungal mycelium and add it to some sterile liquid medium.

2 You can then incubate this for a week at 25 °C and at the end of the time either measure the increase in diameter, or dry it and weigh it to measure the dry mass.

3 You can compare this with the dry mass of a mycelial disc the same size as the one you placed in the nutrient medium, and see how much growth has occurred.

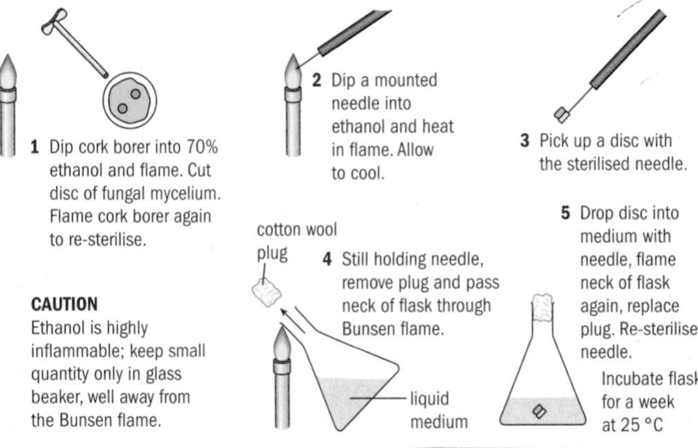

Inoculating a flask with a mycelial disc.

Activity 15.2E

Over the course of a week, a mycelial disc increased in diameter from 5 mm to 15 mm.

Calculate the percentage increase.

Culturing viruses

As viruses can replicate only when inside a host cell many are grown in laboratories in cell cultures. The inside of glassware is coated with a layer of cells. There are certain cell lines, such as human T-cell lines, that are used. Smallpox viruses were grown, to make vaccine, inside the tissues of chicks developing in hens' eggs.

Did you know?

The most well-known cell line was developed in the 1950s when a doctor, in the USA, took some cancer cells from a deceased patient, Henrietta Lacks. Because they were cancer cells, they can keep on dividing and are 'immortal'. Her cell line, HeLa, has been used in labs all over the world and there are now more of her cells than there were in her body. Jonas Salk used them to develop a vaccine against polio. Henrietta Lacks has been recognised for her contribution to science as her cells have been used for research into cancer, AIDS, gene mapping and the effects of radiation and other toxins on cells. However, her cells have also caused some problems. They are so prolific that they have contaminated other cell cultures in some labs.

One biologist, Leigh Van Valen, says that, as Henrietta Lacks's cells have evolved, they should be considered as a separate species of a single-celled life form and could be classified as *Helacyta gartleri*.

Bacteriophage viruses infect bacteria, so you can grow bacteria either as a lawn on an agar plate or in a liquid medium, and then infect the bacteria with bacteriophage viruses. As the viruses attach to bacterial cells and inject their DNA into them, the bacterial cells make many copies of the virus. The bacterial cells then lyse (split open) and release the new viruses to infect many more bacterial cells.

Viral plaque counts on solid media

If this occurs on a bacterial lawn, circular clear areas, called viral plaques, can be seen in the bacterial lawn. Counting the plaques gives an indication of the number of virus particles that were in the original stock bacteriophage culture solution.

The phage virus culture solution has to first be diluted, because if too many plaques are produced on the plate, they cannot be counted accurately. Too few viral plaques are considered unreliable, so we need to have plates with between 30 and 300 viral plaques.

The bacteriophage virus T_2 coliphage infects the bacterium *E. coli*.

The plates used have a layer of hard agar as a base and a mixture of soft agar, bacterial cells and phage virus particles as a top layer.

Worked example

200 viral plaques are counted on an agar bacterial lawn.

The viral stock solution was diluted by a factor of 1 million (10^6); this can also be described as a 1 in 10^6 dilution or a 10^{-6} dilution.

1 cm^3 of the diluted viral solution was added to the plate.

The number of virus particles in 1 cm^3 of diluted viral solution was 200.

So the number of virus particles in 1 cm^3 of undiluted stock viral solution was 200×10^6

$= 2 \times 10^8$

Activity 15.2F

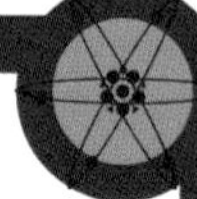

A viral stock solution was diluted by a factor of 10^4.

1 cm^3 of this diluted viral solution was added to several agar bacterial lawn plates. The plates were then incubated for a short time, all at the same temperature.

The plaque counts were as follows:

Plate:	1	2	3	4	5	6
Viral plaques:	250	200	220	270	230	210

Calculate the mean viral plaque count from the above data.

Calculate the number of viral particles in 1 cm^3 of the original undiluted stock.

Explain why setting up six plates can check the reliability of the data.

A suitable method is described here. Of course, aseptic technique is used throughout the preparation of materials and during the practical. Tryptone agar and tryptone broth are selective media as they contain the amino acid tryptone to promote the growth of *E. coli* bacteria.

Each group needs:

- a culture of *E. coli* and a culture of T_2 coliphage virus
- five Petri dishes containing hard tryptone agar
- five test tubes, each containing 2 cm³ soft tryptone agar, maintained in a water bath at 45 °C
- 10 test tubes each containing 9 cm³ tryptone broth
- Bunsen burner, water bath, 1 cm³ sterile pipettes, OHP pen for labelling.

Method

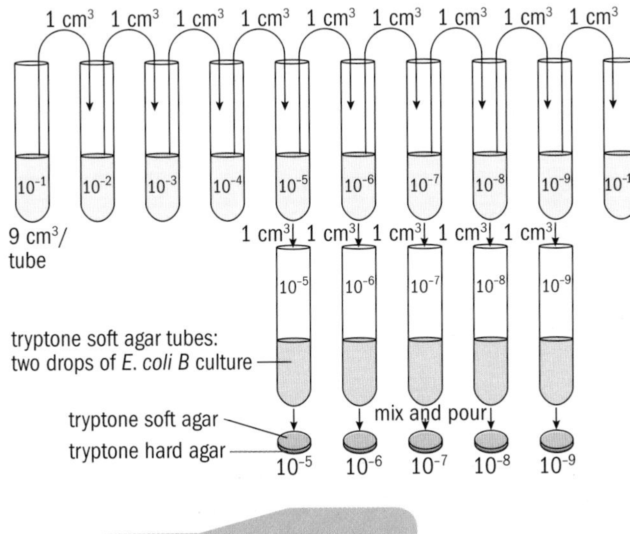

Method for carrying out viral plaque counts.

1 Label the five agar plates: 10^{-5}; 10^{-6}; 10^{-7}; 10^{-8}; 10^{-9}.

2 Label the 10 tryptone broth tubes: 10^{-1} through to 10^{-10}.

3 Transfer 1 cm³ of the phage virus stock solution to the first tryptone broth tube, labelled 10^{-1}. Keeping the tube vertical, roll it between the palms of your hands to mix the contents. This has made a 1 in 10 (×10) dilution.

4 Transfer 1 cm³ from this tube to the 10^{-2} tryptone broth tube. Mix the contents. This makes a 1 in 100 (×100 or 10^{-2}) dilution.

5 Transfer 1 cm³ of this to the next tube and so on. You will now have a **serial dilution** of the phage virus stock solution, from ×10 to ×10^{10}.

6 Add two drops of *E. coli* culture solution to the tube containing soft tryptone agar. Also add 1 cm³ of the 10^{-5} dilution phage virus solution. Roll the tube between your palms and pour its contents onto the hard agar plate labelled 10^{-5}. Replace the lid, swirl the plate and allow it to harden.

7 Use separate pipettes and repeat step 6 for the tryptone broth phage dilution tubes 10^{-6}, 10^{-7}, 10^{-8} and 10^{-9}.

8 Tape the lids and incubate all plate cultures, upside down, for 48 hours at 25 °C.

9 Examine all plates and count the viral plaques, where there are between 30 and 300 plaques. You can divide the plate into four (or more) sections and use a colony counter or a hand tally counter to help you.

10 Calculate the number of viral particles in 1 cm³ undiluted solution.

11 Tabulate your data and share it with the rest of the class. Are the results fairly similar? Are they reliable?

Observing viral activity in liquid media

If you inoculate tryptone broth tubes with 1 cm³ *E. coli* solution and let them incubate at 25 °C for 48 hours, they will be cloudy due to growth of the bacteria.

Label five such tubes 10^{-5}, 10^{-6}, 10^{-7}, 10^{-8}, 10^{-9}.

Add 1 cm³ of the appropriate diluted phage virus stock solution, as prepared in the method above, to five of the tubes containing cloudy solution.

Incubate at 25 °C for another 24–48 hours. Observe and measure transmission in a colorimeter. Use the cloudy solution from tube 6 as **zero transmission**.

Use the results from both methods to calibrate the colorimeter readings you get.

Assessment activity 15.2

1 You are a microbiologist training laboratory technicians. Prepare a short presentation to explain to them the principles underlying aseptic technique – which means explain what aseptic technique is and why each stage has to be carried out. M2

2 You are a trainee technician in a medical microbiology lab. You need to demonstrate that you can carry out some microbiological lab techniques.

   a Under supervision, cultivate microorganisms using aseptic techniques.

   b Inoculate a tube of liquid medium. Incubate it and use colorimetry to measure the turbidity.

   c Pour some agar plates.

   d Inoculate one solid medium plate using the streak plate technique.

   e Inoculate another solid medium plate using the lawn technique.

   f Incubate the plates and observe the results. P3

   You also need to demonstrate competence in measuring growth of fungi.

3 Using aseptic technique throughout and under the supervision of your teacher:

   Inoculate a tube of liquid medium with yeast. Incubate at 25 °C for 24–48 hours and:

   a use colorimetry to measure how much growth has occurred.

   b use a haemocytometer to calculate the number of yeast cells (total count) in 1 cm^3 of the undiluted inoculated and incubated medium.

   Inoculate a flask of liquid medium with a mycelial disc. Incubate the flask for a week and then measure the increase in diameter and the increase in dry mass to measure the growth of the fungus.

   Record all your results and share them with the rest of your class. Do you all get similar results?

   In addition to the skills demonstrated above, you also need to demonstrate competence at doing viral plaque counts in solid and liquid media. P3

4 Under supervision and using aseptic technique throughout, carry out the practicals described on page 295. Record your data and compare them with others in your class to check reliability. P3

5 Write a short report on this practical and explain the principles underlying the techniques involved: aseptic technique; serial dilution; viral plaque counts. M2

Grading tips

For M2 keep your presentation short and to the point. Do not put too much information on each slide. Have the main points there and then explain them to your audience. Remember to explain *why* each step is carried out. Practice so you are fluent. Make sure you will be able to answer any questions. You may also like to prepare a handout to give to your audience.

For P3 follow all instructions carefully. Observe aseptic technique and work safely and competently.

PLTS

Independent enquirer, Effective participator and Self-manager

You will develop your skills as an independent enquirer when researching information; you will also develop skills as an effective participator when giving the presentation and when sharing your results. You will develop your self-management skills as you research information and plan and carry out practical activities.

15.3 Factors that influence the growth of microorganisms

In this section: 

Key terms

Buffer – chemical that resists a change in pH, by accepting or donating protons.

Proton – hydrogen ion.

Viable – living

Sterilisation – killing all the microorganisms in a particular place.

Generation time – time taken for the population to double.

Metabolites – chemicals produced by an organism's metabolic process.

Fermentation technology – large-scale cultivation of microorganisms to produce a commercially useful substance.

Nutrients

All microorganisms need:

- a source of carbon, hydrogen and oxygen to make lipids, carbohydrates and amino acids
- nitrogen to make amino acids (and hence proteins), nucleic acids and ATP
- sulfur to make amino acids, and hence proteins
- phosphorus to make nucleic acids and ATP
- water, as their cytoplasm is about 70% water
- minerals such as calcium, potassium, magnesium and iron, as these help some enzymes to work
- traces of copper, cobalt, manganese, zinc and molybdenum.

Some bacteria can photosynthesise and make their large organic molecules from inorganic ones such as carbon dioxide and water or hydrogen sulfide. They are **photoautotrophs**.

Many bacteria are **heterotrophs** and need organic carbon. They secrete digestive enzymes, break down the large molecules and absorb the products of digestion. The growth medium for these bacteria needs to have a source of organic carbon, such as glucose.

Nutrient broth or agar also contains proteins. Some special growth media will also have certain vitamins.

Temperature

All living organisms have an optimum temperature for their growth. If the temperature is too low, enzyme-catalysed reactions may proceed too slowly to sustain life. If the temperature is too high then proteins may denature and lipids, which make up the cell membranes), will melt.

Most eukaryotic organisms will not be able to exist at very cold or extremely hot temperatures. However, some bacteria can live at the extremes.

Thermophiles

Some bacteria live in hot springs (where the temperature is around 100 °C), compost heaps, peat bogs, hot tubs and some (hyperthermophiles) inhabit thermal oceanic vents, where the water may be above 250 °C. These bacteria have very stable proteins with many disulfide bridges to maintain their 3D structure. Many thermophiles are Archaea (see page 284). Enzymes from thermophiles are used in biotechnology, in washing powders and for the PCR (see page 355).

Most thermophiles thrive at temperatures between 45 and 80 °C. Some can live at lower temperatures.

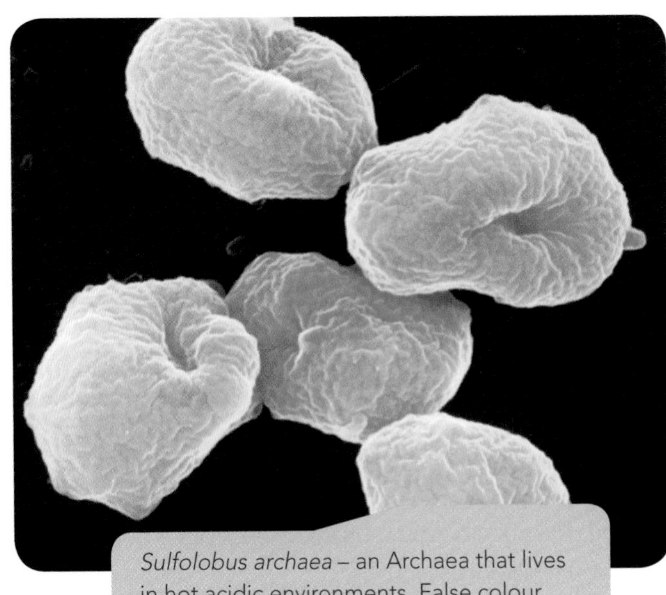

Sulfolobus archaea – an Archaea that lives in hot acidic environments. False colour scanning electron micrograph. Mag: × 29 880.

Mesophiles

These bacteria grow best at moderate temperatures, between 20 and 40 °C. Pathogenic bacteria are mesophiles. Some mesophiles also live in soil. The bacteria used in cheese and yoghurt making are mesophiles.

Psychrophiles

These bacteria, also described as cryophiles, thrive at low temperatures. Their upper limit is 20 °C and some live in fridges, freezers, polar ice caps and deep sea waters. Their lipid cell membranes are resistant to the stiffening usually caused by low temperatures, and they may have special proteins that act as 'antifreeze' to keep their cytoplasm fluid and protect their DNA.

Did you know?

There are some psychrophilic eukaryotes, for example fungi, that live under the alpine snowfields.

Activity 15.3A

1 Why do you think enzymes for washing powder are obtained from thermophilic bacteria?

2 Suggest why pathogenic bacteria are mesophiles.

Oxygen

When life first evolved on Earth there was no free oxygen in the atmosphere. Some Archaea developed photosynthesis. Many used hydrogen sulfide as the source of hydrogen and electrons for this process, but later some used water. This released free oxygen into the atmosphere and killed many species.

There are still some bacteria that are **obligate anaerobes**. These are killed by oxygen. They will use other chemicals, such as sulfur, during their energy metabolism.

Some bacteria are **aerotolerant** – they cannot use oxygen but can live in its presence. Others are **facultative anaerobes** – they can grow without oxygen but use it if it is present. Yeast is a facultative anaerobe.

Some bacteria are obligate aerobes and depend on oxygen for their respiration.

Anaerobic microorganisms can obtain energy from a variety of anaerobic metabolic pathways and we make use of these organisms in biotechnology. Yeast and some bacteria can respire anaerobically and produce ethanol.

Activity 15.3B

When free oxygen was first released into the Earth's atmosphere it killed many of the bacterial species present at the time. Were these bacteria obligate anaerobes, facultative aerobes, obligate aerobes or aerotolerant?

Carbon dioxide

Some bacteria, including many human pathogens, grow well when carbon dioxide levels are raised.

pH

Most bacteria grow best within a range of pH 6–8. Our stomachs produce hydrochloric acid which gives a low pH of about 2. This kills many of the bacteria that enter with our food. We also produce secretions that keep our skin acidic. The vagina is also acidic, which reduces the chance of infection.

Acidophilic bacteria can withstand low pH. *Helicobacter pylori* bacteria can live in the stomach and cause ulcers. Some such bacteria can live at very low pH values of between 0 and 2. *Acetobater aceti* produces acetic acid (vinegar) by oxidising ethanol. Other acidophiles live in deep-sea hydrothermal vents and some live in the volcanic, sulfur-rich hot springs of Yellowstone National Park. They are hyperthermophiles as well as acidophiles. They oxidise sulfur to sulfuric acid. Acidophiles have lipid plasma membranes that are totally impermeable to hydrogen ions or have mechanisms to pump the hydrogen ions (**protons**) out of their cells.

All bacteria will produce waste products that may alter the pH of the medium they are growing in. This is why, when you grow bacteria in the lab, you need to add a **buffer** solution to the medium.

Some fungi can tolerate low pH, but most grow best at pH 5–6. These are important in decomposing organic matter in acidic peat bogs.

Hot spring in Yellowstone National Park, Wyoming, USA. The pool is warmed by volcanic activity and the bright colours come from mineral deposits and thermophilic bacteria.

Water potential

Microorganisms need water, as metabolic reactions in their cells take place in solution. If the water potential of their surroundings is greater than inside their cells, water enters the cells by osmosis. Bacteria have cell walls so they may swell but not burst. However, in microorganisms with fragile cell walls they will undergo lysis.

If the water potential of their surroundings is lower than inside their cells, microorganisms will lose water by osmosis. If the cytoplasm dehydrates, metabolic reactions cannot take place and the organism will die.

Some microorganisms can survive in very salty solutions, such as the Dead Sea. These bacteria are called halophiles (salt-loving). However, most cannot tolerate being in a solution of low water potential, which is why jam-making and salting are traditional methods of food preservation.

Activity 15.3C

1 Explain to a partner how making jam preserves fruit.

2 If antibiotics or antiseptics are not available, wounds can be packed with sugar or honey to prevent infection. Explain why this prevents infection.

Viable counts

You have seen how the haemocytometer can be used to make a total count of microorganisms in a small volume of culture. The problem with that method is that all cells are counted and some of them may be dead. **Viable** counts allow us to find out how many living cells are in a small volume of culture medium. If these are done at intervals, then the rate of growth of a particular bacterium can be measured.

This method involves transferring small amounts of a bacterial culture and spreading it onto nutrient agar plates. Where each bacterium lands on the nutrient medium, it will divide by binary fission to give 2, 4, 8, 16 and so on, until it forms a visible colony of bacteria. These colonies can be counted and each one represents a single bacterium.

The number of colonies indicates the number of bacteria in 1 cm^3 of the nutrient broth. Because the number may be high a serial dilution is carried out first, to give smaller numbers of colonies that can be counted accurately.

Did you know?

Many bacteria can divide by binary fission every half an hour. This means that after a day (24 hours), provided there is enough food, one bacterium can become 130 thousand billion!

Method

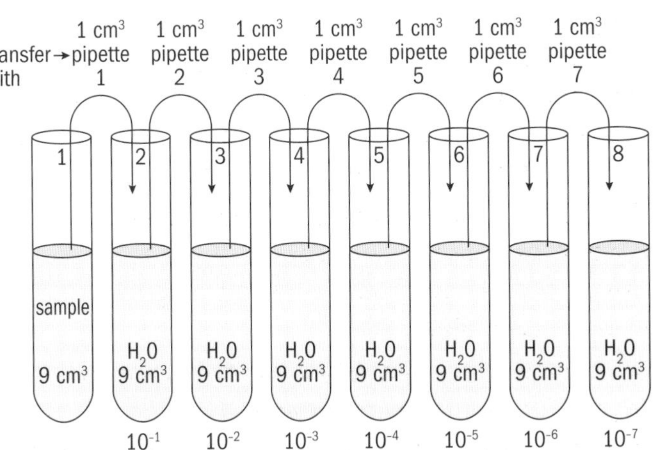

Method used when carrying out a serial dilution.

1 Use aseptic technique throughout.
2 Inoculate a flask of buffered nutrient broth, pH 7, with a loopful of bacteria, for example *E. coli.*
3 Mix the contents and immediately withdraw 1 cm^3 of the solution. Add this 1 cm^3 to a test tube containing 9 cm^3 sterile water. This gives a ×10 dilution.
4 Roll the tube between your palms to mix the contents. Transfer 1 cm^3 from this tube to another tube containing 9 cm^3 sterile water. This gives a ×100 dilution. Roll the tube to mix the contents.
5 Transfer 1 cm^3 of the ×10 dilution to a Petri dish containing nutrient agar (a nutrient agar plate) and spread the solution evenly over the agar.
6 Transfer 1 cm^3 of the ×100 dilution onto another nutrient agar plate and spread it evenly over the agar.
7 Tape the lids onto these plates, label the plates and incubate them upside down, at 30 °C for 24 hours.
8 After 30 minutes, shake the contents of the inoculated flask and remove 1 cm^3. Add this 1 cm^3 to a 9 cm^3 tube of sterile water and mix the contents (10^{-1}). Take 1 cm^3 from this tube and add to another 9 cm^3 sterile water tube, and mix the contents (10^{-2}). Transfer 1 cm^3 from this tube to another 9 cm^3 sterile water tube (×10^{-3}) and mix. Keep repeating this process to make a series of dilutions from ×10^{-3} to ×10^{-7}.
9 Transfer 1 cm^3 from the ×10^{-3}, ×10^{-4}, ×10^{-5}, ×10^{-6} and 10^{-7} dilutions onto five separate nutrient agar plates. Tape, label and incubate them as described for step 7.
10 Repeat steps 7 and 8 at 30 minute intervals, for 4–5 hours.
11 After 24 hours examine the nutrient agar plates and choose the plates that have between 20 and 200 visible colonies.
12 Count the colonies and calculate how many bacteria were present in 1 cm^3 of undiluted nutrient broth.
13 Put your data in a table and graph the data, using logarithmic graph paper.
14 Calculate the generation time for *E. coli* at 30 °C.

Methods for controlling the growth of microorganisms

Temperature

Low temperatures reduce enzyme activity and slow the growth of microorganisms. This is why keeping food in fridges and freezers keeps it fresh for longer. However, some bacteria and fungi still grow at fridge temperatures and some grow very slowly in freezers, so food cannot be kept indefinitely.

You have seen previously that most microorganisms cannot exist at high temperatures. For this reason, heat can be used to **sterilise** surgical instruments or equipment used in food manufacture and storage. Test tubes and Petri dishes that have been used to culture bacteria are also heated under pressure at 121 °C for 20 minutes, to kill bacteria so they can be disposed of safely. A similar procedure, using moist heat, is used for sterilising surgical instruments and equipment where food is processed or equipment used to grow microorganisms for biotechnological processes. Dry heat can also be used but bacteria need to be exposed to higher temperatures for a longer time.

Most bacteria and fungi are killed by temperatures of 60–80 °C but some bacteria form spores that are heat resistant, so the higher temperature is necessary to make sure all bacteria are killed.

Did you know?

The bacterium *Listeria monocytogenes* is found in ice cream, raw smoked fish and soft cheeses. It can grow and multiply at 0 °C so will do so in a fridge. Pregnant women are advised not to eat these foods as *Listeria* bacteria can infect their babies when they are being born and may cause the babies to die from a type of meningitis.

Pasteurisation

This method is used for milk and beer. A temperature of 72 °C is applied for 15–20 seconds. This reduces the numbers of any pathogenic organisms so the drinks will not cause diseases.

Canning

If food is heated to kill bacteria and then placed in a sealed tin to prevent entry of microorganisms, it will not spoil. Tins of bully beef from the First World War were found to be preserved after 80 years. However, if the tins are damaged, then microorganisms can enter and contaminate the food. There have been cases of severe food poisoning, botulism, caused by anaerobic bacteria that were present in tinned food.

pH

Pickling has long been used to preserve food. The low pH of the vinegar prevents the growth of most microorganisms.

Yoghurt is made using certain species of bacteria that are harmless and reduce the pH of the milk. This coagulates milk proteins and renders the resulting yoghurt uninhabitable by many other bacteria or fungi.

Drying

Microorganisms need water to live and reproduce. Drying food makes it impossible for microorganisms to grow in it and spoil it. The dried food must be kept dry. Food can be dried in the sun or in ovens.

Irradiation

The plastic Petri dishes you use to grow bacteria will have been irradiated with gamma radiation, so that they are sterile and there are no bacteria on them to contaminate your cultures. Small doses of ionising radiation, such as X-rays or gamma rays or ultraviolet light, will damage the DNA of bacteria and prevent the organisms from making proteins or dividing. Some food is treated with ionising radiation to kill bacteria on it.

Water potential

Placing food in strong salt or sugar solutions preserves it because water leaves bacterial cells by osmosis and the cells cannot carry out their metabolic reactions.

Activity 15.3D

You are a microbiologist and have to explain to people who work in catering how methods of food preservation work.

Prepare a short PowerPoint presentation to explain *how* various food preservation methods work by killing or preventing the growth of bacteria.

Antiseptics

These are chemical substances that can be used to kill microorganisms on living tissue, such as skin. Before surgery skin should be treated with antiseptics, such as iodine solution, alcohol or hydrogen peroxide. Hot salt water is good as an antiseptic mouthwash.

Antiseptic hand gel.

Case study: Microbiologist in a dairy

I have a microbiology degree and work in a small dairy that makes milk products such as yoghurts. I am responsible for making sure there is a code of conduct so that all equipment and preparation areas are sterilised. I make sure all staff know that they must change into their work clothes before entering the work area and wear special boots. They have to walk through a trough of disinfectant as they enter the workplace. Hair should be covered and no jewellery or nail varnish is allowed. The rules are strict and any staff that do not observe them may get a severe warning. Some activities, like eating in a food preparation area, lead to instant dismissal. Staff with colds or stomach upsets are not allowed to come into work and any cuts or grazes must be covered with special blue plaster. I also run the laboratory where we test the products to make sure there are no microorganisms other than the types of bacteria used to make the yoghurt. If we notice any yoghurt cartons that are domed (the lid is raised) we immediately know that there are unwanted bacteria in those pots (they have produced gas which makes the lids rise) so we have to scrap the whole batch. I enjoy the work as I have a lot of responsibility.

1. **Why are jewellery and nail varnish not allowed in the dairy?**
2. **What is the gas likely to be in yoghurt cartons with raised lids?**

Disinfectants

These are chemicals that kill or inhibit the growth of microorganisms on non-living material, such as surfaces in kitchens, restaurants and hospitals. Bleach, alcohol and potassium permanganate are all disinfectants.

Antimicrobials

These are chemicals that kill or inhibit the growth of microorganisms. They include:

- **antivirals** – chemicals that inhibit the replication of viruses. Examples include aciclovir and zidovudine
- **antibiotics** – chemicals that kill or inhibit the growth of bacteria. Some of them are derived from living organisms such as fungi or bacteria. The chemical sulfonamides are also antibiotics. These chemicals all interfere with some aspect of the bacterial metabolism.

Antibiotic	Mode of action
penicillin cephalosporin vancomycin teicoplanin	prevents the synthesis of cell walls of some types of *growing* bacteria; if the cell walls are not made properly, water can enter the bacterial cells and cause lysis
polymixin	inhibits the function of bacterial plasma membranes
rifampicin	inhibits the transcription stage of protein synthesis
erythromycin chloramphenicol tetracycline streptomycin	inhibits protein synthesis in bacteria
sulfonamides	interfere with metabolic reactions

Bacteria and fungi probably produce antibiotics to reduce competition from other microorganisms. Scientists have developed their use to treat infections caused by bacteria. However, some populations of bacteria have evolved resistance to certain antibiotics and may cause infections that are difficult to deal with. MRSA is a hospital-acquired infection caused by a bacterium, *Staphylococcus aureus*, that is resistant to many antibiotics including a powerful one often kept to use as a last resort, called meticillin. This bacterium causes more deaths per year in the UK than are caused by road accidents.

Some antibiotics are ineffective against certain bacteria so medical microbiology labs may screen bacteria to find which antibiotic is most effective against them.

Antifungals are chemicals that treat fungal infections. Some are described below.

Antifungal	Use
amphotericin	aspergillosis (farmer's lung), candidiasis (thrush), fungal meningitis
clotrimazole	skin and nail infections such as athlete's foot
griseofulvin	nail infections
terbinafine	ringworm (a fungal skin infection)

Activity 15.3E

You are a hospital microbiologist. Make a poster that can be displayed in hospital corridors and wards that explains to hospital staff, patients and visitors why they should use the alcohol hand rubs.

There are many chemicals that are known to have antimicrobial properties, such as lemon juice, garlic, onions and tea tree oil.

Industrial production processes

There are three stages for scaling up a laboratory process to a full-scale industrial production process.

1 Basic screening – a researcher grows the microorganism that produces a useful product, in a small flask, to find out its optimum conditions and requirements for growth. They investigate its:
   - nutrient requirements
   - optimum temperature
   - optimum pH
   - oxygen requirements.

   They also need to know at which stage of the growth cycle the useful product is made.

2 Pilot plant – the microorganism is grown in a small-scale bioreactor (fermenter) to see if the optimum conditions for growth on this scale are any different from those found in the basic screening process.

3 Plant scale – now the microorganism is grown in a massive bioreactor. There will be problems as the motor of the reactor and the bacterial metabolism both generate heat, so a cooling jacket is needed (see page 305). Chemical engineers must take lots of different factors into account when they design the bioreactors.

The growth curve

When microorganisms are first introduced into a growth medium, they need to synthesise enzymes needed to make use of the nutrients in the medium. This takes time, as genes have to be switched on and transcribed so the growth rate is at first slow. This is called the **lag phase**.

They then enter an exponential phase of growth, called the **log phase** or **growth phase**. They grow quickly, with their numbers or biomass doubling possibly every 30 minutes, for some quick-growing bacteria. Because the rate of growth is so fast, log graph paper is used to plot a growth curve as millions of bacterial cells will be produced. The log phase lasts as long as there are plenty of nutrients, space and oxygen and as long as their toxic waste is not building up enough to kill them.

When the bacteria begin to run out of nutrients, space or oxygen, or start to be poisoned by their waste, the rate of cell division equals the death rate, and they are in the **stationary phase**.

When more die than are being produced, the population is in the **decline phase** or **death phase**.

lag phase organisms are adjusting to the surrounding conditions. This may mean taking in water, cell expansion, activating specific genes and synthesising specific enzymes. The cells are active but not reproducing so population remains fairly constant. The length of this period depends on the growing conditions.

stationary phase nutrient levels decrease and waste products like carbon dioxide and other metabolites build up. Individual organisms die at the same rate at which new individuals are being produced. **Note**: in an open system, this would be the **carrying capacity** of the environment.

decline or death phase nutrient exhaustion and increased levels of toxic waste products and metabolites lead to the death rate increasing above the reproduction rate. Eventually, all organisms will die in a closed system.

log (exponential) phase the population size doubles each generation as every individual has enough space and nutrients to reproduce. In some bacteria, for example, the population can double every 20-30 minutes in these conditions. The length of this phase depends on how quickly the organisms reproduce and take up the available nutrients and space.

Population size

Time

Growth curve for a population of microorganisms.

Activity 15.3F

Suggest how you might be able to keep a culture of bacteria permanently in the growth (log) phase.

Primary and secondary metabolites

Microorganisms grown on an industrial scale to make something useful for us produce either primary or secondary **metabolites**.

Microorganisms produce primary metabolites as they divide and grow. This means that if we measure the production of the primary metabolite and plot it on a graph, it has a similar curve to the growth curve although it just slightly lags behind the growth curve.

Microorganisms produce secondary metabolites when they are reaching the end of their growth (log) phase and are entering the stationary phase. Some secondary metabolites are made as the microorganism converts a primary metabolite into a different chemical.

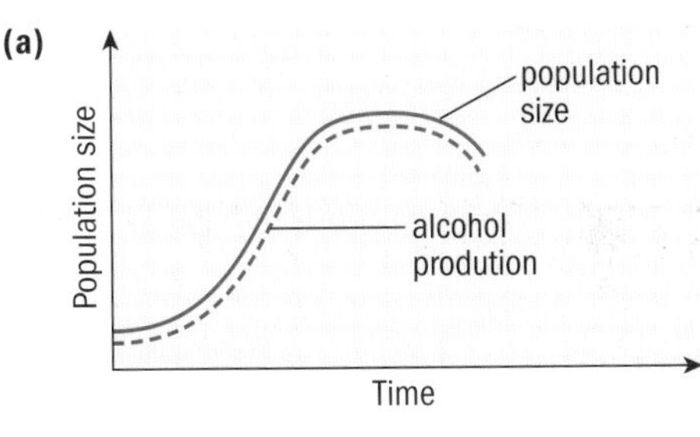

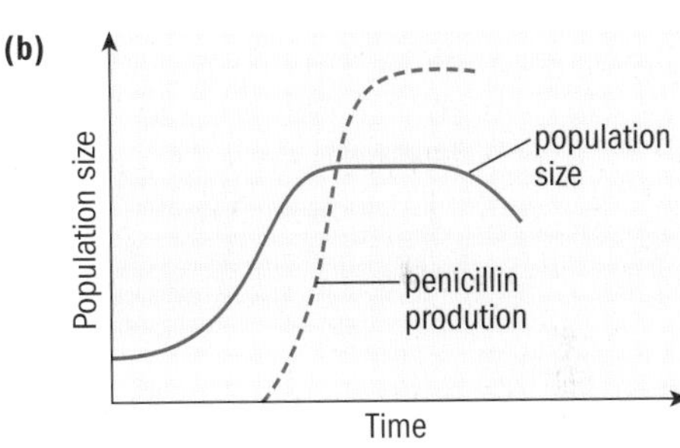

Growth and product curves showing the production of **(a)** alcohol by a population of yeast fungi and **(b)** penicillin by *Penicillium* fungi.

Activity 15.3G

Look at the graphs above.

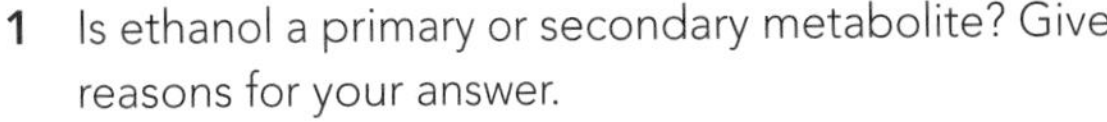

1 Is ethanol a primary or secondary metabolite? Give reasons for your answer.

2 Is penicillin a primary or secondary metabolite? Give reasons for your answer.

If we look at the increase in growth during the log phase we can calculate the time it takes for a particular microorganism to double its numbers, or, in the case of fungi, to double the dry mass of a mycelial disc (see page 293). This time is called the **generation time**.

The growth of bacteria can be measured by examining a culture at intervals over the course of several hours. This could be done by a direct count using a haemocytometer (see page 293) or by a viable count (see page 299) or by colorimetry (see page 292).

Generation time

A graph of population size, or optical density, is plotted against time and the time taken for the population size to double can be read from the graph.

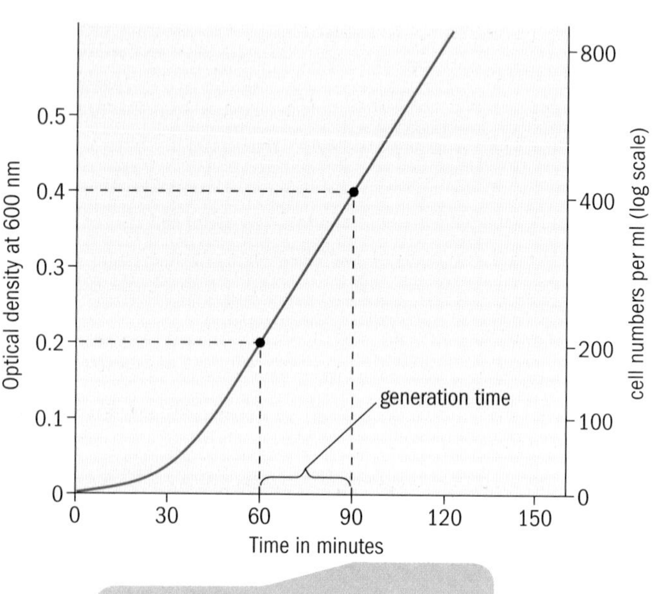

Indirect method for determining generation time.

Measuring growth rates

From the graph above, if you read the number of cells in 1 cm^3 growth medium solution at the beginning of the log phase and the number at any other point during the log phase, you can calculate the increase. If you know the time of the log phase, then the rate of growth is

$$\frac{\text{increase in population size}}{\text{time}} \text{ cells cm}^{-3} \text{ unit time}^{-1}$$

The unit used for time may depend on how quickly the bacteria in question grow. You may calculate an increase per minute or per hour or per day. The growth rate will also depend on the conditions under which the bacteria are growing.

Did you know?

Some bacteria have a circadian (24 hourly) rhythm (as do most other organisms on Earth). These bacteria divide just once a day and it is at night. This reduces the chance of their DNA undergoing a mutation as it is being copied, as there is no UV light at the time of replication. UV light can damage DNA. Faster growing bacteria are usually pathogenic and they are inside another living organism, so they are shielded from UV light anyway.

Industrial microbiology involves **fermentation technology** – the large-scale cultivation of microorganisms, or other single cells, to produce a commercially useful substance.

The dairy, wine-making and brewing industries have used anaerobic fermentation for a long time to make cheese, wine and beer. Companies making biomedical products, such as insulin and human growth hormone from genetically engineered microorganisms, have adapted the earlier fermentation technology. Industrial fermentation is also used by biotechnology companies to make useful products from genetically engineered plant and animal cells. For example Chinese hamster ovary cells are used to grow viruses for making vaccines.

Bioreactors

The large vessels used in industrial fermentation are called bioreactors or fermenters. They are designed with attention to the following areas:

- **Interior surfaces** – must be smooth and polished so there are no rough areas or little cracks that could harbour unwanted microorganisms.
- **Sterilisation** – there is an inlet to allow steam in to sterilise the bioreactor, between different batches of cultures. This ensures aseptic conditions.
- **Aeration** – microorganisms may need oxygen. Air enters through an inlet pipe with a diffuser. The diffuser breaks up the air stream to help the oxygen to dissolve in the growth medium. The air has to be filtered so that the growing culture is not contaminated with any unwanted microorganisms. The air is first heated and then cooled, as heat destroys any bacteriophage viruses. Air leaving the vessel also has to be filtered to sterilise it.
- **pH control** – the specific microorganisms should be grown at their optimum pH and the pH needs to be kept stable. The pH is monitored and acid or alkali added to maintain the optimum pH.

- **Temperature control** – the microorganism needs to be grown at its optimum temperature and this temperature should be kept stable. Microorganisms are living so their metabolism produces heat as a by-product. In a large vessel there is a small surface area compared with the volume, so the heat will not be dissipated. A cooling jacket of water is needed around the vessel.
- **Nutrients** – there is an inlet to add nutrients so that, if the microorganisms have to be maintained in their growth phase, they will not run out of nutrients.
- **Mixing** – bioreactors have baffles and paddles (impellers), driven by a motor, to stir the mixture. This distributes oxygen and nutrients and prevents the microorganisms forming clumps which could block the inlet or outlet pipes.
- **Foaming** – antifoaming agents are added as foaming may block inlets and outlets.
- **Harvesting**– if the microorganisms are to be maintained in the growth phase (if their useful product is a primary metabolite), then the product needs to be removed regularly as it may be toxic to the microorganisms, for example ethanol. There is an outlet where the product can be drained off.

A microbiologist adjusting a bioreactor (fermenter).

Did you know?

The design of bioreactors is very complex. People with degrees in biochemical engineering are involved in designing bioreactors.

Batch and continuous fermentation

Both are methods of aqueous fermentation – where the microorganisms are grown in liquid in a bioreactor.

Batch fermentation uses a **closed** bioreactor. The microorganisms and their nutrient medium are put into the reactor and left for the fermentation to take place. While this is happening, the temperature is regulated but nothing is added or removed from the vessel and at the end of the growth period the product is harvested. The vessel is cleaned and sterilised and a new culture batch set up. It is easy to set up a batch fermenter and to control the environmental factors. The vessels are versatile and can be used for making more than one type of product. If the culture becomes contaminated only one batch is lost.

Continuous fermentation takes place in open fermenters so that nutrients can be continuously added. The pH, oxygen levels and temperature are controlled and the microorganisms are kept in their growth phase. This uses smaller vessels and is more cost-effective than closed fermentation. However, there is more likely to be foaming and if there is a problem there is more wastage.

Safety

Industrial laboratories, as well as research and medical microbiology laboratories, have a very strict set of guidelines:

- All surfaces, including floors, must be smooth and easy to clean.
- There are designated clean zones where personnel must wear full protective clothing and follow the rules to avoid releasing pathogens or contaminating cultures.
- There is a regular cleaning and maintenance routine.
- Work is often carried out in laminar flow cabinets or clean air cupboards, which are sterile working areas and where air is filtered.
- Aseptic technique is always used.

Assessment activity 15.3

1 You are a technician in a medical microbiology lab and you need to grow bacteria.

Plan how you would investigate the following:

- the optimum temperature for the growth of *E. coli* bacteria
- the optimum pH for the growth of *S. albus* bacteria.

You need to think about:

- how you are going to measure growth
- whether you will use liquid or solid media
- what factors you will keep constant and how you will keep them constant
- how many replicates you will do
- what the independent variable is and what the dependent variable is.

When you have written your plan, get it checked by your teacher and then carry out your investigation. Remember that you will need to use aseptic technique throughout.

Record your results and share them with the rest of your class. Are their results similar? P4

If you have grown the bacteria for the same length of time, you can calculate the growth rate for the bacteria under the different conditions of temperature and pH. You can determine growth rate by measuring colonies per day or optical density (light absorption in the colorimeter) during a certain period of time.

Compare the growth rates under different conditions. M3

Explain how data about growth rate and generation time for a particular microorganism would be used to draw conclusions in a biotechnology or medical microbiology lab. D3

2 You are now a microbiologist working for a biotechnology company. You often have to find out about rates of growth of bacteria.

a From the graph shown on page 294, calculate the growth rate of the bacteria, during the log phase.

b How would you expect the growth rate to vary under the following conditions? Give reasons for your answers.

i lower temperature

ii increased oxygen concentration

iii increased nutrients. 

c One of the products the company makes is a new antibiotic. The antibiotic was discovered during routine screening of soil for bacteria. A bacterium has been found that produces a chemical that suppresses the growth of other bacteria.

i Explain why it is important to find out if the antibiotic is a primary or secondary metabolite, before these bacteria are grown on a large scale to produce the antibiotic.

ii What practical investigations would you need to carry out to find the conditions needed for the best growth of this bacterium? D2

3 You now work for the PR department of a biotechnology company. You need to produce some information, for potential customers, about how antibiotics are made.

There are many antibiotics but penicillin is still very important. Penicillin is a secondary metabolite produced by a fungus. The original fungus, investigated by Fleming, was *Penicillium notatum* but this was unsuitable for large-scale production. After much searching, a better strain, *Penicillium chrysogenum*, was found and used. This strain has also been modified by genetic manipulation and selection to give high-yielding strains. Since the 1940s the yield per cm^3 of penicillin by industrial fermentation has increased by over 30 times.

a Make a large labelled flow diagram to show the process of penicillin production on a large scale. Indicate whether it uses batch or continuous fermentation and explain why this method is used. Include information about the growth requirements of penicillin and about the substrate used. Show how the product is harvested and purified for use. Explain why *P. chrysogenum* rather then *P. notatum* is used for large-scale production. **D2**

b Indicate why scientists working in a large-scale production plant need to know about the growth rate of the fungus. **D3**

c Use biology textbooks and the Internet to help you.

d Show your flow diagram to the rest of your group.

Grading tips

For **P4** think carefully about your plan; refer to information given above to help you. Write out your plan clearly, indicating that you have considered all the variables, replicates and method. When you have had it checked, carry out the practical investigations competently and safely. Record your results and share them with the rest of the class to see if your data are reliable.

For **M3** in question 1 use your data and those of the rest of the class to compare growth rates under different conditions.

For **M3** in question 2 remember the log phase is when they are growing fastest, so look at the start and end of this phase. Read figures from the graph carefully. Calculate the increase over a particular time period.

For **D3** in question 1 think about growth rate for bacteria. You would need to know this if growing a bacterial culture to produce something; you would also need to know it for treating the bacteria with antibiotics. Slow-growing bacteria such as TB take a long time to treat as antibiotics only kill bacteria that are dividing. For **M3** in question 3 apply your knowledge about growth rates and growth curves, as well as primary and secondary metabolites.

In question 3, plan your flow diagram carefully for **D2**. Make sure you include all the information specified above in the assessment description.

PLTS

Self-manager, Effective participator, Team worker, Reflective learner, Independent enquirer and Creative thinker

You will develop your skills as a self-manager and effective participator. If you work in groups you will also develop teamwork skills. You will develop your reflective learner skills by applying knowledge to solve problems. You will develop your independent enquiry skills whilst researching the information for the flow diagram and your creative thinking skills to design your flow diagram.

15.4 Identifying microorganisms

In this section:   

Key terms

Virology/virologist – the study of viruses.

Reverse transcriptase – enzyme obtained from/found in retroviruses. It catalyses the reaction of making DNA from RNA.

Taxonomy – the study of the principles of classification.

Mycology – the study of fungi.

Gram positive – bacteria that stain purple with Gram stain. They retain the purple stain as they have a thick cell wall.

Gram negative – bacteria with thin cell walls. They do not retain the purple stain but take up the pink safranin and stain pink when Gram stained.

Bacillus – rod-shaped bacterium.

Coccus – round-shaped bacterium.

Classification of viruses

Because viruses are not properly defined as living or non-living, they are not designated a kingdom like the bacteria, protoctists, fungi, plants or animals. However, they contain nucleic acid with a genetic code that is the same as, although much smaller than, that of all other organisms on Earth.

They do not have cells and are extremely small. They can pass through filters that do not allow bacteria through and can only be seen with electron microscopes.

Scientists do not fully know the origin of viruses. There are theories and these have been discussed on page 286. They are unlikely to have a common ancestor and we cannot be sure of how they are related to each other. For all the living organisms that scientists classify, they follow the rules of **taxonomy** and the classification indicates how the organisms are related and how far back they shared a common ancestor. This does not apply to viruses. There are over 5000 described species of virus and probably more waiting to be discovered. They are all parasites and so are important as they cause diseases of people or of their livestock and crops. Some infect and kill organisms such as algae in the sea and thus contribute to the recycling of nutrients. Because viruses can move between cells of different hosts, sometimes from one species to another, they have contributed to the evolution of their host species as they can alter the host genome (some of our genes are remains of viral DNA) and when they infect a population they cause natural selection of those individuals most resistant to them.

The classification of viruses is based on their:

- morphology (shape – round, long, with or without an envelope)
- type of nucleic acid (see below)
- mode of replication (see below)
- host organisms (bacteria, fungi, protoctist, animal or plant)
- pathology (type of disease they cause).

The Baltimore classification

In 1971, David Baltimore, a **virologist** and Nobel prize-winner, devised the Baltimore classification system. This places viruses into seven groups (I–VII) depending on the type of nucleic acid they have. All viruses depend on the host cell machinery to copy the viral nucleic acid and make their proteins.

Nucleic acids (RNA and DNA) exist in the following forms:

- ds DNA is double-stranded DNA that consists of a coding strand and a template strand
- ss +DNA is the single coding strand of DNA
- ss −DNA is the single template strand of DNA (complementary to the coding strand)
- ss +RNA is messenger RNA. It is a copy of the coding strand of a piece of DNA
- ss −RNA is complementary to a length of +RNA
- ds +/− RNA is double-stranded RNA; one strand is mRNA, the other is complementary to it.

David Baltimore put all viruses into one of the six groups above, according to their type of nucleic acid. Later, a seventh group was described. These viruses, for example hepatitis B, have gapped nucleic acid.

Group	Genome	Examples of viruses	Examples of diseases caused
I	ds DNA	adenoviruses, herpesviruses, poxviruses	meningitis, chickenpox, smallpox
II	ss +DNA	parvoviruses	parvovirus infection
III	ds RNA	reoviruses	gastroenteritis
IV	ss +RNA	picornoviruses, togaviruses	hepatitis A, polio, SARS, foot and mouth, yellow fever, hepatitis C, rubella
V	ss −RNA	orthomyxoviruses, paramyxoviruses, rhabdovirus	influenza, measles, mumps, Ebola, Marburg, rabies
VI	ss +RNA-RT	retroviruses; have **reverse transcriptase**	HIV/AIDS
VII	ds DNA-RT	hepadnaviruses; have reverse transcriptase	hepatitis B

All viruses have to 'trick' their host cells into making viral proteins. Therefore the host cells have to make mRNA that has the genetic code of the viral nucleic acid. The diagram below shows how the viruses of each group do this.

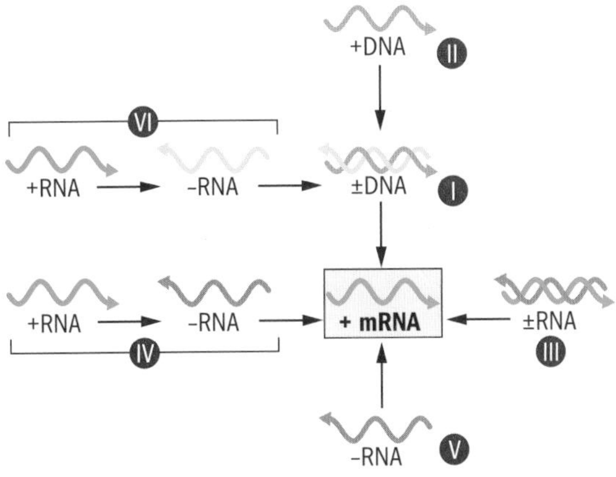

The Baltimore classification system for viruses. Viruses are classified according to the type of nucleic acid they have. All viruses have to cause the host cell to make mRNA.

Activity 15.4A

Discuss with a partner why all viruses need to cause mRNA to be made in the host cell.

The International Committee on the Taxonomy of Viruses

In addition to this broad classification system, there are specific naming conventions following the guidelines set out by the International Committee on Taxonomy of Viruses (ICTV). The International Union of Microbiological Societies (IUMS) has given the ICTV the task of refining and maintaining a universal virus taxonomy.

The taxonomy system for viruses shares certain features of the taxonomic system for living organisms. However, whereas for living organisms the sequence of taxonomic groups is kingdom, phylum, class, order, family, genus, species; virus classification begins with the level 'order'. Each taxon, apart from species, has a suffix:

- Order (~ virales)
- Family (~ viridae)
- Subfamily (~ virinae)
- Genus (plural: Genera)
- Species.

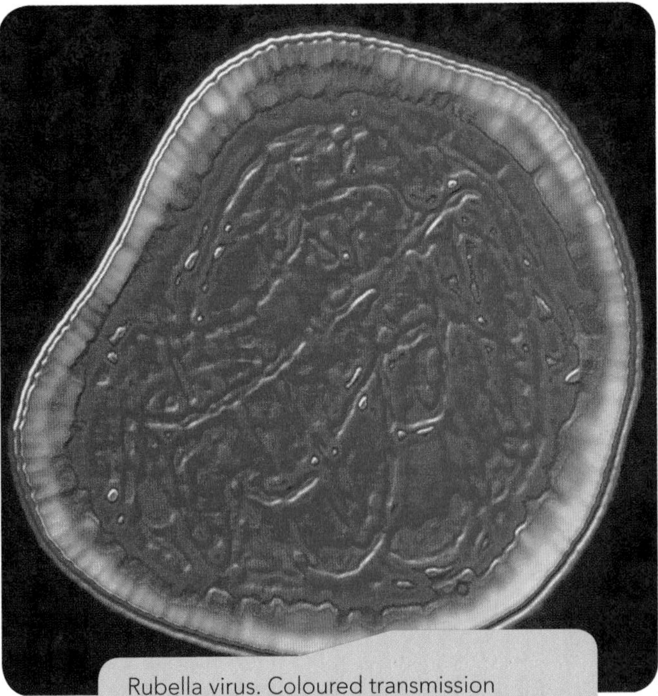

Rubella virus. Coloured transmission electron micrograph. The capsid (coat) is yellow and the single-stranded RNA is shown as pink. The diameter is about 60 nm. Mag: ×486 000.

The ICTV has established six orders:

Order	Genome	Hosts	Examples of diseases caused
Caudovirales	ds DNA	bacteria	T_2 phage virus causes bacteria to lyse
Herpesvirales	ds DNA	eukaryotes	*Herpes simplex* causes cold sores
Mononegavirales	ss –RNA	plants and animals	Ebola virus causes haemorrhagic fever
Nidovirales	ss +RNA	vertebrate animals	SARS (severe acute respiratory syndrome)
Picornavirales	ss +RNA	plants, insects and other animals	strawberry mottle virus
Tymovirales	ss +RNA	plants	potato virus X

Within each order are several families, subdivided into subfamilies, genera and species.

Viroids

These are infecting agents that are even smaller than viruses but have some of their properties. They infect plants. An example is the potato spindle tuber viroid.

Prions

Pronounced '*preeons*'. These are infectious particles made of protein. They do not have any nucleic acids and cannot be inactivated by the procedures, such as heat, that inactivate viruses. Some infect yeast fungi but the best-known ones infect mammals and cause scrapie in sheep, bovine spongiform encephalopathy (mad cow disease), its human equivalent CJD (Creutzfeldt–Jakob disease) and kuru (laughing disease).

Did you know?

There are also subviral agents called satellites. One causes chronic paralysis in bees.

Why is studying diseases of bees so important?

Characteristics of fungi

Microbiologists study some fungi – notably moulds and yeasts. There are about 100 000 identified species of fungi, mostly living on land, although there are a few marine species.

All fungi share a common ancestor so they form a group of related organisms. They are eukaryotic, more closely related to animals than to plants, and have cell walls that contain chitin. This makes the walls hard and stiff, and resistant to drying out. Fungi are the most resistant eukaryotes and some can grow in acid. Others can grow where there is no source of nitrogen. Fungi do not have undulipodia (eukaryotic flagella).

Fungi form spores which germinate to form slender tubes called hyphae. These hyphae are divided by cross walls, called septa, into cells. Some of the septa may break down, leaving 'cells' with many nuclei. Where the septa remain intact, they do not fully separate the 'cells' and cytoplasm can flow from one to another.

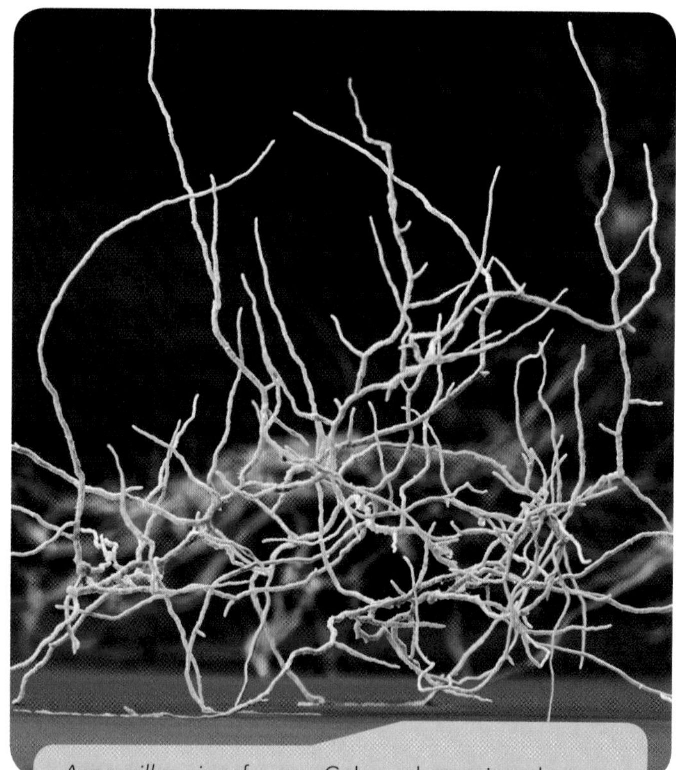

Aspergillus niger fungus. Coloured scanning electron micrograph (SEM) of a network (mycelium) of vegetative filaments called hyphae. *A. niger* grows in household dust, soil and decaying vegetable matter, including stale food. It is one of the most common causes of otomycosis (fungal ear infections) and, if large amounts of spores are inhaled, may result in the serious lung disease aspergillosis.

A large mass of threadlike hyphae forms a mycelium, which is the growing body for most fungi. Often this is below ground and is therefore generally hidden. Sometimes a fruiting body (a spore-bearing structure for reproduction), such as a mushroom, comes above ground.

The nuclei in the hyphae are haploid. When many fungi reproduce sexually they produce haploid gametes which fuse to give diploid zygotes. These then undergo meiosis to give haploid hyphae.

Some fungi do not have any sexual reproduction and produce haploid spores that germinate into haploid hyphae. Others can reproduce both sexually and asexually.

Activity 15.4B

What form of nuclear division will fungi with haploid hyphae use to make haploid gametes?

Most fungi are aerobic (need oxygen to respire) and heterotrophic. They secrete digestive enzymes onto their food substrate. These enzymes break down the complex organic molecules in the food to smaller ones that can then be absorbed, through the plasma membranes, into the hyphae. Fungi play a key role in decomposition of dead matter in soil and in recycling of nutrients.

Some fungi live freely in soil but others form partnerships with trees or other plants. Most orchid seeds need fungi to help them germinate. Many forest trees rely on fungi (mychorrhiza), growing in their roots, to transport minerals from the soil to the tree roots. There are fossil fungi dating from 300 million years ago. In each case the fossilised fungi are associated with fossilised plants. Some scientists have suggested that these plant–fungus associations enabled plants to colonise dry land.

Leaf cutter ants make leaf compost and cultivate a type of fungus to eat. The particular species of fungus is found only in the nests of these ants.

Some fungi are important sources of antibiotics. These are chemicals that the fungi produce to reduce competition with other microorganisms but we have made use of them to help overcome infections.

Some fungi produce toxins, such as alkaloid mycotoxins. Some of these are hallucinogenic and have been used in religious rites and ceremonies. The ergot fungus *Claviceps pupurea* lives in damp rye.

Bread made with infected rye causes people to exhibit a range of symptoms that may have led to the belief that they are bewitched. The symptoms included:

- diarrhoea
- headaches
- nausea and vomiting
- muscle spasms and seizures or convulsions
- hallucinations
- mania, psychosis or delirium
- gangrene and loss of fingers and toes or limbs
- death.

In the Middle Ages this affliction was known as St Anthony's fire. Often a whole community was affected as they would have eaten bread made from the same batch of rye. The condition is now called ergotism. The alkaloid toxin made by this fungus is structurally very similar to the hallucinogenic drug LSD.

Did you know?

Some scientists think that people accused of witchcraft in the Middle Ages may have been affected by eating rye bread infected with ergot fungus. The occurrence of witchcraft incidents coincides with areas where rye bread was eaten during years with higher rainfall, when the fungus would have thrived on the rye. As late as 1952 in Paris, several people were affected by ergotism from a batch of infected bread. Most were taken to lunatic asylums!

Mycotoxins in stale food eaten during the Middle Ages were probably responsible for people thinking they had seen dragons and other monsters. These toxins may also have been used in certain religious ceremonies during pagan times.

- Some enzymes used in detergents are obtained from fungi.
- Yeast fungi are used in making wine, beer, soy sauce and bread.
- Mould fungi are used in cheese-making.
- Quorn™ is a protein obtained from fungi.
- Mushrooms and truffles are important sources of food for humans.

Pathogenic fungi

Although fungi are eukaryotic organisms, the pathogens are also regarded as microorganisms. The study of these fungi is medical **mycology**. Some fungal pathogens infect humans and others infect livestock or crop plants. Hence it is important for scientists to study these fungi.

Activity 15.4C

Fungi that are plant parasites produce the enzyme cellulase, which breaks down cellulose. Suggest why they do this.

Fungus	Disease	Characteristics
Aspergillus niger	farmer's lung, asthma, allergies, ear infections	cough, breathlessness, chest pain
Aspergillus flavus	cancer	toxin made by fungus growing on nuts causes cancer
Cryptococcus	meningitis	usually lives in soil and is not harmful, but if it infects people with weakened immune systems can cause severe meningitis
Pneumocystis	pneumonia	may infect elderly, newborns and those with weakened immune systems
Stachybotrys	chest infection	may occur where people live in very damp houses
Candida	thrush	may infect mouth or vagina
Botrytis	mould infections of plants	may affect grapes and members of the cabbage family
Puccinia	wheat leaf rust	affects wheat and rye
Tinea corporis	ringworm	skin infections in humans and livestock
Tinea pedis	athlete's foot	

Did you know?

The potato blight organism, *Phytophthora infestans*, is not actually a fungus but is a fungus-like protoctist. It caused the potato blight famines in Ireland in the 1840s, which reduced the population greatly. Someone discovered by chance that potatoes growing near an old copper mine did not get infected.

In France, during the same time period, a professor at the University of Bordeaux, studying winemaking, noticed that where grapevines near roads had been sprayed with copper sulphate and hydrated lime, to deter passers-by from eating the grapes, the grapes did not get botrytis. He carried out trials using this mixture, which is still known as Bordeaux mixture. Bordeaux mixture is still used to protect grapes and potatoes from fungus or blight infections.

Identifying bacteria

The term 'bacteria' was originally used to describe all prokaryotic organisms. They were first classified according to similarities of cell structure and metabolism or on differences of cell chemical components.

Modern classification of bacteria depends very much on genetic techniques, such as comparing gene sequences, particularly of the gene for ribosomal RNA. The PCR can be used to increase the amount of bacterial DNA to be tested.

Unit 18: See page 355 for more information on PCR.

The table describes the main groups of fungi.

Group	Characteristics	Examples
Zygomycota	no septa (cross walls); reproduce sexually and asexually	*Rhizopus* – black bread mould
Basidiomycota	reproduce sexually and have club-like fruiting body	mushrooms, toadstools, puffballs, *Mycorrhiza*
Ascomycota	reproduce sexually with asci (capsules containing spores)	yeast, truffles
Deuteromycota	reproduce asexually; hyphae divided by septa (cross walls)	*Penicillium, Candida, Cryptococcus, Aspergillus*

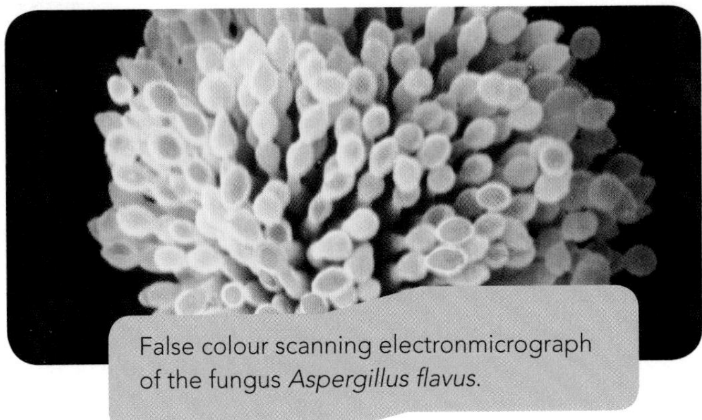
False colour scanning electronmicrograph of the fungus *Aspergillus flavus*.

Scientists have recognised that there are two domains within the prokaryote group. These used to be called Archaebacteria and Eubacteria. These groups are now called Archaea and Bacteria. They evolved from a common ancestor but the Archaea have more in common with eukaryotic organisms than they do with bacteria.

Remember that all classification of living organisms is man-made, based on scientific observations, and, as we learn more about these organisms, their classification may change.

It is very important that medical microbiologists can correctly identify bacteria that are causing infections in people, so that the most effective antibiotic treatment can be given. Because different subgroups of bacteria have often widely differing metabolism, some antibiotics are specific for certain bacteria.

The Gram stain

In 1884, a Danish microbiologist, Hans Christian Gram, developed a staining technique that could distinguish between two groups of bacteria. This Gram stain differentiated between **Gram positive** and **Gram negative** bacteria. It is still an important diagnostic technique as the two groups of bacteria have different cell wall structures and respond differently to antibiotics. Penicillin prevents cell wall synthesis in growing Gram positive bacteria but is not effective against Gram negative bacteria.

The stages of the Gram stain are as follows:

1 Place a drop of sterilised water on a clean microscope slide.

2 Using a sterilised microbiological wire loop, transfer some bacteria from a colony, or from a liquid culture, into the water. Move the loop in a circular motion to spread the bacteria onto the slide.

3 Allow the slide to dry, either in the air or over a beaker of hot water, or by passing it quickly to and fro in a hot Bunsen flame. This last step heat fixes the smear onto the slide.

4 Add one drop of crystal violet and leave for 1 minute. This stains the bacterial cell walls. Avoid skin contact with the crystal violet.

5 Wash off the excess crystal violet with tap water.

6 Flood the smear with Gram's iodine solution. This is a mordant – it intensifies the violet colour of the stain. Avoid skin contact with the iodine solution.

7 Wash off the Gram's iodine with tap water.

8 Add 95% ethanol, drop by drop, until no more crystal violet washes from the smear.

9 Wash with tap water.

10 Counterstain with safranin for 45 seconds. Avoid skin contact with the stain.

11 Wash off with tap water.

12 Gently dry the slide in a clean Bunsen flame.

13 Place a drop of cedar oil on the smear and examine the bacteria, using the oil immersion objective on a light microscope.

14 If the bacteria are Gram positive they will have retained the purple stain.

15 Gram negative bacteria have a thinner wall and two lipid membranes. This allows the ethanol to wash out the crystal violet and then the bacterial walls take up the safranin and stain pink.

Did you know?

A microbiologist in the nineteenth century developed a stain for bacteria. He wrote it down but no one else could get it to work. It turned out he had forgotten to say that he had stirred it with a rusty spoon. The iron oxide from the rust on this spoon was crucial to the stain working. This just shows that you have to write down every step of a scientific method you use because other scientists need to carry out the same tests to make sure the results are the same.

Completing the diagnosis

The shape of the bacteria can also be seen. Rod-shaped bacteria are bacilli (singular: **bacillus**). Round ones are cocci (singular: **coccus**). Using the Gram stain, bacteria can be put into four groups: Gram positive bacilli; Gram positive cocci; Gram negative bacilli; Gram negative cocci.

Some bacteria, such as the *Mycobacteria* that cause TB and leprosy, are neither Gram positive nor Gram negative. They can be identified with an acid-fast stain. However, for the purposes of treatment they respond to antibiotics used to treat Gram negative bacteria.

Special media are also used to help identify bacteria. The media will suppress growth of other bacteria (such as non-pathogenic bacteria in faeces), and promote the growth of pathogens that may be there. Once the causative pathogen has been isolated it can be further investigated and tested to see which antibiotic will be effective against it.

Classifying bacteria

Bacteria make up the bulk of the biomass on Earth. There are many different species but many have probably not yet been identified. Only a small minority of bacteria are pathogenic.

Classification of bacteria is determined by rules laid down by the International Committee on Systematic Bacteriology. For medical microbiology the four groups mentioned above, plus spiral bacteria, are the most important.

Some important Gram positive bacteria:

Gram positive cocci	staphylococci (in clusters); can produce the enzyme catalase which breaks down hydrogen peroxide. *Staphylococcus aureus* is an important cause of hospital-acquired infections
	streptococci (in strings); can't make catalase. *Streptococcus pneumoniae* causes meningococcal meningitis. Other streptococci cause sore throats, tooth decay, endocarditis
Gram positive bacilli	spore-forming: *Bacillus cereus* – food poisoning; *Bacillus anthracis* – anthrax; *Clostridium tetani* – tetanus; *Clostridium botulinum* – botulism; *Clostridium perfringens* – gangrene; *Clostridium difficile* – colonitis
	non-sporing: *Listeria monocytogenes* – food poisoning; *Corynebacterium diphtheriae* – diphtheria; *Propionibacterium acne* – acne

Some important Gram negative bacteria:

Gram negative cocci	*Neisseria gonorrhoea* – gonorrhoea; needs raised carbon dioxide levels to be cultured; bacteria appear in pairs (diplococci)
Gram negative bacilli	*Escherichia coli* – occur normally in the gut – some strains can be harmful; *Salmonella enteriditis* – food poisoning; *Salmonella typhi* – typhoid fever; *Yersinia pestis* – plague; *Pseudomonas aeruginosa* – wound infections – needs oxygen; *Legionella* – legionnaires' disease – needs iron and cysteine to be cultured; *Bordetella pertussis* – whooping cough – needs activated charcoal in culture medium to absorb its toxins
Gram negative curved rod	*Vibrio cholerae* – cholera; *Campylobacter* – food poisoning/gastroenteritis; *Helicobacter pylori* – stomach ulcers
Gram negative spirochetes (spiral)	*Treponema pallidum* – syphilis

Important acid-fast bacteria:

- *Mycobacterium leprae* causes leprosy; cannot be cultured in the laboratory.
- *Mycobacterium tuberculosis* causes TB; very slow growing.
- *Mycobacterium bovis* causes TB in cattle.

PLTS

Independent enquirer, Reflective learner, Creative thinker and Effective participator

You will develop your skills as independent enquirer when finding the information needed for the poster. You will also develop reflective learning and creative thinking skills in designing the poster. You will develop skills as an effective participator by undertaking the practical tasks.

Case study: Clinical microbiologist

I work in a large hospital. I have a degree in medical microbiology and my specialist areas of study were the bacteria, viruses and fungi that cause diseases in humans. I work in the hospital laboratory and have to find out which type of bacterium or fungus is causing an infection. Some of the samples are from patients in the hospital and some are sent in by GPs. For example, if a doctor thinks that a patient has an infection and wants to know the best antibiotic to use, then we need to identify the type of bacterium. This is because not all bacteria respond to all antibiotics. We also need to know if the bacterium is resistant to any antibiotics. I also help educate nurses and doctors about infection control in the hospital.
This involves explaining how and why hand gels and sterile gloves should be used. You may have heard of the bacterium MRSA being a problem in hospitals.

1 **What is MRSA?**

2 **If you were a clinical microbiologist what would you do to prevent its spread in hospitals?**

Assessment activity 15.4

M4 D4 P5 BTEC

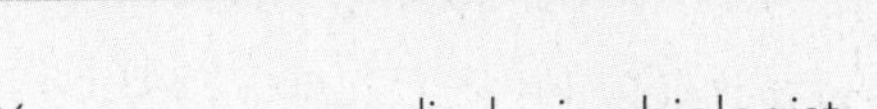

1 You are a virologist. Your laboratory is open to the public on an open day.

a Make a poster to inform people about the importance of viruses, their structure and characteristics and a simple description of how they are classified. **P5**

b Explain how a knowledge of virus structure and classification can help doctors accurately diagnose illnesses. **D4**

2 You are now a mycologist.

a Examine prepared microscope slides showing fungal hyphae and spores. **D4**

b If you can, under the supervision of your teacher, place some dampened bread in a Petri dish and leave covered for a few days. Examine the mould fungus growing on it. You should see the mycelium of hyphae plus aerial hyphae with structures containing spores. Press a piece of sticky tape so the sticky side is lightly touching onto the mould and then place this sticky tape, sticky side down, onto a microscope slide and examine under low and high power. **D4**

c Examine the bread mould on the damp bread, using a binocular microscope. **D4**

d Make a poster to show the characteristics of fungi and of the main subgroups: Ascomycota, Zygomycota, Basidiomycota, Deuteromycota. Include a description of some important examples of each subgroup and explain why these fungi are important to humans and why they need to be studied. **P5**, **M4**

e Describe how studying fungi using a microscope is useful to help identify them. **D4**

3 You are now a medical microbiologist.

a Make a poster describing the main groups of medically important bacteria and show how they are classified according to the Gram stain and their shape. **P5**

b Outline an explanation of how the Gram stain differentiates bacteria according to the structures of their cell walls. **M4**

c Explain why this stain and microscopic examination is an important first step in identifying pathogenic bacteria. What further tests can then be done to make a complete and accurate diagnosis? **D4**

Grading tips

In your answer to question 1a, for **P5**, include in your poster information about the structure, characteristics and importance of viruses. Remember, importance can mean ways in which they are useful and/or ways in which they are harmful. Include a simple outline account of how viruses are classified for **M4**.

To answer question 1b, apply your knowledge of viruses and the diseases they cause for **D4**.

For question 3, you can include *brief* details of the Gram stain here to achieve **P5**. Focus on medically important bacteria and their classification, for example Gram positive bacilli, etc.

To achieve **M4** in question 3 you need to outline an explanation of how this stain works – why it stains some purple and others pink.

For **D4** in question 3 you need to do some research via the Internet and microbiology textbooks to help you with this.

WorkSpace

Amber Lansley

Bacteriologist at the Centre for Emergency Preparedness and Response, Health Protection Agency

I work in a team which researches vaccines and antibiotics for use against bacterial diseases like Anthrax (*Bacillus anthracis*) and Plague (*Yersinia pestis*). I also manage Containment Level 3 laboratories which involves the maintenance of equipment, lab facilities and ensuring staff safety.

We usually work on the development of several methods (assays) at any one time, so time needs to be planned to allow work on all of them. Firstly I plan and organise what work needs to be done for that day with the other members of the bacteriology team.

I then set up the lab with all the equipment that is needed for the experiment. Once I am happy that I have everything we need we start our work. The lab work can take up to several hours working in a MSCIII (high containment microbiological safety cabinet), which is restrictive and tiring, so being part of a trained, well organised team is very important. Every day is different and it is very exciting knowing that I am contributing to drugs which may be used to help people all over the world.

Sometimes not everything goes according to plan. On one occasion the bacteria we were growing were not growing as fast as we had expected. We had to investigate why this was by looking at all the different factors which could impact on the growth, such as incubation temperature, the media the bacteria were grown in, checking that the equipment was working as it should and seeing if it was a problem with the particular strain of bacteria we were working with. Experiments had to be performed to compare growth in different incubators and media to see if these made a difference. This helped us to determine the reason for its slow growth.

Think about it!

1. How would you ensure the safety of yourself and your colleagues when working with bacteria such as those mentioned?
2. Name some types of media in which bacteria can be grown.

Just checking

Work in groups and discuss what you know about microbiology and then answer these questions.

1 Make a table to compare eukaryotic cells, prokaryotic cells and akaryotes. Remember to include similarities and differences.
2 Explain to a partner why it is difficult to decide whether viruses are living.
3 Explain to a partner what serial dilution is and why it is used.
4 Discuss with a partner whether total counts or viable counts are more useful.
5 Explain to a partner how colorimeters can be used to measure bacterial growth.
6 Draw a spider diagram to show all the useful and harmful things that bacteria can do for us.

edexcel

Assignment tips

To get the grade you deserve in your assignments remember to do the following:

- Always use aseptic technique when carrying out practicals.
- Always assess the risks involved, however small.
- Plan and carry out your practicals competently.
- Record your data.
- Research carefully when preparing posters and presentations.
- Make sure you can pronounce all the terms and names.

Some of the key information you'll need to remember includes the following:

- key features of eukaryotic cells, prokaryotes and akaryotes
- use of light microscopes including phase contrast and oil immersion
- principles of electron microscopy
- aseptic technique
- inoculation of liquid and solid media to grow microorganisms
- methods of measuring the growth of microorganisms
- factors that affect the growth of microorganisms
- Gram stain
- how microorganisms are classified.

You may find the following websites useful as you work through this unit

For more information on...	Visit...
the Wellcome Trust	The Wellcome Trust website
virus classification	The Universal Virus Database of the International Committee on Taxonomy of viruses (ICTVdb)
the study and development of biology	The Society of Biology website
bacteria and viruses	The Scientific American website
fungi	Nature Grid – Canterbury Environmental Education Centre, Kent, UK

Other useful sources of information are journals and magazines such as *New Scientist*, *Scientific American*, *National Geographic* and *Focus*.

18 Genetics and genetic engineering

When you study genetics you are studying the basis of life itself. The structure of DNA was determined by Watson and Crick, with help from Franklin and Wilkins, in 1953. Watson and Crick themselves said that they had 'discovered the secret of life'.

In the 1860s Gregor Mendel, a monk in Eastern Europe, carried out some important investigations into the inheritance of characteristics in pea plants. He developed ideas that completely changed the way people had previously understood inheritance but many people did not understand his work and its significance was not appreciated during his lifetime.

Watson and Crick showed how DNA molecules contain coded information that governs the making of proteins in living cells.

In this unit you will develop an understanding of the techniques that are key to modern genetics and molecular biology. You will extract DNA, carry out electrophoresis to obtain a DNA profile, transform bacterial cells and amplify small amounts of DNA using the polymerase chain reaction. You will learn about the applications of genetics in agriculture to improve crops and you will find out how genetics can help scientists to map migrations of animals and to identify any animals illegally obtained from the wild.

You will also examine the impact and ethical implications of gene technology on industry and society.

Learning outcomes

After completing this unit you should:

1 understand the process of protein synthesis
2 be able to investigate the process of cell division in eukaryotic cells
3 understand the principles of Mendelian genetics
4 be able to apply basic techniques of DNA technology.

Assessment and grading criteria

This table shows you what you must do in order to achieve a pass, merit or distinction grade, and where you can find activities in this book to help you.

To achieve a **pass** grade the evidence must show that you are able to:	To achieve a **merit** grade the evidence must show that, in addition to the pass criteria, you are able to:	To achieve a **distinction** grade the evidence must show that, in addition to the pass and merit criteria, you are able to:
P1 compare and contrast the structure of various nucleic acids **See Assessment activity 18.1**	**M1** explain how genetic information can be stored in a sequence of nitrogenous bases in DNA **See Assessment activity 18.1**	**D1** explain the steps involved in biosynthesis of protein including the roles of RNA **See Assessment activity 18.1**
P2 identify the stages of mitosis and meiosis in eukaryotic cells **See Assessment activity 18.2**	**M2** describe the behaviour of chromosomes during cell division using the results of the practical investigations **See Assessment activity 18.2**	**D2** analyse the correlation between observed pattern of dihybrid inheritance and the expected pattern **See Assessment activity 18.3**
P3 carry out practical investigations to record stages of cell division **See Assessment activity 18.2**		
P4 explain how the behaviour of chromosomes leads to variation **See Assessment activities 18.2, 18.3**	**M3** apply principles of modern Mendelian genetics to predict patterns of monohybrid, dihybrid inheritance and variation **See Assessment activity 18.3**	
P5 explain monohybrid and dihybrid inheritance ratios **See Assessment activity 18.3**		
P6 carry out basic DNA techniques **See Assessment activity 18.4**	**M4** describe digestion of DNA by restriction endonucleases and electrophoresis of fragments **See Assessment activity 18.4**	**D3** explain the steps involved in producing a genetically modified organism **See Assessment activity 18.4**
P7 identify applications of genetic engineering **See Assessment activity 18.4**		

How you will be assessed

Some of your assignments will involve carrying out practical investigations and techniques. You need to demonstrate that you can work safely and competently in a laboratory setting and carry out risk assessments where appropriate.
Some assignments will involve posters, presentations and possibly role play. You may also be asked to take part in discussions within your group. The assignments will enable you to develop many skills and you will need to put together a portfolio of evidence.

Moses, 19 years old

I have particularly enjoyed this unit as I would eventually like to work in the field of bioinformatics. I am interested in nutrition and pharmacology and I think that, with the information coming from the Human Genome Project, in the future medicines will be tailored to individuals so that people do not take drugs that are of no benefit or that have nasty side effects. As scientists learn more about foods and how food molecules interact with genes, one day we may be able to overcome the problem of why some people gain weight whilst others eating similar amounts and types of food don't. This is a really exciting time to be studying genetics and this unit has inspired me to aim for a career in genetics. I have really enjoyed the practical work in this unit and it has given me a lot of confidence.

Catalyst

Applications of genetics

Find out about one of the following and share information with the rest of the class:

- the work of Gregor Mendel
- the work of Professor Sir Alec Jeffreys
- the work of Watson and Crick
- the Human Genome Project
- nutrigenomics
- phamacogenomics
- the work of Ian Wilmut
- how DNA is used in archaeology.

18.1 Protein synthesis

In this section:

Key terms

Polymer – large molecule made of repeating smaller molecules, called monomers.

Nucleotide – monomer of nucleic acids; each made of a sugar, organic base and phosphate group.

Nitrogenous base – component of a nucleotide. There are five: adenine, guanine, cytosine, thymine and uracil.

Condensation reaction – reaction between two small molecules that join to form a bigger molecule. A molecule of water is eliminated.

Hydrogen bonds – weak bonds formed by attraction between a negatively charged part of a molecule (such as oxygen) and a positively charged part of another molecule (such as hydrogen).

Gene – a length of DNA that codes for one or more protein(s).

Triplet code – genetic code contained in DNA. Each triplet of bases corresponds to a particular amino acid in a protein, so the sequence of base triplets determines the sequence of amino acids in a protein. This sequence in turn determines the 3D shape of a protein and its ability to carry out its function.

Transcription – making of mRNA that can carry the genetic code from the DNA in the nucleus to the ribosomes where proteins are assembled.

Codon – triplet of bases on a length of mRNA.

Anticodon – triplet of bases on a piece of tRNA. Each anticodon is complementary to a codon.

Translation – assembly of amino acids into a protein at a ribosome, using coded instructions on mRNA.

Amino acid activation – attachment of amino acids to their specific tRNA molecules to be carried to the ribosome for assembly into proteins.

Degenerate code – the genetic code is described as degenerate because each amino acid has more than one base triplet code in the DNA.

Nucleic acids are universal and are polymers

Nucleic acids are found in all living organisms – prokaryotes and eukaryotes, and in akaryotes. Hence, they are universal. Prokaryotes and eukaryotes contain both DNA (deoxyribonucleic acid) and RNA (ribonucleic acid). DNA and RNA contain coded information, called genetic information, to build the organism and govern the biochemical processes that keep it alive.

You will find out more about the DNA and RNA of prokaryotes in Unit 15.

Did you know?

Viruses (akaryotes) have *either* DNA **or** RNA, not both. Prokaryotes and eukaryotes have both.

Both DNA and RNA are **polymers**. This means they are large molecules made of repeating subunits or monomers. The monomers in this case are **nucleotides**. The polymers are polynucleotides.

A nucleotide consists of a pentose (5-carbon) sugar, a phosphate group and a **nitrogenous base**.

The nucleotides join together by **condensation reactions**.

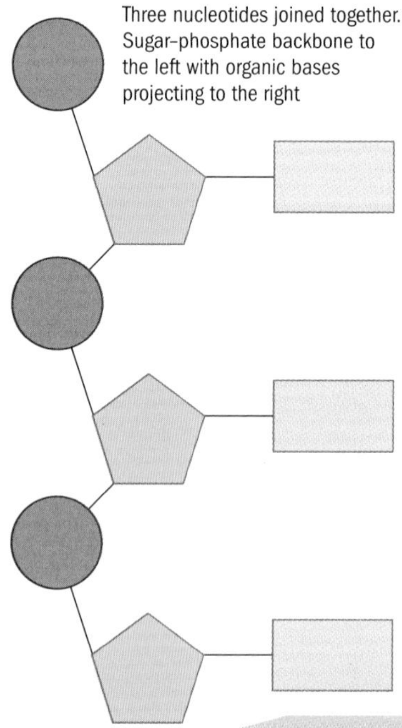

Three nucleotides joined together. The sugars (orange) and phosphates (red) make up the backbone. The organic bases are shown in green.

There are five different nucleotide (also described as nitrogenous or organic) bases:

- purines, which are larger, having a double ring structure: adenine, A; guanine, G
- pyrimidines, which are smaller, having a single ring structure: thymine, T; cytosine, C; uracil, U.

RNA contains the bases adenine, uracil, guanine and cytosine. DNA contains adenine, thymine, guanine and cytosine.

Purines

adenine

guanine

Pyrimidines

thymine

uracil

cytosine

Purine and pyrimidine nucleotide bases.

DNA is a double helix

In eukaryotes some DNA is found in mitochondria and chloroplasts but nearly all of it is in the nucleus, associated with histone proteins and coiled and wound tightly into structures called chromosomes. Each chromosome consists of one huge molecule of DNA.

The DNA molecule consists of two backbone chains of sugars and phosphate groups. The pentose sugar is deoxyribose. The two chains are described as anti-parallel because they run in opposite directions. Pairs of organic bases join the backbones together. The pairs of bases are held together by **hydrogen bonds**.

A purine base always joins with a pyrimidine base so that these 'rungs in the ladder' are the same width. Adenine pairs with thymine using two hydrogen bonds. Guanine pairs with cytosine using three hydrogen bonds. This type of pairing is very specific and is also known as **complementary base pairing**.

The hydrogen bonds make the molecule of DNA very stable and strong, whilst also enabling it to unzip in order to copy itself, during interphase, before the nucleus and cell divide.

The hydrogen bonds also enable part of a DNA molecule (a gene) to unzip so that a piece of complementary RNA can be made to carry the instructions for protein assembly out of the nucleus to the ribosomes, for protein synthesis.

The paired bases are protected by being held within the backbones and this helps reduce corruption of the genetic information.

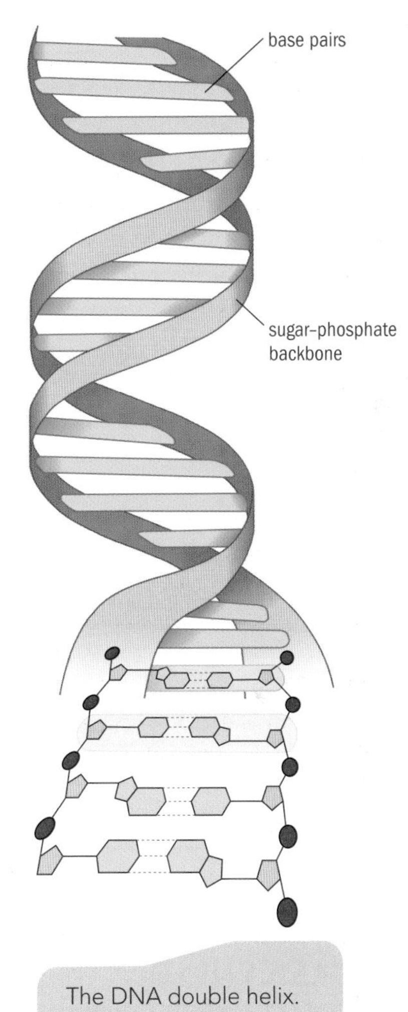

The DNA double helix.

Activity 18.1A

Use the Internet or biology textbooks to find out the names of the scientists who worked out the structure of DNA. When did they do this? When did some of them gain the Nobel Prize?

RNA is different from DNA

There are three types of RNA:

- messenger RNA (mRNA) which carries the genetic code from the DNA in the nucleus to the ribosomes, where the code is read and amino acids are assembled into proteins
- transfer RNA (tRNA) which carries amino acids from a store in the cytoplasm to the ribosomes for assembly into proteins. Each tRNA will only attach to and carry a certain amino acid. Each tRNA has three unpaired bases at one end and this is where the amino acid joins. The sequence of these three bases determines which amino acid joins the tRNA molecule
- ribosomal RNA (rRNA) – ribosomes are assembled in the nucleolus of eukaryotic nuclei and are made of protein and RNA.

RNA is structurally different from DNA in the following ways:

- The pentose sugar in the nucleotides is ribose, rather than deoxyribose.
- The polynucleotide chain is usually single stranded, rather than double stranded.
- The nitrogenous base uracil is found instead of thymine.

Activity 18.1B

Use the Internet and biology textbooks to find out how many genes there are in each human cell nucleus.

The shape of protein molecules is crucial

You may have already seen in Unit 11 that proteins are very important. They make up about 75% of our dry mass and are key components in all tissues, particularly connective tissue, muscle and bone. They are also part of the structure of all cell membranes.

Haemoglobin is a protein that carries oxygen from lungs to cells for aerobic respiration. We also rely on many proteins, such as enzymes and antibodies, to perform specific functions.

Activity 18.1C

Discuss with a partner why (a) enzymes and (b) antibodies are so important to us. Share your thoughts with the rest of your class.

You may remember from Unit 11 (page 221) that proteins are polymers, made of long chains of monomers – amino acids. These amino acids are joined together by peptide bonds, formed during condensation reactions. There are 20 different amino acids and each one may be used more than once in a particular protein. They can be joined into chains of anything from several to well over 100 amino acids. The sequence of amino acids (the primary structure) in the protein chain determines how the protein will fold into its 3D shape. In turn, the 3D shape enables the protein to carry out its specific function. Thus, if the sequence of amino acids is wrong, the protein will not be able to fold into its proper shape and will not be able to carry out its function.

What determines the primary structure of proteins?

A **gene** is a length of DNA that codes for a protein. The sequence of nucleotide bases provides the genetic code that determines the sequence of amino acids in the protein.

- It is a **triplet code**. A sequence of three nucleotide bases codes for a particular amino acid.
- There are four bases, so if they are arranged in groups of three, the total number of different combinations is $4^3 = 64$. This is more than enough to code for the 20 amino acids plus start and stop codes.
- All amino acids except methionine have more than one base triplet that codes for them. Because of this feature, the genetic code is described as a **degenerate code.**

There are lots of genes on each chromosome.

Protein synthesis

Protein synthesis takes place in two stages: **transcription** and **translation**.

Transcription

The instructions for making the protein are copied from the piece of DNA onto a messenger molecule – messenger RNA. Transcription occurs in the nucleolus of the nucleus.

- The part of the chromosome with the relevant gene dips into the nucleolus, which is a dense area of the nucleus containing lots of RNA nucleotides. The gene (length of DNA) unwinds and unzips as the hydrogen bonds between the nucleotide bases break.

- This exposes DNA bases on the nucleotides.
- Free RNA nucleotides line up along one chain of the DNA, known as the template strand, and make temporary hydrogen bonds with their complementary base. Adenine from an RNA nucleotide pairs with thymine on the DNA. Uracil from an RNA nucleotide pairs with adenine on the DNA. Guanine from an RNA nucleotide pairs with cytosine on the DNA template strand and cytosine from an RNA nucleotide pairs with guanine on the DNA template strand. The enzyme RNA polymerase catalyses this reaction.
- Sugars and phosphates of adjacent RNA nucleotides bond together.
- This forms a single polynucleotide chain that is complementary to the template strand of the DNA molecule. It is therefore a copy of the other DNA strand, the coding strand, because it carries a copy of the information held in the DNA coding strand for the assembly of a specific sequence of amino acids to make a specific protein.
- A triplet of bases on the mRNA that codes for a specific amino acid is called a **codon**.
- This piece of messenger RNA can now break away from its DNA template, pass through pores in the nuclear envelope and go to a ribosome.

There are regions within a gene called introns. These are spliced out during transcription so that only the coding regions of DNA, known as exons, are transcribed to RNA.

First position	Second position				Third position
	T	C	A	G	
T	Phe	Ser	Tyr	Cys	T
	Phe	Ser	Tyr	Cys	C
	Leu	Ser	STOP	STOP	A
	Leu	Ser	STOP	Trp	G
C	Leu	Pro	His	Arg	T
	Leu	Pro	His	Arg	C
	Leu	Pro	Gln	Arg	A
	Leu	Pro	Gln	Arg	G
A	Ile	Thr	Asn	Ser	T
	Ile	Thr	Asn	Ser	C
	Ile	Thr	Lys	Arg	A
	Met	Thr	Lys	Arg	G
G	Val	Ala	Asp	Gly	T
	Val	Ala	Asp	Gly	C
	Val	Ala	Glu	Gly	A
	Val	Ala	Glu	Gly	G

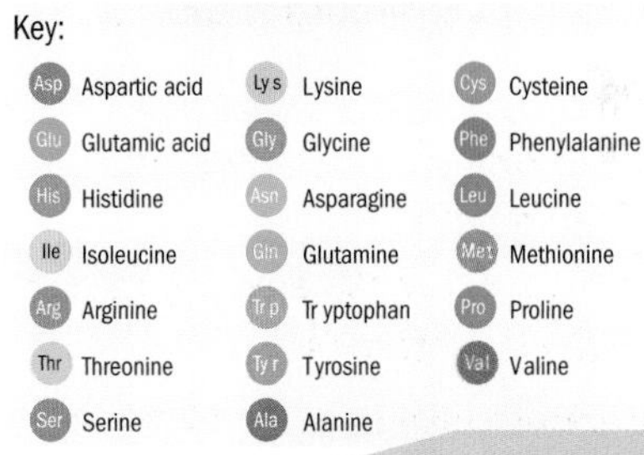

The standard DNA triplet code.

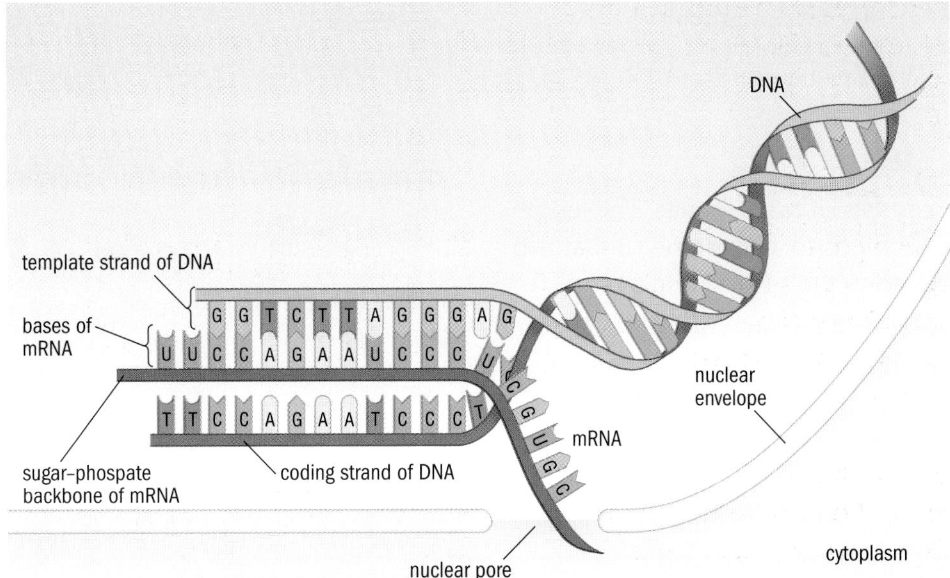

The transcription of a gene. The length of DNA (a gene) unwinds and unzips. Free activated RNA nucleotides pair up and bind temporarily, with hydrogen bonds, to their complementary bases on the unwound template strand of DNA. Bonds form between the sugar and phosphate groups of the RNA nucleotides. The single-stranded piece of RNA is a copy of the coding strand of the unwound DNA. The mRNA leaves the nucleus through a pore in the nuclear envelope.

Activity 18.1D

Discuss with a partner why the genetic code would not work if the DNA bases were read in pairs instead of in triplets.

Amino acid activation

Before translation can happen, specific amino acids are attached to their corresponding transfer RNA molecules. Each reaction is catalysed by a specific aminoacyl synthetase enzyme.

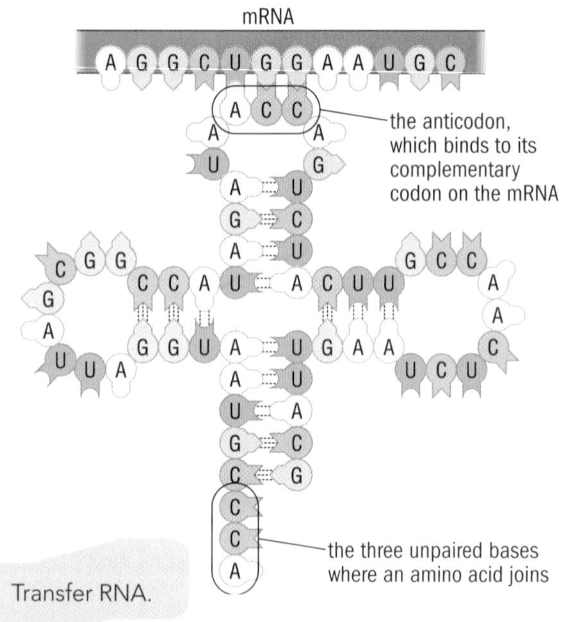

Transfer RNA.

Translation

This happens at ribosomes.

- Each ribosome consists of two subunits. The length of messenger RNA binds to a ribosome so that two codons are attached to the small subunit.
- The first exposed mRNA codon is always AUG. A tRNA molecule, with a corresponding **anticodon** – UAC, forms hydrogen bonds with this codon. Energy from ATP and an enzyme as catalyst are needed for this to happen.
- A second tRNA molecule brings a different amino acid. Remember that the three unpaired bases at the end of the tRNA determine which amino acid joins it. As described above, the enzyme aminoacyl transferase catalyses the reaction. The anticodon on the tRNA binds temporarily to the second codon on the small subunit.
- The anticodon on a tRNA will bond only to its complementary codon on the mRNA. For example, only a tRNA with the anticodon UAC will bind to the mRNA codon of AUG.
- Now, two amino acids are side by side and a peptide bond forms between them, forming a dipeptide. Another enzyme, present in the subunit of the ribosome, catalyses this reaction.
- The ribosome now moves along the mRNA so that codons 2 and 3 are exposed to the subunit.
- A third tRNA brings a third amino acid and a peptide bond forms between it and the dipeptide. The first tRNA molecule leaves and is free to collect and bring another amino acid, of the same type, to the ribosome.

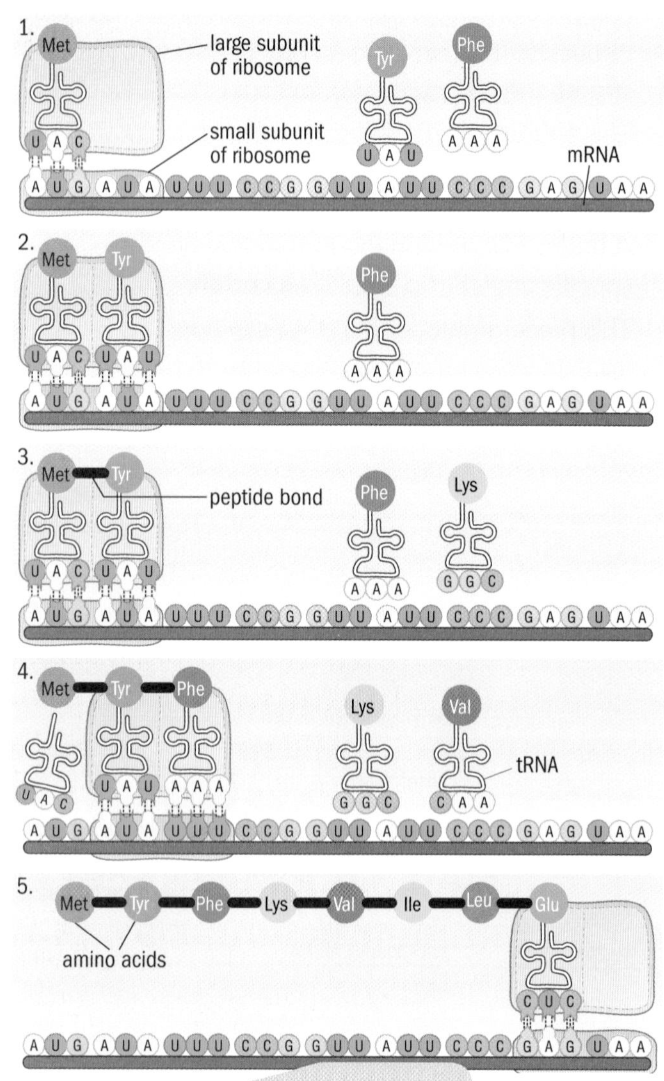

Translation. Stages of assembly of a protein at a ribosome.

Protein synthesis in prokaryotes

Prokaryote DNA is not enclosed in a nucleus. However, the gene is still copied into mRNA and this is read straight away, at the prokaryote's smaller ribosomes, with tRNA bringing the amino acids for assembly.

The genetic code for prokaryotes is almost completely identical to that for eukaryotes, which is very useful for genetic engineering. For example, putting the gene for human insulin into bacteria results in the bacterial ribosomes reading the code and making the same insulin protein.

Activity 18.1E

1 Explain how the instructions for building a protein can be used at ribosomes, although the DNA, which carries the genetic code, does not pass out of the nucleus.

2 Suggest how genes are responsible for synthesising non-protein molecules such as triglycerides.

Did you know?

The standard definition of a gene is that it is a length of DNA that codes for one protein. However, scientists have found that some genes can code for more than one protein, depending on how the RNA is spliced.

Assessment activity 18.1

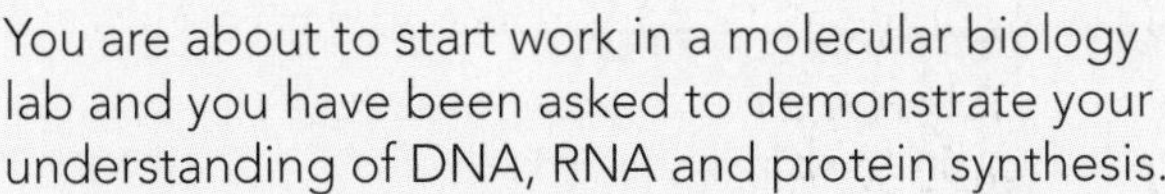

You are about to start work in a molecular biology lab and you have been asked to demonstrate your understanding of DNA, RNA and protein synthesis.

1 Make a table to compare and contrast the structure of DNA with that of RNA. P1

2 Draw labelled diagrams to show the structures of DNA and RNA. P1

3 Make a simple 3D model of a piece of DNA. P1

4 Explain how genetic information can be stored in a sequence of nitrogenous bases in DNA. You may like to make a 3D model or a 3D poster to help you. Or you may wish to give a PowerPoint presentation to the rest of your class. M1

5 Using 3D models that you have made, or a series of annotated diagrams, explain the steps involved in the biosynthesis of protein molecules. Include the roles of the three types of RNA. D1

Grading tips

When making a table for P1 make sure you compare the same feature across a row in the table. For example, when comparing DNA and RNA look at the number of polynucleotide strands, bases present, type of sugar present, how many types of DNA/RNA there are. You have been asked to compare their structures so do not include things like where they are found in cells.

For M1 you can build on the work you did for P1 and use the diagrams or models you made to help you *explain* how a length of DNA contains coded information.

To achieve D1 you need to *explain* the steps – transcription and translation – of protein synthesis.

PLTS

Independent enquirer, Effective participator, Creative thinker

You will develop skills as an independent enquirer as you use information sources to help you. You will also develop skills as an effective participator as you explain the concepts involved to others in your class, and develop creative thinking skills to make your presentation interesting and understandable.

18.2 Cell division in eukaryotic cells

In this section:    

Key terms

Chromosome – thread-like structure, made of DNA , found in nucleus of eukaryote cells.

Chromatid – when the DNA of a chromosome replicates, two identical sister chromatids are produced. These will eventually separate at mitosis or meiosis and will then be chromosomes.

Centromere – region where two chromatids are joined to each other and where they attach to the spindle during cell division.

Autosomes – chromosomes that are not involved in determining the sex of an individual.

Sex chromosomes – chromosomes responsible for determining the sex of an individual. They may also have genes for other characteristics.

Centrioles – structures that produce protein threads that form the spindle for mitosis or meiosis.

Cytokinesis – cleaving of cytoplasm, after mitosis or meiosis, to give two new cells.

Diploid – nuclei with two sets of chromosomes, arranged in pairs. For humans one set (*n*) of chromosomes is 23, so cell nuclei with 23 pairs (2*n*), each pair consisting of a maternal and paternal chromosome, are diploid.

Haploid – nucleus/cell or organism with one set of chromosomes.

Meiosis – type of nuclear division that produces cells with half the number of chromosomes as the parent cell.

Zygote – fertilised egg cell.

Fertilisation – joining of male and female gamete nuclei.

Allele – version of a gene.

Bivalent – pair of homologous chromosomes during meiosis.

Human chromosomes

Chromosomes are the physical bearers of genetic information. All the genetic information of an individual is called the genome. Prokaryotes have one single, circular chromosome plus some small circular molecules of DNA, called plasmids. However, humans, like all other eukaryotes, have the majority of their genome divided into a collection of linear chromosomes. Each chromosome is one molecule of DNA that contains many genes. A small part of our genome is in the mitochondria, as these have their own DNA.

Activity 18.2A

Use the Internet to find out how mitochondrial DNA is used for ancestry tracing. Share what you have found with the rest of the class.

Most of your DNA is in the nucleus of cells. The strands of DNA are wrapped around histone proteins and, when cells are not dividing, this DNA is in a rather shapeless network, called chromatin. The strands of chromatin are unduplicated chromosomes and can be seen with an electron microscope but not with a light microscope.

DNA replication

Before the cell divides, during a part of the interphase known as the S phase (see cell cycle, page 330), each DNA molecule replicates. It does this by a process called semi-conservative replication.

1 The double helix of each DNA molecule unwinds.

2 The hydrogen bonds between the nucleotide bases break, so the molecule 'unzips'.

3 This exposes the nucleotide bases.

4 Free DNA nucleotides in the nucleoplasm bond onto the exposed bases, using hydrogen bonds, and following complementary base-pairing (A with T, and C with G).

5 Covalent bonds form between the sugar of one nucleotide and the phosphate group of the adjacent nucleotide, forming the new backbones.

As a result, two new identical molecules of DNA are made. Each new molecule contains one conserved (old) strand and one new strand, which is why this replication is described as semi-conservative.

Mutations

Very rarely, a mistake is made when DNA is being replicated. Sometimes a nucleotide base is slightly altered and it pairs with the wrong exposed base on the DNA strand. Substances that can alter a base in

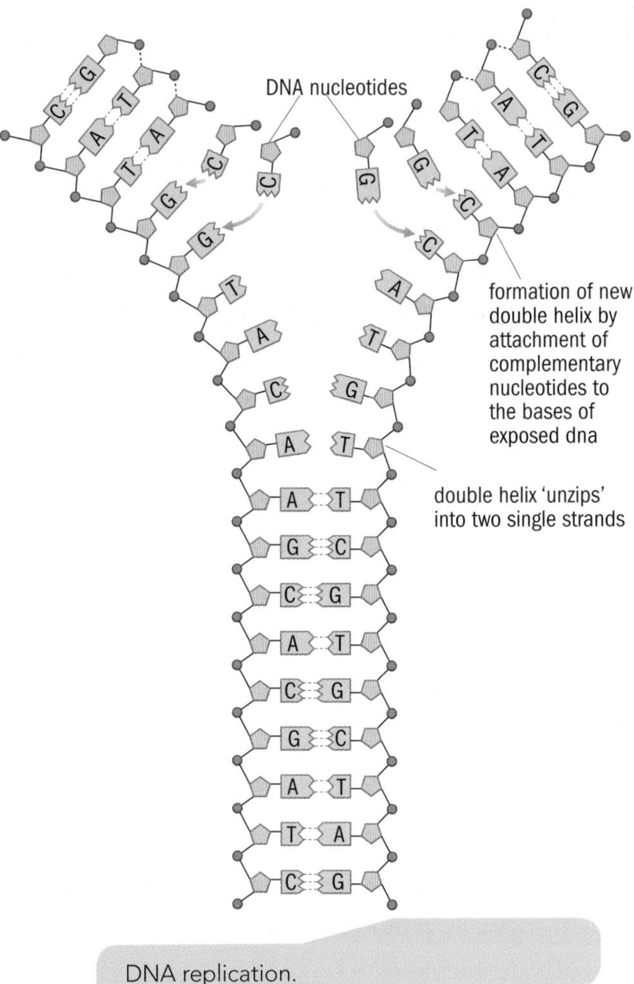

DNA replication.

this way are called mutagens. The resulting mutation is called a **substitution**. This results in the triplet code being altered and may lead to a different amino acid being placed in the protein, which could change the 3D shape of the protein and alter its ability to carry out its function. However, because the DNA code is degenerate, some base substitutions change the triplet but still code for the same amino acid. These are called silent mutations.

For example, TCT is the base triplet code for the amino acid serine. If the third base is substituted by A, C or G, the new base triplet will still code for serine, as TCT, TCA, TCC and TCG all code for serine.

However, if the second base is substituted by T, making the triplet TTT, then this codes for the amino acid phenylalanine.

Sometimes a base pair is lost or an extra one is added. Because the genetic code is in base triplets, adding or deleting a base will cause a **frameshift** where all the succeeding base triplets are altered.

Worked example

ATG CAG CAG CAG TTT TTA CGC AAT CCC codes for a polypeptide with the following amino acids: methionine, glycine, glycine, glycine, phenylalanine, leucine, arginine, asparagine, proline.

If the second T is lost from TTA, the code becomes ATG CAG CAG CAG TTT TAC GCA ATC CC_ which codes for methionine, glycine, glycine, glycine, phenylalanine, tyrosine, valine, threonine.

Mutations that happen during interphase just before mitosis will not be passed to offspring. They may contribute to ageing or to the development of a cancerous tumour.

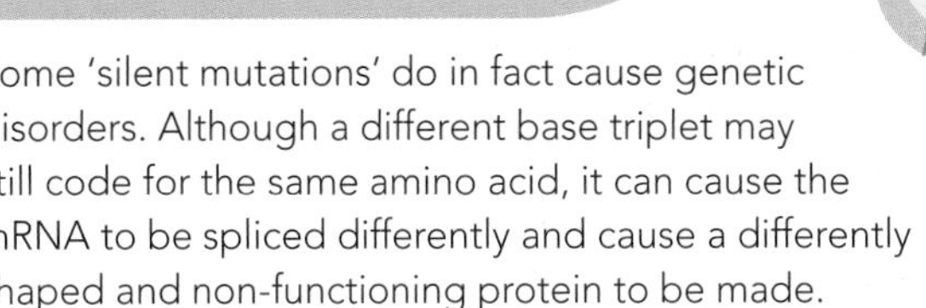

Did you know?

Some 'silent mutations' do in fact cause genetic disorders. Although a different base triplet may still code for the same amino acid, it can cause the mRNA to be spliced differently and cause a differently shaped and non-functioning protein to be made.

Chromosomes coil and condense

As the duplicated chromatin condenses and supercoils into duplicated chromosomes, they take up stains so that you can see them with a light microscope (as well as with an electron microscope). In this state, the chromosomes are short and sturdy enough to be moved around more easily, during the process of cell division. Each chromosome consists of two identical sister **chromatids**, joined at a region called the **centromere**. However, in this supercoiled state, they cannot be transcribed, so cannot express the proteins that they code for.

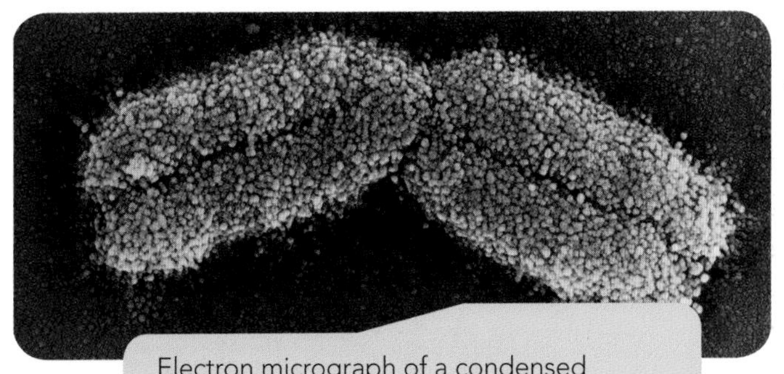

Electron micrograph of a condensed chromosome that consists of two identical sister chromatids. Mag: ×36000.

Activity 18.2B

1 Discuss with a partner the difference between chromosomes and chromatids.

2 Why do you think that each chromosome has to make a copy of itself before the cell divides?

3 Why is it important that chromosomes are not in their supercoiled state for very long?

4 Explain why mutations that happen when DNA copies itself just before mitosis are not passed to your offspring.

How many chromosomes do we have in each cell?

In every cell of your body that has a nucleus there are 46 chromosomes. Remember that you started life as one cell – a fertilised egg. The egg cell nucleus had 23 maternal (from the mother) chromosomes and the sperm cell nucleus had 23 paternal (from the father) chromosomes. The fertilised egg cell divided many times to produce all the cells in your body. In each nucleus you have 23 pairs of chromosomes, each pair consisting of one maternal and one paternal chromosome.

Autosomes

Twenty–two pairs of your 23 pairs of chromosomes are homologous pairs, meaning that the two in each pair are the same size and they contain the same genes; they are called **autosomes**. They have nothing to do with determining what sex we are.

Sex chromsomes

The other pair is the **sex chromosomes**. If you have a large X chromosome and a small Y chromosome, they are not fully homologous and you will be male. If you have two large X chromosomes you will be female. There are other genes on the X chromosome that code for other characteristics, such as blood clotting factors. These genes are described as sex-linked because they are on one of the sex chromosomes.

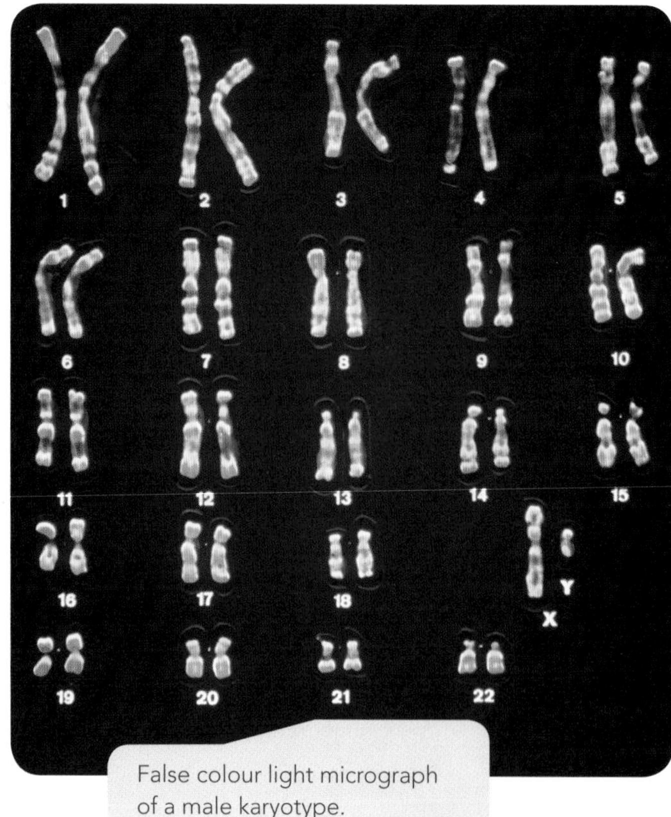

False colour light micrograph of a male karyotype.

Did you know?

Chimpanzees have 24 pairs of chromosomes. The members of pair 1 have fused with members of pair 2 in humans, so we have 23 pairs. We share 98.7% of our DNA with them. We also share 50% of our genes with bananas.

The cell cycle

Cells spend about 5–10% of their time dividing. The rest of the time they are in interphase. Interphase is further divided into G_1, S and G_2 phases.

After a cell has divided, the two new cells are in the G_1 growth phase. They then enter the S, synthesis, phase where all the DNA is replicated. After that there is another growth phase, the G_2 phase. G_1, S and G_2 are all part of the interphase. At the end of interphase, the chromosomes have all duplicated, each cell is full size, organelles have replicated and the cell is ready to divide either by mitoisis or by **meiosis**, as shown by M on the diagram.

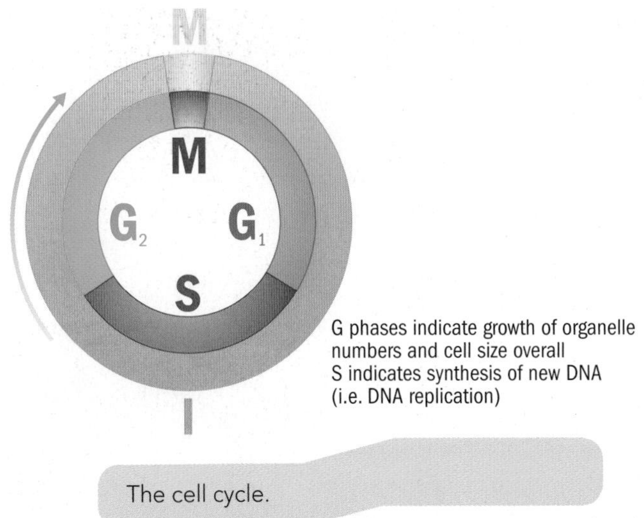

The cell cycle.

Activity 18.2C

How does the number of chromosomes in the nucleus of a cell in G_2 growth phase differ from when the cell is in G_1 growth phase?

Mitosis

Eukaryotic cells can divide by mitosis to produce two daughter cells that are genetically identical to the parent cell and to each other. This type of division is used by some eukaryotes, such as strawberry plants, for asexual reproduction. It is used by us for:

- growth
- repair
- replacement of red blood cells
- cloning of B lymphocytes and T lymphocytes during an immune response.

The nucleus of the cell divides into two genetically identical nuclei and this is followed by the cytoplasm cleaving (**cytokinesis**) to give two new cells.

Mitosis is one type of nuclear division. The whole process of mitosis is continuous but for convenience we describe it in four stages.

Prophase

The chromosomes that have already replicated during interphase now supercoil to become shorter and thicker. You can see, when looking at them using a light microscope, that each one consists of a pair of sister chromatids.

The nuclear envelope breaks down and disappears. The **centriole**, an organelle, divides into two and each daughter centriole is moved to opposite ends of the cell. Threads of protein arise from the centrioles to form a spindle. The shape of the spindle is rather like lines of longitude around an imaginary globe. We call the ends of the spindle the poles. We call the middle part the equator.

Unit 11: See page 199 for more information on the centriole.

Metaphase

The chromosomes are moved to the equator of the spindle. Each chromosome, consisting of two sister chromatids, attaches to the spindle by its centromere.

Anaphase

The centromeres of the chromosomes split and the spindle threads begin to contract. The centromeres are still attached to the spindle threads so, as the two parts of the split centromeres are pulled apart, the sister chromatids are separated. They assume a V shape as they are pulled, centromere first, further away from each other towards the poles.

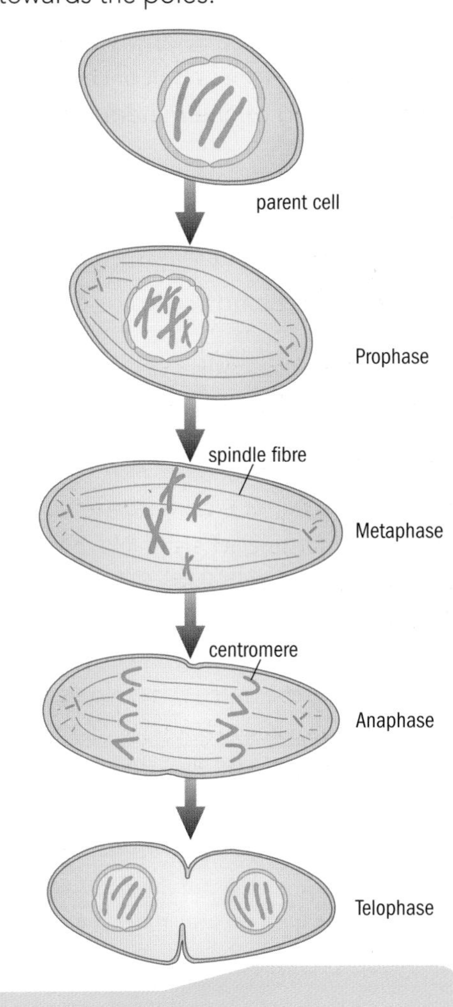

The main events in mitosis.

Telophase

The separated sister chromatids are now called chromosomes. As they reach the poles, a new nuclear envelope forms around each group of chromosomes. The spindle breaks down and disappears. The chromosomes uncoil and you can no longer see them under the microscope.

Cytokinesis

The whole cell now splits into two cells. Each new cell contains the full number of chromosomes, in our case 23 pairs. Cells with the full number of chromosomes are described as **diploid**.

When mitosis gets out of control

Normally cells divide only a certain number of times. This number is around 50 and is called the Hayflick limit. Cells then undergo programmed cell death or apoptosis. There are genes that control and regulate mitosis and apoptosis. If those genes mutate (change) and a cell fails to undergo apoptosis, then the cell keeps on dividing to form a ball of cells, which becomes a tumour. The tumour develops its own blood supply and some cells from it may enter the lymphatic system and be carried to other parts of the body, where they set up secondary tumours. A tumour that can spread in this way is described as malignant, or cancerous. The spreading process is called metastasis.

Did you know?

Prokaryotes (bacteria) do not carry out mitosis. They divide by binary fission and their DNA replicates before this happens. Akaryotes (viruses) do not divide at all but insert their nucleic acids into the host cell's genome and cause the host cell to make lots of copies of the virus.

Activity 18.2D

1 A human skin cell contains 46 chromosomes. How many chromosomes are there in (a) a human brain cell; (b) a liver cell; (c) a mature red blood cell?

2 How many chromatids are present in a human liver cell (a) in the G_1 phase; (b) in the G_2 phase; (c) in metaphase?

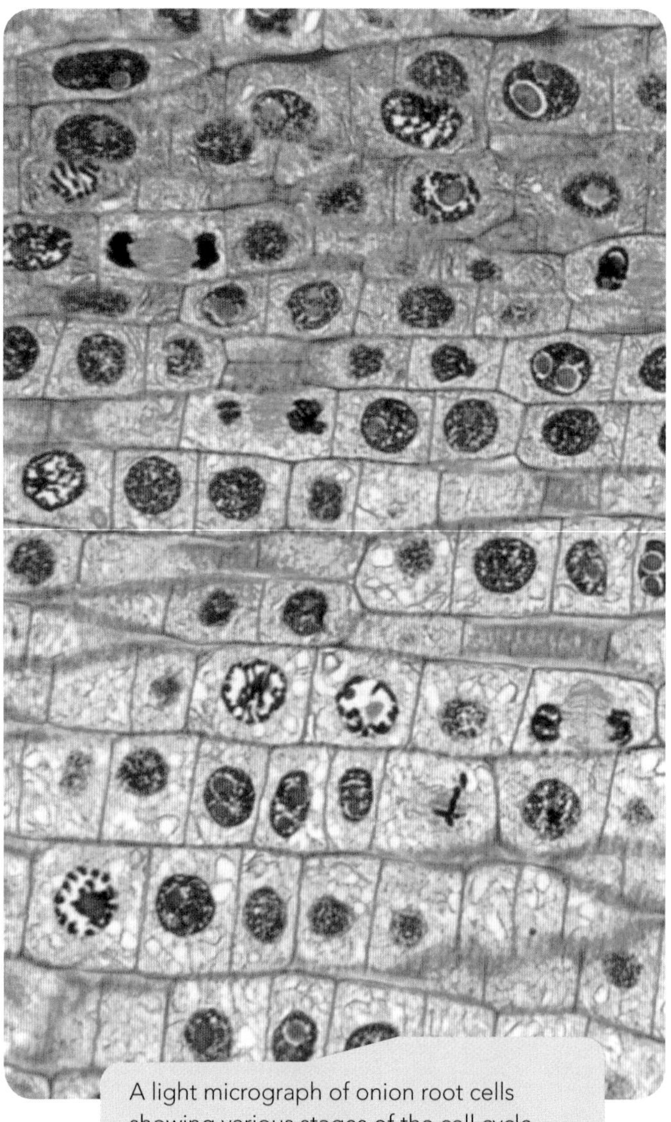

A light micrograph of onion root cells showing various stages of the cell cycle.

Meiosis

Humans reproduce by sexual reproduction. Each parent produces special reproductive cells called gametes, made by a type of nuclear division called meiosis. Our gamete nuclei have half the number of chromosomes as our other cells (one set as opposed to two sets) and are described as **haploid**. When gamete nuclei fuse together at **fertilisation**, a diploid **zygote** is formed. The zygote is genetically different from both parents. This is because it contains genetic information from two (usually unrelated) individuals and also because of certain processes that happen during meiosis.

Meiosis has two separate divisions referred to as meiosis 1 and meiosis 2. Each division has four stages: prophase, metaphase, anaphase and telophase.

During interphase, before prophase of meiosis 1, the DNA replicates and, as a result, each chromosome consists of two sister chromatids.

Meiosis 1

- **Prophase 1**

1 The chromosomes, each consisting of two sister chromatids, supercoil.

2 They shorten and thicken and can take up stains so that you can see them under a light microscope.

3 The chromosomes come together in their homologous pairs to form **bivalents**. A small part of the Y chromosome is homologous to a small part of the X chromosome so they also form a pair. Each pair consists of one maternal and one paternal chromosome. Each member of the pair has the same genes at the same place (locus) on it, but they may have different versions of the gene. These different versions are called **alleles**.

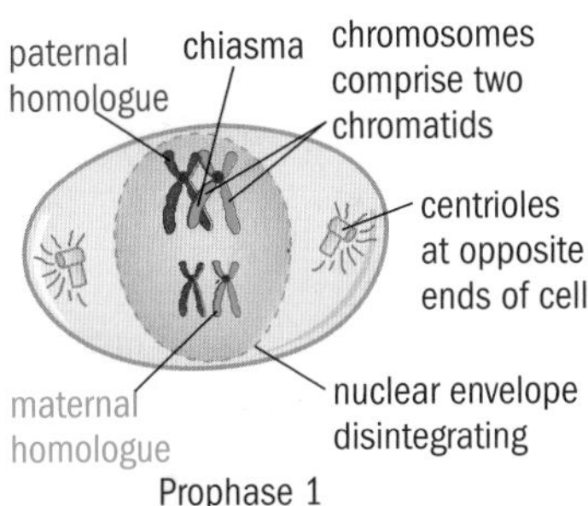

Prophase 1

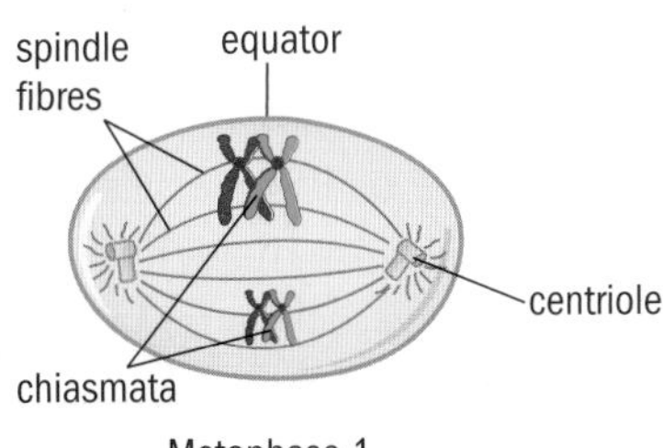

Metaphase 1

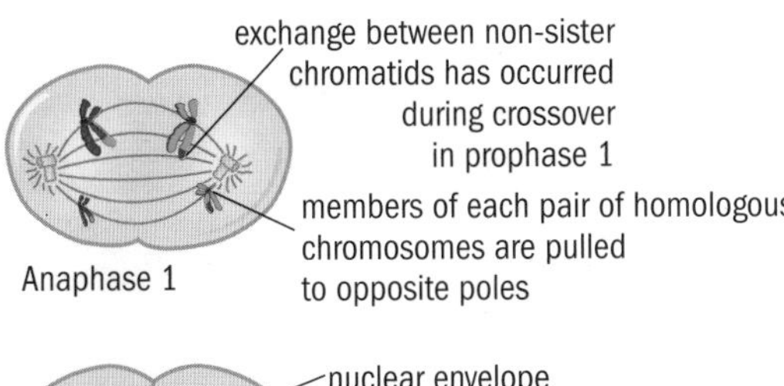

Anaphase 1

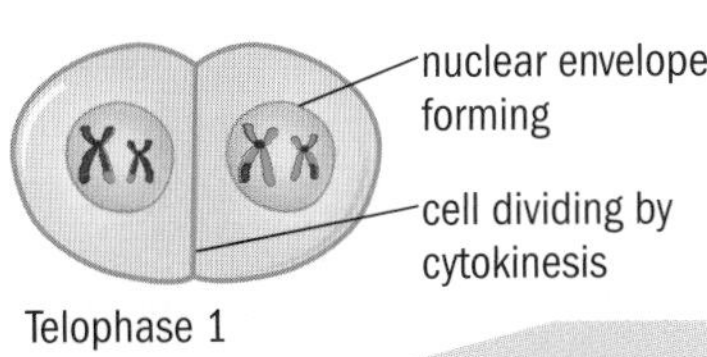

Telophase 1

Stages in meiosis 1.

4 The non-sister chromatids within a pair of chromosomes attach at points called **chiasmata** (singular: chiasma), wrap around each other and may swap sections of chromatids with one another. This process is called **crossing over**.

5 The nucleolus disappears and the nuclear envelope breaks down and disappears.

6 The centrioles divide and migrate to opposite ends of the cell. Protein threads from the centrioles form the spindle, just as they do in mitosis.

- **Metaphase 1**

1 The pairs of homologous chromosomes line up on the equator of the spindle. They join to the spindle by their centromeres. Their chiasmata are still present.

2 The members of each pair of chromosomes are arranged randomly and face opposite poles of the spindle. This allows them to **independently segregate** when they are pulled apart in anaphase 1.

- **Anaphase 1**

1 The members of each pair of chromosomes are pulled apart by the spindle fibres and move to opposite poles.

2 The chiasmata separate and the lengths of chromatid that have been crossed over remain with the non-sister chromatid to which they have become attached. This means the sister chromatids in each chromosome are no longer genetically identical to each other. Their centromeres are still intact.

- **Telophase 1**

1 Two new nuclear envelopes form, one around each set of chromosomes at each pole, and the cell divides.

2 Each new cell is haploid. It has half the number of chromosomes of the original parent cell but each chromosome consists of two chromatids.

3 There may be a brief interphase before the second meiotic division.

Meiosis 2

This division is in a plane at 90° to meiosis 1. Two cells are now dividing.

- **Prophase 2**

1 The nuclear envelopes break down.

2 The centrioles divide and migrate to opposite poles, forming new spindles.

- **Metaphase 2**

The chromosomes arrange randomly on the equator, attached to the spindle by their centromeres.

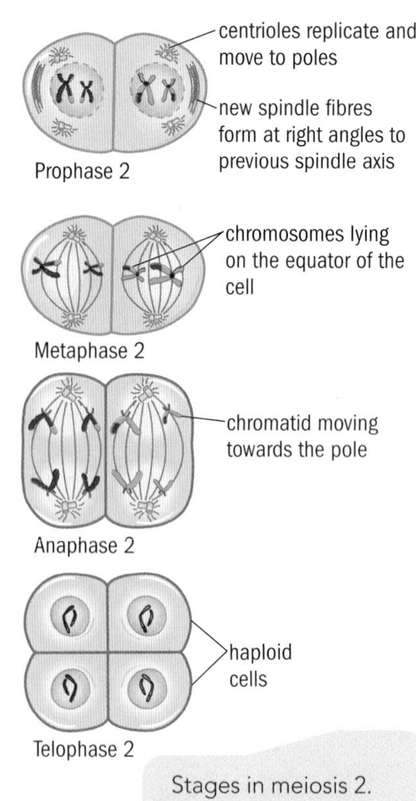

Stages in meiosis 2.

- **Anaphase 2**

The centromeres divide and the chromatids of each chromosome are pulled apart to opposite poles. This segregation of the chromatids is random, according to how they are arranged on the equator.

- **Telophase 2**

1 Nuclear envelopes reform around the four daughter haploid nuclei.

2 The cells divide.

During interphase before meiosis 1, when DNA replicates, it may undergo a mutation. This may also increase genetic variation, but remember that this type of mutation can happen before mitosis as well.

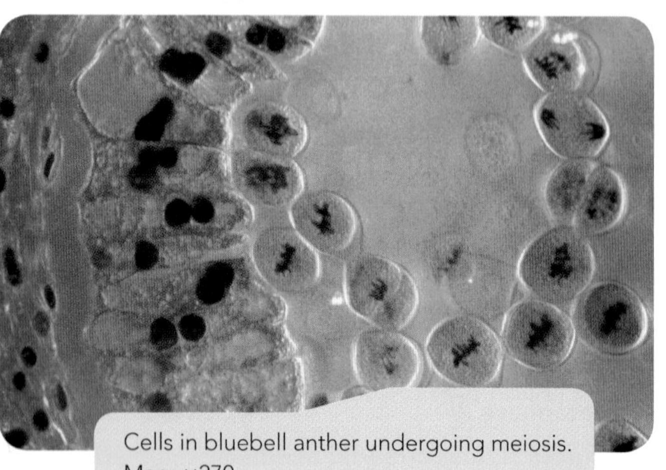

Cells in bluebell anther undergoing meiosis. Mag: ×270.

Chromosome mutations

Sometimes during meiosis, a pair of chromosomes may not segregate. If both chromatids for chromosome 21 went to one pole, then the resulting gamete would have an extra chromosome. It would have 24 instead of 23. If that gamete was fertilised by a normal gamete, the resulting zygote would have 47 chromosomes. It would have three copies of chromosome 21, instead of the normal two. This is called trisomy 21, also known as Down syndrome.

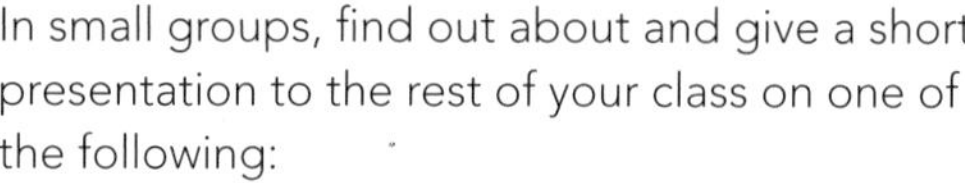

Activity 18.2E

In small groups, find out about and give a short presentation to the rest of your class on one of the following:

- Down syndrome
- Trisomy 13
- Trisomy 18
- Turner syndrome
- Klinefelter syndrome.

How is our sex determined?

Humans have 22 pairs of autosomes and one pair of sex chromosomes. Male sex chromosomes are X and Y; females have two X chromosomes.

Only half of the chromosomes are in the gamete nuclei. All eggs have one X chromosome but half the sperm have an X chromosome and half have a Y chromosome. If the egg released from the ovary is fertilised by a Y-bearing sperm the child will be male. If it is fertilised by an X-bearing sperm, the child will be female. Fertilisation is random and theoretically in a population equal numbers of boys and girls should be produced each year.

PLTS

Effective participator, Self-manager, Reflective learner

When working through Assessment activity 18.2, you will develop your skills as an effective participator and as a self-manager when carrying out the practical examination. You will develop your reflective learning skills as you apply what you know about the stages of cell division, identify these stages in a microscope slide and describe how the chromosomes are behaving.

Assessment activity 18.2

You are about to start work in a molecular biology lab and you have been asked to demonstrate your understanding of cell division.

Look at the micrograph on page 332 showing onion root tip cells in various stages of the cell cycle.

1 Identify and make labelled drawings of cells in the following stages:
   - a interphase
   - b prophase
   - c metaphase
   - d anaphase
   - e telophase. P2

2 Annotate each drawing to describe the behaviour of the chromosomes (describe what they are doing) at that stage. P2

3 If your teacher can supervise you, you may also make slides of onion root tips and observe cells in various stages of mitosis; you may also examine prepared slides of onion root tip squashes. Observe the slides and identify cells in various stages of mitosis. Make labelled drawings of these cells. P3

4 Make 3D models or posters to describe how chromosomes behave in mitosis. Work in groups, with each group responsible for making the models of one stage of the division. Use all the models to make a 3D poster or chart to show the whole process. M2

5 Look at the photomicrograph on page 334. Identify cells in some stages of meiosis. Can you tell easily whether they are in meiosis 1 or 2? P2

6 Observe some prepared microscope slides of either locust testes or lily anthers. Identify cells in various stages of meiosis. Make labelled drawings of the cells and annotate them to describe what the chromosomes are doing. P3

7 Make 3D models or posters to describe how chromosomes behave in meiosis. Work in groups, with each group responsible for making the models of one stage of each division. Use all the models to make a 3D poster or chart to show the whole process. M2

8 Explain how each of the following contributes to genetic variation among offspring:
   - a prophase 1 of meiosis
   - b metaphase 1 of meiosis
   - c metaphase 2 of meiosis
   - d fertilisation. P4

Grading tips

For P2 in questions 1 and 2 make large clear diagrams showing each stage of mitosis. Label the structures and add annotations. These are notes on the diagrams, usually associated with the labels, indicating what the chromosomes are doing at each stage.

To achieve P3 in question 3 you can make your own slides of onion root tips and see how many of the stages of mitosis you can see in the cells. Make sure you follow the instructions carefully as cells are only dividing in a small region just behind the very end of the root tip. If you can't make your own slides or if they are not very successful, examine prepared slides of onion root tip squashes and identify cells in interphase, prophase, metaphase and anaphase. Make a labelled drawing of each one.

Pipe cleaners make good models for chromosomes. You can stick them onto paper to make 3D posters for M2 in question 4. Don't try to have 23 pairs, just two pairs will do to show what happens. For question 7 use different sized pipe cleaners to make two matching pairs of chromosomes in your first cell. Colour them to show how crossing over has occurred. Annotate the 3D drawings to explain what is happening at each stage.

For question 5, it is not always easy to be able to tell if cells dividing by meiosis are in meiosis 1 or 2. As long as you can identify prophase, metaphase, anaphase and telophase and have tried to see if the cells are in meiosis 1 or 2, you will achieve P2.

For P3 in question 6 make labelled drawings of cells in as many stages of meiosis as you can identify. Annotate the diagrams to show what the chromosomes are doing in each stage.

Explain what happens at each stage to produce genetic variation to achieve P4. For example in prophase 1 crossing over occurs. Say how this causes the chromosomes in the daughter cells to be different from those in the parent cell. Hints for the others: think about independent assortment of chromosomes and of chromatids. Think about how fertilisation is random – which sperm fertilises the egg?

18.3 Mendelian genetics

In this section:    

Key terms

Genotype – refers to alleles present in a cell or organism for a particular characteristic.

Phenotype – observable characteristics of an organism.

Dominant – characteristic that is observed in the phenotype even if only one allele for it is present in the genotype.

Recessive – characteristic that is observed in the phenotype only if two alleles for it are present in the genotype, or if there is only one allele for it and no dominant allele is present.

Homozygous – having two identical alleles of a particular gene; true breeding.

Heterozygous – having two different alleles of a particular gene.

Monohybrid inheritance – inheritance pattern for one characteristic.

Dihybrid inheritance – inheritance of two characteristics.

Wild type – describes characteristics of organisms, such as *Drosophila*, as they occur in nature, as opposed to mutant varieties.

Discontinuous variation – variation where there are distinct categories within the population, for example free ear lobes and attached ear lobes.

Continuous variation – variation where there is a range within the population, for example height and skin colour in humans.

Huntingtin – protein needed for the nervous system to function correctly.

Co-dominance – inheritance pattern where, in an individual, both alleles of a gene contribute to the phenotype.

Sickle cell anaemia – genetic disorder where the haemoglobin is abnormal and the red blood cells can become sickled in shape.

Linkage/linked – genes on the same chromosome are linked.

Sex linkage – the gene for a characteristic is on one of the sex chromosomes, usually on the larger X chromosome.

Non-disjunction – failure of a pair of chromosomes or chromatids to separate at cell division.

Trisomy – where there are three copies of a chromosome in a cell, instead of the usual two copies.

Who was Gregor Mendel?

Gregor Mendel.

Gregor Mendel was born in Heinzendorf, which is now in the Czech Republic, in 1822. He was brought up on a farm and during that time he gardened and kept bees. He became a monk and also studied science at the University of Vienna. For eight years, beginning in 1856, he studied variation in plants, using the monastery garden in Brünn to cultivate several generations of pea plants. He published a paper about his findings in 1866. However, few people read it and no one appeared to understand it.

What did he discover?

Before Mendel, most biologists thought that characteristics were inherited via a blending mechanism, giving characteristics in the offspring that were intermediate between the parental types. His research with peas showed otherwise but it was criticised at the time and not accepted. By 1900, other biologists were trying to establish a theory of discontinuous inheritance, rather than blending, and some of them independently duplicated Mendel's work and rediscovered his writings and laws.

They acknowledged his work and other biologists began to establish genetics as a science.

Mendel worked with the edible pea, *Pisum sativum*, as he could easily obtain seeds of lots of varieties.

Pisum sativum – pea plants.

- He established that each variety he used was true breeding, which means that when allowed to self-fertilise, they always produced offspring the same as the parents.
- He selected varieties that had different characteristics, such as round seeds, wrinkled seeds, yellow seeds, green seeds, tall stems and short stems.
- He crossed different varieties, taking pollen (male gametes) from one type and applying it to the stigmas (female parts) of another type.
- He collected the seeds and grew them, producing the F_1 or first filial generation. He looked at their characteristics and found that they all had the characteristics of just one of the parents. For example, when peas with long stems were crossed with peas having short stems all the F_1 generation had long stems. He described long stems as **dominant** and short stems as **recessive**.
- He allowed the F_1 plants to self-fertilise and collected and grew their seeds, producing the F_2 generation.
- He observed the F_2 plants and he did something very unusual for the time – he counted the number of plants of each type. In the F_2 generation there were 787 long-stemmed plants and 277 short-stemmed plants. This gave a ratio of very nearly three long-stemmed (the dominant characteristic) plants for every one short-stemmed (the recessive characteristic) plants.

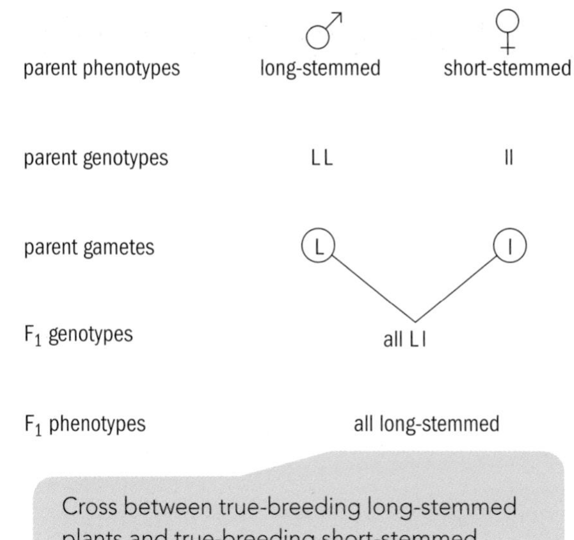

Cross between true-breeding long-stemmed plants and true-breeding short-stemmed plants to produce the F_1 generation.

From these results he postulated that each pollen grain and each egg carry only one determinant (we now call these genes) for each characteristic. He said plants were true breeding if all their gametes carried the same determinant. If a seed inherited one of each determinant, the dominant characteristic would be seen, but not the recessive characteristic.

However, these F_1 plants have both determinants so half their gametes will have a determinant for long stems and half will have a determinant for short stems.

The diagram shows what we expect or predict to happen if random fertilisation occurs – any pollen grain can combine with any egg.

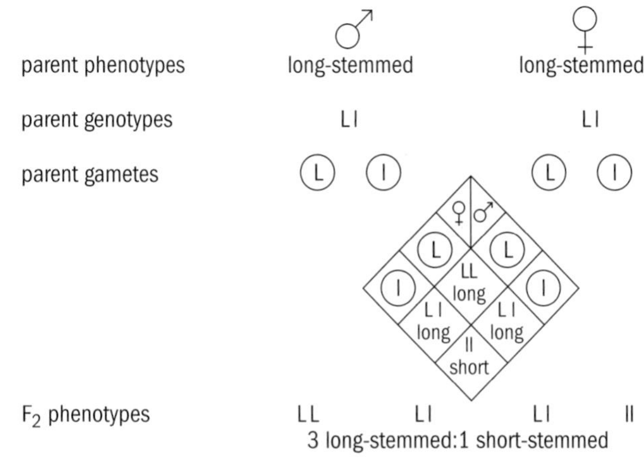

Cross between members of the F_1 generation to produce the F_2 generation, showing the 3:1 ratio of long-stemmed and short-stemmed plants.

Activity 18.3A

Draw a flow diagram showing Mendel's experiment crossing long-stemmed and short-stemmed pea plants.

This ties in nicely with what Mendel observed.

- We say that the F_1 offspring are all **heterozygous**, as they have one of each type of allele in each of their cells.
- Therefore half their gametes have one type of allele and half have the other type. Because they can produce different gametes they would not breed true.
- They all look the same as the tall parent so their **phenotype** is the same.
- However they have a different genetic make-up from the parent (long-stemmed) plant, so they have a different **genotype.**
- This type of inheritance pattern is now described as **Mendelian dominant**.
- It is also an example of **monohybrid inheritance** as it involves studying the inheritance of just one characteristic – in this case, stem length.

Interpreting Mendel's results using our knowledge of genes, chromosomes and cell division

Remember that in each cell we, and peas, have pairs of homologous chromosomes, one of each pair derived from our mother and the other in each pair derived from our father. These matching chromosomes have genes for the same characteristics on the same place along their length.

Slightly different versions of genes, produced by mutation, are called alleles. In this case we say the gene is for stem length and the alleles of that gene may code for either long stems or short stems.

A pea plant that is true breeding for long stems has two alleles for long. It is **homozygous** for the dominant characteristic. If it is crossed with a pea plant that is homozygous for the recessive characteristic (short stems), then we can predict what will happen. As the gametes are produced by meiosis, they will have only half the number of chromosomes and so will have only one allele for each gene. So all the gametes from the long-stemmed plants will have an allele for long stem; all the gametes from the short-stemmed plants will have an allele for short stem.

All the offspring will have one of each allele and will have the phenotype of long stems. However, their genotype is not the same as that of the tall-stem parent. They will not breed true. They are heterozygous.

Dihybrid inheritance

Mendel also did experiments where he crossed peas with two different characteristics. He was investigating **dihybrid inheritance**. He was lucky as he happened to choose characteristics that had genes on different pairs of chromosomes so they could undergo independent assortment at meiosis.

He crossed pea plants that had round and yellow seeds with pea plants that had wrinkled and green seeds. All the F_1 generation had round and yellow seeds. These two characteristics were dominant.

He allowed members of the F_1 generation to self-fertilise, collected and grew the seeds and observed the phenotype (observable characteristics) of the resulting plants in the F_2 generation. He found that the plants in the F_2 generation showed the following characteristics: round yellow seeds, round green seeds, wrinkled yellow seeds and wrinkled green seeds in the ratio of 9:3:3:1.

We can draw genetic diagrams that explain Mendel's results (page 339).

Mendel's laws of inheritance

Mendel deduced two 'laws' from his observations. He used terms that are different from the ones we use today as he did not know anything about meiosis or genes and alleles.

His laws, in modern language, are:

1. **The law of segregation**: the cells of the diploid parent organisms have two alleles for each characteristic but the haploid gametes have only one allele for each characteristic.
2. **The law of independent assortment**: any one of the male gametes can combine randomly with any one of the female gametes. Thus, when we make predictions using genetic diagrams, we show what would happen if each type of male gamete combined with each type of female gamete.

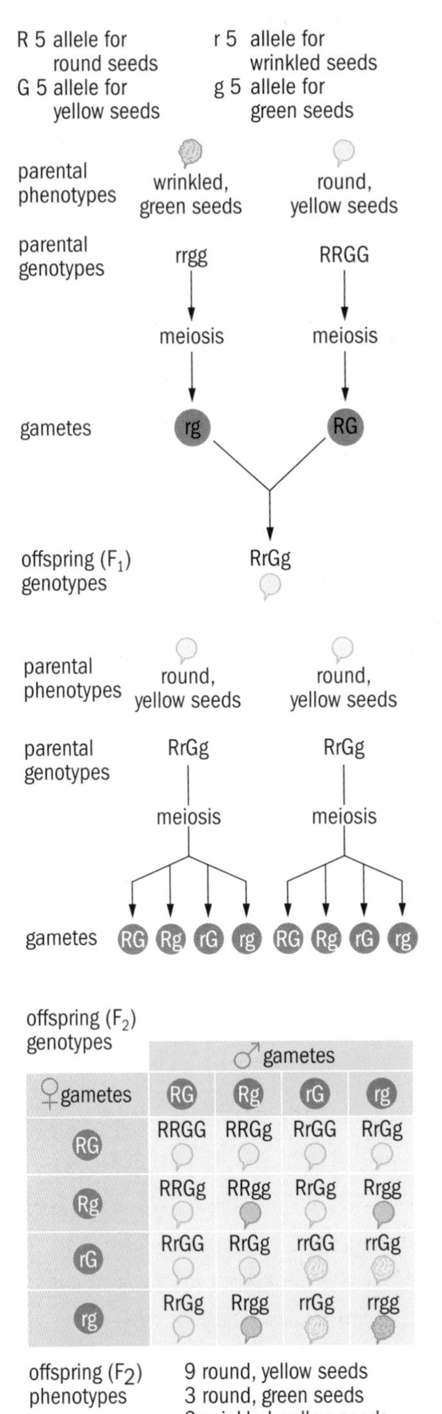

A genetic explanation for Mendel's dihybrid cross between peas with yellow round seeds and peas with green wrinkled seeds.

Drawing genetic diagrams

The diagrams above are called genetic diagrams. They help us explain the results we see when we carry out genetic crosses. Here is a standard way of doing this:

1 Show the parental phenotypes (what each parent looks like).

2 Then show the parental genotypes (the alleles present for a particular characteristic).

3 Then show the gametes that could be produced by these parents. Put circles around the gametes.

4 Use a Punnet square to show the possible combinations of gametes, with the resulting genotypes of the offspring.

5 Indicate the phenotypes of each offspring.

6 Work out the phenotype ratios.

Activity 18.3B

Using the Punnet square above, write down:

a all the genotypes that can produce the phenotype round and yellow seeds

b all the genotypes that can produce the phenotype round and green seeds

c all the genotypes that can produce the phenotype wrinkled and yellow seeds

d all the genotypes that can produce the phenotype wrinkled and green seeds.

In experiments, observed phenotype ratios in offspring do not always match the expected ratios

Mendel was lucky in using plants for his experiments because they produce lots of seeds. The more offspring produced, the nearer the observed phenotype ratios are likely to be to the expected ratio. Sometimes scientists predict a certain ratio of phenotypes among the offspring (the expected ratio) of a cross but the observed ratio does not match. They can do a statistical test, the chi-squared (pronounced *ki-squared*) test, to see if the difference between observed and expected ratios is just due to chance – for some reason some gametes did not get fertilised or the embryos died. If the test says that the difference between observed and expected is not significant and just due to chance, then scientists accept that their predictions were correct and the inheritance pattern they were using is also correct. However, if the statistical test result indicates that the difference between observed and expected results is not due to chance, but it is a significant difference, then the scientists need to think again about the pattern of inheritance for the characteristic they are investigating.

Worked example

In mice, yellow coat is dominant to grey coat. If two heterozygous yellow-coated mice are crossed, we would expect three-quarters of the offspring to have yellow coats and one-quarter to have grey coats. This is a 3:1 ratio as predicted using Mendel's monohybrid inheritance pattern. However, in this case, there are always about two-thirds yellow-coated offspring and one-third with grey coats (a 2:1 ratio). Scientists now know that, for these mice, having two yellow (Y) alleles is a lethal combination and these embryos do not survive.

Activity 18.3C

Drosophila melanogaster, the fruit fly, has been used by geneticists for just over 100 years. The picture below shows a **wild type** fruit fly.

Wild type fruit fly *Drosophila melanogaster.*

These flies are easy to breed in a laboratory. They lay eggs which hatch into larvae that feed on food containing fruit and yeast. The larvae pupate and then hatch into the next generation of adults. At 25 °C, this life cycle takes about 12 days.

There are several mutant varieties. One type has rather shrivelled wings, called vestigial wings. Another type has long wings.

A geneticist crossed true-breeding (homozygous) female long-winged flies with true-breeding (homozygous) male vestigial-winged flies. All the adults of the F_1 generation had long wings.

He allowed some of the F_1 to interbreed and observed their offspring. In the F_2 generation he counted 297 with long wings and 103 with vestigial wings.

a Calculate the ratio of long-winged to vestigial-winged fruit flies in the F_2 generation.

b Which characteristic is recessive – long wings or vestigial wings?

c Draw genetic diagrams, showing the results of the crosses described here. Use the following symbols:

allele for long wings = L

allele for vestigial wings = l

genotype for homozygous long-winged type = LL

genotype for homozygous vestigial-winged type = ll.

Genes and environment

We are all products of both our genes and our environment. Environment can include the food we eat, the amount of exercise we do, the way we are brought up and any diseases or injuries we may suffer.

Consider this: you may be born with genes that can code for certain proteins, including enzymes, which give you a genetic potential to be 6 feet tall. However, if you are neglected as a child and not fed properly, you will not reach that maximum potential height. But if you are fed properly and take exercise you can reach that maximum height. However, if a child had a disability that stopped them from walking and running about, they would not grow to their full potential height. Similarly, a child may be born with the genetic potential to have good intelligence and achieve well at school and beyond. However, if their parents do not talk to them or allow them to play, or do not feed them properly, or if the child does not go to school, the child won't achieve their potential intelligence level. Children who are fed properly, get enough sleep and are given opportunities for stimulating play and whose parents read to them and talk to them can achieve their full intelligence potential.

Eight-year-old suffering from undernutrition and having stunted growth as a result.

Many of us have genes that predispose us to certain illnesses, such as cancer and heart disease. These diseases will usually strike us in later life. They may also need a trigger from the environment. For example, those of us with genes that predispose us to cancer or heart disease will be more likely to develop these diseases if we:

- smoke tobacco
- drink more alcohol than the recommended amount

- eat a high fat diet
- don't eat enough fruit and vegetables
- don't make enough vitamin D
- are overweight
- don't take much exercise
- expose our skin to too much sunburn.

Some people follow a very healthy lifestyle and still become ill with cancer or heart disease. In these cases, they probably have a very high genetic predisposition for these diseases. Some people can smoke tobacco for years and not get lung cancer or heart disease, so they probably have a very low genetic predisposition to these diseases.

Continuous variation

This describes differences between phenotypes in a population where there is a range of variation rather than a few distinct categories. For example, in humans, height, skin colour and intelligence are all examples of **continuous variation**.

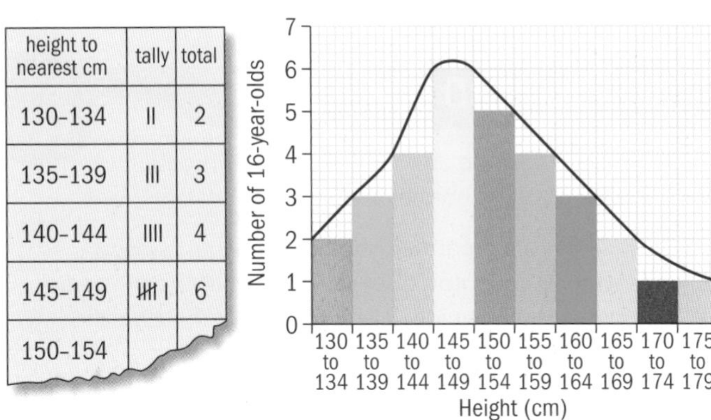

height to nearest cm	tally	total
130-134	II	2
135-139	III	3
140-144	IIII	4
145-149	卌 I	6
150-154		

Height of a class of 16-year-olds.

There are many genes involved in producing this type of variation and the environment can exert quite a large influence on the expression of these genes.

Many plants that humans grow for food have continuous variation. The length of maize (corn) cobs is determined by several genes, each of which has two alleles. So if the alleles of three genes A/a, B/b and C/c each contribute 2 cm/1 cm to the length of a corn cob, plants with the genotype AABBCC will have corn cobs of 12 cm length and plants with the genotype aabbcc will have corn cobs of 6 cm length. However, if any of the plants are grown without enough light, water, minerals or in an unsuitable temperature, then their corn cobs may not reach their full potential length.

Maize corn cobs.

Did you know?

Scientists have discussed something called the nature–nurture debate, to try to establish whether certain traits and behaviours (such as criminality) in humans are determined by genes or environment. However, as they have found out more about how genes work, it seems that both nature and nurture work together to determine many traits and behaviours.

Discontinuous variation

Characteristics that are controlled by just one gene produce phenotypes that are quite distinct. For example, the blood transfusion service categorises donors and recipients according to their blood group. Each person belongs in a distinct category, either group A, group B, group AB or group O.

About half the population can't smell freesia flowers. Their genetic make-up means they don't have receptors in their noses to detect freesia smell molecules. The rest of the population can smell them, so there are two distinct categories of people. This is another example of **discontinuous variation**.

Dominant and recessive genetic disorders

Some genetic disorders have a dominant inheritance pattern but the majority have a recessive inheritance pattern.

Huntington disease – a genetic disorder with a dominant inheritance pattern

This is an autosomal dominant disorder. That means that the mutation is of a gene on one of the autosomes; in this case it is on chromosome 4 (see page 330). It affects on average about 1 in 15 000 individuals of Western European origin and fewer people in other parts of the world.

The gene in question codes for a protein, now called **huntingtin**, that plays an important role in the functioning of nerve cells, but its exact function is still unknown. One end of this gene has a base triplet, CAG, which is repeated from 7 to 35 times. As a result of this repeat, the huntingtin protein has between 7 and 35 glutamine (amino acid) residues at one end.

The mutation that occurs is called an expanding triple nucleotide repeat. This means that the number of CAG triplets increases when the DNA replicates before meiosis. If the number increases beyond 36, then the person will suffer symptoms of the disease. This is because the abnormal form of the protein causes gradual damage to specific areas of the brain.

The greater the number of repeats above 36, then the earlier the symptoms will appear, but usually they appear between the ages of 35 and 44 years.

The symptoms include:

- involuntary (uncontrollable) muscle movements (sometimes called chorea, which is a Greek word that means dancing)
- progressive mental deterioration and loss of mental faculties as brain cells are lost
- personality changes and behaviour problems
- memory loss
- weight loss due to being unable to coordinate eating movements.

Because the abnormal protein actually interferes with nerve function, sufferers need only have one abnormal allele of the gene. This means they produce some abnormal huntingtin protein (as well as some normal protein) and the abnormal protein interferes with nerve function. Sufferers are heterozygous. This means that only one parent needs to have the disorder and any child has a 50% chance of inheriting the disorder and, because the symptoms don't usually appear until the mid- to late-30s, people may well have had children before they realise they have the disorder.

There is a genetic test available for people who have a parent with the disorder to find out if they themselves have it. However, there is no cure and only limited treatment to reduce the severity of some of the symptoms. People are given information from a genetic counsellor, so they can make an informed decision as to whether to be tested.

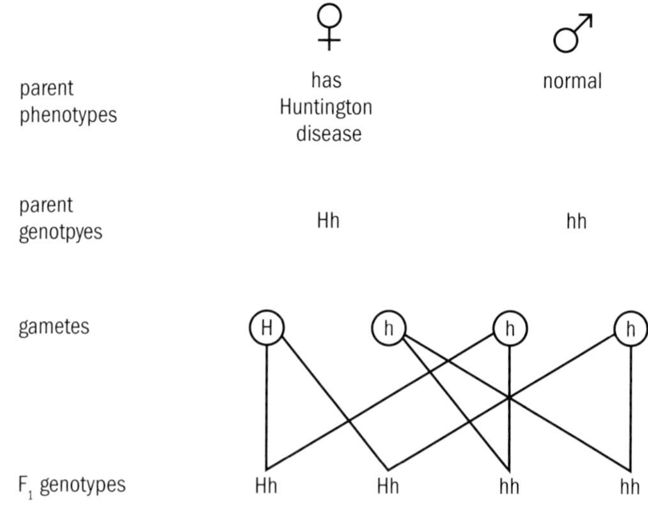

Genetic diagram showing how the dominant condition of Huntington disease is inherited.

Activity 18.3D

1 Explain how the abnormal huntingtin protein will differ from the normal one.

2 Discuss with a partner whether you would want a test for Huntington disease. Think of all the pros and cons. Share your ideas with the rest of the class in a discussion.

3 If you had family members with this disease, would you have a fetus tested? Think of all the ethical issues that such testing raises. Share your ideas with the rest of the class.

Cystic fibrosis – a genetic disorder with a recessive inheritance pattern

This is an autosomal recessive disorder. The gene in question (CFTR, which stands for cystic fibrosis transmembrane regulator gene) is on an autosome, in this case chromosome 7. The gene codes for chloride channel proteins in the plasma membranes of epithelial cells lining the respiratory tract, the gut, the reproductive tract and the sweat glands.

Individuals who are heterozygous and have one normal allele (of the gene) and one abnormal allele can still produce enough normal chloride ion channels so they have no symptoms. However, they are carriers, so if they have children with another carrier there is a 1 in 4 chance, at each pregnancy, of having a child with the disorder.

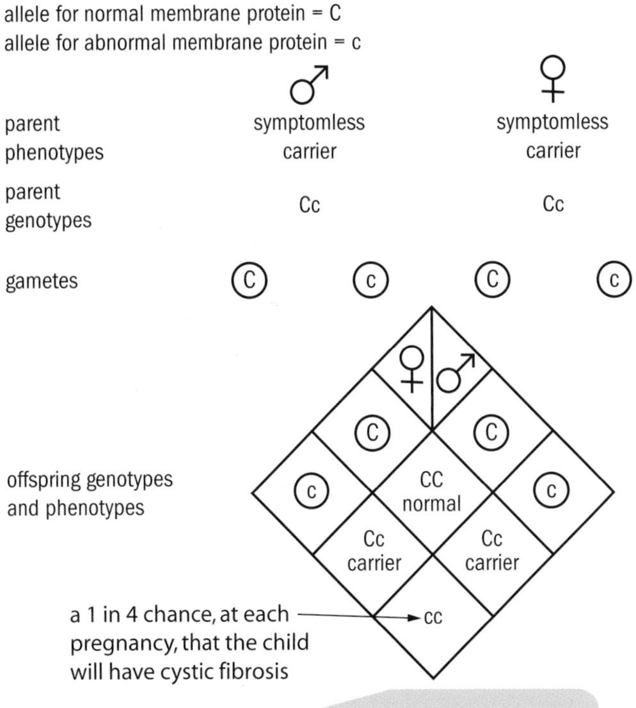

Genetic diagram to show the inheritance pattern for cystic fibrosis. Both parents are symptomless carriers.

Sufferers inherit two abnormal alleles of the gene and cannot make chloride ion channel proteins that function properly. They are homozygous. In their respiratory, digestive and reproductive tracts the mucus produced is thick and sticky.

In the respiratory tract the cilia do not get properly hydrated and cannot beat to remove the thick mucus, so bacteria or fungi trapped in it can multiply and cause infections. Patients take antibiotics to prevent these infections. They also have to beat their chests and cough up the mucus.

The patient's sweat is more salty than usual.

In the gut the thicker mucus stops pancreatic enzymes from reaching the small intestines and reduces the absorption of digested food so a baby suffering from cystic fibrosis fails to thrive due to malnutrition. Patients have to swallow enzyme capsules with their meals.

Some males with cystic fibrosis may be infertile as their sperm ducts are blocked with thick mucus. Some females with cystic fibrosis are infertile as the mucus in their cervix is thicker than normal and prevents sperm from entering the uterus.

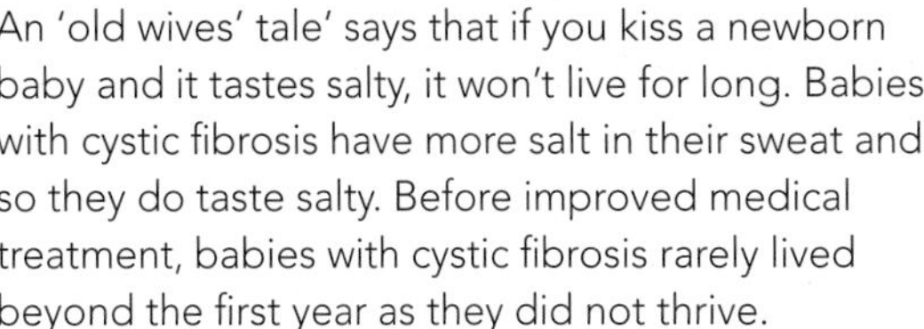

Did you know?

An 'old wives' tale' says that if you kiss a newborn baby and it tastes salty, it won't live for long. Babies with cystic fibrosis have more salt in their sweat and so they do taste salty. Before improved medical treatment, babies with cystic fibrosis rarely lived beyond the first year as they did not thrive.

The most common mutation that causes cystic fibrosis is a loss (deletion) of three nucleotide base pairs (a codon) from the CFTR gene, which leads to a loss of one amino acid (phenylalanine) from the normal 1480 amino acids in the protein.

There is a genetic test so that people can find out if they carry this mutation. However, there are about 600 other mutations to the CFTR gene that cause cystic fibrosis. Some, but not all, of these can be tested for so the test is about 90% accurate.

Co-dominance

Two alleles of the same gene are described as **co-dominant** if they both contribute to the phenotype of the organism.

Blood groups

In humans the most common blood grouping system used, to make sure that donor and recipient bloods are compatible for transfusions, is the ABO system.

Erythrocytes (red blood cells) have proteins (antigens) on their plasma membranes. These antigens are coded for by one gene that has three different alleles, I^A, I^B and I^o.

The allele I^o leads to there being no antigen produced, so it is recessive to the other two alleles. However, an individual with the genotype I^AI^B will have both types of antigen on the plasma membranes of their erythrocytes, and will belong to blood group AB. The alleles I^A and I^B are co-dominant.

The table below shows the possible genotypes for people of different ABO blood groups.

Blood group (phenotype)	Possible genotypes
A	I^AI^A or I^AI^o
B	I^BI^B or I^BI^o
AB	I^AI^B
O	I^oI^o

Sickle cell anaemia

All individuals with **sickle cell anaemia** have the same mutation. Haemoglobin, the protein in erythrocytes that carries oxygen from lungs to cells for respiration, is made of four polypeptide chains. There are two alpha chains and two beta chains. The mutation leads to a change in amino acid number 6 in each beta chain. Instead of the usual glutamic acid at position 6, there is valine.

Just this one small change (two amino acids out of about 600 in the haemoglobin molecule) causes the haemoglobin to be less soluble and more 'stringy'. In homozygous individuals, with two alleles for the abnormal haemoglobin, all their haemoglobin is abnormal and less soluble. This, in turn, leads to their erythrocytes becoming sickle shaped rather than the usual round, disc shape.

Sickled erythrocytes cannot squeeze through tiny capillaries and may block these capillaries, reducing the blood (and oxygen) supply to organs. The sufferer feels this anaemia as a 'painful crisis'. Eventually, organs such as the heart, lungs and kidneys become damaged. Treatment involves blood transfusions about every 3 months.

The presence of the normal haemoglobin prevents their erythrocytes from becoming sickle shaped, so they do not have any symptoms. They are symptomless carriers. In that respect, sickle cell anaemia can also be described as having a recessive inheritance pattern.

Heterozygous advantage

Malaria is caused by a protoctist parasite that enters and destroys erythrocytes. The parasite is carried from infected to uninfected people by female *Anopheles* mosquitoes.

In some regions of Africa, where a particular type of malaria is endemic (always present), people who are heterozygous for sickle cell do not suffer from malaria. This is because some of the haemoglobin in each of the red cells is stringy. This isn't enough to change the shape of their red blood cells but the parasite needs to feed on haemoglobin, so it can reproduce. It cannot digest and get enough energy from the haemoglobin in the red cells of carriers as half of it is stringy, so it cannot survive or reproduce. In these regions there is a higher percentage of such heterozygous people than there is in the rest of the world. This is an example of natural selection.

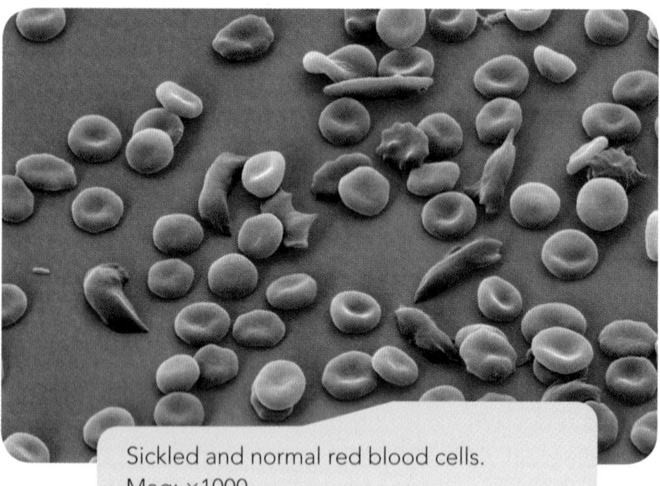

Sickled and normal red blood cells. Mag: ×1000.

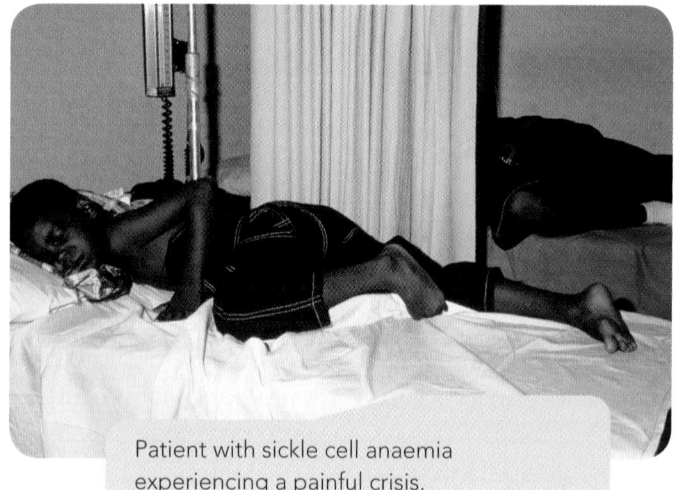

Patient with sickle cell anaemia experiencing a painful crisis.

The genotype of people homozygous for normal haemoglobin can be written as Hb^NHb^N. The genotype of people with sickle cell anaemia is Hb^SHb^S. Heterozygous people have the genotype Hb^SHb^N. Half the haemoglobin in each of their erythrocytes is normal and half is abnormal. So, at this level, both alleles, Hb^N and Hb^S, contribute to the phenotype. They are co-dominant.

Linkage

If two or more genes are on the same chromosome they are **linked**. The alleles of these genes are usually inherited together as they do not independently assort at meiosis (see page 332) unless chiasmata have formed between them during crossing over.

In humans there is a rare genetic disorder called nail patella syndrome. People who suffer from this disease have misshapen finger- and toenails and small or missing kneecaps (patellae). This condition has a dominant inheritance pattern and the gene is on chromosome 9. It is very close to the gene for the ABO blood group system. These two genes are linked, and are on an autosome.

Sex linkage

A characteristic is **sex-linked** if the gene that codes for it is on one of the sex chromosomes, usually the larger X chromosome. On the human X chromosome there are many genes that are not concerned with our gender. These include genes for blood clotting factors, a gene for a muscle protein called dystrophin and a gene for colour vision.

Haemophilia

Haemophila was the first human trait shown to have sex-linked inheritance. There is more than one type of haemophilia and for all of them there is a deficiency in the normal blood-clotting mechanism.

The most significant type of haemophilia is haemophilia A. This is the result of an abnormal allele of the gene for a protein, called factor 8, which is needed for blood to clot when we are injured. An abnormal allele codes for an abnormal factor 8 protein that does not function. The gene for factor 8 is on the X chromosome.

Females have two X chromosomes. If the allele for factor 8 is abnormal on only one of the X chromosomes, they can still produce enough factor 8 for normal blood clotting as the allele for factor 8 on the other X chromosome is normal. However, they may pass the abnormal allele to some of their children, so they are symptomless carriers. If the abnormal allele passes to a male child, he will suffer from haemophilia.

Males have only one X chromosome, so if they inherit one abnormal allele for factor 8, they will suffer from the disorder, haemophilia A. Cuts and grazes aren't usually a problem as they can be treated and wounds stitched if necessary. It is knocks and bruises, which produce internal bleeding, that are more dangerous. Bleeding into the joints leads to chronic damage. Bleeding into the brain is a frequent cause of death in haemophiliacs.

Treatment involves giving whole blood transfusions or factor 8 obtained from donated blood. Since 1985, all blood and blood products for such treatment are screened for HIV. Since 1994 it has been possible to make factor 8 using genetically modified (transgenic) organisms.

Queen Victoria was a carrier of haemophilia. Scientists think that this arose as a spontaneous mutation in her father's X chromosome, during DNA replication before meiosis. Her granddaughter, Alexandra, married Tsar Nicholas II of Russia and their son, Alexei, suffered from haemophilia.

Haemophiliac males cannot pass the disease to their sons as their sons inherit their father's Y chromosome. However all their daughters will be carriers.

The inheritance patterns for haemophilia

We use the following symbols when showing the inheritance patterns for a sex-linked characteristic, such as haemophilia:

H = the allele for normal factor 8

h = the abnormal allele for non-functioning factor 8.

Inheritance of haemophilia from a carrier mother and a normal father.

parental phenotypes: carrier mother, normal father

parental genotypes: X^HX^h, X^HY

gametes

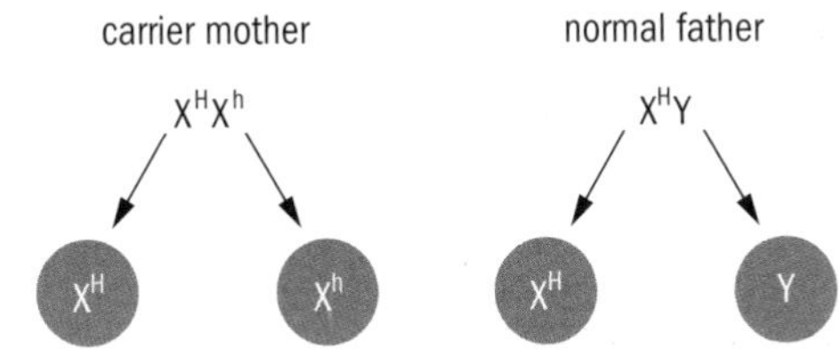

offspring genotypes

	male gametes	
female gametes	X^H	Y
X^H	X^HX^H Normal female	X^HY Normal male
X^h	X^HX^h Carrier female	X^hY Haemophiliac male

Tsarina Alexandra of Russia and her son, Alexei, who had haemophilia. Alexandra was Queen Victoria's granddaughter and was a symptomless carrier.

Chromosome mutations

Chromosomes may undergo changes to their structure or to their numbers. These are both chromosome mutations and can cause genetic disorders.

Changes to the number of chromosomes

In humans the normal number of chromosomes in each body (somatic) cell is 46, with 23 in each gamete (sex cell). Sometimes, during meiosis, the chromosomes of a homologous pair fail to separate at meiosis 1, or the chromatids of a pair fail to separate at meiosis 2. Such a failure to separate is called **non-disjunction**.

Activity 18.3E

1 At which stage of meiosis 1 would homologous chromosomes fail to separate?

2 At which stage of meiosis 2 would chromatids of a pair fail to separate?

Down syndrome

If non-disjunction occurs in chromosome 21, then a gamete with two copies of chromosome 21 can result. If this is fertilised by a normal gamete, containing one copy of chromosome 21, then the resulting zygote has three copies of chromosome 21. This is called **trisomy** 21. The total number of chromosomes in the zygote nucleus is 47. As the zygote divides by mitosis, all the body cells of the individual will have the extra chromosome. The presence of the extra chromosome causes Down syndrome.

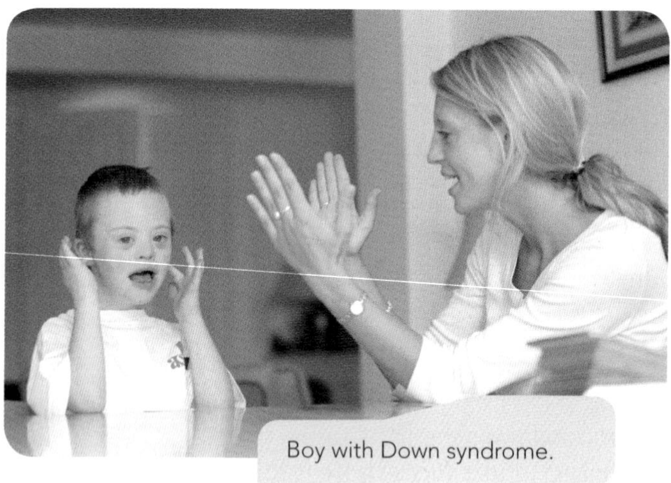

Boy with Down syndrome.

About 1 in 1000 births in the UK are Down syndrome. Individuals with Down syndrome may have some of the following characteristics:

- 'floppy' babies
- adults tend to be short with poor muscle tone
- short, broad hands with a single palm crease
- varying degrees of learning difficulties and increased risk of autism
- flattened face and flattened back of the head
- upward slant of eyes
- an extra skinfold on the eyes
- large tongue
- small ears
- congenital heart defects
- gastrointestinal abnormalities
- increased risk of leukaemia
- increased risk of Alzheimer disease
- thyroid dysfunction
- increased risk of infections

- increased risk of eye cataracts
- hearing problems
- premature ageing.

With improved medical care, the life expectancy of people with Down syndrome is now around 60 years.

The non-disjunction may occur in the father's testes or in the mother's ovaries during meiosis. About two-thirds occur during egg production. The risk of this happening increases with the age of the mother (or father) and pregnant women are usually screened.

The non-disjunction is spontaneous and was not present in the mother's or father's genotypes so, although Down syndrome is a genetic disorder, it is not hereditary.

The most effective and safe method of screening is the combined or triple test carried out between weeks 10 and 14 of the pregnancy. The baby is measured in the uterus with an ultrasound scan and levels of certain chemicals in the mother's blood are measured. The midwife will also take into consideration the mother's age and will calculate her risk of having a baby with Down syndrome. If the risk is high, the mother is offered an amniocentesis test or a chorionic villus test. These tests involve taking some fetal cells from the fluid around the fetus inside the uterus or from the placenta. The cells are grown so they are dividing. They are then placed in distilled water so they swell and burst. This spreads out the chromosomes which can be photographed under a microscope. The photograph is called a karyotype. The chromosomes can be counted.

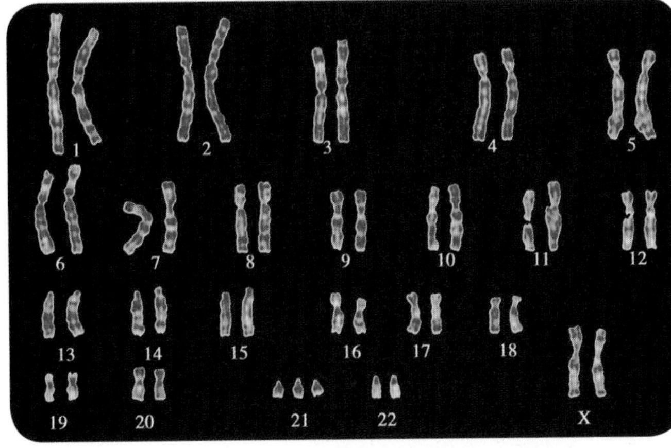

Down syndrome karyotype.

However, these tests are invasive and there is a risk that they could trigger a miscarriage. If the fetus is found to have Down syndrome, the parents can make a decision as to whether or not to terminate the pregnancy.

About 4% of children with Down syndrome do not have trisomy 21. Instead they have a 14/21 translocation. This means that part of chromosome 21 became attached to chromosome 14 in one of their parents. The parent suffered no ill effects from this as he or she still had the correct chromosome quota. However, at meiosis, the extra piece of chromosome 21 will be in a gamete and the baby will have Down syndrome.

About 1% of Down syndrome babies are mosaics. This means that the non-disjunction happened after fertilisation as the cells in the early embryo divided by mitosis. Some cells in the body have 46 chromosomes and some have 47.

Other trisomy abnormalities

Syndrome	Cause	Main characteristics
Patau syndrome	trisomy 13	organ defects and severe learning difficulties
Edwards syndrome	trisomy 18	most die before birth or within first year
Klinefelter syndrome	XXY	affects only males; born with small testes; may have enlarged breasts; no significant learning difficulties
XYY syndrome	extra Y chromosome	affects only males; may grow quickly and be taller than average

Turner syndrome

Just over 50% of females with Turner syndrome have only one X chromosome instead of two. The rest have two X chromosomes but one is abnormal and has parts of it missing. Turner syndrome females are short with a webbed neck and shield-shaped chest. They do not mature sexually and may have heart and kidney defects.

Changes to the structure of chromosomes

Sometimes part of a chromosome is lost. This is called a deletion. The loss of part of chromosome 15 can lead to Prader–Willi syndrome, if it is part of the paternal chromosome that is missing. If the equivalent part of the maternal chromosome 15 is missing, then the child has Angelman syndrome. The symptoms are entirely different.

Fragile X syndrome

At one end of the X chromosome is a gene with a repeating base triplet, CGG. If the number of repeats in this triplet increases to over 230, it makes the end of the X chromosome fragile and the gene does not get transcribed. It does not make its protein. Females with 5–230 repeats are carriers. Their children may inherit fragile X syndrome, which is about 1.6 times more common in males than in females. The characteristics are long narrow face, prominent jaw, forehead and ears, and learning difficulties.

Assessment activity 18.3

BTEC

1 Mendel crossed green-podded pea plants with yellow-podded pea plants. All the F_1 generation had green pods. He allowed these to self-fertilise and grew the seeds. The F_2 generation contained 428 plants with green pods and 152 plants with yellow pods.

   a Which characteristic is dominant? P5

   b What is the ratio of green-podded plants compared with yellow-podded plants? P5

   c Draw genetic diagrams to show how these characteristics are inherited and to explain these ratios. P5

2 When true-breeding fruit flies with grey bodies are crossed with true-breeding fruit flies with black bodies, all the F_1 generation had grey bodies.

   What phenotypes would you expect to see in the F_2 generation if the members of the F_1 generation were allowed to interbreed?

   Use genetic diagrams to explain your predictions. M3

3 In a breed of dog, coat colour is determined by a gene, with two alleles, B/b. The allele B produces a black coat and b produces a white coat. Another gene on a different pair of chromosomes, S/s, produces a short-haired coat or a long-haired coat. S codes for short-haired coat.

   a Two dogs with black, short-haired coats have been crossed several times and produced many litters of puppies. Altogether, they have produced puppies with the following phenotypes:

   18 with black, short-haired coats

   5 with white, short-haired coats

   7 with black, long-haired coats

   2 with white, long-haired coats.

   i Calculate the ratio of the four phenotypes in the offspring.

   ii Use genetic diagrams to explain these ratios. You will need to deduce the genotypes of each parent – use trial and error here. P5

   b Analyse why the observed ratio is not exactly the same as the expected ratio. How could geneticists determine if this difference between observed and expected is due to chance? D2

   c A dog breeder has a white, long-haired female dog. She knows that the genotype of that dog is bbss. She wants to find out the genotype of a black, short-haired (unrelated) male dog of the same breed. She crosses the two dogs.

   i Write down all the genotypes that can give the phenotype black with short hair.

   ii Use genetic diagrams to predict which puppies will be produced if a black, short-haired male dog of each genotype was mated with the white, long-haired female dog.

   iii How could the dog breeder find out if the black short-haired dog was true-breeding (homozygous) for both characteristics? M3

4 You are a genetics counsellor.

   a Explain, by means of annotated drawings and talking (you may like to use role play) to parents, how their child has inherited cystic fibrosis although neither of them have it. P5

   b Use role play to explain to a couple, one of whom has a mother with Huntington disease, the pros and cons of being tested to see if they have the allele for Huntington disease. P5

c The normal gene for CFTR protein codes for 1480 amino acids. Part of the base triplet sequence and the amino acid sequence is shown below:

DNA	AAT	ATC	ATC	TTT	GGT	GTT
Amino acid	aspar-agine	isoleucine	isoleucine	phenyl-alanine	glycine	valine
Position	505	506	507	508	509	510

The most common mutation causing cystic fibrosis is:

DNA	AAT	ATC	AT– ––T	GGT	GTT
Amino acid	asparagine	isoleucine	isoleucine	glycine	valine

Explain why, although bases are lost from two base triplets, only one amino acid is lost and the amino acid at position 507 remains unaltered. D1

5 Draw annotated genetic diagrams to predict the possible genotypes and phenotypes of the children of a couple, both of whom are symptomless carriers of sickle cell anaemia. M3

6 A gene for colour vision is found on the X chromosome. Abnormal alleles of this gene can cause red–green colour blindness. About 4% of males suffer from this. It is very rare in females.

The inheritance pattern is the same as that for haemophilia A.

You are a genetic counsellor. Produce a poster for your clinic that explains to people how colour blindness can be inherited. M3

7 In pairs or singly, find out about one of the subjects below and give a presentation to the rest of your class.

Patau syndrome (trisomy 13)

Edwards syndrome (trisomy 18)

Klinefelter syndrome

Turner syndrome

XYY syndrome

Prader–Willi syndrome

Angelman syndrome

Fragile X syndrome

Amniocentesis

Chorionic villus sampling

Screening tests for Down syndrome. 

Grading tips

P5 For question 1a, explain how you arrived at your answer (because if you guess you have a 50% chance of being correct!). Also show your working for question 1b. For question 1c follow the rules for drawing genetic diagrams given on page 339.

For M3 in question 2 you need to follow the rules for genetic diagrams. Link phenotypes to genotypes, then write sentences giving your answer, but the diagrams show how you have worked this out and you must keep them.

For P5 in question 3, once you have worked out the ratio, think about Mendelian ratios for dihybrid inheritance and you will have an idea of the genotypes of both dogs. Do some rough genetic diagrams to confirm if you are correct. Then draw out your genetic diagrams, following the rules given on page 339.

For M3 in question 3 write down all the possible genotypes that can give a dog with black hair and short coat. Draw a genetic diagram to predict what would happen if each type was mated with the white dog. Now you should be able to tackle part c. Think about what she would do.

For D2 in question 3 refer to the information on the chi-squared test on page 182. You do not need to carry out such a test on these results (you may if you wish) but explain why the observed ratio is not exactly as you would expect (think about the size of dog litters) and explain which statistical test can be used to see if the difference is significant or just due to chance.

For P5 in question 4 you can find the information you need for this in the text and simply set it out in a clear way. Then you can use this and the role play exercise to explain it verbally.

For D1 in question 4 you need to apply what you have learnt and really think about the question you are being asked. Look carefully at the information you are given in the assessment question 3. Think about frameshift mutations and the fact that the genetic code is degenerate.

For M5 in question 5 follow the rules for genetic diagrams. Start with parental phenotypes, then show parental genotypes, gametes and possible offspring genotypes, then finally offspring phenotypes. Make sure that corresponding phenotypes and genotypes line up. Make sure you put circles around the gametes.

18.4 DNA technology

In this section:    D3 M4 P6 P7

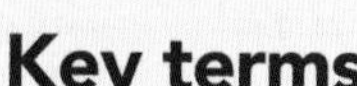

Key terms

Centrifuge – machine that spins very fast and causes fragments to drop due to increased pull of gravity.

Supernatant – the liquid in the tube, after centrifuging, above the fragments that have dropped to the bottom of the tube.

Plasmid – small circular piece of DNA.

Restriction enzyme – endonuclease enzyme that can cut DNA at a specific recognition site.

Gel electrophoresis – method to separate DNA fragments.

Vectors – agents that carry something; in this context we mean something that can carry DNA into a cell to genetically modify the cell.

Marker gene – gene with a known location and a clear-cut phenotype.

Ligase – enzyme that catalyses the joining of sugar and phosphate groups in a strand of DNA.

Screening – process to identify which bacteria have been transformed.

Transformed cells – bacteria that have taken up DNA from another type of cell.

DNA polymerase – enzyme that catalyses replication of DNA.

Primers – short single-stranded fragments of DNA that can bind to flanking regions of the DNA to be copied.

Target region – portion of a DNA molecule to be copied by the PCR.

Flanking region – regions of DNA either side of the target region.

Replication cycle – one cycle of duplication of a piece of DNA .

Gene therapy – addition of a functional allele of a gene into a cell to treat a genetic disorder.

Genetic engineering – manipulation, by means other than sexual reproduction, of an organism's genetic material so that it is altered.

GMO (genetically modified organism) – an organism with DNA altered by manipulation in a laboratory.

Vector – an agent that can carry a gene (DNA) into another cell, which becomes genetically modified as a result; examples include plasmids, viruses, liposomes and bacteria such as *Agrobacterium tumefaciens* – this bacterium infects plant cells and inserts its genes into them, causing crown gall. It can be altered to carry other genes into plant cells.

Genome – all the genetic material in a cell or an organism.

Extraction of DNA

You can extract DNA from any cell. You can use plant cells, such as the very first leaves (called cotyledons) from cress seedlings like the ones you get in a punnet of salad cress.

1 Use scissors to chop the head (cotyledons or first leaves) from a punnet of salad cress, into a mortar.

2 Use a pestle and grind them with a pinch of sand to help break open the cells.

3 Add some buffered detergent solution to break down the cell membranes as detergents emulsify the lipids in cell membranes. The buffer maintains a suitable pH. The DNA from chromosomes, chloroplasts and mitochondria will be dissolved in this detergent. Remove any cell debris from the solution by **centrifuging** (spinning at a very high speed so larger fragments are pulled to the bottom of the tube).

4 Pour the detergent solution with dissolved DNA very carefully down the inside of a test tube half-filled with ice-cold ethanol. The DNA precipitates out where the two liquids meet, because it is insoluble in ethanol.

5 Centrifuge to isolate the pellet.

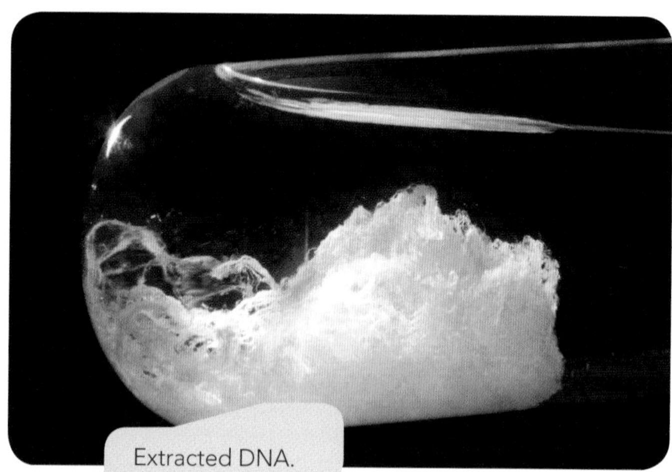

Extracted DNA.

Did you know?

There is a test for DNA. If you heat the extracted DNA to 95 °C in acid, this hydrolyses the molecules and releases the deoxyribose sugar and changes it slightly. If you then add diphenylamine it reacts with this derivative of deoxyribose. A blue colour shows that the substance is DNA.

Sometimes, scientists use protease enzyme to remove any proteins that might be attached to the DNA. They may also add a chemical to bind to calcium ions and stop enzymes from breaking down the DNA.

Extracting DNA from mitochondria or chloroplasts

Both these organelles contain DNA. If tissues, such as leaves or liver, are ground up in ice-cold 2% buffered sucrose solution, and then centrifuged, first of all the nuclei go to the bottom of the centrifuge tube. Then the liquid above the nuclei (the **supernatant** fluid) is taken and put into another tube and spun around in the centrifuge again. Now chloroplasts come down (if the tissue is from plants) or mitochondria come down.

These can then be treated with chemicals to extract the DNA into a solution. If another chemical mixture, such as phenol/chloroform/iso-amyl alcohol, is poured into the tube containing the isolated DNA, the DNA precipitates out.

Activity 18.4A

1 Why do you think the solution used to grind up tissues is kept ice-cold?

2 Why do you think the solution used to extract chloroplasts and mitochondria contains 2% sucrose?

Some scientists use boiling or grinding with glass beads to break open the cells to get the mitochondria or chloroplasts out.

Mitochondrial DNA is used for ancestry tracing. DNA from chloroplasts can be used to help classify plant species.

Extracting DNA from bacterial plasmids

A single colony of bacteria is taken from an agar plate and put into a solution of sodium hydroxide and SDS (a strong detergent) at 37 °C overnight. This causes the cell walls and membranes to burst. The solution is then centrifuged and the cell wall and membrane debris, and the large circular chromosome, go to the bottom of the tube. The liquid supernatant is taken off and treated with a 50:50 mixture of chloroform and phenol. It is then centrifuged again so that any denatured proteins go to the bottom. The clear supernatant is taken off into another tube and ice-cold ethanol is added. This makes the **plasmid** DNA precipitate out.

Scientists need to investigate bacterial plasmids as these circular bits of DNA may have genes for resistance to antibiotics.

Using electrophoresis to separate DNA fragments

Scientists sometimes need to separate lengths of DNA into fragments. The process of electrophoresis separates the fragments and produces a banding pattern, a bit like a bar code. These patterns, or DNA profiles, can be used to see if a suspect's DNA matches that found at the scene of a crime. It may also be used to establish whether someone is the mother or father of a child.

Unit 32: See page 486 for more information on how the banding patterns produced during DNA analysis are used in forensic science.

Gel electrophoresis separates DNA fragments that are of different lengths.

- First the DNA sample to be analysed is incubated at 35 °C for about an hour with a restriction endonuclease enzyme. **Restriction enzymes** are obtained from bacterial cells. Bacteria use them to cut DNA to protect themselves from invading viruses. Each restriction enzyme cuts a length of DNA at a specific base pair sequence, called a **restriction site**. They break bonds between sugar and phosphate groups in the DNA backbone.

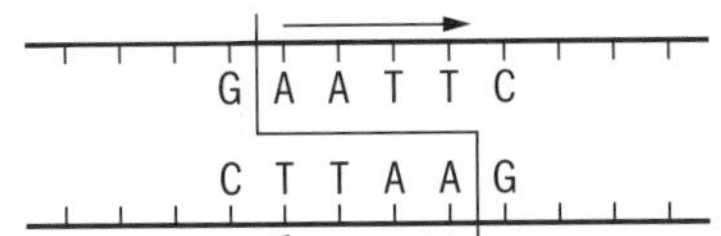

The sequence GAATTC is the recognition sequence for a specific restriction enzyme.

The enzyme makes a staggered cut.

This leaves exposed nucleotide bases or sticky ends

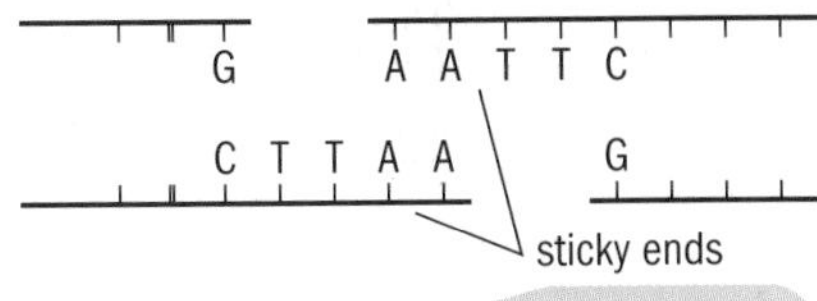

How restriction enzymes cut DNA.

- The restriction enzyme used will recognise a specific base sequence in a length of DNA and make a cut wherever that sequence is. This produces DNA fragments of varying length.
- Once the DNA has been cut by the restriction enzymes, a dye, such as bromophenol blue, is added to make the DNA solution more dense.
- The electrophoresis tank has meanwhile been set up and liquid agarose gel poured into it. There are combs in place and once the gel has set these are removed, leaving wells in one end of the gel.
- A buffer solution is poured into the tanks, so that the gel is covered.
- The DNA samples to be analysed or compared are each added to a well. This has to be done carefully. The micropipette is held above the well and the dyed solution allowed to fall into the well.
- A cathode and anode (electrodes) are placed at either end of the tank, with the anode at the end away from the wells. A power supply is connected to them and left for about 2 hours.
- Because DNA has an overall negative charge, due to its many phosphate groups, the fragments travel through the gel towards the anode, which is positive.
- Shorter fragments of DNA travel further than longer fragments, in much the same way that a large person trying to squeeze through a crowd of people would find it harder than a thin person.
- After about 2 hours, the power supply is disconnected and the buffer solution poured away. A dye is added and this stains the DNA fragments to show the resulting banding pattern.

DNA can be analysed, for example, to identify criminals, to establish someone's innocence in a crime, to establish maternity or paternity, to establish evolutionary relationships between organisms and to find out if someone is carrying an allele for a genetic disorder.

(a) samples are placed in wells cut into the gel at one end using a fine pipette

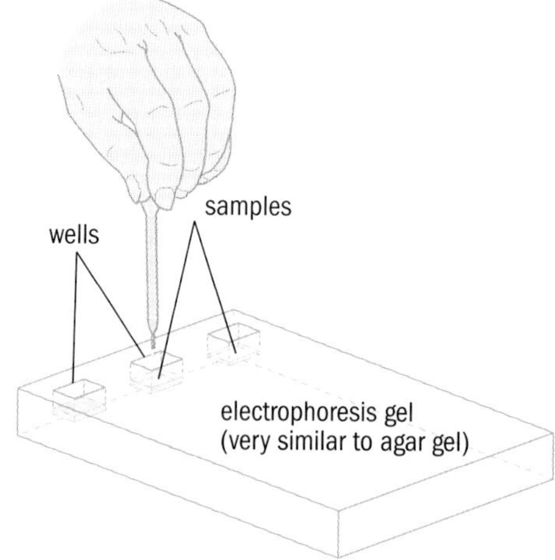

(b) electrophoresis tank

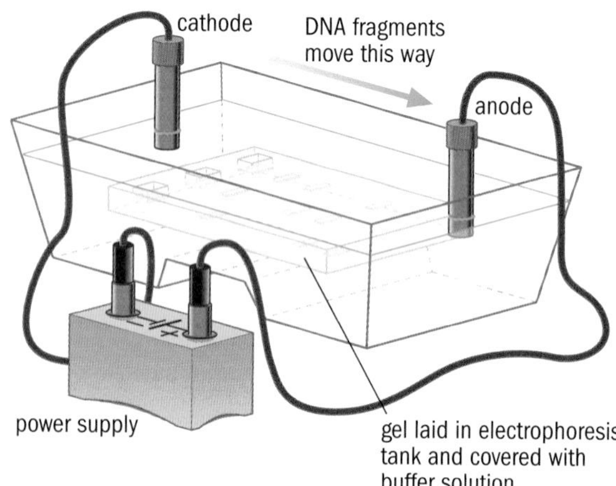

(c) electrophoresis gel showing separated DNA fragments, revealed by flooding with a DNA-binding dye.

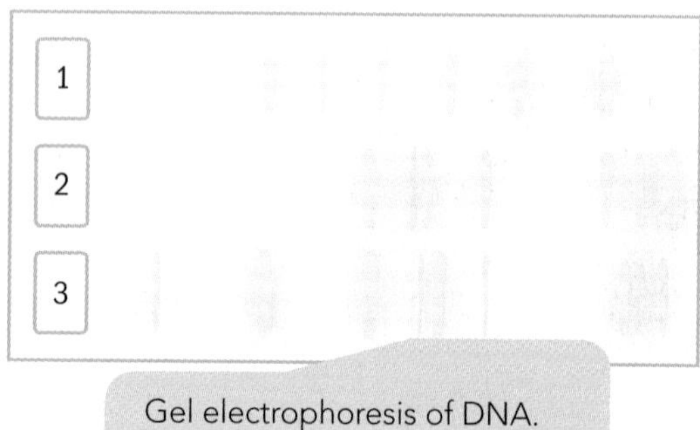

Gel electrophoresis of DNA.

Case study: DNA fingerprinting

In 1984 DNA fingerprinting was used for the first time to catch a murderer and rapist called Colin Pitchfork. His crimes were committed in Leicestershire and the police had heard about the new technique being developed by (the now) Professor Sir Alec Jeffreys at Leicester University.

DNA evidence has since also been used to establish the innocence of people who had wrongly been convicted of murder.

1. **Should there be a database containing DNA samples for all people in the UK? Explain your views.**
2. **How can DNA evidence be used in old criminal enquiries?**

Mitochondrial DNA is used for ancestry tracing as all of your mitochondria come from your mother. This is because the gamete (egg) from your mother contained mitochondria but the mitochondria in your father's gamete (sperm) do not enter the egg at fertilisation.

Transformation of cells

Many bacteria can naturally take up DNA from other bacterial cells. There are three ways they can do this: transduction, conjugation and transformation.

Transduction

Bacteriophage viruses (often called phage viruses), which infect bacteria, can transfer DNA from a donor bacterial cell, which they have infected and used to make many copies of virus particles, to a recipient bacterial cell as a new virus particle infects it.

Conjugation

Some bacteria have, in addition to their circular chromosome, smaller circles of DNA called plasmids. Two bacteria attach to each other and the donor cell passes a copy of its plasmid to the recipient cell.

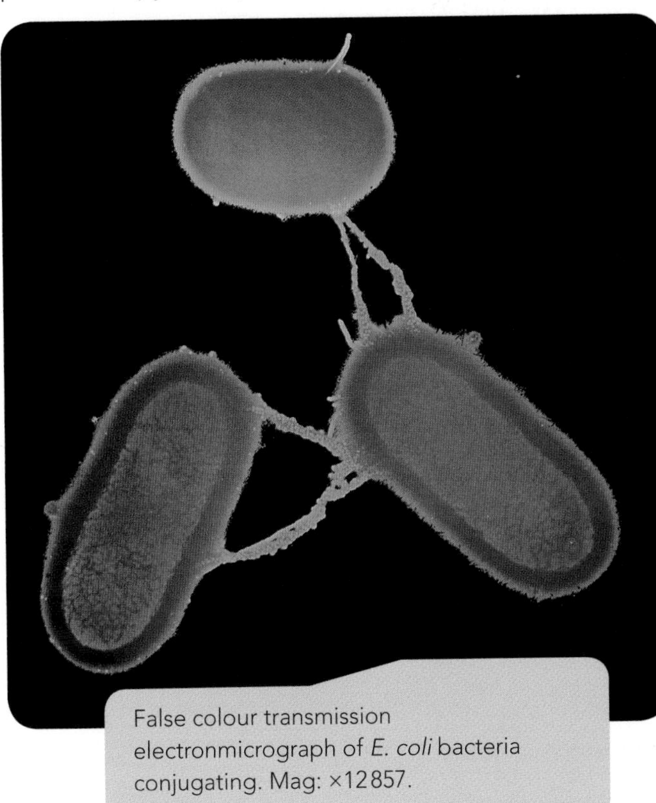

False colour transmission electronmicrograph of *E. coli* bacteria conjugating. Mag: ×12857.

In transduction something carries the DNA from one bacterium to another. This something is called a **vector** (a carrier). You may have heard about mosquitoes being vectors of malaria. In this context scientists use the term to mean something that carries DNA into another cell. Bacteriophage viruses can be vectors. Scientists can use bacterial plasmids as vectors when carrying out genetic engineering.

Activity 18.4B

On plasmids of bacteria there are genes for resistance to antibiotics. When you take antibiotics this sometimes kills many of the bacteria that live in your gut. These are good bacteria and we need them. However, some will have resistance to antibiotics because they have a gene on their plasmid and they will survive and multiply. How might pathogenic bacteria that invade your gut acquire antibiotic resistance from your gut bacteria?

Transformation

Some bacteria can naturally absorb (take up) DNA from their surroundings. For instance they may take up DNA from dead bacteria. A recipient bacterial cell that has taken up DNA, perhaps from dead bacteria around it, is **transformed**.

The recipient cells have to be in a state where their walls are permeable to large DNA molecules.

Transforming *Escherichia coli*

The bacterium *Escherichia coli* does not naturally take up DNA from it surroundings. However, scientists have found ways to make it take up DNA. One method is to treat the bacterium with calcium chloride and then give it heat shock – put it in ice for a minute and then into a waterbath at 40 °C. These two treatments make temporary pores in the cell walls so the bacteria can take up DNA. *E. coli* has been genetically modified to make useful proteins for humans.

Making insulin

- mRNA is obtained from cells in human pancreas tissue that make insulin.
- The enzyme **reverse transcriptase**, obtained from some viruses, is used to make a single complementary strand of DNA from this mRNA.
- The enzyme **DNA polymerase** is then used to make the other complementary DNA strand.
- Now we have the gene for insulin – a length of double-stranded DNA that has the code for making the protein insulin.

- Three unpaired nucleotides are put onto both ends of the molecule. These unpaired nucleotides (with exposed bases) are called **sticky ends**.
- Plasmids are obtained from *E. coli* bacteria.
- A restriction enzyme is used to cut open the plasmid. The enzyme makes a staggered cut (see page 351) and leaves sticky ends. Scientists know which bases are in those sticky ends so the sticky ends they add to the insulin gene are complementary to the plasmid sticky ends.
- Plasmids and genes for insulin are mixed in a small tube (called an Eppendorf tube) and DNA **ligase** enzyme is added. Complementary sticky ends can join, by hydrogen bonds, and DNA ligase catalyses the condensation reaction between the sugars and phosphates in the DNA backbones.
- The plasmid containing a human insulin gene is called **recombinant DNA**.

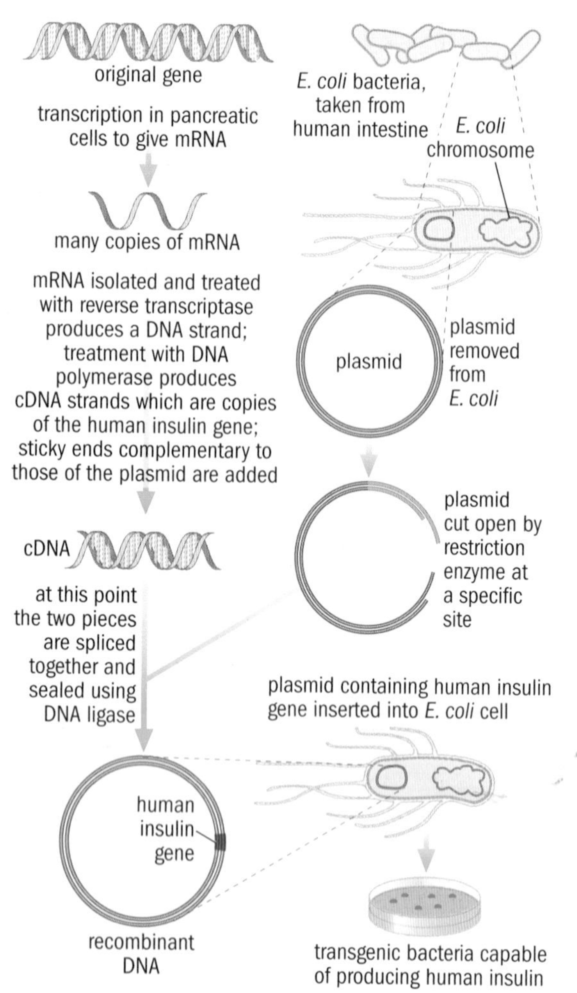

Production of genetically modified bacteria containing the human insulin gene.

- The recombinant plasmids and *E. coli* bacteria are mixed. Calcium chloride is added and they are subjected to heat shock. Some bacteria take up recombinant plasmids and these will be capable of making human insulin.

Screening the bacteria to identify transformed cells

Some of the *E. coli* bacteria will take up plasmids. However, some won't take up plasmids and some will take up a plasmid that does not have an insulin gene in it (the cut plasmid just sealed itself without joining to the insulin gene). The bacteria have to be **screened** to find out which ones are transformed.

- The antibiotic resistance genes on the *E. coli* plasmids are used as **marker genes**.
- The plasmids are cut at a restriction site that is in the middle of a gene for resistance to tetracycline. If the plasmid accepts a human insulin gene, then the tetracycline resistance gene will not work. However, it still has an intact gene for resistance to ampicillin.
- The bacteria are plated out onto agar plates containing ampicillin. Any bacteria that didn't take up a plasmid won't grow. Those that have taken up a plasmid will grow and form visible colonies. However, we need to know which have taken up recombinant plasmids.
- A sterile piece of velvet is placed on the surface of the agar plate and then pressed onto another agar plate that contains tetracycline. This transfers bacteria from colonies on one plate to the other plate.
- On the tetracycline plate only colonies that have a plasmid but no human insulin gene will grow.
- So the scientists now know which colonies on the ampicillin plate they need.
- These bacteria are taken and grown in large culture vessels and their insulin is harvested.

Bacteria with either the recombinant plasmid containing the insulin gene or the original plasmid without the insulin gene will grow on ampicillin agar

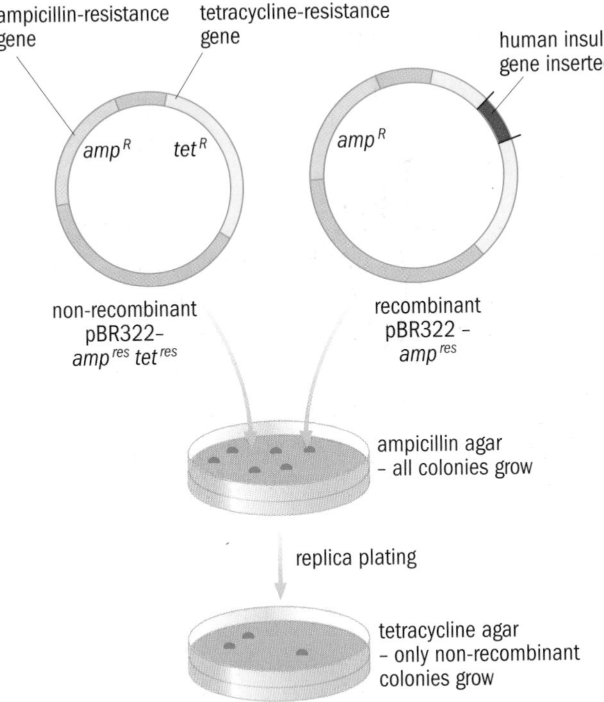

The replica plates are formed by transferring bacterial cells from colonies on one growth medium to another

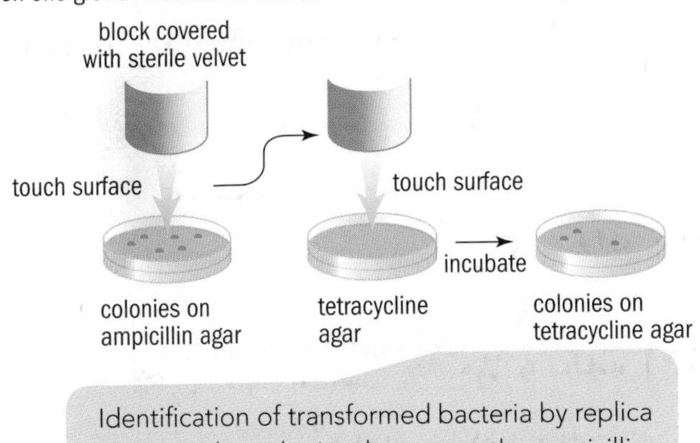

Identification of transformed bacteria by replica plating – the colonies that are on the ampicillin plate but not on the tetracycline plate are the transformed bacteria.

Amplification of DNA: the polymerase chain reaction (PCR) and some of its applications

In 1983 a young biochemist in California, Kary Mullis, realised that, as DNA can make copies of itself in cells, this reaction could be duplicated in a laboratory by the polymerase chain reaction (PCR). He published an article about this in *Scientific American* where he began by saying:

'Beginning with a single molecule of DNA, the PCR can make 100 billion copies of it in an afternoon.'

At first his boss at the biotechnology company thought this was an interesting concept but that it would probably have no practical applications. PCR is now widely used in forensic and medical science to multiply very small samples of DNA and produce enough molecules to analyse it. In 1993 Kary Mullis was awarded the Nobel prize in Chemistry for developing the PCR.

Activity 18.4C

Why do you think there was a time lapse of 10 years between Kary Mullis developing the PCR and receiving the Nobel prize for it?

Share your ideas with the rest of your class.

How the PCR works

The PCR can be used to replicate a small length of DNA, but not whole chromosomes as happens in cells. The portion of a DNA molecule that is to be copied is called the **target region**. The regions of the DNA that flank the target region are sequenced (their nucleotide base sequence is found). **Primers** (small lengths of DNA) are synthesised that are complementary to the DNA that flanks the target regions.

1. The DNA to be copied is heated to 95 °C to separate its complementary strands. The strands separate because the hydrogen bonds that hold the complementary base pairs together break. This gives two single strands of DNA.
2. The mixture is then cooled to 55 °C and the primers are added. They bind to the **flanking regions** of the two single strands by complementary base pairing. The process of them binding is called **annealing**.
3. This gives a region, either side of each single strand of DNA to be copied, that has double strands. The DNA polymerase enzyme can bind to these double-stranded ends.
4. DNA polymerase, obtained from a bacterium, *Thermus aquaticus*, that lives in hot springs, is added to the mixture along with many nucleotides. This enzyme is called **Taq polymerase**.
5. The temperature is raised to 72 °C and the enzyme adds free nucleotides to the unwound DNA. It moves from one particular end of the DNA molecule, called the 3′ end.

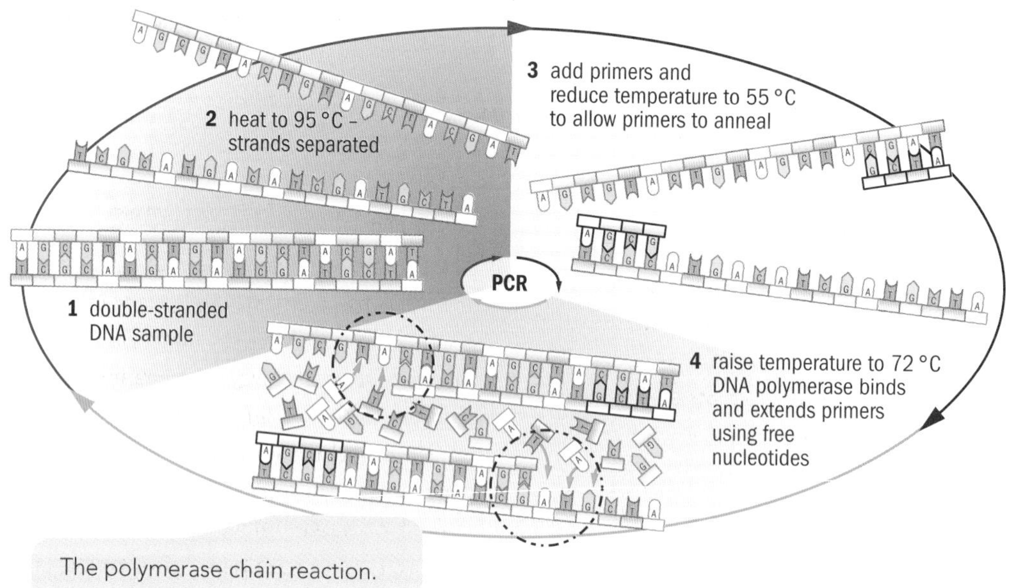

The polymerase chain reaction.

These five steps form one **replication cycle**. This cycle can be repeated many times. Twenty cycles gives one million copies of the original DNA molecule and 30 cycles gives one billion (1 000 000 000) copies.

Activity 18.4D

Show how 20 cycles of replication of a single length of DNA produces one million copies.

When this reaction was first done in a laboratory, a DNA polymerase enzyme that did not withstand such high temperatures was used. This meant that at each cycle the temperature had to be cooled to 35 °C for stages 4 and 5. It takes longer to cool to this lower temperature and then longer to heat up again to 95 °C, so it took many hours for 20 cycles. Using Taq polymerase speeds up the PCR and makes all the tests it is used for cheaper and easier to carry out.

Applications of the PCR

Early diagnosis of cancer

The PCR can be used to detect leukaemia and lymphomas (cancer of the lymph nodes) in the very early stages. The sooner treatment begins the better the outcome (prognosis) for the patient.

Early diagnosis of viral diseases

The DNA (or RNA in the case of some viruses) can be amplified using PCR and analysed to give an early diagnosis so treatment can begin early.

Preimplantation genetic screening

If two parents are both carriers of cystic fibrosis, they may have IVF (*in vitro* fertilisation). The embryos are screened when they are at the eight-cell stage. At this stage, one cell can be taken from an embryo without causing it any harm. The DNA can be extracted from the nucleus of the cell and multiplied using the PCR. It can then be analysed to see if there are two copies of the abnormal allele that leads to cystic fibrosis. The parents can then choose to have only healthy embryos, which do not have cystic fibrosis alleles, implanted into the woman's uterus.

The same sort of screening may be used when both parents carry an allele for other recessive genetic disorders or if one parent is known to be a carrier for a dominant genetic disorder.

Activity 18.4E

Discuss with a partner all the ethical issues raised by this type of screening. Share your ideas with the rest of your class.

Forensic science

In some cases only a trace of DNA is found at the scene of a crime. The PCR can be used to multiply this sample so that it can be analysed and a DNA profile (genetic fingerprint) made.

Analysis of ancient DNA

DNA can be obtained from very old bones of up to tens of thousands of years old. If a small amount of DNA is extracted, then it can be multiplied so there is enough to analyse. Remains of mammoths, Egyptian mummies and the remains of the Romanov family (Tsar Nicholas II and his family, who were executed during the Russian Revolution) have all been analysed using this technique.

Some applications of genetic manipulation

Insulin

You have seen on pages 353–55 how genetically modified bacteria can be used to make large amounts of human insulin to treat diabetics. The advantages of using this method for obtaining insulin, compared with extracting it from pig pancreas tissue, are:

- a lot can be made easily and cheaply
- there is no risk of the insulin being contaminated with a virus or other disease
- there are no ethical issues, such as giving people of certain religious faiths material from a pig
- no one raises ethical objections about bacteria being exploited
- because these bacteria are cultured in large vessels with controlled conditions, they can be grown anywhere in the world – climate does not matter
- the bacteria in question are also modified so that they lack an enzyme and cannot make some of the nutrients they need. Hence they can only survive in the laboratory environment where they are given such nutrients. They cannot escape into the environment and survive there
- there are strict guidelines for people working with these organisms, to minimise any risk of contamination or any other potential biohazards.

Activity 18.4F

Draw an annotated flow diagram to show how insulin is made using genetically modified bacteria.

Human growth hormone (somatotropin)

Some children grow at a slower rate than normal because their anterior pituitary glands do not make enough human growth hormone (hGH). This condition, pituitary dwarfism, can be treated by injecting the children with human growth hormone. This used to be obtained from the pituitary glands of corpses but there have been cases of new variant CJD (Creutzfeldt –Jakob disease), which is caused by and transmitted by an abnormal protein called a prion (pronounced *pree-on)* protein, being transferred to some recipients of the hormone. Genetically modified bacteria can now be used to make hGH. The advantages of using **GMOs** to produce hGH are that it:

- is safer as there is no risk of transmitting CJD
- is easier to produce hGH in large quantities so all children with pituitary dwarfism can be treated
- is cheaper to produce
- may be more acceptable for those being treated.

Activity 18.4G

Find out how human growth hormone promotes growth. Find out about the symptoms of CJD. Suppose you have a child with pituitary dwarfism. Think about which source of hGH you would prefer your child to be treated with. Share your ideas with the rest of your class.

Genetically modified crops

In 1974 the first World Food Conference was held in Rome. At that conference it was predicted that, within 10 years, no child in the world would go to bed hungry. Sadly, the prediction was wrong and today 1 billion people will go to bed hungry.

Between now and 2050 the world population will rise by about 30% but demand for food will increase by 70% and the demand for meat will rise by 100%. There is very little land left to be able to increase that used for farming and in many parts of the world water is very scarce. Added to these problems are climate change and the fact that, for the past 25 years, the governments of many countries have not invested properly in the development of their agriculture.

If world food production is to be increased over the next 40 years, without unlimited land and water, technology will be more important than it was over the last 40 years during the Green Revolution. More efficient fertilisers and pesticides will be needed. One key way to increase yield is to develop genetically modified crops, such as those that:

- need less water
- produce more fruits and seeds
- have short growing seasons and can fruit twice a year
- are resistant to diseases (so pesticides are not needed)
- can fix nitrogen so will need fewer fertilisers
- have been bio-enhanced to contain more nutrients
- have a slower ageing process so can be stored for longer.

Many governments are now paying attention to food security (being able to feed the population) and self-sufficiency (being able to produce all the food needed without having to import any).

In some countries there has been opposition to GM crops, to the extent that the trials have to be conducted at secret places. In the USA, GM crops have been eaten for 10 years with no ill-health effects reported so far, but there are plenty of ill-health effects from eating processed foods containing non-GM foods.

Gene therapy

Gene therapy is being developed and trialled for treating disorders such as cystic fibrosis. If a person inherits two recessive alleles of the CFTR gene (see page 342) they will suffer from cystic fibrosis. Gene therapy involves inserting a functioning CFTR allele into the epithelial cells of the sufferer's respiratory tract. This is not classed as **genetic engineering** because:

- not all of the person's cells are altered
- the alteration is only temporary
- the genetic alteration cannot be passed to the patient's offspring.

However, genetically modified viruses have been used to introduce the CFTR functioning allele into the patient's respiratory tract epithelial cells. These viruses used as vectors (carrying agents) were genetically modified:

- to make sure that they would not cause a disease
- to carry the CFTR allele.

(a) a liposome is an artificial vesicle

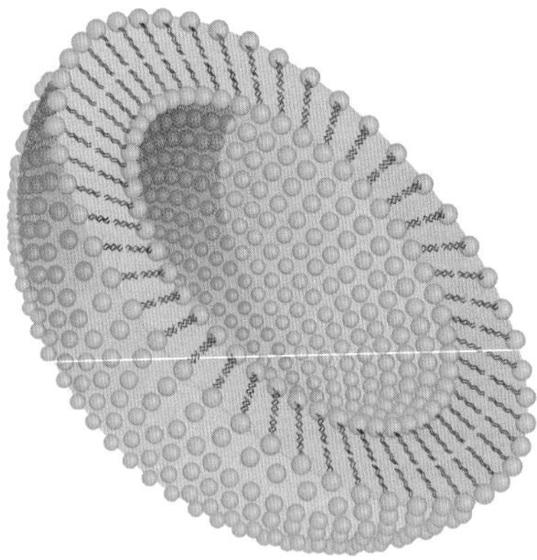

(b) genes can be enclosed in the liposome and so are able to pass through the plasma membrane of the target cell

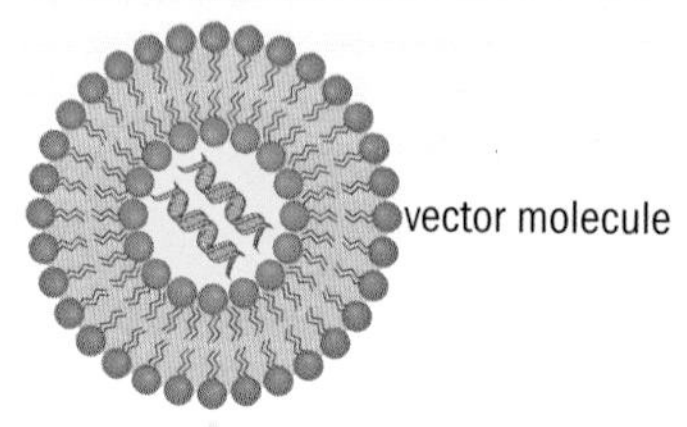

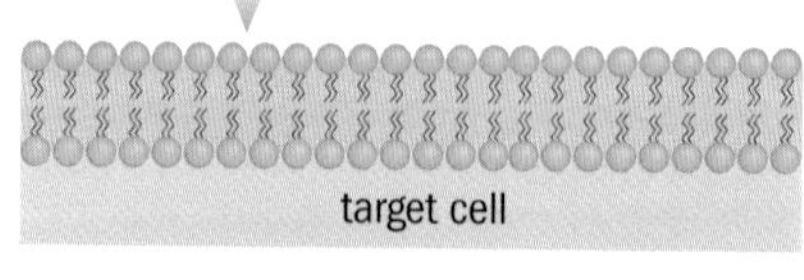

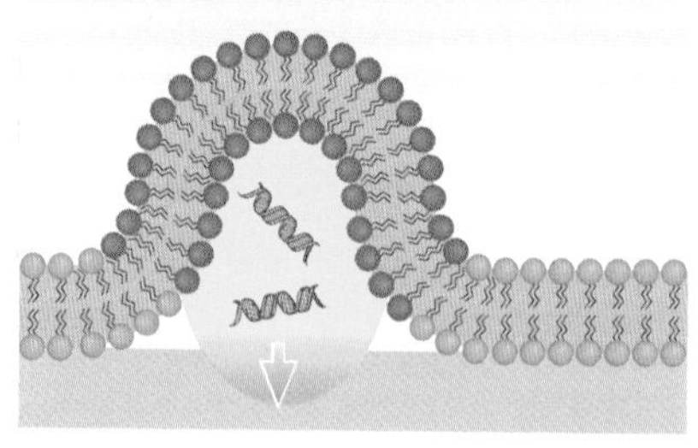

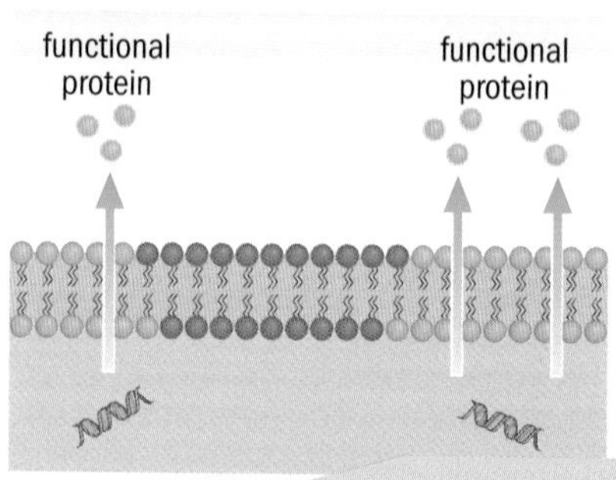

Gene therapy using a liposome vector.

Because the epithelial cells are replaced at regular intervals, the treatment has to be given about every 10 days. However, the patient mounts an immune response to this virus even though it is harmless, so this method of introducing the functioning allele has not been very successful. After the first use, the immune response of the recipient makes the delivery of the gene into the cells difficult, as antibodies coat the viruses and stop them entering the patient's cells.

Liposomes are spherical lipid particles that can enclose the allele. They can dissolve in the lipid layer of cell membranes and pass through the plasma membranes and nuclear envelopes of cells. Once in the nucleus, the allele may insert itself into the **genome** and direct the making of functioning chloride ion channel proteins. The liposomes can be sprayed, in an aerosol, into patients' noses. They are vectors and the recipient does not mount an immune response against them.

Gene therapy cannot at present be used for Huntington disease as it does not involve removing faulty alleles. However, in the future it may be possible to use gene therapy for such dominant genetic disorders using antisense or silencing strands of RNA. These would be:

- complementary to the mRNA transcribed from the faulty huntingtin allele
- able to bind to the mRNA that codes for huntingtin protein
- able to prevent the translation stage of protein synthesis.

Transgenic animals

Genetic engineering of humans is illegal and considered to be unethical. However, transgenic animals have been produced. The gene is inserted into the zygote, so that every cell of the embryo and resulting adult animal is genetically altered. These animals are genetically modified and are described as transgenic. Transgenic sheep can produce human proteins in their milk. These proteins, such as factor 8 to treat haemophilia, are too large to be made in genetically modified bacterial cells. The protein can be obtained and purified from the sheep's milk. The sheep do not suffer and are too valuable to be eaten!

Assessment activity 18.4

1 Under the supervision of your teacher, extract DNA from kiwi fruits, cress cotyledons, onions or your own cheek cells. P6

2 You can also carry out gel electrophoresis on some lambda phage virus DNA or calf thymus DNA, depending on which kit you use. Follow the instructions in the kit carefully. Your tutor should identify the hazards of any chemicals used. P6

3 Write a report of the practicals, describing how DNA is digested by restriction endonuclease enzymes and how the resulting fragments are separated by gel electrophoresis. M4

4 Draw an annotated flow diagram to explain the steps needed to make genetically modified (transformed) *E. coli* bacteria that can be used to produce human insulin to treat people with diabetes. D3

5 Explain the advantages of using genetically modified bacteria, as opposed to extracting insulin from pig or cow pancreas tissue. D3

6 Make a poster to inform people about the process of DNA amplification using the polymerase chain reaction, and to show some of its applications. P7

7 Prepare some ideas so that you can contribute to a class discussion about GM crops:

   a Think about the benefits of using GM crops to feed more people.

   b Consider all the arguments against the use of GM crops and evaluate these in the light of not using them and not being able to feed hungry people.

   c Reach a conclusion.

   d Share your ideas with the rest of the class. P7

8 In pairs or individually, find out about one of the following topics and give a presentation to the rest of your class.

- Golden Rice™
- Flavr Savr™ tomatoes
- frost-resistant potatoes
- pest-resistant strains of crop plants
- GM corn that contains fish oils
- bioenhanced GM sorghum
- GM bananas that are pest resistant
- transgenic mammals.

In your presentation:

a outline the steps used to genetically modify the organisms

b outline the benefits and disadvantages of using the GMO

c discuss the commercial, social and ethical issues associated with the GMO. P7

Grading tips

Carry out the practicals competently and safely. Write your reports in the conventional way.

For D3 in questions 4 and 5 draw a large clear annotated flow diagram. The annotations should include the explanations of why each step is done. Show clearly somewhere on the diagram the advantages of using GMO to produce insulin as compared with extracting it from pig or cow pancreas tissue.

For P7 in question 6 remember to include accurate biology but the poster should also be able to communicate to people with little biological knowledge.

PLTS

Independent enquirer, Reflective learner and Effective participator

You will develop your skills as an independent enquirer when researching information. You will develop reflective learner skills when considering the pros and cons and the commercial, social and ethical issues. You will also be an effective participator when communicating your ideas to the rest of the class.

WorkSpace

Gillian Hamilton

Alzheimer's Society Fellow, Medical Genetics Section, University of Edinburgh

With a degree in genetics, my job includes research, together with teaching and supervising students. I work as part of a team investigating the genes associated with Alzheimer's disease and I carry out functional work: mutating genes to see if this affects how the gene functions.

A typical day can be quite varied. I spend the majority of my time in the lab, where I modify DNA and grow cells. DNA can be modified inside bacterial cells and then introduced into mouse cells using a virus. I can then determine whether the mutation has affected gene function. Some experiments take a few minutes to complete, but it might be a week before you get results from others.

Communication in science is important. I spend time reading journals to find out new results from other labs and also attend meetings where new results are presented. I also prepare results from my own work to be published in journals or presented at seminars. Part of my day may also include lecturing, supervising students and marking exams.

My lab has about 20 people working in it. Although we all work on different projects the experiments we do are mostly the same. I also collaborate with researchers in London, Oxford and Cardiff.

I thoroughly enjoy working through a particular experiment, eliminating possibilities, tackling problems and sharing information with colleagues.

The process of determining how a gene might work often involves looking at what happens when a gene has been removed from the genome. (The genome is the complete recipe book for making an organism and the human genome is composed of over 30000 genes.) We can remove a single gene from the genome and then carry out tests to see how this has affected the organism.

Think about it!

1. Do you think that the field of genetics will be important in the development of treatments for patients with Alzheimer's disease?
2. How much of an effect do you think lifestyle has on the risk of developing Alzheimer's disease?

Just checking

1 Explain how meiosis contributes to genetic variation.
2 Explain how mutations can contribute to genetic variation.
3 Explain how DNA codes for proteins.
4 Describe how DNA copies itself.
5 Explain the following terms: dominant; recessive; genotype; phenotype; co-dominant; allele.
6 Describe how bacteria can be genetically modified to produce a human protein, such as insulin.

Assignment tips

To get the grade you deserve in your assignment remember to do the following:

- Follow instructions for practical work carefully.
- Always assess the risks involved.
- Work competently and safely.
- Record your data.
- Research carefully when preparing posters and presentations.
- Extract relevant information and present it in a clear and succinct way.

Some of the key information you will need to remember includes the following:

- the structure of DNA
- the process of protein synthesis
- the structure of chromosomes
- the behaviour of chromosomes at cell division
- Mendel's experiments on monohybrid and dihybrid inheritance
- alleles and genes, how environment and genotype interact to give phenotype
- genetic disorders such as cystic fibrosis and Huntington disease
- DNA technology such as gel electrophoresis
- applications of genetic engineering.

For information on:	Visit:
The work of the Wellcome Trust	The **Wellcome Trust** website
Mendelian genetics	**Hobart and William Smith Colleges** website
The work of the Society of Biology	The **Society of Biology** website
virtual fly lab	**WKU Biology Department** website
DNA profiling	**How Stuff Works** website

Other useful sources of information are journals and magazines such as *New Scientist, Scientific American, National Geographic* and *Focus*.

Credit value: 10

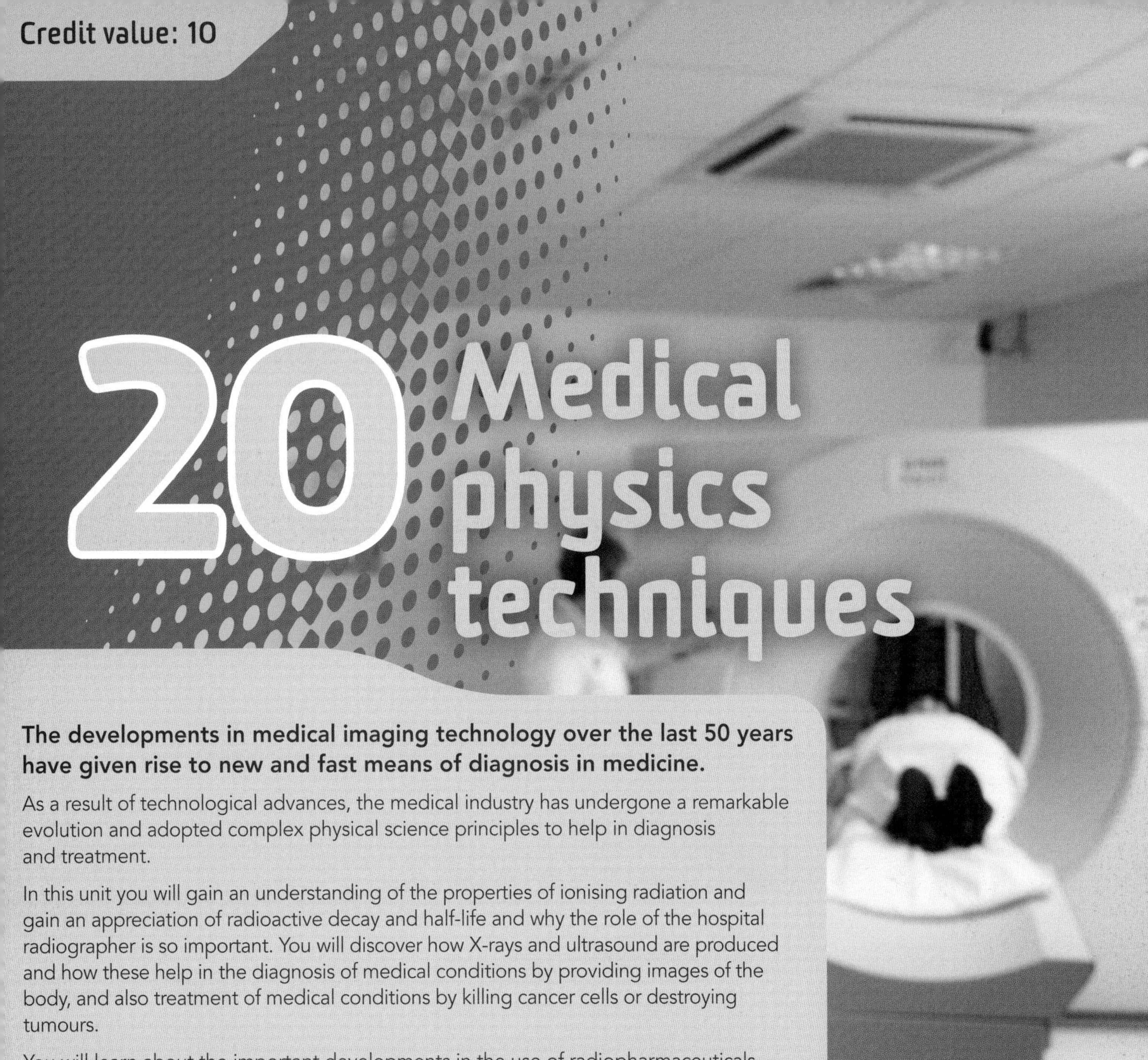

20 Medical physics techniques

The developments in medical imaging technology over the last 50 years have given rise to new and fast means of diagnosis in medicine.

As a result of technological advances, the medical industry has undergone a remarkable evolution and adopted complex physical science principles to help in diagnosis and treatment.

In this unit you will gain an understanding of the properties of ionising radiation and gain an appreciation of radioactive decay and half-life and why the role of the hospital radiographer is so important. You will discover how X-rays and ultrasound are produced and how these help in the diagnosis of medical conditions by providing images of the body, and also treatment of medical conditions by killing cancer cells or destroying tumours.

You will learn about the important developments in the use of radiopharmaceuticals in diagnostic imaging and their detection by a gamma camera. You will have the opportunity to study the process of magnetic resonance imagery, the application of the method and safety aspects.

Finally, you will learn how the continued use of radiation and high-energy rays depends on an understanding of their effects on the body and a strict adherence to safety protocols.

Learning outcomes

After completing this unit you should:

1 know atomic structure and the physical principles of ionising radiation and ultrasound
2 understand how radiopharmaceuticals are used in diagnostic imaging
3 know the basic principles of magnetic resonance imaging
4 understand the importance of radiation safety to the treatment of malignant disease with radiotherapy.

Assessment and grading criteria

This table shows you what you must do in order to achieve a **pass, merit** or **distinction** grade, and where you can find activities in this book to help you.

To achieve a **pass** grade the evidence must show that you are able to:	To achieve a **merit** grade the evidence must show that, in addition to the pass criteria, you are able to:	To achieve a **distinction** grade the evidence must show that, in addition to the pass and merit criteria, you are able to:
P1 describe radioactivity, including atomic structure **See Assessment activity 20.1**	**M1** explain the random nature of decay and how it relates to half-life **See Assessment activity 20.1**	**D1** analyse the effect of the operation and design of the tube/head on a typical X-ray spectrum **See Assessment activity 20.1**
P2 describe the production of X-rays and ultrasound **See Assessment activity 20.1**		
P3 describe the production and detection of radiopharmaceuticals **See Assessment activity 20.2**	**M2** compare the desirable biological properties and radiological properties of radionuclides used for imaging **See Assessment activity 20.2**	**D2** evaluate the choice of radiopharmaceuticals for a range of clinical imaging requirements **See Assessment activity 20.2**
P4 explain the role of pharmaceuticals within the operating principles of the gamma camera **See Assessment activity 20.2**		
P5 outline the process of magnetic resonance imaging including the instrumentation and equipment used **See Assessment activity 20.3**	**M3** explain the factors influencing signal intensity in MRI **See Assessment activity 20.3**	**D3** evaluate the appearance of bone and soft tissue in an MRI scan and a conventional X-ray **See Assessment activity 20.3**
P6 explain the principles and effects of radiation therapy including the equipment used **See Assessment activity 20.4**	**M4** explain how excessive exposure to radiation can cause harm **See Assessment activity 20.4**	**D4** evaluate a range of therapy techniques, types of radiation available and the equipment used **See Assessment activity 20.4**

How you will be assessed

Your assessment could be in the form of:

- diagrams and posters
- presentations
- written reports
- notes taken during site visits.

Yasmine, 17 years old

I really enjoyed this medical physics unit and was fascinated with the many incredible facts I was able to find out. I was particularly amazed at the technology which is used in X-rays, radiopharmaceuticals and MRI scans. Our visit to the radiology department in our city hospital really put our class work into context.

The section on magnetic resonance imaging was very detailed and quite complicated but I managed to understand the principles and explain how it works.

I have had X-rays myself and have known people who have needed treatment for cancer. This unit has helped me to understand what actually happens to patients in the radiology department and that I shouldn't be afraid of this kind of technology.

Catalyst

Making radiation treatment clear

We use highly dangerous types of radiation in the treatment of cancer and other illnesses.

When some serious illnesses are diagnosed, the treatment can be as worrying to people as the illness itself. To most, even the words used to describe the diagnosis and subsequent treatment are very difficult to understand.

How would you explain to a patient forms of diagnosis and treatment such as X-rays, ultrasound, radionuclides, radiotherapy and magnetic resonance imagery?

Research these terms in groups and present them to your class in as simple an explanation as possible.

20.1 Radioactivity, X-rays and ultrasound

In this section:

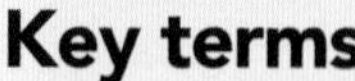

Key terms

Half-life – the time taken for the number of active nuclei in a radioactive isotope to fall to half its initial value ($t_{\frac{1}{2}}$).

Radioactive decay constant – a measure of how radioactive an isotope is. It is inversely proportional to the half-life. If an isotope has a long half-life, it will have a low decay constant; if it has a short half-life, it will have a high decay constant.

Activity – the number of atoms that decay in one second (s^{-1}). The becquerel (Bq) is also used as a unit for activity (1 decay per second).

Photon – a tiny 'packet' of electromagnetic energy.

Piezoelectric material – a substance that can produce an electric current when squeezed.

Resonance – if an outside forcing frequency matches the natural frequency of any system, the vibrations produced are very large.

Acoustic impedance – how substances affect sound waves passing through them, defined as the ratio of sound pressure to particle velocity at a given point.

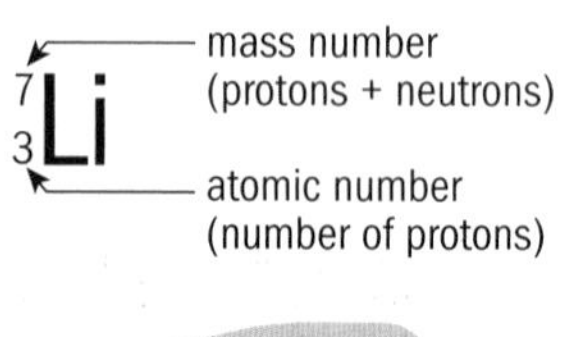

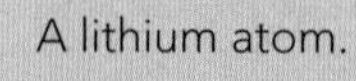

A lithium atom.

Radiation warning sign.

Radioactivity involves a process of change taking place in the nucleus of an atom. It can be explained as an attempt by an unstable nucleus to become more stable. The lithium atom shown above, for example, contains three protons and four neutrons. This imbalance in the nucleus is not enough to make it radioactive. When the number of neutrons reaches five or more (in heavier lithium isotopes) the nucleus is very unstable and radioactive. Radioactivity is not like a chemical reaction for a number of reasons:

- It takes place in the nucleus, unlike a chemical reaction.
- It does not involve the atom's electrons.
- The changes are spontaneous and unaffected by pressure and temperature.
- It releases much more energy than a chemical reaction.

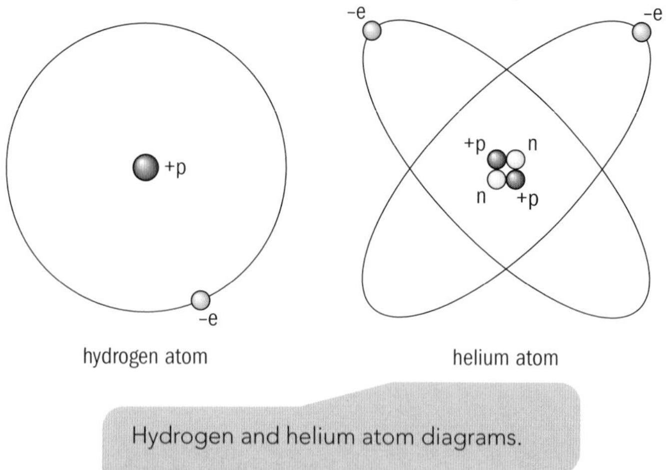

Hydrogen and helium atom diagrams.

Decay law

Radioactive decay is a completely random process whereby nuclei disintegrate. When this happens, the atom is changed according to the type of radiation emitted. The disintegration continues through a number of stages until a stable atom is formed. The original atom is called the **parent** and the atom formed when radiation is emitted is called the **daughter**.

How active a radioactive substance is depends on the number of disintegrations of nuclei in a given time. This is called the **activity** and, as it corresponds to the rate of change of the number of undecayed nuclei with time, the proportion can be calculated using a mathematical constant, the **radioactive decay constant** (λ).

A radioactive substance decays exponentially with time. The graph (page 367) shows a radioactive decay curve. You can see that the **half-life** is the time taken for the number of active nuclei in a source at any given time to fall to half its value. There are other examples of such curves including: a capacitor discharging, a shock absorber damping and the cooling of hot water.

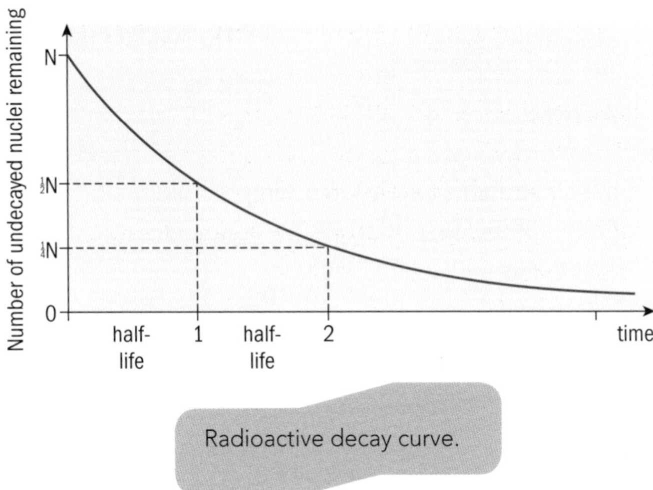

Radioactive decay curve.

Alpha decay

The alpha particle is made up of two protons and two neutrons (identical to a helium nucleus). If a radioactive atom loses an alpha particle it therefore loses four mass particles. For example:

radium → *radon*

$^{226}_{88}\text{Ra} \rightarrow {}^{222}_{86}\text{Rn} + {}^{4}_{2}\alpha$

Beta decay

Explaining this type of disintegration is more difficult. It involves complex energy changes and the creation of other sub-atomic particles. Simply, if a β^- particle is emitted, a neutron converts into a proton and the overall charge on the atom becomes positive. If a β^+ particle is emitted, a proton converts into a neutron and the overall charge on the atom becomes negative. Example of β^- emission:

sodium radioisotope → *magnesium*

$^{24}_{11}\text{Na} \rightarrow {}^{24}_{12}\text{Mg} + {}^{0}_{-1}\text{e}^-$

Gamma radiation

When radioactive nuclei disintegrate, they emit either an alpha or beta particle. When this happens, the nucleus is in a changed energy state. This surplus energy is lost as a gamma ray and the nucleus achieves some stability. There is no alteration to the proton or neutron number when gamma rays are emitted.

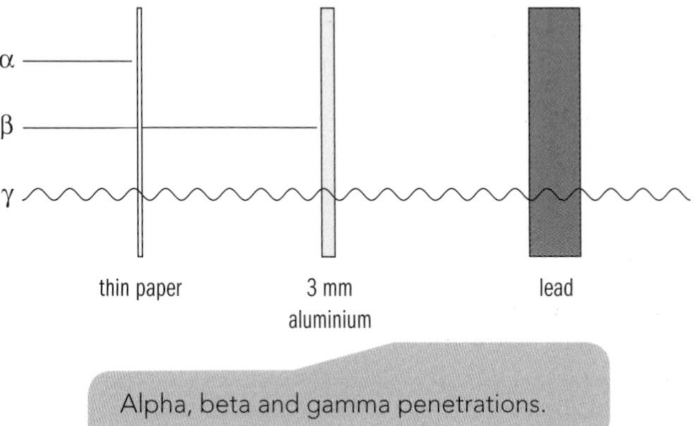

Alpha, beta and gamma penetrations.

Radiation	Penetration	Other characteristics
alpha α	Alpha particles have a range of a few centimetres in air. Can be stopped by skin, a sheet of paper or thin sheet of aluminium foil.	Positively charged particles deflected by an electric and magnetic field. Identical to a helium nucleus. Interact with most atoms in their path and therefore cause strong ionisation of particles per millimetre in air and are least penetrating. Detected by photographic film, cloud chamber, spark counter and thin-window Geiger–Müller tube.
beta β	High velocities mean that these electrons have higher energies. They can be stopped by a few millimetres of aluminium.	Negatively charged particles deflected by electrical and magnetic fields. An electron moving at high speed. Emitted from a source at a range of speeds causing slightly weaker ionisation than alpha per millimetre in air. Detected by photographic film, cloud chamber and a G–M tube.
gamma γ	Varying energy levels. Gamma radiation is not stopped by an absorbing material but its intensity is reduced (typically reduced by 50% in 10 cm of lead).	An electromagnetic wave that is not deflected by an electric or magnetic field. Causes lower ionisation per millimetre in air than alpha and beta radiations and travels at the speed of light. Detected using photographic film, a cloud chamber and a G–M tube.

Activity 20.1A

Use the example above to predict the atomic number and mass number of the element astatine (At) formed when polonium ($^{218}_{84}Po$) loses a beta particle (β^-) during decay. Write your answer in the same way as the example shown.

Radioactive decay formulas

$A = \lambda N$

$N = N_0 e^{-\lambda t}$

A = the total activity of the sample in decays per second

N = number of radioactive particles remaining after time t

N_0 = the initial number of radioactive particles in the sample

t = time

λ = decay constant

e = 2.718

Worked example

Find the number of atoms remaining in a radioactive sample after 6 hours. The half-life ($t_{\frac{1}{2}}$) of the sample is 10 hours.

Answer

$A = A_0 e^{-\lambda t}$

Initial number of atoms in a sample from experiment, $N_0 = 10^{19}$

The decay constant is found by:
$-\ln 2 \div 10 \times 60 \times 60 = 1.9 \times 10^{-5}$.

Convert the time (6 hours) into seconds = 21 600 s

$N = 10^{19} \times e^{-1.9 \times 10^{-5} \times 21600}$

$= 10^{19} \times 0.663$

$= 6.63 \times 10^{18}$ atoms remaining.

Did you know?

Alexander Litvinenko (said to be a former Russian spy) died of radiation poisoning in 2006. It is thought that he was given polonium-210, an alpha emitter, in some food or drink.

Uses of radioactivity

Application	Description
radioactive tracers	*Medical tracers*: iodine-123 is used to detect problems in kidneys; technetium-99 is injected into the body and emits gamma rays; it has a half-life of 6 hours. *Industrial tracers*: pistons in engines may contain radioactive elements so that if the metal wears away the particles can be detected; pipeline maintenance is carried out using a beta-emitter isotope injected into the oil or gas flow to locate the leaks.
sterilising	*Gamma rays*: used to kill bacteria in food that is ready for retail sale and requires a long shelf-life; for sterilising hospital equipment, especially plastic syringes, which would melt under heat treatment.
radioactive dating	Uranium-238: this element decays very slowly into lead with a half-life of 4.5 million years. By measuring the amount of lead in rocks it is possible to determine their age. Carbon-14: has a half-life of 5700 years and is continuously replenished, and so present in a constant amount, in animals and plants when living. On death, the element decays. This can be used to date any organism or object which has absorbed carbon (radioactive isotope carbon-14).
other uses	Gamma rays are used to control the regular thickness of thin materials such as tin or aluminium in motor car manufacture. Americium-241 is used to ionise air in smoke detectors to conduct electric current. If particles of smoke enter the detector, the interference in current flow sets off the alarm. Gamma rays are used to detect flaws in welds for construction. Gamma rays are used to destroy cancer cells.

X-rays

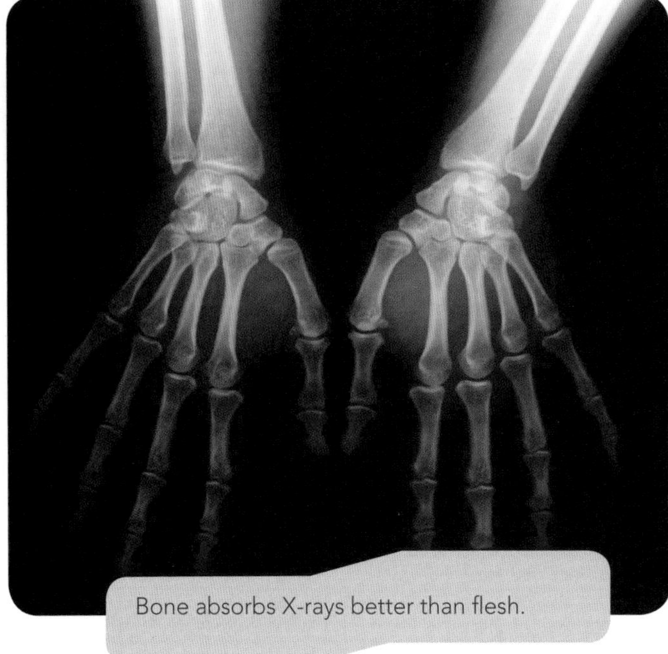
Bone absorbs X-rays better than flesh.

These are waves produced by the sudden stopping of very high-speed electrons when they hit a target plate. They have a typical wavelength of 10^{-10} m. They were given the name X-rays because, when they were discovered in 1895 by a physicist named Wilhelm Röntgen, nothing was known about them.

Today, X-rays have many applications mainly using their penetrating power. These include:

- **Medicine** – X-rays are used for diagnosis, and for treatment of some cancers. Care is taken to reduce exposure to patients because of the damage that X-rays can do to healthy tissue. In X-ray diagnosis, the energy of the photons is controlled such that they pass through flesh but are absorbed by denser diseased tissue and bone. The areas where X-rays are absorbed showing white on photographic film; other areas show up dark. To reduce **absorption** of low-energy **photons** by healthy tissue, X-rays are **filtered** through a thin metal plate before going through the patient. Otherwise the tissues would heat up and become damaged.
- **Industry** – castings and welds can be inspected using X-rays just like bone fractures. Machines can be inspected without being dismantled.
- **X-ray crystallography** – this is the study of crystal structure. This method has been used to analyse simple crystal structures but has also revealed complicated structures such as DNA.

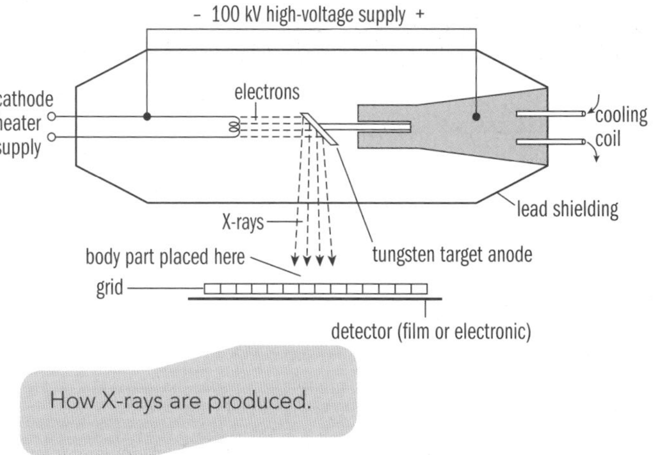

How X-rays are produced.

X-rays are produced with an evacuated tube. This is how it works:

1 A filament is heated by the flow of an electric current.

2 Electrons, negative in charge, are given off (thermionic emission) and accelerated towards the anode, which is given a positive charge of p.d. up to 100 kV.

3 The electrons' acceleration increases their kinetic energy.

4 When the electrons hit the target (usually a tungsten or copper block) a fraction of the kinetic energy from the majority of electrons is transformed into X-rays. Most of their kinetic energy is transformed into heat.

5 The anode is kept cool using circulating oil or water.

6 X-rays pass out through a small window in the lead shielding.

Changing X-ray intensity

- Increase the filament current – more electrons are produced and collide with the anode, so more X-ray photons are produced.

Changing X-ray penetration (*quality*)

- Increase the p.d. of the anode – the speed of the electrons is increased and the frequency of the X-rays is increased.

X-ray characteristics

Like all electromagnetic waves, X-rays spread out as they travel. Their intensity drops with the square of the distance. In other words, if the distance to an object is doubled, the intensity of the X-rays falls to a quarter of their original value. This is called an inverse square relationship.

X-rays are also absorbed or scattered by the presence of other atoms. When X-rays are absorbed their intensity decreases. The thicker the absorbing material, the lower the X-ray intensity. Experiments have shown that the relationship follows an exponential curve like that of radioactive decay.

$$I = I_0 e^{-\mu x}$$

where I = intensity after absorption

I_0 = intensity before absorption

μ = linear absorption coefficient

x = thickness of absorber

When the intensity of the X-rays is reduced to half its original value by an absorbing material, the thickness of the material is called the **half-thickness** ($x_{\frac{1}{2}} = \frac{ln2}{\mu}$), which is found using various thicknesses of metals and a G–M tube.

X-rays penetrating matter experience scattering to various extents depending on the atomic composition of the matter and the wavelength of the X-rays. In medical diagnosis, X-rays can be scattered when entering the body, allowing low-energy X-rays to be absorbed in the patient's body. To prevent this they are filtered using combinations of aluminium, copper and tin plates. Scattered X-rays exiting the body are passed through a lead grid to ensure that they do not reduce the contrast image. The grid is placed between the patient and the detector.

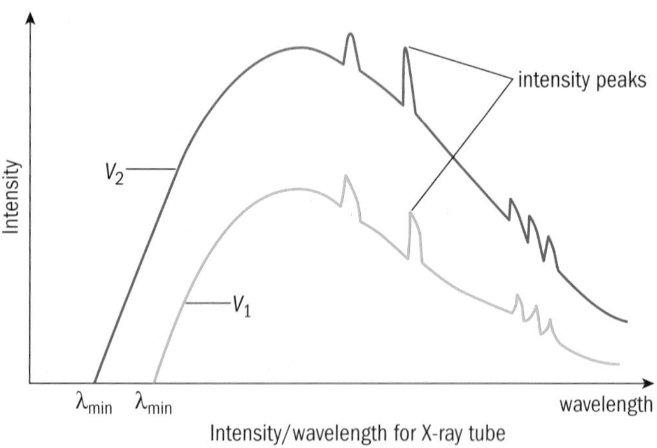

Graph of wavelengths of X-rays against intensity.

In the graph of wavelength vs intensity, shown above, V_1 and V_2 are the anode voltages. An X-ray machine emits a 'spectrum' of wavelengths. The intensity peaks are a function of the target metal. Low-energy wavelengths are filtered out by a metal filter to stop unnecessary harm being caused to the patient.

Activity 20.1B

Give two ways of reducing the intensity of X-rays.

Ultrasound

When sound waves hit a solid object in air, they bounce off (they are reflected). This is called an echo and also applies to sound waves moving through liquids. Measuring the time taken for the sound pulse to travel to the object and back, and knowing the speed of sound in both media, it is possible to calculate the distance to the object using the formula: distance = speed × time.

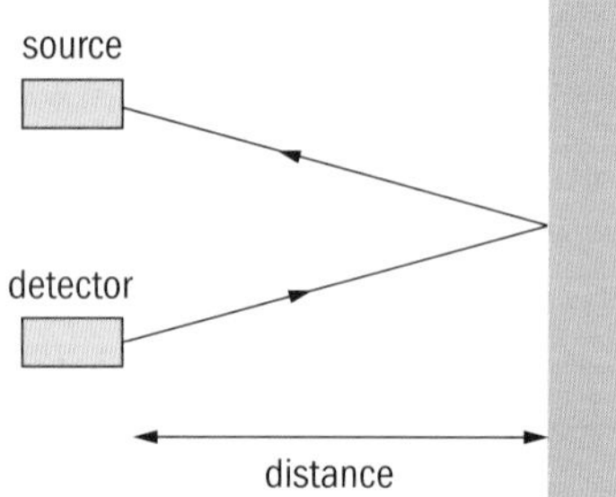

Sound waves with a frequency more than 20 000 hertz (20 kHz) are called *ultrasonic* or **ultrasound**. In medicine frequencies of several MHz are used and they have much shorter wavelengths than the sound waves that we can hear.

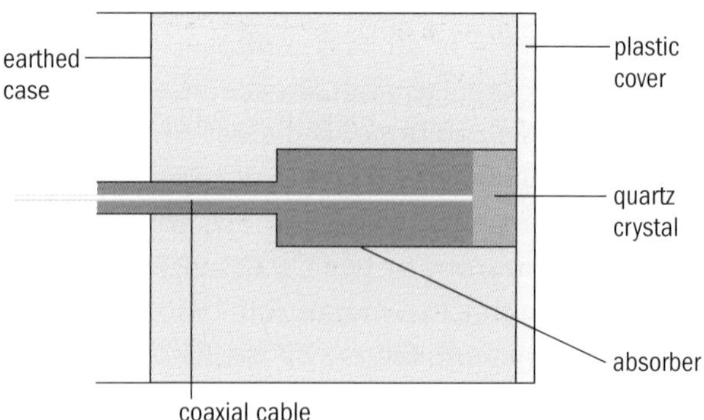

Production of ultrasound using the quartz crystal method. In modern medical ultrasound, other crystals are now used such as synthetic ceramics or lead zirconate titanate.

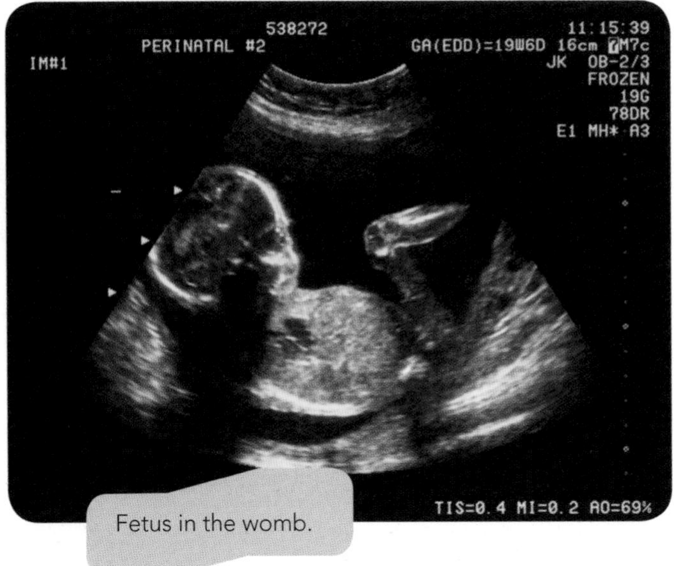

Fetus in the womb.

Production of ultrasound

An ultrasonic transducer converts electrical energy into ultrasound when an alternating voltage is applied. A quartz crystal has a special property – it is **piezoelectric**, which means that it vibrates regularly when stimulated by electricity. If the applied frequency is equal to the natural frequency of the crystal, it **resonates**. This produces very large vibrations and ultrasonic waves.

The frequency is in the megahertz range (MHz). Pulses of waves pass through an object and are partly reflected by boundaries within the material (such as between muscle and bone in the human body). These are reflected back to the transducer, which converts the ultrasound to electrical energy. The image can then be displayed on a monitor screen. Unlike X-rays, ultrasound is quite safe and gives very good contrast images for soft tissues.

- **Acoustic impedance** in medical ultrasound

When sound travels through matter, some is transmitted and some is reflected. The amount that is reflected depends on the density characteristics of the matter and the velocity of ultrasound through it. Bone, for example, has four times the acoustic impedance of muscle and will reflect more ultrasound waves.

Calculating the fraction of ultrasound reflected from the perpendicular:

$$\propto = \frac{(z_2 - z_1)^2}{(z_2 + z_1)^2}$$

z_2 and z_1 are two different acoustic impedance values

If the resulting figure is large (compared with a table of known values) then there is a strong echo. If the figure is small, this corresponds to a weak echo.

- **Refraction**

When the ultrasound pulse encounters a boundary or interface between two mediums, two tissues for example, most of the pulse is reflected but a significant amount is refracted. This means it passes through the boundary into the second medium until encountering another boundary. When this happens, some of the pulse is again reflected and some refracted. This will produce another echo slightly later than the first. A series of echoes will build up, representing different depths of boundaries.

- **Intensity**

As ultrasound passes through body tissue, it is weakened by reflection and absorption and some of its penetrating ability is lost. This is called **attenuation**. The penetrating ability of the ultrasound depends on intensity (in medical diagnosis, measured in watts per square cm, W cm^{-2}) and the attenuation is a change in this intensity as the energy moves through the different media. Differences in intensities are recorded in decibels (dB). As the decibel scale is logarithmic, a measurement of –20 dB is 100 times less intense than –10 dB, a measurement of –30 dB is 1000 times less intense than –10 dB and so on.

Ultrasonic scanning

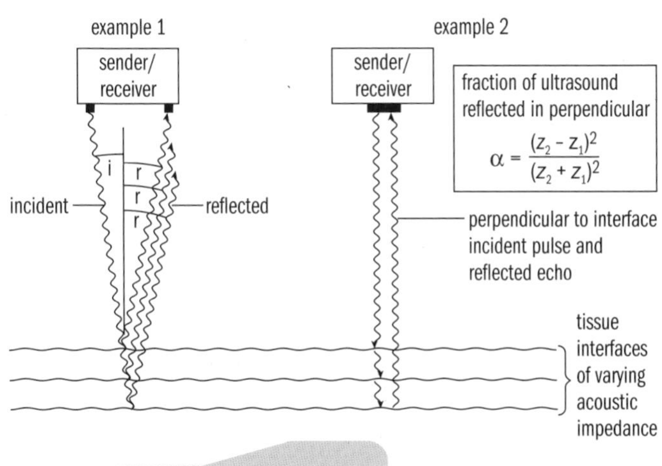

Reflection principles.

Displaying ultrasonic scans

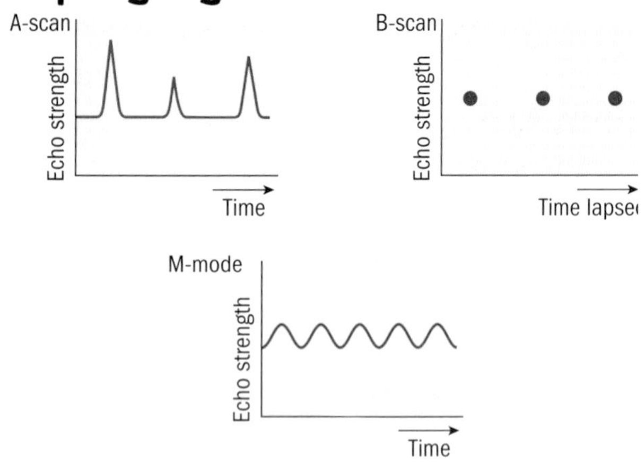

A-scan: A single-dimension pulse. Echoes are plotted as peaks and are a function of depth into the body.

B-scan: Displayed as bright spots where the brightness is a measure of signal strength. In a two-dimensional B-scan, a probe allows movement around the body to develop a cross-sectional image.

M-mode scan: The M in M-mode stands for motion. This is a moving scan image produced by a fast-moving series of B-mode scans over the body. This allows the image to resemble the true motion of organs and fetuses.

Safety and hazards

Safety note on use of ultrasound: Medical organisations regularly review their policies on the use of ultrasound. The British Medical Ultrasound Society produces safety guidelines on the use of ultrasound which are regularly updated.

The Doppler effect

You have all heard the **Doppler effect** in action. When an ambulance passes by, the sound of its siren changes in pitch when it comes towards you and then moves away from you. The sound waves are 'bunching up' as it approaches and 'spreading out' as it goes past.

This change in frequency can help us to work out the velocity of an object or substance and whether it is moving towards us or away from us.

In medicine, the use of the Doppler effect can indicate how fast or slow the blood is moving. This provides useful information about restricted or blocked blood flow in arteries and possible blood clotting. An echocardiogram (use of ultrasound Doppler effect on the heart) can also determine the direction of blood flow over heart valves.

Did you know?

Navies around the world regularly use ultrasound for detection and communication. It is thought that many animals such as whales and dolphins that have been washed up on beaches worldwide may have been affected by these activities.

Uses of ultrasound

Application	Description of use
medicine	fetal scans during pregnancy; high-intensity ultrasound treats kidney stones and benign and malignant tumours; help in the actions of antibiotics in killing bacteria; dental hygiene – cleaning teeth and regenerating tooth enamel; therapeutic ultrasound as a muscle relaxant (this form of treatment has not been fully proven to work)
industry	ultrasonic testing used to detect flaws in materials; increasing heat transfer in liquids, e.g. ethanol production; cleaning equipment by activating microscopic bubbles which collapse at the surface of an object; locating position using badges which respond to ultrasound and send back location signals; sonochemistry – increasing reaction rates
nature	Many animals either produce or make use of ultrasound, including bats, insects, dolphins, whales, fish and horses.
research, marine and military	Sonar – used in range finding where the ultrasonic wave echo reflected from objects is timed and distances calculated.

Assessment activity 20.1

BTEC

Your work as a junior technician in the radiography section of a large hospital involves working with other highly qualified personnel, talking to patients undergoing therapy and periods of personal study. You must show that you have a clear understanding of the terms used and an understanding of the basic science principles involved in your department.

1. Draw sequences which show what happens to radioactive elements when they lose: an alpha particle; a beta particle. What happens to an atom when gamma rays are emitted? P1
2. Draw a fully labelled diagram demonstrating the principles of: an X-ray tube; production of ultrasound. P2 Use a suitable diagram to analyse what happens to an X-ray spectrum when the tube voltage is changed. Show some known X-ray peaks in your diagram. What do these peaks tell you? D1
3. Using graph paper, show a decay curve and mark on:
   - the axis showing the fraction of undecayed nuclei remaining
   - the axis showing time
   - half-life intervals
   - fractions of original number of undecayed nuclei remaining. M1

Grading tips

Include labels of protons and neutrons in your answer and at least two element sequences for each decay to achieve P1. To achieve M1 you could add a simple demonstration set of results using dice to illustrate the random aspect.

PLTS

Self-manager

You will develop your skills as a self-manager when organising and listing new terminology and definitions for use in the unit.

Functional skills

Mathematics

You will develop your skills in mathematics when carrying out calculations for radioactive decay.

20.2 Radionuclide and diagnostic imaging

In this section:

Radiation warning sign.

Key terms

Radiopharmaceuticals – the name given to pharmaceuticals containing radionuclides.

Sterile – free from living microorganisms.

Pyrogen-free – conditions or substances which do not raise body temperature.

Electronvolt (eV) – the amount of energy carried per electron, a very small number (1 eV = 1.6×10^{-19} joules). 1 eV is the energy gained by an electron when moved through a potential difference of 1 V.

Radionuclides, *radioactive isotopes* and *radioisotopes* are different terms to describe the same thing. It depends on which one you prefer to use. For the purpose of this text they will be referred to as **radionuclides**.

In simple terms, the nucleus of an atom is made of protons and neutrons. If the number of neutrons and protons is similar, the atom is usually stable. Atoms of the same element with different numbers of neutrons are called *isotopes*. All atoms in the periodic table have isotopes, for example:

Oxygen

$^{16}_{8}O$	$^{17}_{8}O$	$^{18}_{8}O$
8 protons	8 protons	8 protons
8 neutrons	9 neutrons	10 neutrons

Whilst most isotopes in nature are stable, many are unstable. These are radioactive isotopes (radionuclides). As they are unstable, the nucleus breaks up into smaller parts; this is radioactive decay (page 366). When this happens the atom emits either an alpha particle, a beta particle or gamma rays.

It was discovered that radionuclides can be made by smashing alpha particles into the nuclei of other atoms. Since then, these radionuclides have been artificially made on a large scale and have many uses in industry, manufacturing and medicine.

Natural radionuclide formation

$$^{14}_{7}N + ^{1}_{0}n \rightarrow ^{14}_{6}C + ^{1}_{1}H$$

In this example, cosmic ray energy from space removes neutrons (n) from atoms in the atmosphere, which then collide with nitrogen (N) atoms. This forms the radionuclide carbon-14, which is taken in by plants and trees during photosynthesis. The carbon-14: carbon-12 ratio is used to date materials that have carbon in them.

Artificial radionuclide formation

$$^{27}_{13}Al + ^{4}_{2}He \rightarrow ^{30}_{15}P + ^{1}_{0}n$$

In this example, aluminium (Al) collides with alpha particles (He nucleus) and a nuclear reaction produces the phosphorus (P) radionuclide.

Radionuclide generators

Radionuclides are produced for medical purposes in a specialist generator. This is basically a method in which a parent radionuclide with a long half-life is allowed to decay to a short half-life daughter radionuclide. The daughter is then separated chemically into solution with suitable chemical compounds. This addition of chemicals to the radionuclides produces **radiopharmaceuticals**. There are more than 60 types currently used in medicine; examples of the more commonly used are given in the following table:

Radiopharmaceutical name	Element and general description
^{123}I	iodine, ready to use
^{67}Ga citrate	gallium, ready to use
^{99m}Tc	technetium, available in MDP (methylene diphosphonate), DTPA (diethylenetriaminepenta-acetic acid) and MAA (macroaggregated albumin) kits
^{201}Tl chloride	thallium, ready to use
^{133}Xe	xenon gas, ready to use

Did you know?

It is estimated that at least 50% of people currently living in Western civilisations will experience some form of radiopharmaceutical treatment.

An effective radionuclide generator should:

- be portable and easily transported between hospitals
- provide easy separation in **sterile**, **pyrogen-free** form
- produce a high yield of radiopharmaceutical.

An effective radiopharmaceutical should:

- emit pure gamma rays (to reduce radiation doses and produce good images), with energies between 100 and 250 keV (kilo**electronvolts**)
- produce lots of photons (reduces imaging time)
- have no toxic effects on patients
- be quick to administer with no complicated equipment
- be easily available
- have a half-life that is long enough to complete the diagnosis and short enough to minimise the radiation dose to the patient.

Activity 20.2A

As a technician in a radiography unit of a busy general hospital, from the two examples shown below, which would you choose as the more suitable radiopharmaceutical? Use information in the previous notes to explain your answer.

Patient scan no. 1 (bone scan)
Procedure time = 4 hours
Radiopharmaceutical = ^{99m}Tc-MDP
Half-life = 6 hours

Patient scan no. 2 (bone scan)
Procedure time = 4 hours
Radiopharmaceutical = sodium fluoride F18
Half-life = 1 hour 50 minutes

The gamma camera

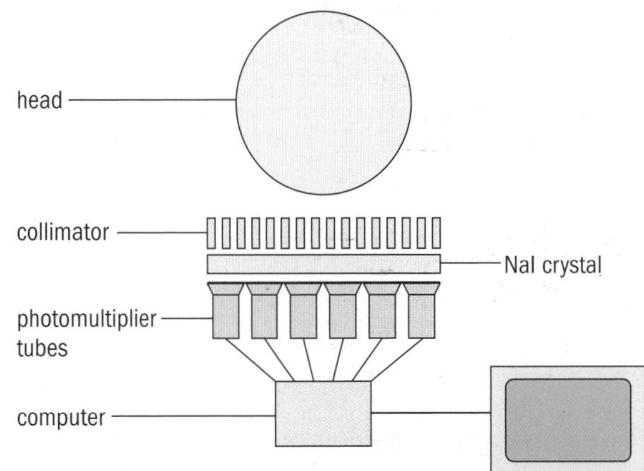

A simplified gamma camera. The patient's head is shown by the circle. ***Collimator*** – made of lead and collimates gamma rays so the image is clear. ***Photomultiplier tubes*** – activated by radiation, coming from a specific location in the patient's body, which is then converted to an electrical signal.

How it works

- Radiopharmaceuticals are ingested by the patient as a *tracer*.
- These radioactive chemical compounds are absorbed by bone, tissues and other body organs (e.g. lungs, heart, brain, kidneys...).
- Gamma rays are emitted as the substance decays.
- These gamma rays are collimated by the lead collimator.
- Gamma rays strike a detection crystal and converted into electrical pulses by the photomultiplier tubes.
- The stronger the electrical signal in one particular position, the more localised the position of the radiopharmaceutical.
- This information is fed into a computer which works out where a gamma ray originated and provides an image.

Activity 20.2B

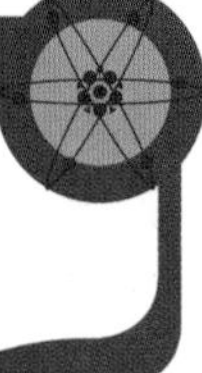

What does a strong electrical signal mean to the medical analyst when examining the image of a patient from a gamma camera?

Diagnosis examples from a bone scan

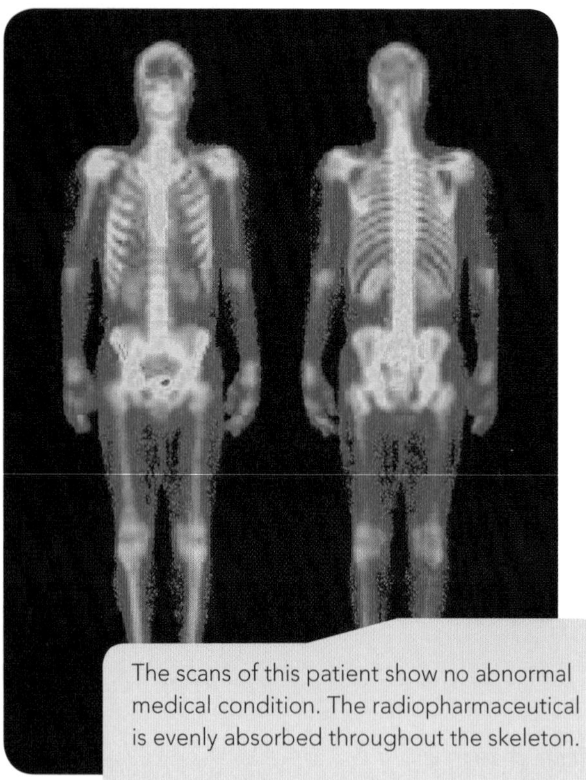

The scans of this patient show no abnormal medical condition. The radiopharmaceutical is evenly absorbed throughout the skeleton.

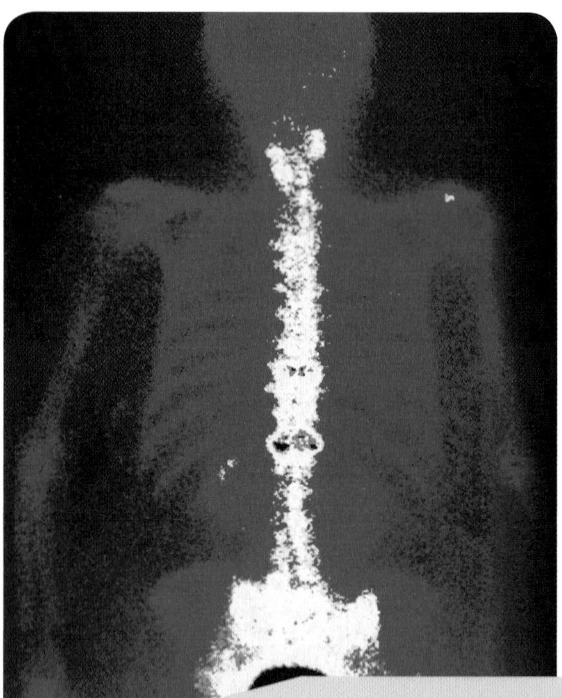

Gamma camera image of an elderly patient. The radioactive tracer used is technetium-99 and shows accumulation in areas of bone which are brightly coloured red. This indicates osteoporosis, bone fracturing and degeneration.

Why use radionuclide imaging?

Good points	Bad points
Shows the functions and processes of the internal parts of the body and is low radiation risk to patients.	Images are not very sharp and some people cannot use them, e.g. pregnant women or those with an underlying heart or lung problem.

Associated imaging techniques

- PET – positron emission tomography – mainly used for cancer diagnosis
- SPECT – single photon emission computed tomography – primarily used in blood flow analysis
- MUGA (or RNA) – multiple gated acquisition scan (radionuclide angiography) – concerned mainly with diagnosis of heart complaints

Quality control in the workplace

When we are working with radiopharmaceuticals there are a number of important points to consider to prevent serious side-effects from the treatment. Training in this industry is intensive and rigorous, and workers take part in regular quality-control training programmes. This ensures that important checks are made, for example:

- prevention of cross-contamination
- correct waste disposal
- sterile equipment – clothing, syringes, etc.
- maintenance of apyrogenic conditions
- carrying out of specialist work in controlled areas
- daily checking of radioactivity dose calibrators for accuracy
- maintenance of the radiopharmaceuticals at a suitable pH and chemically pure.

Assessment activity 20.2

BTEC

As a recent addition to the technical and nursing staff of a large city hospital, you must show that you are familiar with the radiopharmaceuticals used and the way in which they are detected within the body of a patient.

1 Make a list of the most common radiopharmaceuticals used in medicine, describe how they are produced and briefly describe what happens when these substances enter the body. P3
2 Explain how the gamma camera works using a fully labelled diagram. P4
3 Using your list for P3, provide details of what qualities you are looking for when choosing a suitable radiopharmaceutical. Remember that patients have to inhale or be injected with these substances. M2
4 Use information in this chapter and your own research to evaluate which radiopharmaceuticals are best for a given purpose. D2

Grading tips

You should include the formulas for your radiopharmaceuticals in your answer for P3 and what the images received by the gamma camera tell us for P4. Additional research is necessary for M2 and D2, which should provide more information on the choices made by doctors for particular radiopharmaceuticals in specific parts of the body. Health of the patient is vital and the image produced is very important.

PLTS

Creative thinker

You will work as a creative thinker when deciding which radiopharmaceuticals are best for a given set of conditions.

Functional skills

ICT

You will use your ICT skills when searching a wide range of sources for information on radiopharmaceuticals.

20.3 Magnetic resonance imaging

In this section:   P5

Field coils – turns of conductive wire that carry the electrical energy necessary to produce a powerful magnetic field in an electromagnet.

Spin – natural property of sub-atomic particles.

Photons – 'packets' of energy associated with electromagnetic waves.

Precession – the movement of a rotating object that does not spin on a perfectly vertical axis – similar to a spinning top which loses energy and begins to slow down.

Nuclear magnetic resonance (NMR)

NMR is a very complicated but extremely detailed way of producing an image of the inside of a patient's body without the need for surgery or ionising radiations. It is painless, can be used on pregnant women and has no known side effects, but can be quite noisy when the procedure is running. For this reason patients are given ear plugs. Patients enter the large circular-shaped scanner and lie down on a sliding platform known as a **flight path**. The technologist or radiologist operating the system must make sure that all iron-containing items are removed from the patient and surrounding area before the procedure is carried out (see MRI safety). Once all the safety aspects have been taken care of, all the patient must do is keep very still!

Different body tissues contain different amounts of hydrogen atoms. The protons in these atoms are spinning, and this spin can be disturbed by electromagnetic waves of a certain frequency. The nuclei can resonate as a result and produce a radio wave signal. This is detected and the information is analysed. NMR produces a very detailed picture of body tissue without the need for X-rays.

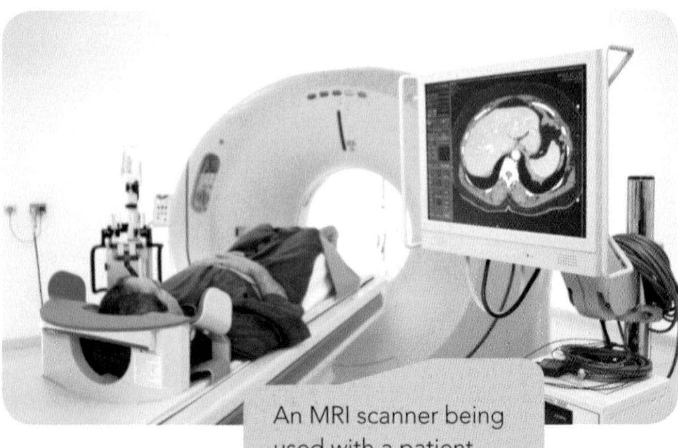

An MRI scanner being used with a patient.

Particles in spin

Just like electrical charge or mass, **spin** is a property of nature. Electrons, protons and neutrons all have this property which comes in multiples of ½ (+ ½ and – ½). If particles pair up with opposite spins, they can cancel out the observed spin effect.

In general, almost every element has an isotope with a net spin, in other words not zero.

Here are some examples:

Nuclei	Net spin
^{1}H	$\frac{1}{2}$
^{2}H	1
^{23}Na	$\frac{3}{2}$
^{13}C	$\frac{1}{2}$
^{19}F	$\frac{1}{2}$

Particles which have spin behave like tiny magnets when they are placed in a uniform magnetic field. They show a north and south pole which will line up with the poles of the magnetic field.

This behaviour allows the particle to absorb energy from incoming electromagnetic waves called **photons** provided that the frequency of the wave is of a particular value.

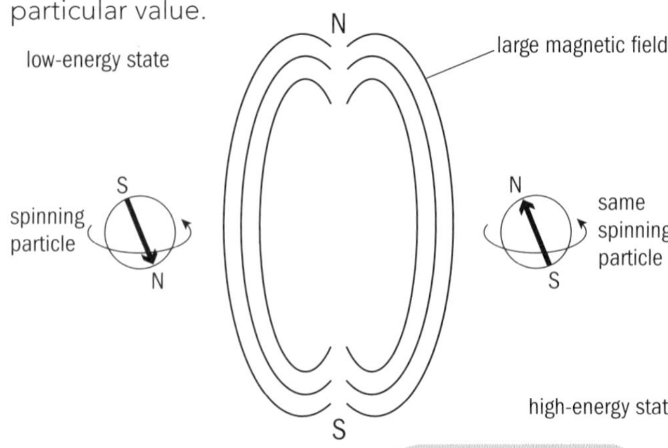

Large magnetic field and two particles in spin showing two energy levels.

Activity 20.3A

A pair of spin particles have opposing spins of equal value. One has a $+\frac{1}{2}$ spin and one has a $-\frac{1}{2}$ spin. What will be the net spin value?

Energy levels

Spinning particles can 'jump' between the energy states shown in the diagram above. If energy of an exact amount is given to it, the particle can change its spin axis so that the poles line up from N–S S–N to N–N S–S. (The spin is not perfect, but resembles the rotation of the Earth about its axis and rotates with a slight wobble. This kind of motion is called **precession**.) The particle has moved from a low-energy state to a high-energy state. You can relate this to holding two magnets so that their north and south poles are together; it takes effort on your part to keep them there. The energy value of the photon that can do this is directly linked to its frequency and in NMR is called the resonance frequency or **Larmor frequency.**

Typically, in magnetic resonance imaging (MRI) for hydrogen scans, the resonant frequency is in the range of 12→85 MHz.

The particle does not remain in the high-energy state for long and returns to the low-energy state, oscillating between the two states for a time after the electromagnetic pulse of energy is applied. The average length of time that a nucleus is in the high-energy state is called its **relaxation time**. Energy is emitted as a radio frequency. The signal detected can remain for a few milliseconds or even seconds and can provide vital information about the viscosity (flowing properties) of body tissue.

What affects signal intensity?

Factor
the density of protons within a volume of tissue sample and the relaxation times inside these volumes
the contrast of tissue types in the body and the need to introduce a contrast medium for comparison
the strength of the magnetic field produced by the primary magnet and the field strength of the gradient coil; the sequences of RF pulse that excite the protons and the timings of the receiving antenna

Did you know?

Nuclear magnetic resonance does not make use of any ionising radiation and is usually referred to as simply 'magnetic resonance'. It is thought that, this way, public perception of the procedure will be more positive.

Activity 20.3B

What is meant by the term relaxation time?

Image resolution

This is simply how sharp the image is and is quite complicated in MRI. In brief, the resolution can be improved by:

- reducing voxel size (a volumetric measure of substance similar to TV pixel)
- reducing the signal-to-noise ratio (SNR)
- reducing the field of view (FOV)
- reducing the slice thickness for cross-section
- increasing the speed of the signal sample.

Nuclei spin in magnetic field

↓

Incoming electromagnetic pulse wave of frequency = Larmor frequency at 90° to permanent magnetic field

↓

Nuclei absorb energy and move from low to high-energy state and oscillate

↓

Alternating wave is emitted as a signal and detected

↓

Nuclei return to low-energy state

↓

Signal decays

MRI scanner

The magnetic resonance imaging (MRI) scanner is a very large and heavy piece of equipment that uses the most up-to-date technological understanding and complex computer systems. Most of its weight comes from the permanent magnet and electromagnets which provide the very strong magnetic fields needed to polarise nuclei. These can weigh up to 100 tonnes. The MRI scanner is constructed of:

- **Main magnet** – originally a large conventional permanent magnet but now being replaced with superconducting electromagnets, cooled by liquid helium to –269 °C (4 K). They are costly but produce a high field strength of high stability.
- **Gradient field coils** – a resistive electromagnet which produces a gradient field where different magnetic strengths over the body help to pinpoint the signals. When scans are being carried out, a knocking sound can be heard when the gradient field coils are switched on and off.

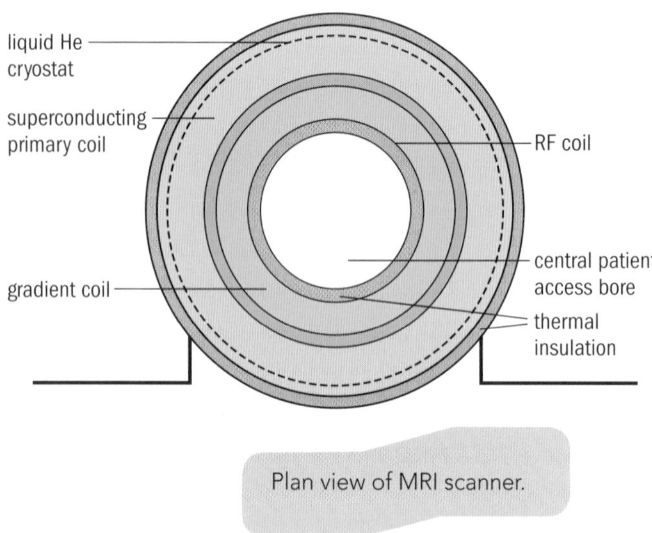

Plan view of MRI scanner.

- **Radio frequency (RF) coils** – produce the input radio wave which excites the nuclei and results in the NMR signal. They need to sustain a power output of up to 1 kW.

MRI uses

- brain scans
- spine scans
- joints (knee, hip, shoulder, wrist, etc.)
- pelvic region scans
- blood vessels
- heart conditions
- breast diseases
- abdominal issues
- other organs and tissues
- detects presence of abnormal body water, e.g. swelling, infection, bleeding and cysts.

MRI safety

- Remove implants and foreign bodies, metallic objects, metal clothing pieces and jewellery.
- All ferromagnetic materials must be removed.
- Hearing aids should be removed, and pacemakers should have been removed prior to examination at an earlier date.
- Dental fillings can distort the MR image if the head is the part of the body under investigation. The radiologist can make allowances for this if the information is made known.
- Quenching – in the event that the MRI scanner needs to be shut down immediately for safety reasons, the liquid helium which cools the primary coil to superconductivity can be evaporated either automatically or by pressing an emergency button.

Note: MRI is safer than CT and X-ray.

Case study: MRI operator

I am a technologist working as a MRI operator for a university hospital.

My job involves a busy schedule of health and safety checks, equipment monitoring and patient care throughout the day, which can be quite daunting.

I work with many highly qualified specialists and have learned a lot about how MRI works, even though physics was not one of my best subjects at school.

I enjoy helping patients to understand the machine and talking through the treatment procedure to make them feel more relaxed about what is happening. This is a very important part of the job because a calm patient will help to provide good scanning images.

Why will a calm patient produce better images than a nervous patient?

MRI versus X-ray image

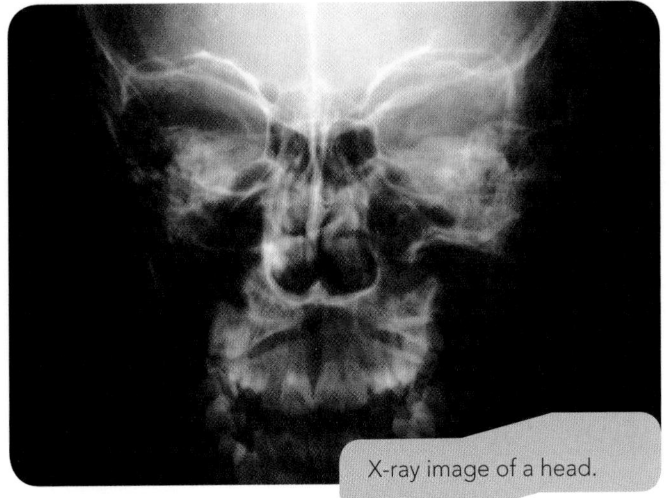
X-ray image of a head.

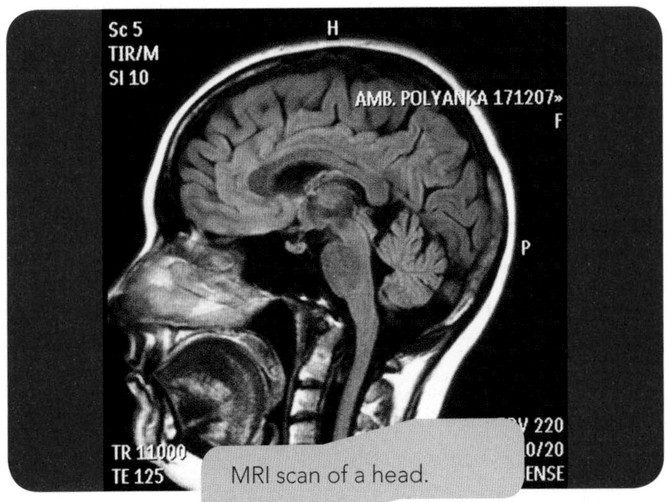

MRI scan of a head.

MRI	X-ray
clear differences in soft tissue types	clear contrast between soft tissue and bone
good contrast between tissues	outline tissue is not well defined
high resolution providing detail	little information concerning muscle, tendons or joints
	definition of tissue interfaces is good
bright and dark areas exposed	bright to dark gradient image obscures tissue
very small tissue changes can be detected	small blood clots or tumours may not show up
	bone density differences cannot be easily differentiated early in the patient's treatment

PLTS

Creative thinker

You will use your creative thinking skills when investigating how to change the intensity of signal of MRI.

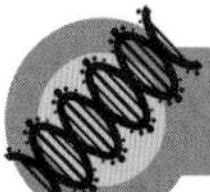

Functional skills

English

You will need to use your English skills when writing documents that analyse the differences and similarities of images.

Assessment activity 20.3

BTEC

You are called upon to provide an explanation of the procedure of an MRI scan to a patient as part of your duties as a technician within the radiology department of a major hospital.

1. Describe how the MRI scanner works in simple terms and list the main components with a brief description of each. P5
2. Provide an explanation of the principles of nuclear magnetic resonance and how different factors change the signal intensity. M3
3. Use a variety of images of the same body parts to evaluate the similarities and differences between X-ray and MR images. D3

Grading tips

Include some mention of proton spin and what happens to particles in a magnetic field in your answer for P5 with a more detailed explanation for M3. To complete D3, you will need to put yourself into the role of an image analyst and become familiar with the detail of individual images.

20.4 Radiation effects and safety

In this section:

Key terms

Malignant – when a disease, such as cancer, can affect other cells and cause death, it is said to be malignant.

Ionising – when an atom gains or loses one or more electrons it is said to be ionised. This can happen when more energy is given to the atom.

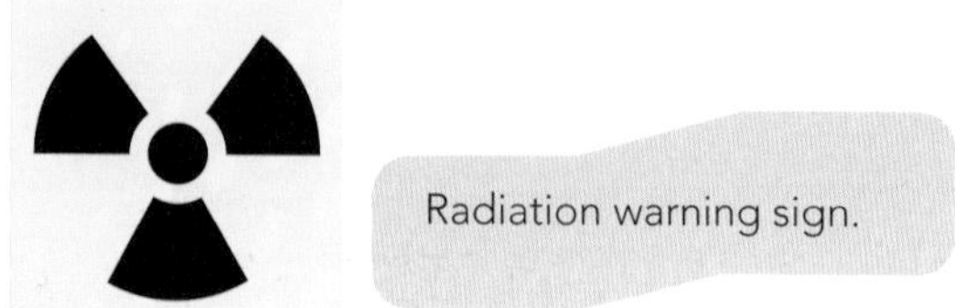
Radiation warning sign.

Effect of X-rays

Ionising radiation such as X-rays is used in the treatment of **malignant** medical conditions such as cancer. This treatment does not make the patient themselves radioactive and is painless. The benefits of using this treatment far outweigh the potential risks and the procedure is continually developing to become ever more efficient at destroying cancer cells.

The malignant tissue in the patient is called the 'target' and radiation used to destroy target cells often damages or destroys cells around the target tissue area. DNA within the cells, for example, can be damaged, affecting the ability of the cells to grow and divide.

The treatment of malignant cancer is a fine balancing act. Doctors target the cancerous cells as best they can and try to limit the number of healthy cells that are damaged. Most healthy cells can recover from this treatment if they are allowed sufficient time. This is why doses of radiation therapy are given in short bursts over many weeks. The following outlines what can happen to cells after varying doses of radiation:

1 Cells may be damaged but repair themselves and operate normally.
2 Cells are damaged but repair allows them to operate abnormally.
3 Cells die.

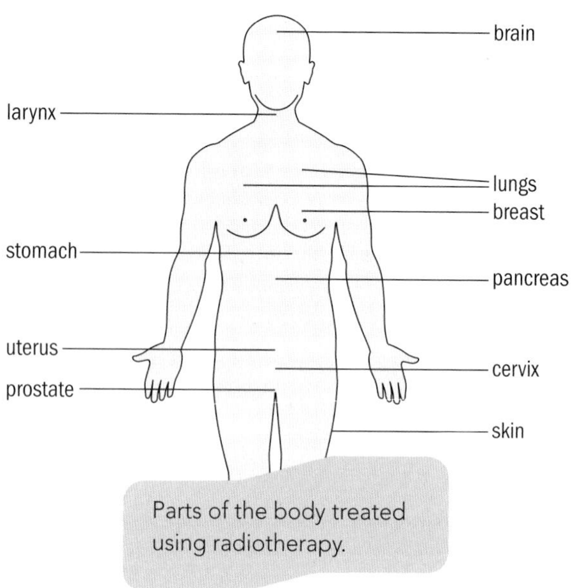

Parts of the body treated using radiotherapy.

Radiation dose

The amount of radiation given to patients in diagnosis is dependent on how close vital organs and tissues are to the malignant tumour. There are two terms commonly used by scientists when dealing with radiation doses:

- *Absorbed dose* – the amount of energy (in joules) received by a mass of tissue, which is measured in kilograms (kg). It has the unit J/kg and is called the gray (Gy).
- *Effective dose* – if the ionising radiation types are compared using the same amounts of energy, alpha particles cause much more biological damage, 20 times more damage than X-rays. In medicine, radiation affects different tissues and organs in different ways and so each tissue or organ has a number which is used as a quality factor. The absorbed dose is multiplied by this number to give the figure for effective dose – also measured in J/kg but called the sievert (Sv).

Did you know?

The average dose of radiation per person per year in the UK is 2.5 mSv. Most of this radiation comes from rocks. A small proportion comes from medical treatment and other sources.

Activity 20.4A

Why are patients given small but regular doses of radiotherapy over a long time period?

Radiotherapy

Radiotherapy (or radiation therapy) is the treatment of mainly malignant diseases carried out using either X-rays, gamma rays or electrons, although there are other forms of this treatment which use protons or neutrons to disrupt the cancerous cells. Patients are given a calculated dose of radiation over a precisely-calculated period of time. The dose is administered by a radiation oncologist who is highly trained in this area of science. There are numerous types of radiotherapy in use at present, for example:

- superficial system – treatment of skin tumours on the skin surface (50–150 kV)
- megavoltage system – using photons (6–22 MeV) for deep-seated tumours, and electrons (6–22 MeV) for superficial tumours on the skin and thyroid glands.

Linear accelerator (Linac)

This is a complex machine that produces X-rays by the acceleration of electrons. It allows the patient to be positioned suitably to receive treatment from multiple beams of X-rays that can be rotated around the patient. The X-rays are 'shaped' using wedges or compensators which allows the tumour to be better targeted.

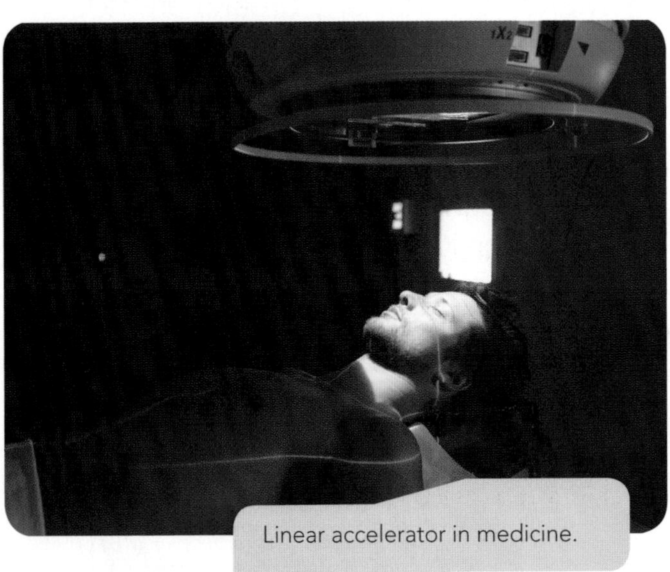

Linear accelerator in medicine.

When equipment such as X-ray machines is used in medical treatment, it is vital that the X-ray beam can be focused accurately enough to be able to destroy the malignant cancer cells whilst causing very little damage to the surrounding cells. Doctors need to work out where to direct the beam, how large it needs to be and what shape is required. To help in directing the beam, some physical devices are used, including:

- compensators – these are made of metals, uniquely manufactured to the patient's tumour size and shape to allow the dose to be distributed normally whilst evening out the skin surface contours
- wedges – shaped like door wedges to progressively lower the beam intensity over the treatment area.

Radiation safety

The table below shows the general effects of radiation doses on humans

Effect	Dose
blood count changes	very low
vomiting occurs	low
some deaths can occur	low – medium
$LD_{50/60}$ with some supportive care	medium
$LD_{50/60}$ with good medical treatment	medium – high
100% chance of death even with treatment	high

Note – $LD_{50/60}$ means that, with these dose levels, 50% of the population will die within 60 days unless medical attention is given.

You will also see doses referred to in the following categories:

- acute dose – a large dose of radiation received in a short time period
- chronic dose – small doses received over a longer period of time.

Not all people suffer the same degree of illness when they are exposed to radiation. Just as with illnesses and infections, much depends on age and general health.

Cells and sensitivity

Some cells react more sensitively to radiation doses than others. Cells which are specialised and do not generally divide quickly are much less sensitive to exposure than non-specialised, rapidly dividing cells.

MOST SENSITIVE

blood-forming cells (e.g. bone marrow)
stomach and intestinal cells
reproductive organs
skin cells
bone and teeth cells
muscle cells
central nervous system

LEAST SENSITIVE

Working with radiography in the medical industry brings employees into contact with more radiation than most people would otherwise experience. The Government has developed specific guidelines on exposure for workers in this type of industry, which are detailed in the Ionising Radiations Regulations (**IRR**) paper from 1999. The recommended maximum dose levels have also been agreed by an advisory group of international experts – the International Commission for Radiological Protection (**ICRP**).

Whole body doses for workers must be no more than 20 mSv averaged over 5 years, and they must not be exposed to more than 50 mSv in any single year.

Activity 20.4B

Which cells are more sensitive to radiation exposure, stomach cells or muscle cells? Why is this the case?

Dosimeters

Employees in the medical industry working with **ionising** radiation must wear sensitive radiation-detecting equipment and follow strict operational guidelines for their own safety and that of others. This equipment must be visibly worn on the body and checked regularly. It is available in two general forms although there are many variations in use:

- film badges – these hold a thin strip of aluminium oxide, which is sensitive to radiation and traps excited electrons, and are measured by a device after every working shift. They are worn on the body between the neck and waist.
- ring badges – these contain lithium fluoride crystals, which also trap excited electrons within them until the badge is heated to very high temperatures. The energy given off as visible light is proportional to the radiation dose. This is called **thermoluminescence**. They are worn on the hand.

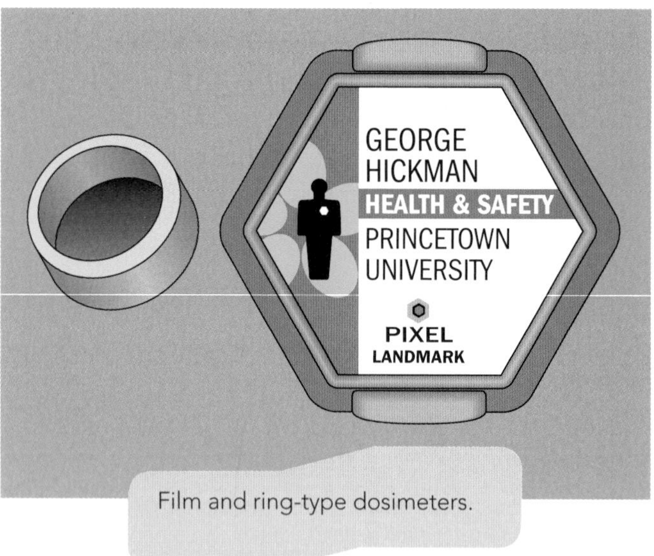

Film and ring-type dosimeters.

REDUCE YOUR RADIATION EXPOSURE!

- Keep a close check on your time in an area of radioactive materials.
- Make sure that any breaks are not taken in areas where radioactive materials are stored.
- Keep your distance from sources and check your dosimeters periodically.
- Shield yourself from exposure using the correct procedures.
- Notify your supervisor if you are pregnant.

Assessment activity 20.4

Working in the X-ray department of a busy hospital means that you will need to attend regular specific additional training sessions for health and safety as part of your continued professional development.

1. Provide a slide demonstration explaining the way in which X-rays are used to treat malignant disease. Provide a brief explanation of equipment which may be used. P6
2. Explain the physical effects of being exposed to a lot of radiation. M4
3. Evaluate the various types of radiotherapy practices that are currently in use and explain the function of equipment that allows these kinds of treatment. D4

Grading tips

List the components of the equipment which focus X-rays onto the target with a simple explanation of how they work and highlight what can happen to cells during radiotherapy for P6. Link the doses of radiation to the symptoms of radiation exposure and comment on preventative measures to achieve M4. You will need to include specific radiation types used for particular diseases for D4 and the equipment used to produce and target the radiation.

PLTS

Independent enquirer

You will use your independent enquirer skills when questioning the safety aspects of radiotherapy.

Functional skills

ICT

You will use your ICT skills when producing a PowerPoint for information on radiotherapy.

WorkSpace

Rebecca Noble

Pre-registration Radiotherapy Physicist

I work as a medical physicist within the radiotherapy department at University Hospital of North Staffordshire. I am in the advanced stages of my training.

My role is to ensure that radiation doses used to treat cancer patients are delivered accurately, safely and in the most effective way.

As a physicist, on a day-to-day basis, I work alongside therapeutic radiographers, technicians, oncologists and nursing staff.

My routine work follows a rota. One week I could be performing quality control tests on the treatment machines to ensure they are performing correctly, the next week I could be working on the planning of patient treatments, or administering a radioactive capsule for an inpatient treatment. Physicists are also on-hand to give advice on complex or unusual treatments.

My work involves research and development, as we are constantly looking at ways to improve the service and bring in new techniques to increase the likelihood of curing the patients.

Knowing that I am using my physics knowledge to contribute to the treatment of cancer patients is very rewarding. There is daily variety in my work and I enjoy the fact that problem solving is key to the role.

We use treatment planning software to calculate and optimise the dose delivered to the patient for a given technique. The software has a new algorithm to calculate dose which I have been involved in testing recently, before its clinical use. Part of this has involved performing simple calculations and physical measurements to establish that the algorithm correctly calculates the dose that the treatment machine delivers, and that it accurately models the underlying physics. I also had to re-calculate a number of plans for different treatment types in order to assess the extent of any differences between the old and new algorithms.

Think about it!

1 What skills do you think would be required to work well within such a multi-disciplined team?
2 What advice might physicists give on complex or unusual treatments?

Just checking

1 Give a six-step guide to how X-rays are produced.
2 What use is the Doppler effect in ultrasound imaging?
3 What are the characteristics of an effective radiopharmaceutical?
4 What are the purposes of the collimator and photomultiplier tubes in a gamma camera?
5 Why is MRI safer than other forms of medical diagnosis?
6 Explain the purpose of the main magnet in MRI.
7 What forms of radiotherapy are used in medical treatment?
8 Name two devices used to detect radiation dose and explain how they work.

edexcel

Assignment tips

To get the grade you deserve in your assignments remember to do the following:

- Make a summarised note of the important aspects of radioactivity and ultrasound. Diagrams will be of value in this and can help you to visualise quite difficult aspects of the physics involved. Much of the unit deals with the equipment used in medical physics and so block diagrams can be used to show their operation.
- Begin by completing all the required tasks for the Pass criterion. For some of these in this unit a suitable list and brief description will be enough, but make sure that the list covers all the key points discussed in your tutor sessions and the assignment tasks.
- To achieve Merit and Distinction criteria you should include explanations where asked. Use a variety of websites and textbooks for the topic and compare the information to be sure that they agree. Make a note as a reference for websites which you have used. Many sites are not as useful as they look. The processes involved in medical physics are very complicated so try to deal with one aspect at a time in a step-by-step approach to your learning.

Some of the key information you'll need to remember includes the following:

- Radioactivity and characteristics of the electromagnetic spectrum are a natural phenomenon which can be put to good use through an understanding of their basic principles.
- The use of radiopharmaceuticals in medical diagnosis is a valuable means of highlighting problems in human physiology and combines the properties of radioactive isotopes and pharmaceutical compounds.
- Diagnosing diseased tissue and organs has developed to an advanced level with the introduction of nuclear magnetic resonance. Images produced of the inside of the human body are now extremely detailed and are becoming more detailed as the technology develops.
- Radiotherapy is an effective method to target and destroy malignant diseases. There are some aspects which need consideration – protection of healthy cells and limiting radiation doses.
- Working in this field of medicine involves a clear understanding of the health and safety implications and adherence to strict codes of practice.

You may find the following websites useful as you work through this unit.

For information on...	Visit...
ionising radiation and ultrasound	National Physical Laboratory website Health and Safety Executive The radiology information resource for patients HSC online
radiopharmaceuticals and gamma camera	SVH School of Nuclear Medicine Technology
nuclear magnetic resonance	Michigan State University Department of Chemistry
radiation effects and safety	University of South Carolina School of Medicine Department of Radiology

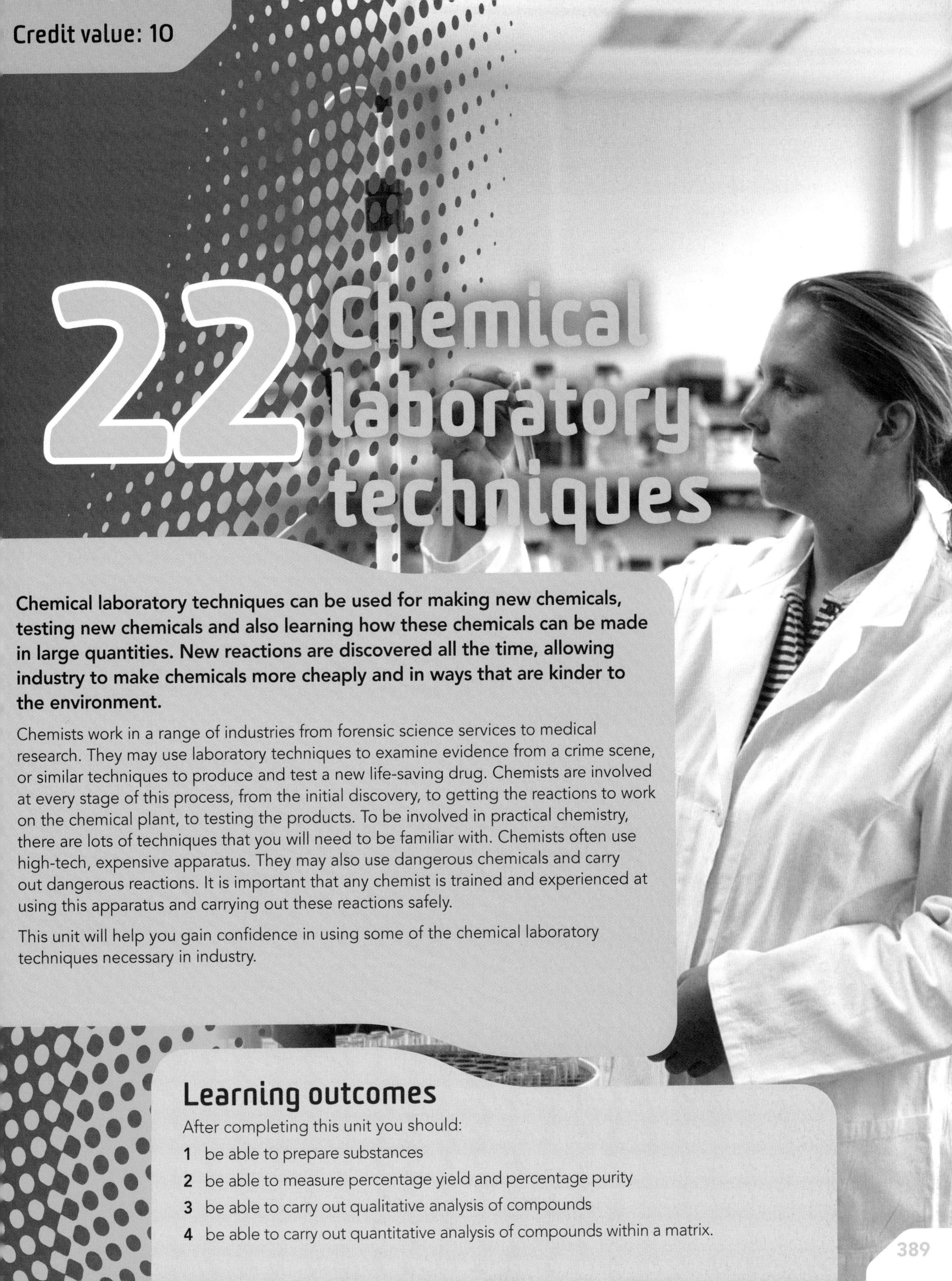

Credit value: 10

22 Chemical laboratory techniques

Chemical laboratory techniques can be used for making new chemicals, testing new chemicals and also learning how these chemicals can be made in large quantities. New reactions are discovered all the time, allowing industry to make chemicals more cheaply and in ways that are kinder to the environment.

Chemists work in a range of industries from forensic science services to medical research. They may use laboratory techniques to examine evidence from a crime scene, or similar techniques to produce and test a new life-saving drug. Chemists are involved at every stage of this process, from the initial discovery, to getting the reactions to work on the chemical plant, to testing the products. To be involved in practical chemistry, there are lots of techniques that you will need to be familiar with. Chemists often use high-tech, expensive apparatus. They may also use dangerous chemicals and carry out dangerous reactions. It is important that any chemist is trained and experienced at using this apparatus and carrying out these reactions safely.

This unit will help you gain confidence in using some of the chemical laboratory techniques necessary in industry.

Learning outcomes

After completing this unit you should:

1 be able to prepare substances
2 be able to measure percentage yield and percentage purity
3 be able to carry out qualitative analysis of compounds
4 be able to carry out quantitative analysis of compounds within a matrix.

Assessment and grading criteria

This table shows you what you must do in order to achieve a pass, merit or distinction grade, and where you can find activities in this book to help you.

To achieve a **pass** grade the evidence must show that you are able to:	To achieve a **merit** grade the evidence must show that, in addition to the pass criteria, you are able to:	To achieve a **distinction** grade the evidence must show that, in addition to the pass and merit criteria, you are able to:
P1 follow a range of procedures to obtain substances by reaction and extraction **See Assessment activity 22.1**	**M1** describe the scientific principles behind the key steps in the preparative methods for substances **See Assessment activity 22.1**	**D1** explain how the yield, purity and atom economy of the substances prepared may be affected by changes to the methods used **See Assessment activity 22.2**
P2 follow methods to determine % yield, % purity and atom economy of prepared substances **See Assessment activity 22.2**	**M2** describe the main problems with each of the preparative methods used **See Assessment activity 22.1**	
P3 use test-tube reactions and infrared spectroscopy to identify functional group compounds **See Assessment activity 22.3**	**M3** explain how the test-tube reactions and infrared spectra allowed the functional groups of each substance to be identified **See Assessment activity 22.3**	**D2** evaluate whether the qualitative analysis carried out was conclusive **See Assessment activity 22.3**
P4 carry out chemical tests to identify inorganic substances **See Assessment activity 22.3**	**M4** explain the bases of the chemical tests used to identify inorganic substances **See Assessment activity 22.3**	
P5 perform quantitative analysis of commercial/natural substances by following given methods **See Assessment activity 22.4**	**M5** identify sources of error and uncertainty in the quantitative analyses carried out **See Assessment activity 22.4**	**D3** evaluate the reliability of the analyses of the commercial/natural compounds **See Assessment activity 22.4**

How you will be assessed

Your assessment is likely to be in the form of:

- practical tasks
- written accounts of practical tasks
- explanations of results from practical tasks.

Ashley, 18 years old

This unit allowed me to gain confidence in handling equipment. I was able to work independently to make and purify substances. I began to understand how substances are made industrially and how they are tested in the laboratory. Here are some of the techniques I used.

I carried out a titration on chemicals of unknown concentration. I used a burette to accurately measure an acid into an alkali where I knew the concentration of the acid but not the alkali. I knew the reaction was complete when the solution changed colour. By carrying out calculations using the amounts of acid and alkali I used, I could then say how concentrated the alkali was.

I also had to extract the pigment from some leaves. I used a technique that was new to me, called reflux, where the reactants are heated at a constant temperature and the vapours condensed and returned to the reacting mixture.

I then had to use some new methods of separating chemicals. Distillation I had used before, but I also carried out vacuum and gravity filtration. In fact gravity filtration is just normal filtration I had used when I was at school, but I had not heard it called that before. For vacuum distillation we used running water to create a vacuum that pulled the liquid through quickly leaving the solid behind.

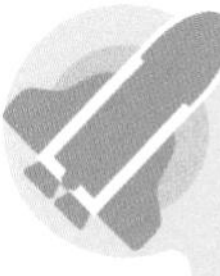

Catalyst

Why is practical chemistry important?

Practically everything we use in our everyday lives has been made or extracted by processes using chemistry – semiconductors inside our computers, pharmaceuticals, plastics, textiles, foods. Chemists need to know how to make and extract substances, to measure the yield and to measure how pure substances are. Nowadays we are more concerned than ever about the environment so we also need to compare different ways of making the same products to see which methods use fewer raw materials (atom economy). We should not ignore the amount of energy used in manufacturing processes and transportation of raw materials.

Work in groups to discuss products made using chemical processes.

List all the things you use in everyday life that have been made using chemical processes. What processes might have been used?

How do you think companies ensure chemical products are pure or fit for purpose?

22.1 Preparing substances

In this section:   

Key terms

Organic compound – substance whose molecules contain one or more carbon atoms, which can be in the form of long carbon chains (including alkanes, alkenes and alcohols).

Inorganic compound – any substance in which two or more chemical elements other than carbon are combined, as well as some compounds containing carbon, but lacking carbon–carbon bonds (e.g. CO_2).

Titration – volumetric analysis where one reagent is added to another from a burette until an end point is reached. The volume added can be used to calculate the concentration of the unknown reagent.

Precipitation – technique used to produce a suspension of solid particles in a liquid by chemical reaction.

Filtration – technique to separate solids from the liquid they are suspended in.

Evaporation – separation technique where liquid is heated until it becomes vapour, leaving any dissolved solid behind.

Recrystallisation – a method of purifying a chemical product. The product and any impurities are dissolved in a solvent which is warmed and then cooled. The pure product will recrystallise out of solution leaving the impurities dissolved.

Reflux – a method using a condenser that will keep a reaction mixture boiling for as long as it needs to complete.

Distillation – separation of liquids from other liquids or from solids due to differing boiling points.

Preparative chromatography – a method of separating and identifying components of a mixture by their differential movements through a two-phase system. This is due to their differing solubilities within the solvent.

Chemical plant – where industrial chemical processes are carried out on a large scale.

van der Waals forces – an attractive force between atoms or molecules. They are much weaker than the forces within bonds and get weaker the further the atoms or molecules are from each other.

As a chemist working in a laboratory you will need to be confident in preparing a variety of different chemical substances. These may be **organic** or **inorganic**. Some preparations may have only one step where you add one chemical to another. Other preparations may have several steps and use complex equipment. You will already have used some preparative techniques in earlier courses. You may have looked at how **distillation** is used to separate different fractions from crude oil to get petrol or diesel, etc. You have probably also done some **preparative chromatography**, although you may have just called it chromatography. You may have tested the ink from several pens to see if the ink has the same components. You do this by placing a spot of each ink on a line near the bottom of a piece of chromatography paper. You place the paper in a solvent such as ethanol so that the spots sit just above the solvent. As the solvent is absorbed by the paper it will rise up the paper taking the different components of the ink with it. The more soluble components will move the furthest. You can then compare the positions of the spots. If they are in the same place then you know the two inks are the same.

Problems with preparative techniques

When making different products, a chemist has to be aware of the difficulties the different techniques used can present. You have to consider any safety issues. You also have to take into account whether the technique will give you a good yield, will it give a pure product, are there any issues with safety with any of the impurities or other chemical made in the reaction. You also need to consider how expensive the technique is and how easy it is to do.

Preparing metal salts

This is often a simple one-step procedure. Acids react with reactive metals and metal compounds, for example:

zinc + hydrochloric acid → zinc chloride + hydrogen

magnesium oxide + nitric acid → magnesium nitrate + water

potassium hydroxide + sulfuric acid → potassium sulfate + water

calcium carbonate + hydrochloric acid → calcium chloride + water + carbon dioxide

Preparing a metal salt and using filtration to get a pure dry product

Copper(II) sulfate can be made by reacting copper(II) carbonate with an acid. This will neutralise the acid to form a salt.

Safety note: Eye protection must be worn. Use a pipette filler. Avoid skin contact with the acid and with the copper compounds and crystals.

Method

1 Pipette 20 cm^3 of 1 mol dm^{-3} sulfuric acid into a 100-cm^3 beaker.

2 Add copper(II) carbonate solid until the solution no longer fizzes. Copper(II) sulfate is produced.

3 Filter the copper(II) sulfate solution into a pre-weighed **evaporating** basin. The excess copper carbonate will be left on the filter paper.

4 Leave the solution to evaporate. Crystals of $CuSO_4.5H_2O$ will begin to form.

5 After about a week, reweigh the evaporating basin.

Filtration equipment.

The method for calculating the percentage yield of your preparation process is described on page 400.

Activity 22.1A

In the process for making copper sulfate described above:

- why does the solution fizz?
- what colour do you expect the crystals to be?

Precipitation reactions

A product can be formed by a **precipitation** reaction. This is where a solid is formed out of a solution. For example, when a solution of silver nitrate is reacted with a solution of sodium chloride then silver chloride and sodium nitrate are formed.

$$AgNO_3(aq) + NaCl(aq) \rightarrow AgCl(s) + NaNO_3(aq)$$

The silver chloride is insoluble in the water and so forms a solid precipitate. This can then be filtered off and dried to get the product.

Preparing a metal salt by titration

Sodium chloride can be made by reacting sodium hydroxide solution with hydrochloric acid solution. In order to find the exact proportions in which the two solutions react, you need to do a **titration** with an indicator (see page 416). Once you have found the amounts you need, you will then prepare the correct mixture in an evaporating basin without indicator.

Safety note: Eye protection must be worn. Use a pipette filler. Avoid skin contact with the acid and alkali. Beware of 'spitting' of hot chemicals in step 11 and splashing of chemicals at any stage; do not heat the evaporating basin directly with a Bunsen burner; use a steam bath.

Method

1 Pour 0.1 mol dm^{-3} sodium hydroxide into a burette, filling it to the 0 mark.

2 Using a pipette, accurately measure 25 cm^3 0.1 mol dm^{-3} hydrochloric acid into a conical flask.

3 Add phenolphthalein as an indicator to the flask.

4 Dropwise add the sodium hydroxide to the flask by slowly opening the tap on the burette. Keep swirling the flask.

5 The reaction is complete when you see the indicator turn from colourless to red.

6 Use a table like this to record your final volume of sodium hydroxide used and repeat the titration five times:

initial burette reading (cm^3)					
final burette reading (cm^3)					
titre (cm^3)					

Case study: Testing solutions

Frank is a laboratory technician in a sixth form college. He prepares all the chemicals ready for the chemistry practicals carried out by the learners. When they are going to do a titration experiment, he prepares and tests the solutions they are going to use.

1. **Why does Frank test the solutions he has made?**
2. **How does Frank test the solutions he has made?**

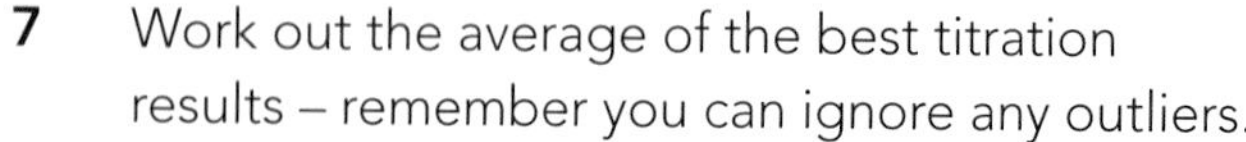

7. Work out the average of the best titration results – remember you can ignore any outliers.
8. Weigh an evaporating basin.
9. Pipette 25 cm^3 of 0.1 mol dm^{-3} hydrochloric acid into the evaporating basin.
10. Add the correct amount of 0.1 mol dm^{-3} sodium hydroxide to neutralise the acid; that is, the average of the best titration results from step 6.
11. Reduce the volume of the solution by about half by evaporating it on a steam bath as instructed and leave until the following week to crystallise.

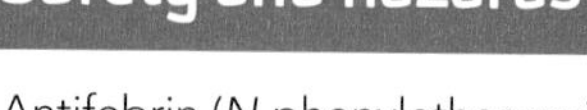

Safety and hazards

Antifebrin (*N*-phenylethanamide) is harmful and must be handled with care. When carrying out any practical procedure a risk assessment should be carried out. You can use a template from your tutor or they are available in textbooks. The following website has a downloadable risk assessment template that could be useful:

The Physical and Theoretical Chemistry Laboratory, Oxford University

In some workplaces, risk assessments have already been carried out and these should be followed. Any new procedure will need a new risk assessment and this may be your responsibility.

Preparing organic compounds

Antifebrin

The pharmaceutical industry is worth millions of pounds. It is important because of the life-saving and life-enhancing drugs that are produced. Over the years, the pharmaceutical industry has tried to produce more effective drugs at lower costs.

Did you know?

In the nineteenth century, antifebrin was used for its analgesic and fever-reducing properties. It was soon established that the drug was toxic. Much later, it was shown that its therapeutic properties were due to its metabolism to produce paracetamol, and its toxic properties to the production of aniline. It is, however, used as a precursor in the manufacture of other pharmaceutical products.

Aniline and ethanoic anhydride react according to the equation:

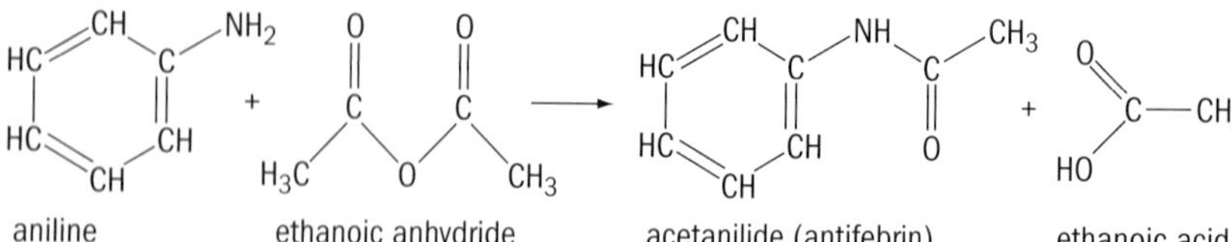

Safety note: Eye protection must be worn. Avoid skin contact with any of the chemicals and avoid inhaling the fumes from them.

Method

1. Pour 50 cm^3 of 0.5 mol dm^{-3} hydrochloric acid into a 100-cm^3 conical flask.
2. Add 2.5 cm^3 of aminobenzene (aniline) (TOXIC) (density 1.02 g cm^{-3}) and swirl to mix.
3. If the solution is coloured, add a small amount of decolorising charcoal, and swirl for a minute.
4. Filter off the charcoal using a fluted filter paper. (Filter into a similar conical flask.)
5. In a 100-cm^3 beaker, dissolve 4 g of sodium ethanoate in 10 cm^3 of water.
6. Warm the sodium ethanoate gently on a steam bath.

7 When the temperature of the sodium ethanoate solution reaches about 50°C, add 3 cm^3 ethanoic anhydride (CORROSIVE) (density 1.08 g cm^{-3}) and swirl.
8 Add the sodium ethanoate solution, containing ethanoic anhydride, to the flask and swirl to mix.
9 Cool the flask in an ice bath for 20 minutes.
10 Filter off the crystals under vacuum using a Büchner funnel.
11 Wash the crystals with a small quantity of iced distilled water. (To do this, detach the suction. Pour on the water and replace the suction.)
12 To **recrystallise** the product, scrape the crystals into a clean conical flask or beaker and add about 10 cm^3 of distilled water.
13 Warm the flask on a hotplate until the water has just reached boiling point.
14 If the crystals have not dissolved, add another 5 cm^3 of water and bring the mixture back to the boil.
15 Repeat the additions of water until the crystals have just dissolved.
16 Set the mixture aside to cool. Purer crystals will form.
17 Filter off the crystals under vacuum using a Büchner funnel.
18 Air dry the crystals before measuring the melting point. (Pure acetanilide (HARMFUL) melts at 115°C.)
19 Use infrared or ultraviolet spectroscopy to determine the purity if possible.

PLTS

Effective participator

Carrying out the practical activities in this unit will show you are an effective participator.

Activity 22.1B

Why is a steam bath used instead of a Bunsen burner in step 6 of the method to prepare antifebrin?

Extracting organic compounds

Did you know?

Many plants synthesise substances that are useful. Some can be used as drugs, dye, fragrances, cosmetics or adhesives. The effective ingredient in aspirin is found in willow bark and has been used in a pain-relieving tea for centuries.

Chemical companies recognise that many plants contain important and useful chemical substances. It is important that they can extract these substances as effectively and efficiently as possible. Understanding how chemicals are bonded can help us decide how the substances can be extracted.

In Unit 1, you studied types of chemical bonding. Covalent bonding and ionic bonding may be regarded as two extremes. Most compounds exhibit behaviour that lies between these two extremes. One way of thinking about the way substances behave is in terms of polarity. Wax and petrol have fairly pure covalent bonding. They are non-polar. Molecules in wax and petrol have only weak forces of attraction between them; that is, only **van der Waals forces**. The compounds have relatively low melting points and boiling points because the forces holding the molecules together are weak.

Compounds that contain atoms with high electronegativity (they attract a positive charge) are usually polar. For example, a water molecule contains an oxygen atom. Oxygen is highly electronegative.

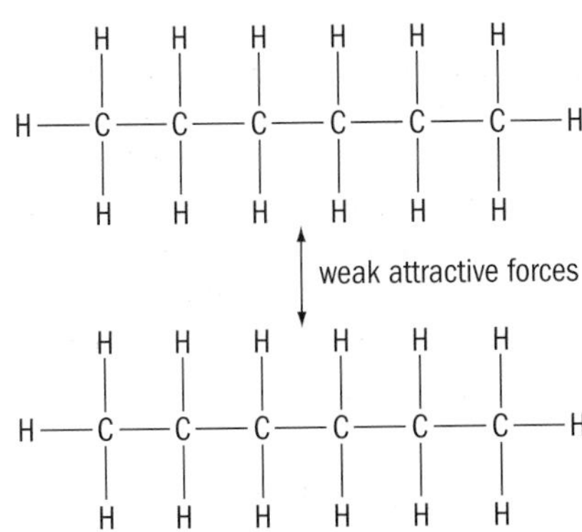

van der Waals forces between adjacent hexane molecules.

The oxygen has a bigger share of the shared pairs of electrons than the two hydrogen atoms have. As a result, the oxygen end of the molecule is slightly negative (and will attract a positive charge), and the hydrogen ends are slightly positive. So, in addition to the weak van der Waals forces present in hexane, there is also an electrostatic attraction between the oxygen end of one water molecule and the hydrogen end of another water molecule. This attraction is known as 'hydrogen bonding'. It is responsible for water having a much higher boiling point than, say, methane which is a similar size.

Hδ+ Hδ+
O δ- ········ H δ+ — O δ-
Hδ+

Hydrogen bond between adjacent water molecules.

The type of bonding between the molecules in solvents leads solvents to be called 'non-polar', in the case of hexane, or 'highly polar' in the case of water. There is a spectrum of polarity in solvents, as shown in the diagram.

Highly polar

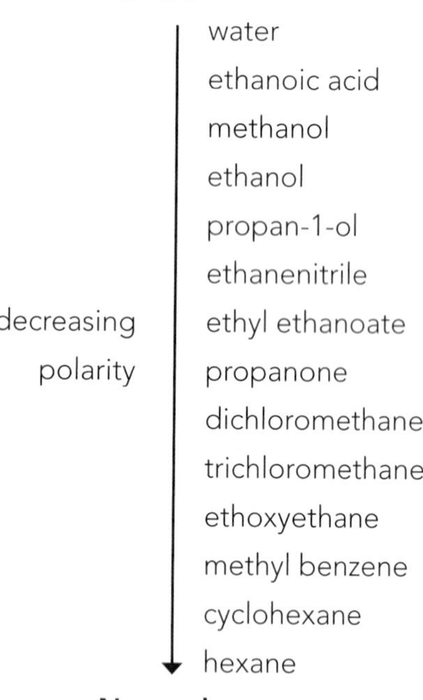

Substances tend to dissolve in solvents whose polarity is similar.

- Wax, which is a non-polar hydrocarbon, will dissolve easily in hexane which is also a non-polar hydrocarbon.
- Sodium chloride, an ionic compound, may be regarded as being highly polar. It dissolves easily in water which is also highly polar.

Note that for something to dissolve in water, the very strong forces of attraction between adjacent water molecules must be overcome. The situation of salt being dissolved in water must provide even stronger attractions between particles than salt on its own and water on its own.

Using reflux for extracting pigments from leaves or herbs

Soxhlet apparatus provides an efficient method of extraction that is continuous and uses pure solvent. This procedure uses a type of **reflux** equipment. The leaves or herbs are placed in the porous thimble. If you do not have access to Soxhlet apparatus, the extraction may be carried out in a beaker (as outlined overleaf).

Rotary evaporation

The solvent may be removed from the extracted chemicals by rotary evaporation.

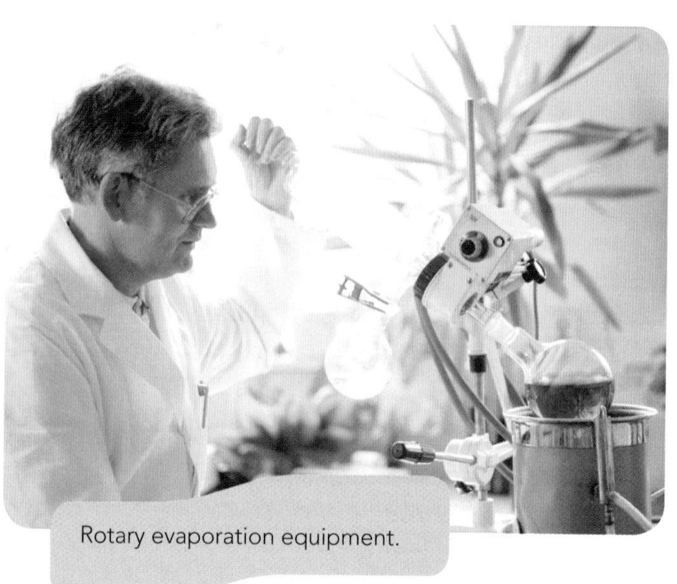

Rotary evaporation equipment.

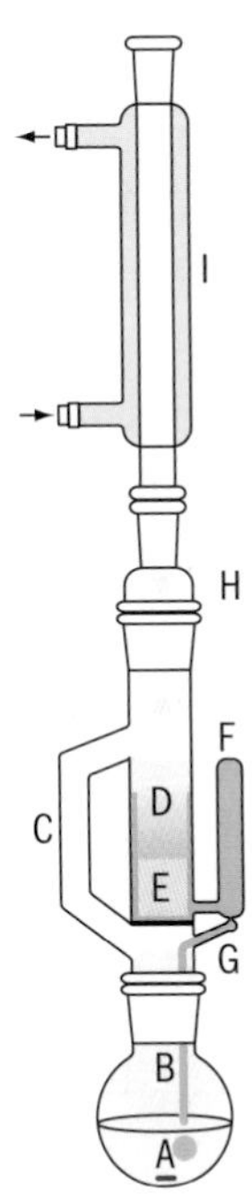

- The substance, E, from which something is to be extracted is placed in a porous thimble, D, and placed in the extractor.
- Solvent is placed in the round-bottomed flask, B, which contains anti-bump chips or a stirrer bar, A.
- A condenser, I, is fitted to the top of the extractor with an expansion adaptor between, H.
- The apparatus is assembled and heat from an isomantle is applied.
- The solvent boils. Solvent vapour travels from the flask, up the side tube, C, to the condenser where it is turned back into a liquid.
- The pure liquid drips on to the herbs, extracting some of the pigment.
- The solvent level in the extractor rises until the level reaches the same height as the top of the siphon tube, F, at which point the solvent siphons down, G, into the flask.
- The process repeats itself many times. Each time, pure solvent extracts more pigment.
- Finally, the extract will be colourless, indicating that no more pigment is being extracted.
- The apparatus may then be cooled and the extract analysed. This is an efficient method of extraction which is continuous and which uses pure solvent.

Soxhlet apparatus.

- The flask is connected to a vacuum pump and, because of the low pressure, the solvent evaporates easily.
- The flask is placed in a hot water bath to ensure that there is enough energy to promote evaporation.
- The extract in the flask is rotated very quickly.
- The extracted solvent is condensed and collected in another flask.

A simple extraction method

This can be used to find out which solvent is optimum for extracting pigments from herbs or leaves. You could use it to investigate the following solvents: water, ethanol, ethyl ethanoate, propanone and petroleum ether (a mixture of hydrocarbons including, and chemically similar to, hexane).

Safety note: Eye protection must be worn. This procedure must be carried out in a fume cupboard as most of the solvents are highly flammable and the fumes they give off should not be inhaled.

Petroleum ether – extremely flammable, harmful, danger to the environment.

Ethanol – highly flammable.

Ethyl ethanoate – highly flammable, irritant.

Propanone – highly flammable, irritant.

Method

1 Weigh 1 g of dried herbs into each of five 100-cm^3 beakers.
2 Place the beakers in the fume cupboard.
3 Using a measuring cylinder, add 20 cm^3 of a different solvent to each beaker.
4 Swirl each beaker in an identical way.
5 Place a watch glass on top of each beaker and leave the beakers for at least 1 hour.
6 Label and weigh five evaporating basins.
7 Filter the contents of each beaker, in turn, into an evaporating basin.
8 Leave the evaporating basins for the solvents to evaporate.
9 Weigh the evaporating basins. (This can be done outside the fume cupboard as the solvents are no longer present.)
10 Calculate the mass of pigment extracted by the solvent and comment on the size of the values you obtained.
11 Write a report on the procedure, explaining each step that you have used and describing any problems with this method.

PLTS

Reflective learner

Making judgements based on your results will help you develop your skills as a reflective learner.

Extracting paracetamol

Some laboratory technicians work in quality control. This is an important role, as a quality controller tests all substances made to ensure they contain the right sort and amount of chemical. They also make sure that the chemical does the job for which it is intended, whether it is a drug, a dye or a cleaning product, for example.

Substances are sampled and tested throughout the manufacturing process and before delivery to customers. The drug paracetamol is sold as tablets that contain substances in addition to paracetamol (acetaminophen). A '500 mg' tablet does not weigh 500 mg (0.5 g) – it weighs more than that; it contains 500 mg of paracetamol, and the rest of the tablet consists mainly of a 'binder', often some sort of starch.

The tablets are sampled and tested to ensure that they leave the factory in the best possible condition. Part of that process is to extract the paracetamol from the tablets to test that the drug has not degraded.

Safety note: Eye protection must be worn. Work in a fume cupboard to avoid inhaling any fumes from the propanone.

Method

1 Weigh two paracetamol tablets using a top-pan balance.

2 Put the two tablets into a mortar and grind them with a pestle.

3 Transfer the powder to a 100-cm^3 flask and add 20 cm^3 of propanone.

4 Warm the flask in a water bath at 60°C in a fume cupboard. Once the solution starts to boil, remove it from the heat.

5 Allow the solution to cool and then filter it (using a glass funnel and filter paper) into a pre-weighed evaporating basin with your name on it.

6 Leave it to one side in the fume cupboard for a week. The propanone will then have evaporated and white crystals of paracetamol will have been left behind.

7 Weigh the evaporating basin.

8 Scrape out the evaporating basin and empty the contents into a 100-cm^3 conical flask. Put a tiny amount in a labelled sample tube for a melting point measurement.

Unit 4: See page 101 for more information about melting point measurements.

9 Recrystallisation: add 5 cm^3 of water to the paracetamol in the flask. Holding the flask in a peg, heat it on a hot plate until it comes to the boil. If the paracetamol has not all dissolved, add a few more cm^3 of water and bring it to the boil. Add more water if it has not all dissolved. Careful, there may still be some filler that is never going to dissolve! Leave the flask on the hotplate while you collect your hot funnel for the next part.

10 You are now going to use the technique of hot **filtration** which removes insoluble impurities, but leaves the product (paracetamol) in solution. Take a clean 100-cm^3 beaker to the oven in the prep room. Collect a hot funnel with a piece of cotton wool inserted and sit it over the beaker.

11 Take this back to your bench and filter the hot solution through it. Let the solution cool. Crystals of paracetamol should form in the beaker.

12 Set up a Hirsch funnel in a clamp beside a vacuum pump. Put two pieces of filter paper on the Hirsch funnel. Dampen them ever so slightly with distilled water and turn on the vacuum pump. Filter off your paracetamol crystals. Allow air to suck through the crystals for a few minutes to dry them. This is vacuum filtration.

13 Put a piece of scrap paper on the bench with a piece of filter paper on top of it. Tip the crystals from the Hirsch funnel on to the filter paper and dry them with another piece of filter paper on top.

14 Label a sample tube and put it on a balance. Set the balance to zero. Remove the tube, add the crystals and then weigh the mass of crystals you have obtained.

Recrystallisation

Selection of the solvent in step 9 is very important. If the substance is soluble in a solvent when it is cold, the crystals will not come back out of solution. You need to select a solvent in which the substance is soluble when the solvent is hot – but not when the solvent is cold. In step 10, any solid impurities are removed by 'hot filtration' – filtering the mixture while it is hot. This requires a hot filter funnel. It is often convenient to use cotton wool to trap the impurities. As the solution cools in step 11, crystals of pure paracetamol form. The more slowly they form, the more regular the crystals will be and the less likely they will be to trap impurities as they form.

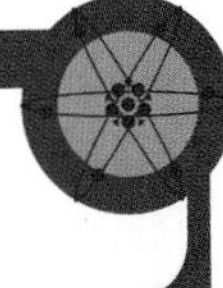

Activity 22.1C

Working with a partner, explain how all the techniques used to extract the paracetamol work.

Assessment activity 22.1

1 You are working as a laboratory technician in a sixth form college. The learners will be carrying out titration experiments on 0.1 mol dm^{-3} sodium hydroxide and 0.1 mol dm^{-3} hydrochloric acid. The tutor has asked you to carry out the experiment first to ensure that the chemicals are the strength he expects and that the learners will be able to get the correct results.
   - a Carry out the experiment, following the correct safety procedures, and write a report describing the key steps in the preparation method you would use and the scientific principles behind these steps. P1, M1
   - b Describe any problems there may be with this method. M2

2 You are working for a pharmaceutical company and have been asked to carry out the preparation of antifebrin in the laboratory so that you can help to explain some of the problems in the **chemical plant**. Following the correct safety procedures, carry out the preparation and write a report for your supervisor describing the process you have used. P1

   What have you learned that would affect how the process was carried out on a commercial scale on the pharmaceutical plant? M2

3 You are a research and development technician, working for a company trying to optimise solvent extraction for one of its preparative processes. You have extracted the pigments from dried herbs using a selection of solvents.

   Write a report explaining which is the best solvent to use. P1, M1

4 You are a quality control technician for a pharmaceutical company. The packaged paracetamol tablets that are sold to the public need to be sampled and tested to ensure that they leave the factory in the best possible condition. Part of that process is to extract the paracetamol from the tablets to test that the drug has not degraded.

   Write a report for your supervisor describing the scientific principles behind the key steps of the procedure you will carry out, explaining each step and describing any problems. P1, M1, M2

Grading tips

For task 1, remember to discuss the type of reaction used. Also explain the method for extracting the sodium chloride crystals. P1

To achieve P1 in task 3, you will need to write enough about every step to show that you have carried out the procedure.

For M1 in these tasks, make sure you clearly relate which scientific principle corresponds to which step of each procedure.

For M2 in these tasks, remember to cover all the problems for each of the procedures.

Functional skills

Mathematics

When you consider the problems that might arise, you will draw conclusions and provide justifications, using your mathematical skills.

English

Writing the reports for these activities will develop your communication skills.

22.2 Measuring percentage yield and percentage purity

In this section:  

Key terms

Percentage yield – the amount of product produced compared to the actual amount of product that it is possible to produce in a reaction, expressed as a percentage.

Purity – the degree to which a substance is undiluted or unmixed with another substance. A 100% pure substance has no other substances present.

Carrying out the reactions to prepare a new substance is only the first step in producing a chemical substance. It is important that the substance made is as pure as possible. Imagine the effects on your health if you took an aspirin which was not pure enough. You would be swallowing other chemicals which might be harmful and not getting enough of the drug required. It is also important to get as high a yield as possible. If the yield is too low then it may be too expensive to make the new substance. So chemists must be able to measure both the **percentage yield** and the **purity** of a new substance.

Calculating percentage yield

The formula for calculating the percentage yield is:

$$\text{percentage yield} = \frac{\text{actual number of moles}}{\text{expected number of moles}} \times 100\%$$

Measurement of mass

This first step in working out the percentage yield is to measure accurately the mass of product you have obtained. How accurate your measurements are may depend on the equipment you have, but should be measured to at least 2 decimal places. You should be able to use a top-pan balance for this. If you are using very small quantities or if you want more accurate measurements, you may use a chemical balance, which will measure to 3 decimal places.

Calculating the actual number of moles

The first stage in calculating the percentage yield is to work out how many moles of product you have made. For this you need to know the mass of the product you have made and the mass of one mole of the product.

Unit 1: See page 10 for an explanation of moles.

$$\text{actual number of moles} = \frac{m}{M_r}$$

where m = mass of product (in g) and M_r = mass of one mole (in g).

The preparation of crystals of copper(II) sulfate is described on page 393. If you weighed the product when you did this preparation, you will be able to calculate your percentage yield from this.

For $CuSO_4.5H_2O$, for example, the mass of one mole is 249.68 g.

So the number of moles made, $n = m/249.68$

Assuming that the mass of the product was 4.0 g (you could put your own value in here), this would give

$$n = \frac{4.0}{249.68} = 0.016 \text{ mol (2 s.f.)}$$

Unit 6: See page 151 for a definition of s.f. (significant figures).

Calculating the expected number of moles

The second stage of the calculation is to work out the number of moles of copper(II) sulfate you would expect under perfect conditions. The equation for the reaction is

$$H_2SO_4 + CuCO_3 \rightarrow CuSO_4 + CO_2 + H_2O$$

From this equation, you can see that one mole of sulfuric acid will produce one mole of copper(II) sulfate.

The formula that relates the expected number of moles to the concentration and volume used in the preparation is given by:

expected number of moles = cV

where c = concentration (in mol dm^{-3}) and V = volume (in dm^3).

Looking at the values from the preparation process described on page 393, the sulfuric acid solution had a concentration (*c*) of 1 mol dm^{-3}. The volume used was 20 cm^3. To convert this volume to dm^3, divide by 1000. This makes the volume (*V*) 0.02 dm^3.

So for this reaction, the number of moles you would expect to make under perfect conditions, is given by

expected number of moles = 1 × 0.02 mol = 0.02 mol

Calculating the percentage yield

Remembering the formula for percentage yield:

$$\text{percentage yield} = \frac{\text{actual number of moles}}{\text{expected number of moles}} \times 100\%$$

$$= \frac{0.016}{0.02} \times 100\%$$

$$= 80\%$$

Activity 22.2A

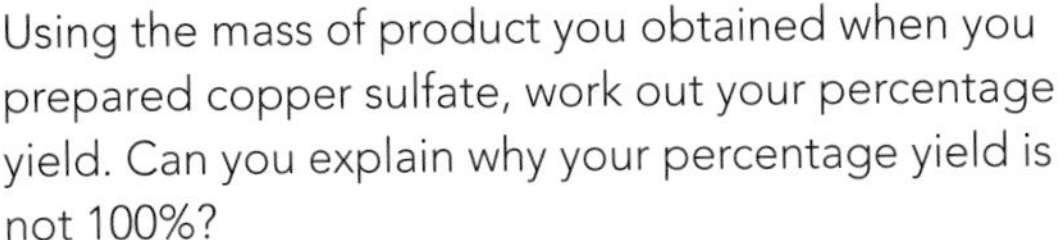

Using the mass of product you obtained when you prepared copper sulfate, work out your percentage yield. Can you explain why your percentage yield is not 100%?

Estimating the purity of a substance

A chemist can test a substance to estimate how pure it is in a variety of ways. These include:

- measuring the melting point
- measuring the boiling point
- chromatography, e.g. thin layer, paper, gas or high performance liquid chromatography.

Measuring the melting point

Checking the melting point of the substance will show if it is pure. A pure substance will melt at an exact temperature, not over a range of temperatures. However, this only shows if the substance is pure or not – it gives an estimate of purity.

Method

The preparation of antifebrin is described on page 394. If you have done this preparation, you can test the melting point of your product using the following procedure. (Pure acetanilide melts at 115°C.)

Safety note: Acetanilide is harmful and skin contact should be avoided. Disposable gloves must be worn.

1 Check the melting point of your product using melting point apparatus.

Unit 4: See page 101 for a detailed procedure on finding the melting point of a substance.

Activity 22.2B

Over what range of temperatures did your acetanilide sample melt? Explain how you could change your method of preparation to improve the purity of your product. (You may need to look back to page 395 to review the process you used.)

Measuring the boiling point

Measuring boiling point is more problematic than measuring melting point. The following method can be used.

Safety note: Eye protection must be worn.

The liquid to be tested is placed in a small test tube, and a short piece of melting point capillary tubing (sealed at one end) is dropped in with the open end down. The test tube should then be attached to a thermometer by a rubber band at the top of the test tube. This is then placed in a Thiele tube containing oil. If you do not have a Thiele tube, a beaker can work, but it is harder to get the temperature constant. The rubber band should be placed well above the level of the oil in the Thiele tube to ensure it doesn't soften and break in the hot oil. Then heat the Thiele tube with a small flame. There will be an *initial* stream of bubbles as dissolved air is expelled and then, a little later, a rapid, continuous stream of bubbles will be seen coming out of the capillary tube. At this point stop heating. The stream of bubbles will slow down and stop. When they stop, the liquid will enter the capillary tube. The temperature at which the liquid enters the capillary is the boiling point of the liquid. This can then be compared to known data in the same way you would compare a melting point. Note that air pressure will have an effect on this result.

Using thin-layer chromatography (TLC)

See Units 4 and 13 for more information on TLC.

In all types of chromatography, there is a mobile phase, in this case a solvent, and a stationary phase, in this case silica (SiO_2) gel on a plate. Silica gel is polar. If a pigment is strongly **adsorbed** to the gel, it will require a highly polar solvent to move it up the plate. If a pigment is weakly adsorbed to the gel, it will move easily up the plate. If the solvent dissolves all the components easily, all the components of the pigment mixture will move up the plate. If the solvent does not dissolve the components easily, only the pigment which is most weakly adsorbed will move.

The solvent chosen as the mobile phase must be suitable to move all the pigments to some extent. Pure propanone is not suitable as a solvent as it moves all the pigments to the top of the plate. Petroleum ether is not suitable either. It only moves adsorbed carotene up the plate weakly. A mixture of 70% (by volume) petroleum ether and 30% propanone is usually the optimum solvent for a separation of components. The component pigments that you might expect to separate are shown in the table.

Leaf pigments	Colour
carotenes	golden
pheophytin	olive green
chlorophyll a	blue green
chlorophyll b	yellow green
lutein	yellow
violaxanthin	yellow
neoxanthin	yellow

If a substance is pure it will only contain one component. If more than one component is shown through chromatography then this indicates that the substance is not pure. Chromatography can also be used to estimate how pure a substance is. For example, the size of the spots on the chromatogram can give you an idea of how many impurities are in the sample. Gas chromatography (GC) can show how pure a substance is: as each component leaves the GC column, a detector measures that component, giving us a good estimation of the purity of the substance.

Method

If you obtained extracts from leaves/herbs using different solvents (see page 397), you can test the purity of these pigments.

Safety note: Eye protection and disposable gloves must be worn. Propanone is highly flammable and an irritant. Avoid skin contact and do not breath the vapour.

1 Add a few drops of propanone to each extract to redissolve it. Carry out the following procedure for each pigment.

2 Obtain a TLC plate. Make a very faint pencil line, 1.5 cm from the bottom of the plate. Make a small and concentrated spot of leaf extract in the middle of the line on the plate. (Dip the spotting tube into the pigment. Touch the tip on the place you want it to go until you see a tiny spot. Lift up the tube. Let the spot dry. Re-spot in the same place. Repeat the spotting procedure about 20 times.)

3 Put your initials in pencil faintly at the top of the plate.

4 Obtain a TLC kit and place the plate in the holder as instructed.

5 Pour 10 cm^3 of the solvent mixture into the small reservoir and stand the holder in the solvent so that the solvent can rise up the plate.

6 Leave the chromatograms to run.

7 Once the solvent front is near the top of the plate, remove the plate and use a pencil to mark the position of the solvent front.

8 Allow the plate to dry and then make a scale drawing of the plate.

9 Measure the distance moved by the solvent, x, and the distance moved by each spot, y. For this you will have to estimate the centre of each spot.

10 Calculate the retention factor, R_f, for each spot using the equation:

$$R_f = \frac{y}{x}$$

11 Try to name some of the pigments. You can do this by calculating the R_f values as shown below and comparing them to data tables.

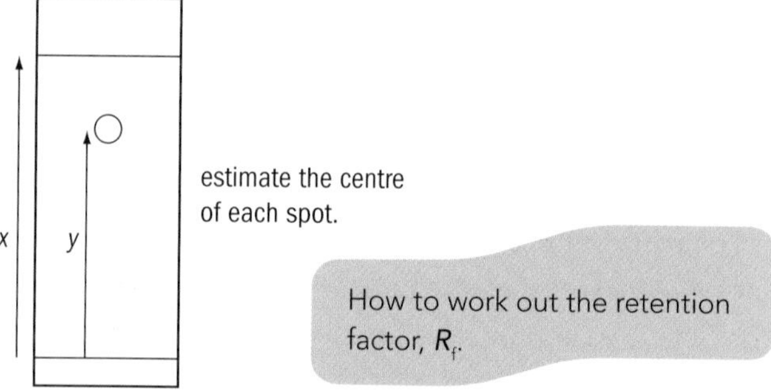

How to work out the retention factor, R_f.

If you have access to a UV/visible spectrophotometer, you could run the visible spectrum of each extract and comment on the results. A spectrophotometer measures light intensity. Light is transmitted through the substance being analysed and the intensity of this light is measured once it has passed through the substance and produces a spectrum that can be compared with reference material to show what is present in the substance.

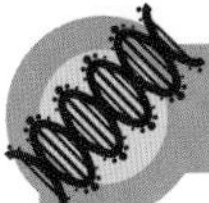

Functional skills

Mathematics

Carrying out the calculations shows you can use the correct mathematical methods to solve them.

Activity 22.2C

You have been given a chromatogram with only two spots on it. What does this say about the purity of the sample? How might the solvent used affect the number of spots seen on the chromatogram?

Measuring the purity of a substance

The methods we've discussed so far only give an estimate of purity. It is often important to know exactly how pure a substance is. For example, a pharmaceutical company must know how pure their medicines are.

Measuring the purity of a substance by titration

Carrying out a titration is one way to show exactly how pure a substance is. The copper(II) sulfate crystals you made earlier (page 393) can be tested for purity by titrating with EDTA.

EDTA (short for ethylenediaminetetraacetic acid) is a large molecule that can form soluble complexes with many metals in a 1:1 ratio. Usually the disodium salt of the acid is used. There are several possible forms of the EDTA in solution, depending on the pH. It is usual to adjust the pH in some way to keep things simple. Often this involves the use of a buffer solution or addition of ammonia. Many EDTA titrations are done at a high pH. However, a pH above 12 poses the risk of the precipitation of insoluble metal hydroxides.

The disodium EDTA molecule.

The indicators for EDTA titrations are 'metallochrome' indicators. These form a complex with a small quantity of the metal ion to be titrated as soon as they are added to the solution. The solution is coloured. Once enough EDTA has been added to complex with all the rest of the metal ion in solution, the next quantity of EDTA added extracts the only remaining metal ions in the flask – those in the indicator. The indicator changes to a different colour when it loses its metal ions.

Safety note: Eye protection must be worn. Avoid skin contact with the chemicals used. The concentrated ammonia solution is corrosive and gives off toxic ammonia gas which irritates the eyes and lungs. Make sure the room is well ventilated and avoid inhaling ammonia fumes.

Method

You could use the crystals you made earlier (page 393) to carry out this investigation into the purity of copper(II) sulfate crystals.

1. Grind the copper sulfate crystals in a mortar with a pestle.
2. You will need to use a top-pan balance accurate to 4 decimal places. Set the top-pan balance to read zero with a weighing bottle on top.
3. Weigh out about 0.6 g of $CuSO_4.5H_2O$ in the weighing bottle.
4. Now set the top-pan balance to zero with nothing on it and reweigh your bottle with the copper sulfate in it. Make sure you read the mass of the bottle with the copper sulfate accurately to 4 decimal places.
5. Add the contents of the bottle to a 250-cm^3 standard flask, through a powder funnel.
6. Reweigh the empty bottle and accurately record its mass to 4 decimal places.
7. Calculate the mass of crystals transferred to the flask.

8 Half fill the flask with water and shake the copper sulfate until it dissolves. Make the volume of liquid in the flask up to the mark with distilled water. Remember that the lower part of the meniscus should touch the line. Use a teat pipette when you get close to the line.

9 Pipette $25\,cm^3$ of the copper solution into a conical flask.

10 Add $100\,cm^3$ of distilled water, $5\,cm^3$ of concentrated ammonia solution and 5 drops of Fast Sulfon Black indicator solution. (Beware, this will stain skin and clothing.)

11 Fill a burette with standard EDTA solution ($0.01\,mol\,dm^{-3}$) and titrate the solution until the indicator changes from deep blue/purple to green. (It may look like a blue-green.)

Repeat the procedure from step 9 until results of the required accuracy have been obtained. You should have final volume readings to an accuracy of $0.05\,cm^3$. You should have at least two titres within $0.5\,cm^3$ of each other.

Calculating the percentage purity from results of titration

The percentage purity is calculated from the formula:

$$\%\ purity = \frac{\text{mass worked out from titration}}{\text{mass weighed out}} \times 100\%$$

Activity 22.2D

Use the values you obtained from your earlier procedures (pages 393 and 403) to work out the percentage purity of the copper(II) sulfate crystals you prepared.

Comment on this value. Does it make sense? If not, why not?

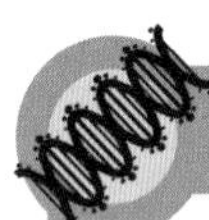

Functional skills

Mathematics

Carrying out the calculations shows you can use the correct mathematical methods to solve the problem.

Worked example

Mass of bottle plus about 0.6 g of copper sulfate = 10.75 g

Mass of bottle after copper sulfate emptied = 10.13 g

Mass of copper sulfate added to flask = 0.62 g

Results from titrations:

initial volume (cm^3)	0	20	0	20
final volume (cm^3)	18.05	38.10	18.10	38.05
titre (cm^3) = final − initial	18.05	18.10	18.10	18.05

Average titre = $18.08\,cm^3$ = $0.018\,dm^3$ (divide by 1000)

Number of moles EDTA in average titre,

$n_{EDTA} = c_{EDTA} \times V_{average\ titre}$

where

c_{EDTA} = concentration of EDTA (in $mol\,dm^{-3}$) = $0.01\,mol\,dm^{-3}$

$V_{average\ titre}$ = volume of average titre (in dm^3) = $0.018\,dm^3$

So $n_{EDTA} = c_{EDTA} \times V_{average\ titre}$
$= 0.01 \times 0.018 = 0.00018\,mol$

Copper ions and EDTA react in the ratio 1 : 1

This (n_{EDTA}) is the number of moles in $25\,cm^3$ which is one tenth of the volume of the flask.

Number of moles in $250\,cm^3$ $n_{copper\ sulfate} = 10 \times n_{EDTA}$

$n_{copper\ sulfate} = 10 \times 0.00018 = 0.0018\,mol$

How many grams is this equivalent to?

$m_{copper\ sulfate} = n_{copper\ sulfate} \times M_r$

where M_r is the mass of one mole of copper sulfate = 249.68 g

$m_{copper\ sulfate} = 0.0018 \times 249.68\,g = 0.45\,g$

So, the percentage purity is given by:

$$\%\ purity = \frac{\text{mass worked out from titration}}{\text{mass weighed out}} \times 100\%$$

$$= \frac{0.45}{0.62} \times 100\%$$

$$= 73\%$$

Green chemistry

Green chemistry, also known as sustainable chemistry, is the design of chemical products and processes that reduce or eliminate the use or generation of hazardous substances. Green chemistry affects all aspects of a chemical product including its design, manufacture and use. Green chemistry technologies provide a number of benefits, including reduced waste, safer products, and reduced use of energy and resources.

Activity 22.2E

Many governments try to get the chemical industries in their country to focus on Green Chemistry. Work in pairs: discuss why you think governments are encouraging the use of Green Chemistry in industry.

Worked example

Consider these two reactions.

In the first reaction:

zinc oxide + hydrochloric acid → zinc chloride + water

$ZnO + 2HCl \rightarrow ZnCl_2 + H_2O$

mass:

71 g	73 g	136 g	18 g

$$\% \text{ atom economy} = \frac{136}{(71 + 73)} \times 100\%$$
$$= 94\%$$

In the second reaction:

zinc carbonate + hydrochloric acid → zinc chloride + water + carbon dioxide

$ZnCO_3 + 2HCl \rightarrow ZnCl_2 + H_2O + CO_2$

mass:

125 g	73 g	136 g	18 g	44 g

$$\% \text{ atom economy} = \frac{136}{(125 + 73)} \times 100\%$$
$$= 69\%$$

Atom economy

The atom economy of a chemical reaction is a measure of how much of the starting materials becomes useful product. Inefficient, wasteful processes have low atom economies. Efficient processes have high atom economies. In industry, it is important to have as high an atom economy as possible because this is also usually the most cost efficient.

The atom economy of a reaction is calculated using the following formula.

$$\% \text{ atom economy} = \frac{\text{molar mass of product}}{\text{molar mass of reactants}} \times 100\%$$

As you can see from the worked example, making zinc chloride from zinc oxide and hydrochloric acid has a higher atom economy than zinc carbonate and hydrochloric acid. However, there may be other reasons why another method may be preferred, e.g. cost of raw materials or safety considerations.

What is the difference between yield and atom economy?

Yield is the amount of product made. Sometimes it is measured as percentage yield, which is the amount made compared to the amount that it is possible to make. Atom economy is slightly different. If a reaction has high atom economy it means that most if not all of the atoms in the reactants are used to make the useful product from the reaction. If a reaction has only one product then all the atoms in the reactants are used to make the product, and so this has the highest atom economy possible.

Atom economy can be calculated using the following equation

% atom economy = mass of desired product from equation/total mass of reactants from equation × 100

Activity 22.2F

Ethanol is a useful product made from glucose by the following reaction.

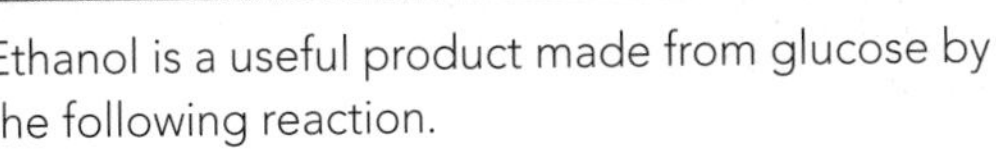

$C_6H_{12}O_6 \rightarrow 2CH_3CH_2OH + 2CO_2$

1 Do you think this reaction has a high atom economy? Why?

Work out the relative molecular masses for each reactant and product using the periodic table to help you. Then calculate the % atom economy for this reaction.

2 Do you still agree with your answer to **1**? Why?

Combinatorial chemistry

Traditionally, research chemists made and tested one new substance at a time. This was a very slow process. Many chemists now use combinatorial chemistry where large numbers of structurally distinct molecules may be synthesised at a time. A variety of reactants are combined using the same reacting conditions in the same reaction vessels to produce multiple new molecules. This is obviously a much quicker process as more than one new substance is made at a time.

Assessment activity 22.2

BTEC

1 As a quality control manager working for a pharmaceutical company, you need to find out what percentage of a sample tablet is paracetamol. It is important that the amount given on the tablet box is correct.

You will need to use the mass you weighed from the earlier procedure (page 398) to carry out this activity.

- a Calculate the % by mass of the tablets that you have extracted as paracetamol. P2
- b Measure the melting point of the paracetamol sample you set aside before recrystallisation and the melting point of the recrystallised paracetamol. Comment on what you find. P2
- c Produce a presentation of your results. Suggest some other tests that the quality control technician might perform on the paracetamol and explain why. P2 D1

2 You work for a chemical company that uses a large amount of sodium chloride in its chemical preparations. You are looking at the best way to make sodium chloride to be used by the company. You have been asked to write a report that explains which method should be used and why. D1

- a From the following list, select substances that would react together to produce the salt, sodium chloride:

  nitric acid sulfuric acid hydrochloric acid sodium oxide

  sodium hydroxide sodium carbonate sodium metal zinc metal
- b Write down word equations and balanced symbol equations for four reactions that would make sodium chloride.
- c Calculate the atom economy of each method. P2
- d Which of the equations in part **b** is the method described for preparing sodium hydroxide in section 22.1?
- e Only one of these reactions would give sodium chloride solution easily. Identify this reaction and explain why the other three reactions would not be satisfactory. D1

Grading tips

For task 1c, discuss how changing the method can affect the yield and purity. D1

In task 2, use the method described on page 393 to help you. Make sure you relate your answers to the preparation you have carried out. P2 Explain how the yield, purity and atom economy may be affected by the different methods used. D1

PLTS

Reflective learner

Making judgements based on your results will show you are a reflective learner.

Functional skills

Mathematics

When you draw conclusions and provide justifications, you will develop your mathematical skills.

English

Writing the presentation will develop your communication skills.

22.3 Qualitative analysis of compounds

In this section:

Key terms

Alkane – a saturated hydrocarbon with the general formula C_nH_{2n+2}.

Qualitative analysis – the testing of a substance to determine its chemical constituents.

Functional group – an atom, or group of atoms, that replaces hydrogen in an organic compound. It defines the structure of the organic compound and determines its properties.

Long before chemists had sophisticated instrumental equipment, they were still able to make discoveries, based on their observations. Early chemists learned how to identify the presence of positive ions (cations) and negative ions (anions). They were also able to identify the class to which an organic compound belonged.

Nowadays, it is sometimes still useful to carry out a quick experiment to identify a substance. This is referred to as **qualitative analysis**. The results of the analysis may then be supported by use of an instrumental technique, e.g. infrared spectroscopy in the case of organic compounds.

Identifying organic compounds

Did you know?

There is no limit to the number of carbon atoms that can be linked together to form an alkane molecule.

Functional groups

The chemistry of **alkanes** is straightforward.

- They burn.
- They can react with halogens in the presence of ultraviolet light.

However, if in place of one of the hydrogen atoms there is another type of atom or group of atoms, a **functional group**, the chemistry is much more extensive.

Look at the molecule below:

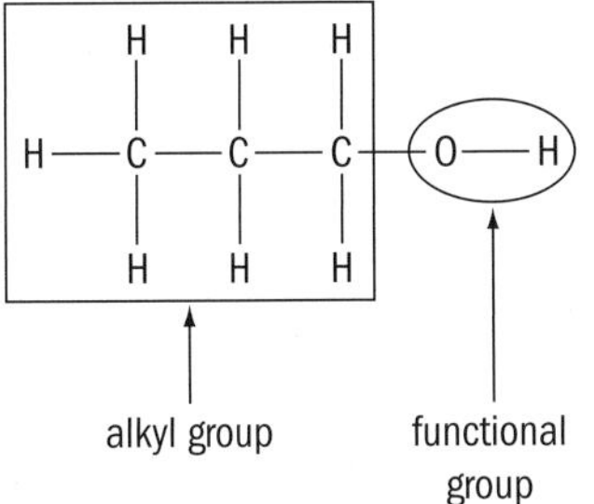

The interesting part, the functional group, in the above molecule is O–H, called the 'hydroxyl' group. Molecules with this type of functional group are alcohols.

The rest of the molecule is the 'alkyl group' or alkyl chain. This is an alkane with a hydrogen atom missing. In the molecule shown, the alkyl group is three carbons long and is called propyl (propane with a hydrogen missing). Propyl could also be written $CH_3CH_2CH_2$–. Other alkyl groups are shown in the table.

Alkyl group	Name
CH_3–	methyl
CH_3CH_2–	ethyl
$CH_3CH_2CH_2CH_2$–	butyl
$CH_3CH_2CH_2CH_2CH_2$–	pentyl

Activity 22.3A

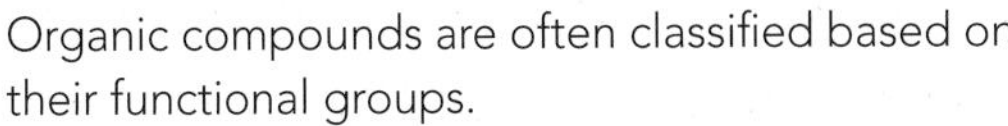

Organic compounds are often classified based on their functional groups.

Copy and complete the table below by drawing the functional group of each organic series.

Series	Functional group
Alkene	
Haloalkane	
Alcohol	
Aldehyde	
Ketone	
Carboxylic acid	
Ester	

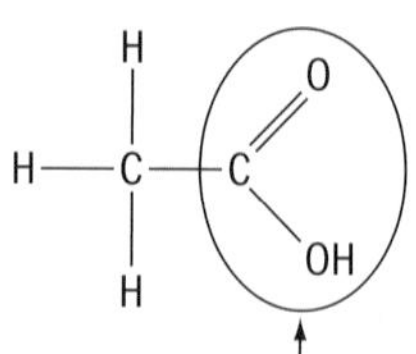

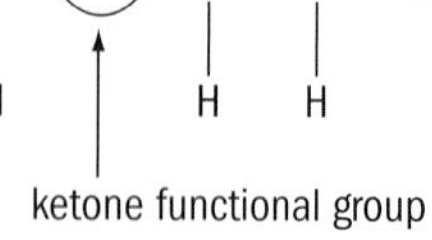

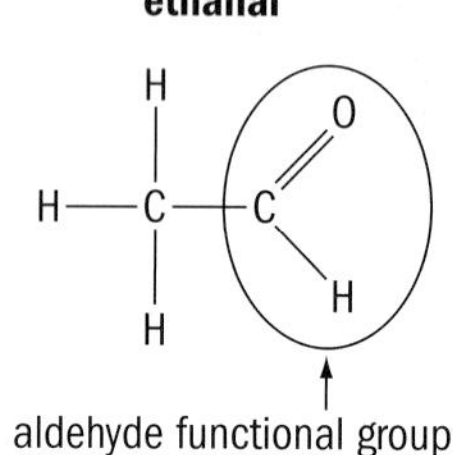

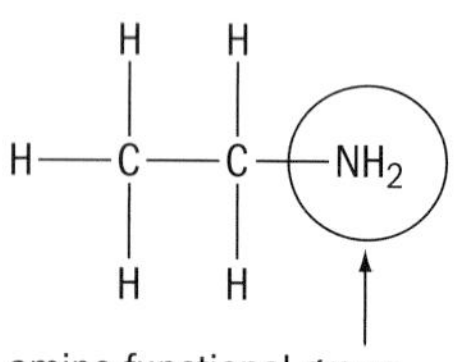

propan-2-ol

secondary alcohol

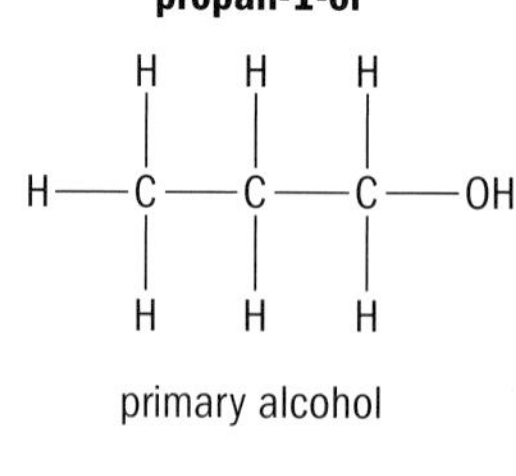

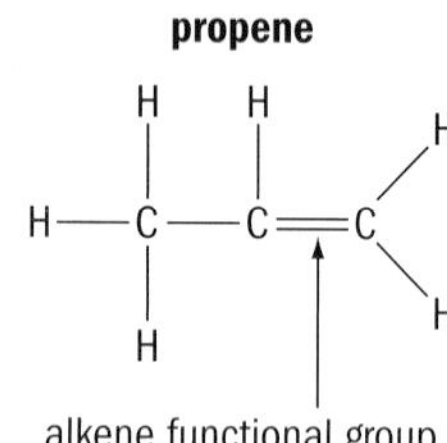

The functional groups of various organic compounds.

Suppose you are given a sample of each of six types of organic compound:

- alkane
- alkene
- primary alcohol
- carboxylic acid
- ketone
- aldehyde.

Each one has a simple molecule with only one functional group per molecule, except for the alkane which has no functional groups. By carrying out simple chemical tests, you can work out which compound is which.

Once you have carried out all the tests, you could then obtain a copy of the infrared spectrum of each compound and explain whether it provides evidence to support your conclusions.

See page 411 for more information on infrared spectra.

PLTS

Effective participator, Creative thinker and Reflective learner

When you carry out this practical you will show that you are an effective participator and a creative thinker. Making judgements based on your results will show you are a reflective learner.

Solubility in water and pH

When presented with a number of different organic compounds to identify, the first step is to consider their solubility in water and their pH. Alkanes and alkenes, being very non-polar, are insoluble in water. The longer the organic chain of the alkene, the less polar and the more like an alkane the molecule becomes. Few long-chain compounds will be soluble in water. The table shows the typical pH of various water-soluble organic compounds.

Water-soluble compound	Acidic, alkaline or neutral solution?
small-chain aliphatic acids	strongly acid, pH < 7
small-chain alcohols	neutral pH = water pH
small-chain aldehydes	neutral or slightly acid due to oxidation
small-chain ketones	neutral pH = water pH
small-chain amines	weakly alkaline, pH > 7

Safety note: Eye protection must be worn. Avoid skin contact with all reagents and products.

Method

1 Add a few drops of the compound to water in a boiling tube.

2 Flick the tube in order to dissolve the compound if it is soluble.

3 If the substance dissolves, test the liquid with a calibrated pH probe or with Universal indicator paper and a glass rod. Ensure that the probe is rinsed well, by dipping it into a beaker of distilled water between solutions.

4 Record your results in a table.

Compound	Soluble in water?	pH if soluble
alkane		
alkene		
primary alcohol		
carboxylic acid		
ketone		
aldehyde		

You can see from this that the two insoluble compounds are the alkane and the alkene. They form two distinct layers in the boiling tube when each is mixed with water.

Looking at the pH, you will see that there is only one acidic compound, the carboxylic acid. However, the test may also show the aldehyde as acidic, due to oxidation. Therefore another test must be carried out to differentiate between these.

Testing with carbonates or hydrogen carbonates to identify acids

When carbonates and hydrogen carbonates react with acid, carbon dioxide is produced. This means you will see effervescence (fizzing). This will identify the carboxylic acid.

See previous Safety note.

Method

1 Add a small quantity of each substance to 1 cm^3 of sodium hydrogen carbonate or sodium carbonate solution in a test tube.

2 Note which substance produces effervescence, i.e. fizz.

Using Brady's reagent to identify aldehydes and ketones

Brady's reagent contains 2,4-dinitrophenylhydrazine. This gives orange crystals when mixed with aldehydes or ketones.

Safety note: Eye protection and protective gloves must be worn. Brady's reagent is very toxic by inhalation and contact. Avoid all skin contact.

Method

1 Add a few drops of each compound to 1 cm^3 of Brady's reagent in a test tube.

2 Note which compounds give orange crystals.

Benedict's test for reducing properties, used to identify aldehydes

Some classes of compound are easily oxidised, in other words they have reducing properties. Compounds that react with Benedict's (or Fehling's A and B) reagent to give a red-brown precipitate of copper(I) oxide have reducing properties. The only class of compounds from the six listed above that definitely reacts with Benedict's reagent is the aldehyde.

Safety note: Eye protection must be worn. Avoid skin contact with the reagents and products.

Method

1 Add about 0.5 cm^3 of each compound to a 1-cm depth of Benedict's reagent (or mixed equal volumes of Fehling's A and B) in a test tube.

2 Holding the test tube in a test-tube clip, warm the mixture in a steam bath.

3 Note which solution gives a red-brown precipitate.

Reaction with warm acidified potassium dichromate(VI) solution, used to test for alcohols and aldehydes

Acidified dichromate(VI) is an oxidising agent.

- Primary alcohols are oxidised to aldehydes and then to carboxylic acids.
- Secondary alcohols are oxidised to ketones.
- Tertiary alcohols are not easily oxidised.
- Alkenes react slowly – in fact, many things are slowly oxidised by warm acidified dichromate(VI).

The orange dichromate ion, $Cr_2O_7^{2-}$, is reduced to blue-green Cr^{3+}. So when you test the compounds with acidified dichromate(VI) you should find that the alcohol and aldehyde are easily oxidised; the solution will change from orange to blue-green.

Safety note: Eye protection must be worn. Avoid skin contact with the reagents or products. Potassium dichromate(VI) is very toxic.

Method

1 Add about 0.5 cm^3 of each compound to a 1 cm depth of acidified dichromate solution in a test tube.

2 Holding the test tube in a test-tube clip, warm the mixture in a steam bath.

3 Note any changes that take place.

Using bromine water to test for alkenes

Alkenes rapidly decolourise bromine water. Other compounds with a double bond will also react in this way. Aldehydes and ketones react with bromine water, producing fumes of hydrogen bromide that you can test for with damp blue litmus paper.

Remember that the solubility test you first carried out will have identified the alkane and the alkene. This test will therefore identify the alkene from these two.

Safety note: Eye protection must be worn. Keep bromine water in the fume cupboard at all times. Avoid skin contact with bromine water and avoid inhaling its vapour, which is very toxic.

Method

1 In a fume cupboard, add about 0.5 cm^3 of each compound to 1 cm^3 of bromine water in a test tube.

2 Tap the tube to mix the reagents.

3 Note which compounds turn the bromine water from brown to colourless.

4 Test any vapour that is produced with damp blue litmus.

5 Note which compound(s) produce(s) a vapour that turns the litmus red.

Identifying amines

Amines that are soluble in water result in a solution that will turn litmus paper blue and have a pH of more than 7. Some amines do not dissolve in water, but will do so if dilute hydrochloric acid is added. Both of these tests can be carried out to identify an organic compound as being an amine.

Summary

Compound	Solubility	pH in water	Reaction with sodium carbonate	Brady's reagent	Benedict's test	Warm acidified potassium dichromate	Bromine water
alkane	no		none	none	none		stays brown
alkene	no		none	none	none	slow reaction	goes colourless
carboxylic acid	yes	<7	fizzing	none	none		
aldehyde	yes	7 (<7)	none	orange crystals	red-brown precipitate	blue-green	acidic fumes
ketone	yes	7	none	orange crystals	none		acidic fumes
amine	yes	>7	none	none	none		
alcohol	yes	7	none	none	none	blue-green	

Identifying aldehydes and ketones from melting points of derivatives

The orange crystals produced by reacting an aldehyde or a ketone with Brady's reagent can be used to identify the compound tested. The crystals must be extracted using filtration and recrystallisation techniques described earlier (page 395). The melting point of the crystals is then measured using melting point apparatus. The melting point can be compared to data tables and the corresponding aldehyde or ketone can be identified.

Infrared spectroscopy

This can also be used to identify functional groups. Infrared radiation makes the bonds in molecules bend and stretch. The table shows the characteristic wavenumbers (1/wavelength in cm) of important peaks in the infrared spectra of some classes of organic compounds.

Vibrating bond	Wavenumber (cm^{-1})
C–H (stretch)	3000–2850
C–H (bend)	1450 and 1375
C=C	1680–1600
C=O aldehyde	1740–1720
C=O ketone	1725–1705
C=O carboxylic acid	1725–1700
C–O alcohols, ethers, carboxylic acids, esters	1300–1000
O–H alcohols	3600–3200
O–H carboxylic acids	3400–2400
N–H (stretch)	3500–3100
N–H (bend)	1640–1550
C–N (amines)	1350–1000

Did you know?

The ranges for wavenumber are shown in descending order because the infrared spectra are plotted with the highest wavenumber on the left and the lowest wavenumber on the right.

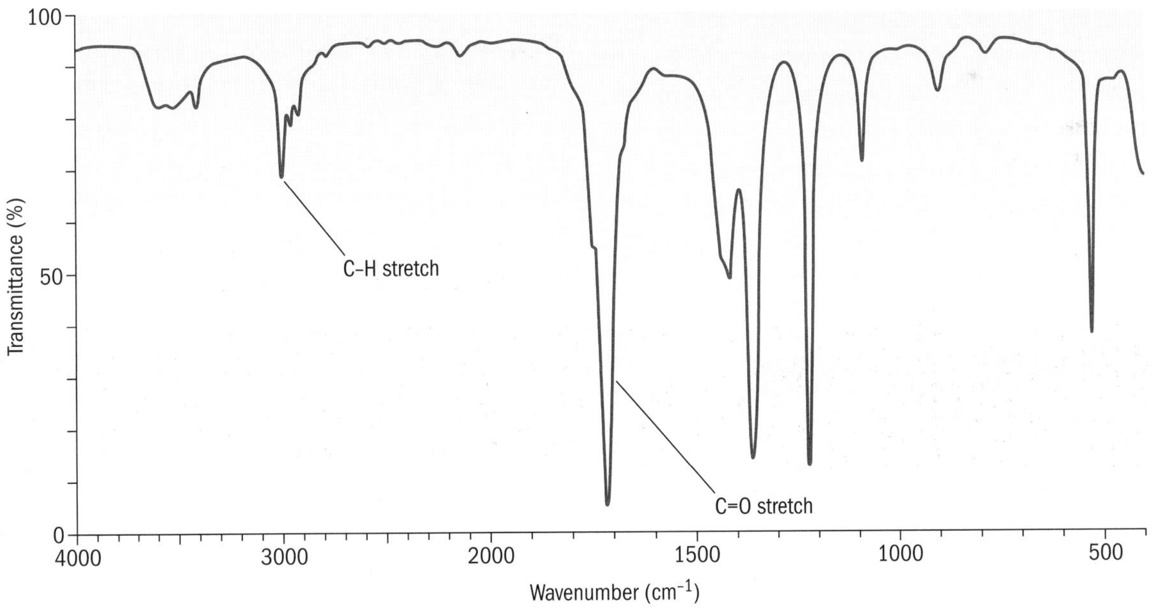

Infrared spectrum for propanone. Note the peaks that correspond to the C–H and C=O bonds.

Activity 22.3B

An infrared spectrum shows peaks at around 3400 cm^{-1}, 3000 cm^{-1}, 1100 cm^{-1}. What functional groups do you think are present in the compound? Can you suggest which compound this is the spectrum for?

Identifying inorganic compounds

It is possible to identify inorganic compounds by carrying out chemical tests. Usually two types of test are needed: one to test for the cation and one to test for the anion.

Flame test for positive ions (cations)

Positively charged metal ions are called cations. When metal ions are placed in a flame, the heat can do several things:

- atomise the substance
- allow the metal to pick up the electrons it needs to become an atom again
- excite the outer electrons of the metal into higher unoccupied electronic levels (excited states).

In the last example, when the electrons drop back down to the ground state again, energy is emitted. The energy associated with this process is in the visible region of the electromagnetic spectrum. As a result, putting a small sample of a metal compound into a flame causes the sample to glow with a colour typical of the metal involved (see table below). Sometimes the flame is looked at through a blue glass to filter out the yellow of sodium.

Substance	Flame colour
potassium bromide	lilac
calcium chloride	brick red
lithium chloride	pink
copper(II) chloride	blue-green
sodium chloride	bright orange

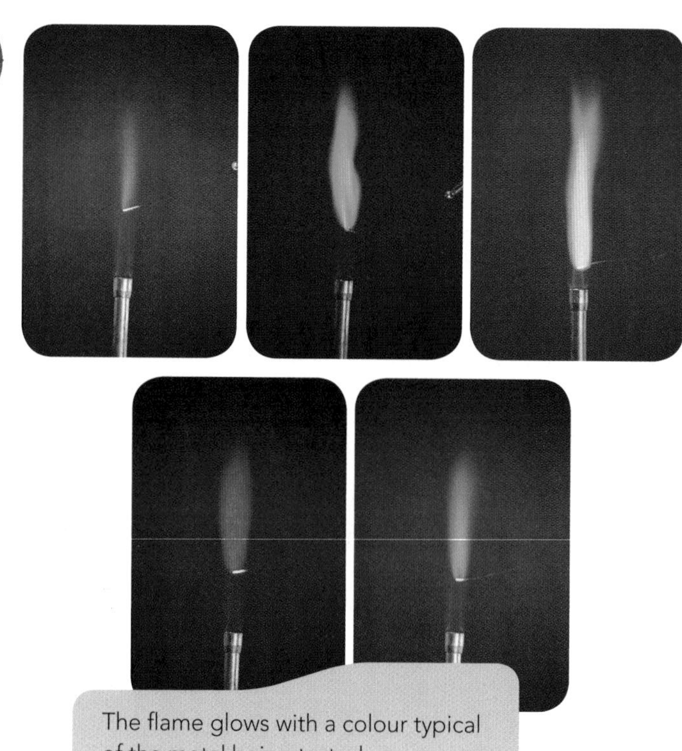

The flame glows with a colour typical of the metal being tested.

When we view the flame colour we see all the emission spectrum colours at the same time. They will appear to be one colour. It is possible to view the separate colours of the emission spectrum by using a spectroscope, which bends light of different energies differently. Low-energy red light is bent the most, and high-energy violet the least. This allows us to see the various distinct colours of the emission spectrum of a sample.

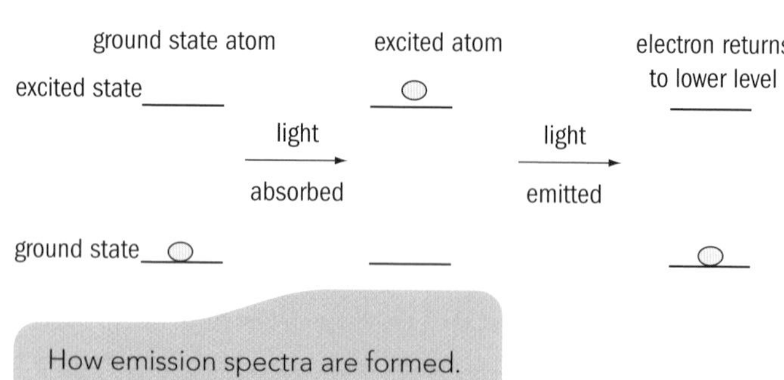

How emission spectra are formed.

Here are the emission spectra of some metal atoms.

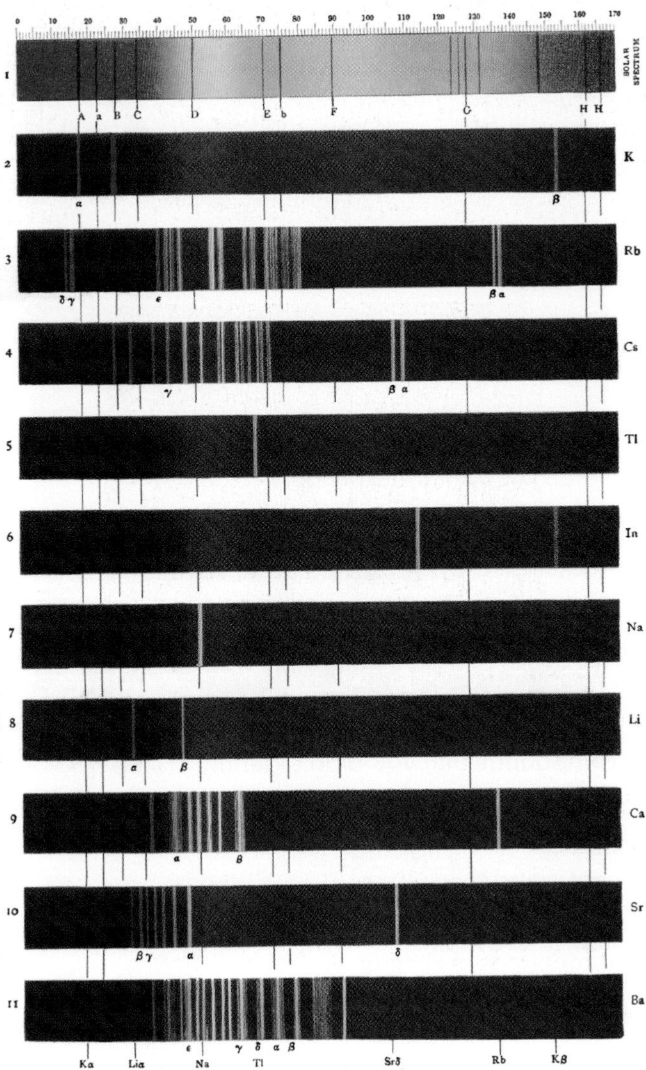

The emission spectra of various metals.

Safety note: You must wear safety goggles and remove any rings, bracelets or watches before you start. Concentrated hydrochloric acid is corrosive, will burn your skin and its fumes will irritate your respiratory system. If you get a splash on your skin, wash it off with lots of cold water immediately and tell your tutor. The acid may splatter from the wire in the flame.

Method

1 Collect a mat and a large Bunsen burner, a piece of blue glass (if available), a dimpled tile, and a flame test wire. (The wire is held in a cork or has a long handle so that you don't burn your fingers.)

2 Ensure there is a test-tube rack with a test tube containing concentrated hydrochloric acid in the fume cupboard. There should also be a small Bunsen burner in the fume cupboard.

3 Put a small amount of the first solution you are testing in one dimple on the tile.

4 Light your large Bunsen burner.

5 Clean the flame test wire by dipping it into the acid (in the fume cupboard) and then clean it by holding it in the flame of the small Bunsen burner until it stops glowing. Return to your own work station.

6 Dip the wire into the solution you are testing. Hold the wire in the flame. Note the colour of the flame.

7 Record your result in a table.

8 Repeat steps 5 and 6 but this time look at the colour through blue glass and through a spectroscope (if available). Record what you see.

9 Repeat steps 5–8 for all the compounds you are testing. Remember to wash the tile carefully and dry it in between each test.

PLTS

Creative thinker

Carrying out this practical will show you are a creative thinker.

Activity 22.3C

Flame tests are used to identify cations. Read the method and the safety measures given above.

- Why must safety goggles be worn?
- Why is the loop dipped in HCl?
- Why do we look through a blue glass?

Testing for negative ions (anions)

Anions react in specific ways with specific chemicals. The following table gives you the reaction used to identify each anion.

Anion	Reaction
chloride	gives a white precipitate with silver nitrate solution, which is soluble in dilute ammonia
bromide	gives a pale yellow precipitate with silver nitrate solution, which is soluble in concentrated ammonia
iodide	gives a yellow precipitate with silver nitrate solution, which is only partially soluble in concentrated ammonia
carbonate	produces bubbles of carbon dioxide when acid is added (note that hydrogen carbonate would do this as well)
nitrate	produces ammonia when reacted with sodium hydroxide, then with aluminium foil, and warmed carefully; ammonia gas will have pH > 7
sulfate	gives a white precipitate with barium chloride solution
sulfite	sulfur dioxide is given off on reaction with hydrochloric acid; this will be a colourless acidic gas with a choking smell

Safety note: Eye protection must be worn. Silver nitrate solution is corrosive, will blacken skin and destroy tissue. Avoid skin contact and wash any splashes off with lots of cold water. Avoid skin contact with other reagents and products and avoid inhaling any gases given off.

Method

This method assumes you are testing seven compounds to decide which is the chloride, bromide, iodide, carbonate, nitrate, sulfate and sulfite.

1 Set up a test-tube rack with one test tube for each compound being tested.
2 Start by dissolving a small amount of each unknown compound in water before testing: add half a spatula-ful of each compound to the test tubes (one per tube), then add a 2-cm depth of distilled water to each tube and dissolve the solid.
3 Add a few drops of silver nitrate solution to each of the tubes. You will be able to identify the chloride, bromide and iodide from the colour of the precipitates that form.
4 Using two clean test tubes, repeat step 2 for the compounds that you didn't identify in step 3.
5 Add a few drops of barium chloride solution to the test tubes. A precipitate will form in the tube containing the sulfate.
6 Make up a fresh solution of the remaining compounds, as in step 2.
7 Collect a test tube containing a few cm^3 of lime water and a teat pipette. Add a few drops of hydrochloric acid solution to the test tube containing the unknown compound. Test any gas produced by bubbling it into lime water using a teat pipette. If the lime water turns milky, you have identified a carbonate.

Case study: Water quality

Liz works for the Environment Agency. She works on a team that checks pollution levels in the rivers around Britain. She has to carry out qualitative tests to see what chemicals may be present in the water and then compares them to tables that list the toxicity of the chemicals. She then reports to her manager who will make decisions on how to deal with the pollution if there is any.

1 What tests might Liz carry out to identify the cations in the river water?

2 What tests might Liz carry out to identify the anions in the river water?

3 What tests might Liz carry out to identify any organic pollutants?

8 Put half a spatula of each of the remaining compounds into a test tube. Add a few drops of sodium hydroxide, then a small piece of aluminium foil, and warm carefully. Test any gas with damp red litmus paper. If nitrate is present the litmus paper will turn blue.

9 Check the remaining compound by placing half a spatula into a test tube. Add a few drops of dilute hydrochloric acid. Test any gas with damp blue litmus paper. If the gas is colourless and turns the litmus paper red a sulfite is present.

Assessment activity 22.3

BTEC

1 You are working as a technician for a large chemical company that makes a whole range of chemical products. You are responsible for controlling the stock of chemicals in the store room. Very occasionally labels can fall off chemical containers. This is very dangerous as it is important that each chemical is treated, stored and used correctly.

Produce a leaflet that will show another lab worker the best way to safely identify six unknown organic chemicals (alkane, alkene, primary alcohol, aldehyde, ketone and carboxylic acid) using both test-tube reactions and infrared spectroscopy. P3 Your leaflet should explain how infrared spectra enable functional groups to be identified M3 and should have a summary evaluating the effectiveness of each test-tube reaction you describe. D2

2 In your role as a technician in task 1, you need to be able to identify unknown inorganic compounds.

Produce a leaflet that will show another lab worker the best way to safely identify unknown inorganic chemicals. This should include testing for cations, and the anions: carbonate, sulfate, nitrate, sulfite, chloride, bromide and iodide. P4 You should explain how emission spectra are formed and give the basis of each chemical test you have described. M4

Grading tips

For task 1, use your experience from actually carrying out the tests yourself to create a flow chart. This should show how the various test-tube reactions enable the organic compounds to be identified, step by step. P3
Make sure you refer to functional groups in your explanations. M3 To achieve D2 you should clearly explain whether each test you have suggested provides conclusive proof of each compound.

In task 2, to achieve P4 your description should show that you have carried out the tests yourself. To achieve M4 you should explain the origin of emission spectra and should be able to explain the anion reactions.

Functional skills

English

Writing the leaflets will develop your communication skills.

22.4 Quantitative analysis of compounds

In this section:   

Key terms

Quantitative analysis – the testing of a substance to determine the amounts and proportions of its chemical constituents.

Volumetric analysis – quantitative analysis using accurately measured titrated volumes of standard chemical solutions.

Spectroscopic analysis – analysis of a spectrum to determine characteristics of a substance, e.g. its composition.

Qualitative analysis tells us what is present in a substance. **Quantitative analysis** tells us how much is present in a substance. Quantitative analysis techniques can be used in many ways; for example, to see if a river is polluted, to ensure the concentration of bleach in a kitchen cleaner is the correct strength, to make sure the vinegar on your chips tastes good.

Chemists can use **volumetric analysis**, e.g. titrations, or **spectroscopic analysis**.

- Titrations are used to prepare and test solutions of known concentration. These solutions are then used to analyse substances also by titration.
- Spectroscopy is the use of light, sound or particle emissions to study a substance. The emissions can provide information about the properties of the substance.

Volumetric analysis

Preparing solutions of known concentrations

Unit 4: See page 85 for information on solutions of known concentrations.

To carry out titrations and other procedures you will need solutions of known concentrations. These can be prepared by adding a known mass of the solute into 100 ml of the solvent in a 1000-ml volumetric flask. This flask is shaken to ensure the solute has dissolved. The flask is then filled with the solvent to the 1000-ml mark and shaken carefully several more times. By knowing the mass of the solute you can work out how many moles of the solute are in the solution, and so can state what the molarity of the solution is. The concentration of the solution can be confirmed by carrying out a titration on it.

Acid–base titrations

Vinegar can be analysed by using an acid–base titration (see Worked example below). This may be done to test the quality of the vinegar being produced.

Worked example

A 10-cm^3 sample of vinegar was titrated with 0.1 M NaOH. The titration was repeated until **concordant** results were gained. The results are shown in the table:

Titre	1	2	3	4
final volume (cm^3)	14.50	30.00	45.00	60.00
initial volume (cm^3)	0.0	14.50	30.00	45.00
total volume of NaOH used (cm^3)	14.50	15.50	15.00	15.00

Titres 3 and 4 are concordant so a value of 15.00 cm^3 is used.

15.00 cm^3 of 0.1 M NaOH is used to neutralise 10 cm^3 vinegar:

$15.00/1000 \times 0.1 = 1.50 \times 10^{-3}$ mol NaOH

The equation for this reaction is

$CH_3COOH + NaOH \rightarrow CH_3COONa + H_2O$

So, 1 mole of sodium hydroxide reacts with 1 mole of ethanoic acid.

10 cm^3 of ethanoic acid must therefore contain 1.50×10^{-3} mol:

$10/1000 \times M = 1.50 \times 10^{-3}$

Concentration of ethanoic acid = 0.15 M.

Sodium hydroxide is used to neutralise the acetic acid in the vinegar. Bromothymol blue is used as an indicator. It will change from blue to pale yellow at the end point of the reaction.

Aspirin can also be quantitatively analysed using an acid–base titration. It is important that the right dosage of aspirin is in each tablet, as too little will be ineffective and too much can be dangerous to your health. The active ingredient in an aspirin tablet is acetylsalicylic acid (this is the chemical term for aspirin). This can also be titrated against sodium hydroxide but in this case phenolphthalein is used as the indicator. It will go from colourless to red at the end point.

Unit 4: See page 86 for more information on acid–base titrations.

Redox titrations

In redox reactions electrons are transferred. Redox titrations rely on oxidation or reduction of the **analyte**. In this case, in order to calculate concentration, we have to work out the number of moles reacting and multiply that by the volume. The number of moles of electrons involved must also be taken into account.

Calculating the mass of sodium hypochlorite in bleach

In industry, it is important to know the concentration of the active ingredient in a product. Sodium hypochlorite is the active ingredient in bleach, and a bleach manufacturer would carry out a redox titration to assess the concentration of this in their product. This is a slightly more complicated procedure than the acid–base titrations described above.

First of all, the sodium hypochlorite must be used to produce iodine. Adding potassium iodide and hydrochloric acid to the bleach will form iodine. One mole of iodine is produced for each mole of sodium hypochlorite present. The iodine produced is determined by adding sodium thiosulfate until the reaction is complete. Starch can be used as an indicator, as it is blue when iodine is present and colourless once all the iodine has reacted. The starch must be added near the end of the reaction as it will otherwise form a complex with the iodine, removing it from solution and giving a false result. The calculations carried out to find the number of moles of iodine will give an equivalent value for the concentration of sodium hypochlorite.

Worked example

10 cm^3 of bleach is reacted with excess aqueous potassium iodide. 0.5 M sodium thiosulfate is used to titrate this solution; 6.0 cm^3 of this is needed to react completely.

Answer

Number of moles of sodium thiosulfate in 6.0 cm^3 of 0.5 M solution = 6.0/1000 × 0.50 = 0.003 mol

1 mole of iodine reacts with 2 moles of sodium thiosulfate

Number of moles of iodine present = 0.003/2 = 0.0015 mol

So, 0.0015 mol of sodium hypochlorite are present in 10 cm^3 of bleach.

M_r of sodium hypochlorite (NaClO) = 74.5

1 mol of sodium hypochlorite has mass of 74.5 g

So, mass of sodium hypochlorite in 0.0015 mol = 0.0015 × 74.5 = 0.11 g

So, mass of sodium hypochlorite in 1 dm^3 of bleach is 0.11/10 × 1000 = 11 g

Calculation of iron content of tablets

Iron tablets are an important supplement for many people, especially growing children and pregnant or menstruating women. So again, it is important that the tablets can be analysed to ensure the correct composition. A redox titration is carried out on the iron tablet using acidified potassium manganate (VII).

$MnO_4^- + 5Fe^{2+} + 8H^+ \rightarrow Mn^{2+} + 5Fe^{3+} + 4H_2O$

Worked example

5 g of iron tablets (5 tablets weighing 1 g each) were titrated against 12.5 cm^3 of 0.01 M $KMnO_4$.

12.5 cm^3 contains 0.000125 moles of MnO_4^-.

Five moles of Fe^{2+} react with every mole of MnO_4^-.

Number of moles of Fe^{2+} reacting is 0.000125 × 5 = 0.00625 moles of Fe^{2+} in 5 g of iron tablets.

One mole of Fe^{2+} is 56 g so 0.00625 moles is 0.35 g.

Divide this by 5 to find out mass of Fe^{2+} in each tablet:

0.35/5 = 0.07 g in each tablet.

The tablets must be weighed, crushed and dissolved in sulfuric acid. The solution will be vacuum filtered to remove any impurities like the tablet coating which may not dissolve. The solution then has distilled water added to it to a known volume (250 ml would be appropriate). This solution is then titrated with acidified potassium manganate (VII). The end point is when the solution stays a faint pink.

Activity 22.4A

List the different ways a product can be analysed. Classify each method as quantitative or qualitative.

Analysis of silver nitrate by a redox titration

Silver ions react with halide ions, like chloride ions, to give a precipitate of silver chloride:

$Ag^+(aq) + Cl^-(aq) \rightarrow AgCl(s)$

Adding silver nitrate solution to a solution of chloride ions will remove all the chloride ions from solution. This forms the basis of a titration method. There are different ways to indicate the end point. The simplest method is to add several drops of potassium chromate (yellow solution – toxic!) to the solution. Once all the chloride ions have been used up, the chromate reacts with the silver ions to form a precipitate and the colour changes to a pale brick-red.

Silver nitrate is not a primary standard. It 'goes off'. Before it is used, it must be standardised by titrating it against a known volume of a known concentration of potassium chloride solution (potassium chloride is very pure and does not 'go off' easily).

Safety note: Eye protection must be worn. Silver nitrate is corrosive and will blacken and damage skin. Potassium chromate is toxic and all contact should be avoided.

Activity 22.4B

Standardisation method

Accurately weigh out 1.900 g of potassium chloride in a weighing bottle. (Weigh it roughly first.) [1.900 g of potassium chloride is needed to make 250 cm^3 of a 0.1 mol dm^{-3} solution.]

Tip the potassium chloride through a funnel into a 250-cm^3 volumetric flask and reweigh the bottle.

Wash round the funnel with distilled water and roughly fill the flask with distilled water. Shake the flask to dissolve the potassium chloride before making up to the mark with distilled water. Shake well to mix the solution.

Pipette 25 cm^3 of the potassium chloride solution into a 250-cm^3 conical flask.

Add 10 drops of potassium chromate indicator.

Fill a burette with silver nitrate solution.

Titrate the potassium chloride solution with the silver nitrate solution.

Repeat the titration until concordant results are obtained. (Strictly, you should do a titration of distilled water only to determine how much silver nitrate the indicator reacts with. This volume is then subtracted from all the titres.)

Results of standardisation and calculation (write your own results in this way)

Calculation of concentration of potassium chloride

Mass of bottle plus potassium chloride = g

Mass of bottle after tipping out potassium chloride = g

Mass of potassium chloride used = g

Number of moles of potassium chloride used, $n = m/M_r$

= /74.551 mol

= mol

Volume of solution, $V = 250\ cm^3 = 0.25\ dm^3$

Concentration of the potassium chloride, $c_{KCl} = n/V$

= /0.25 mol dm^{-3}

= mol dm^{-3}

Silver nitrate titration results:

initial volume (cm^3)				
final volume (cm^3)				
titre (cm^3)				

Average titre = cm^3 = dm^3 (you have to ÷ 1000)

Volume of KCl, $V_{KCl} = 25\ cm^3 = 0.025\ dm^3$

Concentration of KCl, c_{KCl} = mol dm^{-3}

Number of moles of KCl, $n_{KCl} = c_{KCl} \times V_{KCl}$ = × 0.025 mol

= mol

Number of moles of $AgNO_3$, $n_{silver\ nitrate} = n_{KCl}$ = mol

Volume of $AgNO_3$ (titre), $V_{silver\ nitrate}$ = dm^3

Concentration of $AgNO_3$, $c_{silver\ nitrate} = n_{silver\ nitrate}/V_{silver\ nitrate}$

= / mol dm^{-3}

= mol dm^{-3}

Spectroscopic analysis

A spectrophotometer records the spectrum of light emitted (or absorbed) by a given substance, and the light can be used to determine the chemical composition of a substance because of signature spectral lines emitted by known elements.

A colorimeter measures the absorbance of different wavelengths of light in a solution. It is used to measure concentration.

Different chemical substances absorb different wavelengths of light. When the concentration of the solute is higher, it absorbs more light in a specific wavelength.

Infrared spectroscopy

This can be used to analyse gases or liquids. You may have a chance to analyse a liquid using an infrared spectrophotometer.

Two small discs of salt called plates are used to sandwich a small drop of the sample to be analysed. If the sample is a solid it must first be crushed and mixed with a liquid called Nujol®.

Both the sample and a reference have to be measured. This is because the measurement can be affected by the light source or the instrument used. Usually no sample is put in the machine and air is used as the reference. Sometimes pure water is used.

Infrared light is beamed through the sample and the amount of energy absorbed at each wavelength is measured. A spectrum is produced showing the absorption as peaks and troughs on a graph. These peaks can then be measured. Their position on the graph shows the frequency at which the infrared light has been absorbed. When the frequency of the infrared light is the same as the vibrational frequency of a bond, absorption will occur and so we can see which bonds are present. This only works on covalent bonds. A data table can be used to see which types of bonds have which vibrational frequency, and so we can state what is present in the sample, e.g. C=C bonds or C-O bonds.

Assessment activity 22.4

D3 P5 M5 BTEC

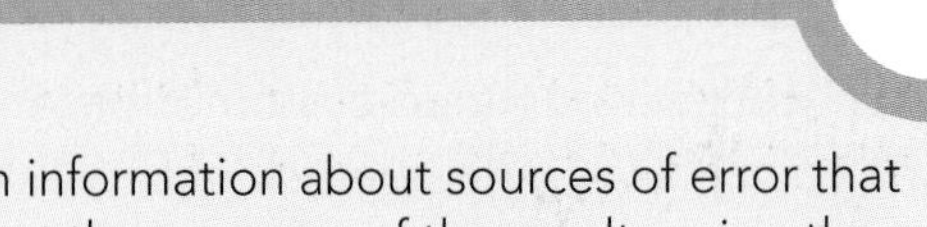

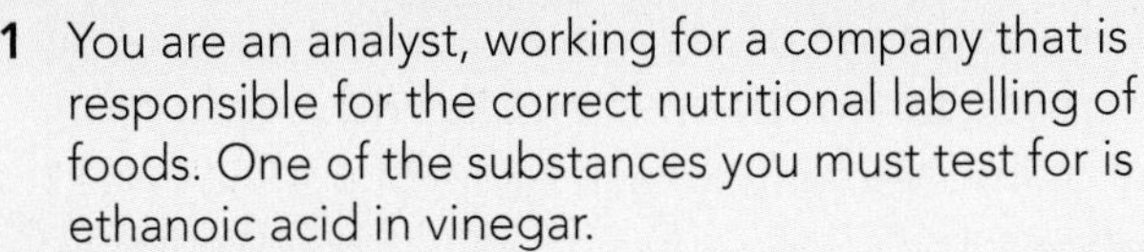

1 You are an analyst, working for a company that is responsible for the correct nutritional labelling of foods. One of the substances you must test for is ethanoic acid in vinegar.

Carry out a quantitative analysis of a sample of vinegar. Write a report explaining how you carried out the quantitative analysis of the amount of ethanoic acid in a particular vinegar. **P5** What sources of error may there be in the analysis? **M5** Explain how the analysis could be made more accurate, and evaluate the degree of reliability to which the result has been found. **D3**

Safety note: Ethanoic acid is corrosive and can cause severe burns. Eye protection must be worn if the acid itself is used rather than a very dilute solution of vinegar.

2 You are a journalist working for a Sunday newspaper magazine that has a weekly feature on supplement tablets.

Write an article explaining how you determined the iron content of iron supplements and interpret the results in the context of the RDA for iron and the claims on the packs. **P5** Use the Internet to research information about sources of error that may affect the accuracy of the results using the method you have described. **M5** Evaluate how reliable your result is, and suggest ways of making the result more accurate. **D3**

Grading tips

For **P5** you must carry out the analyses. Recording your results in a table (see page 416) will help you calculate your titres correctly. Use the worked examples to help you.

In task 1, you must discuss sources of error within the procedures to help you meet **M5** and **D3**. These are not concerning mistakes made by you, but refer to the precision and accuracy of equipment, or reliability of the procedure.

When evaluating a procedure, as in task 2, ensure you discuss the strengths and weaknesses of the procedure as well as giving possible improvements to help you meet **M5** and **D3**. You may look at the use of significant figures.

WorkSpace

Rachel Slater

Senior Scientist, Almac Sciences

I work in the field of organic chemistry carrying out solid phase peptide synthesis and purification. Customers will request a variety of synthesis services. My role involves the production of peptides and proteins for the pharmaceutical industry, for universities, and for research and development (R&D) companies.

In the laboratory, a typical day would involve the synthesis of peptides by SPPS (solid phase peptide synthesis), purification by high pressure peptide liquid chromatography (HPLC) and analysis of the peptides by analytical HPLC and mass spectrometry. Experiments are also carried out in order to develop routes to more novel/ challenging peptide products.

I am also required to prepare quotation requests and costings for customers, together with updating customer-focused progress reports. There are about 10-12 scientists working in the lab, within which there are project-specific teams. Part of my role also focuses on health and safety issues as a member of the on-site safety team.

We work with a varied customer base, such as internal departments (Quality Assurance (QA), analytical), university research groups, small R&D companies and large pharmaceutical firms. As a team we could have four or five different projects running simultaneously, which makes my job interesting and diverse. Some customers want a great deal of involvement at each stage of the job; others simply want to see the results once the job is complete.

In the process of making a peptide (a chain of amino acids) we might reach the end of the chain and some amino acids may not have coupled together properly. This can be incredibly frustrating so we analyse the process at regular intervals to ensure that all is progressing according to plan. Some customers require modifications to the peptide chain, for example the addition of dyes, which mean we need to run trial experiments to develop a suitable addition procedure.

Think about it!

When working in a chemical laboratory:

1. What safety precautions would you need to take?
2. What types of glassware are used?

Just checking

1 Working in groups, list all the different preparative techniques you have used. Against each technique write what type(s) of substance you could prepare with it. Circle all the techniques that are used for extraction.
2 What formula would you use to calculate percentage yield?
3 Draw a diagram of titration equipment and label it to show how you control the accuracy of a titration investigation.
4 Draw a spider diagram showing how you would qualitatively and quantitatively analyse inorganic and organic compounds.
5 Consider the different products you have made. How have you ensured they are pure? How can you check them for purity?

edexcel

Assignment tips

- You will be carrying out several practical preparations and analyses. It is important that you follow the methods carefully in order to get the correct results. You must also consider safety when carrying out the practical work.
- If you are writing a report of your findings, make sure it is clear what you were hoping to find out and what you actually found out. Use your results to back up and explain your conclusions.
- When writing an evaluation, ensure you consider the strengths and weaknesses of the procedure. You should also give some possible improvements and explain why the improvements will help produce better results. Do not just state that you should have been more careful or that you should have done it more times: this is not an evaluation of the procedure but just an evaluation of how you have worked, and this is not what is asked for.

You may find the following websites useful as you work through this unit.

For information on...	Visit...
titrations	Virtlab®: A Virtual Laboratory
laboratory techniques	About.com: Chemistry
risk assessment template	The Physical and Theoretical Chemistry Laboratory, Oxford University

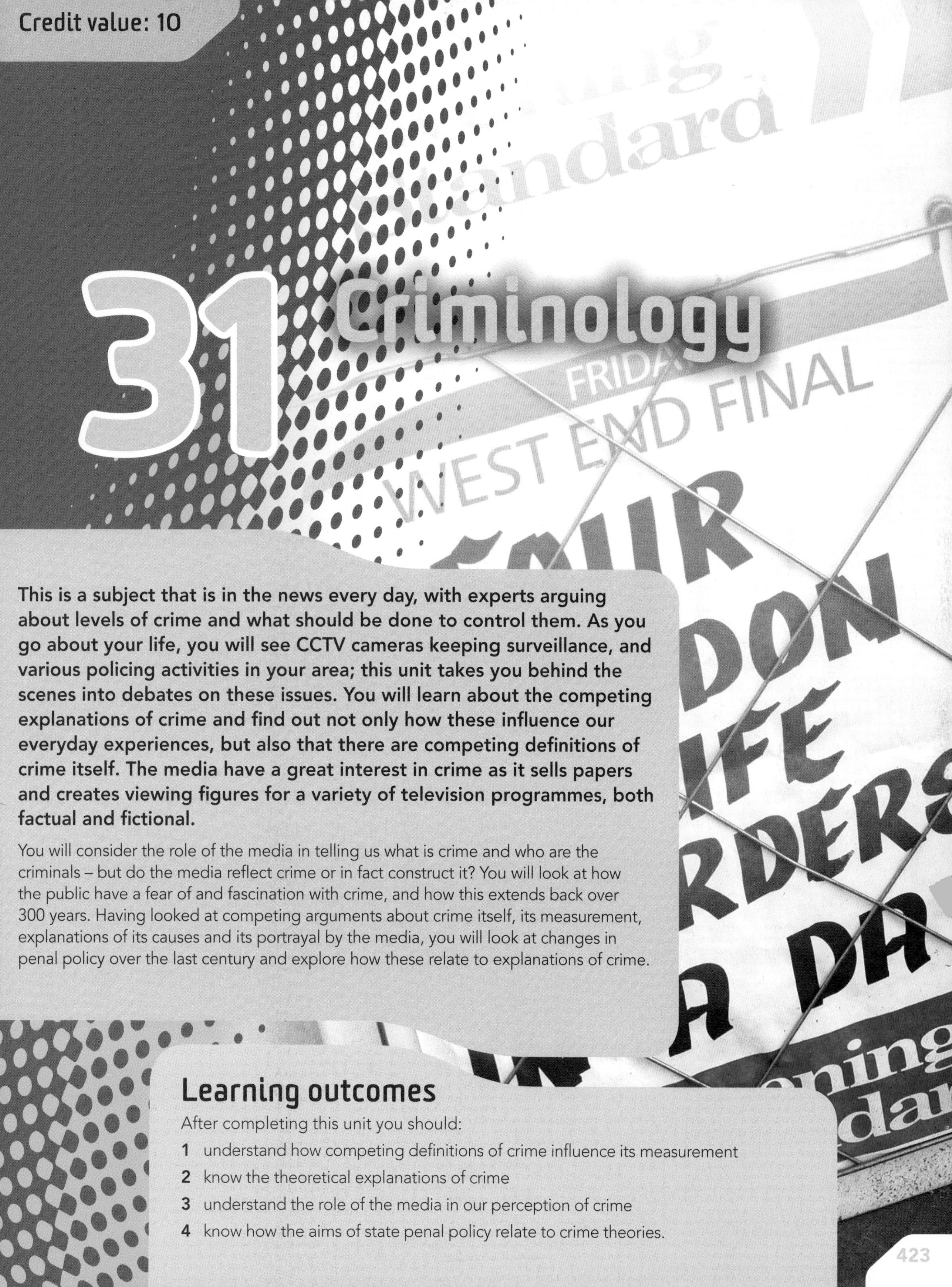

Credit value: 10

31 Criminology

This is a subject that is in the news every day, with experts arguing about levels of crime and what should be done to control them. As you go about your life, you will see CCTV cameras keeping surveillance, and various policing activities in your area; this unit takes you behind the scenes into debates on these issues. You will learn about the competing explanations of crime and find out not only how these influence our everyday experiences, but also that there are competing definitions of crime itself. The media have a great interest in crime as it sells papers and creates viewing figures for a variety of television programmes, both factual and fictional.

You will consider the role of the media in telling us what is crime and who are the criminals – but do the media reflect crime or in fact construct it? You will look at how the public have a fear of and fascination with crime, and how this extends back over 300 years. Having looked at competing arguments about crime itself, its measurement, explanations of its causes and its portrayal by the media, you will look at changes in penal policy over the last century and explore how these relate to explanations of crime.

Learning outcomes

After completing this unit you should:

1 understand how competing definitions of crime influence its measurement
2 know the theoretical explanations of crime
3 understand the role of the media in our perception of crime
4 know how the aims of state penal policy relate to crime theories.

Assessment and grading criteria

This table shows you what you must do in order to achieve a **pass**, **merit** or **distinction** grade, and where you can find activities in this book to help you.

To achieve a **pass** grade the evidence must show that you are able to:	To achieve a **merit** grade the evidence must show that, in addition to the pass criteria, you are able to:	To achieve a **distinction** grade the evidence must show that, in addition to the pass and merit criteria, you are able to:
P1 describe competing definitions of crime **See Assessment activity 31.1**	**M1** draw conclusions from their analysis with consideration of how the figures are produced **See Assessment activity 31.1**	**D1** explain how the difference in crime figures relate to the reported/ recorded crime figures and the definition of crime used by the agency **See Assessment activity 31.1**
P2 analyse the crime figures produced by Home Office official statistics and the British Crime Survey **See Assessment activity 31.1**		
P3 describe theoretical explanations of crime **See Assessment activity 31.2**	**M2** explain the ways in which the theoretical explanations have influenced current political policies on crime control **See Assessment activity 31.2**	**D2** explain the challenge of realism which replaced positivism as the major criminological theory in the late 20th century **See Assessment activity 31.2**
P4 identify the influences that the theoretical explanations of crime have had on current crime control policies **See Assessment activity 31.2**		
P5 describe how the media influence our understanding of crime **See Assessment activity 31.3**	**M3** report on factors which influence fear of crime **See Assessment activity 31.3**	**D3** outline the theoretical explanation of moral panics developed by S. Cohen and use it to illustrate a current crime problem **See Assessment activity 31.3**
P6 explain the difference between the fear of crime and the actual risk of crime **See Assessment activity 31.3**		
P7 outline ways in which criminological theory influenced penal policy over the last century **See Assessment activity 31.4**	**M4** explain how crime theories have influenced penal policy **See Assessment activity 31.4**	**D4** analyse the influence of positivism and realism on penal policy over the last century **See Assessment activity 31.4**

How you will be assessed

Your assessment could be in the form of:

- a presentation showing your knowledge of the definitions of crime and how crime figures are produced
- a newspaper article
- carrying out research and surveys.

Dinesh, 21 years old

After studying criminology, I became very interested in penal policy and applied for a job with the prison service. Once my training is complete, I will be working with offenders, and as a prison officer will work to ensure safe custody for all offenders. This involves carrying out detailed procedures for moving prisoners around the establishment for them to do work in trades or education, in my category C prison. Many offenders need help with basic skills of reading and writing, and education provision helps them to develop these skills which will help them to gain employment on their release from prison. An important part of the prison officer role is to ensure that all prisoners are treated with respect, and we are given training to understand diversity issues. If society does not treat prisoners with respect during their sentence then how can we expect them to treat the public with respect when they are released?

Catalyst

What is crime?

In small groups, discuss what crime means to you, and prepare posters listing the crimes you would identify. In a class discussion, ask if everyone shares the same definition of 'crime'. In this unit you will be studying crime, and it is a good idea at an early stage to think about what exactly that means. If you don't know how to define 'crime' how can you measure it?

31.1 The measurement of crime

In this section:    

Key terms

Crime – the definition is contested but two main definitions are used here, legal and normative.

Quantitative data – a set of data that can be expressed in numbers and is objective.

Qualitative data – a set of data that is collected by interview or questionnaire and is subjective.

The nature and extent of **crime** in the UK is always in the news, and you will have heard politicians quoting Home Office statistics and arguing that crime figures are either falling or increasing. In this section, you will see that there are differences in how statistics are gathered that relate to how the different measures define crime, and you will explore some of the complexity in crime reporting and recording.

Did you know?

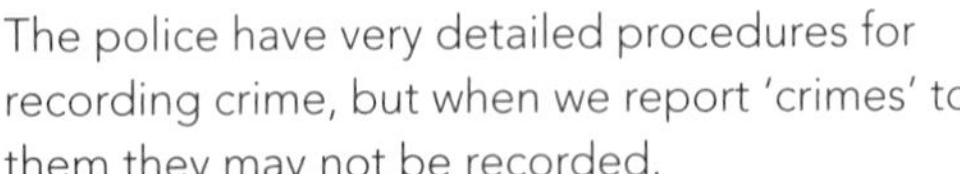

The police have very detailed procedures for recording crime, but when we report 'crimes' to them they may not be recorded.

What is crime?

In order to measure crime it is important to actually define the concept, and that is not as simple as it sounds.

For the police, crimes are 'intentional acts which break the law of the land (see Tappan reference, page 448)' – so unless there is a law prohibiting an act then it cannot be dealt with by the police. This is a legal definition of crime.

Activity 31.1A

Can you think of any crimes today that did not exist in the past? Discuss within a small group some of the crimes in the past that are legal today. Also discuss: does 'crime' change over time?

Did you know?

In the 1930s alcohol consumption was banned in America. The major result of this was gang warfare to control the very profitable supply of illegal alcohol.

The legal definition has been criticised because it means that unless there is a law against an act, then no matter how harmful it may be there cannot be a crime. Criminologists have argued that a more helpful definition is: 'crimes are acts which break or contravene a set of formal or informal norms or codes'. This is known as a normative definition.

Activity 31.1B

Imagine seeing an individual committing an act you feel is wrong. Which definition would you be using: legal or normative? Now, if you were concerned enough to report the act to the police and they can only use the legal definition, would you expect the act to be recorded as a crime?

Measurement of crime

The Home Office gathers information on crime using two different methods:

- Home Office official statistics
- British Crime Survey (BCS) data.

For both these measures, you will look at how the data are produced and what are considered to be crimes.

Home Office official statistics

For this measure of crime, a criminal act must be reported to or by the police, and then it must be formally classified and recorded by the police. When you see discussions of crime statistics, politicians or journalists quote both numbers of crimes committed and also different categories of crime. With this rigorous system of classifying and defining crime, acts recorded by any police force in the UK will be based on a common understanding of crime. This means that the data given in the Home Office official statistics are **quantitative** and objective; later in the unit we will

Activity 31.1C

In groups, think about and debate which definition of crime is used for each of the measures of crime outlined above. Then look at the data provided in the tables below, which tell you that the crime totals for 2008/09 are 10.7 million for the BCS and 4.7 million as recorded by the police. How would you explain the difference? Note that both figures are correct in their own right, and that valuable information can be gained by considering each separately over time.

Number of crimes and risk of being a victim, 2007/08 and 2008/09 BCS

Numbers (000s) and percentage change **BCS**

	2007/08	2008/09	Percentage change and significance[1]	
	Number of incidents (000s)			
Vandalism	2,695	2,769	3	
Burglary	737	744	1	
Vehicle-related theft	1,508	1,514	0	
Bicycle theft	444	540	22	**
Other household theft	1,066	1,184	11	
Household acquisitive crime	3,756	3,982	6	
All household crime	**6,451**	**6,751**	**5**	
Theft from the person	581	725	25	**
Other theft of personal property	987	1,096	11	
All violence	2,200	2,114	-4	
with injury	*1,063*	*1,116*	5	
without injury	*1,137*	*998*	-12	
Personal acquisitive crime	1,883	2,094	11	
All personal crime	**3,768**	**3,936**	**4**	
All BCS crime	**10,219**	**10,687**	**5**	
Risk of being a victim of any BCS crime[2]	**22.2**	**23.4**		**

1. Statistical significance for change in all BCS crime cannot be calculated in the same way as for other BCS figures (a method based on an approximation has been developed).
2. Risk is defined as the proportion of the population being a victim of any BCS crime once or more.

Taken from **Research, Development and Statistics – Home Office** Crime in England and Wales 2008/2009, Chapter 2, Extent and trends.

Number of crimes recorded by the police in 2007/08 and 2008/09

Numbers (000s) and percentage change **Recorded crime**

Offence group	2007/08	2008/09	Percentage change
	Number of offences[1] (000s)		
Violence against the person	961.2	904.0	-6
Violence against the person – with injury[2]	*452.4*	*421.2*	-7
Violence against the person – without injury[2]	*508.8*	*482.8*	-5
Sexual offences	53.5	51.5	-4
Most serious sexual crime[3]	*41.4*	*40.8*	-2
Other sexual offences	*12.0*	*10.7*	-11
Robbery	84.8	80.1	-5
Burglary	583.7	581.4	0
Domestic burglary	*280.7*	*284.4*	1
Other burglary	*303.0*	*297.0*	-2
Offences against vehicles	656.4	592.1	-10
Other theft offences	1,121.1	1,080.7	-4
Theft from the person	*101.7*	*89.7*	-12
Fraud and forgery	155.4	163.3	5
Criminal damage	1,036.2	936.7	-10
Drug offences	229.9	242.9	6
Miscellaneous other offences	69.4	71.1	3
Total recorded crime	**4,951.5**	**4,703.8**	**-5**

1. Numbers given in this table are the latest available and may differ slightly from provisional figures published previously.
2. See Table 2.04 for the full list of offences included in violence against the person with/without injury.
3. Most serious sexual crime comprises rape, sexual assault, and sexual activity with children.

look at categories and trends in crime produced by this measurement. See the tables in Activity 31.1C showing different types of crime figures.

British Crime Survey data

This measure of crime uses information collected from a random sample of the public who self report crime; these data are **qualitative**. It is useful to note that in this measure, if you were one of the sample chosen, everything you reported as a crime would appear in the data. Now ask yourself if you are aware of all the categories of crime as defined by Tappan (1947) and used by the police. Very few people outside of the criminal justice system will have the knowledge to use the legal definition of crime – so people in the sample will use their view of what crime is. From Activity 31.1B, your discussion in class may well have found big differences in people's views of crime. This illustrates why data presented in the BCS are described as subjective, as they depend on individuals defining crime in their own way.

Look at the difference between reported crime and crime actually being recorded by the police (see tables in Activity 31.1C). It is possible that the same crime committed by different individuals in different circumstances would lead to reports of crime being treated differently by the police. This is because police officers have discretion to take circumstances into account. So every crime reported to the police will not necessarily be recorded by them and may not appear in the Home Office official statistics.

Factors influencing Home Office official statistics

Most crimes are reported through observation by the police, stop and search, and policing duties, so it is reasonable to assume that if there is an increase in the number of police officers year on year then crime figures will be affected. More police officers will lead to more crimes being recorded.

You have seen that the police and agencies of the Criminal Justice System use the Tappan legal definition of crime, so if the state introduces new laws then there will be more crimes that can be recorded. This area will be developed further in section 31.3.

The criminologist Mike Maguire (see reference, page 448) has argued that the increasing availability of mobile phones makes it easier for people to report crimes to the police, and that the requirements of insurance companies mean that people need to have a crime number in order to make a claim for a loss. He also argues that previously hidden crimes like domestic violence are now much more likely to be reported than in the past because of changes in society.

Factors influencing BCS data

Home Office researchers Mirrlees-Black *et al.* (see reference, page 448) have undertaken research on why people do not report crimes to the police. They found that:

- 44% of victims felt that the incident was not sufficiently serious to report

Case study: What do your parents do?

Ten sixth-form boys from a public school are seen playing football on a playing field where there is a by-law prohibiting the activity. They have just finished end-of-year exams and are very boisterous. Local people report them to the police, a police car arrives quickly on the scene, and the police officers interview the boys. They give their addresses and parents' jobs to the police with other contact details, and explain that they are celebrating a term of hard work before going to university.

In a nearby town a similar group of 10 boys from the local school are also reported for playing football on a restricted playing field. They too have completed exams and are enjoying a lively game. The police interview them and, again, addresses and parental details are taken by the officers, the boys pointing out that they were doing no harm and were just celebrating the end of exams.

1. **In view of the evidence, do you feel that the two groups would receive the same treatment by the police?**
2. **Would you expect any effect due to class or parental influence?**
3. **Is there a possibility that different social classes may receive different treatment in terms of police discretion?**

- 33% claimed that the police would not be able to do much about it
- 22% thought that the police would not be interested
- 11% felt that they would be better placed to deal with the matter instead of reporting it to the police.
- They also noted that some people chose not to report crimes to the police for fear of reprisals.

These researchers identified that different types of crime were reported at different rates:

- 97% of motor vehicle thefts were reported to the police
- 85% of burglaries were reported to the police
- 45% of woundings were reported to the police
- 26% of vandalism to cars was reported to the police.

Categories of crime

Within the different categories of crime there is a focus on street crime. Criminologists argue that there are other types of crime that are also harmful and dangerous but that do not appear in crime statistics; for example, white collar and corporate crime, and crimes by the state.

Assessment activity 31.1

P1 M1 D1 P2 BTEC

You are a trainee prison officer and you have been asked to prepare a presentation for the Prison Governor on the different types of crime and how they are defined and reported.

Describe the competing definitions of crime in the first part of your presentation. **P1** Carry out research on the Home Office website to find data on the reporting of crime for 2007–2008, and also the recording of crime. Include information in your presentation on the differences you see across the categories of crime. **M1** **D1** **P2**

Grading tips

To help you gain **M1** look at the competing definitions of crime and reflect on what effect these have on the figures produced.

For **D1** think about the following categories of crime: burglary, thefts against the person (including vehicles and bicycles), vandalism, violence, wounding and robbery. Explain how the differences in the crime figures relate to whether they are reported or recorded, and ensure that you bring the different measures of crime into your explanation.

PLTS

Independent enquirer and Team worker

When undertaking the activities and answering the questions in the case study, you will have the opportunity to demonstrate independent enquiry and team-working skills.

Functional skills

ICT, Mathematics and English

You will have the opportunity to use IT systems, and to use ICT to find and select information. In your preparation of reports and presentations you will develop, present and communicate information.

When looking at statistical data from the Home Office and BCS you will demonstrate mathematical skills.

English skills will be developed in reading, writing, and speaking and listening when you research information, work in groups and give presentations.

31.2 Theoretical explanations of crime

In this section:    M2 D2 P3 P4

Key terms

Voluntarism – the argument that offenders choose to commit crime by free will as seen in the classicist and right realist theories.

Positivism – research method based upon scientific observation to find the causes of crime. Once the causes are known then the criminal can be cured.

Determinism – the argument that offenders are forced to commit crime through circumstances beyond their control – seen in positivist theoretical explanations.

Rehabilitation – the process of helping offenders get back into law-abiding society through individual programmes of expert help and support.

Longitudinal studies – in this research method, groups of people are studied over long periods of time and the work often includes different generations of the same family. Then researchers can see whether the criminal behaviour of a grandparent passed down to a parent, then a son and/or daughter.

Syndrome – a medical term for a package of symptoms causing a disease or condition.

Taking crime seriously – this phrase was a key message from realist theorists to attack the way in which they saw crime being excused by positivist criminologists – and it gained massive public acceptance.

Why is theory important?

Have you ever looked through glasses with coloured lenses? If you look through blue lenses then everything you see will be blue, but if you change to red lenses then everything you see will be red. If we apply this to a theory, once you accept a particular theory then everything will be seen from the arguments of that theory, and that can include policies to control crime or to sanction criminals.

In this section you will explore a range of theoretical explanations of crime. Some of them not only aim to explain criminal behaviour but also set out ways to control such behaviour. These explanations are always related to a society at a particular time because, in order to explain criminal behaviour, the theorist must provide evidence to convince people that their argument is sound and provides better understanding than competing theories. As indicated earlier, theoretical explanations can have a direct link to how a society sets out to control crime. Evidence for this link with policy will be explored in section 31.4.

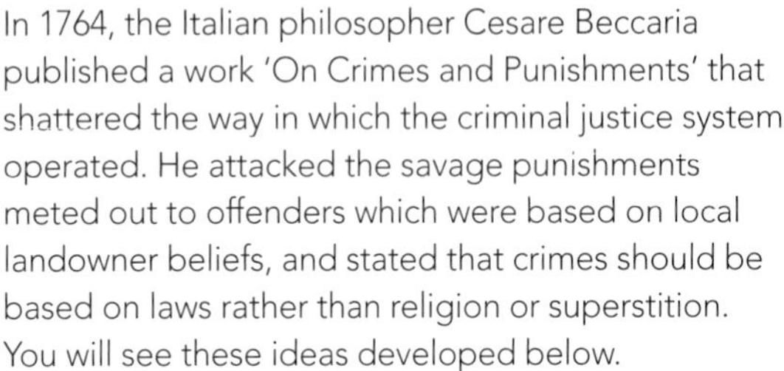

Did you know?

In 1764, the Italian philosopher Cesare Beccaria published a work 'On Crimes and Punishments' that shattered the way in which the criminal justice system operated. He attacked the savage punishments meted out to offenders which were based on local landowner beliefs, and stated that crimes should be based on laws rather than religion or superstition. You will see these ideas developed below.

Theoretical explanations are always based on what is happening in societies at the time they are developed, and Beccaria saw unfairness in what counted as a crime, which could be based on religious laws or even superstition, whilst punishments varied massively for the same offence.

Criminal justice was designed to control the working classes. It was applied by the rich minority in societies and the sanctions could be extremely harsh (in the UK over 200 crimes attracted the death penalty at the time of Beccaria). With this in mind, it is worth considering the ideas Beccaria set out in his theoretical explanation.

- Human intelligence gives us the ability to distinguish right from wrong, so everyone can make rational decisions on whether or not to commit a crime (that is human free will or agency). This argument is called **voluntarism**.
- In making such decisions the individual will look at the benefits and costs of their actions and then decide whether or not to commit the crime.
- Criminal justice should be based only on laws that must apply equally to everyone in society: rich and poor alike.
- Punishment should fit the crime and should be the least possible to deter criminal behaviour.

- Sanctions should be applied soon after the offence in order to deter offenders.

This theory was called classical theory and was adopted widely because it was seen to be a vast improvement on the previous piecemeal system. In this approach, because an individual has the intelligence to make a rational choice between offending or not, it is appropriate for the state to punish the offender. In contrast to the earlier harsh and unreasonable punishments, Beccaria argued that individuals would be deterred by minimal but fair sanctions applied promptly.

Activity 31.2A

In groups, discuss the ideas of Beccaria and contrast them with what was happening in society at that time. Do you agree with the idea that you and other learners have the intelligence to make rational decisions? How would you feel if you were given a harsh punishment for committing an act for which someone else was only told off? If you were charged with an offence would you be deterred in the future by a prompt but fair sanction?

In many societies the rich were not always supportive and the argument about everyone being treated equally did not gain approval from all. In addition, there were criticisms that the sentencing was too lenient and the law was soft on criminals. (These arguments were going on over 200 years ago, and you can see exactly the same arguments in the press today.)

For the first half of the nineteenth century in the UK this theory was accepted and applied, with judges giving minimal sentences to deter offending behaviour. Even at that time, broadsheet newspapers were distributed that described crimes in the local area and, of course, the more sensational the crime the more papers were sold (you will return to this theme in section 31.3). In the nineteenth century there was mass movement of the population from the countryside to the cities driven by the industrial revolution. This gave rise to concern among the ruling classes – large numbers of working class people living in close proximity to each other was thought to create a risk of revolution. There were regular reports of crime and disorder in the big cities with respectable citizens fearing to go out at night. This increased pressure on judges to give harsher sentences to deter criminals.

Did you know?

In 1982, Geoffrey Pearson, Wates Professor of Social Work at the University of London, published a fascinating book *'Hooligan, a History of Respectable Fears'*. He was interested in the way in which groups of individuals were portrayed by the media and undertook research to see if such labelling was a twentieth century phenomenon – instead he found that right back to 1700, groups were given negative labels if they were seen to be a threat to the upper classes. In one fascinating example he graphically described the emergence of a 'new crime' in 1860s London – garrotting – in which the victim was choked and robbed. The reporting of these incidents by *The Times* and the use of cartoons by the magazine *Punch* (showing people wearing special clothing to protect themselves) created a climate of fear. (You will return to this theme in section 31.3.)

The rise of positivism: a scientific criminology

In the nineteenth century great advances in medicine and sciences, such as physics, chemistry and biology, were based on a research method involving careful observation and experiments. Theorists argued that such methods could be applied to the study of criminal behaviour in approaches that became known as **positivism**. Unlike the classicists, positivists argued that human nature was basically good and therefore criminal behaviour must be caused by factors outside of the individual's control – this claim rejects voluntarism. If these factors could be discovered by use of this new research method, then criminals could be cured in the same way that illnesses and disease had been cured. Positivism is a structural approach, as it looks for factors outside of the individual's control that cause criminal behaviour, in order to cure them. This approach is called **determinism**. This argument leads to the conclusion that if an individual has no choice in committing crime then it is wrong to punish that individual for actions outside of their control. Positivists rejected the idea of deterrence and instead said that individuals must be given expert help to resolve their problem; this policy is called **rehabilitation**.

There are three main positivist approaches:

- biology, including genetic
- family influences, a psychological approach
- culture.

Each of these approaches will be outlined, but all share the key claim that criminal behaviour is determined by factors outside of the individual's control, i.e. if you can find the cause then you can cure the criminal.

Biology

In 1876, an Italian physician called Lombroso studied the body shapes and sizes of convicted murderers and attempted to prove that this group was physically different from law-abiding citizens. Note at this time the ideas of Darwin about natural selection were very topical. From his observations, Lombroso argued that there were significant differences in the group of murderers, with primitive skull features, and that these people were biological throwbacks to more primitive times. So the cause of crime was claimed to be a physical or biological abnormality. Society could now identify people with such skull features and act to prevent them going on to commit murder.

Any research or theory is tested by other scientists, and Lombroso's claim was discredited. However, it does illustrate how ideas at a particular time can gain credibility. It is interesting to note that even 200 years after the Lombroso argument that criminals look different from law-abiding citizens, people still make judgements on physical stereotypes. An example is given opposite where people were shown photographs and asked to say which person committed certain crimes (discussed in an academic paper by Bull and Green in 1980; see reference, page 448).

Genetic explanations of crime

In the twentieth century, research on biological factors focused on the transmission of characteristics by genes, and many researchers found that identical twins demonstrated similar tastes, talents and patterns of behaviour – including criminal behaviour. Research by Mednick and colleagues (see reference, page 448) continued the genetic argument that they had been able to discover a pattern of inherited **autonomic nervous system** characteristics among known offenders, which meant that they did not have the same constraints against anti-social behaviour present in non-offenders.

Activity 31.2B

In groups, look at the photographs below and individually match faces A, B and C against the crimes outlined. Compare your views with those of the class on who might have committed the various crimes: fraud, rape, mugging, soliciting, robbery with violence, taking and driving away a car, illegal possession of drugs, gross indecency. Was there any pattern in the class analysis?

From these different biological explanations, the common factor is that the criminal has no choice but to commit crime due to the genetic factor(s) indicated in the research. Scientific research has considerable impact on both the general public and policy makers, and these ideas still have credibility today.

Psychological explanations of family influences

In this approach, social psychological theory is used to investigate whether problem families transmit criminal behaviour across generations. Researchers used **longitudinal studies**, where the criminal behaviour of individuals within particular families was studied over generations and their crimes recorded. These studies have been carried out in America as well as in the UK, and in both countries it was found that there was a strong correlation between criminal offenders and their criminal families.

As an example, in 1994, following longitudinal research in the UK, Farringdon set out two claims (see reference, page 448).

- He argued that criminal offending is part of a larger **syndrome** of anti-social behaviour (a syndrome is the medical name for a set of symptoms indicating the existence of a condition or problem). Here, the syndrome is a collection of anti-social dispositions and patterns of behaviour.
- His second claim is that anti-social or criminal behaviour begins at an early age and develops into a long and serious criminal career.

From research findings he concluded that criminal offenders tend to exhibit socially disruptive behaviour within their families at an early stage, moving to truancy, school failure, family violence, unemployment, and when they have families of their own they show poor parenting skills and their children follow the behaviour of their parents, and so the cycle continues. This pattern was shown across several generations of problem families.

Here, individuals in problem families inherit the criminal syndrome and have no choice but to follow a career in crime, then passing the syndrome on to their children. Have you seen such arguments in the media or tabloid press in your area? Do you see certain family names appearing in local crime descriptions? These arguments about anti-social behaviour have been taken up in legislation on anti-social behaviour orders (ASBOs), where individuals are labelled as a problem and crimes are defined due to interaction between these individuals and the people in contact with them.

If these claims are valid and family influences cause criminal behaviour, then the way to cure crime is for the state to intervene and change problem family behaviour. Note that here families are labelled as a problem, and that the influence of these families creates a cycle of criminal behaviour.

Psychological explanations of cultural pressures

Discussion of this subject emerged from academic research on American juvenile delinquents in large American cities. The researchers collected quantitative data on where young male offenders lived in several big cities, and found that they lived in the inner city areas which had low rents and poor surroundings and facilities. In these areas, they found that the population tended to move into better areas as soon as their circumstances allowed, and there was little continuity of population. The researchers argued that community structures in these areas were not settled and therefore the normal controls of anti-social behaviour were not working. It is argued that in a small community everyone knows everybody else and what people are doing, so individuals realise their actions are observed and will control their behaviour. To investigate this claim, researchers collected lots of hard data on offender location and the types of area in which they lived. They discovered that the offenders did live in inner city locations where people did not stay long enough to know each other.

The research project also used qualitative research by observing and talking to young offenders socially and in gangs to find out what crime meant to them. This qualitative evidence allowed the young offenders to give their view of life in their locality, and the results showed that they formed a sub-culture that did not share mainstream society's values. From this viewpoint, crime was seen as acceptable and was something the gangs enjoyed. Their subculture was labelled as a problem and their behaviour reinforced the negative label.

From this point of view, the young offenders were forced into crime because normal control measures did not apply in the areas in which they lived, and they developed values that allowed crime to be seen as a worthwhile activity, rather than something wrong.

In summary, according to the positivist argument, individuals are forced into crime through no fault of their own, either by biology/genetic influence, or by psychological factors arising from family influence or cultural pressures. These structures in society are the problem and to cure them, the individuals must be given expert help to enable them to reject the life of crime and become valued citizens. Because we are all different as individuals, the treatment programmes devised must be tailored to the individual. Rehabilitation, not deterrence, is the policy for positivists.

Positivism was the dominant criminological theory from the end of the nineteenth century, and after the Second World War the introduction of the National Health Service meant that there was a wide range of experts available to provide rehabilitation for offenders. Because each offender is an individual and, like all of us, unique, criminologists argued that sentencing should take account of this, unlike the classicist argument outlined above. A typical example of what happened when an offender was convicted was that the judge would put the offender in the care of a nominated expert, such as a psychologist, where they would remain under treatment until cured.

Sentences for the same crime could last for 6 months or 6 years depending on the individual and the expert. This has been criticised by competing theorists, such as the classicists, and it was argued that indeterminate sentences were an infringement of the offender's human rights. (Note that indeterminate sentencing has returned to the UK in the twenty-first century, with certain categories of offender who will only be released when experts are convinced they pose no threat to society.) The numbers of offenders given indeterminate sentences by the courts is increasing rapidly in the UK.

The test of any theory is the efficiency of its policy, and classicism went out of favour in the nineteenth century because crime rates kept increasing. The positivist policy of rehabilitation involves open-ended treatment for offenders. It is resource-demanding and uses experts' time. An important measure of effectiveness is the re-offending rate of people who have completed their rehabilitation, and this is evidence which is readily available from prison records. In 1974, researchers Clarke and Sinclair concluded that there was little reason to believe that any one of the widely used methods of treating offenders was much better at preventing re-offending than any other. These findings were supported by Martinson's analysis of 231 studies of particular rehabilitation treatment programmes in America, which concluded that, with few exceptions, the programmes had no appreciable effect on re-offending.

We have already seen that changes in society influence theory, and in both America and the UK there were changes in the political mood, with public opinion moving from left-wing policies implemented by the Labour government to right-wing policies supported by the conservative arguments of US President Ronald Reagan and Prime Minister Margaret Thatcher in the UK in the early 1980s. In general, conservative policies do not support state intervention, and the arguments of positivists about rehabilitation with its widespread use of experts and high cost were going out of favour in both countries. Political leaders develop links with academic theorists in many policy areas, and in crime and justice James Q. Wilson was a key advisor to Ronald Reagan and a key theorist in right realism.

The advance of right realism

Right realist criminologists argue that we must take crime seriously and, instead of looking after the offender, the state should look after the victims of crime and replace the soft liberal criminal justice system with strong penal sanctions (see also section 31.4). The argument about '**taking crime seriously**' was presented as a contrast to the positivist approach, which focused attention on the offender not the crime committed by them. A fundamental argument of right realists is that offenders make choices by free will; this voluntarist approach is the complete opposite of the positivist argument that offenders have no choice but to commit crime. You will see similarities between the early classicist approach and this realist explanation, but there are major differences in their policies to control crime.

In 1975, James Q. Wilson, who was US President Ronald Reagan's adviser on crime and justice, argued that crime exists because wicked people exist, and that most crimes come about because the offender decides that the benefits of the crime outweigh the risks. He argued that the most serious crimes were street crimes committed by a readily identifiable criminal class of repeat offenders – labelled as a major problem. In order to control crime there needs to be tough and swift punishment for the targeted criminal class – you see here a difference compared to the classicist argument for minimal sanctions. Slogans such as 'Prison Works' and 'Three Strikes and you are Out' showed that locking up a criminal meant that they could not then commit further crimes; these are examples of the right realist approach which used imprisonment to prevent repeat offending – a policy known as selective incapacitation. This policy was viewed enthusiastically by right-wing parties in both the UK and America, and the press and public in general were enthusiastic about these simple direct ideas to cut crime.

The theoretical ideas of Wilson were taken up in the law-and-order drive of conservative politicians in the UK and America with phrases such as 'The War on Crime' being very popular. A quote from Michael Howard, UK Home Secretary from 1993 to 1997 – 'We aim to take the handcuffs off the police and put them back on the criminals where they belong' – indicated the aims of the new criminal justice approach.

In order to have their explanation of crime accepted, right realists attacked the positivist theory from several directions:

- research in both the UK and America has shown that rehabilitation does not work
- the costs of rehabilitation are unacceptable
- it is an infringement of an offender's human rights not to know when their sentence will end
- there are so many potential factors influencing offender behaviour that it is impossible to design programmes to cure them all
- liberal arguments about social factors, such as poverty and unemployment, causing crime were wrong, because in the 1950s great improvement in social conditions resulted in a rise in crime rather than a decrease
- individuals commit crime through choice – voluntarism; it is a lack of self control and responsibility that leads to crime
- crime is a symptom of declining moral standards and permissiveness; family breakdown, single parents and poor parenting have led to a criminally motivated underclass
- instead of searching for causes of crime the state needs to reduce opportunity for crime.

These arguments were effective in discrediting positivism and the right realist approach gained dominance in both the UK and America. With this new theory came methods of crime control based on deterrence and reducing the opportunity for crime.

Activity 31.2C

In groups, carry out a survey in your local area to count the number of CCTV cameras you can see. This will involve collecting information from shops, street-mounted cameras and speed cameras, and then discussing the findings in your class. Do you see evidence of security systems in shops, banks or from personal experience in homes? Does this evidence suggest that in the UK there have been initiatives to prevent crime by surveillance measures that have meant that the criminal would be observed and therefore risk arrest?

Situational crime control

The basic argument of right realism is that the offender will make a choice between committing a crime or not, depending on benefit/risk analysis. From this position the state must reduce crime by increasing surveillance and in the design of new housing estates. For example, if you look around your local area you will see CCTV cameras and this is one aspect of surveillance that has expanded dramatically since the 1980s. It is argued that if potential criminals see CCTV cameras then they will make a choice not to commit crime. Do you see any problems with this claim? If the claim is correct then the criminal will not commit the crime within camera vision, but could they then move on to commit crime elsewhere? Wilson and Kelling argued that it is important to repair damage from vandalism and improve rundown areas, because if damage is left unrepaired then a message is sent out that no one cares, and the problem will escalate. Here, maintaining a good environment will contribute to crime control.

Individual responsibility is another part of the right realist argument. You may have seen campaigns in the media to encourage householders to secure their property and car owners to lock their cars to prevent theft – often with thieves portrayed as predatory animals. It is sometimes argued that these campaigns shift responsibility for crime from the offender to the victim – how do you feel about that argument? Right realists argue that low-level crime, if not controlled, will inevitably lead to more serious offences. This has led to a recommendation of zero-tolerance policing, where even the most minor infringement is punished. In recent years, anti-social behaviour has been criminalised, and this also fits with the argument that unpunished crimes will encourage offenders into more serious crimes.

Activity 31.2D

From your experience and research, discuss in class the benefits and/or problems with surveillance, as well as the fairness of transferring blame for crime from the criminal to the victim. What do you think about the way in which ASBOs are used? Do you see evidence of them being used across all social classes and age groups?

It would be good practice to summarise points in a poster/presentational format.

Challenge to the right realist domination by left realism

For two decades, the right realist explanation and its policies were dominant in both the UK and America, with conservative ideas in the ascendency. This meant that left-wing theorists were left with no explanation to challenge the right realist arguments and therefore no chance to influence policy. In the 1980s in the UK, former Prime Minister Tony Blair (then leader of the so-called New Labour party) used the left realist theory of criminologist Jock Young as a base for his crime and justice policy, in the same way that Ronald Reagan had used the ideas of James Q. Wilson. The left realist arguments set out very different views on crime control policy to those described above.

Left realism

This theoretical approach agreed crime had to be taken seriously because the general public accepted that argument, but then argued that the majority of street crime was carried out by members of the working class against other members of the working class. In order to understand the impact of crime, the use of victim surveys was recommended, and it was agreed that the causes of crime must be investigated as, in this view, the right realist policies of imprisonment were not working. A key claim for this approach is that crime is caused by relative deprivation where the expectations of people are not being met by real opportunities. In order to combat this problem, the aim was to build a more equal society. These ideas were developed in the Left Realist Social Crime Control policy.

Activity 31.2E

In groups, discuss how behaviour is regulated in your school or college, then think about what control measures are applied in your neighbourhood – do you see neighbourhood watch schemes, or big houses with security gates? Discuss the idea that police in your area should work with the local community to tackle issues concerning the locality, rather than have to carry out central government policies and meet state targets.

Social crime control

The policy recommendations of left realism include more police accountability to local communities, meaning that communities should influence policing which would be fairer than centrally directed policy. Government policy should aim to reduce inequalities in society, as it is perceived inequality that causes crime. There should be a return to the informal community control structures that prevent crime, as the offenders are known in their communities, with the involvement of families, schools and social services. Here, the neighbourhood watch approach is favoured as it involves local people looking out for each other and regular meetings with local police community officers. Although left realists disagree with right realist theory, to gain public acceptance for their ideas they supported the policies of CCTV and environment maintenance that provided reassurance to communities.

Assessment activity 31.2

BTEC

As a junior reporter on your local newspaper, the Editor has asked you to explain to local residents why there has been a big increase in the number of CCTV cameras, and to report on recent council work to refurbish and repair a run-down estate in the area. In order to explain the changes over the last 30 years, you have agreed that the reasons for changes in crime control policies must be included to stimulate discussion of how budget cuts could be achieved.

1 **Write an article for the newspaper where you describe how criminologists explain why people commit crime.** P3

2 **Add a section to your article in which you explain how the different theories have had an effect on how crime is dealt with in your local area.** P4 M2

3 **Add another section in which you explain how the more recent criminological theory of realism has replaced positivism, and the challenges this presents.** D2

Grading tips

For M2 you need to explain how changes in crime control policies have been influenced by the theoretical explanation of crime described in this section.

For D2 you need to explain to the Editor why the dominant scientific theory of positivism was challenged and displaced by realist criminology, and then indicate the impact on crime control policy.

31.3 The role of the media in our perception of crime

In this section:    

Key terms

Interactionism – an explanation of crime which asks who has the power to define a person as deviant, and what impact does that have on the person.

Labelling – this approach by Becker also looked at the influence of power in putting the label of deviant on a person or group.

Deviance – this term refers to activities or behaviours that are not crimes but offend people – in the recent past anti-social behaviour has become the modern term for it.

Power – power operates when an individual does something they would otherwise choose not to do, consciously or unconsciously.

Moral panic – a theoretical explanation by Cohen to explain the influence of the media in constructing crime.

Fear of crime – the argument that people can live in fear of crime, even if crime figures indicate that there is no problem in their area.

Media campaigns – these are stories begun by the media about actions by a group that are portrayed as deviant or anti-social, and the media dramatise events to provoke action against the target group.

Risk from crime – the argument that some people can be at great risk from crime but do not accept that this is the case.

The terms **interactionism** and **labelling** changed the focus from the causes of crime to question the process of criminalisation. The ideas of the most famous exponent of labelling, the American sociologist Howard Becker, were set out in a paper called 'Outsider: studies in the sociology of deviance' in 1963. Concise summaries of this work can be found by typing his name into an Internet search engine. This is an important work, as he argues that society creates outsiders by labelling certain groups. The label is always negative, and his research revealed that it has two effects: first to make the group different to the norm and then to make members live up to the label. A key factor here is that Becker departed from positivism's search for causes of crime, and instead argued that '**deviance** is not a quality of the act a person commits – but rather a consequence of the application by others of rules and sanctions to the offender'. A deviant therefore is one to whom that label has been successfully applied and this involves another interesting concept for scientists: **power**.

Activity 31.3A

In groups, research how the media get information about the nature and extent of crimes committed in your area. Reading local newspapers and collecting information about reports from magistrates' courts would be useful starting points. Share your findings with the class and move on to investigate the types of crime that are common in the area. If it is possible to visit a magistrates' court, you will see a key role for the police in presenting evidence against the accused person. Remember: the police will only bring a case if there has been an offence committed that breaks a law – so the legal definition of crime applies here.

Presentation of findings by poster or report will demonstrate your understanding of the relationship between the police and the media.

Activity 31.3B

How do you find out about crime?

In a class discussion, collect evidence of how the group has accessed crime stories or programmes in the last week. Research television schedules for the week to see how many programmes are focused on crime and what types of programme are involved – factual, documentary, fictional detective series, police series, forensic science programmes. What is the coverage of crime stories in the local newspaper(s)? Does crime feature strongly in the paper(s)? Do you enjoy reading about crime or viewing crime programmes? Do adults in your family watch crime programmes or discuss articles about crime?

Activity 31.3C

In groups, research tabloid newspapers to see if you can find emotive headlines and capture these terms on posters. They will be useful when you look at the arguments of Stan Cohen about moral panic. For starters have you ever seen 'mindless thugs ...' 'vandals destroy ...'?

Once you have collected the evidence, can you see any trends as to the nature of the people labelled by the media? Aspects to consider here could include social class, powerful groups, young working class people, immigrants, travellers.

In this section you will be looking at how the media portray crime, and from the activities you will now be aware that much of the information on the type and nature of crime collected by the media comes from police sources. The police have the power to apply their definition of crime to us, and make arrests and give cautions where they have evidence to support their actions.

All media survive by making profits, so when you look at the media you will tend to see more reporting of negative events than positive events. For example, a story about young people running wild in an inner city area and causing disturbances is much more likely to feature than young people raising money for charities.

The more horrendous the actions described by the media, the more interest is aroused in their audience – and of course follow-up articles on the original story will sell more copies. The portrayal on television of policing in a range of serials such as *The Bill* and *The Sweeney* attracts big audiences, and new technology as shown in programmes like *Cracker* can be fascinating for people, as these programmes give an insight into the work of parts of the criminal justice system. Reality programmes such as *Crimewatch* not only fascinate audiences, but can also provide useful leads for the police and help to catch criminals featured in their case studies. There are also programmes that portray life in prison; for example, the comedy *Porridge* provides some realistic aspects of prison regimes that show prison officer power in the actions of characters. If you look at television peak-time viewing schedules you will see a wide range of programmes such as *CSI*, *NCIS*, *Unsolved Mysteries* and *Crimes That Shook Britain*, which give an indication of public interest in crime as entertainment. Fictional detective programmes such as *Poirot*, *Miss Marple* and *Sherlock Holmes* attract audiences who are fascinated by the painstaking care needed to collect evidence and lead to the arrest of the criminal(s). There are numerous reality programmes about motorway policing – viewing car chases involving stolen vehicles with police commentary on the actions of the criminals provides entertainment for viewers. Recently, there has been the development of reality programmes looking at the work of 'scene of crime' teams and these benefit from accurate research by the programme makers.

These sources of entertainment show that the public are fascinated by crime, and when you look at television schedules coverage of crime features strongly. The question now is, whilst we are fascinated by crime, do you think this constant exposure to crime may also influence our **fear of crime**? For example, the programme *Crimes That Shook Britain* is specifically designed to cover sensational and horrific crimes and as such may well cause viewers to fear crime.

You have seen how the media gain information about the nature and extent of crime, and investigated how you find out about crime and also about crime as a source of entertainment. Two related theoretical approaches, interactionism and labelling, have alerted you to arguments about the criminalisation of individuals and groups. Now we move on to a famous social science investigation by Stan Cohen and his theory of **moral panic**.

Moral panic theory

Cohen in 1973 (see reference, page 448) investigated reported disturbances by mods and rockers in Clacton on Easter Monday 1964. These were groups of young people who went around together, with differences in their 'uniforms' and bikes. The mods rode scooters whilst the rockers rode motorbikes and there was rivalry between the two groups.

Newspaper headlines after the event described a 'Day of Terror', with Clacton being invaded by a mob hell-bent on destruction and two easily recognisable rival gangs deliberately causing the troubles. Cohen argued that the media's over-reaction to the events forced the police to intervene more strongly in subsequent disturbances, and later that year headlines a bit like the fictional ones (page 439) showed how the media were amplifying their reporting.

THE DAILY CLARION

WILD YOUTHS CAUSE TROUBLE IN SEASIDE TOWN

60 ARRESTED DURING FIGHTS

[illegible]

Violence

[illegible]

[illegible]

[illegible]

Anger

[illegible]

Youths fighting in the market square

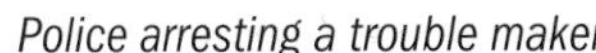

Police arresting a trouble maker

In his research, Cohen gathered qualitative evidence from interviews with people in the affected towns, and also interviewed a sample of mods and rockers who had been charged by the police. His findings presented a different picture to that portrayed by the media – there was no evidence of structured gangs and rivalries were regional, whilst owners of scooters and motorbikes were a minority of the young people.

Police records showed that typical offences were not assault or malicious damage, but were threatening behaviour; this is quantitative evidence. From this evidence, Cohen set out his theory of a deviancy amplification spiral, where the problem escalates until new legislation is put in place to control it.

In the Cohen argument the first stage is to label an act as deviant.

1 An act or group is identified as causing problems – now this stage requires power. If you tell your local newspaper that someone you know is annoying you and is therefore deviant, it is extremely unlikely that they will publish your complaint.
2 The labelled act must be seen as newsworthy by the media and something which could cause widespread concern – your localised complaint would be unlikely to be of interest across the UK so would not be taken up by the media.
3 Through the choice of language and headings in the press the deviant act is given prominence and headlines to attract readers.
4 The reporting of the deviance is so alarmist that the police change priorities to enable them to allocate more officers to control this deviance. This leads to more arrests and the next stage.
5 The law-and-order campaign by the media is now in full swing. More reports of problems and more arrests get the attention of politicians – media demands for crackdowns and harsh legislation continue.
6 The final stage is the introduction of harsh new laws to control the initial deviant act, now classed as a crime.

This explanation shows that media over-reporting and sensationalism force the police to intervene, which leads to more arrests and confrontation. This is again over-reported, and eventually members of parliament are pressurised by their constituents to pass a law against the behaviour of the young people. Note that people in the areas affected by the mod and rocker disturbances did not accept the media's overblown descriptions.

Media campaigns

You have looked at the way in which mods and rockers were labelled as a problem by the media; the process of reporting involved a **media campaign**. Here, the media identify a 'problem' for the general public, label the group (e.g. mods/rockers, skinheads, punks) and simplify the causes, e.g. a decline in moral standards, lack of discipline, etc. Then, the use of emotive language, e.g. in the Clacton extract the use of the phrase 'a new battlefield', imply a war is happening, and the media demand action to stop the problem. Eventually, the government responds to this public demand, fuelled by the media, and harsher legislation results. Can you remember seeing headlines about a government crackdown on problems?

The benefits for the media from running such campaigns include:

- an increase in sales or viewing figures
- the end result of new legislation to control the 'problem' identified by the media is reported as justifying their actions; they then move on to the next campaign.

Fear of crime

One effect of media campaigns and moral panics is to make the public fearful of crime. The groups identified by the media as causing the problems are usually young working class people or travellers/immigrants who can be portrayed by the media as different to the law-abiding majority. These groups do not usually have power in society.

In addition to moral panic reporting, there is increased reporting in the media of domestic violence, and recently of knife crime, where random attacks by strangers on innocent people can cause fear of crime.

In the 1990s, two types of research were combined to test the relationship between the fear of crime and the **risk from crime**. Hough and Mayhew used qualitative methods to collect answers from a random sample of 10,905 people to the question:

'How safe do you feel walking alone in this area after dark?'

The researchers used quantitative research to establish the actual frequency of street crimes against the individuals in their survey, and arranged the data for men and women in three age groups: 16–30, 31–60 and 61+.

From this research, evidence was collected about how people in these groups felt about their risk from street crime and also about their actual experience of street crime. The findings showed that in all age groups females felt more at risk than males, with the over 61 group being the most fearful for both sexes. However, the actual risk from street crime was much greater for the 16–30 group for both males and females. For males, 1% of the 16–30 group felt very unsafe but had a 7% risk of being a victim; in contrast, 7% of the 61+ group felt very unsafe but their risk was 0.6%. For females, 16% of the 16–30 group felt very unsafe, but their risk from crime was 2.8%, whilst 37% of the 61+ group felt very unsafe but their risk was 1.2%.

This evidence is important as it indicates that there is a big difference between the perception of crime and the risk from it. The combination of both qualitative and quantitative research has provided an informative result.

Activity 31.3D

From the information you have read, do you feel that the media have contributed to this situation? From reading about moral panics and media campaigns, do the media report crime or actually construct new crimes and new groups to fear?

In groups, think about the Hough and Mayhew research question. Record what you understood by the question 'How safe do you feel ...' and discuss. This activity could give you insights into the challenges in using qualitative research.

Assessment activity 31.3

You are on a visit to your local newspaper and have an opportunity to speak to the crime reporter, to collect evidence for a school newspaper article on fear of crime.

1 Find out how the reporter gets information about crime and which types of crime readers find the most interesting. Describe how his/her reports influence the way readers view criminal behaviour. P5
2 Carry out research and surveys to gain information about local fear of crime covering different ages/ethnic groups. Explain how this different from the actual risk of crime. P6

3 Research factors that influence fear of crime. M3
4 Research a recent moral panic and use the argument of Cohen to present your evidence. D3

Grading tips

For M3 it is relevant to look at how the public gain knowledge about crime in society and the popularity of crime in the media, e.g. it may be helpful to refer to television schedules.

Cohen argues that new crimes are created by media reporting methods and media campaigns; for D3 research a recent media campaign to collect evidence for or against this argument.

PLTS

Creative thinker and Reflective learner

By completing the assessment activity you will demonstrate your creative thinking and reflective learning skills.

31.4 Relating the aims of state penal policy to crime theories

In this section:   

If you look at any society in the modern world you will see that crime exists to some extent, and all societies over time have set out to control crime by using a penal policy on offenders.

History of the UK penal system

Earlier in this unit, you read about harsh sentences that were meted out at the whim of local landowners or magistrates. In the eighteenth century, prisons in the UK focused on two key aims: custody and coercion. In this approach, accused people were detained in prison before trial and then coercion made sure that they paid any fines before release. This approach rested on voluntarism.

Activity 31.4A

Research the meanings of **deterrence**, **reform**, rehabilitation, **punishment** and **incarceration**. Then in groups discuss what you think penal policy should be.

Capture your findings on posters and share with the class.

Did you know?

Only the state can deprive an individual of their liberty by locking them away in prison. Therefore, there is an argument that the state is responsible for the safe custody of all prisoners and that responsibility should not be delegated. This may indicate that a primary aim of any penal policy must be safe custody.

In the eighteenth century, there was no separation of men and women in prisons. The conditions were dreadful, with a lack of light, sanitation and washing facilities, and prisons could be run for profit. These were places where diseases such as typhus were rife, the gaolers could charge inmates for bedding, food and alcohol, and the state had little input.

The famous reformer John Howard (1726–1790) campaigned against these dreadful conditions, and in 1779 the Penitentiary Act brought central government into the operation of prisons for the first time. The word 'penitentiary' has significant meaning as it involved prisoners having to do hard labour whilst serving their sentence – here, punishment is involved.

Full state centralisation and nationalisation of prisons came about with the passing of the Prison Act in 1877. The aims were to introduce a standard regime across the country based upon the principle of deterrence by punishment rigidly and efficiently applied. Under this penal policy, hard unproductive work was used to enforce obedience and to discipline prisoners rather than to teach prisoners any useful skills. Two of the mechanical devices used were the treadmill and the crank, with the latter used for prisoners sentenced to hard labour. In order to cater for different prisoner strengths, the crank had a system that could regulate the pressure needed to turn the device, and this could be used to increase or decrease the effort required.

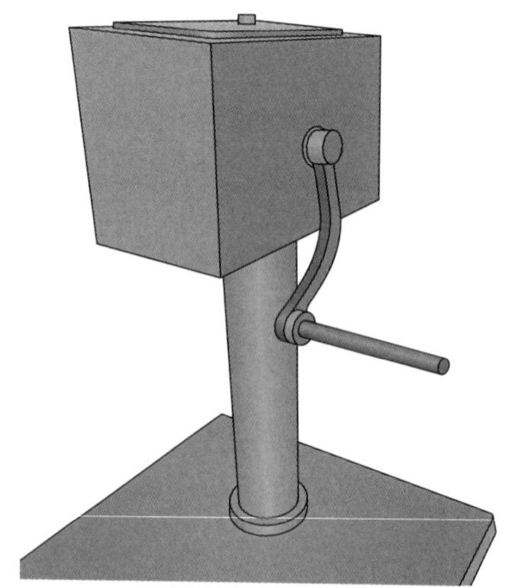

A crank.

Influence of the classicist crime theory

In the early nineteenth century, the classicist theory of criminology was dominant. Its key argument was that individuals had the intelligence and free will to make rational decisions on whether to commit crime or not: a voluntaristic approach. It also argued that in order to make individuals decide not to commit crime the punishment should be prompt and minimal, and should apply equally to all individuals in society. The punishment should be only sufficient to deter the offender – deterrence.

Activity 31.4B

In groups, discuss the classicist approach and capture on posters or in presentations its strengths and weaknesses.

Do you feel that everyone in society can make rational judgements? Do you feel that age or mental capacity might be a factor? Can you think of any group(s) in society who may not support the classicist policies? Prior to this approach, generally only working class and poor people were targeted by the criminal justice system.

The policy of minimal sanctions to deter offenders came under threat during the garrotting panic of 1862, and the Recorder of Birmingham, Matthew Davenport Hill, strongly argued against the presumption of innocence – for him, hanging the offenders on the basis of suspicion would solve the problem. In 1861 a Member of Parliament, Mr Pilkington, had been the victim of a garrotting attack in Pall Mall, London. This led to the Garotters' Act of 1863, which brought back flogging, and was followed by legislation to introduce a minimum of 5 years of penal servitude for a second offence.

See the 'Did you know?' feature on page 431 for more on the garrotting panic of 1862.

In the mid-nineteenth century, therefore, a key part of the classicist policy was abandoned and the idea of minimal sentencing to deter offenders was discredited.

Did you know?

In the upper classes, who dominated Parliament and offices of state, the tradition of public school education involved flogging and bullying of pupils, whilst in the army and navy floggings were a standard form of punishment. It is recorded that the floggings that were carried out on garrotters were not extended to equally violent crimes carried out by members of the upper classes. So the classicist ideal of the law applying equally to all in society was not being practised.

The positivist approach

The evidence as to whether policies are working invariably comes from crime statistics. The harsh punishments of the second half of the nineteenth century did not, in fact, result in crime rates falling. So, the opportunity to advance a new approach to the explanation of crime and methods of crime control was taken by the positivist theorists.

They argued that scientific investigation of the causes of crime would enable policies to be introduced to cure the criminals, and therefore society would be a better, safer place for all. You have read in section 31.2 about different arguments, based on biology, family and culture, as to the causes of crime, and that if these factors could be controlled then there would be no crime. These ideas had considerable appeal, particularly because they appeared to be based on hard scientific evidence, and there was similarity to the arguments that were so successful in the fields of medicine, chemistry and physics, where discoveries had greatly transformed society.

Activity 31.4C

Individually collect evidence from watching television adverts or from adverts in magazines or newspapers to see if you can find 'scientific' claims for products and/or the use of white-coated 'scientists' telling us why their product(s) are superior. Preparation of posters and/or presentations to the class could lead to discussion of the extent of such advertising, and also whether companies and advertising agencies would set out to use 'science' if it did not work.

Rehabilitation

In 1895, the Gladstone Report commissioned by the government stated that deterrence was not working and that penal policy should include reform measures, with rehabilitation given priority. This report was strongly influenced by the positivist approach that was gaining dominance in criminology at the time, showing that theory directly influenced penal policy.

So, at the beginning of the twentieth century the argument was that we are all basically good people, and if an individual commits a crime then there must be some factor that forced them to do so. Following on from that argument, if we have no freedom of choice in our actions then punishment is not appropriate. This means that the policy of deterrence could not work as it does not remove the cause of criminal behaviour. Instead, the offenders need expert help to analyse their problems and devise treatment to cure their behaviour – and as we are all different as individuals then penal policy must recognise that in giving treatment designed for the offender. This policy is called rehabilitation because it aims to cure the causes of crime and rehabilitate offenders into society as law-abiding citizens.

In Britain after the Second World War the introduction of the Welfare State meant that a wide range of professionals and experts became state employees. This created resources that could be applied to the removal of the causes of crime. So, you had a mixture of penal policy which demanded that offenders be given individual treatment and support to bring them back into society, and an expanding welfare system employing professionals and experts who could undertake these treatments.

For two decades, the penal system ensured that imprisonment combined containment with rehabilitation and this policy had an effect on sentencing policy. If each offender requires an individual treatment programme to cure them, then magistrates and judges cannot give fixed prison sentences.

Activity 31.4D

In the rehabilitation argument, each offender will come from different circumstances that forced him or her to commit the crime for which they were imprisoned. So each individual must be assessed by experts to design a programme to cure them.

Have a class discussion on the strengths of this approach – in your class do you see yourselves as a group of individuals or are you all the same? Do you feel that it is a good thing to try to cure criminal behaviour?

Should offenders be helped to become law-abiding citizens or should we just lock them up and throw away the key?

In any area of criminology there are always arguments for and against a policy – so now think about possible problems. (Trigger ideas: are all experts equally effective? How would you feel if a judge told you that your sentence could be 3 months or 5 years, and depended on the judgement of an expert?)

Incapacitation

By the mid-1970s there were challenges to the positivist policy of rehabilitation, with critics arguing that the high costs of rehabilitation were not preventing reoffending on release. Right realist arguments that the criminal justice system should protect victims and not support criminals gained public acceptance – 'taking crime seriously'. Now the argument was that the majority of crimes were committed by a small hardcore of wicked individuals who freely chose careers in crime for profit, so if this group were imprisoned then they could not commit crime. This claim was reported as 'Prison Works'. If this hard core of criminals are kept in prison for longer sentences then they cannot commit crime during this time. This approach is known as incapacitation because prisoners cannot commit further crimes, and the longer the sentences the more effective the incapacitation. However, this approach was challenged as crime rates did not fall and there was pressure on prisons due to overcrowding, so alternatives were investigated.

Restorative justice

Twenty-first century penal policy has aimed to make greater use of alternative sanctions to imprisonment such as community service orders or probation, because there is evidence that prisoners can learn new criminal techniques whilst in prison and on release commit more serious crimes. An important new initiative involves the use of panels where the offender and victim discuss the crime committed and work together to change the offending behaviour. This approach is called restorative justice and has had some success to date.

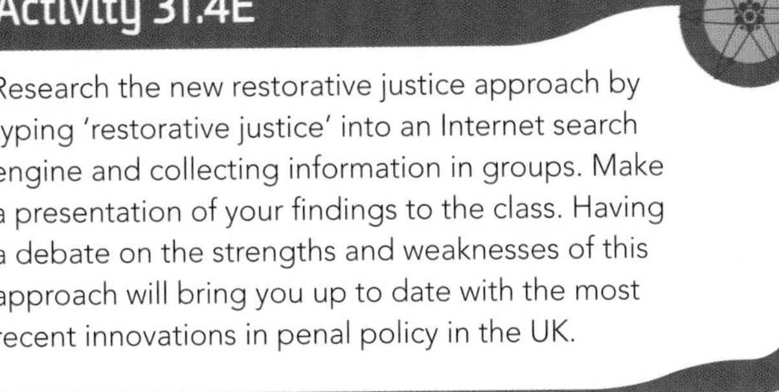

Activity 31.4E

Research the new restorative justice approach by typing 'restorative justice' into an Internet search engine and collecting information in groups. Make a presentation of your findings to the class. Having a debate on the strengths and weaknesses of this approach will bring you up to date with the most recent innovations in penal policy in the UK.

Did you know?

The prison system in the UK has different categories of prison and levels of security to which prisoners are allocated. Details of the regime in England and Wales and also in Scotland and Northern Ireland for different categories of prisoner can be found on the Internet.

For the category A prisons in England and Wales, the security level is the highest possible; for example, HMP Frankland has high-security perimeter fencing and anti-helicopter netting across the prison estate to prevent escapes.

Assessment activity 31.4

You are visiting a local prison to make a presentation to the prison staff.

1 Outline the different aims of penal policy over the last century. P7
2 To show the prison staff your understanding of criminology, explain how penal policy has been influenced by crime theories. M4
3 You find that prison staff are interested in the idea that crime theory can influence penal policy, so provide an analysis of how positivism and realism influenced penal policy in the twentieth century. D4

Grading tips

For M4 you need to explain how different views on the causes of crime have influenced decisions on what penal approach is used. Here, voluntarism and determinism need to be taken into consideration.

The strength of the positivist theory in the early twentieth century directly influenced policy. To help you gain D4 indicate why this approach was discredited and replaced by realist policy.

PLTS

Effective participator

In this assessment you will have the opportunity to demonstrate your effective participation skills.

WorkSpace

Lillian Smith

Trainer for prisoners in a Category C training prison

I am responsible for providing work experience and training for prisoners. In my area, prisoners learn basic skills in reading, writing, maths and art. I must comply with all the security regulations within the prison and am employed by an FE College.

Training in prisons is provided by FE Colleges under contract to the Justice Ministry and my college looks after a range of prisons from high security Category A prisons to open prisons at Category D. My colleagues all work for the college and teach in different subject areas. We are helped by prison staff who escort prisoners to and from the education department and carry out security searches before they come into class and at the end of the class.

The education classes are highly structured. Prisoners are collected from the residence wings and escorted to our separate education block and signed in and out on a daily register. This is a crucial part of the prison regime.

The satisfaction of teaching prisoners is to see them produce good work. Learning new skills can help them when they are released from prison not only to get work but also to fit back into society. It is great to see a prisoner develop reading and writing skills as this gives them more independence and confidence.

Work experience and learning new skills are crucial for prisoners, as being in prison clearly carries a negative image making it difficult to get work on release. If the prisoner can show possible employers that they have voluntarily studied courses in prison it shows good motivation and self-improvement. For many prisoners our courses have given them new horizons and enabled them to carry out day-to-day activities which we take for granted. They can write letters to their families and read letters sent to them – previously they had to rely on other prisoners or prison staff to write/read for them.

Think about it!

1 Do you think prisoners are allowed access to the Internet? If not, why not?

2 How do the media portray prisons? Is this an accurate image?

Just checking

1 What are two competing definitions of crime?
2 Identify the differences between the two official measures of crime: Home Office official statistics and the British Crime Survey.
3 Name two important theories of crime and explain how these have influenced crime control policies.
4 Give an example of the influence of the media on our understanding of crime.
5 Why might those most at risk from crime fear it least, while those who are most fearful are in fact at least risk?
6 Give an example of how criminological theory has influenced penal policy in the last 100 years.

Assignment tips

- You need to read the information provided carefully and ensure that you look at competing arguments.
- In order to demonstrate understanding of crime measurement you need to consider the differences between recording of crime by the police and self-reporting of crime.
- When you are looking at different theories of crime it is useful to remember that any theory sets out to explain crime at a particular time and place. Moreover, for a theory to be successful it must attempt to discredit competing arguments.
- The influence of major crime theories on policy is one reason why there is such competition between different explanations of crime.
- You will read evidence of the public being fascinated by crime and see the different ways in which we find out about crime. This will give you evidence to describe how the media influence our understanding of crime.
- The types of research used to show differences between fear of crime and risk from it will help your understanding of this aspect of crime, but also tie into other units you will meet in this qualification.
- You have read about debates on what is crime and arguments about the causes and control of crime; in regard to penal policy you will see how the aims have changed over time and are still changing.
- For Merit and Distinction grades it is important to be able to use evidence to support your arguments, and also to recognise that there are two sides to every aspect of criminology.
- The ability to link arguments to demonstrate understanding will gain credit. You will see in the unit description of the use of different types of research; the ability to use these effectively will contribute to higher grades.
- If you are able to explain the strengths and weaknesses of quantitative and qualitative research and evidence, this would contribute towards Distinction grades.

You may find the following references/websites useful as you work through this unit.

For information on...	Read/Visit...
definitions of crime	Tappan, P. W. (1947) Who is the criminal? *American Sociological Review* **12**, 100.
factors influencing crime statistics	Maguire, M. (1997) Crime statistics, patterns and trends; changing perceptions and their implications. In: Maguire, M., Morgan, R. and Reiner, R. (eds) *The Oxford Handbook of Criminology* (2nd edn). OUP, Oxford.
unreported crime	Mirrlees-Black, C., Budd, T., Partridge, S. and Mayhew, P. (1998) *The 1998 British Crime Survey*. Home Office, London.
the relationship between physical appearance and criminality	Bull, R. and Green, J. (1980) The relationship between physical appearance and criminality. *Medicine, Science and Law* **20**, 79.
genetic factors in the cause of crime	Mednick, S., Moffitt, T. and Stark, S. (eds) (1987) *The Causes of Crime: New Biological Approaches*. Cambridge University Press, New York.
criminal careers in families	Farringdon, D. (1994) Human development and criminal careers. In: Maguire, M., Morgan, R. and Reiner, R. (eds) *The Oxford Handbook of Criminology*. OUP, Oxford.
moral panics	Cohen, S. (1973) *Folk Devils and Moral Panics*. Paladin, St Albans.
crime, general	Home Office website
research, development and statistics	Research, Development and Statistics – Home Office
legislation	Ministry of Justice
penal policy	HM Prison Service

Credit value: 10

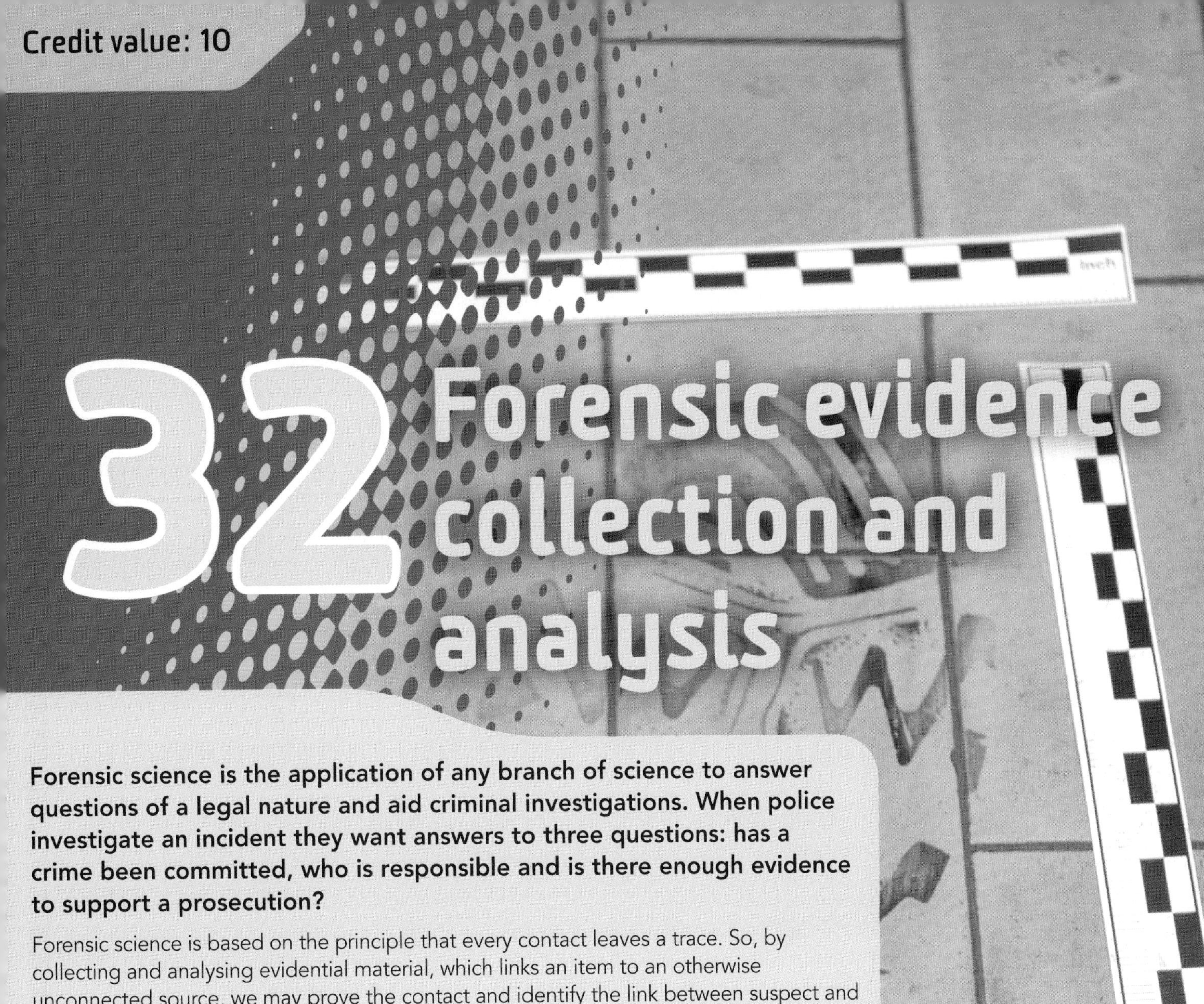

32 Forensic evidence collection and analysis

Forensic science is the application of any branch of science to answer questions of a legal nature and aid criminal investigations. When police investigate an incident they want answers to three questions: has a crime been committed, who is responsible and is there enough evidence to support a prosecution?

Forensic science is based on the principle that every contact leaves a trace. So, by collecting and analysing evidential material, which links an item to an otherwise unconnected source, we may prove the contact and identify the link between suspect and crime scene, and suspect and victim.

Scene of crime officers process the crime scene and recover the evidence, so you will learn how to examine a simulated crime scene, and search for and collect different types of forensic evidence. Forensic scientists then apply scientific techniques to analyse biological, physical and chemical types of evidence, so you will learn how to do this in the laboratory. Both of these experts provide the results and interpretation of their examination in a report for police investigators and the courts, and they may give verbal evidence in court as an expert witness. You will also learn how to document, interpret and present your forensic evidence.

Learning outcomes

After completing this unit you should:

1 be able to gather forensic evidence from a simulated crime scene using appropriate methods
2 be able to use chemical techniques to analyse evidence from a simulated crime scene
3 be able to use physical techniques to analyse evidence from a simulated crime scene
4 be able to use biological techniques to analyse evidence from a simulated crime scene
5 be able to report the analysis of evidence from a simulated crime scene.

Assessment and grading criteria

This table shows you what you must do in order to achieve a **pass, merit** or **distinction** grade, and where you can find activities in this book to help you.

To achieve a **pass** grade the evidence must show that you are able to:	To achieve a **merit** grade the evidence must show that, in addition to the pass criteria, you are able to:	To achieve a **distinction** grade the evidence must show that, in addition to the pass and merit criteria, you are able to:
P1 carry out a forensic examination of a simulated crime scene, using appropriate methods to gather biological, physical and chemical evidence **See Assessment activity 32.1**	**M1** describe the procedures used to gather evidence from a simulated crime scene **See Assessment activity 32.1**	**D1** justify the procedures used to gather evidence from a simulated crime scene **See Assessment activity 32.1**
P2 outline the main techniques used to analyse chemical evidence **See Assessment activity 32.2**	**M2** explain the main techniques used to analyse chemical evidence **See Assessment activity 32.2**	**D2** evaluate the techniques used to analyse chemical evidence **See Assessment activity 32.2**
P3 carry out practical work used to analyse chemical evidence gathered from a simulated crime scene **See Assessment activity 32.2**	**M3** present valid conclusions drawn from the analysis of chemical evidence gathered from a simulated crime scene **See Assessment activity 32.2**	**D3** justify the choice of techniques used to analyse the chemical evidence gathered **See Assessment activity 32.2**
P4 outline the main techniques to analyse physical evidence **See Assessment activity 32.3**	**M4** explain the main techniques to analyse physical evidence **See Assessment activity 32.3**	**D4** evaluate the techniques used to analyse physical evidence **See Assessment activity 32.3**
P5 carry out practical work used to analyse physical evidence gathered from a simulated crime scene **See Assessment activity 32.3**	**M5** present valid conclusions drawn from the analysis of physical evidence gathered from a simulated crime scene **See Assessment activity 32.3**	**D5** justify the choice of techniques used to analyse the physical evidence gathered **See Assessment activity 32.3**
P6 outline the main techniques used to analyse biological evidence **See Assessment activity 32.4**	**M6** explain the main techniques used to analyse biological evidence **See Assessment activity 32.4**	**D6** evaluate the techniques used to analyse biological evidence **See Assessment activity 32.4**
P7 carry out practical work to analyse biological evidence gathered from a simulated crime scene **See Assessment activity 32.4**	**M7** present valid conclusions drawn from the analysis of biological evidence gathered from a simulated crime scene **See Assessment activity 32.4**	**D7** justify the choice of techniques used to analyse the biological evidence gathered **See Assessment activity 32.4**
P8 report on a chemical, physical and biological forensic examination **See Assessment activity 32.5**	**M8** justify the conclusions drawn in the report **See Assessment activity 32.5**	**D8** evaluate their findings, including aspects of probability **See Assessment activity 32.5**

How you will be assessed

- You will be asked to process a simulated crime scene where you will have to demonstrate the correct examination stages, and record, collect and package the forensic evidence. You will need to describe and justify your examination technique in a written report.
- You will then be asked to scientifically analyse the biological, chemical and physical evidence, and draw conclusions from your findings.

Marie, 18 years old

I am studying for a BTEC National Diploma in Forensic Science. The unit I've most enjoyed on the course is the forensic collection and analysis unit. This has given me firsthand experience of how SOCOs (Scene of Crime Officers) search a crime scene and Forensic Scientists analyse the evidence. I have learnt how to approach and analyse a crime scene during the theory lessons in the laboratory, and in the practicals I have been given crime scenarios so I can practise what I have learnt.

For the forensic assignment, I was given a scenario telling me that an unknown male was found dead in a flat, and I had to work out whether the victim had committed suicide or had been murdered. I had to assess the crime scene and draw a plan to scale, then photograph the scene from different angles (and this was before I could even enter the scene!). After that I had to take photographs of the evidence and then collect it. It was very interesting and I was given personal protective equipment worn by real SOCOs that protected me completely and stopped me from contaminating the crime scene. I then had to analyse the evidence as a forensic scientist, from blood samples using blood typing tests, to using chromatography for ink samples.

The final part of the assignment was keeping records throughout my analysis and writing up my results in the form of an expert witness statement. I then had to draw conclusions from my evidence and decide whether the results proved suicide or murder.

Catalyst

Who contributes to a forensic investigation?

A forensic investigation involves a number of different professionals who each play a unique role and contribute a different area of expertise to the case. There is a large difference between the job of the scene of crime officer (SOCO) and the job of the forensic scientist.

Forensic scientists specialise in a specific area of science; for example, DNA analysis or forensic toxicology. Any type of science can be used in a criminal investigation, from simple blood analysis (forensic serology) to more unusual pollen or soil analysis (forensic botany).

What is the difference between the roles of the SOCO and forensic scientist?

List the types of job a SOCO and forensic scientist would carry out during a forensic investigation. What other areas of science can be applied to a forensic investigation?

32.1 Gathering evidence from a crime scene

In this section:   

Key terms

Common approach path (CAP) – a pathway from the police cordon to the focal point of the crime scene. This is used by all authorised personnel to access the site of the crime and minimises contact with the scene and evidence.

Contamination – unwanted transfer of material from a source to a piece of physical evidence.

Cross-contamination – unwanted transfer of material between two or more sources of physical evidence.

Gel lifter – thick, flexible, low-adhesive gelatine layer designed to lift fingerprints, footprints, dust marks and other trace evidence from most surfaces.

Chain of continuity – complete documentation that accounts for the progress of an item of evidence throughout the investigation, from the crime scene to the courts.

Latent – present but not obvious.

Every criminal investigation starts at the crime scene, where the Scene of Crime Officer (SOCO) must search for and recover the relevant evidence. There are a number of basic stages of a crime scene investigation that the SOCO must complete in order to collect the evidence correctly.

Crime scene initial activities

First attending officer

The first member of personnel to arrive at the scene is called the first attending officer (FAO); this is usually a police officer. Their first task is to assess the scene and establish whether assistance is necessary. The FAO will also search for and arrest any suspects if they are still present at the crime scene, ascertain if there are any injuries or witnesses, and may also help to preserve any potential forensic evidence from weathering, e.g. protect from rain.

Restriction of access

When a SOCO first arrives at the crime scene, there are a number of security and safety issues that need to be dealt with before the scene and evidence can be processed. Anybody who enters the crime scene could potentially destroy or contaminate evidence so access must be controlled. Only authorised personnel should be allowed entrance to the crime scene via a suitable **common approach path**. A log must be kept recording who has had access to the scene, their role and their time of entry and exit. To preserve the crime scene and protect the evidence the SOCO must identify the boundaries of the crime scene and then isolate the scene. This can be done using barrier tape, vehicles and guards to create a police cordon enclosing the crime scene. Personnel allowed to enter the crime scene and individuals who are excluded are listed in the table below.

Authorised access	Unauthorised access
SOCOs & crime scene manager	general public
forensic scientists	media
police officers & investigators	family of victim
paramedic staff	witnesses
fire brigade staff	non-essential personnel, e.g. police, paramedics and firefighters

Did you know?

SOCOs are employed by the 43 police forces in the UK. They also work for regional police forces, the British Transport Police, the Ministry of Defence Police, and forensic science agencies such as the Forensic Science Service. SOCOs normally work for about 40 hours a week in shift patterns and may be called out to a crime scene at any time of the day or night or on the weekends. SOCOs are usually based at police stations.

Prevention of contamination

Contamination is the unwanted transfer of material, which must be avoided at all costs at the crime scene. Individuals can contaminate the scene and evidence at any stage of an investigation, but unwanted transfer of material can also occur between two or more sources of evidence (**cross-contamination**). The table below shows a number of prevention measures that can be used to reduce the potential for contamination. Contaminated items of evidence cannot be used in a court trial and they may also bring into question the integrity of the other items of evidence in a criminal investigation.

How can evidence be contaminated?	How can contamination be prevented?
SOCO leaving fingerprints, footprints, hairs, fibres or DNA at the scene	wearing full personal protective equipment: SOCO suit, gloves, mask, overshoes. No food, drink or personal items taken onto the crime scene
unsealed packaging	choosing the correct type of packaging and sealing carefully
placing two items of evidence into one evidence bag	packing items of evidence into separate bags
using old, dirty equipment at a number of different crime scenes	using new, sterile, disposable equipment at every scene, e.g. tweezers, swabs and scissors

Use of personal protective equipment

To protect the scene from contamination by the SOCO, full personal protective equipment (PPE) is put on before entry to the crime scene. This disposable protective clothing and equipment worn by SOCOs also reduces the risk of injury and protects the SOCO from harm at the crime scene.

Activity 32.1A

Find an image of a SOCO, label the clothing and identify why it must be worn.

Health and safety

Once the scene is secure, the SOCO determines whether it is safe to enter the crime scene by assessing the hazards and controlling the risks. The most important thing is health and safety. The risks and hazards will be different for different types of crime scene (see table).

Activity 32.1B

You are a SOCO and the police have requested you attend the crime scene of a violent sexual assault. Make a list of the potential hazards that may be present at the scene and describe the control measures you will put in place to reduce the risks.

Crime scene	Risks and hazards	Control measures
murder	potential infectious materials, e.g. blood or body fluids contaminated with HIV or hepatitis; sharp or dangerous weapons, e.g. knife, gun	vaccinations, PPE, careful handling of sharps (needles, knives etc.) and guns, packaging in hard container, correct waste disposal and disinfection techniques
burglary	broken glass, trip hazards, e.g. overturned furniture, suspect still present at scene	lift heavy furniture correctly to avoid injury, carefully handle sharps and package in hard container, police check scene
arson	electrical hazards, toxic chemicals and fumes, structural damage to buildings	all electrical supplies switched off at mains, eye protection and face masks, structural engineer to check stability of building

Health and safety regulations

Work at the crime scene is regulated by a number of pieces of legislation that serve to protect employees and other individuals. SOCOs and their employers must follow certain requirements to ensure everybody is kept safe and healthy.

Unit 2: See page 50 for further information on the key health and safety government act that applies to crime scene investigation: Health and Safety Act 1974, Control of Substances Hazardous to Health (COSHH) Regulations 1996.

Activity 32.1C

You are a SOCO manager and one of your roles is to ensure the health and safety of your SOCO employees at work. You have decided to produce a health and safety information leaflet to help train new SOCOs. The leaflet must describe the different laws which protect them and the actions that must be carried out by both the SOCO and yourself to maintain high standards of health and safety. Using the Internet, identify and research three key pieces of legislation to create an appealing training leaflet.

Scene and evidence documentation

Observation and recording of the scene

When the crime scene has been made safe, the SOCO can enter and make a general survey, identifying obvious pieces of evidence and potential points of entry and exit that may have been used by the offender(s). At this point, the SOCO may protect any evidence to stabilise the scene and prevent damage or deterioration to the evidence. For example, a tent may be used to protect the victim's body from the weather at an outdoor murder scene.

The SOCO then has to record the crime scene in its original state, before any evidence is moved or collected. There are a variety of methods for documenting the scene.

Note-taking. Notes should be specific, detailed, accurate, organised and legible:

- name and signature of SOCOs
- date and times of examination
- crime case number – police reference
- location of crime scene
- detailed description of crime scene
- location of items of evidence
- description of evidence including evidence item numbers, time of discovery and sketch of evidence
- description of evidence collection and packaging techniques including evidence bag number and time of collection
- storage and transportation information.

Sketching. Rough sketches and measurements are made and the location of items of evidence recorded. The sketch should be labelled and include a key and north line showing direction.

Photographs. Overview photos of the entire scene and surrounding area, including points of entry and exit, should be taken. Items of evidence should be photographed to show their position. Close-up photos of evidence are then taken to record detail. A measuring scale is included in the photo to show the size of each item.

Video. May be used in addition to still photography.

Recording the scene is a continuous process, and once the search for evidence starts the SOCO continues to document all new items of evidence as they are located.

Continuity and chain of evidence

For each piece of evidence submitted to court there must be thorough and complete documentation of the evidence (notes, sketches and photographs) recording the procedures and methods used. In addition, a log must be kept showing who was in possession of the evidence at every stage of the criminal investigation. Together, these maintain the **chain of continuity** or custody.

Scene search and evidence recovery

Search for material of potential evidential value

Once the crime scene has been recorded, the SOCO can start to intensively search the scene for evidence. There are a number of ways to search the scene, following different search patterns:

- quadrant search pattern
- link method
- strip or line search
- grid method
- spiral search method
- wheel or ray method.

Activity 32.1D

Using the Internet, identify what is involved with each search pattern. Include a diagram of each pattern to show the pathway the SOCO would follow. Also identify any advantages and disadvantages that each search pattern may have. Finally, explain which pattern you would follow and why.

PLTS

Independent enquirer

Using the Internet will develop your skills as an independent enquirer. Remember not all information that you find on the Internet is valid.

Targeting and recovery of evidence

When an item of evidence is discovered it is documented: before it is collected it is photographed and details of the item, including a description, the location and time of discovery, are recorded. Each item of evidence must be given a unique evidence number, usually the initials of the SOCO followed by a sequential number. The evidence is then collected. There are a number of different techniques the SOCO can use depending on the type of evidence; three of them are shown in the table below.

Activity 32.1E

Research the techniques below that are also commonly used by forensic scientists to collect evidence:

- casting
- vacuuming
- hand picking
- swabbing.

Method of evidence collection	Description of technique	Types of suitable evidence
shaking	gently shake item over large piece of paper and collect loose particulate material that falls off	to recover trace evidence, e.g. glass fragments, paint chips, hairs, fibres
brushing	brush surface with clean tooth/paint brush and collect debris on paper/container	to remove trapped particles from surfaces like shoes, pocket linings, suspect or victim's hair, e.g. gunpowder residue, soil, pollen, hairs, fibres
taping	apply strips of clear sticky tape or **gel lifters** to surfaces to pick up trace evidence; sequentially, pull off strips of tape, stick down onto clear plastic acetate sheets and examine with microscope	to recover fingerprints, fibres and hairs from clothing, car seats, window ledges, edges of broken glass at point of entry, any dry surface

Packaging and labelling

When the evidence is collected, suitable packaging is used to protect it and keep the item secure. There are a range of types of packaging available to the SOCO, depending on the type and size of evidence, and any health and safety hazard. A common type of packaging used is the tamper evident bag, a plastic bag imprinted with a unique evidence bag number and an evidence details label that is completed by the SOCO at the scene. The bag has a sticky strip to seal it – if tampered with in any way the seal distorts.

Soft packaging:

- **small and large plastic tamper evident bag** for small items of clothing, cans, bottles, mobile phones and other small items
- **small grip-seal bags** for trace evidence, e.g. hairs, fibres, drugs, glass particles
- **big brown paper evidence bags** for large items of clothing, bed sheets, curtains.

Hard packaging:

- **metal air-tight tins** for small, trace items, e.g. bullets, soil and hairs; used for volatile liquids, e.g. petrol or paraffin, to prevent evaporation into air
- **weapons tube** for sharp items to prevent injury e.g. knives, syringes and scissors
- **cardboard boxes** to protect small and large types of evidence, e.g. guns and mobile phones.

The packaged evidence must be labelled accurately and completely with the details of the item, to maintain the chain of continuity. The following pieces of information should be recorded on the evidence packaging:

- evidence number
- description and location of evidence
- time and date of collection
- crime case number
- name and signature of SOCO.

Correct packaging and labelling of evidence is essential. Evidence packaged incorrectly may cause harm or deteriorate. Packaging that appears to have been tampered with may be dismissed from court.

Activity 32.1F

You are a SOCO and are processing the scene of a drugs raid. The offender is suspected of producing and selling illegal drugs. Make a list of a range of types of evidence you might find at this crime scene and describe how each item should be appropriately packaged.

Characterisation and comparison

In the forensic laboratory the evidence will be analysed using an identification and comparison process. However, to carry out a meaningful analysis the laboratory needs a control, or reference, sample to compare to the crime scene evidence. For example, a fingerprint found at the scene must be accompanied by suspect fingerprints to allow a comparison of the two to take place.

Marks and impressions

There are a range of types of marks and impressions that can be found at the crime scene. It is important to remember to photograph and record these types of evidence before collection.

Fingerprint marks

When a finger touches a surface, an impression of the fingerprint is left as a two-dimensional layer of sweat. This **latent** sweat print is invisible to the naked eye, but can be visualised using fingerprint powders or enhancement chemicals. Finger marks can also be visible, if the finger that leaves the mark was dirty or covered in a coloured substance, e.g. blood, ink, mud, make-up, etc. Latent and visible fingerprint marks may be recovered using the taping technique.

Fingerprint impressions

If a finger touches a soft surface, a three-dimensional negative impression of the fingerprint pattern may be left in the surface, e.g. putty, Blu-Tack®, chocolate. These 'plastic' fingerprint impressions can be recovered using putty-type casting materials.

Footprint marks

Two-dimensional footprint marks may be left on hard surfaces, especially when shoes are wet or dirty, e.g. with mud, blood, paint. Footprint marks can be recovered using the taping technique.

Footprint impressions

Three-dimensional footprints can be left in soft surfaces, e.g. cement, mud, sand, snow. Footprint impressions can be recovered using plaster-casting material, e.g. Crown stone.

Tool-mark impressions

These are three-dimensional marks left in soft and hard surfaces; for example, impressions left in a wooden window frame or front door while breaking in using a screwdriver. Tool-mark impressions can be recovered using putty-type casting materials.

Tyre-print impressions

Three-dimensional track impressions can be left by the wheels of a car in soft surfaces, e.g. cement, mud, sand, snow. Tyre-print impressions can be recovered using plaster-casting material.

Teeth impressions

Three-dimensional tooth marks can be left in soft surfaces; for example, bite marks in apples, chocolate, the skin. Dental impressions can be recovered using putty-type casting materials.

Many of these types of evidence are left by items that change over time, and these changes make the marks and impressions unique. On use, objects gain individual 'wear and tear', so a screwdriver impression can be linked back to the specific screwdriver used both by make and by the individual tool-mark pattern that it produces. Damage also occurs to the shoes that we wear, so that shoe prints can be linked to an individual by size, make, and wear and tear.

Footprint impression.

Case study: Jessica Chapman and Holly Wells

On 4th August 2002, Ian Huntley murdered 10-year-old schoolgirls Jessica Chapman and Holly Wells. Huntley was a caretaker at Soham College in Cambridgeshire. Holly and Jessica lived nearby and had gone to buy sweets in the early evening. On their return they walked past Huntley's house and he invited them inside. He killed them in his bathroom and transferred their bodies to a ditch near Lakenheath airbase 17 miles away. When the girls were discovered missing a police investigation and search was conducted. Huntley was interviewed by the police and media journalists and originally he was not suspected. Two weeks later the half-burned remains of Holly and Jessica's shirts were found in a storage building at Soham College and on 17th August Huntley and his girlfriend Maxine Carr were arrested. On the same day the bodies of the girls were discovered. Ian Huntley was convicted with two counts of murder on the 17th of December 2003 and he was sentenced to life imprisonment.

1. **Identify the types of evidence that the SOCOs would have looked for at the scene of the murders, on and in Huntley's car, and at the location where the bodies were discovered.**
2. **What were the health and safety issues associated with each crime scene?**

Storage and transmission to laboratory

When the evidence has been collected it must be safely and securely transferred to the forensic laboratory. The SOCO takes the evidence back to the police station and sends it, along with the paperwork, to the laboratory for analysis using a secure and reliable courier company. The storage of evidence at all times must be secure to prevent unauthorised personnel from removing or tampering with the evidence.

While in storage it is important that the evidence is stored under suitable conditions. The evidence must be preserved and protected from deterioration and contamination. For example, blood swabs must be stored at 4°C or –20°C, to prevent the blood sample from decaying.

Assessment activity 32.1

You are a SOCO working for the police. While you are on call a suspected suicide is reported, and the police ask you to attend the home of the victim to examine the crime scene and collect the forensic evidence. The victim has died from a gunshot wound and there is a gun and blood present at the crime scene. A suicide note has been found at the scene which may contain fingerprints. A suspected drug or poison has also been located at the scene.

1. Make a detailed list of the methods that you would use to gather evidence at the crime scene and carry some of them out. **P1**
2. Produce a plan of the examination procedures that you would follow at the crime scene. Include a description of each stage of the search process from arrival to transportation of the evidence. **M1**
3. Justify why you have chosen the procedures described in your answer to Question 2. **D1**

Grading tips

To achieve **P1** you should make sure that you consider all types of evidence that may be at the crime scene, i.e. biological, physical and chemical.

To achieve **M1** the procedures that you describe should be related to the simulated crime scene detailed in the question.

To achieve **D1** you must include in your justification of the procedures what could happen if the procedure had not been followed. You must explain how and why the procedures you described would allow you to recover all the relevant forensic evidence effectively without degradation.

32.2 Using chemical techniques to analyse evidence

In this section:   D2   

Key terms

Transparent – capable of transmitting light, easily seen through.

Cutting agents – a variety of cheap chemicals added to drugs of abuse to increase the quantity and value. They decrease the concentration of the drug and may be potentially harmful to health.

Porous – absorbent, full of pores through which gas or fluids may pass.

Stationary phase – in chromatography, the part of the apparatus that does not move with the sample, e.g.filter paper.

Mobile phase – in chromatography, the gas or solvent liquid, e.g. water or alcohol, that carries the components of the mixture.

Solvent front – in chromatography, the leading edge of the moving solvent as it progresses along the stationary phase. The distance the solvent moves is measured from the sample baseline to the point where the solvent front stops.

Sensitivity – amount by which substance being tested may be diluted and still give a reaction in a presumptive test.

Specificity – the likelihood that a presumptive test gives a positive reaction ONLY when exposed to the substance being tested (high to low).

Carcinogen – any natural or artificial substance that can cause cells to become cancerous by altering their genetic structure, e.g. asbestos, UV radiation, X-rays and some chemical substances.

Asphyxia – a condition caused by the inability to breathe, where there is a large decrease in the amount of oxygen in the body and an increase in carbon dioxide. This usually results in loss of consciousness and can lead to death.

Many types of evidence have a chemical nature and a number of chemical analytical techniques are used in the forensic laboratory to analyse evidence.

Spectrometry

Spectroscopy is the term for an analytical laboratory technique that uses electromagnetic radiation to analyse chemical substances. Many chemical materials absorb some wavelengths of electromagnetic radiation and reflect or transmit others to produce a specific spectrum. Different substances and solutions can be characterised by the spectra they produce. There are a range of spectroscopic techniques, discussed below.

The basic components of a spectrophotometer, the instrument used to analyse electromagnetic spectra, are the same regardless of whether the instrument is designed to measure absorption of ultraviolet, visible or infrared light, although the sample preparation varies. The components include:

- a light source
- a monochromator or frequency selector
- a sample holder
- a detector to convert electromagnetic light into an electrical signal
- a recorder to produce a record of the signal.

Unit 4: See pages 103–105 for more information about spectroscopic techniques.

Colorimetry

Visible white light is composed of a range of colours, red through to violet. To see an object, light has to reflect off the object; the colour of the object is determined by the colour of light reflected. Colour is therefore a visual indication that objects absorb certain colours of visible light and reflect or transmit others. For example, when white light is shone onto a red material, all colours of light are absorbed by the material except for red, which is reflected back by the material to our eye.

Substances absorb a particular frequency or colour of light according to their specific energy requirements. Different substances have different energy requirements and therefore absorb light at different frequencies. For example, a red material absorbs blue, green and yellow light (400–650 nm) and reflects red light (650–800 nm), whereas a blue material absorbs red and green light and reflects blue light.

Colorimetry measures the type of visible light radiation a material absorbs as a function of wavelength or frequency. This is unique for a given substance, and the information can be used to identify the substance. Colorimetry is a simple tool for determining the probable identity of a coloured material, but the technique does not provide a definitive result.

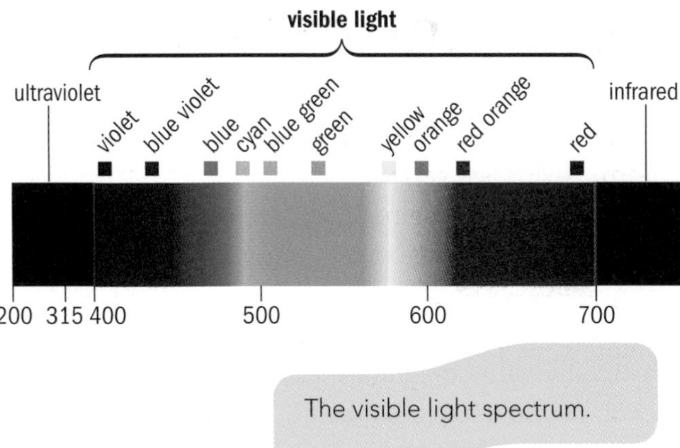

The visible light spectrum.

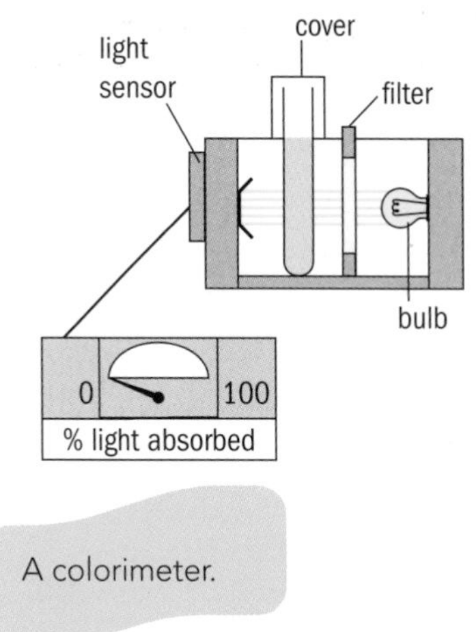

A colorimeter.

Colorimetry can also be used to determine the concentration of a sample by establishing the quantity of light absorbed by a substance or solution, and the amount of light transmitted through. How much light a substance absorbs depends on the concentration of the material and is defined by the relationship known as the Beer–Lambert law, which states that: 'absorbance is directly proportional to the concentration of a solution, i.e. as the concentration increases the amount of light absorbed increases'. For example, very weak diluted solutions are light in colour and **transparent**; they do not absorb much light and transmit most light through the solution. As the concentration increases the number of molecules in the solution increases – more molecules can absorb more energy.

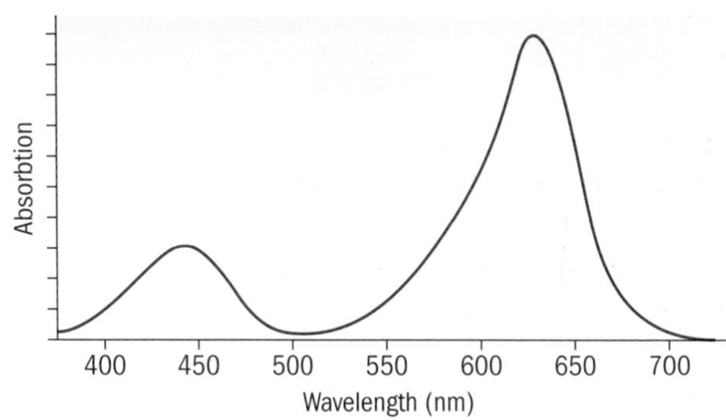

The absorption spectrum for the dye malachite green isothiocyanate/MeCN.

How does it work?

1 The substance or solution is placed into the sample holder of the colorimeter.
2 A beam of white light containing a wide range of frequencies is passed through the material.
3 The atoms of the material absorb light at particular frequencies and reflect or transmit at other wavelengths.
4 The instrument detects which wavelengths have been absorbed and records the information.
5 An absorption spectrum graph can then be produced for a particular material, which plots the fraction of visible light absorbed at each wavelength.

The graph usually shows one or more typically broad absorption bands; every type of substance or solution has a unique absorption spectrum. The results can be converted into energy which can be used to determine information about the atomic structure of the material.

UV spectroscopy

Ultraviolet (UV) light radiation lies just above the violet region of visible light on the electromagnetic spectrum. It has shorter wavelengths than visible light but longer wavelengths than X-ray radiation. The UV spectrophotometer uses a beam of UV light instead of white light. The UV light is absorbed by molecules of organic compounds and it excites the electrons, which move to a higher energy level and state of vibration. The excited molecules collide with each other and this causes them to lose energy. The electrons drop to their original ground states releasing the rest of the energy as low-energy fluorescent light. Molecules drop to different vibrational energy levels and the emitted fluorescent light will have different frequencies. The wavelength and amount of fluorescent light emitted is characteristic of the molecules in a material.
The UV spectrophotometer records the intensity of fluorescence emitted, producing an excitation

spectrogram of fluorescent light intensity against UV wavelength. Analysis of the different frequencies helps to identify forensic substances.

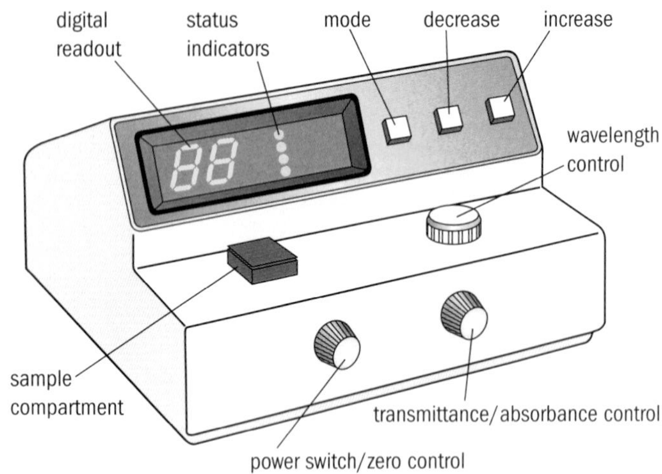

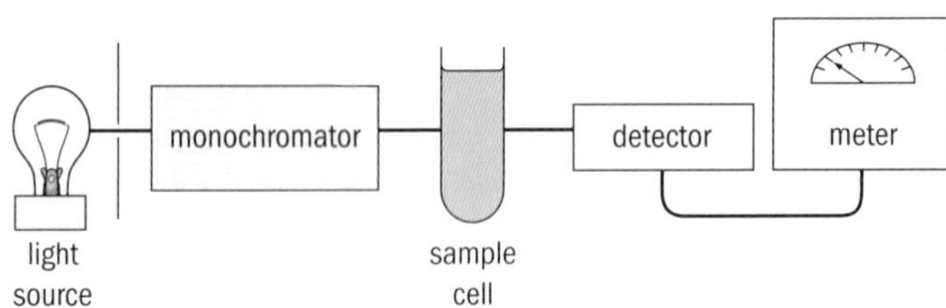

A UV spectrophotometer.

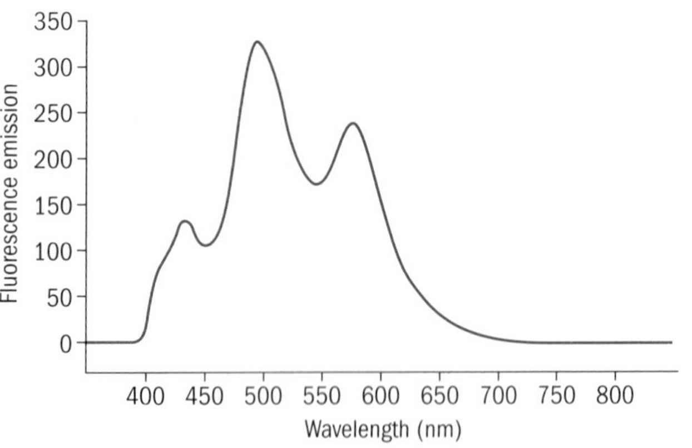

Fluorescence emission spectrum of textile fibre with 365 nm excitation wavelength.

Infrared spectroscopy

Infrared (IR) light is a type of electromagnetic radiation that lies between the visible light and microwave regions of the electromagnetic spectrum. IR light has wavelengths of between about 1 mm and 750 nm. We cannot see infrared radiation, but we can sometimes feel it as heat. IR spectroscopy is used to analyse forensic evidence samples – to gather information about a substance's structure, assess its purity and to identify it. Covalent bonds in organic substances are not rigid but continuously vibrate by bending and stretching. The molecules vibrate at a unique frequency which falls within the infrared spectrum. When an individual molecule absorbs IR light energy that matches its vibrational frequency, the radiation energy kicks the molecule into a higher state of vibration. The amount of energy it needs to do this will vary from bond to bond, and so each different type of bond in a substance will absorb a different frequency and wavelength (and hence energy level) of IR radiation.

The radiation source of an IR spectrophotometer emits light at different frequencies in the IR region of the electromagnetic spectrum and produces an IR spectrum – a graph plotting wave number against percentage of radiation transmittance. The absorption in the IR region provides a more complex pattern than UV or visible spectroscopy, with enough characteristics to specifically identify a substance. Different materials always have distinctively different IR spectra; therefore each IR spectrum is equivalent to a 'fingerprint' of the substance. IR spectroscopy can be used to analyse a range of types of forensic evidence:

- layers of paint in chips from the clothes of a victim involved in a hit-and-run accident (the specific make, model, and year of vehicle can be identified from the paint)
- synthetic fibre
- drug evidence.

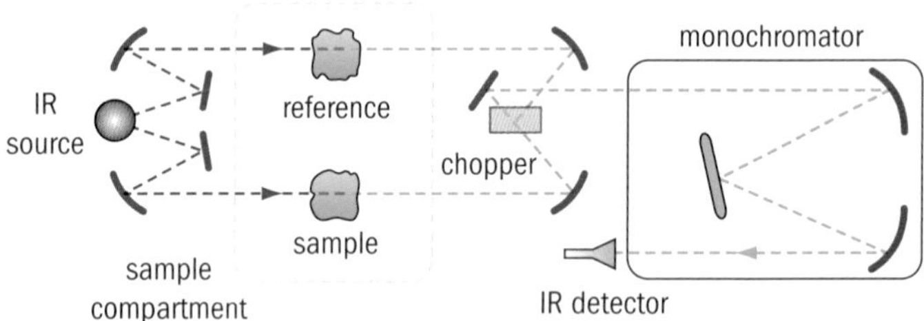

The parts of the IR spectrophotometer.

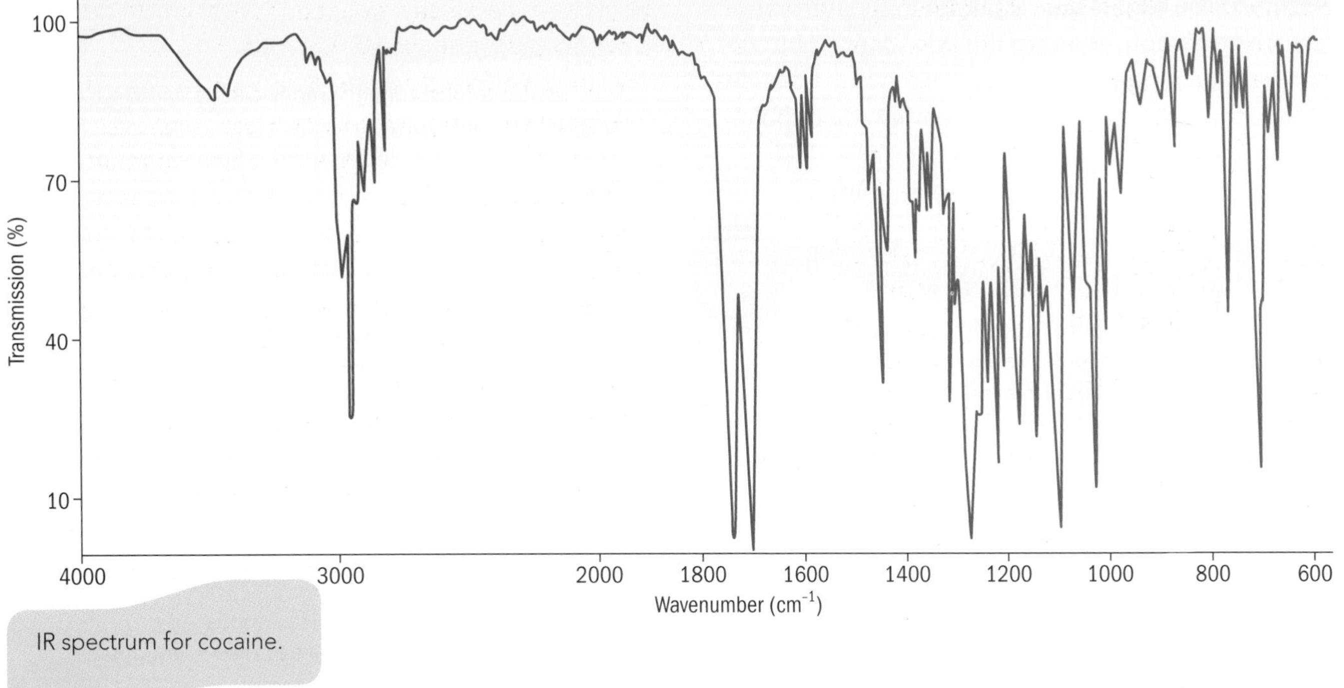

IR spectrum for cocaine.

Mass spectrometry (MS)

This is a powerful analytical technique for identification of the structure and chemical properties of solid, liquid and gas molecules. It is utilised in the forensic laboratory to quantify and identify the chemical structure of unknown substances, usually in conjunction with a chromatographic technique. MS can be used to analyse a range of types of forensic evidence in a variety of criminal investigations; for example:

- to detect illegal dumping of poisons and toxins in river water
- to determine whether food or drink samples have been spiked with drugs and poisons
- to detect and identify the use of steroids in athletes.

The MS technique consists of four main stages.

Stage 1: Ionisation

In order to measure the characteristics of individual molecules, a mass spectrometer converts them to ions so that they can be moved about and manipulated by external magnetic fields. An 'ion source' (e.g. a heated filament) produces a high-energy beam of electrons. These bombard the sample molecules, fragmenting them and ionising the atoms by knocking off one or more electrons to give positive ions (cations). Because ions are very reactive and short-lived, their formation and manipulation must be conducted in a vacuum.

Stage 2: Acceleration

The ions are accelerated so that they all have the same kinetic energy, and are finely focused into a single beam.

Stage 3: Deflection

Once the atoms have been turned into a single beam of electrically charged ions, they pass through the mass analyser. This contains a strong external magnetic field perpendicular to the ion beam's direction of motion. This bends the ion beam and deflects the ions. Two factors affect the amount an ion deflects. Different ions are deflected by the magnetic field by different amounts according to the mass of the ions. The lighter they are, the more they are deflected. The amount of deflection also depends on the number of positive charges on the ion, i.e. how many electrons were lost. Ions with two (or more) positive charges are deflected more than ones with only one positive charge. These two factors are combined into the mass/charge ratio, given the symbol *m/e*. The deflection therefore separates the different ions in the sample according to their mass and charge.

Stage 4: Detection

The separated beams of ions passing through the machine are then detected electrically and the information is stored and analysed in a computer. By varying the strength of the magnetic field, ions of different mass can be focused on the detector fixed at the end of the curved mass analyser tube. The separated ion fragments' mass and abundance are measured. The results are displayed in the form of a mass spectrograph, showing the abundance of the ions against a range of *m/e* ratios. No two substances produce the same fragmentation pattern, and the mass spectrum pattern produced is an individual 'fingerprint' of the substance being examined.

Activity 32.2A

You are a forensic scientist working in the chemistry department of the Forensic Science Service. You have been asked to go to court to explain the spectroscopic technique to the jury. This must involve a presentation, which you can use as a visual aid for the jury.

In small groups, create a colourful, clear and informative presentation describing the different types of spectroscopic techniques and the types of evidence they can be used to analyse.

PLTS

Team worker

Working in groups will develop your skills as a team worker.

Chromatography

Instrumental techniques

Chromatography is a collective term for a number of similar laboratory techniques that are used to separate mixtures into individual pure components. There are a number of reasons why a forensic scientist would want to separate a mixture.

- They may want to analyse and examine liquid or gas mixtures, and the relationship of their components to one another.
- They may need to identify a mixture and its components, or purify and separate components to isolate one of interest for further investigation.
- They can also quantify (determine the amount of) a mixture and its components in a sample.

There is a range of chromatographic techniques (see the table on page 464) that can be used to analyse different forensic samples; e.g. inks, dyes, pens, markers, highlighters, lipstick, clothing dyes, food colouring (e.g. in sweets), coloured pigments in plants, flowers, drugs and accelerants.

See Unit 4 (pages 96 and 102) and Unit 13 (pages 248–250) for more about chromatographic techniques.

Did you know?

The technique of chromatography can be applied to analyse different types of mixtures for a range of criminal investigations and is used in many different types of forensic laboratory.

- Forensic toxicology laboratories use chromatography to detect alcohol and drugs in victim or offender blood samples.
- Forensic environmental laboratories use chromatography to determine the level of illegal pollutants in a lake or river.
- Forensic chemistry laboratories use chromatography to determine the purity of a drug sample and identify the drug and **cutting agents**.
- Forensic questioned-document laboratories use chromatography to compare components from pen ink or printer ink samples.
- Forensic fire-investigation laboratories use chromatography to analyse samples of accelerant used to start fires at arson scenes.

Paper chromatography

In paper chromatography, a small sample of a mixture is placed on a **porous** surface (**stationary phase**), e.g. filter paper. The paper is placed in contact with a liquid solvent (**mobile phase**) which dissolves the mixture sample (e.g. ink). The molecules present in the mixture distribute themselves between the stationary and mobile phase, depending on how well the components dissolve in the solvent. The solvent moves through and up the paper and components of the mixture are carried with the solvent a certain distance. Individual components that dissolve in the solvent are carried along quickly and move far up the filter paper.

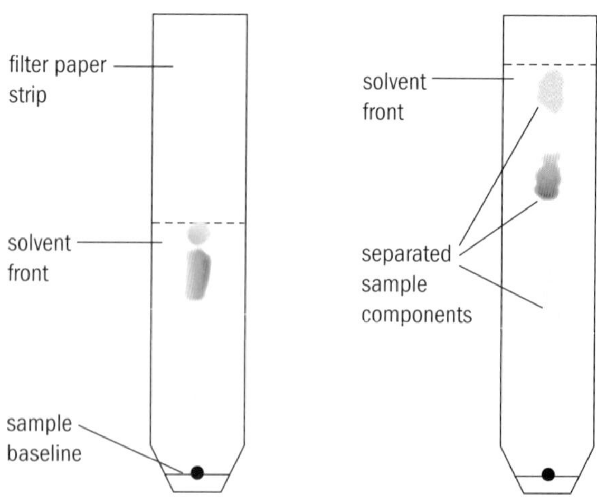

Chromatographic separation of black pen ink.

Components with a stronger attraction to the paper than the solvent move more slowly and not as far in the same amount of time. The mixture therefore becomes separated, and by this technique forensic scientists can identify what chemicals are present in various mixtures.

Chromatogram analysis

Molecules in mixtures travel at different speeds when pulled along the paper by the solvent and so travel different distances in the same given time. The retention factor (R_f) can be determined for samples separated on a chromatogram to give a measure of a component's properties in a mixture. This calculates how far a component has moved compared to how far the solvent moved (the **solvent front**), and provides a quantitative method for identifying components in a mixture.

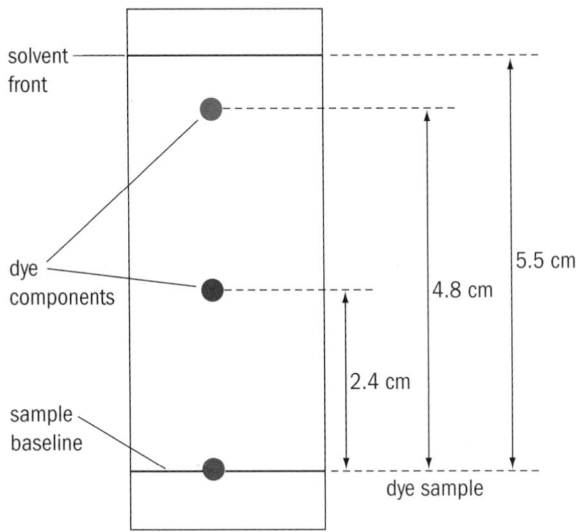

How to determine the R_f value of component samples.

For substances that are soluble in the solvent, the R_f will be close to one – the substance will move quickly up the paper. For substances that are less soluble in the solvent, the R_f value will be closer to zero and the substance will move slowly up the paper, not moving far. Mixtures will produce the same pattern on the paper as long as the paper, solvent, and time are kept constant. Therefore, the R_f values will be identical for the same components in different mixtures. Using different solvents or shortening the time of development would change the patterns, so it is important when comparing crime scene and reference samples that all variables are kept constant. Known reference samples are usually separated on the same chromatogram to compare with unknown crime scene samples. This allows the forensic scientist to compare R_f values and identify unknown samples.

Activity 32.2B

Calculate the R_f values for the known reference samples and evidence components, and compare them to determine the identity of drug B and any cutting agents present. The solvent front moved a distance of 10.2 cm.

Known pure reference sample results

Component	Distance moved (cm)	R_f value
flour	1.9	
cellulose	4.3	
rat poison	5.5	
LSD	7.0	
heroin	8.8	
anabolic steroids	9.2	

Drug B evidence sample results

Component	Distance moved (cm)	R_f value	Identification
1	4.3		
2	5.7		
3	8.8		

Functional skills

Mathematics

You will develop your mathematics skills when carrying out the R_f calculations.

Thin-layer chromatography (TLC)

This technique also separates dried liquid samples using a liquid solvent (mobile phase). However, instead of paper as the stationary phase, it uses an absorbent material; for example, a thin layer of silica gel dried onto a flat glass, plastic or metal plate. The sample mixture is placed at the bottom of the plate and the plate is inserted into the solvent. The solvent moves across the plate, and this causes the mixture components to travel up the porous gel at different speeds. It separates components in the same way as paper chromatography. These can then be compared with known standards. TLC is used in forensic science to analyse the dye composition of fibres, inks and paints. It can also be used to detect pesticide or insecticide residues in food.

Types of chromatography	Types of sample separated	Mobile phase	Stationary phase	Uses
paper	dried liquid samples	liquid solvent	filter paper strip	one of the most common types of chromatography; to analyse pen inks, lipsticks, food & fabric dyes, etc.
TLC	dried liquid samples	liquid solvent	TLC sheet – glass/plastic plate covered with a thin layer of silica gel	to analyse the dye composition of fibres, inks and paints; to detect pesticide or insecticide residues in food
liquid e.g. high performance liquid chromatography (HPLC)	liquid samples that may incorporate insoluble molecules	liquid solvent	column composed of silica or alumina gel powder or suspension of solid beads in a liquid	to test water samples to look for pollution in lakes and rivers; to analyse metal ions and organic compounds in solutions; to analyse blood found at a crime scene
gas e.g. pyrolysis gas chromatography (PGC)	vaporised samples and gas mixtures	carrier gas, e.g. nitrogen, hydrogen or helium, is used to move gaseous samples	column composed of a liquid or of absorbent solid beads	to detect bombs in airports; to analyse fibres on a person's body; to test for the presence of accelerants in arson cases and residue from explosives (PGC); to analyse body fluids for the presence and level of alcohol and illegal substances

Fingerprint chemical enhancement

The surface of fingerprint skin is covered in ridges which create our fingerprint patterns. On the top of fingerprint ridges are sweat pores: small openings in the skin from which sweat, produced in the sweat gland, is secreted onto the ridge surface. Sweat is mainly composed of water, but also contains a number of other substances, including amino acids, proteins, salts and sugars. When a finger touches a surface the sweat is transferred to the surface, producing a two-dimensional mirror impression of the fingerprint ridge pattern. These sweat marks are called latent fingerprints and are usually invisible to the naked eye. They must be enhanced, giving them colour, so that they can be photographed and examined.

There are a number of chemical treatments that can be used at the crime scene and in the forensic laboratory to enhance latent fingerprints. Chemical reagents react with different component substances of sweat causing a colour change. Usually a number of chemical enhancement techniques are used in sequence, to achieve the best, or optimum, enhancement. The technique used depends on the surface that the latent mark is on, as there are different chemical reagents for porous and non-porous surfaces.

Remember: that the aim of enhancement is to provide contrast between the fingerprint and the surface it is on. If the invisible fingerprint was on a black door, what colour chemical do you think you would use to make it visible?

Safety and hazards

Many chemical enhancement reagents pose a health and safety risk, e.g. superglue and iodine fumes are toxic and strongly corrosive. A risk assessment must be carried out and control methods put in place to reduce the hazards. Only trained staff should chemically enhance fingerprints using Home Office standardised operating procedures.

See page 488 for more information on fingerprints.

Ninhydrin

Ninhydrin is a commonly used reagent for developing latent prints on porous items. It reacts with the amino acid and water-soluble components of sweat, producing a bluish-purple or pink visible fingerprint mark. The blue-coloured substance (Ruhemann's purple) is formed by the reaction of ninhydrin with its reduction products, hydrindantin and ammonia. Ninhydrin solution can be applied to the surface of the item that the latent print is on by spraying, painting or dipping. The reaction may be quite slow; marks may take between several hours and several weeks to fully develop, but the process can be speeded up by heating the finger mark using an incubator.

Ninhydrin-enhanced fingerprint marks.

Iodine

Iodine fuming is the oldest technique used to develop latent fingerprints but is now not commonly used in the forensic laboratory. The technique works best on fresh prints, usually no more than a few days old, and can be used on both porous and non-porous surfaces. The technique is non-destructive and can be used in sequence before other treatments, e.g. ninhydrin treatment of paper.

Fingerprint marks enhanced with iodine.

Activity 32.2C

There are many more fingerprint enhancement techniques used by forensic scientists; for example:

- amido black
- superglue fuming (cyanoacrylate fuming)
- silver nitrate.

You have been asked to make a leaflet to help train new forensic scientists. It should include information about the fingerprint enhancement techniques listed above. You could include pictures of the equipment used to enhance the print. Identify which types of fingerprints it could enhance and from which surfaces. You could also get pictures to show how the results would look after using each technique. Also, you must use books and the Internet to describe the science behind the techniques. Remember to make this leaflet eye-catching and informative.

Chemical presumptive tests for body fluids

Presumptive tests are chemical reagent based, colour change tests capable of detecting the presence of blood, semen, saliva and other body fluids. They are sensitive, quick, simple to use and cheap. They can be used at the crime scene and in the laboratory. Any fluid or stain thought to be a body fluid should be tested with a presumptive test at the crime scene before collection. A sequence of chemicals is applied to suspected stains, which change colour in any area where the body fluid is present. Presumptive tests are not 100% accurate; they only indicate the presence of a body fluid and the results must be confirmed in the laboratory using conclusive analytical techniques.

Presumptive tests always give a positive result in the presence of the specific fluid being tested for. However, they can also give false positive results – giving a positive reading or indication when the fluid is not actually present.

Blood presumptive tests

Presumptive blood tests react with haemoglobin (Hb) present inside red blood cells. Hb is an iron-containing metalloprotein found in the red blood cells of mammals and other animals. Its role is to transport oxygen from the lungs to the rest of body. The tests are based on the ability of the Hb to catalyse the oxidation of reagents.

- Leucomalachite green (LMG): the LMG test is a non-toxic, low-**sensitivity** presumptive blood test with high **specificity**. It has a number of false positives and identifies 1 ml of blood in 10 000 ml of water.
- Luminol: this is a presumptive test chemical reagent that releases blue chemiluminescent light when mixed with an oxidising agent. When blood stains are suspected of being present but are invisible to the eye, luminol reagent can be sprayed over a large area and examined in the dark. Luminol is very sensitive and detects minute traces of blood, so may indicate the presence of blood after an offender attempts to clean up and wash away the evidence.

Did you know?

Luminol is a hazardous, toxic chemical and is considered a possible **carcinogen**. When in use at the crime scene, full PPE should be worn, including gloves and a face mask. A new chemiluminescent reagent has recently been developed, called BLUESTAR®, which has replaced the use of luminol at the scene. It is non-toxic, very sensitive, and easy to prepare and use; it does not require complete darkness and it has a stronger and longer-lasting luminescence. It is a more effective presumptive test than luminol, without many of its drawbacks.

Semen presumptive tests

Semen evidence may be submitted to the forensic laboratory from rape, sexual assault and abuse cases, and provides important evidence of sexual contact. Semen is commonly recovered from bed sheets, underwear, clothes, or from vaginal, anal or oral swabs taken during the medical examination of a victim.

Semen contains high levels of the acid phosphatase enzyme secreted by the prostate gland. Semen and semen stains at the crime scene and on items of evidence may be presumptively tested for using the acid phosphatase (AP) test. When the AP reagent is applied to semen, the AP enzyme catalyses the conversion of sodium-naphthylphosphate to free naphthyl, which converts brentamine dye from a colourless to a purple stain. The AP enzyme is also present in a number of other biological fluids, e.g. blood, saliva, urine and vaginal secretions, although in much smaller quantities, and these will react weakly with the AP test. A positive result must be confirmed in the laboratory using microscopic examination to visually identify the presence of sperm cells.

Saliva presumptive tests

Saliva is another commonly encountered biological sample analysed in forensic laboratories. It is often found on the rim of drinking glasses, cans and bottles, on food items, bite marks, cigarette butts, stamps, envelopes, chewing gum, balaclavas, and on the skin in sexual contact cases. Saliva is a watery fluid secreted by the cells of the salivary glands and carried to the mouth via tube-shaped salivary ducts. It is 99% water, and the other 1% consists of organic molecules (glycoproteins, lipids), electrolytes (e.g. calcium, phosphates), mucus, enzymes and antibacterial compounds. Saliva also contains epithelial cells shed from the inside surfaces of the cheeks and therefore contains DNA.

Saliva contains high levels of the salivary amylase enzyme, which digests and breaks down starches in food. The Phadebas® presumptive test identifies the presence of amylase activity and is used in forensic laboratories to indicate the presence of saliva. Other body fluids contain low levels of amylase, including blood, semen, sweat and urine, although they usually do not have sufficient quantities to react with this presumptive test.

Activity 32.2D

Produce a leaflet for trainee SOCOs about body fluids they may be faced with when working. It could include:

- types of body fluid found at crime scenes
- a presumptive test for each type of body fluid
- advantages and disadvantages of presumptive tests including false positives.

Functional skills

ICT

You will develop your ICT skills when producing the training leaflet.

Toxicology

The science of poisons

Toxicology is the scientific study of poisons and drugs of abuse – their source, properties, effects and antidotes. Forensic toxicologists analyse blood, urine and other tissues in the laboratory in a range of criminal investigations. A poison is defined as a substance that, when taken into or forming in a living organism, causes injury, illness or death. There are many different types of poison (see table), which cause different effects on the human body, and they can be classified in a number of ways. For example, they may be grouped into classes depending on their chemical structure, composition and properties, their effects on the human body, or how they are analysed.

Types of poison	Description	Examples
anions	some anions are poisonous; they are harmful or deadly to living organisms	weed killers, insecticides and bleaching agents; one of the most poisonous anions is cyanide ion
corrosive poisons	corrosive poisons are acids (e.g. hydrochloric acid), alkalis (e.g. potassium and sodium hydroxide), and other substances (e.g. heavy metal salts and strong detergents) that cause physical destruction or 'burning' of the skin and body tissues through direct contact; they are commonly ingested causing severe surface and tissue damage to the mouth, oesophagus, stomach and intestines	bleach
gaseous and volatile poisons	a number of gases are poisonous to living organisms and can cause injury and death if inhaled; different gases affect the body in different ways; some gases are corrosive or irritant; chlorine and mustard gas cause blisters and attack the eyes and lungs; nerve gases break down the action of the nervous and respiratory system; other gases, e.g. carbon monoxide and hydrogen cyanide, cause **asphyxia** – they affect the bloodstream, restricting the body's ability to absorb oxygen	chlorine and mustard gas, carbon monoxide and hydrogen cyanide also volatile poisons include: cigarette lighter refills, hair sprays, deodorants, air fresheners, cleaning products, nail-varnish removers
metal and metalloid poisons	a number of metals are poisonous to humans; can enter the body via ingestion, inhalation or absorption through the skin and mucous membranes and be stored in the body's soft tissues, where they compete with other ions and bind to proteins; results in damage to organs throughout the body and multi-organ system failure leading to death; symptoms of metal poisoning include vomiting, diarrhoea, abdominal pain, muscle cramps and paralysis	lead (Pb), lithium (Li), mercury (Hg), thallium (Th), arsenic (As)
pesticides	the broad term 'pesticide' is defined as any chemical or natural substance used to control, reduce and kill organisms considered to be pests; ingestion, inhalation or skin contact with pesticides causes a range of symptoms depending on the type of pesticide, but commonly including headache, runny nose and eyes, increased saliva, vomiting, diarrhoea, sweating, general weakness, seizures, shallow breathing, dizziness, abdominal pain or cramps, convulsions	insect killers (insecticides), mould and fungi killers (fungicides), weed killers (herbicides), rat and mouse killers (rodenticides)
toxins	toxins are naturally occurring poisonous substances produced by living organisms – plants, animals, fungi, microorganisms; some alkaloids have medicinal properties and others, e.g. cocaine and heroin, are drugs of abuse	plant alkaloids, snake and insect venom, carcinogenic aflatoxins produced by the *Aspergillus* fungus, rat poison

Activity 32.2E

Use the Internet and books to research the poison cyanide. In pairs, produce a list of facts about the poison and find two documented crimes that may have involved cyanide.

Activity 32.2F

In pairs, see if you can find any criminal cases that involved arsenic. Find out what happened, how it was used, symptoms experienced and make a storyboard about the crime.

Drugs of abuse

Drugs of abuse are defined as illegal drugs, or prescription or over-the-counter drugs used for purposes other than those for which they were made for. The forensic scientist investigates a range of drug evidence samples to identify and quantify the drug present, as UK legislation prescribes different charges and sentencing for different drugs and amounts of drugs. There are two key pieces of legislation that are used to control drugs of abuse. The Misuse of Drugs Act 1971 is aimed at preventing unauthorised use of drugs of abuse. It categorises drugs into three groups (A, B and C; see table below) based on the harm the drug can cause if misused – category A are the most dangerous drugs, category C are the least dangerous. These categories are used in court to determine the penalty for misuse of a drug. The second piece of legislation is the Misuse of Drugs Regulations 1985, which was amended in 2001.

These regulations categorise drugs into five groups and describe the requirements for the legitimate prescription, distribution, production, record keeping and storage of drugs. Drugs of abuse can be broadly classified according to their effect on the central nervous system (CNS) and their impact on the activity of the brain. For example, stimulants like cocaine and amphetamines arouse and stimulate the CNS, depressants such as heroin, alcohol and tranquilisers have a depressing effect on the CNS and inhibit brain activity, and hallucinogens (e.g. cannabis, ecstasy and magic mushrooms) alter perception and mood without stimulating or depressing the CNS.

Instrumental analysis

Poisons and drugs of abuse are analysed in the forensic laboratory to determine their identity, composition and quantity of their components. A range of tests can be used, and usually there is a sequence of testing to confirm the identity of a substance.

Visual examination

The first stage of analysis involves observing the physical properties of the sample with the naked eye or microscope. Characteristics such as colour, shape, dimensions and markings can provide valuable clues about a drug's identity.

Presumptive tests

Poisons and drugs of abuse can be tested using chemical presumptive tests, which are similar to the body fluid presumptive tests described earlier in this section. They indicate the presence of certain chemical substances. A small amount of evidence sample is tested with chemical reagents that produce

Drug	Examples	Penalties for possession	Penalties for dealing
Class A	ecstasy, LSD, heroin, cocaine, crack, magic mushrooms (whether prepared or fresh), methylamphetamine (crystal meth), other amphetamines if prepared for injection	up to 7 years in prison or an unlimited fine, or both	up to life in prison or an unlimited fine, or both
Class B	cannabis, amphetamines, methylphenidate (Ritalin), pholcodine	up to 5 years in prison or an unlimited fine, or both	up to 14 years in prison or an unlimited fine, or both
Class C	tranquilisers, some painkillers, GHB (gamma-hydroxybutyrate), ketamine	up to 2 years in prison or an unlimited fine, or both	up to 14 years in prison or an unlimited fine, or both

characteristic colours on reaction with the substance of interest, indicating a positive result. Presumptive tests for drugs and poisons are not specific as they also produce false positive results. They provide the forensic scientist with information that allows the scientist to choose the correct type of confirmatory analytical test.

TLC

Evidence samples, e.g. blood and urine, that contain trace levels of drugs or poisons are mixtures, and can therefore be separated and analysed using chromatography. Different drugs and poisons appear as different coloured spots on a TLC plate after visualisation; they also travel different distances and therefore have different R_f values. This makes TLC more discriminative than presumptive tests – it can distinguish among a larger number of drugs. However, it is not a confirmatory test and should be backed up with other analytical tests.

Immunoassay

Immunoassays are highly sensitive tests used widely to identify the presence of a drug or poison in ante-mortem and post-mortem samples. The technique utilises the antigen–antibody interaction, in a similar way to the blood typing technique. A specific antibody is used to detect a drug or poison within a body fluid. The substance of interest acts as the antigen. If the substance is present in the sample, the antibody selectively binds with the drug or poison to form an antigen–antibody complex. As the reaction is so specific, this technique is able to confirm the identity of a substance.

GC-MS and HPLC-MS

There are a number of chemical analytical techniques that can be used to confirm the identity of a sample. Which technique is used usually depends on a number of factors, including the nature of the drug or poison in question, the nature of the substrate the drug or poison is in, and the concentration and amount of sample.

To analyse trace levels of organic drugs and poisons, gas chromatography (GC) or high performance liquid chromatography (HPLC) is normally used to separate the substance in question from the mixture it is in and provide information on its identity and quantity. The GC or HPLC apparatus may be linked to a MS instrument, which provides more detailed information (see pages 461 and 462).

UV and IR spectroscopy

Many drugs and poisons absorb light in the UV–visible and IR region of the electromagnetic spectrum, the wavelength of light absorbed is a unique characteristic of the substance. UV and IR spectroscopy can be used to quantify and identify the amount of a questioned substance present in a sample. By comparing spectrograms of an unknown questioned sample to a known drug or poison, the identity of the sample can be determined (see pages 459 and 460).

Specimen collection (ante and post mortem)

Many types of poison and drugs of abuse samples are analysed in the forensic laboratory. These types of evidence are either bulk samples, large enough to be weighed, or trace samples that are so small they cannot be weighed. A large range of bulk samples are analysed, including: seized illegal drugs in the form of powders, tablets, plant material, etc.; legal tablet and capsule drugs; poisons and suspicious materials (e.g. alcohol suspected of being spiked with drugs). Many other types of sample are analysed for the presence of trace levels of poisons or drugs; for example, drug-taking paraphernalia, including syringes, wrapping materials and ashtray contents, and items that have been in contact with drugs (e.g. containers, clothing, cutlery, crockery, bank notes). Other trace items that may be analysed include food or drink suspected of containing poison or drugs, and items used in illegal drug laboratories (e.g. glassware, solvents).

Activity 32.2G

List as many biological samples that you can think of that are taken from living individuals for toxicology, and make another list of biological samples taken from deceased individuals that may be of use for toxicologists.

PLTS

Independent enquirer, Self-manager and Team worker

Researching forensic applications of analytical techniques, completing work to standard and deadline, and working as a team to analyse forensic evidence will help you develop these respective skills.

Assessment activity 32.2

BTEC

You are a forensic scientist working for the chemistry department of the Forensic Science Service. A SOCO has submitted a number of items of evidence from a suspected suicide case. The police have asked you to analyse the evidence using chemical techniques of analysis. You must establish whether the victim killed themselves or if their death is more suspicious and could have been murder.

Types of evidence collected from the crime scene:

- suspected drug of abuse or poison
- suicide note and pen
- fingerprints located on the gun handle and suicide note
- suspected blood stains.

1 Make a list and give a brief description of the techniques you would use to analyse the chemical evidence recovered from the crime scene. Carry out some of the practical techniques you have listed (make sure you check the safety notes for each technique you use). P2 P3

2 Explain each of the chemical techniques of analysis. Present valid conclusions from your practical work. M2 M3

3 Discuss the advantages and disadvantages of the chemical techniques of analysis. D2 D3

Grading tips

To achieve P2 and P3 you must be able to briefly describe the main points of the chemical techniques used to analyse forensic evidence. When you carry out the techniques, make sure you complete the appropriate laboratory documentation, and accurately present your results.

To achieve M2 and M3 you must be able to explain fully the chemical analysis techniques, and how they were used to analyse the forensic evidence collected from the crime scene. You should draw conclusions from the results of your chemical analysis, and support your conclusions with evidence and reasoning; suggest also how to further analyse the evidence.

To achieve D2 and D3 you must be able to evaluate the scientific techniques of chemical analysis, and explain why the techniques used were suitable for analysis of the forensic evidence from the crime scene. You must also identify a range of types of evidence that can be analysed using the techniques, and how they could be applied to different types of criminal investigation.

Functional skills

ICT and English

You will be developing your ICT skills by selecting and using a variety of sources of information to research and create a portfolio of analytical techniques with forensic applications; and when you enter, develop and format information, to suit its meaning and purpose, in building a case file on a simulated crime.

When reading information on analytical techniques you will develop your English skills, in comparing, selecting, reading and understanding texts and using them to gather information, ideas, arguments and opinions. When writing a portfolio on techniques used in a forensic laboratory and conclusions from the results of analytical work, and building a case file, you will be practising your English skills, in communicating effectively and persuasively.

32.3 Using physical techniques to analyse evidence

In this section: P4 M4 D4 P5 M5 D5

Any object, mark or impression that originates from a non-living object can be classified as physical evidence, and a wide range of types of evidence falls into this category; for example, ballistic evidence, questioned documents, footprints and tyre prints.

Ballistics

Firearms

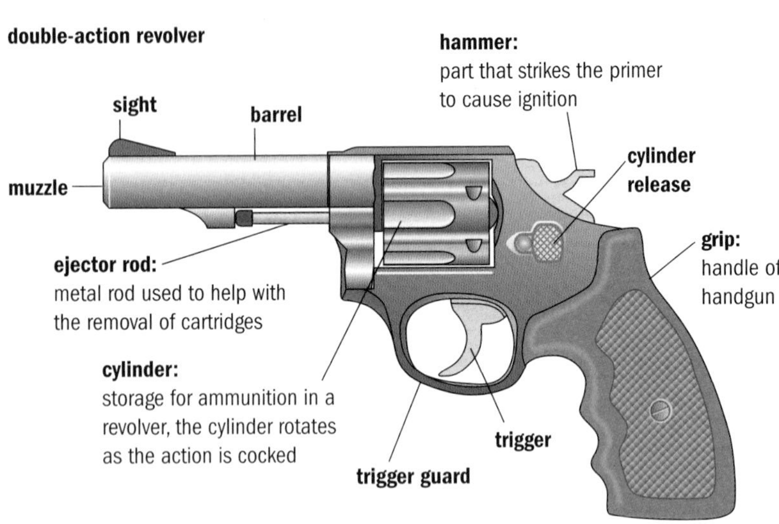

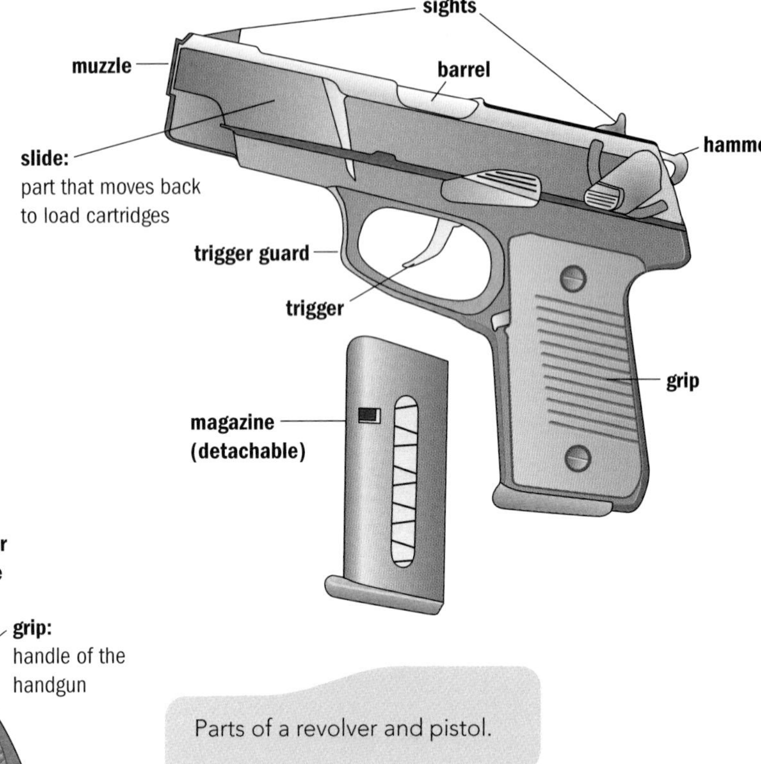

Parts of a revolver and pistol.

A firearm is any lethal weapon that uses a burning propellant to discharge one or many projectiles. Modern firearms are manufactured in a range of shapes and sizes for a variety of purposes, and many of these different types are associated with criminal activity and may be analysed in a forensic laboratory. The guns most commonly used for crime fall into four categories: the pistol or revolver, both hand guns, and the rifle or shotgun, which are both long guns.

Handguns

Handguns are small, light, compact, concealable weapons that are often used for short-range firing, e.g. for self defence. They have short barrels (the metal tube that the bullet projects from) and are designed to be fired using one or two hands. They usually use less powerful ammunition than rifles or shotguns.

There are two different types of handgun. The first is the revolver, which has a revolving cylinder (or magazine) that contains a number of firing chambers (usually six). Each chamber in the cylinder is loaded with a single cartridge, and after every shot the cylinder revolves bringing the next cartridge into line with the barrel. The spent, empty cartridge case remains in the cylinder or is manually ejected from the gun. The revolver is cheap, reliable and accurate, and is simple in design and use. However, it is limited to six shots, is relatively slow to reload, and there is a lot of resistance when the trigger is pulled.

The second type of handgun is the pistol, which has a single fixed firing chamber in the rear of the barrel, and a removable magazine so they can be used to fire more than one round. Pistols can be single-shot or self-loading mode, and magazines can hold up to 10 ammunition cartridges.

Rifles

Rifles differ from handguns in the length of their barrel. The rifle barrel is much longer and is designed to be fired from the shoulder. Rifles have a shoulder stock so it is possible to hold the rifle comfortably when firing. As rifles are large they are harder to carry and are poorly concealable. However, they are much more accurate and use more powerful cartridges than handguns.

Shotguns

Shotguns have a similar external appearance to rifles, they have a long thin barrel, are large and heavy, and are also fired from the shoulder. They are powerful firearms and are easier to hit a target with, often used for hunting game in flight or on the ground. There are many types of shotgun; they are normally single or double barrelled and can be reloaded using a number of mechanisms, including single shot, manual pump action and self-loading (semi-automatic).

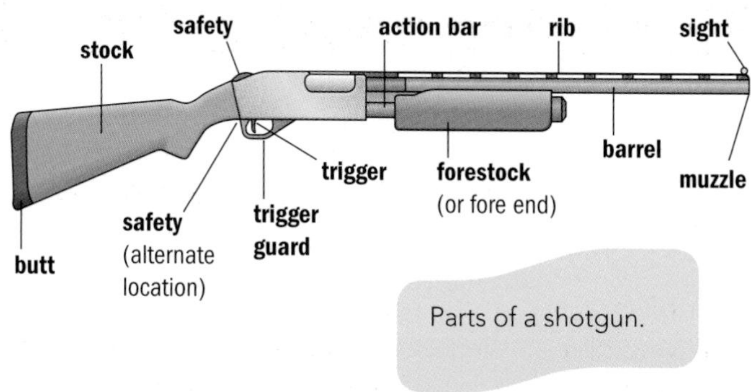

Parts of a shotgun.

Bullets and cartridges

The ammunition loaded into rifle and handgun magazines for firing is in the form of cartridges. The cartridge is composed of a shell containing the bullet, gunpowder and primer.

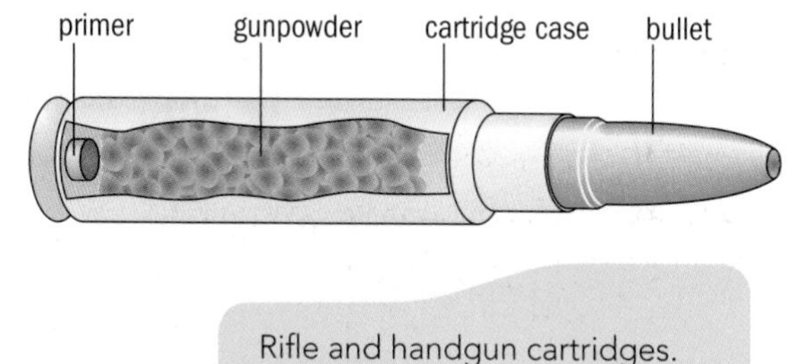

Rifle and handgun cartridges.

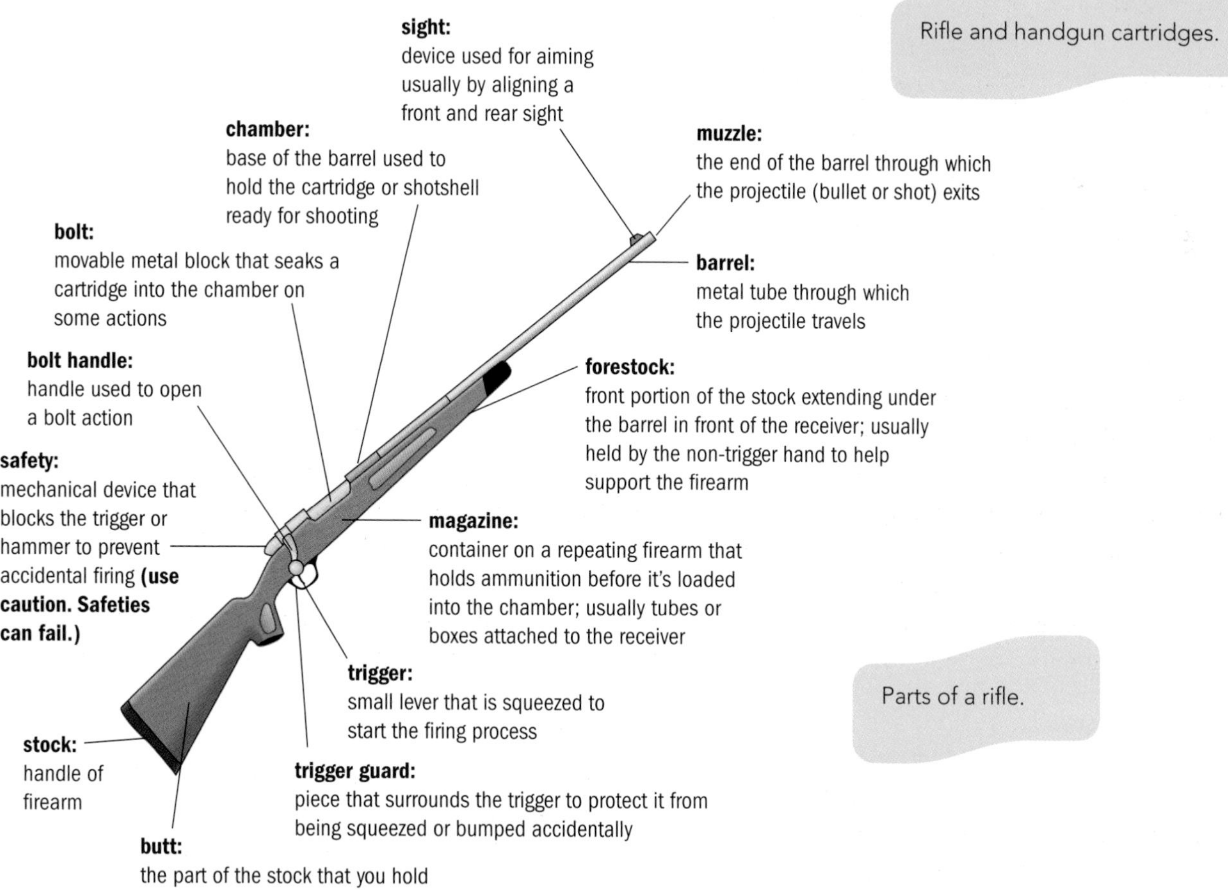

Parts of a rifle.

The cartridge case or shell is a cylindrical case, made to precisely fit the firing chamber of a firearm. It is usually made of brass, steel or aluminium. The base of the case is imprinted with a headstamp. These markings are punched into the bottom of the case during manufacture and identify the manufacturer, the calibre size and for military ammunition the year of manufacture.

Cartridge headstamp.

The bullet is the projectile, which is forced out of the barrel and impacts the target. The bullet itself does not contain any explosives, but damages the target by impacting and penetrating, causing external and internal wounding. Bullets can be made of lead, brass, copper, steel or tungsten, rubber or plastic, and may have a harder 'jacket' covering, usually made of copper, nickel, aluminium or steel. These metals are softer than those used in the manufacture of gun components, and therefore the cartridge case and bullet become scratched by the interior of the gun during use.

In contrast to the handgun and rifle, the shotgun does not use bullets, and instead is designed to discharge pellets, or shot. Shot is usually made from lead or steel, and the size and weight can vary. The shotgun cartridge case may contain one large projectile, called a slug, a few pellets of large shot, known as buckshot, or many tiny pellets, called birdshot. There may be as many as several hundred small spherical shot pellets in a shotgun cartridge. The higher the number of pellets shot on discharge, the larger the spread and therefore likelihood of hitting the target, causing maximum damage to the target. The wad is usually made of plastic or paper and separates the gunpowder from the shot as well as holding the shot together as it moves through the barrel of the gun. It is projected from the barrel of the gun along with the shot when a shotgun is fired.

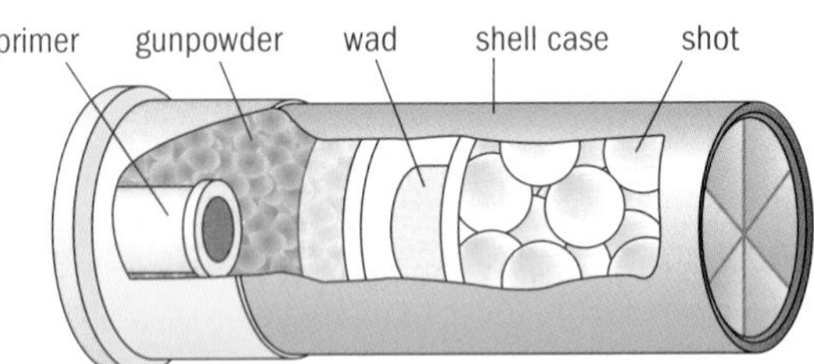

Shotgun cartridge.

Propellants

All firearms work according to the same basic principles. When the trigger of a gun is pulled the weapon's firing pin is released. The firing pin strikes the primer, which produces a spark and ignites the propellant. The propellant material contained within the cartridge case is a volatile and highly explosive chemical that, on ignition by the primer, burns, creating large volumes of gas in the small space between the cartridge case and the bullet. The gas generates pressure behind the bullet which forces the projectile bullet down and out of the barrel of the gun at high velocity, giving the bullet speed, direction and stability in flight. The barrel of a gun is very strong and sealed to hold the expanding gases and prevent them from escaping, to ensure the bullet gains optimum velocity as it is thrust by the gases. The spent cartridge case is pushed backwards with equal force and is ejected manually or automatically from the gun.

Rifling

The insides of both rifle and handgun barrels are cut with a number of twisting spiral grooves, which cause the bullet to spin as it travels through and out of the barrel. These grooves are called rifling and give the bullet stability in flight. Firearms can be manufactured with any number of grooves, usually 4–16, spiralling to the left or right. They vary in shape and twist rate.

The raised areas between the grooves are termed 'lands'. However, the interior of the shotgun barrel is completely smooth and does not have rifling; hence the name 'smoothbore shotgun'.

Bullets fired from rifled firearms are 'engraved' with rifling marks from the grooves, and this can provide information on the gun that fired the bullet. For example, the calibre, make and model of suspect rifled firearms can be determined from the direction of the rifling, the width of the lands and grooves, the depth of the grooves, and the angle and pitch of the rifling on a bullet. Every bullet that is fired from a specific type of weapon will carry the same rifling impressions. Rifling data for many makes and models of firearms are well documented, and a database of the information can be used to help identify a suspect weapon.

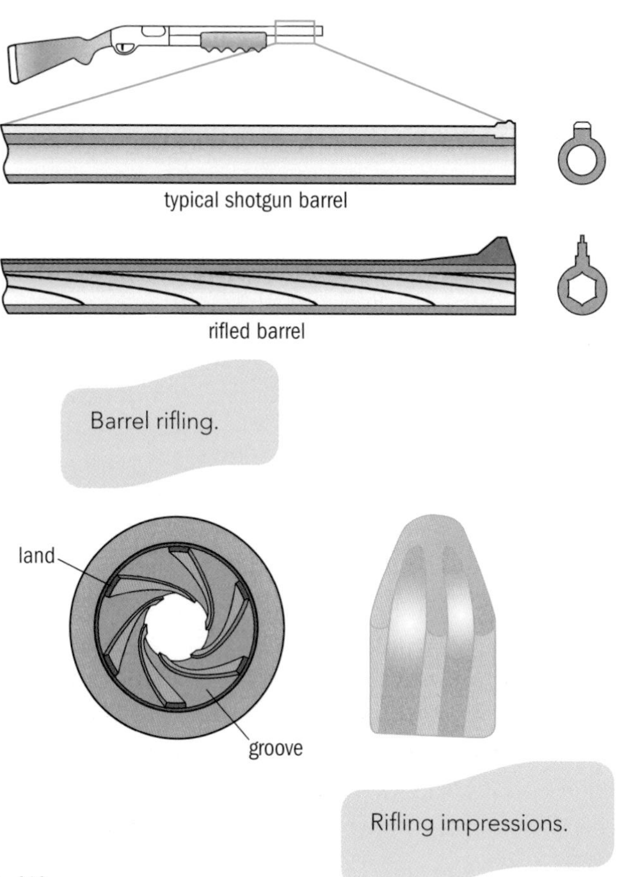

Barrel rifling.

Rifling impressions.

Calibre

The calibre of a weapon is the internal diameter of the gun barrel, or the bore, measured between opposite lands for rifled barrels. Handguns and rifles are manufactured in a range of different calibres. Shotguns aren't classified by calibre but come in different gauges. The gauge of a shotgun is determined by the number of round lead balls of bore diameter that it takes to equal one pound in weight. The calibre is also used to specify the type and size of cartridge and bullet designed to be used with a weapon. Calibre is normally recorded in hundredths of an inch or millimetres. For example, in the US the imperial system of measurement is used, and a .30 Winchester cartridge, which is 30 hundredths of an inch in diameter, is called a 30-calibre bullet. In the UK the metric system is used, and a .30 Winchester would be referred to as a 7.62-mm bullet.

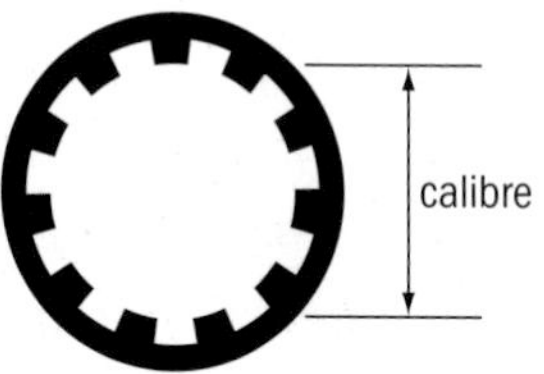

Calibre measurement of a rifled barrel.

Activity 32.3A

Use the cartridge charts on these websites:

pistols: **Pistol Cartridge Lists**

rifles: **Centrefire Metric Rifle Cartridges**

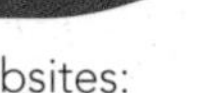

to determine the cartridge case type, bullet diameter and total cartridge length, type of primer and year the cartridge was introduced for the following weapon calibres:

- 9-mm Winchester Magnum pistol
- 44 Remington Army pistol
- 7-mm Mauser rifle
- 15-mm French carbine rifle.

NB All measurements in the cartridge charts are in inches.

Ballistic fingerprinting

As well as rifling marks, the bore of a firearm can leave other unique markings on the bullets it discharges. All firearm bores have slight irregularities and imperfections in the metal caused by production machines when the gun is made. In addition, over time and with regular use, wear and tear will occur to the bore of the firearm, changing the bore by creating new shallow impressions. These slight variations are individual characteristics and create distinct imprints and striations as they scratch the surface of the bullet during projection through the barrel.

The specific pattern of marks on a bullet is called the ballistic fingerprint, and can be used to match a bullet recovered from a victim or crime scene to the particular gun it was fired from. If the weapon has been recovered, it is used to produce a reference sample for comparison to the recovered bullet. However, if the projecting force causes a bullet to deform on impact with a target, the individual striations and marks on their surface are destroyed and the evidential information is lost.

Ballistic fingerprint analysis cannot be carried out on shotgun projectiles, as shotgun cartridges containing shot are covered with a plastic wrapping or sleeve that protects the projectile. This stops it from touching the barrel and prevents the barrel from making impressions and striations on the projectile. In addition, as the barrels of shotguns are smoothbore, they do not leave rifling marks on the projectile.

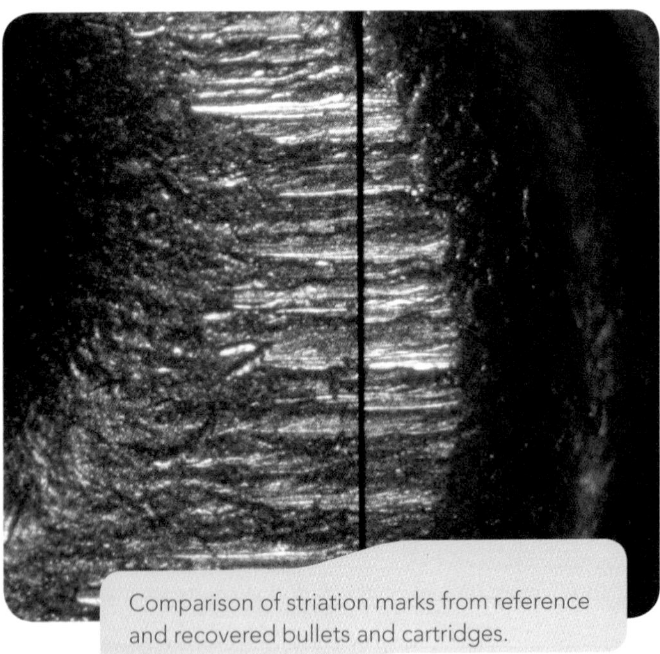

Comparison of striation marks from reference and recovered bullets and cartridges.

Microstamping

Microstamping is a technique that imprints information about a gun on each bullet casing it ejects. A laser is used to etch the firearm's make, model and a unique serial number onto the tip of the gun's firing pin. When the gun is fired, the firing pin strikes the casing of the bullet, imprinting the information on discharged cartridge cases. The microstamp is too small to read with the naked eye, and a microscope is used to identify the details which can be cross-referenced with a database of registered weapons. This allows forensic scientists to identify and trace a firearm from stamped bullet casings recovered from a crime scene. However, if a criminal picks up his or her shell casings, the microstamp will not be recoverable.

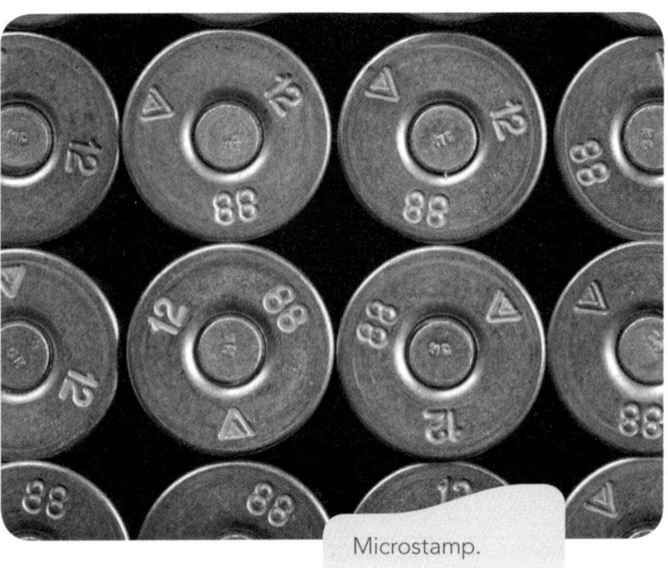

Microstamp.

Discharge residue

When a gun is fired, a number of types of gases, vapours and particulate matter are formed by different parts of the firearm. These deposits are known as firearm discharge residue (FDR), or gunshot residue. The particles can be embedded in or deposited on the surface of a target, and when detected, identified and quantified by a forensic scientist they provide valuable and significant evidence.

FDR released from a pistol when fired.

Any component of the cartridge or the firearm can contribute to the formation of the residues, which are a complex mixture of organic and inorganic components; however the main sources are the bullet, primer and propellant. FDR is normally a combination of gunpowder residues and lead residues from the primer and bullet. Some modern ammunition has been produced that contains virtually lead-free compounds to reduce the toxicity levels. The amount of residue emitted from a gun can vary slightly from shot to shot.

When the trigger is pulled and the firing pin strikes the cartridge primer, the primer ignites producing gases. Primers commonly contain lead styphnate, barium nitrate and antimony sulphide compounds, which produce a vaporous cloud of lead residue that is expelled from the muzzle of the firearm.

The primer ignites the propellant which burns and changes from a solid to produce more gases. Not all of the propellant will burn, and partially unburnt propellant granules form organic particulate matter called gunpowder residue. This also contains the carbonaceous soot from completely burned gunpowder. Both smokeless and black gunpowder contain nitrate compounds that form nitrite-based residues when they burn. When the nitrite particles are emitted from the muzzle of a firearm, they will strike a nearby target and either be embedded in the target's surface or leave a deposit of nitrite residue.

When the propellant burns to gas, pressure is created in the cartridge and the bullet is projected down the barrel of the gun. The friction of movement produces heat and the hot gases in the chamber melt the base of the bullet which releases metal and lead vapours. This mixes with the lead vapour residues from the primer which condense to particles on cooling. The primer lead vapours also coat the gunpowder particles and form lead particulate deposits. As the large bullet travels down the bore, the hard metal of the bore rifling shaves and strips metal from the sides of the bullet, forming particulate matter. This residue of minute lead particles has more mass than vaporous lead residue and travels greater distances from the weapon.

A number of analytical techniques can be used in the forensic laboratory to detect and identify FDR. Colour change tests, similar to presumptive tests, have been developed as a simple, reliable, quick and inexpensive method for indicating the presence of FDR. A number of chemical tests can be used that react with different components of gunshot residue.

- The **modified Griess test** detects the presence of nitrite residues, the by-product of the combustion of smokeless gunpowder, which are expelled from the muzzle of a firearm on discharge.
- The **dermal nitrate test** detects the presence of unburned gunpowder and nitrate residues on the hands of a suspect. The hands are covered in hot paraffin or wax, which is removed once it has hardened. Residues are transferred from the hands to the paraffin, which is tested with the chemical diphenylamine; this reacts with nitrates and turns blue.
- The **sodium rhodizonate test** is used to determine if lead gunshot residues are present on a target surface. The surface is sprayed with sodium rhodizonate solution, which reacts with any lead present on the surface and changes in colour to bright pink. This is followed by hydrochloric acid solution, which turns the colour to blue and confirms the presence of lead.

A more conclusive technique that has been developed is the **particle analysis method**, which can confirm the presence and identity of FDR and distinguish between environmental sources. This technique utilises a scanning electron microscope linked to an X-ray analyser, which enables the forensic scientist to examine the unique morphology (shape and size) of individual residue particles and analyse their elemental composition. This technique has some disadvantages: the equipment is very expensive, and the analysis is time-consuming and can only be carried out by trained staff.

Wound patterns

The wound patterns created on the surface of the target of impact by a projecting bullet have very specific characteristics. Gunshot wound patterns can be analysed to help establish the circumstances when the firearm was fired, for example the distance of the gun from the target victim, the type of firearm, the location of the shooter and the track of the projectile.

When a bullet strikes a body with high velocity, the skin is first pushed in and then perforated while it is in a stretched state. Entrance wounds usually have a rim of abrasion, an 'abrasion collar', surrounding the wound, caused by the projectile dragging the surrounding skin surface into the wound slightly and abrading it as it passes through.

A bullet may come to rest inside the body of the victim, for example if it hits bone, or it may pass completely through and exit the other side of the body. The bullet leaves the body via an exit wound, where the projectile penetrates through the skin and pushes it outward.

The degree of damage caused by a bullet depends on the energy of the projectile at the time of impact. The greater the energy of the bullet at the moment of impact, the greater the tissue destruction caused and likelihood of death. The impact energy of a projectile is determined by the product of its mass or weight multiplied by the square of its velocity.

The distance from which a shot was fired can be estimated by examining the wound pattern. As the firing distance (the length between the muzzle of the gun and the target surface) increases, the characteristic features change. Identifying these wound features allows the forensic scientist to establish an approximate firing distance.

Wounds created by shotguns are different to those from other firearms, as shotguns fire multiple pellets rather than a single projectile, creating a pattern of entry holes. When a shotgun is fired, the shot exits the barrel and begins to spread out into a pattern that increases in diameter the further the shot moves away from the barrel.

Gunshot wounds are associated with a very specific type of blood-stain pattern. When a bullet impacts a body, blood is projected from the entry wound created. This 'back spatter' leaves the body and moves backwards towards the gun and shooter, landing on the floor and sometimes on the gun itself, or the hands, arms and clothes of the shooter. The back-spatter blood stains are very fine small spatters, with a low density of staining, and are only seen at close range of fire.

If the bullet passes through and exits the body, creating an exit wound, a second pattern of staining is produced. Blood projecting from an exit wound is called 'forward spatter', and it moves forward from the source of the blood in the same direction as the projectile. Forward spatter also has a fine, atomised distribution, but the staining is more dense than back spatter. It is normally found on nearby surfaces, objects, people, and especially on the wall behind a victim, and can be seen at any range of fire. If there is no exit wound there will be no forward spatter staining present.

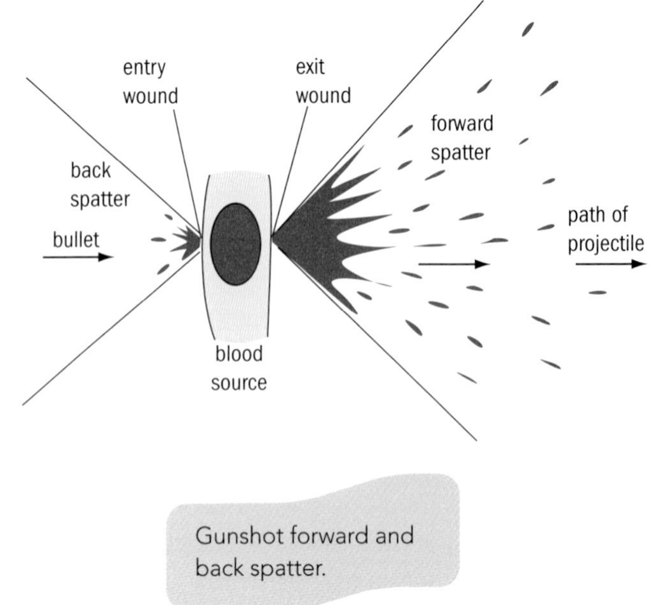

Gunshot forward and back spatter.

Trajectory

The trajectory of a moving object is the path the object follows when moving through space. When a bullet is fired from a gun, it is only briefly powered by the pressures of the gases forming behind it. After this, its flight is governed by the laws of classical mechanics and follows a typical trajectory pattern. External ballistics is the application of the science of mechanical physics to study the behaviour of a bullet during its flight, after it exits the barrel and before it hits the target. Knowledge of a bullet's trajectory allows the forensic scientist to reconstruct the crime and establish where the shooter was located when the gun was discharged.

When in flight, the main forces acting on a bullet are gravity and air resistance. As soon as the bullet leaves the barrel of the gun, gravity imparts a downward acceleration on the bullet and it immediately begins to drop. In addition, air resistance progressively slows the flight of the bullet. If the barrel of a gun is horizontal to the ground when the gun is fired the bullet never rises above the barrel, and gravity causes it to immediately descend. Therefore, to fire an accurate 'horizontal shot' that impacts the target in a straight line from the eye, the barrel of the gun is usually tilted upwards slightly, so that when the bullet exits the barrel it rises upwards slightly in an arc shape and then begins to descend. This is called a parabolic trajectory.

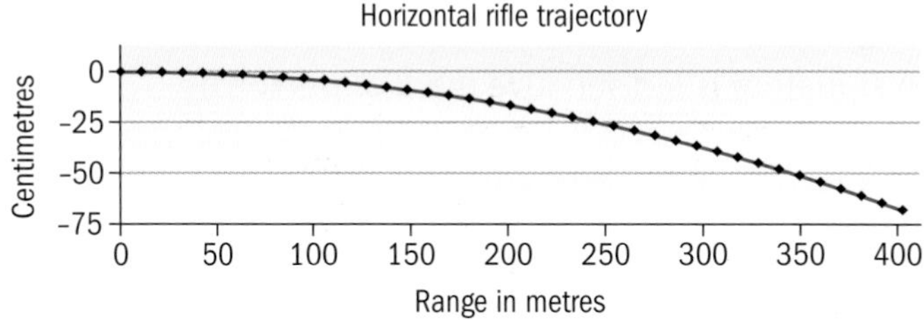

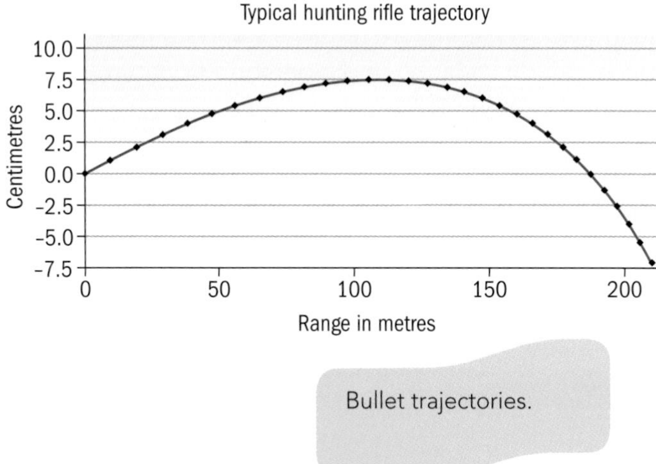

Bullet trajectories.

Document examination

Documents are examined in a forensic laboratory to determine their origin and history – to find out how a document was made, where it came from and what happened to it since it came into existence. Criminals sometimes attempt to forge important documents such as wills, cheques and deeds, either by altering the document or creating a false document. Forensic examination of questioned documents can identify forgeries, identify the author, date documents, and detect and interpret deletions, alterations and additions to a document. In legal terms, a 'document' can be anything with writing on it, and a 'questioned document' contains any signature, handwriting, typewriting, or other mark or symbol whose source or authenticity is in dispute or in doubt. A large range of types of documents can be analysed as evidence for many different types of criminal investigation. For example, forged cheques, passports or loan applications may be analysed in a fraud case, a ransom note may be analysed in a kidnapping investigation, and graffiti may be analysed in a vandalism case.

The forensic document examiner uses a number of techniques to investigate various properties and characteristics of a questioned document. Their role is to:

- identify handwriting and signatures
- determine whether a document is a forgery
- identify typewriters, photocopiers and laser or ink-jet printers used to produce the document
- detect alterations, additions, deletions or substitutions to a document
- decipher alterations and erasures
- identify and decipher indented writing and other impressions on a document
- identify and compare inks, and identify the type of writing instrument
- identify and compare the type of paper the document is made of.

Handwriting analysis

Most work carried out by forensic document examiners is connected with the analysis of handwriting. Handwriting identification is based on the principle that handwriting is unique to an individual. Handwriting has individual features and characteristics that distinguish one person's writing from that of another. Handwriting analysis involves examination of the construction, proportion and shape of writing on a questioned document, and comparison to known pieces of writing. Handwriting is very distinctive to individuals, and even disguised handwriting will exhibit some of the person's individual characteristics.

There are three different types of handwriting: block capitals which are upper-case un-joined letters, cursive writing which is lower-case joined-up letters, and script which is lower-case un-joined letters. Normal handwriting of most individuals contains a mixture of cursive and script. When carrying out a handwriting examination, the forensic document examiner compares the questioned document to a reference sample of an individual's handwriting. The type of handwriting must be the same in both samples; for example, cursive writing on a questioned document can only be compared with cursive writing on a reference sample. The reference sample can be a piece of writing by the individual from before the beginning of an investigation or can be dictated text written at the time of the investigation. The document examiner can then examine and compare the two pieces of handwriting in order to identify whether they were written by the same individual and to distinguish forgeries. They closely analyse the individual's handwriting to highlight the particular ways in which they write, and look for similarities and dissimilarities between the questioned and reference samples.

(a) THE CASE OF JOHN WHITE

(b) The case of John White

(c) The case of John White

Block capital, cursive and script handwriting.

Signatures

A signature is a distinctive mark of personal identification made by an individual. A signature is usually used to sign a written document by hand to signify knowledge, approval and acceptance of the document. Because signatures are used so often, people usually write them automatically, and although an individual's signature remains standard there is much natural variation and a person's signature is never identical on two different occasions.

Characteristics for comparisons

The guidelines in the following table are used when comparing questioned and known reference handwriting samples to determine whether the handwriting or signature on a questioned document is forged.

Characteristic	Points of comparison
line quality	Do the letters flow? Are they irregular, laborious and shaky, with a lack of rhythm?
size/proportion	Is the ratio of letter height and width the same? Are the letters larger, wider, higher, inconsistent? Is the spacing of words and letters consistent? Is there different spacing after capital letters?
pen lifts	Are capital letters and lower case letters connected? Are there frequent lifts off the paper? Are there any retracing strokes?
angle/slant	Are the letters slanted to the left or the right? Is the angle slight or very pronounced? Is there a greater than 5-degree change?
pen pressure	Is the pressure used heavier than usual?
loop formation	Are loops circular or more teardrop or egg-shaped? Are they made in a clockwise or anti-clockwise direction? Is there wider spacing between loops? Are there more squared, shorter or broken loops?
stroke formation	Are the Ms and Ws wider? Are there more squared or wedge-shaped stokes? Are there any unusual letter formations?
alignment	Is there a change in baseline habits? Do the letters stay even on the baseline? Are there more downward slants from the baseline?
diacritics	How are the letters t crossed and i dotted? Are there position placement changes? Are some t crossings and i dots heavier than others?

Activity 32.3B

Ask two different people to write the following sentences in their normal handwriting:

'I saw the suspect leaving the scene of the crime. He was wearing a baseball cap, a red jumper and black trousers.'

Compare the handwriting samples using the table above. Are there obvious differences? Is it easy to tell them apart?

Now ask the two individuals to rewrite the sentences by copying each other's handwriting. How easy is it for them to forge each other's handwriting? Can you tell the difference between the original and forged pieces of writing? Use the table above to identify how they differ.

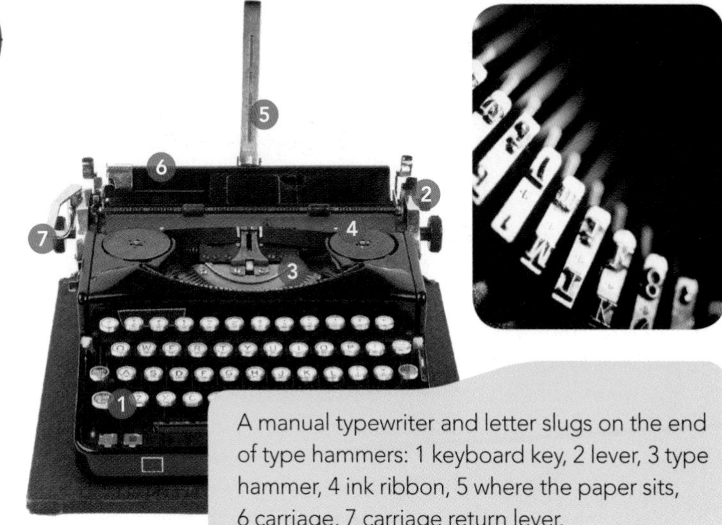

A manual typewriter and letter slugs on the end of type hammers: 1 keyboard key, 2 lever, 3 type hammer, 4 ink ribbon, 5 where the paper sits, 6 carriage, 7 carriage return lever.

Printed documents

The way documents are produced in the office and at home has changed dramatically over the last 150 years. The first types of machines that were used for written communication were manual typewriters. These were first developed in the nineteenth century, and they became so commercially successful that almost every office owned a typewriter before computers became popular in the 1980s. When computer-based word-processor systems were introduced, the use of typewriters declined and printers used to print word-processed documents became commonplace. Modern technology has also led to the development of photocopiers which can mass-produce copies of documents. Forensic questioned-document examiners are often required to examine documents produced by typewriters, printers and photocopiers, to determine the make and model of the machine that produced a document and to link a document to a specific individual machine.

Typewriters

A manual typewriter works using a complicated arrangement of levers and springs and is completely mechanical. When a key is pressed on the keyboard, a lever attached to the key swings a second lever, a type hammer, up towards the paper held in the carriage at the back of the typewriter. The type hammer has a metal block with a raised letter at the end, called a slug. This hits a piece of ink ribbon between the type slug and the paper, creating a printed impression of the letter on the paper.

Although typewriters are not as popular as they used to be, they are still used and may be linked to a questioned document in a criminal investigation. With time and use, typewriters become worn and damaged so that the printing they produce deteriorates. If the slugs become damaged or accumulate dust or dirt then the letters may not print accurately or correctly on the page. The letters can appear out of alignment if the type hammer or levers are damaged, producing type that is too much to the right or left, or too high or low on the line. Also, letters may not be inked uniformly, appearing fainter or heavier on one part of the letter. These defects are unique to an individual typewriter and may allow a forensic document examiner to identify which specific machine produced a particular document, or whether the same machine produced two different documents.

Photocopiers

For most businesses a photocopier has become a standard piece of equipment. It was originally developed in America in the 1940s and has become one of the most important office inventions of the 20th century. It allows many copies of a document to be made very quickly.

The forensic document examiner may be asked about the origin of a photocopy and to identify the photocopying machine used to create the photocopied document. There are two types of mark that may appear on a photocopy that can be used to link a photocopied document to a specific photocopier. The first type is called trash marks; these are dots or spots on a document produced by dust particles or debris on the glass surface of a photocopying machine where the original document is placed for copying. However, these are only temporary markings, because if the glass surface is cleaned the machine will stop creating them. If there are a number of documents with exactly the same trash markings, this indicates that they have been produced by the same photocopying

Case study: Harold Shipman

Harold Shipman was a doctor who became an infamous serial murderer in the UK. He came under suspicion as there was a high death rate amongst his patients. In June 1998, one of Shipman's patients, Kathleen Grundy, was found dead at her home after a visit from Shipman. Grundy's daughter went to the police when she discovered that her mother's will had been changed and Shipman had been left £386000. Kathleen's body was exhumed and re-examined. Traces of the drug diamorphine were found in her system and Shipman was arrested. When the police searched Shipman's home they found an old typewriter which was forensically analysed and compared to Kathleen's will. The forensic scientist proved that Shipman had forged the will using his typewriter and this evidence helped to convict Shipman of murder. Shipman was eventually found guilty of the murder of 15 of his patients; however, it is believed that he may have been responsible for the death of up to 250 patients. Shipman was sentenced to life imprisonment in 1999 and he committed suicide in his cell a year later.

machine. The second type is marks caused by the drum or mechanism of the photocopier. These appear as regularly spaced marks on the photocopied document that correspond to damage present on the rotating drum or within the mechanism of the machine.

Laser printers

Laser printers were developed to print computer-processed documents in the 1970s and rely on the same technology as photocopiers. They produce high-quality documents and it is usually very difficult, if not impossible, to identify which specific printer has produced an individual document. However, if a fault occurs on the printer drum it may produce drum marks in the same way that a photocopying machine does, which could be used to link a document to a printer.

Inkjet printers

Inkjet printers were first produced in the 1980s as a cheaper alternative to laser printers. They contain a large number of nozzles inside a print head which spray small drops of ink from an ink cartridge directly onto the printing paper. Although a forensic scientist may be able to establish that an inkjet printer was used to create a document, as these types of printers tend to produce images with slightly blurred outlines, it is as difficult to link an individual inkjet printer to a document as it is with a laser printer.

Paper

The type of paper a questioned document is made of can be analysed to aid a criminal investigation. For example, paper analysis can determine whether one or more pages have been added to a multi-page document, or to prove whether a document was really produced at a certain time. Paper can be forensically examined to establish a number of characteristics; for example, the size, thickness, colour, brightness, density and finish of paper. These features can be used to identify a specific paper type and to compare paper types between documents.

Activity 32.3C

Collect as many different types of A4-sized paper from a range of different manufacturers as you can.

- Feel the paper between your fingers – does it feel rough or smooth? Are the pages coated?
- Weigh the paper – do the different types weigh the same?
- Hold the paper up to the light – can you see a regular pattern? Is there a watermark present?
- Place the paper on a page containing bold-type writing – can you see the type through the paper?
- Tear a corner off and place the page on a dark background. Look at the torn edge of the paper using the magnifying glass – can you see any of the fibres that make up the paper? If you can see any fibres, are they thick or thin, long or short?
- Place a drop or two of water on the surface of a corner of the paper – does it stay on the surface or is it absorbed into the paper? Tug at the wet corner of the paper with your fingers – is the paper still strong, or did the water weaken it?

Some types of paper, for example, good quality writing paper, art drawing paper and envelopes, contain a watermark – a design or pattern put into paper during its production that can be seen when holding the paper up to the light. Usually, they show the manufacturer's name and a geometric design or image. They are created by making thinner (line or *wire* watermarks) or thicker (shadow watermarks) on the layer of pulp during the manufacturing process when the paper is still wet (hence the name watermark). The purpose of watermarks is to identify the origin and date of manufacture of the paper, or as a security measure to avoid forgery of important documents such as bank notes and passports.

Ink

Ink is an important type of forensic evidence in a range of criminal investigations, including fraud cases, embezzlement, kidnapping, theft and drug dealing. Ink is used on a range of documents, and many of these are submitted to the questioned-documents laboratory for analysis; for example, fake driving licences and passports, ransom notes, threatening letters, suicide notes.

Pen inks are mixtures, which can be separated using chromatography. Two pen-ink samples can be separated and compared to determine whether they contain the same components. For example, the ink from a suspect's pen can be compared with ink on a forged document to establish whether there is a match, indicating the guilt of the suspect. There are three main categories of ink ingredient, although ingredients may vary widely according to the type of printing process and print application:

- pigments – natural or synthetic substances that produce the colour of the ink, blended together in specific proportions to create a desired colour
- vehicle – substance in the ink mixture that carries the pigment and binds it to the printed surface
- modifiers and additives – change the properties of the ink so that it can be used appropriately for different types of print process and application.

Assessment activity 32.3

BTEC

You are a forensic scientist working for the physical sciences department of the Forensic Science Service. A SOCO has submitted a number of items of evidence from a suspected suicide case. The police have asked you to analyse the evidence using physical techniques of analysis. You must establish whether the victim killed themselves or if their death is more suspicious and could have been murder.

Types of evidence collected from the crime scene:

- gun
- suicide note written on white paper
- victim's diary showing samples of handwriting.

1. Make a list and describe very briefly the techniques you would use to analyse the physical evidence recovered from the crime scene. Carry out some of the practical techniques you have listed (make sure you check the safety notes for each technique you use). P4 P5
2. Explain each of the physical techniques of analysis. Present valid conclusions from the analysis of your practical work. M4 M5
3. Discuss the advantages and disadvantages of the physical techniques of analysis. D4 D5

Grading tips

To achieve P4 and P5 you must be able to briefly describe the main points of the physical techniques used to analyse forensic evidence. You must also be able to carry out the techniques, completing the appropriate laboratory documentation and accurately presenting your results.

To achieve M4 and M5 you must be able to explain fully the physical analysis techniques and how they were used to analyse the forensic evidence collected from the crime scene. You should draw conclusions from the results of your physical scientific analysis, and support your conclusions with evidence and reasoning; suggest also how to further analyse the evidence.

To achieve D4 and D5 you must be able to evaluate the scientific techniques of physical analysis and explain why the techniques used were suitable for analysis of the forensic evidence from the crime scene. You must also identify a range of types of evidence that can be analysed using the techniques and how they could be applied to different types of criminal investigation.

PLTS

Independent enquirer, Self-manager and Team worker

Researching forensic applications of analytical techniques, completing work to standard and deadline, and working as a team to analyse forensic evidence will help you develop these respective skills.

32.4 Using biological techniques to analyse evidence

In this section:    P6 M6 D6 P7 M7 D7

Key terms

Antigen – substance present on surface of red blood cells that can stimulate an immune response, e.g. formation of antibody.

Antibody – protein in blood serum, produced by body's immune system, that reacts with antigens on red blood cells.

Biological evidence comes from living organisms – humans, animals and plants. This type of evidence has a cellular structure, and if nuclei are present within the cells they will also contain DNA material. DNA is unique and is used to identify individuals, so is of high evidential value in a criminal investigation.

Blood group analysis

Blood is the most common type of body fluid found at crime scenes and is usually associated with violent crimes, e.g. murder, sexual or violent assault. Blood may also be present at 'volume' crime scenes; for example, at burglaries, muggings and vehicle theft scenes. Forensic serology involves the examination and analysis of blood found at a crime scene or on an item of evidence. There are a number of different laboratory techniques the forensic scientist can use.

- Presumptive colour tests may be utilised to get a preliminary indication that the sample is blood.
- Blood typing methods can then be used to identify the blood type.
- Blood may yield DNA evidence which can provide information on the identity of the blood source.

Blood is a circulating tissue composed of fluid plasma and different cells. It moves around the body in blood vessels and is circulated by the heart. The most common type of blood cell is the red blood cell (RBC), or erythrocyte; the role of the RBC is to deliver oxygen to body tissues via the haemoglobin protein molecules inside red cells. Mammalian RBCs are flat, circular shaped and depressed in the centre. They lack a nucleus and organelles and are approximately 6–8 μm in diameter.

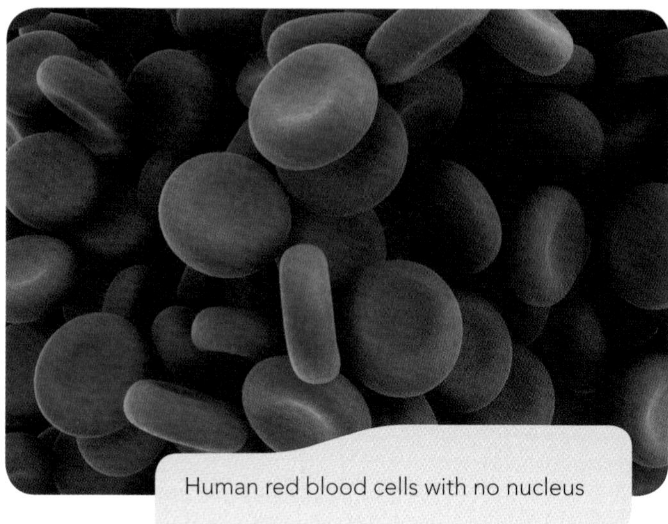

Human red blood cells with no nucleus

The RBC membrane is composed of lipids (~44%), proteins (~49%) and carbohydrates (~7%).

Activity 32.4A

Using books and the Internet, find a labelled diagram of the red blood cell membrane.

You need to find out and make notes on the function of each structure below:

- lipid
- protein
- carbohydrate
  - glycoprotein
  - glycolipid.

Red blood cell antigens

Human and animal red blood cells have **antigens** – extracellular glycoprotein structures, located within and sticking out from the RBC membrane. These surface antigens give a specific blood-type characteristic to the cells, which determines a person's blood group. Variations in the antigens on RBC membranes creates different blood group types. There are only a small number of different blood group types, and the population can be divided into groups of people who share the same blood types. The identification of blood group cannot identify an individual, but can help eliminate those who possess other blood group types from the inquiry. For example, if a blood sample is recovered from a crime scene and laboratory analysis establishes the type is group AB, it can only be linked to any person who has type AB, which is a large group of people. However, if the suspect is blood type O, they can be eliminated from the investigation as they could not be the source of the AB type blood.

The original blood typing system, the ABO system, was identified by the Austrian biologist Karl Landsteiner in 1901. He categorised human blood into four groups, A, B, AB and O, based on the presence or absence of either or both antigen 'A' and antigen 'B' on the surface of RBCs (see table, right).

- An individual with only A antigens present on the surface of their RBCs is categorised as blood type A.
- Those who have only B antigens present are categorised as type B.
- If A and B are present the blood type is AB.
- If neither antigen is present the blood type is group O.

Since the determination of the ABO blood typing system, 15 different antigen systems based on blood groups have been identified. These are inherited independently of each other, and an individual may have any combination of blood antigens.

Another blood typing system is based on the rhesus (Rh) antigen, discovered by Landsteiner and Alexander Wiener in the 1940s. This antigen was originally identified in rhesus monkeys; subsequent serological testing demonstrated that approximately 85% of the human population have the Rh antigen on their RBC surfaces. Individuals with the Rh antigen are termed rhesus positive, and individuals who lack the antigen are rhesus negative (see table below).

Forensic blood typing analysis determines which antigens are present on the RBC surface and therefore identifies the blood group of an evidential blood sample.

Blood group	Antigens	Reaction with anti-A antibody	Reaction with anti-B antibody
A	A present B absent	Yes	No
B	B present A absent	No	Yes
O	A & B absent	No	No
AB	A & B present	Yes	Yes

Blood group	Rh antigen	Reaction with anti-Rh antibody
Positive	Present	Yes
Negative	Absent	No

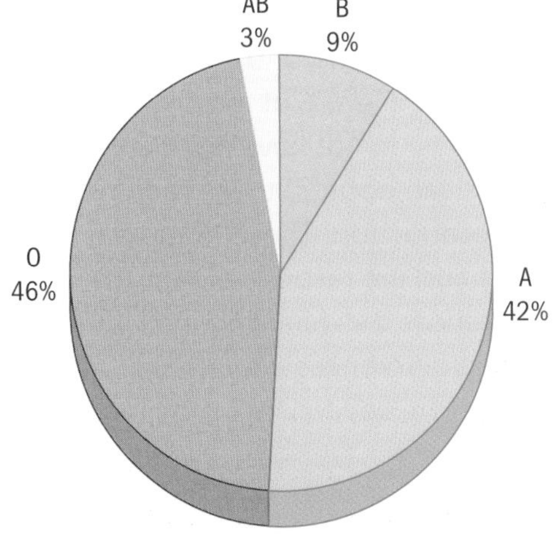

Proportion of ABO blood groups in the population.

blood type (genotype)	type A (AA, AO)	type B (BB, BO)	type AB (AB)	type O (OO)
red blood cell surface proteins (phenotype)	A agglutinogens only	B agglutinogens only	A and B agglutinogens	no agglutinogens
plasma antibodies (phenotype)	B agglutinin only	A agglutinin only	None	A and B agglutinin

The ABO blood type system.

Activity 32.4B

A violent fight takes place between two gangs of youths in the street and the police are called. On the arrival of the police, some of the offenders flee the scene of the fight. The police barricade the street and SOCOs attend to examine the crime scene for forensic evidence. They identify a number of blood stains which they recover for forensic analysis. The following day, police arrest four suspects and their blood is taken for reference. In the laboratory, blood group analysis is carried out on the crime scene and suspects' blood samples. Examine the results of the analysis below; establish each blood group and determine whether any of the suspects were involved in the fight.

	Reaction with anti-A antibody	Reaction with anti-B antibody	Reaction with anti-Rh antibody	Blood group type?
Crime Scene: Blood sample 1	–	–	–	
Crime Scene: Blood sample 2	+	+	–	
Suspect 1: Blood sample	–	+	+	
Suspect 2: Blood sample	–	–	+	
Suspect 3: Blood sample	+	+	–	
Suspect 4: Blood sample	+	–	+	

Key: + = positive result, precipitation formed; – = negative result, no precipitation.

The serological technique consists of testing **antibodies** and antigens present in the blood. Antibodies are produced in the blood as an immune response to antigens. The tables above show the different blood group antigen–antibody reactions.

Genetics – DNA sequencing and genetic fingerprints

Deoxyribonucleic acid (DNA) is the genetic material that carries hereditary information from one generation to the next. It is a large molecule tightly packaged into chromosomes in the nucleus of a human cell. DNA provides the genetic information for each individual and determines their physical characteristics, e.g. eye colour, hair type and height. Every cell containing DNA within an individual has the same genetic information and every individual's DNA is unique (apart from identical twins who originate from the same fertilised egg). Therefore DNA can be used to identify individuals.

The genetic information is stored in DNA in a genetic code – genes are made up of a sequence of nucleotide bases that determine the amino acid sequence. Different sequences of amino acids code for different proteins in the human body. These protein-coding regions, or exons, are very similar in different individuals. However, not all of the DNA code is used to produce proteins – most of the DNA sequence is in fact 'junk' DNA. Although human DNA is 3.3 billion base pairs in size, only a fraction of this, ~2–3%, is used to produce specific proteins, and non-coding intron junk DNA makes up 97–98% of the DNA code. These non-coding regions do not have any genetic function and mutations can occur in these regions that do not affect protein production. These mutations alter the genetic code and are responsible for the small amount of differences between two individuals' DNA sequences. These differences are termed 'polymorphism' and are the key to DNA profiling.

DNA profiling

The restriction fragment length polymorphism (RFLP) technique was developed in 1984 and was the first type of DNA profiling method used to analyse evidence samples in forensic laboratories. It involves the use of

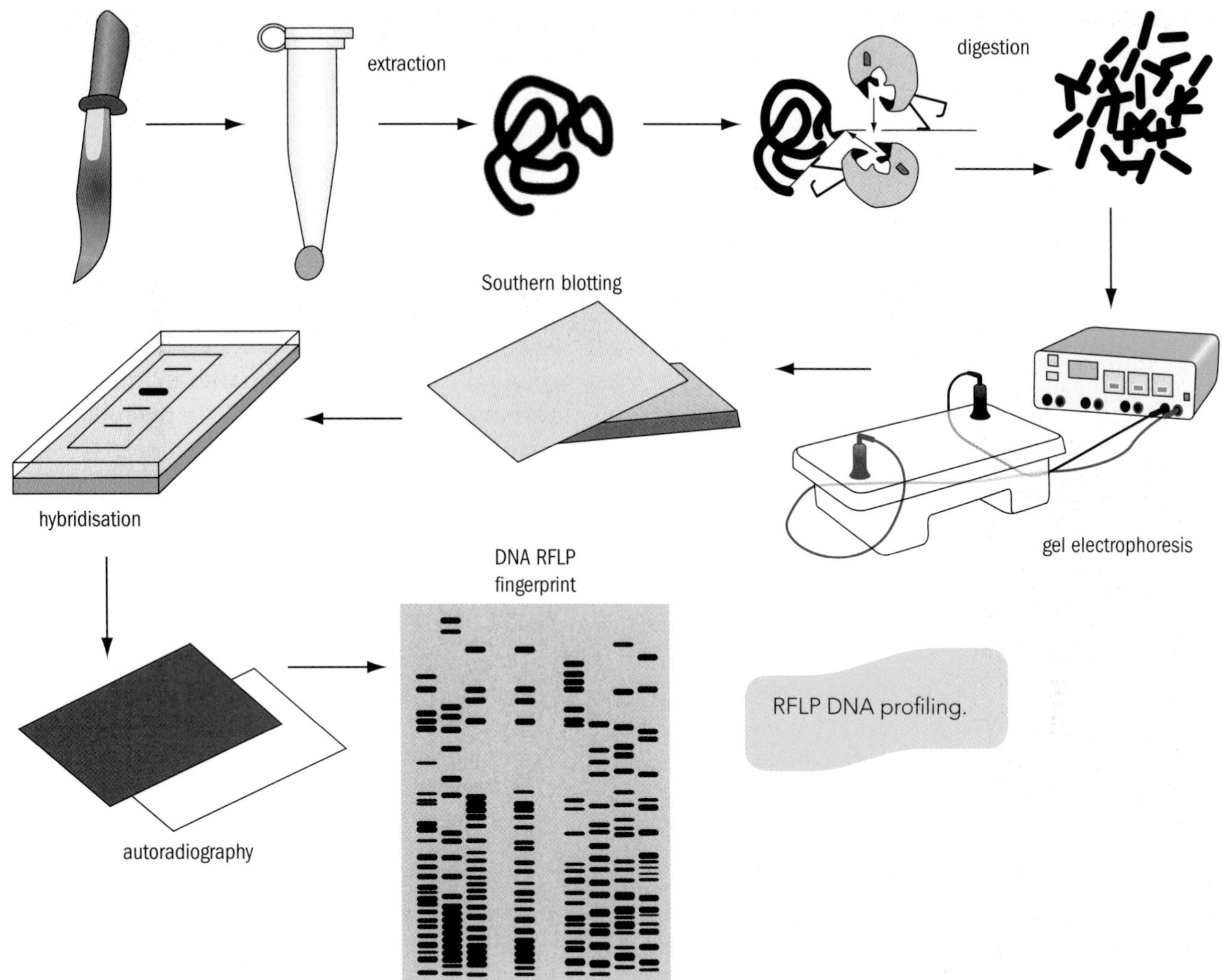

RFLP DNA profiling.

restriction enzymes that digest and cut the extracted DNA to produce fragments of repeating DNA sequence. The small DNA fragments can then be separated using gel electrophoresis and their size analysed.

The most common DNA samples analysed in the forensic laboratory are blood, semen, epithelial skin cells and muscle tissue. These all contain DNA except for red blood cells which do not have a nucleus; however, white blood cells do contain DNA and can be profiled. The first stage of DNA profiling is the extraction and isolation of the DNA from the nucleus of the cell.

Stage 1: DNA extraction

To extract DNA, the cell membranes, structural materials, and proteins and enzymes are destroyed by an alkaline solution at high temperature (50–100°C) so the DNA can be released and safely separated from the rest of the cell components.

Stage 2: DNA digestion

DNA-cutting restriction enzymes are added to the extracted DNA, which digest the DNA into hundreds of small fragments. The enzymes are extremely specific and only cut across the DNA molecule in certain locations when they recognise a particular sequence of nucleotides, called a restriction site.

Stage 3: DNA separation

Once the DNA has been digested into many small fragments, the DNA pieces must be separated so they can be analysed and measured. This is done using the gel electrophoresis technique, a method that separates large molecules, e.g. nucleic acids or proteins, on the basis of the size and electric charge of the molecule. The difference in rate of movement causes the different sized pieces of DNA to move different distances and therefore separate on the gel. DNA fragments of the same size move to the same position on the gel and form a pattern of DNA bands.

Stage 4: DNA visualisation

It is not possible to see the DNA fragments on the electrophoresis gel and an enhancement technique must therefore be used to visualise the DNA band pattern. First, the Southern blotting technique is used, where a nitrocellulose membrane is laid on top of the gel and the DNA bands transfer to its surface. The membrane is then soaked in a solution containing radioactively labelled pieces of DNA, called probes, which recognise and bind to specific sequences of DNA on the membrane, a process known as hybridisation. The excess probe is washed away and the membrane is exposed to an X-ray film, a method called autoradiography. The radioactively labelled DNA fragments leave an image on the film that, when developed, shows the location of each band of DNA fragments; this pattern is known as the DNA fingerprint.

Stage 5: DNA analysis

To measure the DNA fragments, a standard marker is run alongside the DNA samples on the electrophoresis gel, which appears next to the DNA fingerprint on the film. This marker is called a DNA ladder and is composed of a number of pieces of DNA of known length. By comparing the distances the unknown pieces of DNA move in relation to the DNA ladder fragments of known size, it is possible to establish the size of the unknown fragments. Usually in the forensic laboratory, unknown crime scene DNA samples are separated next to suspect reference DNA samples, which allows direct comparison of the DNA patterns. If the two DNA fingerprints have the exact same DNA band pattern with DNA fragments of the same size, then the suspect is the source of the crime scene DNA sample. If the pattern does not match exactly, then the DNA did not originate from the suspect.

A new technique of DNA profiling has since been developed based on the same principles as RFLP profiling, except it uses very short fragments of DNA sequence, called short tandem repeats (STR). This method involves the use of the polymerase chain reaction (PCR) technique instead of enzyme digestion, and allows the analysis of very small quantities of DNA, e.g. from a single cell.

Fingerprints

Fingerprint patterns

Fingerprints are the most common type of evidence found at crime scenes and are associated with a wide range of crimes. Fingerprints are biological characteristics that are unique to every individual and are formed during fetal development. The primary goal of a criminal investigation is identification, so fingerprints have high evidential value as they can uniquely associate a person with a crime scene or evidentiary item. Fingerprints are one of the last features to be lost from the skin during decomposition and in certain circumstances can be used to identify dead bodies months or even years after death.

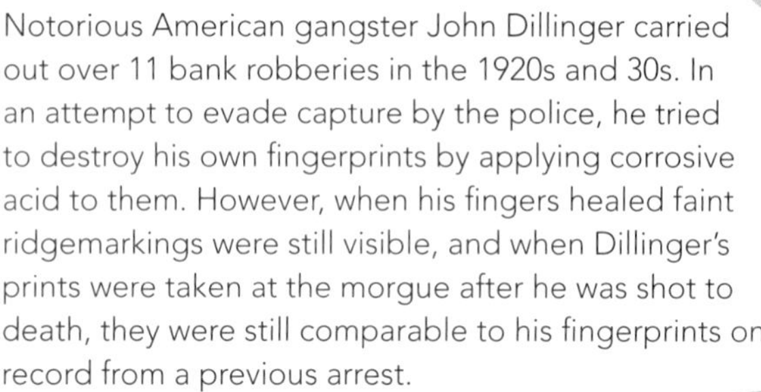

Did you know?

Notorious American gangster John Dillinger carried out over 11 bank robberies in the 1920s and 30s. In an attempt to evade capture by the police, he tried to destroy his own fingerprints by applying corrosive acid to them. However, when his fingers healed faint ridgemarkings were still visible, and when Dillinger's prints were taken at the morgue after he was shot to death, they were still comparable to his fingerprints on record from a previous arrest.

Fingerprinting is a system of identification based on skin ridge patterns on fingertips. Friction skin ridges are a series of elevated lines of skin of different sizes and formations called ridges (or hills), the long deep grooves in between are called furrows (or valleys). Friction ridge skin is present on the surface of palms, palm side of fingers and thumbs, and soles of feet and is designed by nature to provide us with a firm grip and to resist slippage.

The flow of the skin ridge surface on a fingertip determines the fingerprint pattern. According to Henry's System of Classification, there are four types of fingerprint patterns. Each pattern type contains sub-groups that possess the same basic characteristics and similar differences among patterns.

- **Loop** – this is the most common type of pattern where the fingerprint ridges flow in from one side of the finger, curve round and exit on the same side of the finger. Loops are categorised according to the slope of the ridges; the loop can slope down to the left or the right.

- **Arch** – these are the simplest and rarest types of prints, where the ridges flow in from one side of the finger, rise in a wave formation and exit on the other side of the finger.
- **Whorl** – These are the most complex type of prints, where the fingerprint ridges curve round in a complete circuit.
- **Composite** – any combination of two or more fingerprint patterns or unusual patterns that do not fit in any other group; sometimes categorised as accidental whorls.

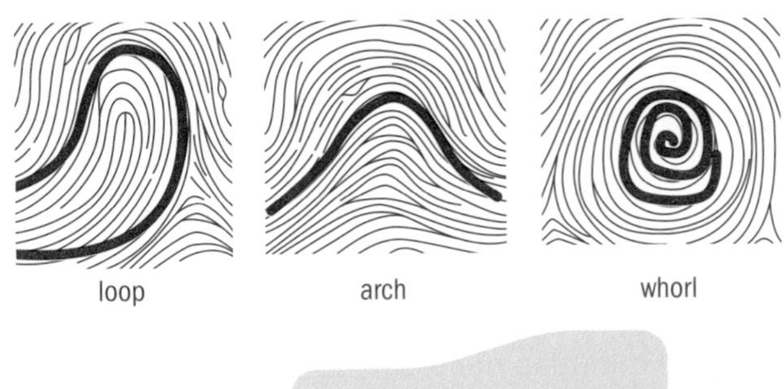

Fingerprint patterns.

Ridge counting

An easy way to distinguish between two different loop patterns is by counting the number of fingerprint ridges present within the pattern. To do this you need to be able to identify three key features of a loop pattern:

- core
- type lines
- delta.

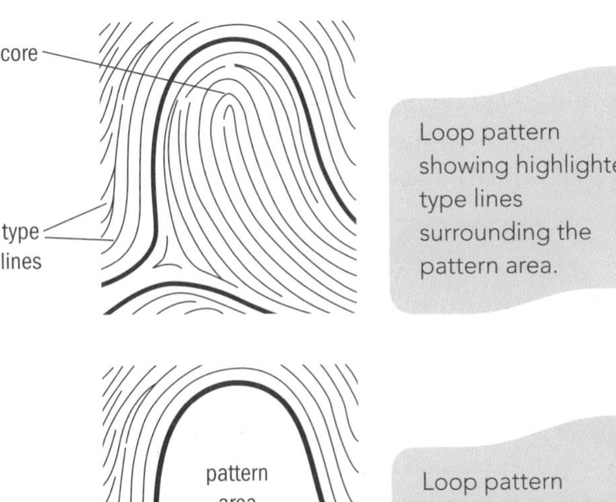

Loop pattern showing highlighted type lines surrounding the pattern area.

Loop pattern showing location of delta at divergence of type lines.

Once these features have been identified on a loop pattern it is possible to calculate the ridge count. To do this, an imaginary line must be drawn between the delta and the core of the loop pattern. The ridge count can then be established by counting from the delta the number of ridges that cross the line. Do not include the core or delta in the count.

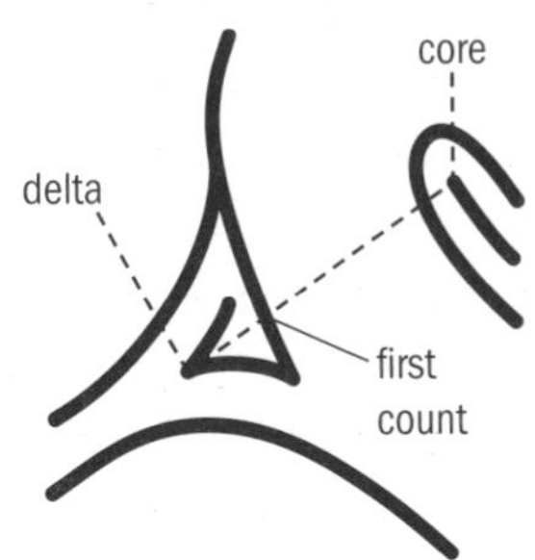

Diagram of a simplified loop pattern showing the type lines, delta, core and imaginary line used to determine the ridge count. Two fingerprint ridges cross the line excluding the delta and core; therefore the ridge count is 2.

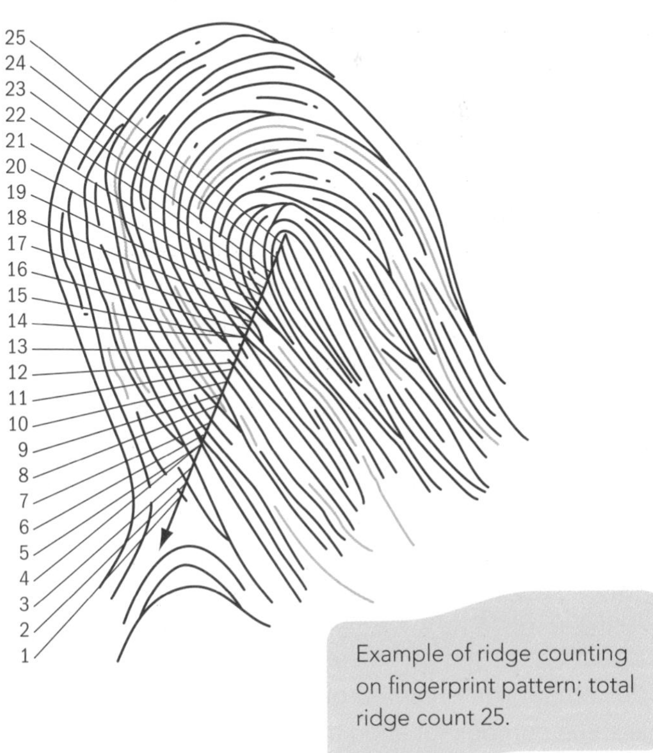

Example of ridge counting on fingerprint pattern; total ridge count 25.

All of these fingerprint features can be used to help the forensic scientist classify an individual's fingerprint pattern. For example, if only one delta is identified within a pattern the print can be categorised as a loop, if none are present the print is an arch and if two are identified the pattern is a whorl. Characteristic features of each fingerprint pattern and their proportions in the population are shown in the table below. As there are only four different types of fingerprint pattern, every individual's fingerprint patterns fall within one of these groups, and large numbers of the population have the same pattern. This is a class characteristic and by examining the pattern alone will lead to many comparison matches. Loops are the most common pattern; therefore many people could potentially match a loop fingerprint mark found at a crime scene. In contrast, arch patterns are much less common and only 5% of the population could potentially be the source of the fingerprint mark. This makes finding an arch pattern of higher evidential value with it being easier to identify the individual.

Fingerprint pattern	Proportion in population (%)	Type-lines	Delta	Core
Loop	60–65	Yes	One	Yes
Arch	5	No	None	No
Whorl	30–35	Yes	Two	Yes
Composite	Rare	Yes	At least two	Yes

Fingerprint minutiae

Although the fingerprint ridge pattern is a class characteristic, the number, location, flow and formation of individual ridges within the pattern makes a fingerprint individual. Fingerprint ridges have a number of different types of characteristic, or minutiae, that can be used to identify and compare fingerprints. For example, where each ridge starts and stops is termed a ridge ending, and a bifurcation is where one ridge splits into two. Some typical ridge characteristics are shown below. When a fingerprint mark is recovered from the scene and a suspect's recorded prints are available, a side-by-side comparison is carried out to determine whether the same number, location and type of minutiae are present in each mark. Until recently, a forensic scientist had to find 16 matching minutiae in two fingerprints to identify them as from the same source. However, this led to many fingerprints being declared as not matching even when the fingerprint examiner was convinced it was a match. Today, there is no minimum number of minutiae necessary to identify a match, and the decision is left to the discretion of the trained and experienced forensic scientist.

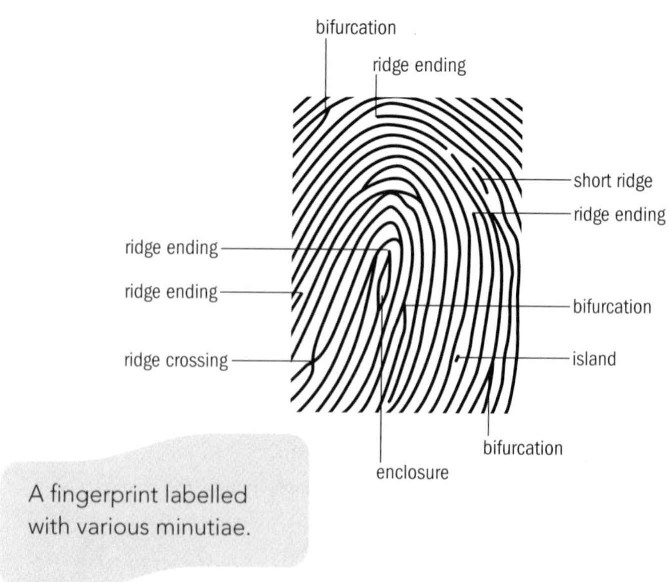

A fingerprint labelled with various minutiae.

Activity 32.4C

A car theft is reported to the police and later that day a car matching the victim's description of the stolen vehicle is located abandoned in a car park. A SOCO examines the car and recovers a partial fingerprint from the interior. The fingerprint is sent to the forensic laboratory for analysis where it is scanned into the Fingerprint Database. This database contains millions of records of fingerprints from criminals convicted of a crime. A computer compares the scanned image to the stored fingerprint records and matches a previous offender's prints to the crime scene print. You must confirm the match by identifying the fingerprint patterns, and the location and type of 10 minutiae present in both of the fingerprint patterns below.

Unknown crime scene fingerprint and suspect's reference fingerprint.

Bones and skeleton physiology

Forensic anthropology is the study of bones or human remains. Forensic anthropologists assist in searching, locating, excavating and recovering suspicious buried or surface remains at the scene of a crime, and are often called to scenes of mass disasters and war graves to identify bone fragments and teeth and identify victims. Forensic anthropologists specialise in the analysis and identification of human skeletal remains for legal and humanitarian investigations. In many murder, suicide, accidental and mass disaster cases, when a body is recovered a while after death has occurred, the tissues of the body have decomposed so traditional means of identification, for example visual identification of the victim or a forensic post mortem of the body, are not possible. Bones often survive the process of decay by many years and provide an important form of identification after death. The forensic anthropologist is a bone specialist who applies standard scientific techniques developed in physical anthropology to identify decomposed, mummified, burned or dismembered human remains. They examine skeletons to identify victims; by establishing the biological profile of the skeleton, they can identify:

- gender
- age at death
- height
- racial ancestry.

The forensic anthropologist may be able to determine the cause and manner of death, by identifying fracture patterns and trauma to the bones, and may be able to estimate the time of death.

The human skeletal system is the rigid frame that supports the body and soft organs and is composed of 206 separate named bones in the adult skeleton. Bones perform several other important functions, including protection of the organs, mineral storage, the formation of blood cells, and allowing movement of the body by acting as levers for the muscles. Bones come in an assortment of shapes and sizes, depending on their specific functions. The adult human skeleton is divided into two parts.

- The axial skeleton forms the long axis of the body. It includes the bones of the skull, vertebral column and rib cage. It is involved in protecting, supporting, or carrying other body parts.
- The appendicular skeleton attaches limbs to the axial skeleton. It includes the bones of the upper and lower limbs and the girdles (shoulder bones, hip bones), which aid movement and manipulation of the environment.

When the forensic anthropologist is sent human remains for forensic investigation, there are a number of key questions they must answer.

- Is the sample human or animal?
- Is the bone of forensic value?
- What bones are present?
- What is the minimum number of individuals (MNI)? There may be bones from more than one individual present and the forensic anthropologist must determine how many victims there are, e.g. by counting the number of skulls present.
- What is the biological profile, for example, the sex, age, ancestry, stature?
- What individual features can be found, for example, trauma to the bone or anatomical anomalies that can help to identify the victim?

Determination of sex

There are a number of key obvious differences between the male and female skeleton, and most of these differences can be seen in the size of the skeleton and the shape of the skull and pelvic girdle. Firstly, the skeletons of males are generally larger in size and more robust than females. Secondly, the skulls of males usually have a wider jaw, squarer chin, more sloping forehead and more pronounced eyebrow bones than females. Finally, differences in shape can be seen in the male and female pelvis, as the female pelvis is designed to be able to give birth.

Activity 32.4D

Forensic scientists can use the pelvis to help identify the sex of human remains. Your job is to research the differences between the male and female pelvis.

Find diagrams and pictures of the male and female pelvis, study them, label them and record any differences you find in a table.

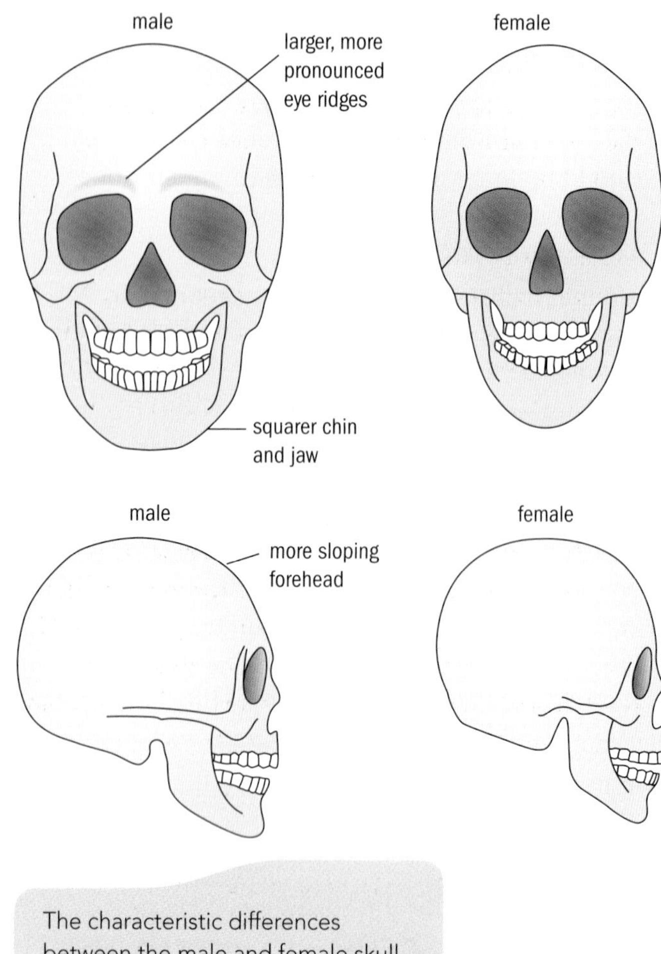

The characteristic differences between the male and female skull.

Determination of stature

The most common method for estimating living stature (height and build) from a skeleton is to measure the lengths of the long limb bones, as long bone length is proportional to height. Ideally, the forensic anthropologist will have the six upper and six lower long bones; however, height can still be determined when the remains of only one long limb bone are present. Precise measurements are obtained using an osteometric board and the measurements are input into mathematical formulas developed through gender research to produce a fairly accurate estimate of stature, with a small standard of error.

Determination of age

As individuals age their bones grow, change and eventually deteriorate, and the typical changes that occur over time can be used to determine the age of a victim when they died from their skeletal remains. The techniques used to age child and adult skeletons are different as the bones of children are still growing, and methods for determining age of children at death are therefore based on the growing skeleton and changes to the teeth. In contrast, adult bones are fully grown and weaken with time, and when examining adult skeletal remains the changes related to deterioration are used to estimate age.

In addition to bone growth, the teeth of children are also growing and changing as they get older and these dentition changes can be used to age child skeletons. Teeth contain enamel, one of the hardest substances in the human body, which makes teeth very strong and resistant to degradation. In fact, they are the last part of the body to decompose, and can withstand extremely high temperatures in a fire, so they are often a useful source of identification in murder or mass disaster investigations.

Hair and fibre evidence identification and analysis

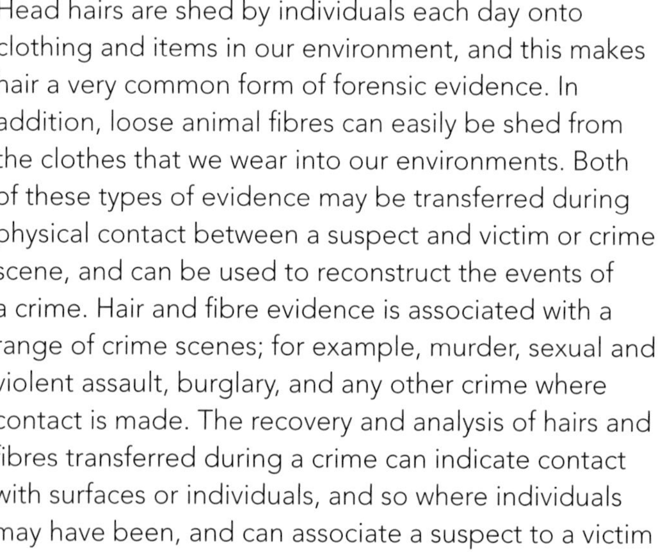

Did you know?

You shed approximately 100 head hairs a day.

Head hairs are shed by individuals each day onto clothing and items in our environment, and this makes hair a very common form of forensic evidence. In addition, loose animal fibres can easily be shed from the clothes that we wear into our environments. Both of these types of evidence may be transferred during physical contact between a suspect and victim or crime scene, and can be used to reconstruct the events of a crime. Hair and fibre evidence is associated with a range of crime scenes; for example, murder, sexual and violent assault, burglary, and any other crime where contact is made. The recovery and analysis of hairs and fibres transferred during a crime can indicate contact with surfaces or individuals, and so where individuals may have been, and can associate a suspect to a victim or a crime scene.

Animal hairs are natural fibres and these are also analysed in the forensic laboratory for criminal cases such as theft of animals and furs, or illegal breeding, and may be an additional source of linking evidence in any human case where the victim or suspect has a pet.

Hair and fibre evidence can be transferred between suspect and crime scene or victim in one of two ways.

Hairs can be transferred directly from the region of the body where they are growing. This is called primary transfer and occurs when hairs fall from the head onto an item of clothing, seat cover or floor. In secondary transfer, hairs do not transfer directly from the area of growth, but from a location where hair has already fallen. For example, from an item of clothing onto the floor or from a seat cover onto an item of clothing. This means that a suspect could carry on their clothes a hair from an innocent person (primary transfer) and transfer it to the crime scene (secondary transfer), making the innocent person look guilty. When assessing hair evidence, the possibility that hair is present at a scene due to secondary transfer must be considered.

Did you know?

Hair evidence can only be used to identify individuals using DNA analysis if a living root is attached to the hair. As the hair shaft itself is composed of dead, keratinised cells that do not contain DNA, the root is the only part of the hair that has living cells containing a nucleus and DNA, which are necessary for DNA profiling. However, hair can be used to help reduce the number of suspects by looking at the characteristics the hair possesses.

Different animal species, including humans, possess hair with distinguishing characteristics; for example, the length, colour, shape, root appearance and internal microscopic features. The same basic structures and morphology are shared by every individual, although there is a degree of variability in the arrangement, distribution, and appearance of some of these microscopic hair characteristics between individuals. These variations allow a forensic scientist to differentiate between hairs from different individuals and animal species. Hair evidence is valuable; it can provide strong corroboration for placing a suspect at a scene. However, morphological characteristics cannot be used to individualise the hair, and an individual can only be associated with hair evidence and not identified by it.

Each hair fibre consists of three main microscopic structures:

- the **cuticle**, a hard outer layer that protects the hair shaft
- the **cortex**, surrounding the **medulla**, which is the main bulk of the hair
- the soft medulla at the centre of the hair.

Did you know?

We can compare the structure of the hair to a pencil: the cuticle would be the paint, the cortex would be the wood, the medulla would be the lead.

The cuticle is a clear, protective, outside covering that provides resistance to chemicals, and enables hair to retain its structural features over long periods of time. The cuticle is not continuous, but formed by separate overlapping scales that always point towards the tip end of the hair. At the hair tip there may be no scales left or fragments may have chipped off edges causing hairs to split. This damage is usually seen in dry hair or hair that has not been trimmed for a while, and is caused by brushing, drying and the environment. There are three basic types of scale structure or pattern, and these patterns are an important feature for species determination – distinguishing human hair from animal and distinguishing between animal species.

Did you know?

The hair cuticle scales point towards the tip of the hair. You can feel the scales against your fingers if you pull a hair between your fingers from the tip to the root end. This pushes the scales in the opposite direction to which they lie, and feels rough in texture. Pulling from the root to the tip pushes the scales flat and feels much smoother.

Comparing hair evidence

The forensic analysis of hair evidence usually involves the use of comparison microscopy, where the structure and characteristic features of questioned hairs are compared to known hair samples. A number of the characteristic microscopic features are examined, and a comparison made between the hairs to determine whether the features are similar or different. If the questioned and known hairs have the same microscopic characteristics, this points to them originating from the same source, which suggests that the suspect's hairs were transferred to the crime scene. If the crime scene hairs are microscopically different, they cannot be associated to the source of the known hairs, the suspect. If there are similarities and slight differences observed between the questioned and known hair, no conclusion can be reached as to whether the questioned hair originated from the same source as the known hairs.

A large number of hairs must be taken from the suspect from all areas of the head, to ensure there is a representative control sample. This is because there is some variation between hairs from the same individual (intrapersonal variation), as not every hair on the same head is exactly the same, which is one reason why hair fibre evidence cannot be used for identification. However, when compared with an adequate number of reference control samples, hair can provide strong circumstantial evidence.

The forensic hair examiner can answer the following questions from morphological examination of hair, including points for comparison.

- What is the cuticle pattern?
- What is the degree of medullation and medullary index?
- What colour is the hair? What is the size, distribution and concentration of pigment granules?
- Is a root present? Did hair fall out or was it pulled?
- What is the condition of the hair and tip? How was the hair cut, treated or damaged?
- Is the hair human or animal?
- If human, child/adult? Male/female? Which racial group? What part of the body is the hair from?
- If animal, which species?
- Does questioned hair recovered from scene compare to reference hair from known individual?

Activity 32.4E

Why not have a look at your own hair and maybe other hairs from people in your class. Collect hair samples by plucking out your own hairs, or searching your hair brush or in and around your room for some hair. Mount onto a microscope slide, and use different magnifications to view your hair. See if you can then answer the questions above.

Look at the root, tip and hair shaft. Draw what you see, and add labels. See if you can identify your medulla and if so what type is it? Also have a look at more than one of your own hairs and see if there are differences between them.

Microscopy

Comparative microscopy

The comparison microscope is a specialised forensic microscope usually used to analyse hair evidence and other types of trace evidence. It consists of two connected compound light microscopes that allow two samples of hair, or other microscopic types of forensic evidence, to be viewed next to each other at the same time. For example, a glass microscope slide holding questioned hairs from a sexual assault crime scene is positioned on the stage of one microscope, and a slide with known reference hairs from a suspect is positioned on the stage of the other microscope. The microscopic characteristics of the two samples can be examined and compared directly in one field of view, allowing feature differences and similarities between the known and questioned hairs to be easily observed. The range of magnification used is approximately 40× to 400×.

Measurement

Measurement is a very important tool in forensic science. It allows the forensic scientist to compare, for example, hair fibres, pollen grains, and bullet and cartridge indentations. Trace evidence can be measured under the microscope if the microscope has an eyepiece graticule – an accurate measuring scale in the lens of the microscope. Items under the microscope are aligned with the scale in the field of view and the ruler is used to measure the size of the object. This provides accurate measurements that can be used for comparison purposes against suspect samples.

Assessment activity 32.4

P6 M6 D6 P7 M7 D7 BTEC

You are a forensic scientist working for the biological department of the Forensic Science Service. A SOCO has submitted a number of items of evidence from a suspected suicide case. The police have asked you to analyse the evidence using biological techniques; for example, blood, DNA and fingerprint analysis. You must establish whether the victim killed themselves or if their death is more suspicious and could have been murder.

Types of evidence collected from the crime scene:

- fingerprints on gun and suicide note
- suspected blood stains
- hairs found on the clothes of the victim.

1. Make a list and describe very briefly the techniques you would use to analyse the biological evidence recovered from the crime scene. Carry out some of the practical techniques you have listed (make sure you check the safety notes for each technique you use). P6 P7
2. Explain each of the biological techniques of analysis. Present valid conclusions from the analysis of your practical work. M6 M7
3. Discuss the advantages and disadvantages of the biological techniques of analysis. D6 D7

Grading tips

To achieve P6 and P7 you must be able to briefly describe the main points of the biological techniques used to analyse forensic evidence. You must also be able to carry out the techniques, completing the appropriate laboratory documentation and accurately presenting your results.

To achieve M6 and M7 you must be able to explain fully the biological techniques, and how they were used to analyse the forensic evidence collected from the crime scene. You should draw conclusions from the results of your biological analysis, and support them with evidence and reasoning; suggest also how to further analyse the evidence.

To achieve D6 and D7 you must be able to evaluate the scientific techniques of biological analysis and explain why the techniques used were suitable for analysis of the forensic evidence from the crime scene. You must also identify a range of types of evidence that can be analysed using the techniques and how they could be applied to different types of criminal investigation.

PLTS

Independent enquirer, Self-manager and Team worker

Researching forensic applications of analytical techniques, completing work to standard and deadline, and working as a team to analyse forensic evidence will help you develop these respective skills.

Functional skills

ICT and English

You will be developing your ICT skills by selecting and using a variety of sources of information to research and create a portfolio of analytical techniques with forensic applications; and when you enter, develop and format information, to suit its meaning and purpose, in building a case file on a simulated crime.

When reading information on analytical techniques you will develop your English skills, in comparing, selecting, reading and understanding texts and using them to gather information, ideas, arguments and opinions. When writing a portfolio on techniques used in a forensic laboratory and conclusions from the results of analytical work, and building a case file, you will be practising your English skills, in communicating effectively and persuasively.

32.5 Reporting the analysis of evidence from a crime scene

In this section:

Key terms

Probability – the likelihood that a given event will occur.

Like any normal scientist, a forensic scientist must document their scientific investigations, interpret and assess their results, and report their findings. As an expert witness, the forensic specialist must also present a written statement and may be called to give verbal evidence in court.

Documentation

Throughout a forensic criminal investigation, various documents are generated that follow the life of the evidence, from the crime scene to the laboratory to court. The documentation records every action taken by the SOCO and forensic scientist throughout the investigation, and their findings and the conclusions they have drawn from the evidence. The pieces of documentation together make up the forensic case file which is submitted to court for the trial.

Crime scene examination

As discussed earlier (page 454), a number of documents are completed during the crime scene examination to record the scene and the evidence. The documents clearly describe the stages that the SOCO followed and how the evidence was collected and packaged. These include an entry log, crime scene sketch, evidence log and paperwork, and a photograph log, which records each photograph taken at the scene. Each item of packaged evidence is clearly labelled, and this information must accurately match the recorded information on the evidence paperwork, ensuring the chain of continuity. This is to prevent the evidence being called into question in court and being disregarded at trial. All of these documents are submitted as part of the investigation case file.

Laboratory analysis

During the laboratory analysis, a number of forms are used to document the evidence and the procedures carried out, as well as the results of the scientific examinations and analysis tests. Standard forms are used by the forensic scientist to ensure that all of the necessary information is recorded. It is essential that the paperwork is completed accurately, completely and in detail, and maintains the chain of continuity for each item of evidence. The paperwork is examined and signed by a second forensic scientist who checks that the analysis procedures have been carried out correctly and competently. The laboratory analysis forms are submitted as part of the case file documentation.

Expert witness statement and report

The SOCO who processes the crime scene and the forensic scientist who analyses the evidence during a criminal investigation are required to write an expert witness statement that describes their role in the investigation. For the forensic scientist, this statement is in the form of a report on their scientific examination and analysis results. In many criminal cases, the scientist may not be called to give evidence in court, and the report that is admitted for use in court must therefore be comprehensive, descriptive, accurate and organised. The report must contain all the relevant details of the analysis carried out by the scientist and a clear explanation of the results and conclusions drawn. It must be written using 'laymen's terms'; this means that it is written in very simple language without scientific jargon, which allows non-scientists, for example members of the jury, the police, lawyers, judges and magistrates, to understand the contents of the report.

Individualisation

The aim of any criminal investigation is identification, of a suspect or victim. According to Locard's principle, every contact leaves a trace. The SOCO and the forensic scientist collect and analyse evidence from a crime scene to source and identify the origin of the trace.

Some types of evidence have individual characteristics and can be sourced back to a specific individual with an extremely high degree of **probability**. For example, fingerprints are said to be unique; no two people have the same fingerprint. It has been estimated that the odds of two individual fingerprints being the same are 1 in 64 billion. Therefore, a fingerprint mark can be linked to and identify a single person. In addition, DNA is individual (except in the case of identical twins) and can be used in a criminal investigation for identification.

Whether a piece of evidence has individual or class characteristics impacts on the significance of the evidence. If we can identify an individual from an item of evidence, then that item has high evidential value and is very useful in a criminal investigation. If we can only link an item of evidence to a group of people then this provides less useful information and the evidential value is limited. How limited the information is often depends on population statistics. For example, if blood type O is found at a crime scene then 46% of the population could have left that evidence, which is an extremely large group of people. However, if blood type AB is found, then only 3% of the population could be the source of the blood, a much smaller group to investigate. Therefore, the rarer the blood type, the higher the evidential value and significance of the evidence.

Activity 32.5A

Make a list of different chemical, biological and physical types of evidence. For each type of evidence establish whether the evidence has class or individual characteristics. Consider which types of evidence are the most and least significant in criminal investigations, and place the types of evidence in order of their evidential value.

Interpretation

The aim of forensic science is to answer questions of a legal nature and to use forensic evidence to reconstruct a crime. The purpose is to establish whether the evidence corroborates, or confirms, the victim, suspect or witness statements given to the police. Therefore, as an expert witness, one of the roles of the SOCO and forensic scientist is to not only state the facts of their investigation, but also to provide their opinion on what the evidence and the results of their investigations actually mean. They must draw rational and balanced conclusions from their observations at the crime scene or from the laboratory test results.

When analysing and interpreting an item of evidence, we often use probability to give us an indication of the value of the evidence. Probability is a measure of how likely it is that an event has occurred. The expert's opinion is designed to provide the court with the most probable or likely explanation of the evidence to assist the court in reaching a decision on an individual's guilt or innocence. Using probability estimates provides the court with a quantitative description of how likely the expert's conclusions are to be true, and allows the court to determine the significance of the evidence.

Forensic analysis often involves the comparison of a crime scene evidential sample and a reference sample taken from a victim or suspect. We usually express the results of the comparison analysis using probability estimates. For example, the DNA profile of an unknown crime scene blood sample is compared to the DNA profile of a suspect, and the profiles match. The forensic scientist may calculate the likelihood that the match has occurred by chance and give a probability estimate of 1 in 1 billion. This means that there is a 1 in 1 billion chance that a random member of the population's DNA would match the DNA profile, and therefore it is likely that the suspect is the source of the blood stain.

Presentation

In addition to writing a statement reporting their findings, the SOCO and forensic scientist may be called to give verbal evidence in court as an expert witness for either the prosecution or defence. They are specialists in their field and must explain to the jury and judge in very simple terms how they contributed to the criminal investigation, which procedures they followed and what results they obtained. They must remain unbiased and impartial in their testimony to ensure that the evidence they provide is fair. Their role is different from that of an ordinary witness, as they do not simply repeat the facts revealed by their investigations and scientific tests, but must also offer their interpretation of the findings in the context of the criminal case. As well as giving factual evidence on any issue that falls within their area of expertise, they must give their opinion on what the evidence means.

Activity 32.5B

One of the roles of the expert witness is to explain complicated scientific techniques to the jury, who usually have very little or no scientific knowledge. The expert witness has to learn to describe the analytical techniques without using scientific terminology.

To appreciate how difficult this is, you need to imagine that you are an expert witness who has to try and explain everyday processes to an alien, who does not know anything about Earth. In small groups, try to explain the following procedures in the simplest terms possible:

- making a piece of toast
- ironing clothes
- sending a text
- making a cup of tea.

Assessment activity 32.5

BTEC

You are a forensic scientist working for the Forensic Science Service. You have been involved in the laboratory analysis of the evidence from a suspected suicide investigation. The police have requested you to complete an expert witness statement report describing your analysis, results and conclusions, which will then be submitted to court for the trial. Your report will make up part of your investigation case file along with your crime scene and laboratory documentation. You have also been called to give verbal evidence in court as an expert witness.

1 Describe your crime scene examination, and biological, chemical and physical analysis of the forensic evidence. Identify your results and conclusions. P8
2 Discuss your results and conclusions and explain the evidential value of the evidence. M8
3 Discuss the strengths and weaknesses of your results and conclusions. D8

Grading tips

To achieve P8 you must produce a written report describing your crime scene examination and chemical, physical and biological analysis of the forensic evidence. Your report must include the results and conclusions you have drawn from your investigations. The report must be submitted along with your crime scene and laboratory analysis documentation as part of your investigation case file. You must also give verbal evidence in a mock court explaining your role in the criminal investigation and your findings.

To achieve M8 you must include in your report and verbal evidence an explanation of how you interpreted the evidence collected from the crime scene in the context of the criminal case, justifying the conclusions you have drawn from your scientific analyses in this investigation. You must assess the significance of the evidence and determine the evidential value.

To achieve D8 you must evaluate your results and conclusions, discussing the strengths and weaknesses of your investigation. You must be able to explain why your interpretation of the evidence is the most likely explanation using probability estimates.

PLTS

Effective participator, Reflective learner and Self-manager

Acting as expert witness in role-play activities, using several sources to reach a conclusion, and completing work to standard and deadline will help you develop these respective skills.

Functional skills

ICT and English

You will be developing your ICT skills when presenting information as an expert witness in ways that are fit for purpose and audience; and when you enter, develop and format information, to suit its meaning and purpose, in building a case file on a simulated crime.

By delivering an expert witness statement or questioning the information presented by an expert witness, you will be practising your English speaking and listening skills. When building a case file, including writing an expert witness statement, you will be practising your English writing skills, in communicating information, ideas and opinions effectively and persuasively.

WorkSpace

Kate Morgan

Crime Scene Investigator (CSI), Lancashire Constabulary

As a Crime Scene Investigator (CSI), I need to know how to plan and carry out a scene examination and to understand what certain evidence can prove. I also support police officers in their investigations and need to be able to explain the processes and their implications to a jury.

When I am called upon to attend a crime scene, I collate as much information as possible before examining it. I use Personal Protective Equipment (PPE) and my Health and Safety knowledge to avoid contamination and I carefully package and label evidence to ensure its integrity. I regularly recover footwear marks and tool impressions and being able to do this successfully has led to many detections.

I regularly write statements for court, which must be professional and clear, with a full description of my scene search, techniques and findings. As an expert witness I am regularly called to give evidence in court, so an understanding of my role within the Criminal Justice System is essential, as well as knowledge of different court types and procedures.

For me, to be able to deduce what has happened at a crime scene based on the evidence found is fascinating. I thoroughly enjoy working with state-of-the-art equipment as it helps me to reach conclusions about the course of events.

My understanding of underlying scientific methods of analysis and the techniques available helps me to determine the appropriate methods to use. For example, knowing which components of fingerprints can be chemically enhanced allows me to recover appropriate items for further examination. Also, by knowing what substances DNA can be recovered from, and analysis options, allows me to swab items appropriate to the crime. Having studied blood spatter and entomology, I have used this knowledge to interpret evidence and understand how, when and where violent crimes have occurred.

Think about it!

1. As a CSI, what steps would you take in the investigation of a burglary?
2. How would these steps differ if an assault had taken place?

Just checking

1 What must the SOCO do during the initial assessment of the crime scene?
2 Describe the stages of a crime scene examination.
3 What are the differences between chemical, physical and biological evidence?
4 Describe a range of techniques used to analyse chemical, physical and biological evidence.
5 Identify the documents that make up a criminal investigation case file.
6 What is the difference between class and individual evidence?
7 What is the role of the expert witness?

Assignment tips

To get the grade you deserve in your assignment remember to do the following...

- The role of the forensic practitioner follows the evidence from the crime scene to the laboratory and finally to court. These are the key steps in this process, which you will need to remember.
- The SOCO searches for and collects evidence from the scene of a crime.
- The forensic scientist examines the evidence collected and submitted to the laboratory to provide information previously unknown or to corroborate information already available.
- They provide the results of their examinations in a report to enable the investigator to trace an offender or to back up other evidence to provide a case for presentation to court.
- They present verbal evidence in court to describe their examinations and explain their interpretations of the findings.

Glossary

absorption – gamma or X-ray radiation being taken into the tissue or medium it is passing through

addition rule – rule for mutually exclusive events, sometimes called the OR rule: P(A OR B) = P(A) + (B)

adsorbed – attract and hold to a surface

aeration – addition of air/oxygen

aerotolerant – organisms that cannot use oxygen but can tolerate it

allocation of public money – government decisions made as to where best to spend money earned through taxation

analyte – a sample being analysed

annealing – combining of two strands of DNA from different sources

antibiotics – chemicals made by some bacteria or fungi that kill or prevent the growth of other bacteria or fungi

antifungals – chemicals that kill fungi/treat fungal infections

antivirals – chemicals that inhibit viral replication

arterioles – blood vessels of small diameter that branch out from an artery and lead to capillaries

asbestos – a naturally occurring silicate fibrous material; it has a number of insulating properties that made it popular as a building material in the past; however, the asbestos fibres cause serious lung diseases such as asbestosis when breathed in by people

attenuation – the drop in energy of a wave as a result of absorption or scattering

autonomic nervous system – nerves that automatically control some normal bodily functions

base (units) – the internationally agreed units from which all other units are derived; there are seven in total

batch fermentation – method of growing microorganisms in small closed fermenters; nothing is added or removed until the end of the fermentation when the product is harvested in batches rather than continuously

bile – dark green/yellowish brown fluid made in liver and stored in gall bladder; contains bile salts, water and hydrogencarbonate ions; neutralises chyme and emulsifies fats

bimodal – a distribution that has two distinct peaks

blinds – a strategy to reduce bias in investigations, such as testing drugs, when the experimenter does not know which patient is receiving which drug until the trial is finished

blood – liquid tissue consisting of red cells, white cells and platelets suspended in plasma

body mass index – a measurement of a person's mass relative to their height

boiling point – the temperature at which a liquid turns to gas forming bubbles in the liquid

bone – rigid tissue formed from ossified cartilage

cartilage – type of connective tissue found at joints

chemical digestion – enzymic digestion of food molecules into smaller molecules, by hydrolysis

chemical energy – energy of a substance stored in the bonds that can be released by a chemical reaction

chemical species – generally, the name given to chemically identical molecular substances

chiasmata – area on a pair of homologous chromosomes where crossing over produces an exchange of parts between non-sister chromatids

chute – a sampling device which separates using chutes

class intervals – statistics – a range of data when grouping large sets of figures

closed – refers to fermentation vessels for batch fermentation

complementary base pairing – DNA bases pair up: adenine with thymine; guanine with cytosine

conclusion – a judgement formed based on the evidence or results obtained

conditional probabilty – when the probability depends on a previous event that has happened

confidence limits – the degree to which you are sure that the measurement could not have occurred by chance

continuous fermentation – microorganisms grown in open containers where substances are added and removed throughout the fermentation process; this maintains the organisms in the log (exponential) phase of growth

control variable – factors which must be kept constant in an investigation whilst two other variables are observed

crossing over – exchange of genetic material between homologous chromosomes

cuvette – a container made of transparent material designed to hold liquid for spectroscopic analysis

daughter – the atom after radioactive emission

death phase – last stage of growth curve, where bacteria run out of nutrients and die

decline phase – phase of growth curve where numbers dying exceeds numbers being produced

dependent variable – a factor in an investigation which changes as a result of a change in another factor, e.g. temperature change with time

derived (units) – units of measurement made up from two or more of the seven base units, e.g. units for speed from metres and seconds

deterrence – initially a policy by classicist criminologists where minimal sanctions were designed to prevent offenders choosing to commit crime; used in late twentieth century by realists in a different sense where, it was argued, very harsh sentences deter offenders

Doppler effect – a change in frequency detected when a source of waves is coming towards or moving away from an observer

elastic arteries – arteries with many elastin and collagen fibres in the wall

electrical energy – energy associated with the flow of electric charge (electrons) through a conductor

error – the amount to which a measured value can differ from the correct/true value

evaluation – a judgement made about the quality of the investigation procedure involving considering all the strengths and weaknesses, as well as alternative methods

event – something that may or may not happen

expected value – the value you expect to measure based on an assumption

facultative anaerobes – organisms that can respire aerobically if oxygen is present but can respire anaerobically in the absence of oxygen

faeces – waste from digestive tract; contains undigested food, dead cells and bacteria, plus breakdown products from bile

filtered – allowing of particular wavelengths to pass through (in the case of waves)

flight path (MRI) – the area of space which can come under the influence of the strong magnet in MRI diagnosis

foaming – foam produced at outlet valves in a fermenter

frameshift – effect of an addition or deletion mutation in a length of DNA; the subsequent base triplets all change

growth phase – log phase

half-thickness – the term used to describe the thickness of a material required to reduce X-ray intensity to half of its original value

harvesting – removing the useful product from a fermenter

heterogeneous – samples of substances which are not uniform in size or composition throughout

heterotrophs – organisms that obtain energy from breaking down organic carbon-based molecules

ICRP – International Commission for Radiological Protection – group of international experts in the field of radiology

ileum – small intestine

immunosuppression – a means of restraining the body's ability to fight infection by using specialist drugs during bone marrow transplants for example

incarceration – the process whereby offenders are locked up for a long time during which they cannot commit further crimes

independent – two events if the outcome of one event does not affect the outcome of the other event

independent variable – a factor in an investigation which is actively changed by the investigator

independently segregate – random assortment of chromosomes or chromatids at meiosis (metaphase/anaphase)

ingestion – take in a substance, usually through the mouth

intellectual property – the material or findings produced by a person as a result of their personal work and ideas

interior surfaces – inside surfaces

intersection – the shaded region in a Venn diagram

ion – a changed atom, an atom that has either lost or gained an electron

IRR – Ionising Radiations Regulations document 1999

isokinetically – taking a gas sample which moves through the sampler at a constant rate

isotopes – atoms with the same number of protons but different numbers of neutrons

lag phase – first stage of microbial growth curve; it is slow as genes are being switched on and enzymes are being synthesised

laminar flow cabinet – safety cabinet for carrying out work with microorganisms

Larmor frequency – the frequency of precession of a nucleus (e.g. proton) in a magnetic field

light energy – energy associated with electromagnetic radiation, such as light which is visible to the human eye

log phase – exponential growth phase

lumen (biology) – cavity or channel within a tubular structure

main magnet – large, superconducting electromagnet in an MRI scanner

maximum absolute uncertainty – same as error

mechanical digestion – physical breaking of food into smaller parts

mechanical energy – sum of kinetic energy and potential energy in a system

Mendelian dominant – characteristic that is dominant and has Mendelian inheritance pattern

mercury – a silver-coloured liquid metal element; it is highly toxic

mixing – movement of liquid culture medium, usually by stirring, to distribute nutrients and oxygen

multiplication rule – rule that applies only to independent events: P(A AND B) = P(A) x P(B)

muscular arteries – muscular elastic vessels that carry blood away from the heart

mutally exclusive – events that cannot happen at the same time

negative containment – method of prevention of escape of microorganisms to protect personnel from exposure to pathogens; the cabinet in which they are housed has negative air pressure

negative correlation – when an increase in one quanity appears to cause another quantity to decrease

nuclear energy – energy released from splitting atoms of radioactive materials

nutrients – chemicals that organisms need to live and grow

obligate anaerobes – organisms that fail to grow in the presence of oxygen

observed value – the value you have measured in an experiment

oxidation – gaining oxygen

pancreatic juice – digestive juice, containing enzymes, produced by cells in pancreas

parallax – the apparent change in the position of an object resulting from a change in the viewer's position; it can lead to errors reading instruments with pointers

parent – the original atom before radioactive decay

peer review – an assessment process carried out by intellectual equals and experts in science

periodic table – table listing all the known elements in terms of their atomic structure and properties

peristalsis – wavelike muscular contractions that move substances through a tube

pH control – keeping the pH constant or within a set range

photoautotrophs – organisms that produce their own food (organic carbon-based molecules) from inorganic molecules and light energy

polarimeter – a device which can measure the extent to which polarised light is rotated when passing through a substance

polypeptides – chain of amino acids joined by peptide bonds; the chain may be coiled and folded into a tertiary structure (3D shape)

population – a collection of items
porosity – extent to which a material will allow solids, liquids or gases through it
positive containment – use of aseptic technique to prevent contamination of a microbiological sample
positive correlation – occurs when a increase in one quantity causes an apparent increase in another quantity
precipitation reaction – a chemical reaction involving solutions which results in an insoluble substance
primary data – data you collect yourself
primary structure (protein) – sequence of amino acids
probable – the approximate and realistic estimate of error which may occur in a reading
proteins – chains of amino acids essential for growth and body repair composed of hydrogen, carbon, oxygen and other elements
pulmonary circulation – part of the circulatory system that carries blood from heart to lungs and back to heart
punishment – the argument that offenders should be subject to punishment whilst in prison to show them that crime does not pay – an early example was the crank – hard unproductive labour
quadrat – a square frame for environmental sampling
quantum computing – idea for computers based on quantum mechanics using the 'qubit' or quantum bit, which can hold an infinite number of values
quaternary structure (protein) – protein made of more than one polypeptide chain
radio frequency coils (MRI) – conductive coils producing radio waves of frequencies required to excite the hydrogen nuclei in a patient
radionuclides – a radioactive isotope
random errors – errors which occur as a result of limitations in the measurement technique and may be identified when repeat measurements produce different values
recombinant DNA – DNA that has joined with another piece of DNA, usually from another organism
redox – the simultaneous gaining and losing of oxygen in a chemical reaction
reducing sugars – sugars which can reduce copper sulphate on heating
reduction – losing oxygen
reform – an argument where penal policy is aimed to change offender behaviour; for example, offenders doing productive but repetitive work such as sewing mailbags would gain employment skills and discipline
refraction – bending of light when it passes from one medium to another
refractive index – a figure which provides the ratio of the speed of light in a vacuum to its speed in other mediums
refractometer – a physical device which can measure refractive index
regression – a statistical method used to determine the equation that describes a line of best fit
relaxation time – average length of time that a nucleus remains in its high energy state
reliable – results of an investigation produced using good laboratory practice and procedure
repeatable – being able to achieve the same results when doing an investigation a second time
residue – each time two amino acid molecules join together to form a peptide, a water molecule is eliminated; the remaining part of each amino acid is referred to as an amino acid residue
restriction site – area on a length of DNA that is recognised by (fits active site of) and cut by a restriction enzyme
retention factor (R_f) – distance travelled by a compound divided by the distance travelled by the solvent
sample – a selection of items from a given population
sample space – list of all possible outcomes
secondary data – data that you didn't collect yourself but obtained from someone else
secondary structure (protein) – peptide chain coiled into an alpha helix or folded into a beta pleated sheet
selective growth medium – a nutrient-rich substance designed to identify a particular bacterium
serial dilution – stepwise dilution of a sample or culture of bacteria; adding 1 ml to 9 ml distilled water this makes a 1 in 10 dilution, or a dilution of x 10; if 1 ml of that is added to 9 ml distilled water, the dilution is now x100; the dilution factor at each step is the same giving logarithmic dilution
solute – a substance dissolved in a solvent
sound energy – energy carried by a sound wave as particles vibrate
spinning riffler – a sampling device which spins samples and separates them
starch – a natural glucose polymer which stores energy – carbohydrate
stationary phase – phase of growth curve where numbers being produced are equal to the numbers dying
sticky ends – unpaired nucleotide bases at the end of a piece of cut DNA
strong correlation – when there appears to be a relationship between two quantities
substitution – type of DNA mutation; one base pair is substituted for another
systematic circulation – part of the cardiovascular system that carries blood from heart to body and back again
systematic errors – errors produced by faulty equipment or incorrect calibration
Taq polymerase – enzyme, obtained from bacterium that lives in hot springs, used in polymerase chain reaction
temperature control – keeping the temperature constant
tertiary structure (protein) – folded protein which assumes a 3D configuration
thermal energy – energy in the form of heat due to molecular motion within a material
thermoluminescence – energy given off as visible light in proportion to the dose of radiation it receives
total lung volume – vital capacity plus residual air volume
ultrasound – wave frequencies in excess of 20 000 hertz
unimodal – a distribution that has one maximum
vacuum line – chemistry apparatus which allows experiments to take place in a vacuum; the experimental apparatus usually connects to the vacuum source via the vacuum line via joints which provide a good seal
veins – blood vessels that carry blood from tissues to heart
venules – blood vessels of small diameter that carry blood from capillary bed to vein
weak correlation – when there does not appear to be a relationship between two quantities

Index

Page numbers in bold show key term definitions

D

E